Atlas of the Developing Rat Nervous System

SECOND EDITION

Atlas of the Developing Rat Nervous System
SECOND EDITION

George Paxinos
School of Psychology
University of New South Wales
Sydney, Australia

Ken W. S. Ashwell
School of Anatomy
University of New South Wales
Sydney, Australia

Istvan Törk[†]
School of Anatomy
University of New South Wales
Sydney, Australia

[†]Deceased

ACADEMIC PRESS

San Diego   New York   Boston   London   Sydney   Tokyo   Toronto

Academic Press, Inc.

A Division of Harcourt Brace & Company
525 B Street, Suite 1900, San Diego, California 92101-4495

United Kingdom Edition published by
Academic Press Limited
24-28 Oval Road, London NW1 7DX

Library of Congress Cataloging-in-Publication Data

Paxinos, George, date.
    Atlas of the developing rat nervous system / by George Paxinos,
Ken W. S. Ashwell, Istvan Törk. — 2nd ed.
        p.   cm.
    Rev. ed. of: Atlas of the developing rat brain / George Paxinos
... [et al.]. c1991.
    Includes bibliographical references (p.   ) and index.
    ISBN 0-12-547610-8 (paper)
    1. Rats—Nervous system—Growth—Atlases. 2. Embryology—Mammals-
-Atlases. I. Ashwell, Ken W. S.   II. Törk, Istvan.   III. Atlas of
the developing rat brain.   IV. Title.
QL737.R666P39   1994
599'.033—dc20   94-17100
                                                               CIP

PRINTED IN THE UNITED STATES OF AMERICA
94   95   96   97   98   99   EB   9   8   7   6   5   4   3   2   1

To Jenny, Christine, and Peter

Contents

Acknowledgments

The *Atlas of the Developing Rat Nervous System,* Second Edition, includes all plates and diagrams of the *Atlas of the Developing Rat Brain* and therefore we incorporate below the Acknowledgments for that book. The present book contains an additional 95 plates and 95 diagrams and we acknowledge here the intellectual and technical support we received.

We thank Oscar Scremin for substantive assistance on the delineations of the vascular system. We thank Luang Ling Zhang for speedy and professional assistance in sectioning tissue and for constructing the final drawings on Adobe Illustrator. We thank Paul Halasz for constructing the index electronically and Xu-Feng Huang for taking most of the new negatives. We thank David Tracey (School of Anatomy) for availing photographic equipment for this project and Kevin McConkey (School of Psychology) and Arthur Toga (Department of Neurology, UCLA) for availing computers for the graphics part of the project. We thank the Academic Press staff, Dr. Graham Lees, Ms. Anne Schermer, and Ms. Linda Shapiro for ensuring high quality of the production of figures and a timely publication.

We owe a special thanks to Elly Paxinos for constructing figures outside of her own professional duties.

Research for this atlas was supported by the National Health and Medical Research Council of Australia.

During work on this project we experienced the professional and personal loss of our third author, Professor Istvan Törk. Although he was not able to contribute to the present book his input on the delineations of peripheral features in the earlier book was enormous and for this reason we have retained him as co-author on the present book.

Acknowledgments to the First Edition

We wish to express our gratitude to Jack D. Barchas, in whose laboratory the project commenced, while one of us (G.P.) was on sabbatical (1988). Jack's long-term view of research created an atmosphere in which we could embark on a project without immediate reinforcement.

Among the people attracted to the congenial environment which Jack Barchas had created in his laboratory was Alicia Fritchle, a graphic artist with a strong background in the fine arts. Alicia urged us to proceed with the construction of the atlas even when the project was one of the more remote possibilities under consideration. She commenced with the staining of the tissue for the atlas and continued with the photography of sections, finally coming to Sydney to enter our pencil drawings on a Macintosh computer. Photographing developing tissue proved difficult and it was due to her perseverance that the project was completed within three years of its commencement. Alicia dedicates her work in this book to the memory of her father, Frank P. Fritchle.

Diane and Pete Ralston (Anatomy Department, University of San Francisco) were essential to the construction of this atlas because they allowed us generous access to their department's Nikon Multiphot, Durst enlarger, and paper processor for the production of nearly all the photographs.

Joseph Altman attended closely to some brain regions of ages E14 and E19 and he generously made available to us for inspection his outstanding collection of embryonic brains cut in the three cardinal planes.

We are also pleased to acknowledge assistance with delineations provided by Elizabeth Taber-Pierce, John Pintar, Ian Curthois, Dzung Vu, Liz Tancred, Richard Robertson, Bill Mehler, David Rapaport, and Christine Gall. We thank Paul Halasz and Annette Dorfman for organizing lists of abbreviations and index and Joan Hunter and Antonia Milroy for technical assistance.

Financial support for work done in Australia came from NH & MRC grants.

Introduction

The publication of the *Atlas of the Developing Rat Nervous System,* Second Edition, affords us the opportunity to expand the previous publication, the *Atlas of the Developing Rat Brain* (Paxinos, Törk, Tecott, and Valentino, 1991), in order to provide a more comprehensive guide for those interested in the developing nervous system of this species. In the present book the entire nervous system (and to a lesser extent the entire body) is delineated in a number of ages and thus the name has been changed to *Atlas of the Developing Rat Nervous System.* All the old atlas plates are reproduced and they are supplemented by 95 plates and 95 diagrams depicting Coronal E14, Coronal and Sagittal E16, and Sagittal E19 brains.

The following table presents the number of plates and diagrams found in this publication:

	Coronal		Sagittal	
	Plates	Diagrams	Plates	Diagrams
E14	20	20	8	8
E16	2 × 22*	2 × 22*	16	16
E17	—	14	—	—
E19	49	49	15	15
P0	74	74	—	—

* Each plate/diagram of coronal E16 presents photographs/diagrams of two anterior–posterior levels.

We realized the need for a developmental atlas during the course of our own investigations of normal embryos as well as embryos that had sustained pathological changes following experimental manipulations (Ashwell, 1987, 1991, 1992; Ashwell and Waite, 1991; Ashwell and Zhang, 1992; Tecott *et al.,* 1989; Waite *et al,* 1992). Our work was impeded because the topography of the anlagen of the approximately 700 nuclei that exist in the adult rat brain had not been comprehensively established.

While the usual copyright restrictions pertain for reproduction of figures of this atlas, Academic Press and the authors will respond promptly to written requests to reproduce specific sets of figures (no request has been denied previously). We have one request to make of users of this atlas: in the interest of facilitating communication, please consider the suitability of our system of nomenclature and abbreviations for your work. Unfortunately, neuroscience communities concerned with different systems have developed identical abbreviations for completely different structures; for example, SO stands for both supraoptic nucleus and superior olive, SC for suprachiasmatic nucleus and superior colliculus, and IC for inferior colliculus and inferior olive. In dealing with the entire nervous system (as increasingly more researchers do) these parochial abbreviation schemes become impossible to implement. An additional complication arises when homologous structures are nonetheless named or abbreviated differently in different species. We have made an effort to establish homologies and are using the same abbreviations for homologous structures in atlases of the mouse (Franklin and Paxinos, in press), monkey (Paxinos *et al.,* in press), and human (Paxinos *et al.,* 1990; Paxinos and Huang, in press; Mai *et al.,* manuscript in preparation). The abbreviations used in the present and our other work were developed using the following principles:

1. The abbreviations represent the order of words as spoken in English (e.g., DLG = dorsal lateral geniculate nucleus).

2. Capital letters represent nuclei, and lowercase letters represent fiber tracts. Thus the letter 'N' has not been used to denote nuclei, and the letter 't' has not been used to denote fiber tracts.

3. The general principle used in the abbreviations of the names of elements in the periodic table was followed wherever appropriate: the capital letter representing the first letter of a word in a nucleus is followed by the lowercase letter most characteristic of that word (not necessarily the second letter; e.g., Mg = magnesium; Rt = reticular thalamic nucleus).

4. Compound names of nuclei have a capital letter for each part (e.g., MD = mediodorsal thalamic nucleus).

5. If a word occurs in the names of a number of structures, it is usually

given the same abbreviation (e.g., Rt = reticular thalamic nucleus; RtTg = reticulotegmental nucleus of the pons). Exceptions to this rule are made for well-established abbreviations such as VTA.

6. Abbreviations of brain regions are omitted where the identity of the region in question is clear from its position (e.g., Arc = arcuate hypothalamic nucleus, not ArcH).

7. Arabic numerals are used instead of Roman numerals in identifying (a) cranial nerves and nuclei (as in the Berman, 1968, atlas), (b) cortical layers, (c) cerebellar folia, and (d) layers of the spinal cord. While the spoken meaning is the same, the detection threshold is lower, ambiguity is reduced, and they are easier to position in small spaces available on diagrams.

8. The abbreviations of the neuroepithelium of nuclei are indicated by lowercase letters (consistent with the smaller size of the primordia when compared to the fully developed structures).

Dating Embryos and Other Nomenclature Questions

Timed Pregnancies

Animals (3 females:1 male) were paired overnight from 6 PM to 8 AM. [While we followed this pairing procedure, we recommend a shorter pairing period (2 hr) if increased dating accuracy is required (see below).] The presence or absence of sperm in a vaginal smear was determined the next morning. Pregnant rats were killed late in the morning of the indicated day. Mating often occurs immediately upon pairing (experienced males) but at times as late as the early morning hours. Sperm is noted in the oviduct within 1 hr of copulation (Blandau and Money, 1944), and 90% of ova are fertilized within 3 hr of copulation (Odor and Blandau, 1951).

The most vexing question we faced was whether to adopt the E0 or E1 convention for the first 24 hr after placing the rats together. We believed that the advent of a new atlas provides the opportunity to assist in the standardization of terminology and for this reason we wrote to 60 developmental neurobiologists seeking advice and attempting to discern whether there was an emerging consensus. We received 31 replies with 22 respondents favoring E0 and 9 respondents favoring E1 for the first 24 hr following pairing. For the day of birth, the vote was also 22 in favor of P0 and 9 in favor of P1. (Three of the respondents who preferred E0 opted for P1 and another three who preferred E1 opted for P0.) A survey of articles published in *Developmental Brain Research* in 1989 and the first 4 months of 1990 revealed that there were 13 articles using E0 to 10 articles using E1 and a 26 to 11 preference of P0 over P1.

We consider the split schemes of E0/P1 and E1/P0 to be in insufficient usage and further burdened by a logical inconsistency. We consider the E0/P0 scheme to be as logical as the E1/P1 scheme. We believe that much as the argument as to whether there should have been a year zero, the argument on E0/P0 versus E1/P1 should be solved by usage. We decided, therefore, to follow the majority of our advisors and call the first 24 hr following the commencement of pairing E0 and the first 24 hr following birth P0. We reproduce below an edited version of the advice given to us by Marie-Claude Bélanger and Raymond Marchand for dating of embryos in the belief that some readers may find the explicit descriptions of these scientists useful.

The first and most important point in dating embryos is to mate the rats for as short a time as possible. With the short interval allowed for breeding, the pups are generally born at the end of the 22nd day, that is E21 plus several hours. The rats are paired, for example, from 4 PM to 6 PM. The end of this period is considered as $E0_0$ (0 day and 0 hour). During the next 24 hr, the embryo is in its first day of gestation, but not yet one day old. As a convention it is proposed that when the word day precedes the cardinal number, this

indicates that the age is at least that many days. For instance, an E12 embryo is at least 12 days of age, but an embryo of the 12th day of gestation is aged anywhere between E11 and E12 and consequently called $E11_x$ (where x is the number of hours beyond the completed 11 days). The zygote formed 3 hr after the breeding is not 24 hr old; it is rather at the onset of its first day and this is why it should be called an $E0_3$ embryo.

Names of Structures

The terms presented in *The Rat Brain in Stereotaxic Coordinates* (Paxinos and Watson, 1986) were employed wherever appropriate. In accepting a particular developmental term we were guided by the suggestions of our advisers and by the following principles: (a) terms can be ratified by usage; (b) a viable 'systematic' term is preferable to a slightly more frequently used but misleading synonym; (c) eponymous terms can be dispensed with when viable alternatives exist; (d) excessive subordination of terms ("babooshka doll" nomenclature) should be avoided.

The Bases of Delineation of Structures

The Postnatal Day 0 (P0) brain is very similar to that of the adult rat. Nearly all structures can be recognized on the basis of some familiarity with the adult equivalents. The delineations for this age, therefore, were on the same basis as given for the delineations in *The Rat Brain in Stereotaxic Coordinates* (1986). We have benefited by reading descriptions of nuclei given by the various contributors to *The Rat Nervous System*, Second Edition (Paxinos, 1995).

The cortical delineations for *The Rat Brain in Stereotaxic Coordinates* were based on the comprehensive work of Zilles on the cortex of the adult rat (Zilles, 1985). The cortical areal maps given for E19 and P0 rats are not based on direct observations but on extrapolations from the work of Zilles (1985) and Zilles and Wree (1995) on the adult rat; given the immaturity of the cortex at birth, these delineations must be considered as mere suggestions.

Diagrams for E14, E16, and E19 feature delineations of the cranium (ganglia, nerves, bones, foramina, muscles, arteries, veins). One of our most enjoyable tasks was to trace nerves and arteries as they follow lengthy courses to their target structures. The close spacing of plates in P0 at times affords dramatic views of structures. For example, Figure 197 captures the pterygopalatine artery as it passes between the crurae and the base of the stapes in the middle ear. The delineation of cranial structures was based nearly exclusively on the comprehensive work of Greene (1959). Articles by Youssef (1966, 1971) as well as by Ambach and Palkovits (1979) and Scremin (1995) were also consulted.

The delineation of the neuroepithelium of E15, E16, and E19 and the identification of structures that had not yet assumed an approximation of the mature form were made by comparisons with fate maps published by Altman and Bayer (1982, 1984, 1985a,b, 1986, 1988a,b,c; Bayer and Altman, 1995a,b).

In the first edition we misidentified two blood vessels. The errors are as follows: The caudal rhinal vein (crhv) was misidentified as the superior petrosal sinus (spets). We made this error because we followed the otherwise excellent work of Greene (1959) and we were alerted to the error by Oscar Scremin. He also informed us of the fact that, unlike in the human, the large vessel that appears at rostral levels is the olfactory artery (olfa) and not the frontopolar artery (frp).

Photography

Photographs of Nissl-stained sections were taken with a Nikon Multiphot macrophotographic apparatus on 4"×5" Kodak Technical Pan film. Agfa multigrade or grade 4 paper was used for printing.

Drawings

The drawings were entered into a Macintosh computer using Adobe Illustrator 1988 and 1990. It was thought that the drawings would be more informative if they were not stylized, and for this reason artistic license was rarely taken. When part of an atlas section was missing or severely distorted, it was often drawn in after consideration of adjacent sections.

In general, fiber tracts were outlined by solid lines and nuclei by dashed lines. Usually, the abbreviations are placed in the center of structures. The ventricles are filled in with solid black. E19 Sagittal blood vessels are colored in black and the oral and nasal cavity and vestibulocochlear system are grey.

Histology

Embryonic Day 14 Coronal (E14 Cor)

A whole female fetus was fixed by immersion in Bouin's fixative, dehydrated in ascending ethanol concentrations, cleared in histolene, and subsequently embedded in paraffin. It was cut with a rotary microtome at 10 μm in the coronal plane and every fifth section was mounted and stained with hematoxylin/celestine blue and eosin, according to the usual methods for this technique. Shrinkage from fresh dimensions was estimated to be 25% by reference to the dimensions of the frozen sectioned sagittal plates at this age. Knife angle and cutting speed were adjusted to minimize the compression effects common with paraffin material. The plates used constitute a regular series at intervals of 200 μm. Note that block orientation was adjusted after the first plate to ensure optimal symmetry of subsequent plates. It was not possible to accurately assess symmetry until the ventricles were reached.

Embryonic Day 14 Sagittal (E14 Sag)

A whole female embryo was immersed for one day in Zamboni's fixative (15% saturated aqueous picric acid, 4% paraformaldehyde in 0.1 M phosphate buffer, pH 7.4) and subsequently frozen on dry ice. It was cut at 20 μm in the sagittal plane. It was difficult to obtain sections without unacceptable distortions from this embryo. As a consequence, the plates used are simply those with tolerable distortion and do not constitute a regular progression through the entire body. Sections were stained with cresyl violet according to already published procedures (Paxinos and Watson, 1982). The Nissl-stained sections presented were examined in conjunction with sections stained for acetylcholinesterase (AChE) (Paxinos and Watson, 1982) and cytochrome oxidase (Wong-Riley, 1978).

Embryonic Day 16 Coronal (E16 Cor)

A whole female fetus was fixed by immersion in Bouin's fixative, dehydrated in ascending ethanol concentrations, cleared in histolene, and subsequently embedded in paraffin. It was cut and stained as for the E14 Coronal series. As with the E14 Coronal, knife angle and cutting speed were adjusted to minimize compression, but this type of distortion was more noticeable with this fetus. The plates used are a regular series at intervals of 100 μm. To optimize symmetry, block orientation was also adjusted once the ventricles were visible.

Embryonic Day 16 Sagittal (E16 Sag)

A whole male fetus was fixed, sectioned, and stained as for the E14 Coronal and E16 Coronal series. The left side of the fetus was sectioned first, and once the midline was reached, the orientation of the block was adjusted to ensure that subsequent sections were truly sagittal. Consequently, the plates depicted are of the right side of the fetus. While a regular series was obtained, the plates shown have been chosen to provide maximal coverage of structures close to the midline (E16 Sag 1 to 13 at intervals of 50 μm), while the last 3 plates (E16 Sag 14 to 16 at intervals of 150 μm) show structures in the lateral parts of the cerebral hemispheres and brainstem.

Embryonic Day 17 Coronal (E17 Cor)

An embryo of unknown sex was decapitated and the head postfixed for one day in Zamboni's fixative. It was then frozen on dry ice and sectioned at 16 μm. We failed to obtain sections without unacceptable distortions from this animal and as a consequence the intervals presented are simply the best available and their progression through the brain is not regular. Only diagrams are depicted for this age because our tissue displayed an unacceptable number of tears and distortions. This embryo was erroneously dated as E16 in the first edition.

Embryonic Day 19 Coronal (E19 Cor)

An embryo of unknown sex was decapitated and the head postfixed for one day in Zamboni's fixative at the end of the 18th day of gestation. It was frozen on dry ice and stored at −80°C until cutting (a few weeks later). The tissue was not blocked, but was positioned on the cryotome chuck so that when sectioned through the diencephalon it would give a plane approximating that given by coronal sections of adult tissue. The thickness at which the tissue was sectioned was not recorded; it was probably 16 μm. In each sequence ten sections were collected and four discarded. The first and the seventh sections were used (in 38 out of the 49 plates), except when these were distorted and adjacent sections were used. This spacing gives a distance between plates of 96–128 μm. There are two exceptions to this spacing. First, Figures 147–152 which cover the olfactory bulb and peduncle, are spaced much further apart. Second, plates in the caudal medulla (Figures 194 and 195) are not at regular intervals because many sections were lost when the block fell off the stage. One set of the sections in the series was stained for acetylcholinesterase (AChE) and another for cytochrome oxidase. The remaining sections were stained for Nissl and plates of the atlas depict only Nissl-stained sections.

Embryonic Day 19 Sagittal (E19 Sag)

The head of a fetus of unknown sex was fixed as for the other paraffin-embedded specimens. The specimen was sectioned at 25 µm and stained with cresyl violet. Sections are of the right side of the head. Plates depicted were chosen at intervals of 75 µm for E19 Sag 1 to 9 and 150 µm for the remaining plates.

Postnatal Day 0 (P0)

The head of a newborn female pup was obtained and frozen on dry ice without fixation. The tissue was not blocked but was positioned on the cryotome chuck in an orientation that produced a section through the diencephalon approximating that of the adult coronal plane.

The tissue was cut at 25 µm thickness and four sections were taken for staining (three for Nissl and one for AChE) while another four were discarded at each interval. The best Nissl-stained section of the available four was used as a plate and this was usually the first or the second section taken at each interval (67 out of the 72 plates). This procedure produced an average interval of 200 m. At levels rostral to Figure 68 only an occasional section was taken. At levels between Figures 128 and 145, a number of sections were lost due to a necessary adjustment to the tissue block to correct asymmetry. Three of the sets of sections were stained for Nissl and one for AChE.

References

Altman, J., and Bayer, S.A. (1982). Development of the cranial nerve ganglia and related nuclei in the rat. *Adv. Anat. Embryol. Cell Biol.* **74**, 1–90, Springer-Verlag, Berlin.

Altman, J., and Bayer, S.A. (1984). The development of the rat spinal cord. *Adv. Anat. Embryol. Cell Biol.* **85**, 1–166, Springer-Verlag, Berlin.

Altman, J., and Bayer, S.A. (1985a). Embryonic development of the rat cerebellum. II. Translocation and regional distribution of the deep neurons. *J. Comp. Neurol.* **231**, 27–41.

Altman, J., and Bayer, S.A. (1985b). Embryonic development of the rat cerebellum. III. Regional differences in the time of origin, migration, and settling of Purkinje cells. *J. Comp. Neurol.* **231**, 42–65.

Altman, J., and Bayer, S.A. (1986). The development of the rat hypothalamus. *Adv. Anat. Embryol. Cell Biol.* **100**, Springer-Verlag, Berlin.

Altman, J., and Bayer, S.A. (1988a). Development of the rat thalamus: I. Mosaic organization of the thalamic neuroepithelium. *J. Comp. Neurol.* **275**, 346–377.

Altman, J., and Bayer, S.A. (1988b). Development of the rat thalamus: II. Time and site of origin and settling pattern of neurons derived from the anterior lobule of the thalamic neuroepithelium. *J. Comp. Neurol.* **275**, 378–405.

Altman, J., and Bayer, S.A. (1988c). Development of the rat thalamus: III. Time and site of origin and settling pattern of neurons of the reticular nucleus. *J. Comp. Neurol.* **275**, 406–428.

Ambach, G., and Palkovits, M. (1979). The blood supply of the hypothalamus in the rat. In P.J. Morgane and J. Panksepp (Eds.), *Handbook of the Hypothalamus,* Marcel Dekker, New York, pp. 267–377.

Ashwell, K.W. (1987). Direct and indirect effects on the lateral geniculate nucleus neurons of prenatal exposure to methylazoxymethanol acetate. *Dev. Brain Res.* **35**, 199–214.

Ashwell, K.W.S. (1991). The distribution of microglia and cell death in the fetal rat forebrain. *Dev. Brain Res.* **58**, 1–12.

Ashwell, K.W.S. (1992). The effects of prenatal exposure to methylazoxymethanol acetate on microglia. *Neuropathol. Appl. Neurobiol.* **18**, 610–618.

Ashwell, K., and Waite, P.M.E. (1991). Cell death in the developing trigeminal nuclear complex of the rat. *Dev. Brain Res.* **63**, 291–295.

Ashwell, K., and Zhang, L.-L. (1992). The ontogeny of afferents to the fetal rat cerebellum. *Acta Anat.* **145**, 17–23.

Bayer, S.A., and Altman J. (1995a). Development: Neurogenesis and neuronal migration. In G. Paxinos (Ed.), *The Rat Nervous System,* 2nd ed., Academic Press, San Diego.

Bayer, S.A., and Altman J. (1995b). Development: Principles of neurogenesis, neuronal migration, and neuronal circuit formation. In G. Paxinos (Ed.) *The Rat Nervous System,* 2nd ed., Academic Press, San Diego.

Blandau, R.J., and Money, W.L. (1944). Observations of the rate of transport of spermatozoa in the female genital tract of the rat. *Anat. Rec.* **90**, 255–260.

Berman, A.L. (1968). *The Brainstem of the Cat: A Cytoarchitectonic Atlas with Stereotaxic Coordinates.* University of Wisconsin Press, Madison.

Cristy (1964). Developmental stages in somite and post-somite rat embryos, based on external appearance, and including some features of the macroscopic development of the oral cavity. *J. Morph.* **114**, 263.

Franklin, K., and Paxinos, G., *The Mouse Brain in Stereotaxic Coordinates,* Academic Press, San Diego, in press.

Greene, E.C. (1959). *Anatomy of the Rat.* Hafner, New York.

Mai, J.K., Assheuer, J., and Paxinos, G., *The Human Brain: Topography, Topometry, and a Myeloarchitectonic Atlas.* Manuscript in preparation.

Odor, D.L., and Blandau, R.J. (1951). Observations of fertilization and the first segmentation division in rat ova. *Am. J. Anat.* **89**, 29–63.

Paxinos, G., Ed. (1985). *The Rat Nervous System,* 2nd ed., Academic Press, San Diego.

Paxinos, G., and Huang, X.-F., *The Human Brainstem: A Cyto- and Chemoarchitectonic Atlas,* Academic Press, in press.

Paxinos, G., Huang, X.-F., and Toga, A., *The Monkey Brain in Stereotaxic Coordinates,* Academic Press, San Diego, in press.

Paxinos, G., and Watson, C. (1982). *The Rat Brain in Stereotaxic Coordinates,* Academic Press, Sydney.

Paxinos, G., and Watson, C. (1986). *The Rat Brain in Stereotaxic Coordinates,* Second Edition, Academic Press, San Diego.

Paxinos, G., Törk, I., Halliday, G., and Mehler, W.R. (1990). Human homologs to brainstem nuclei identified in other animals as revealed by acetylcholinesterase activity. In G. Paxinos (Ed.), *The Human Nervous System,* Academic Press, San Diego, pp. 149–202.

Scremin, O.U. (1995). Cerebral vascular system. In G. Paxinos (Ed.), *The Rat Nervous System,* 2nd ed., Academic Press, San Diego.

Tecott, L.H., Rubenstein, L.R., Paxinos, G., Evans, C.J., Eberwine, J.H., and Valentino, L.L. (1989). Developmental expression of proenkephalin mRNA and peptides in rat striatum. *Dev. Brain Res.* **49**, 75–86.

Waite, P.M.E., Lixin, L., and Ashwell, K.W.S. (1992). Developmental and lesion-induced cell death in the rat ventrobasal thalamus. *NeuroReport* **3**, 485–488.

Wong-Riley, M. (1978). Changes in the visual system of monocular sutured or enucleated cats demonstrable with cytochrome oxidase histochemistry. *Brain Res.* **171**, 11–28.

Youssef, E.H. (1966). The chondrocranium of the albino rat. *Acta Anat.* **64**, 586–617.

Youssef, E.H. (1971). The chondrocranium of *Hemiechinus auritus aegyptius* and its comparison with *Erinaceus europaeus. Acta Anat.* **78**, 224–254.

Zilles, K. (1985). *The Cortex of the Rat: A Stereotaxic Atlas.* Springer-Verlag, Berlin.

Zilles, K., and Wree, A. (1995). Cortex: A real and laminar structure. In G. Paxinos (Ed.), *The Rat Nervous System,* 2nd ed., Academic Press, San Diego.

List of Structures

Names of the structures are listed in alphabetical order. Each name is followed by the abbreviation of the structure.

A

A5 noradrenaline cells   A5
A7 noradrenaline cells   A7
A8 dopamine cells   A8
abducens nerve or its root   6n
abducens nucleus   6
accessory abducens nucleus   Acs6
accessory facial nucleus   Acs7
accessory olfactory bulb   AOB
accessory trigeminal nucleus   Acs5
accumbens nucleus   Acb
accumbens nucleus, core   AcbC
accumbens nucleus, shell   AcbSh
adrenal gland   Adr
agranular insular cortex   AI
alar orbital bone   AOrb
alisphenoid bone   ASph
alisphenoid foramen   ASphF
allantois   Al
alveus of the hippocampus   alv
ambiguus nucleus   Amb
ampulla of anterior semicircular duct   AntA
ampulla of horizontal semicircular duct   HorA
ampulla of posterior semicircular duct   PostA
amygdala   Amg
amygdalohippocampal area   AHi
amygdaloid fissure   AF
amygdaloid neuroepithelium   amg
amygdalopiriform transition area   APir
amygdalostriatal transition area   AStr
angular thalamic nucleus   Ang
ansa lenticularis   al
anterior amygdaloid area   AA
anterior amygdaloid area, dorsal part   AAD
anterior amygdaloid area, ventral part   AAV
anterior cerebellum   ACb
anterior cerebral artery   acer
anterior commissural nucleus   AC
anterior commissure   ac
anterior commissure, anterior part   aca
anterior commissure, posterior part   acp
anterior cortical amygdaloid nucleus   ACo
anterior facial vein   afv
anterior hypothalamic area, anterior part   AHA
anterior hypothalamic area, posterior part   AHP
anterior hypothalamic neuroepithelium   ah
anterior hypothalamic nucleus   AH
anterior inferior cerebellar artery   aica
anterior lacerated foramen (foramen orbital rotundum)   ALF
anterior lobe of the pituitary   APit
anterior medial preoptic nucleus   AMPO
anterior olfactory nucleus   AO
anterior olfactory nucleus, dorsal part   AOD
anterior olfactory nucleus, lateral part   AOL
anterior olfactory nucleus, medial part   AOM
anterior olfactory nucleus, posterior part   AOP
anterior pituitary anlage   apit
anterior pretectal nucleus   APT
anterior semicircular duct   Ant
anterior spinal artery   aspina
anterior tegmental nucleus   ATg
anterior thalamic neuroepithelium   ath
anterior thalamus   ATh
anterior transitional promontory   ATP
anterobasal nucleus   AB
anterodorsal thalamic nucleus   AD
anteromedial thalamic nucleus   AM
anteroventral thalamic nucleus   AV
aorta   aorta

aortic arch   aa
aortic valva   aortval
apex of the cochlea   Apex
aqueduct (Sylvius)   Aq
arcuate eminence of the petrosal part of the temporal bone   ArcE
arcuate hypothalamic nucleu   Arc
arcuate nucleus neuroepithelium   arc
arytenoid swelling of the larynx   Ary
ascending facial nerve   asc7n
atlas (C1 vertebra)   Atlas
atlas (C1 vertebra)   Atlas
auditory neuroepithelium   aud
auriculotemporal nerve   aute
axis (C2 vertebra)   Axis

B

Barrington's nucleus   Bar
basal nucleus of Meynert   B
basal plate neuroepithelium   bp
basal telencephalic neuroepithelium   btel
basal telencephalic plate, anterior part   bta
basal telencephalic plate, intermediate part   bti
basal telencephalic plate, posterior part   btp
basal telencephalon   BTel
basal vein   basv
base of stapes   BStapes
basilar artery   bas
basioccipital bone   BOcc
basisphenoid bone   BSph
basocochlear fissure   BCF
basolateral amygdaloid nucleus   BL
basolateral amygdaloid nucleus, anterior part   BLA
basolateral amygdaloid nucleus, ventral part   BLV
basomedial amygdaloid nucleus   BM
basomedial amygdaloid nucleus, anterior part   BMA
bed nucleus of the accessory olfactory tract   BAOT
bed nucleus of the anterior commissure   BAC
bed nucleus of the stria terminalis   BST
bed nucleus of the stria terminalis, intraamygdaloid division   BSTIA
bed nucleus of the stria terminalis, lateral division   BSTL
bed nucleus of the stria terminalis, lateral division, juxtacapsular part   BSTLJ
bed nucleus of the stria terminalis, lateral division, posterior part   BSTLP
bed nucleus of the stria terminalis, lateral division, ventral part   BSTLV
bed nucleus of the stria terminalis, medial division, anterior part   BSTMA
bed nucleus of the stria terminalis, medial division, posterior part   BSTMP
bed nucleus of the stria terminalis, medial division, posterointermediate part   BSTMPI
bed nucleus of the stria terminalis, medial division, posterolateral part   BSTMPL
bed nucleus of the stria terminalis, medial division, posteromedial part   BSTMPM
bed nucleus of the stria terminalis, ventral division   BSTV
bile duct   biled
blood vessel   bv
brachial plexus   bplex
brachiocephalic trunk   brctr
brachiocephalic vein   brcv
brachium of the inferior colliculus   bic
brachium of the superior colliculus   bsc
buccal nerve   buccn

C

C1 adrenaline cells   C1
CA1 field of the hippocampus   CA1
CA2 field of the hippocampus   CA2
CA3 field of the hippocampus   CA3
canal of the facial nerve   7canal
cardiac ganglion   Card
cardiac plexus   cardpx
carotid body   CtdB
carotid canal   Ctd
carotid plexus   cplex
caudal linear nucleus of the raphe   CLi
caudal rhinal vein   crhv
caudate putamen (striatum)   CPu
caudate putamen neuroepithelium   cpu
caudoventrolateral reticular nucleus   CVL
cavernous sinus   cav
celiac artery   cela

N

nasal cavity   Nasal
nasal epithelium   NasalE
nasal septum   NSpt
nasal turbinate   NasT
nasopalatine nerve   npal
nasopharyngeal cavity   NasoPhar
nerve   n
nerve of the pterygoid canal   ptgcn
nerve of vomeronasal organ   vno
neuroepithelium   ne
notochord   noto
nuclear transitory zone   ntz
nuclei of the diagonal band   DB
nuclei of the lateral lemniscus   LL
nucleus of Darkschewitsch   Dk
nucleus of Roller   Ro
nucleus of the ansa lenticularis   AL
nucleus of the brachium of the inferior colliculus   BIC
nucleus of the fields of Forel   F
nucleus of the horizontal limb of the diagonal band   HDB
nucleus of the lateral olfactory tract   LOT
nucleus of the lateral olfactory tract, layer 1   LOT1
nucleus of the lateral olfactory tract, layer 2   LOT2
nucleus of the lateral olfactory tract, layer 3   LOT3
nucleus of the optic tract   OT
nucleus of the posterior commissure   PCom
nucleus of the solitary tract   Sol
nucleus of the stria medularis   SM
nucleus of the trapezoid body   Tz
nucleus of the vertical limb of the diagonal band   VDB

O

occipital artery   occ
occipital bone   Occ
occipital cortex   Oc
occipital sinus   occs
oculomotor nerve or its root   3n
oculomotor nucleus   3
oculomotor nucleus, parvocellular part   3PC
olfactory artery   olfa
olfactory bulb   OB
olfactory bulb neuroepithelium   obn
olfactory epithelium   olfepith
olfactory nerve   olf
olfactory nerve layer   ON
olfactory tubercle   Tu
olfactory ventricle (olfactory part of lateral ventricle)   OV
olivary pretectal nucleus   OPT
olivocerebellar tract   oc
omental bursa   Omen
ophthalmic artery   opha
ophthalmic nerve of trigeminal   5ophth
ophthalmic vein   ophv
optic chiasm   ox
optic fiber layer   OF
optic foramen   OptF
optic nerve   2n
optic nerve layer of the superior colliculus   Op
optic recess of third ventricle   OptRe
optic stalk   os
optic tract   opt
oral cavity   Oral
orbital artery   orb
orbital cavity   Orbit
orbital cortex   Orb
organ of Corti   Corti
oriens layer of the hippocampus   Or
otic ganglion   Otic
otic vesicle   Oticv
oval foramen   Oval
oval paracentral thalamic nucleus   OPC
oval window   OvalW

P

palate   Palate

palatine artery   pala
palatine bone   Pal
palatine nerve   pal
pancreas   Panc
parabigeminal nucleus   PBG
parabrachial nuclei   PB
parabrachial pigmented nucleus   PBP
paracentral thalamic nucleus   PC
paracollicular tegmentum   PCTg
parafascicular thalamic nucleus   PF
paraflocular cavity (subarcuate fossa)   PFlCv
paraflocculus   PFl
paraflocculus, dorsal part   PFlD
paraflocculus, ventral part   PFlV
paralemniscal nucleus   PL
paramedian raphe nucleus   PMR
paramesonephric duct   pmesd
paranigral nucleus   PN
parapyramidal reticular nucleus   PPy
parastrial nucleus   PS
parasubiculum   PaS
paratenial thalamic nucleus   PT
paratrigeminal nucleus   Pa5
paratrochlear nucleus   Pa4
paraventricular hypothalamic nucleus   Pa
paraventricular hypothalamic nucleus, anterior parvocellular part   PaAP
paraventricular hypothalamic nucleus, posterior part   PaPo
paraventricular thalamic nucleus   PV
paraventricular thalamic nucleus, anterior part   PVA
paraventricular thalamic nucleus, posterior part   PVP
parietal bone   Parietal
parietal cortex, area 1   Par1
parietal cortex, area 2   Par2
parietal plate   ParietalP
parotid gland   Par
parvocellular reticular nucleus   PCRt
parvocellular reticular nucleus, alpha part   PCRtA
pedunculopontine tegmental nucleus   PPTg
periaqueductal grey neuroepithelium   pag
pericardium   PCard
perifacial zone   P7
perifornical nucleus   PeF
perilymph   Perilymph
peripeduncular nucleus   PP
perirhinal cortex   PRh
peritoneal cavity   Periton
peritrigeminal zone   P5
periventricular hypothalamic nucleus   Pe
petrous part of the temporal bone   Petrous
phrenic nerve   phrn
pigment epithelium of retina   Pig
pineal gland   Pi
pineal recess of the third ventricle   PiRe
piriform cortex   Pir
platysma muscle   Platysma
pleural cavity   Pleural
polymorph layer of the dentate gyrus   PoDG
pons   Pons
pontine flexure   PnF
pontine migration (corpus pontobulbare)   pnm
pontine nuclei   Pn
pontine nuclei neuroepithelium   pn
pontine reticular nucleus, caudal part   PnC
pontine reticular nucleus, oral part   PnO
portal vein   portv
posterior cerebellum   PCb
posterior cerebral artery   pcer
posterior commissure   pc
posterior communicating artery   pcoma
posterior cortical amygdaloid nucleus   PCo
posterior cupula of the nasal capsule   PCNC
posterior dorsal recess of nasal cavity   PDR
posterior hypothalamic area   PH
posterior hypothalamic neuroepithelium   ph
posterior hypothalamic neuroepithelium, dorsal part   phd
posterior hypothalamic neuroepithelium, ventral part   phv
posterior inferior cerebellar artery   pica
posterior intralaminar thalamic nucleus   PIL
posterior isthmal recess   PIs
posterior limitans thalamic nucleus   PLi
posterior lobe of the pituitary   PPit

posterior pituitary   ppit
posterior pretectal nucleus   PPT
posterior semicircular duct   Post
posterior superior alveolar artery   psa
posterior thalamic neuroepithelium   pth
posterior thalamic nuclear group   Po
posterior thalamic nuclear group, triangular part   PoT
posterior thalamus   PTh
postermedian thalamic nucleus   PoMn
postero-orbital follicle   pof
posterodorsal tegmental nucleus   PDTg
posterolateral cortical amygdaloid nucleus (C2)   PLCo
posteromedial cortical amygdaloid nucleus (C3)   PMCo
precerebellar nuclei neuroepithelium   pcb
precommissural nucleus   PrC
predorsal bundle   pd
premammillary nucleus, dorsal part   PMD
premammillary nucleus, ventral part   PMV
premaxilla   Premaxilla
preoptic area neuroepithelium   poa
preoptic recess of the third ventricle   P3V
prepositus hypoglossal nucleus   PrH
prerubral field   PR
presphenoid bone   PSph
presphenoid wing   PSphW
presubiculum   PrS
pretectum   PTec
pretectum neuroepithelium   ptec
primordia of ear ossicles   audos
principal sensory trigeminal nucleus   Pr5
principal sensory trigeminal nucleus, dorsomedial part   Pr5DM
principal sensory trigeminal nucleus, ventrolateral part   Pr5VL
pterygoid bone   Ptg
pterygoid canal   PtgC
pterygopalatine artery   ptgpal
pubis   Pubis
pulmonary trunk   pult
pulmonary vein   pulv
Purkinje cell layer (cerebellum)   Pk
putamen   Pu
pyramidal cell layer of the hippocampus   Py
pyramidal decussation   pyx
pyramidal tract   py

R

raphe magnus nucleus   RMg
raphe obscurus nucleus   ROb
raphe pallidus (postpyramidal raphe) nucleus   RPa
Rathke's pouch   Rathke
recess of the inferior colliculus   ReIC
recurrent laryngeal nerve   rln
red nucleus   R
red nucleus, magnocellular part   RMC
red nucleus, parvocellular part   RPC
Reichert's cartilage   Reichert
reservoir of migrating cells   Res
respiratory epithelium   respepith
reticular protuberance   rp
reticular thalamic nucleus   Rt
reticulotegmental nucleus of the pons   RtTg
retina   Retina
retinal ganglion cell layer   RGn
retroambiguus nucleus   RAmb
retrochiasmatic area   RCh
retroethmoid nucleus   REth
retrolemniscal nucleus   RL
retrorubral field   RRF
retrorubral nucleus   RR
retrosplenial agranular cortex   RSA
retrosplenial granular cortex   RSG
reuniens thalamic nucleus   Re
rhinal fissure   RF
rhomboid thalamic nucleus   Rh
rib   Rib
rib 1   Rib1
rib 2   Rib2
right atrium   RAtr
right lobe of liver   RLiver

right superior vena cava   rsvc
right ventricle   RVent
roof plate of spinal cord neuroepithelium   rfp
rostral interstitial nucleus of medial longitudinal fasciculus   RI
rostral linear nucleus of the raphe   RLi
rostral periolivary region   RPO
rostroventrolateral reticular nucleus   RVL
round window   RoundW
rubrospinal tract   rs

S

saccular macula   SMac
saccule   Sacc
sagulum nucleus   Sag
scaphoid thalamic nucleus   Sc
scapula   Scap
second pharyngeal arch   Arch2
semilunar valves   SLun
sensory root of the trigeminal nerve   s5
septal neuroepithelium   spt
septofimbrial nucleus   SFi
septohippocampal nucleus   SHi
septohypothalamic nucleus   SHy
septum   Spt
sinoauricular valve   sinoaur
solitary tract   sol
sphenoid nucleus   Sph
sphenopalatine artery   spa
sphenopalatine ganglion   SphPal
spinal accessory nerve   11n
spinal cord   Spinal
spinal nerve   spn
spinal trigeminal nucleus   Sp5
spinal trigeminal nucleus, caudal part   Sp5C
spinal trigeminal nucleus, interpolar part   Sp5I
spinal trigeminal nucleus, oral part   Sp5O
spinal trigeminal tract   sp5
spinal vestibular nucleus   SpVe
squamous part of the temporal bone   Squamous
stapedius muscle   Stapedius
stapes (ossicle)   Stapes
stellate ganglion   StGn
sternomastoid muscle   StM
sternum   Sternum
stigmoid hypothalamic nucleus   Stg
stomach   Stom
straight sinus   sts
stratum radiatum of the hippocampus   Rad
stria medullaris of the thalamus   sm
stria terminalis   st
strial part of the preoptic area   StA
striohypothalamic nucleus   StHy
stylomastoid foramen   StyMF
subbrachial nucleus   SubB
subclavian artery   subcla
subclavian vein   subclv
subcoeruleus nucleus   SubC
subcommissural organ   SCO
subcommissural organ neuroepithelium   sco
subfornical organ   SFO
subgeniculate nucleus   SubG
subiculum   S
subincertal nucleus   SubI
sublingual gland   SLG
submammillothalamic nucleus   SMT
subparafascicular thalamic nucleus   SPF
subparafascicular thalamic nucleus, parvocellular part   SPFPC
subpeduncular tegmental nucleus   SPTg
substantia innominata   SI
substantia nigra   SN
substantia nigra, compact part   SNC
substantia nigra, lateral part   SNL
substantia nigra, reticular part   SNR
substriatal area   SStr
subthalamic nucleus   STh
subventricular cortical layer   SubV
sulcus limitans   sl
superficial gray layer of the superior colliculus   SuG

superficial plate neuroepithelium   sp
superficial temporal vein   stempv
superior cerebellar artery   scba
superior cerebellar peduncle (brachium conjunctivum)   scp
superior cerebellar peduncle, descending limb   scpd
superior cervical ganglion   SCGn
superior colliculus   SC
superior colliculus neuroepithelium   sc
superior glossopharyngeal ganglion   S9Gn
superior mesenteric artery   smes
superior mesenteric ganglion   SMes
superior neuroepithelial lobule of diencephalon   snl
superior oblique muscle   SOb
superior olive   SOl
superior ophthalmic artery   sopha
superior paraolivary nucleus   SPO
superior rectus muscle   SRec
superior sagittal sinus   sss
superior thalamic radiation   str
superior vagal (jugular) ganglion   S10Gn
superior vena cava   svc
superior vestibular nucleus   SuVe
suprachiasmatic nucleus   SCh
supramammillary decussation   sumx
supramammillary nucleus   SuM
supraoculomotor central gray   Su3
supraoptic decussation   sox
supraoptic nucleus   SO
supraoptic nucleus, retrochiasmatic (diffuse) part   SOR
supraspinal nucleus   SSp
supratrigeminal nucleus   Su5
suspensory ligament of lens   SuspLig
sympathetic plexus   splex
sympathetic trunk   Symp
synovial cavity   Syn

T

tail   Tail
tectospinal tract   ts
tegmentum   Tg
telencephalon   Telen
temporal cortex, area 1 (primary auditory cortex)   Te1
temporal cortex, area 2   Te2
temporal cortex, area 3   Te3
temporal muscle   Temp
tenia tecta   TT
tensor tympani muscle   TensT
terete hypothalamic nucleus   Te
terminal nerve   term
thalamic neuroepithelium   th
thalamus   Th
third ventricle   3V
thymus gland   Thym
thyroid cartilage   ThyrCart
thyroid gland   Thyr
tongue   Tong
trachea   Trachea
transverse fibers of the pons   tfp
transverse sinus   trs
transverse sinus of the pericardium   trspc
trapezoid body   tz
trigeminal ganglion   5Gn
trigeminal nerve   5n
trochlear nerve or its root   4n
trochlear nucleus   4
tuberomammillary nucleus   TM
tympanic branch of glossopharyngeal nerve   ty9
tympanic bulla   TyBu
tympanic cavity   TyC
tympanic membrane   TyM

U

umbilical artery   umba
umbilical vein   umbv

ureter   ureter
urethra   urethra
urinary bladder   Blad
urogenital mesentery   urgmes
urogenital tract   urgen
utricle   Utr
utricular macula   UMac

V

vagal ganglion   10Gn
vagus nerve   10n
vascular organ of the lamina terminalis   VOLT
vein (unidentified)   v
ventral amygdalofugal pathway   vaf
ventral cochlear nucleus   VC
ventral cochlear nucleus, anterior part   VCA
ventral cochlear nucleus, posterior part   VCP
ventral commissure of spinal cord   vc
ventral funiculus   vf
ventral funiculus of the spinal cord   vfu
ventral hippocampal commissure   vhc
ventral horn   VH
ventral lateral geniculate nucleus   VLG
ventral lateral geniculate nucleus, magnocellular part   VLGMC
ventral lateral geniculate nucleus, parvocellular part   VLGPC
ventral median fissure of the spinal cord   VMnF
ventral neuroepithelial lobule of diencephalon   vnl
ventral nucleus of the lateral lemniscus   VLL
ventral pallidum   VP
ventral periolivary nuclei   VPO
ventral posterolateral thalamic nucleus   VPL
ventral posteromedial thalamic nucleus   VPM
ventral reuniens thalamic nucleus   VRe
ventral root   vr
ventral spinocerebellar tract   vsc
ventral tegmental area (Tsai)   VTA
ventral tegmental area, rostral part   VTAR
ventral tegmental nucleus (Gudden)   VTg
ventral third ventricle   V3V
ventricular zone of the retina   Vent
ventrolateral thalamic nucleus   VL
ventromedial hypothalamic nucleus   VMH
ventromedial hypothalamic nucleus, central part   VMHC
ventromedial hypothalamic nucleus, dorsomedial part   VMHDM
ventromedial hypothalamic nucleus, ventrolateral part   VMHVL
ventromedial thalamic nucleus   VM
vertebra   Vert
vertebral artery   vert
vestibular area neuroepithelium   ve
vestibular ganglion   VeGn
vestibular root of the vestibulocochlear nerve   8vn
vestibulocochlear nerve   8n
vitreous of the eye   Vitr
vomer   Vomer
vomeronasal cavity   VNC
vomeronasal organ   VNO

X

xiphoid thalamic nucleus   Xi

Z

zona incerta   ZI
zona incerta, dorsal part   ZID
zona incerta, ventral part   ZIV
zonal layer of the superior colliculus   Zo
zygomatic arch   Zyg

Index of Abbreviations

The abbreviations are listed in alphabetical order. Each abbreviation is followed by the structure name and the numbers of the figures on which the abbreviation appears.

xx

E14 Coronal Section Plan

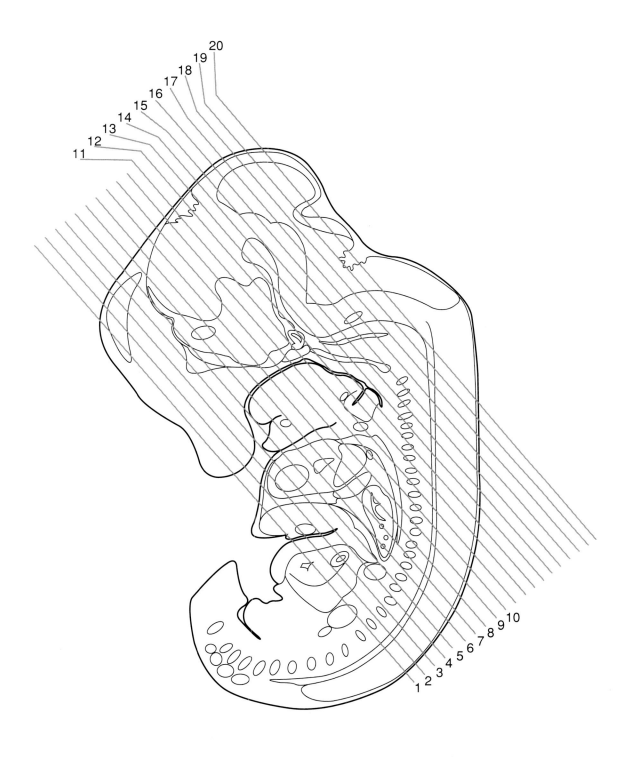

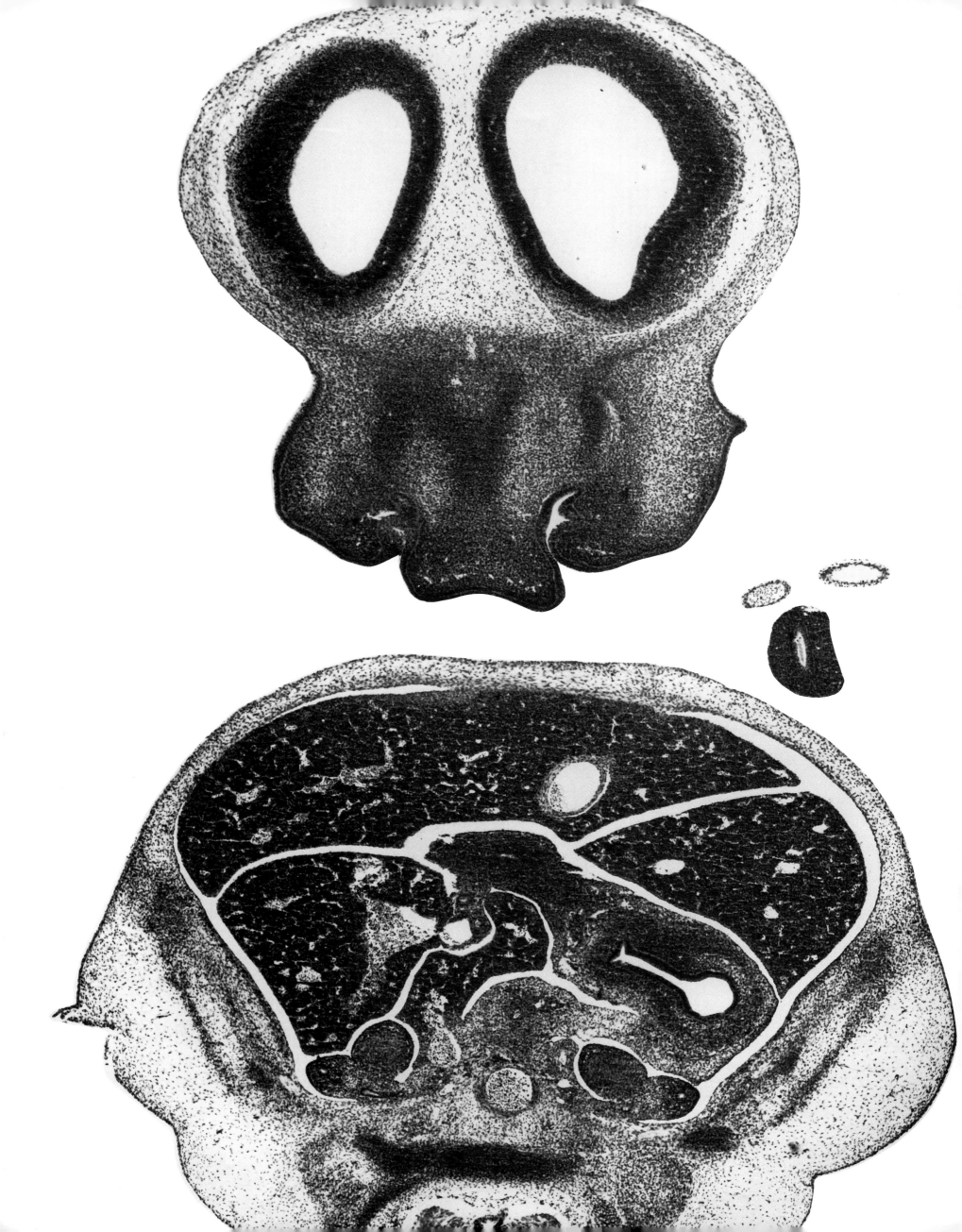

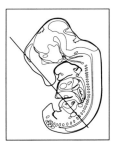

Figure 1
E14 Coronal 1

10n   vagus nerve
aorta   aorta
biled   bile duct
bta   basal telencephalic plate, anterior
cela   celiac artery
Celiac   celiac ganglion
cx   cortical neuroepi
DMes   dorsal mesentery
DRG   dorsal root ganglion
dven   ductus venosus
Gon   gonad
icn   intercostal nerve
IntZ   intermediate zone of spinal gray
LLiver   left lobe of liver
loment   lesser omentum
LV   lateral ventricle
MesN   mesonephros
midgut   midgut in physiological herniation
MLiver   middle lobe of liver
Nasal   nasal cavity
ne   neuroepithelium
noto   notochord
obn   olfactory bulb neuroepi
Omen   omental bursa
pmesd   paramesonephric duct
RLiver   right lobe of liver
sss   superior sagittal sinus
Stom   stomach
Symp   sympathetic trunk
Telen   telencephalon
umba   umbilical artery
umbv   umbilical vein
urgmes   urogenital mesentery
VH   ventral horn
Vomer   vomer

sss
cx
Telen
bta   LV
obn

Vomer

Nasal

umbv   umba
midgut

MLiver   dven
loment
biled   LLiver
RLiver   10n
Omen   Stom
cela   DMes
pmesd   Gon   Celiac
MesN   aorta
icn   Symp
noto   urgmes
VH   DRG
IntZ
ne

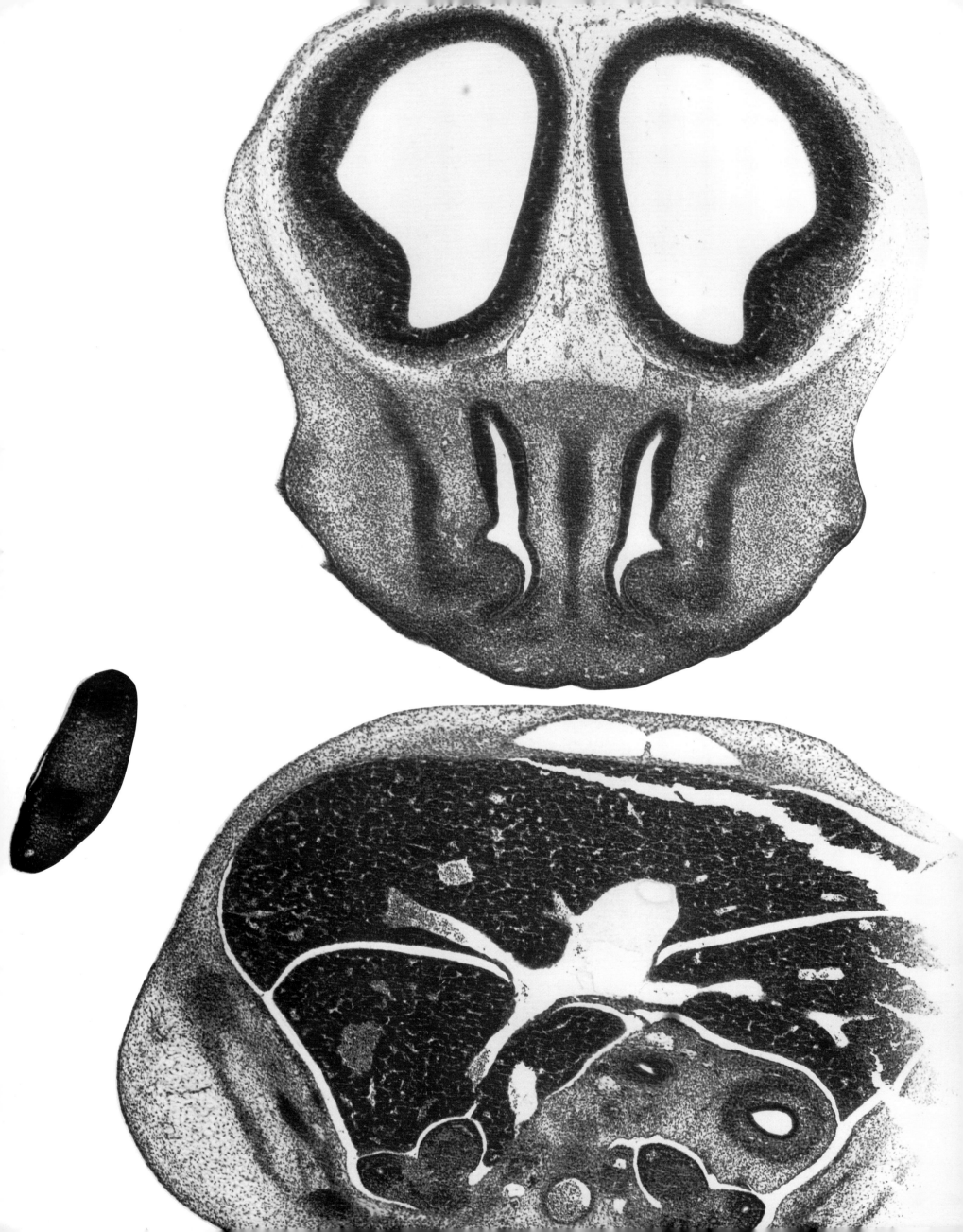

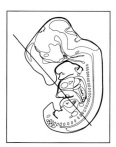

Figure 2
E14 Coronal 2

10n   vagus nerve
acer   anterior cerebral artery
Adr   adrenal gland
aorta   aorta
bta   basal telencephalic plate, anterior
BTel   basal telencephalon
cx   cortical neuroepi
DRG   dorsal root ganglion
dven   ductus venosus
FLimb   forelimb
Gon   gonad
gsac   greater sac of abdominal cavity
icn   intercostal nerve
LLiver   left lobe of liver
LV   lateral ventricle
Maxilla   maxilla
MaxT   maxilloturbinate
mcer   middle cerebral artery
MesN   mesonephros
MLiver   middle lobe of liver
Nasal   nasal cavity
noto   notochord
OB   olfactory bulb
olf   olfactory nerve
olfepith   olfactory epithelium
Omen   omental bursa
PCard   pericardium
PDR   posterior dorsal recess of nasal cavity
pmesd   paramesonephric duct
Premaxilla   premaxilla
respepith   respiratory epithelium
RLiver   right lobe of liver
sss   superior sagittal sinus
Stom   stomach
Symp   sympathetic trunk
Telen   telencephalon
term   terminal nerve
urgmes   urogenital mesentery
VH   ventral horn
Vomer   vomer

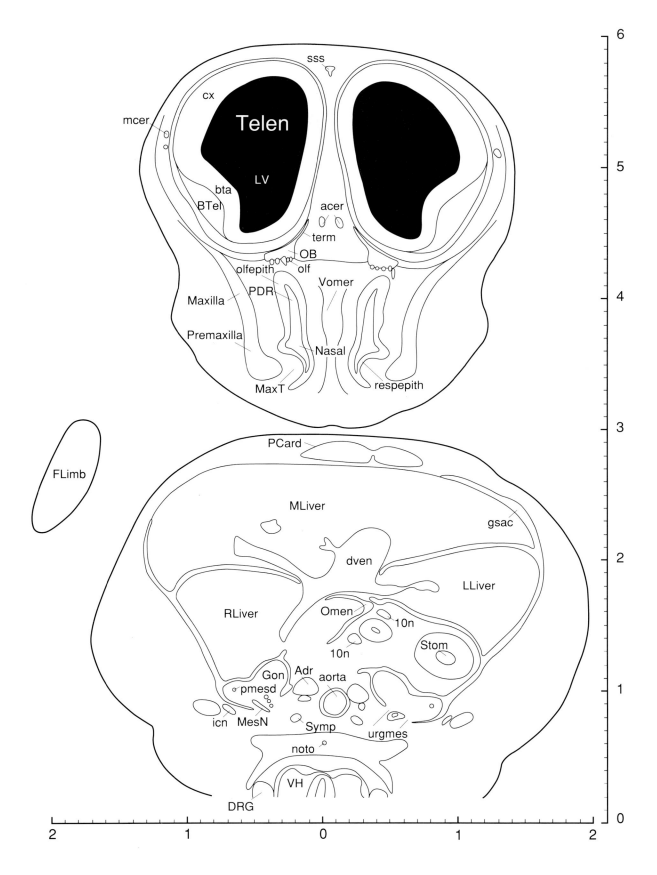

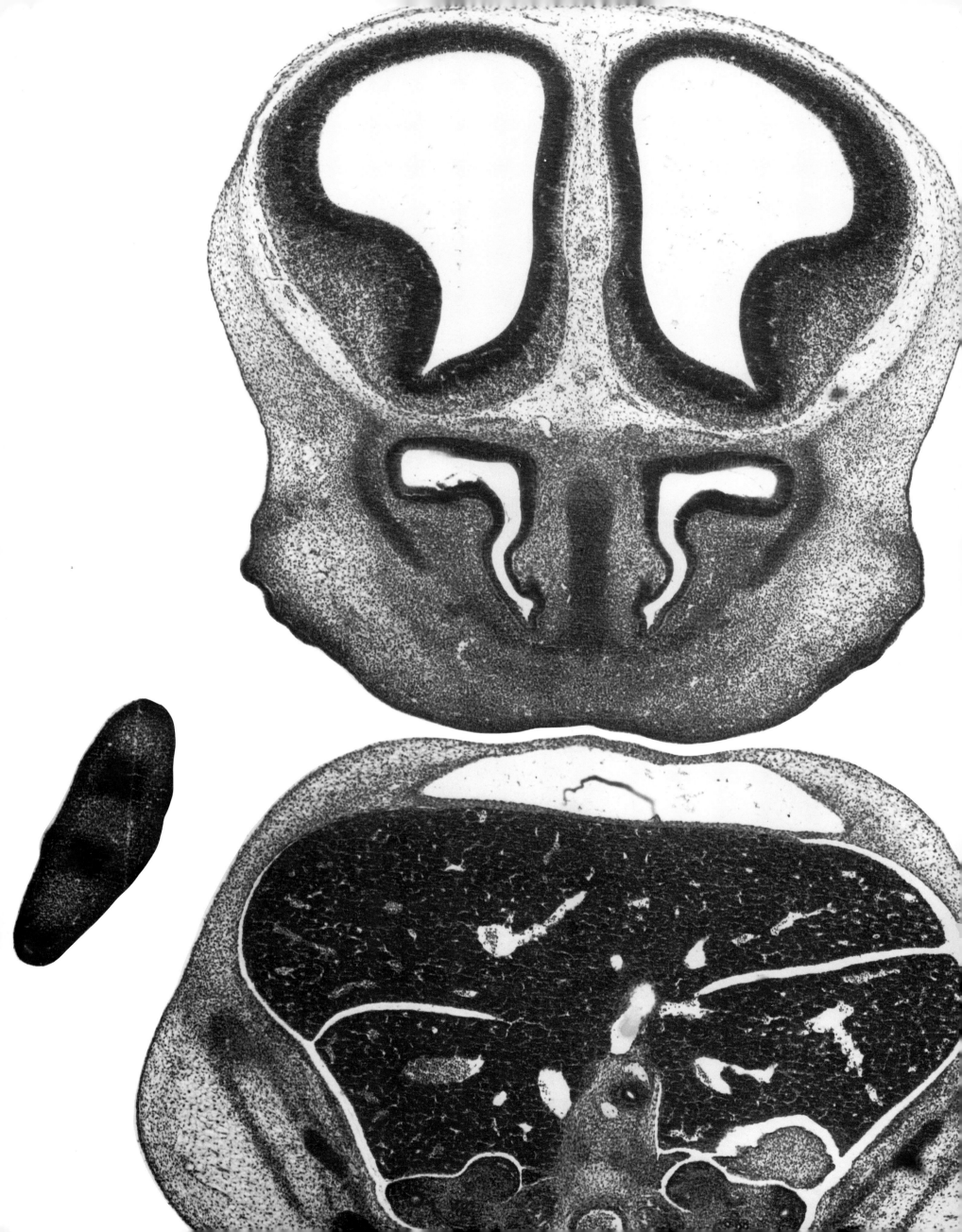

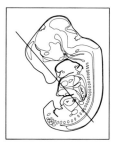

Figure 3
E14 Coronal 3

5max   maxillary nerve
10n   vagus nerve
acer   anterior cerebral artery
Adr   adrenal gland
aorta   aorta
BTel   Basal telencephalon
bta   basal telencephalic plate, anterior
cx   cortical neuroepi
Diaph   diaphragm
Ecto1   ectoturbinate 1
EndoII   endoturbinate II
Eso   esophagus
FLimb   forelimb
Gon   gonad
hi   hippocampal formation neuroepi
icn   intercostal nerve
inf   infraorbital branch
LLiver   left lobe of liver
LV   lateral ventricle
mcer   middle cerebral artery
MesN   mesonephros
MLiver   middle lobe of liver
MS   medial septal nucleus
olf   olfactory nerve
olfa   olfactory artery
PCard   pericardium
PDR   posterior dorsal recess of nasal cavity
pmesd   paramesonephric duct
Rib   rib
RLiver   right lobe of liver
spt   septal neuroepi
Stom   stomach
Telen   Telencephalon
v   vein (unidentified)
Vomer   vomer

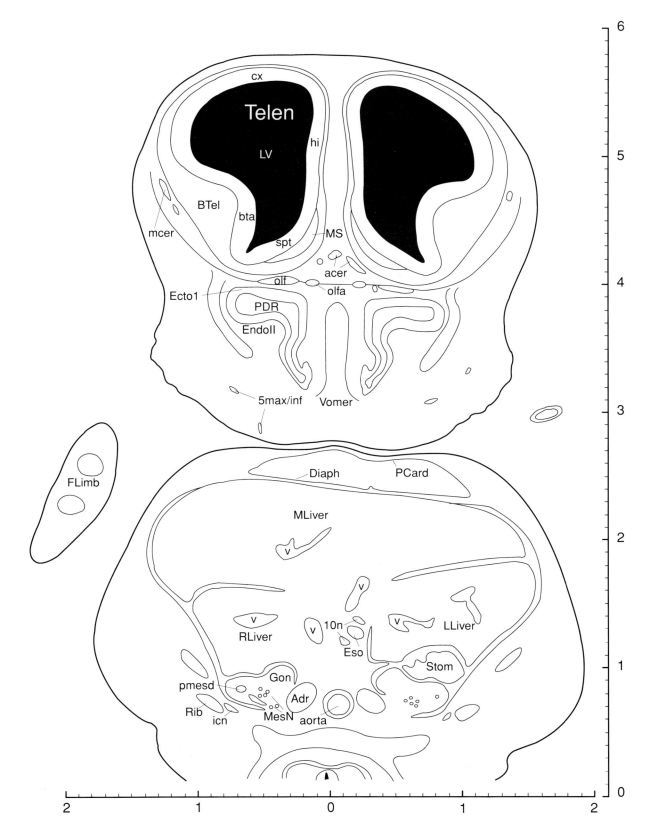

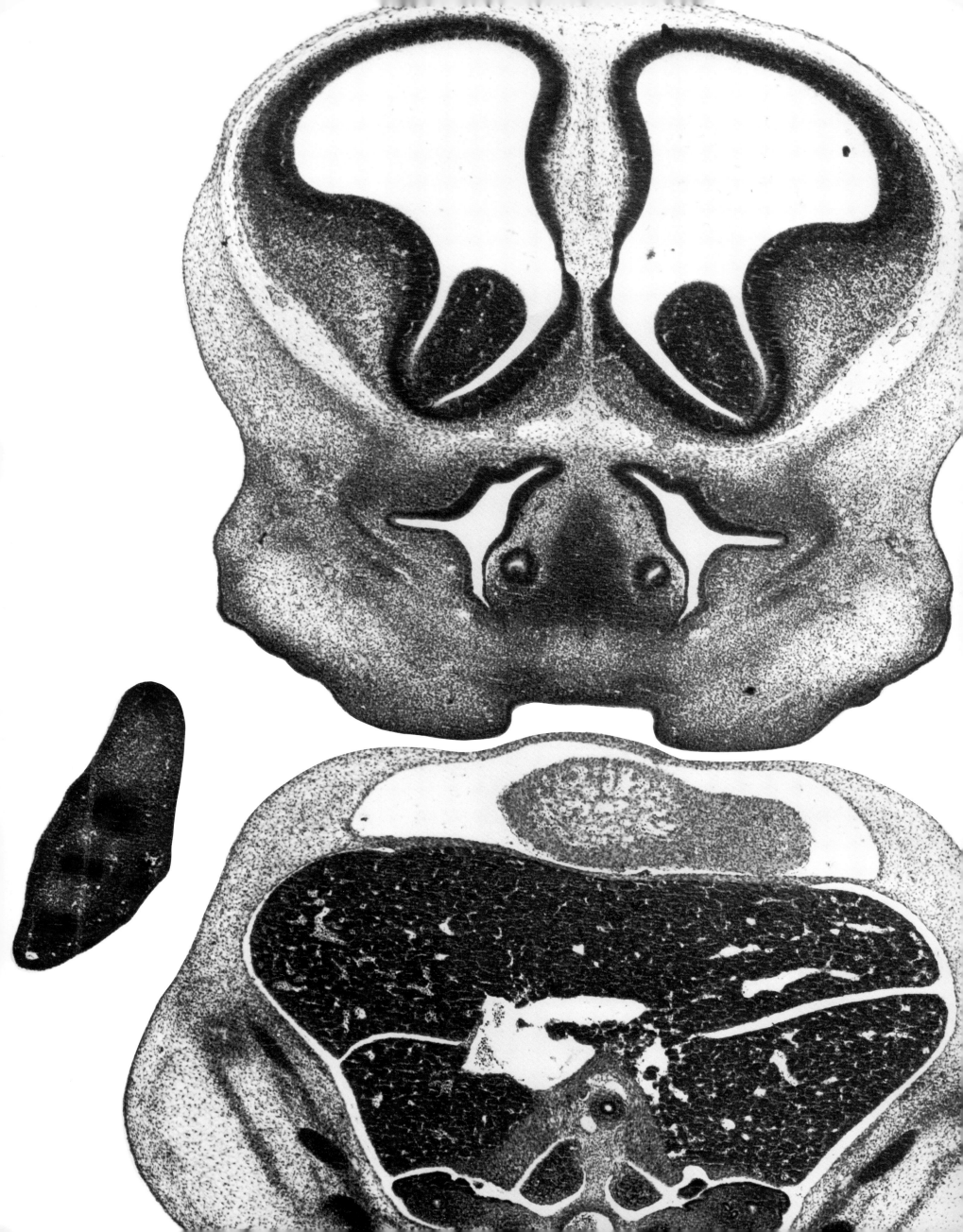

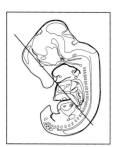

Figure 4
E14 Coronal 4

5max   maxillary nerve
10n   vagus nerve
acer   anterior cerebral artery
Adr   adrenal gland
aorta   aorta
bta   basal telencephalic plate, anterior
BTel   basal telencephalon
bti   basal telen plate, intermediate
cx   cortical neuroepi
Diaph   diaphragm
Ecto1   ectoturbinate 1
Ecto2,2'   ectoturbinates 2, 2'
EndoII,II'   endoturbinates II, II'
Eso   esophagus
FLimb   forelimb
hi   hippocampal formation neuroepi
ibf   interbasal plate fissure
icn   intercostal nerve
inf   infraorbital branch
ivc   inferior vena cava
LLiver   left lobe of liver
LV   lateral ventricle
mcer   middle cerebral artery
MesN   mesonephros
MLiver   middle lobe of liver
MS   medial septal nucleus
olf   olfactory nerve
olfa   olfactory artery
PCard   pericardium
PDR   posterior dorsal recess of nasal cavity
pmesd   paramesonephric duct
RLiver   right lobe of liver
RVent   right ventricle
Spt   septum
spt   septal neuroepi
sss   superior sagittal sinus
Symp   sympathetic trunk
Telen   telencephalon
VNO   vomeronasal organ
vno   nerve of vomeronasal organ
Vomer   vomer

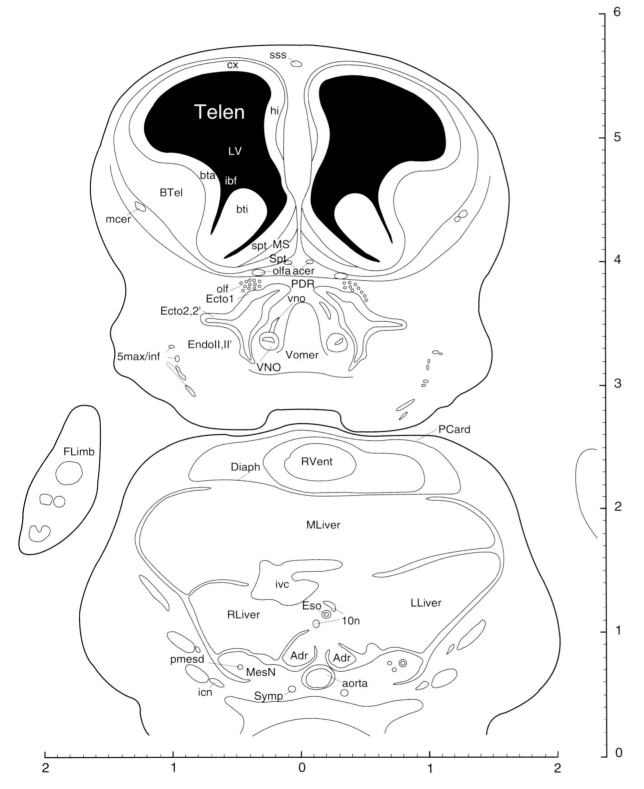

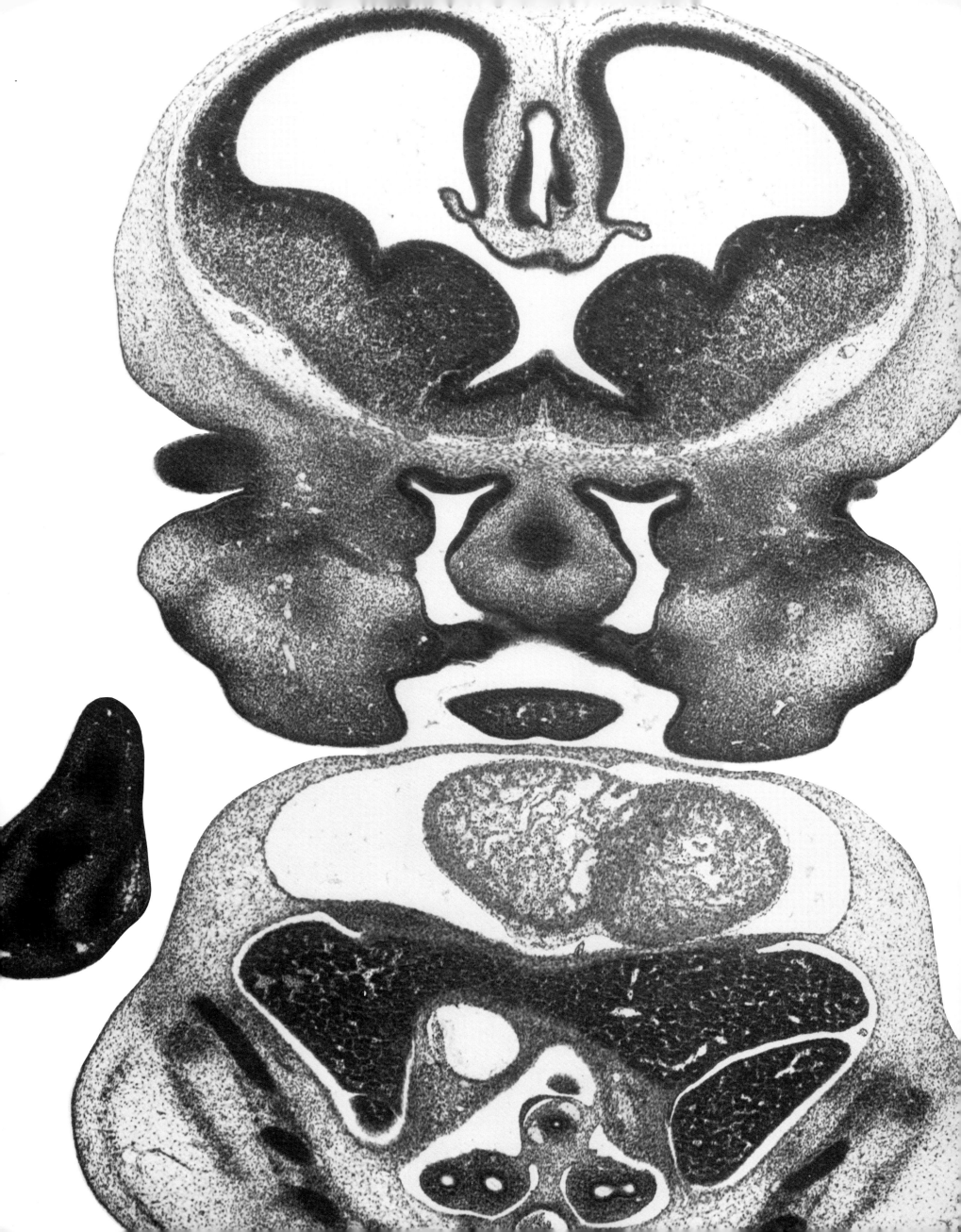

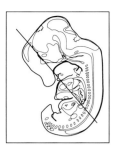

Figure 5
E14 Coronal 5

3V   third ventricle
5max   maxillary nerve
10n   vagus nerve
acer   anterior cerebral artery
aorta   aorta
bta   basal telencephalic plate, anterior
BTel   basal telencephalon
bti   basal telen plate, intermediate
chp   choroid plexus premordium
cx   cortical neuroepi
DB   nuclei of the diagonal band
Diaph   diaphragm
Ecto1   ectoturbinate 1
Ecto2,2'   ectoturbinates 2, 2'
Eso   esophagus
FLimb   forelimb
hi   hippocampal formation neuroepi
ibf   interbasal plate fissure
icn   intercostal nerve
inf   infraorbital branch
ivc   inferior vena cava
IVF   interventricular foramen
Jaw   jaw
LLiver   left lobe of liver
LTer   lamina terminalis
Lung   lung
LV   lateral ventricle
LVent   left ventricle
mcer   middle cerebral artery
MLiver   middle lobe of liver
Nasal   nasal cavity
PCard   pericardium
phrn   phrenic nerve
Pi   pineal gland
RLiver   right lobe of liver
RVent   right ventricle
snl   superior neuroepith lobule of dien
spt   septal neuroepi
sss   superior sagittal sinus
Symp   sympathetic trunk
Telen   telencephalon

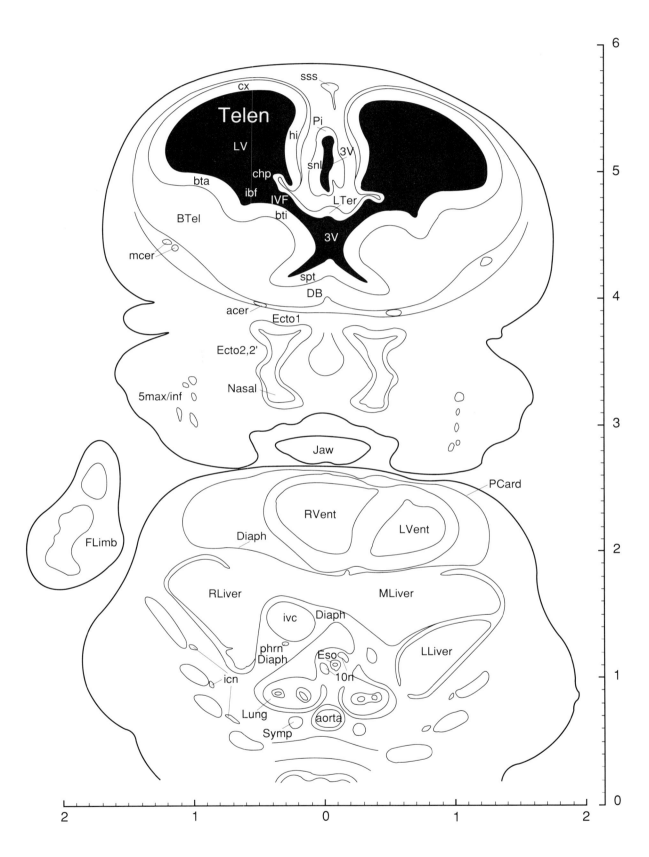

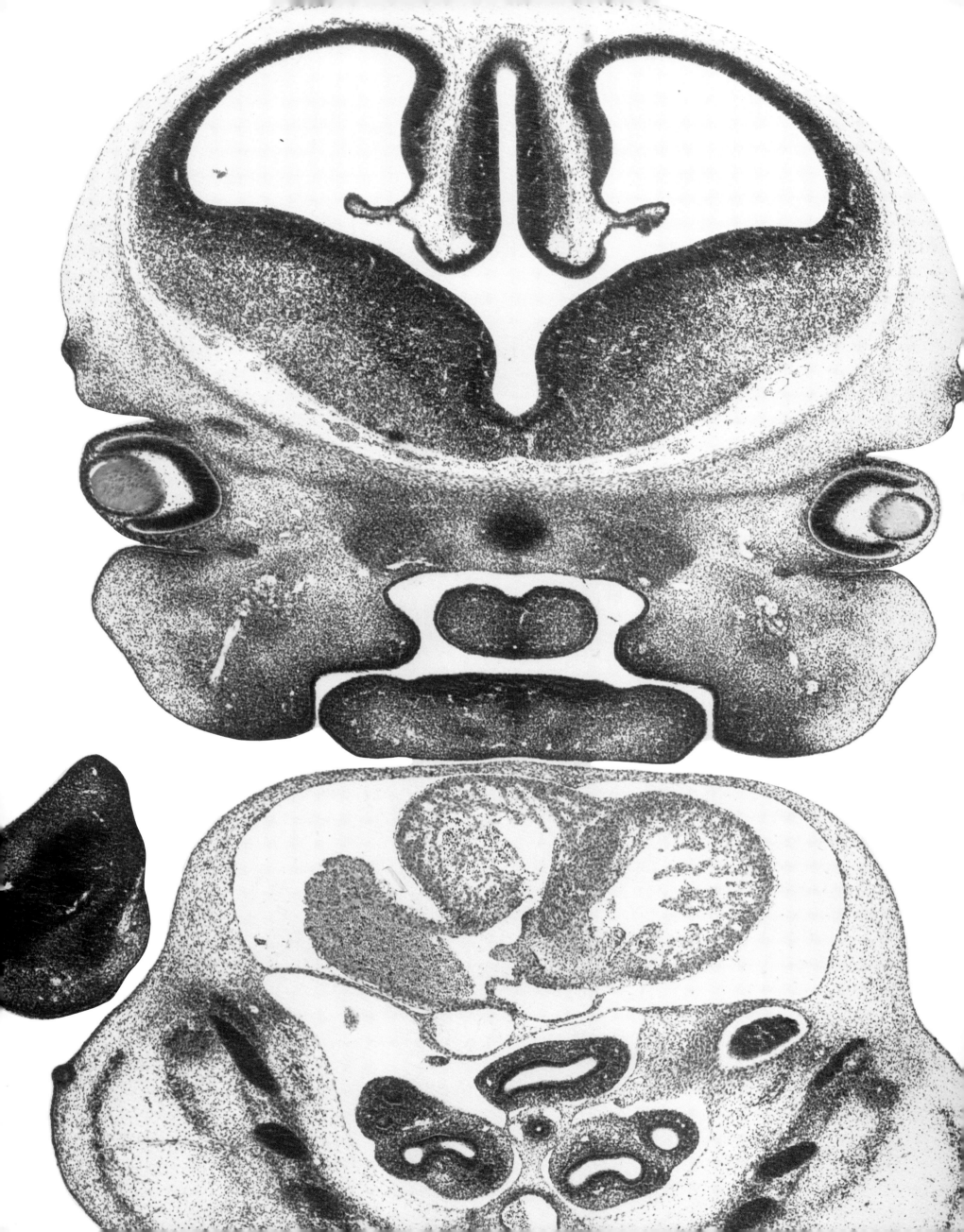

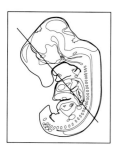

Figure 6
E14 Coronal 6

5max   maxillary nerve
10n   vagus nerve
acer   anterior cerebral artery
aorta   aorta
ATh   anterior thalamus
BTel   basal telencephalon
bti   basal telen plate, intermediate
btp   basal telencephalic plate, posterior
chp   choroid plexus premordium
Cornea   cornea
cx   cortical neuroepi
D3V   dorsal third ventricle
Diaph   diaphragm
Dien   diencephalon
DRG   dorsal root ganglion
ectring   ectodermal ring
Eso   esophagus
FLimb   forelimb
hi   hippocampal formation neuroepi
ialvn   inferior alveolar nerve
ibf   interbasal plate fissure
icn   intercostal nerve
inf   infraorbital branch
ivc   inferior vena cava
IVF   interventricular foramen
Jaw   jaw
Lens   lens
Liver   liver
lsvc   left superior vena cava
Lung   lung
LV   lateral ventricle
LVent   left ventricle
mcer   middle cerebral artery
noto   notochord
PCard   pericardium
phrn   phrenic nerve
Pi   pineal gland
Pig   pigment epithelium of retina
Pleural   pleural cavity
poa   preoptic area neuroepi
RAtr   right atrium
Retina   retina
RVent   right ventricle
sm   stria medullaris of the thalamus
snl   superior neuroepith lobule of dien
Symp   sympathetic trunk
Telen   telencephalon
Tong   tongue
V3V   ventral third ventricle
vnl   ventral neuroepith lobule of dien

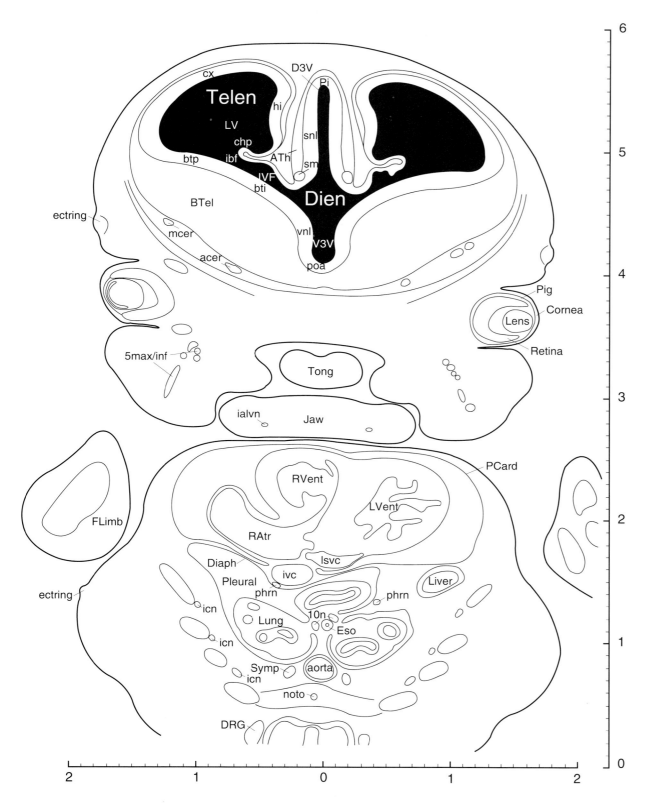

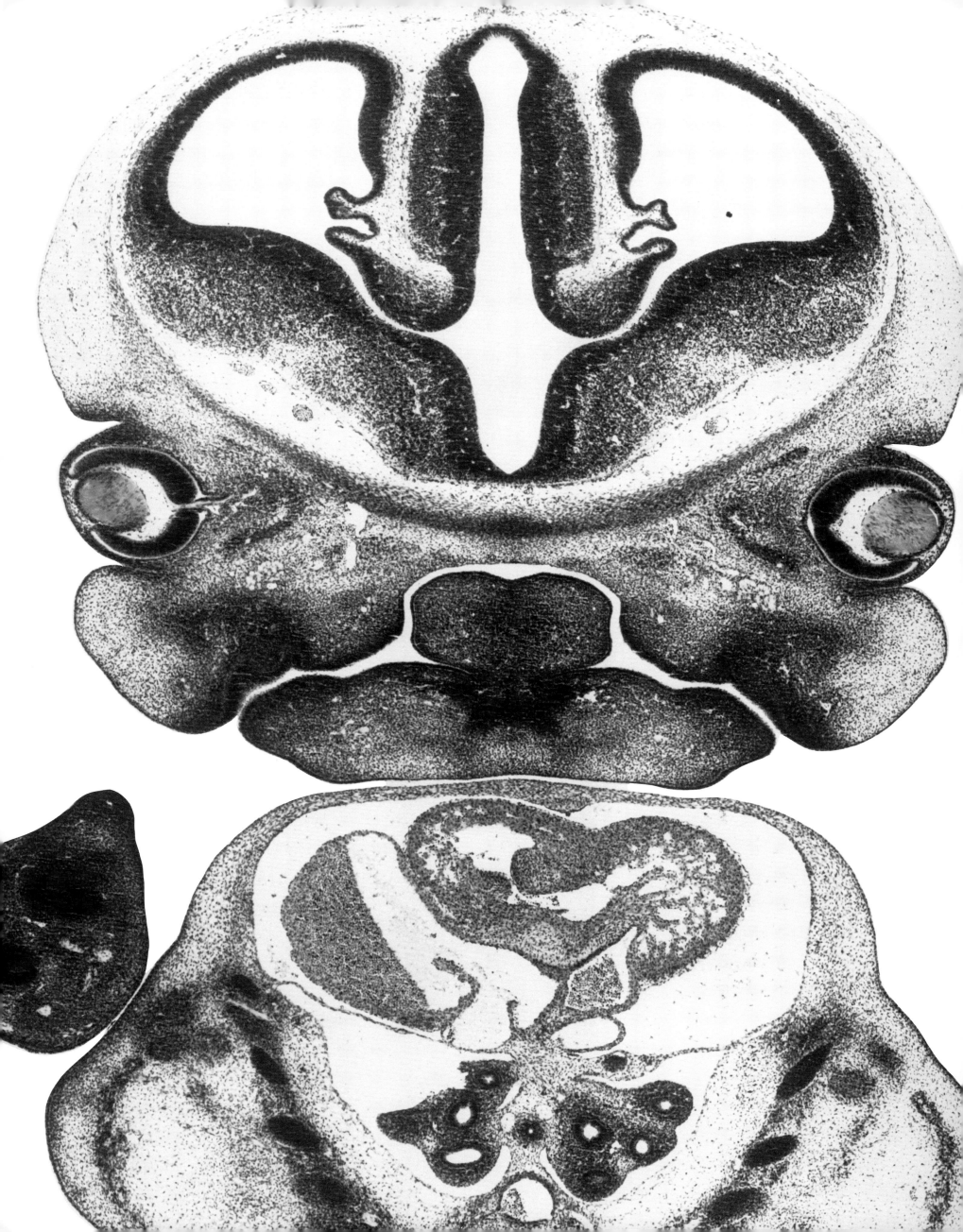

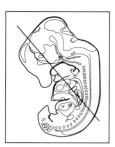

Figure 7
E14 Coronal 7

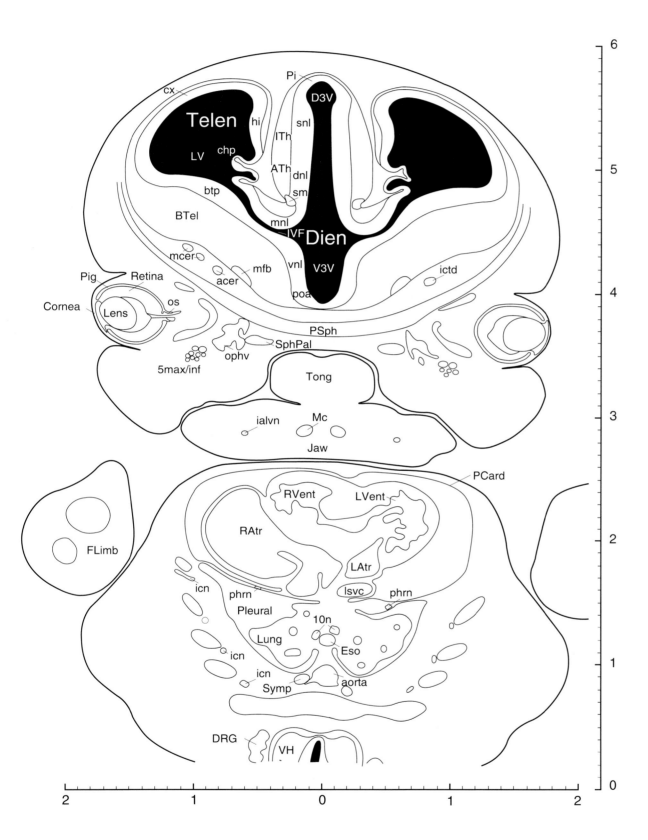

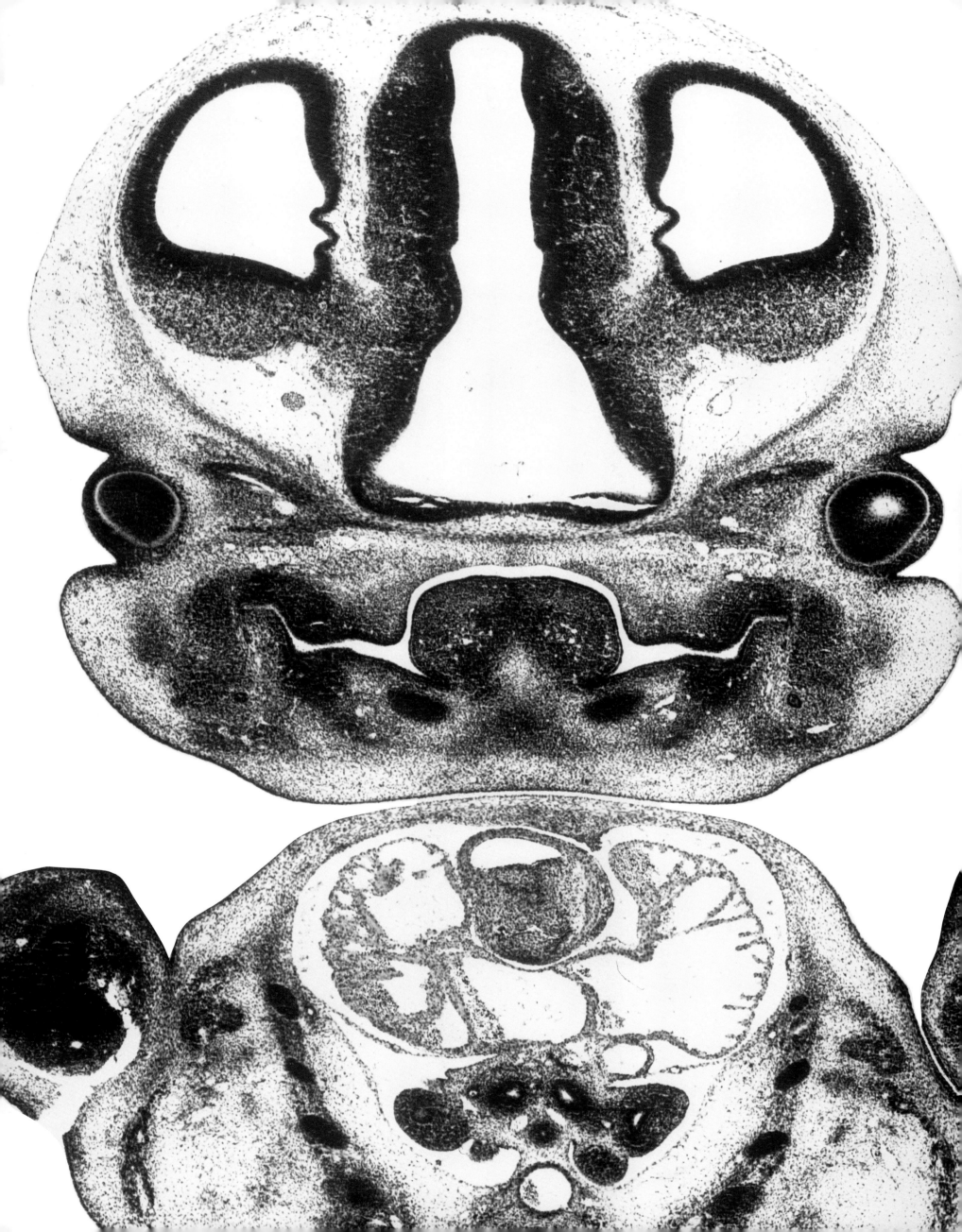

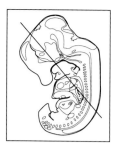

Figure 8
E14 Coronal 8

3n   oculomotor nerve or its root
5max   maxillary nerve
10n   vagus nerve
ah   anterior hypothalamic neuroepi
Amg   amygdala
aorta   aorta
aortval   aortic valva
ATh   anterior thalamus
BTel   basal telencephalon
btp   basal telencephalic plate, posterior
cardpx   cardiac plexus
chp   choroid plexus premordium
cx   cortical neuroepi
D3V   dorsal third ventricle
Dien   diencephalon
dnl   dorsal neuroepith lobule of dien
EP   entopeduncular nucleus
Eso   esophagus
fr   fasciculus retroflexus
GP   globus pallidus
hi   hippocampal formation neuroepi
Hum   humerus
ialvn   inferior alveolar nerve
icn   intercostal nerve
ictd   internal carotid artery
inf   infraorbital branch
inl   inferior neuroepith lobule of dien
IRec   inferior rectus muscle
ITh   intermediate thalamus
Jaw   jaw
LAtr   left atrium
LHb   lateral habenular nucleus
lsvc   left superior vena cava
Lung   lung
LV   lateral ventricle
Mc   Meckel's cartilage
mfb   medial forebrain bundle
mhb   medial habenula neuroepi
mnl   middle neuroepith lobule of dien
ophv   ophthalmic vein
OptRe   optic recess of third ventricle
PCard   pericardium
phrn   phrenic nerve
Pig   pigment epithelium of retina
Pleural   pleural cavity
PTh   posterior thalamus
pult   pulmonary trunk
RAtr   right atrium
Retina   retina
rsvc   right superior vena cava
snl   superior neuroepith lobule of dien
SphPal   sphenopalatine ganglion
SRec   superior rectus muscle
Symp   sympathetic trunk
Telen   telencephalon
Tong   tongue
trs   transverse sinus
V3V   ventral third ventricle
vnl   ventral neuroepith lobule of dien
ZI   zona incerta

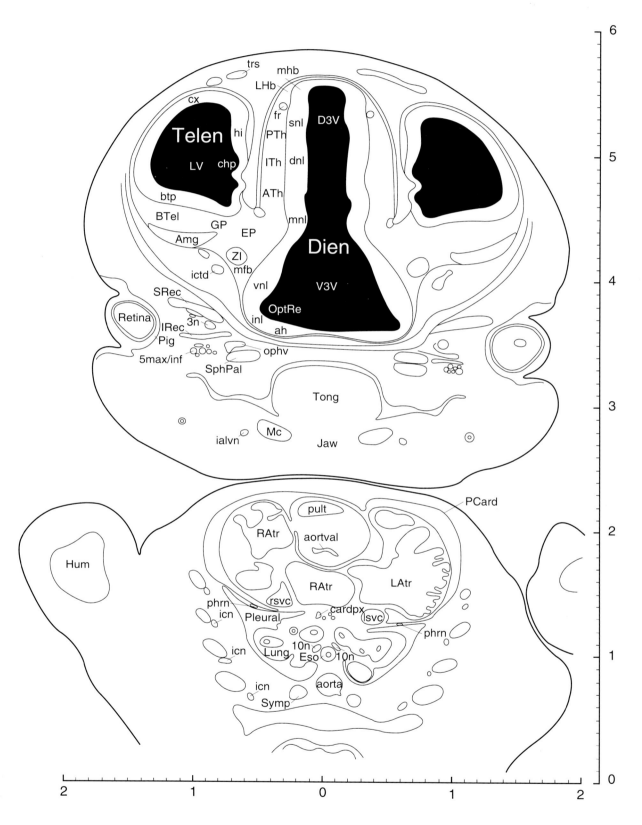

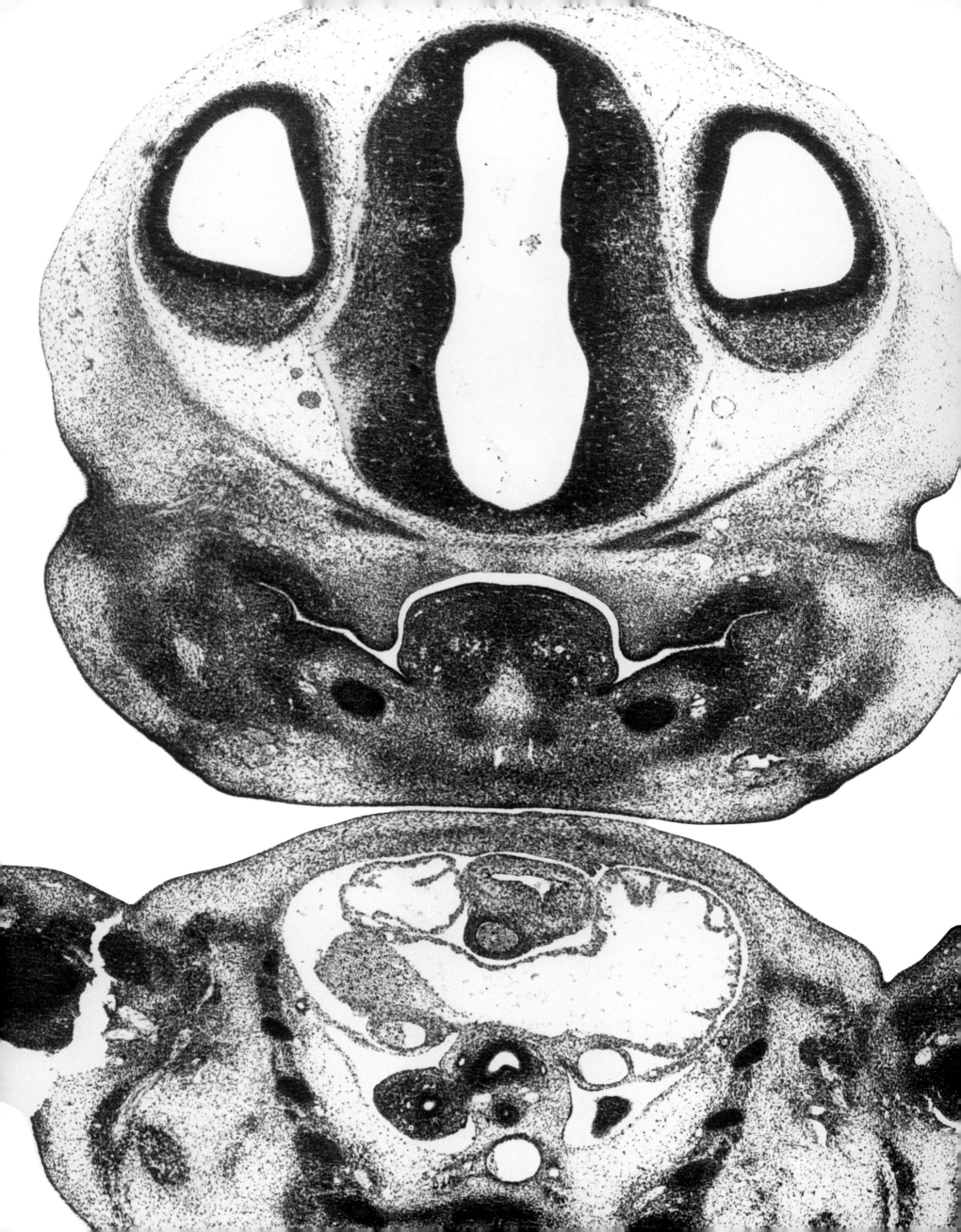

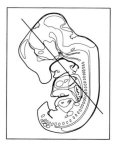

Figure 9
E14 Coronal 9

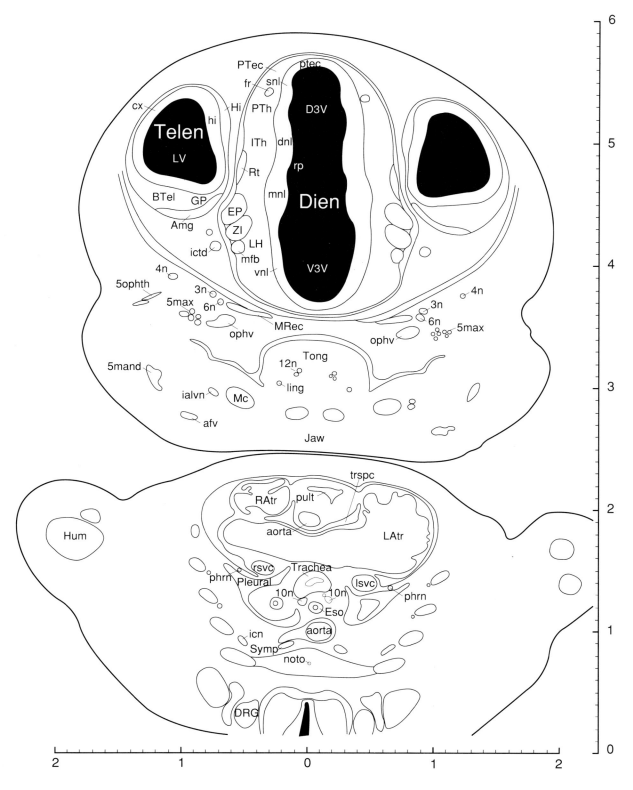

3n   oculomotor nerve or its root
4n   trochlear nerve or its root
5mand   mandibular nerve
5max   maxillary nerve
5ophth   ophthalmic nerve of trigeminal
6n   abducens nerve or its root
10n   vagus nerve
12n   hypoglossal nerve or its root
afv   anterior facial vein
Amg   amygdala
aorta   aorta
BTel   basal telencephalon
cx   cortical neuroepi
D3V   dorsal third ventricle
Dien   diencephalon
dnl   dorsal neuroepith lobule of dien
DRG   dorsal root ganglion
EP   entopeduncular nucleus
Eso   esophagus
fr   fasciculus retroflexus
GP   globus pallidus
Hi   hippocampal formation
hi   hippocampal formation neuroepi
Hum   humerus
ialvn   inferior alveolar nerve
icn   intercostal nerve
ictd   internal carotid artery
ITh   intermediate thalamus
Jaw   jaw
LAtr   left atrium
LH   lateral hypothalamic area
ling   lingual nerve
lsvc   left superior vena cava
LV   lateral ventricle
Mc   Meckel's cartilage
mfb   medial forebrain bundle
mnl   middle neuroepith lobule of dien
MRec   medial rectus muscle
noto   notochord
ophv   ophthalmic vein
phrn   phrenic nerve
Pleural   pleural cavity
PTec   pretectum
ptec   pretectum neuroepi
PTh   posterior thalamus
pult   pulmonary trunk
RAtr   right atrium
rp   reticular protuberance
rsvc   right superior vena cava
Rt   reticular thalamic nucleus
snl   superior neuroepith lobule of dien
Symp   sympathetic trunk
Telen   telencephalon
Tong   tongue
Trachea   trachea
trspc   transverse sinus of the pericardium
V3V   ventral third ventricle
vnl   ventral neuroepith lobule of dien
ZI   zona incerta

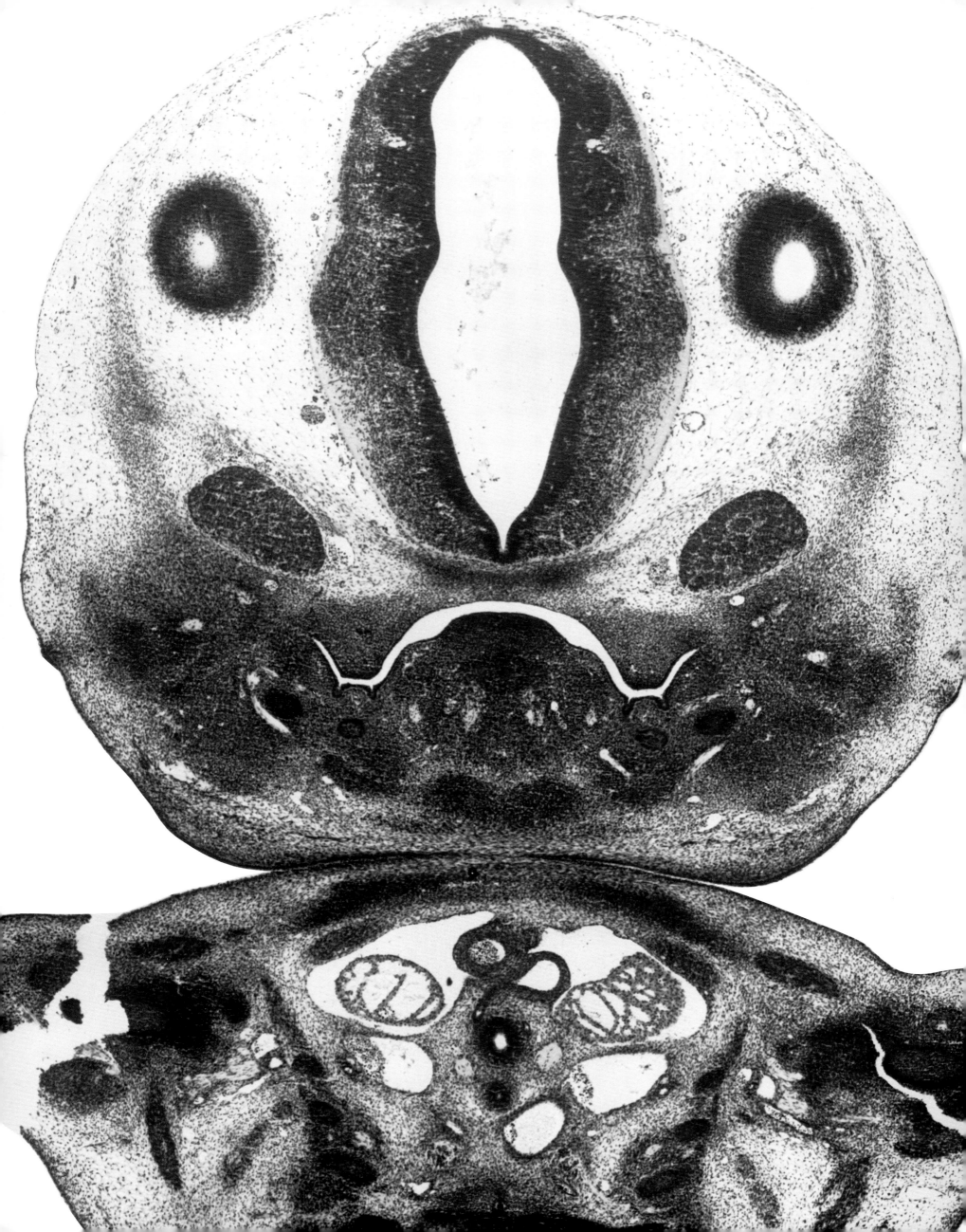

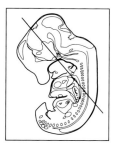

Figure 10
E14 Coronal 10

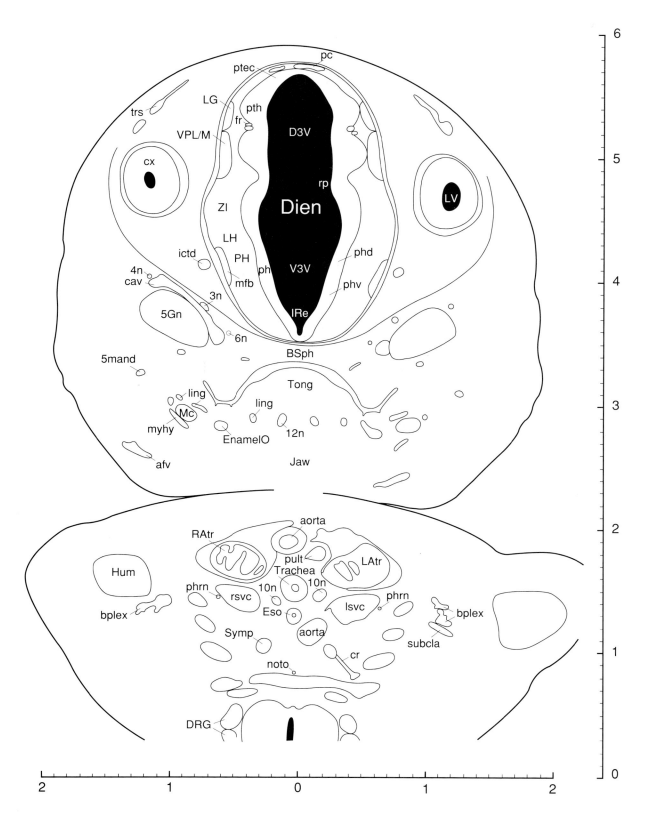

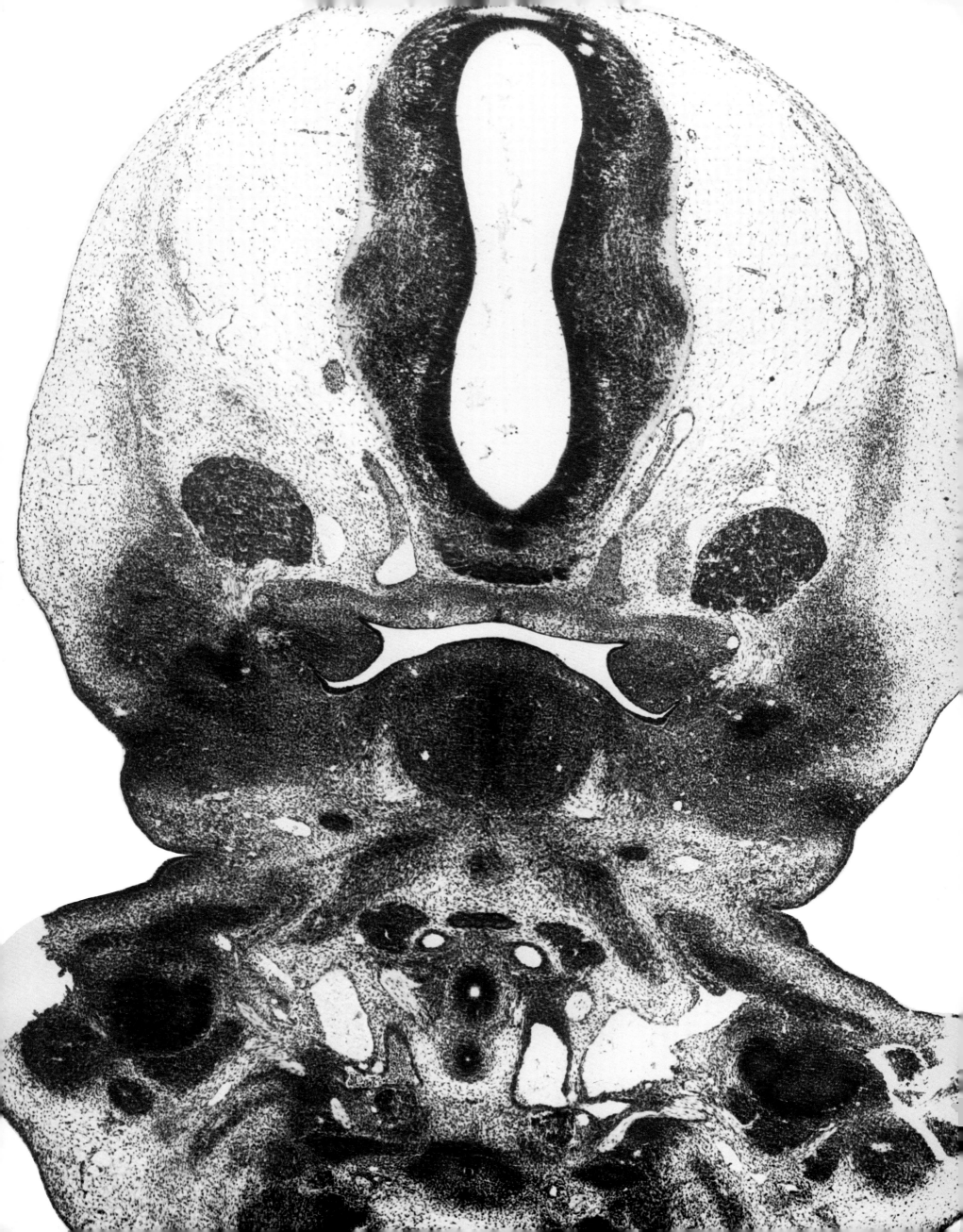

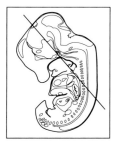

Figure 11
E14 Coronal 11

3n   oculomotor nerve or its root
4n   trochlear nerve or its root
5Gn   trigeminal ganglion
5mand   mandibular nerve
6n   abducens nerve or its root
7n   facial nerve or its root
10n   vagus nerve
12n   hypoglossal nerve or its root
aa   aortic arch
apit   anterior pituitary anlage
bplex   brachial plexus
BSph   basisphenoid bone
cav   cavernous sinus
cctd   common carotid artery
D3V   dorsal third ventricle
Dien   diencephalon
DRG   dorsal root ganglion
Eso   esophagus
fr   fasciculus retroflexus
Hum   humerus
Hyoid   hyoid bone
ictd   internal carotid artery
InfS   infundibular stem
LG   lateral geniculate nucleus
LH   lateral hypothalamic area
lsvc   left superior vena cava
Mc   Meckel's cartilage
mfb   medial forebrain bundle
MG   medial geniculate nucleus
mtg   mammillotegmental tract
noto   notochord
pc   posterior commissure
pcoma   posterior communicating artery
ph   posterior hypothalamic neuroepi
phd   posterior hypoth neuroepi, dors
phv   posterior hypoth neuroepi, vental
ppit   posterior pituitary
PTec   pretectum
rln   recurrent laryngeal nerve
rp   reticular protuberance
rsvc   right superior vena cava
spn   spinal nerve
StGn   stellate ganglion
subcla   subclavian artery
subclv   subclavian vein
Thym   thymus gland
Thyr   thyroid gland
Tong   tongue
Trachea   trachea
trs   transverse sinus
V3V   ventral third ventricle
ZI   zona incerta

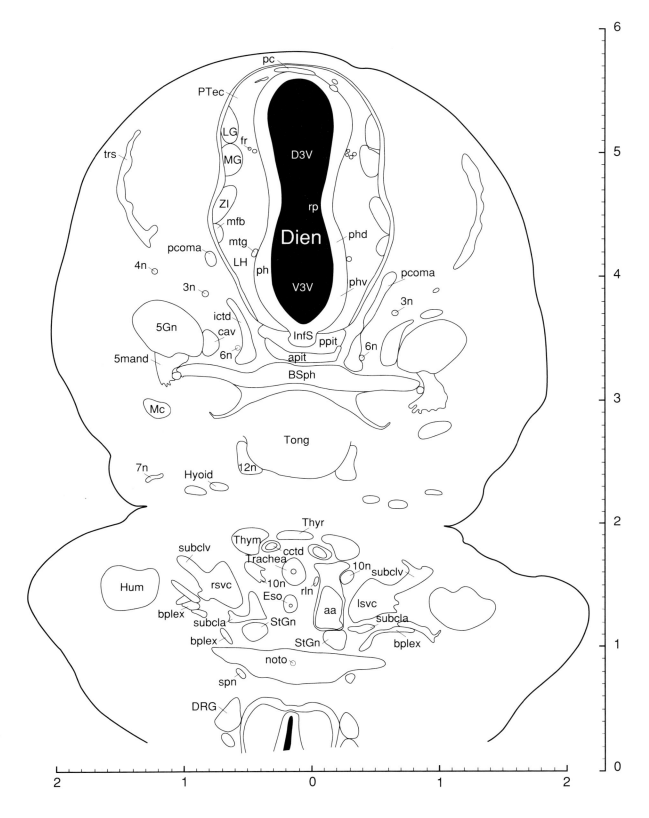

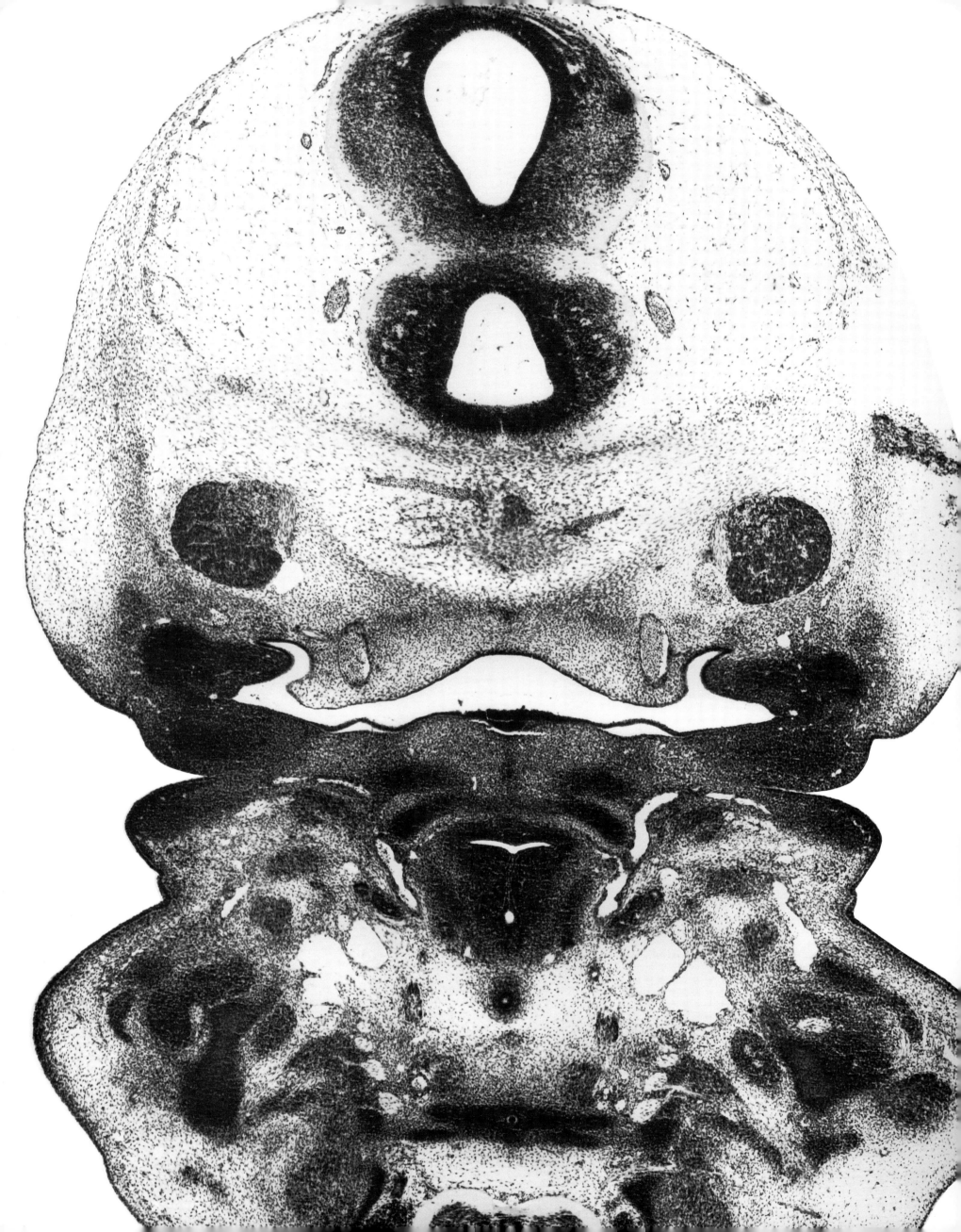

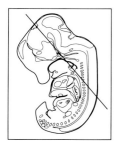

Figure 12
E14 Coronal 12

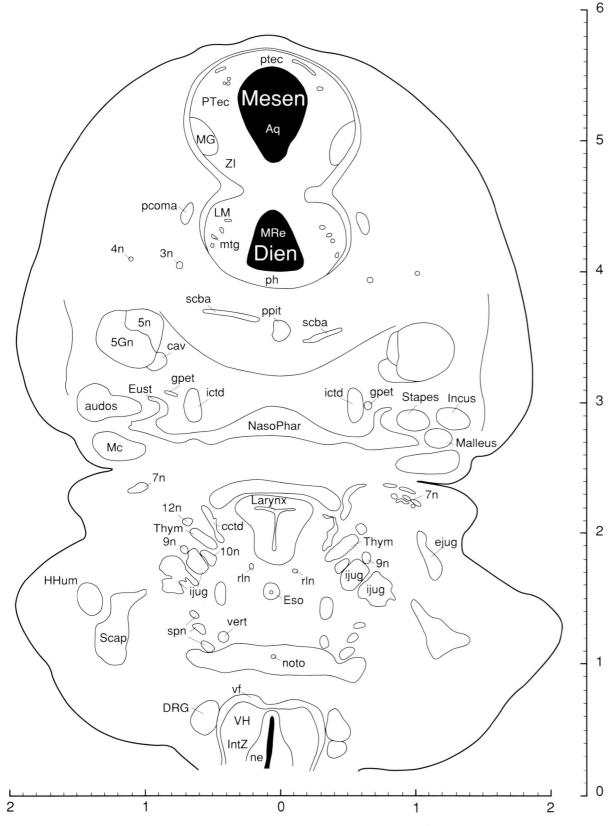

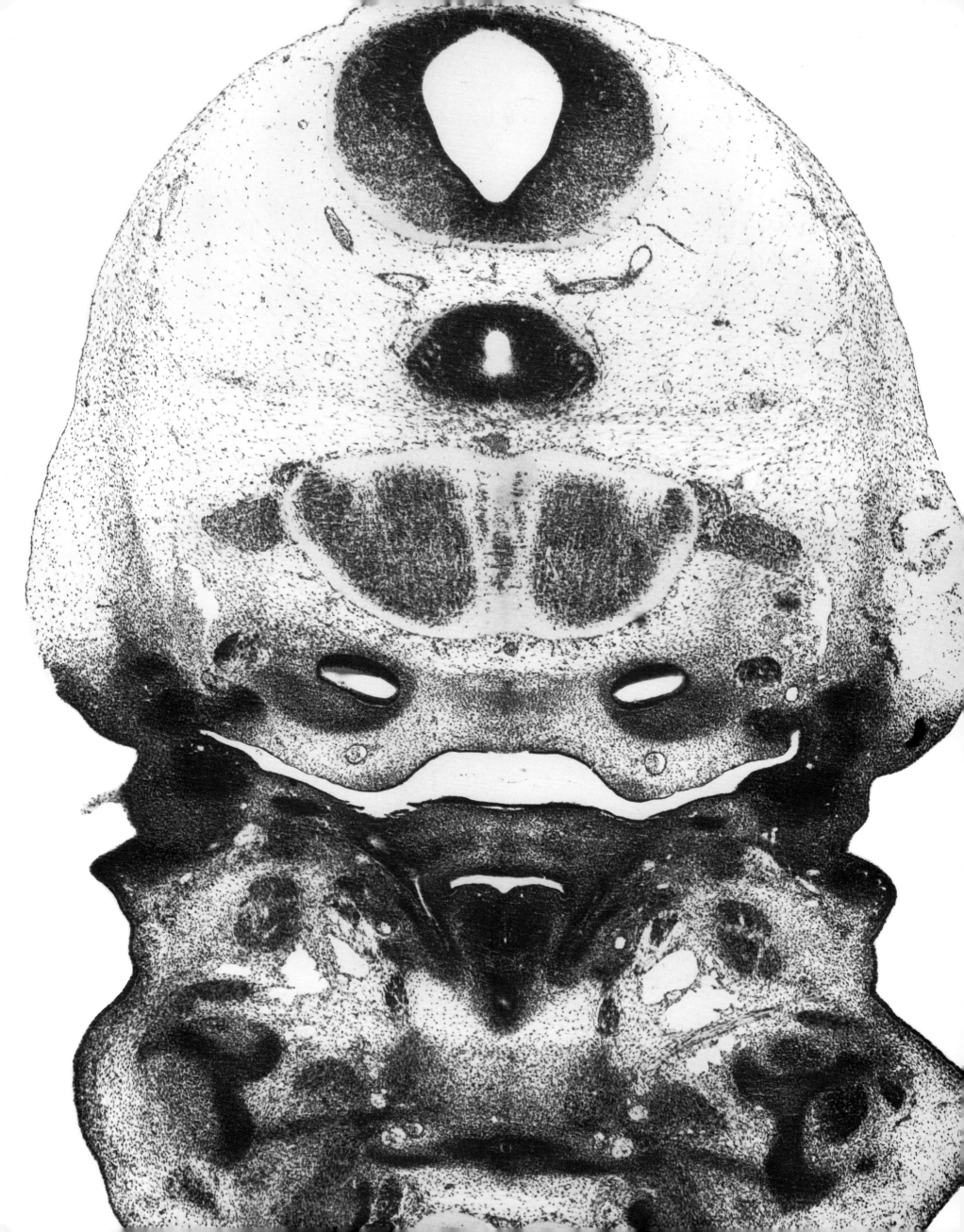

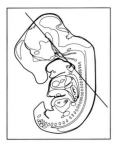

Figure 13
E14 Coronal 13

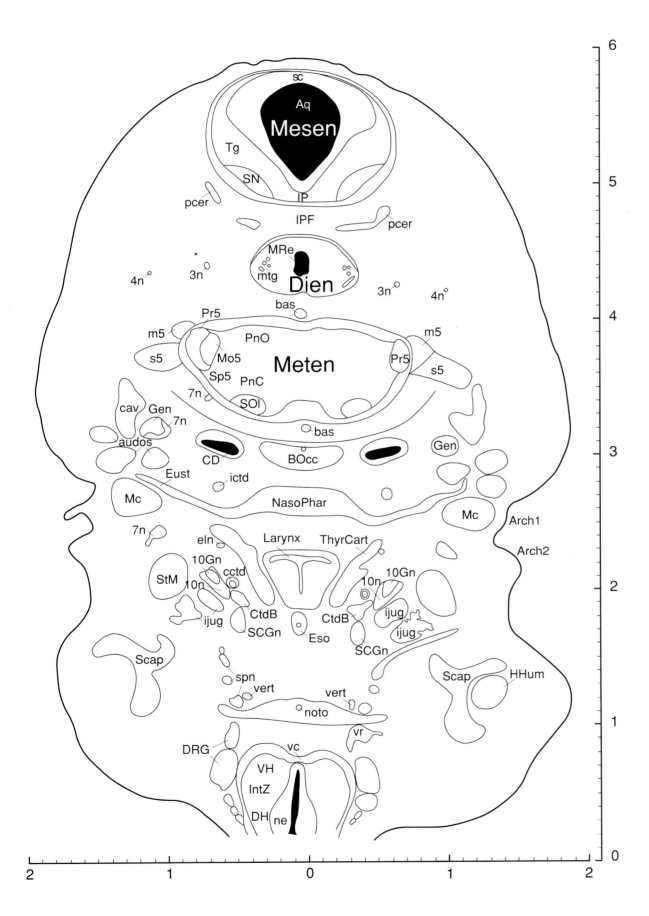

3n   oculomotor nerve or its root
4n   trochlear nerve or its root
7n   facial nerve or its root
10Gn   vagal ganglion
10n   vagus nerve
Aq   aqueduct (Sylvius)
Arch1   1st pharyngeal arch (mandibular portion)
Arch2   2nd pharyngeal arch
audos   primordia of ear ossicles
bas   basilar artery
BOcc   basioccipital bone
cav   cavernous sinus
cctd   common carotid artery
CD   cochlear duct
CtdB   carotid body
DH   dorsal horns of spinal cord
Dien   diencephalon
DRG   dorsal root ganglion
eln   external laryngeal nerve
Eso   esophagus
Eust   Eustachian tube
Gen   geniculate ganglion
HHum   head of humerus
ictd   internal carotid artery
ijug   internal jugular vein
IntZ   intermediate zone of spinal gray
IP   interpeduncular nucleus
IPF   interpeduncular fossa
Larynx   larynx
m5   motor root of the trigeminal nerve
Mc   Meckel's cartilage
Mesen   mesencephalon
Meten   metencephalon
Mo5   motor trigeminal nucleus
MRe   mammillary recess of the 3rd ventr
mtg   mammillotegmental tract
NasoPhar   nasopharyngeal cavity
ne   neuroepithelium
noto   notochord
pcer   posterior cerebral artery
PnC   pontine reticular nu, caudal part
PnO   pontine reticular nu, oral part
Pr5   principal sensory trigeminal nu
s5   sensory root of the trigeminal nerve
sc   superior colliculus neuroepi
Scap   scapula
SCGn   superior cervical ganglion
SN   substantia nigra
SOl   superior olive
Sp5   spinal trigeminal nucleus
spn   spinal nerve
StM   sternomastoid muscle
Tg   tegmentum
ThyrCart   thyroid cartilage
vc   ventral commissure of spinal cord
vert   vertebral artery
VH   ventral horn
vr   ventral root

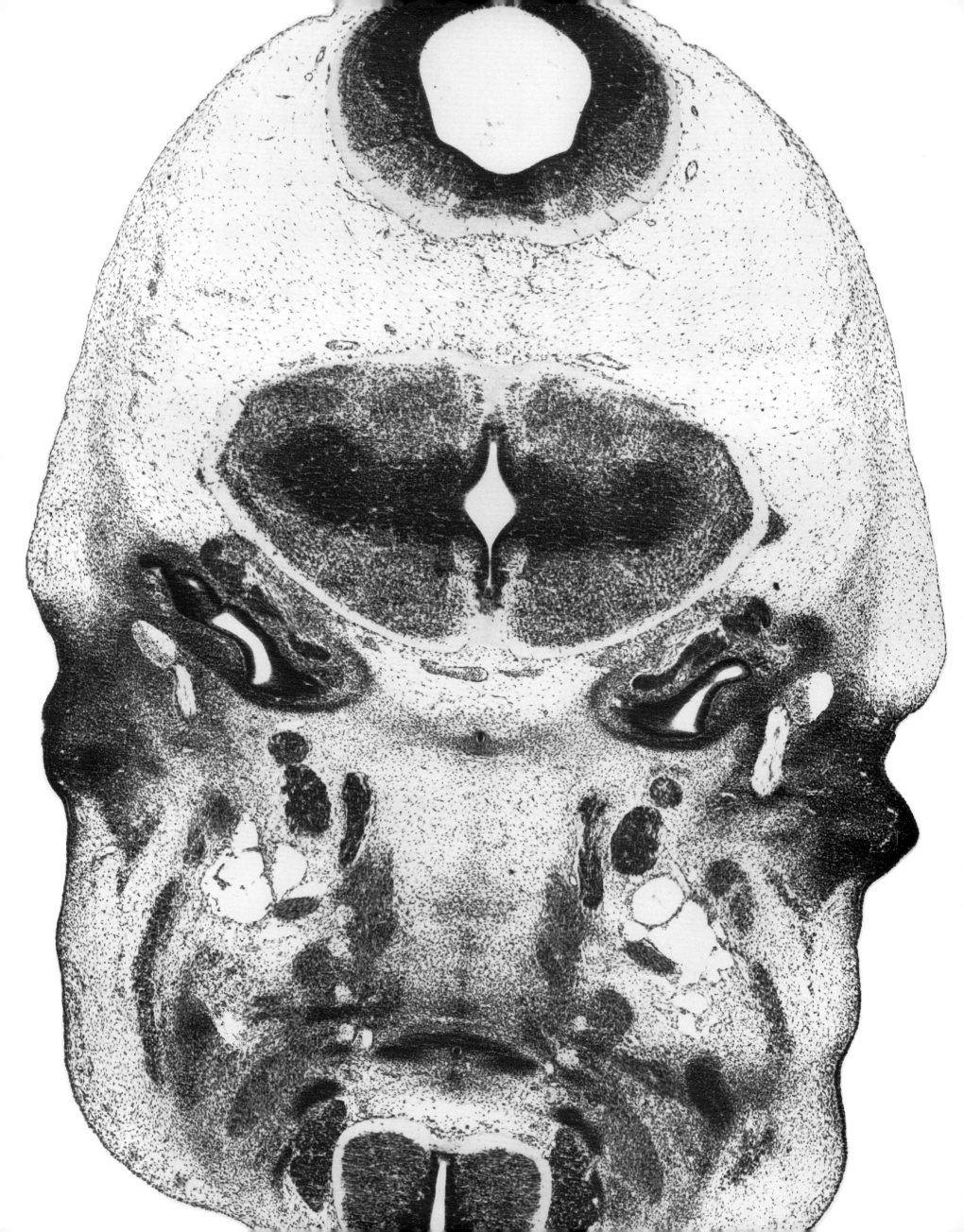

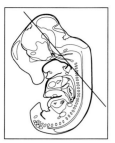

Figure 14
E14 Coronal 14

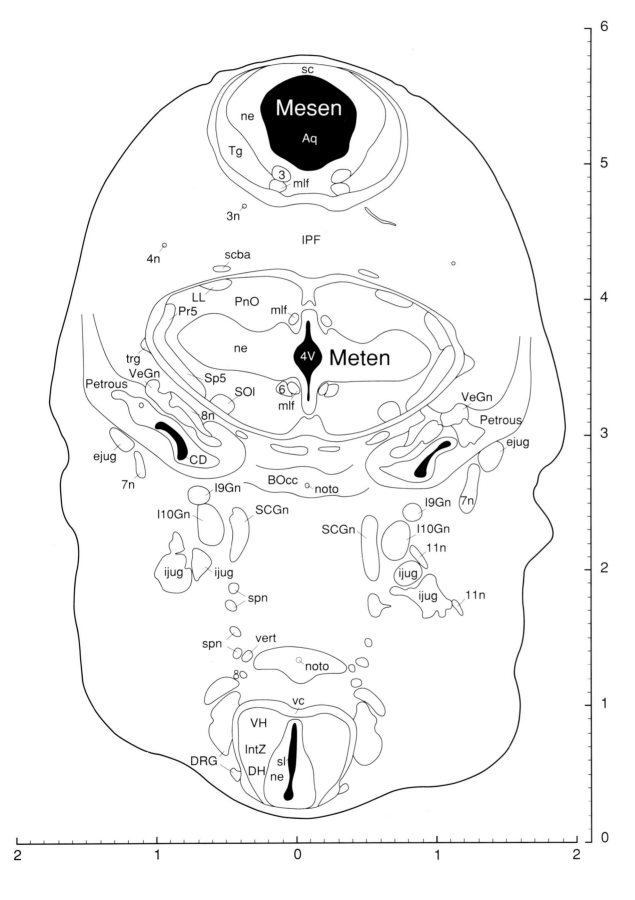

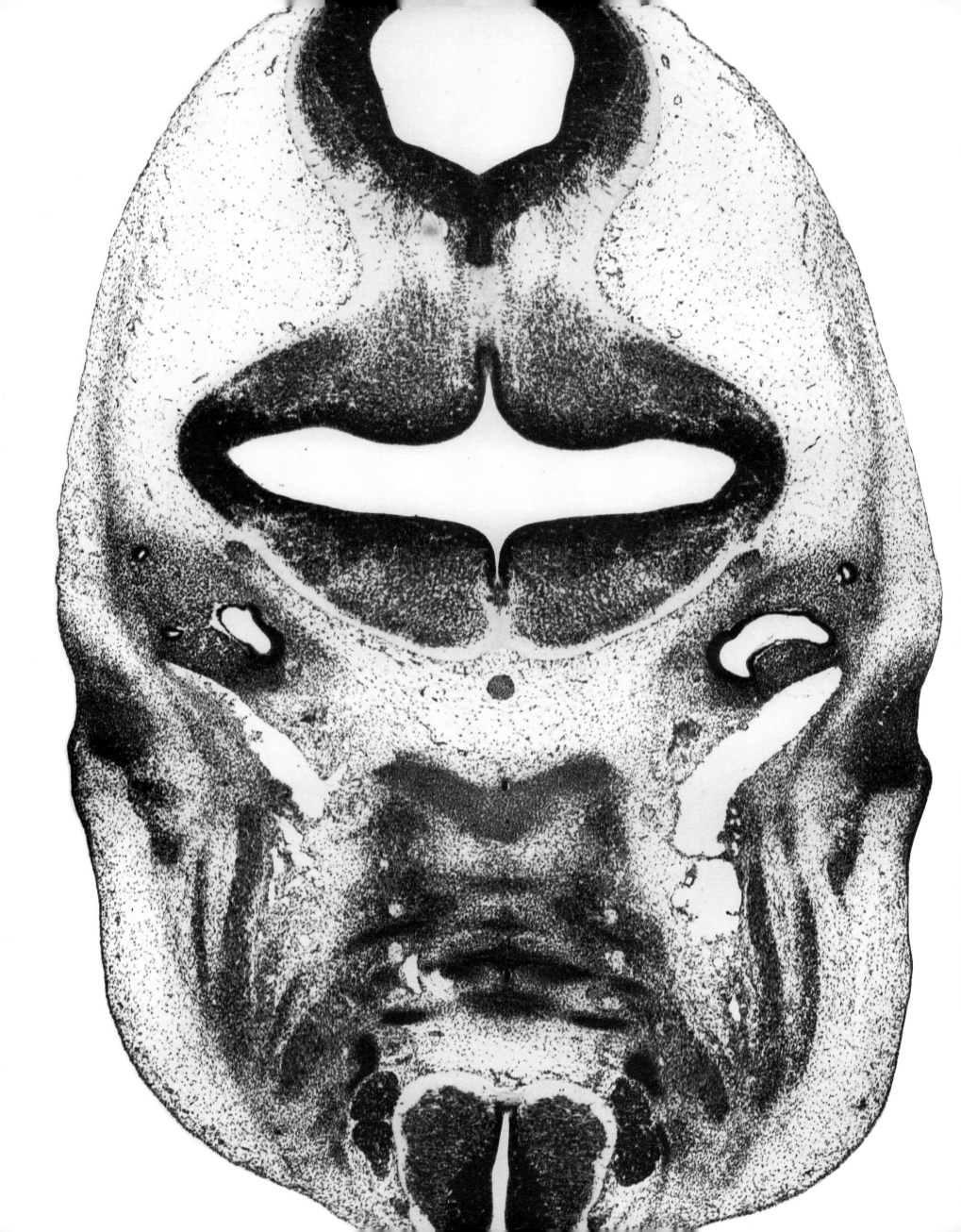

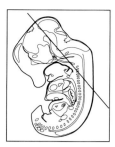

Figure 15
E14 Coronal 15

3   oculomotor nucleus
4n   trochlear nerve or its root
4V   fourth ventricle
7   facial nucleus
8n   vestibulocochlear nerve
10Gn   vagal ganglion
Ant   anterior semicircular duct
Aq   aqueduct (Sylvius)
bas   basilar artery
BOcc   basioccipital bone
CD   cochlear duct
ctz   cortical transitory zone of Cb
DH   dorsal horns of spinal cord
DRG   dorsal root ganglion
fp   floor plate of spinal cord neuroepi
Gi   gigantocellular reticular nucleus
ijug   internal jugular vein
IntZ   intermediate zone of spinal gray
Is   isthmus region
JugF   jugular foramen
lf   lateral fissure
LR4V   lateral recess of the 4th ventr
Md   medulla
Mesen   mesencephalon
Meten   metencephalon
mlf   medial longitudinal fasciculus
ne   neuroepithelium
noto   notochord
ntz   nuclear transitory zone
Petrous   petrous part of the temp bone
rp   reticular protuberance
sc   superior colliculus neuroepi
scba   superior cerebellar artery
sl   sulcus limitans
Sp5   spinal trigeminal nucleus
spn   spinal nerve
SuVe   superior vestibular nucleus
Tg   tegmentum
vc   ventral commissure of spinal cord
ve   vestibular area neuroepi
VH   ventral horn
vr   ventral root

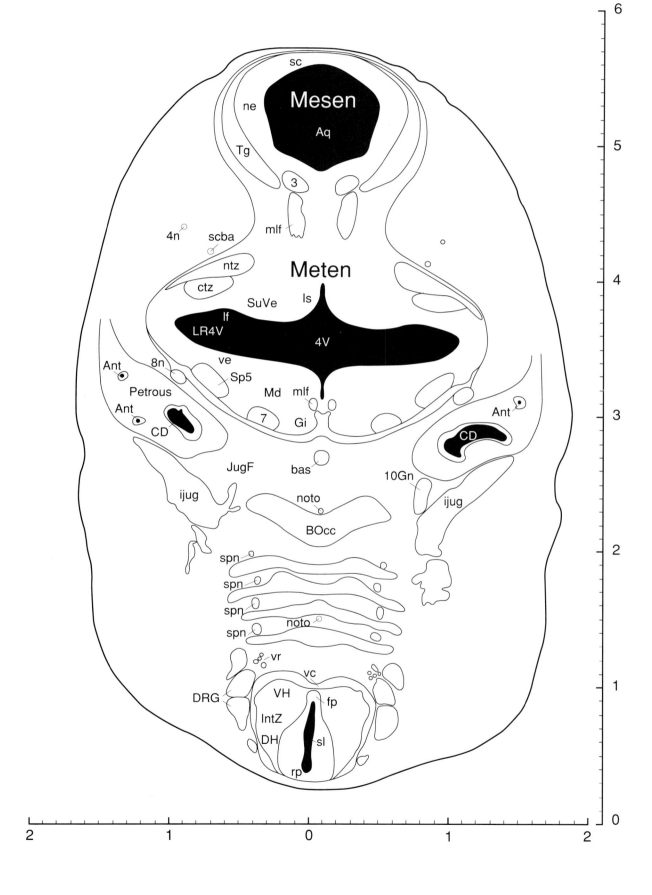

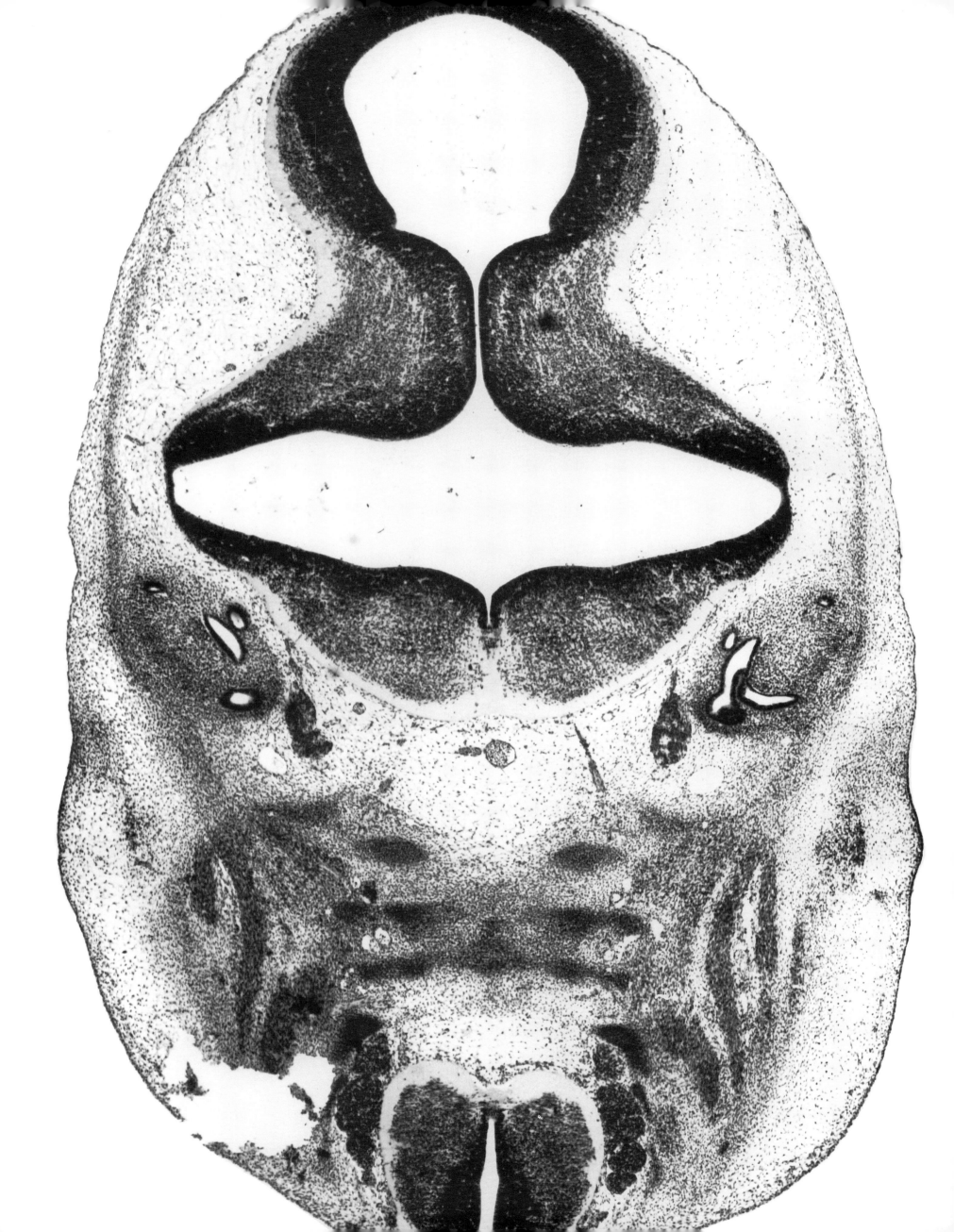

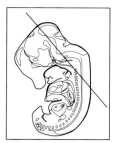

Figure 16
E14 Coronal 16

4n trochlear nerve or its root
12n hypoglossal nerve or its root
Ant anterior semicircular duct
Aq aqueduct (Sylvius)
bas basilar artery
ctz cortical transitory zone of Cb
DH dorsal horns of spinal cord
dr dorsal root
DRG dorsal root ganglion
eld endolymphatic duct
FovIs fovea of the isthmus
Gi gigantocellular reticular nucleus
ijug internal jugular vein
IntZ intermediate zone of spinal gray
IRt intermediate reticular zone
Is isthmus region
Lat lateral (dentate) cerebellar nu
lf lateral fissure
LR4V lateral recess of the 4th ventr
ma migration a of Rüdeberg
ma1 migration a1 of Rüdeberg
ma2b1 migration a2b1 of Rüdeberg
Md medulla
me5 mesencephalic trigeminal tract
Mesen mesencephalon
Meten metencephalon
mlf medial longitudinal fasciculus
Myelen myelencephalon
noto notochord
ntz nuclear transitory zone
Petrous petrous part of the temp bone
S10Gn superior vagal (jugular) ganglion
Sacc saccule
sc superior colliculus neuroepi
scba superior cerebellar artery
sl sulcus limitans
Sp5 spinal trigeminal nucleus
SuVe superior vestibular nucleus
Tg tegmentum
trg germinal trigone
Utr utricle
vc ventral commissure of spinal cord
ve vestibular area neuroepi
VH ventral horn
vr ventral root

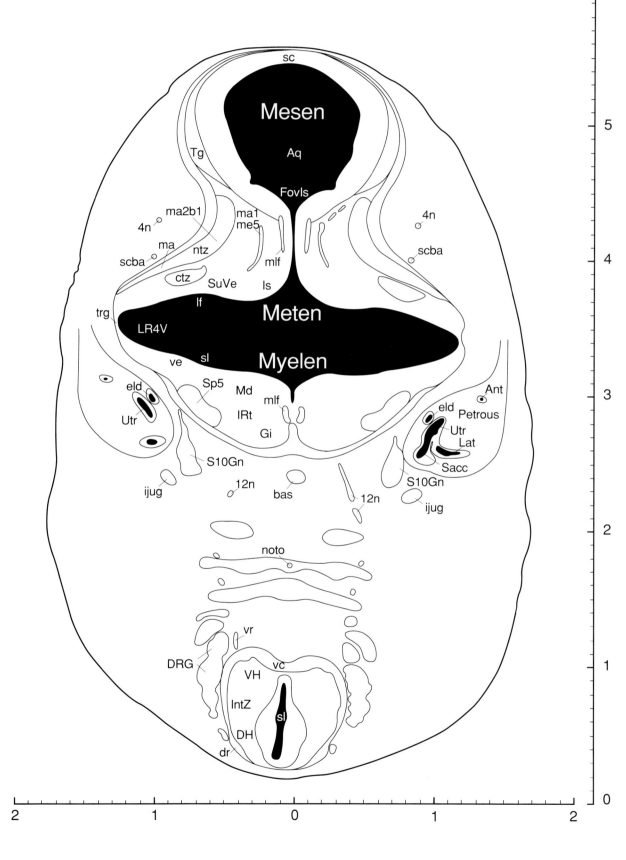

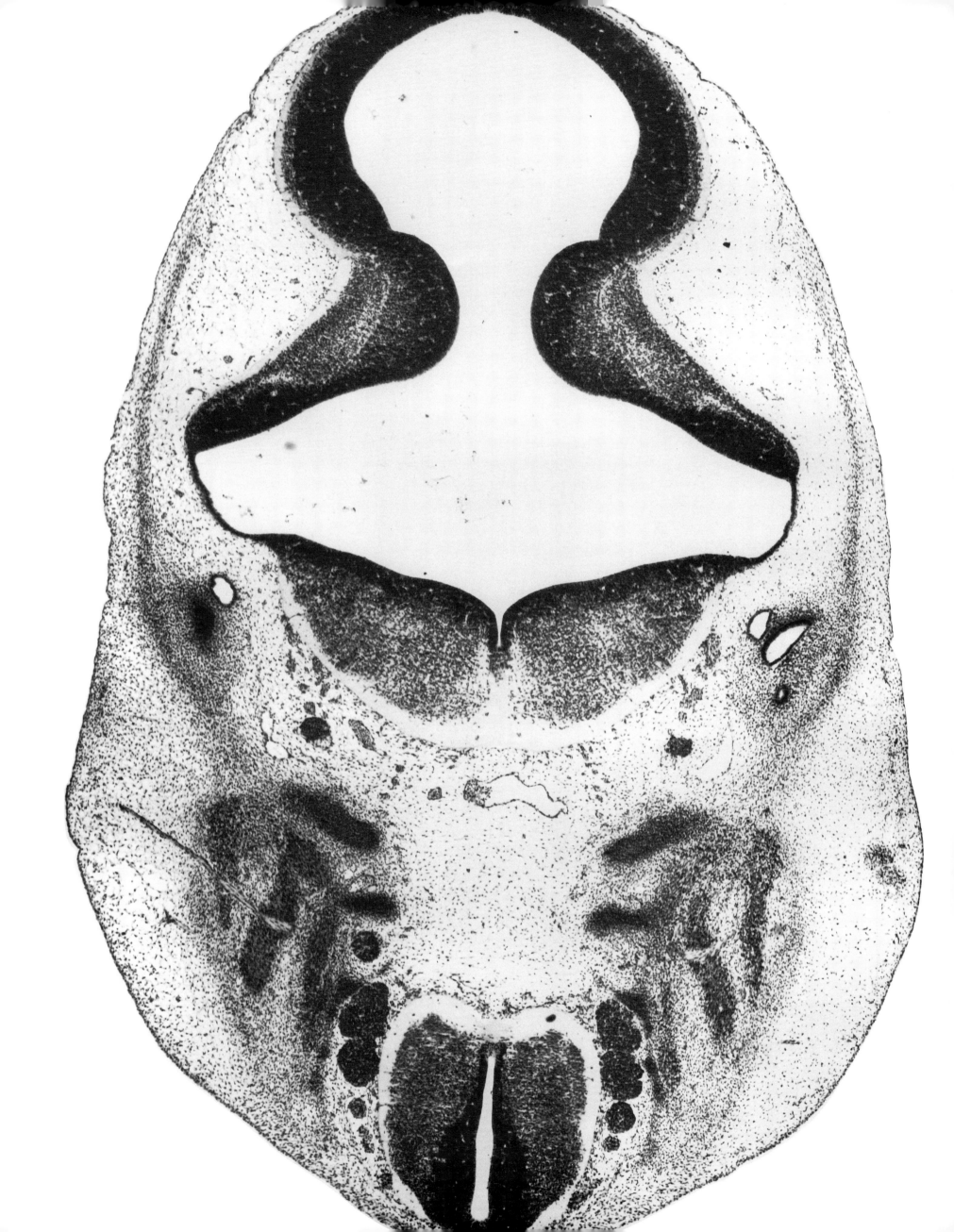

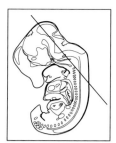

Figure 17
E14 Coronal 17

4n   trochlear nerve or its root
4V   fourth ventricle
9n   glossopharyngeal nerve
10Gn   vagal ganglion
10n   vagus nerve
Aq   aqueduct (Sylvius)
bas   basilar artery
BOcc   basioccipital bone
ctz   cortical transitory zone
DH   dorsal horns of spinal cord
dr   dorsal root
DRG   dorsal root ganglion
eld   endolymphatic duct
Gi   gigantocellular reticular nucleus
ijug   internal jugular vein
IntZ   intermediate zone of spinal gray
IRt   intermediate reticular zone
Is   isthmus region
lf   lateral fissure
LR4V   lateral recess of the 4th ventr
ma   migration a of Rüdeberg
ma1   migration a1 of Rüdeberg
ma2b1   migration a2b1 of Rüdeberg
mb   migration b of Rüdeberg
Me5   mesencephalic trigeminal nu
Mesen   mesencephalon
Meten   metencephalon
mlf   medial longitudinal fasciculus
Myelen   myelencephalon
ne   neuroepithelium
ntz   nuclear transitory zone
pag   periaqueductal grey neuroepi
Petrous   petrous part of the temp bone
Pons   pons
Post   posterior semicircular duct
sc   superior colliculus neuroepi
sl   sulcus limitans
Sp5   spinal trigeminal nucleus
spn   spinal nerve
trg   germinal trigone
Utr   utricle
vc   ventral commissure of spinal cord
ve   vestibular area neuroepi
VH   ventral horn

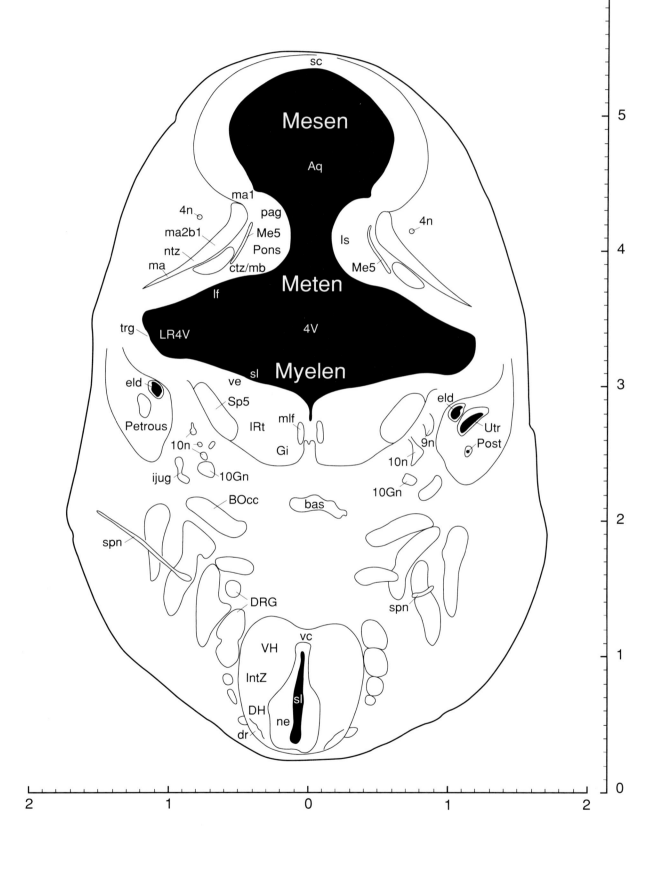

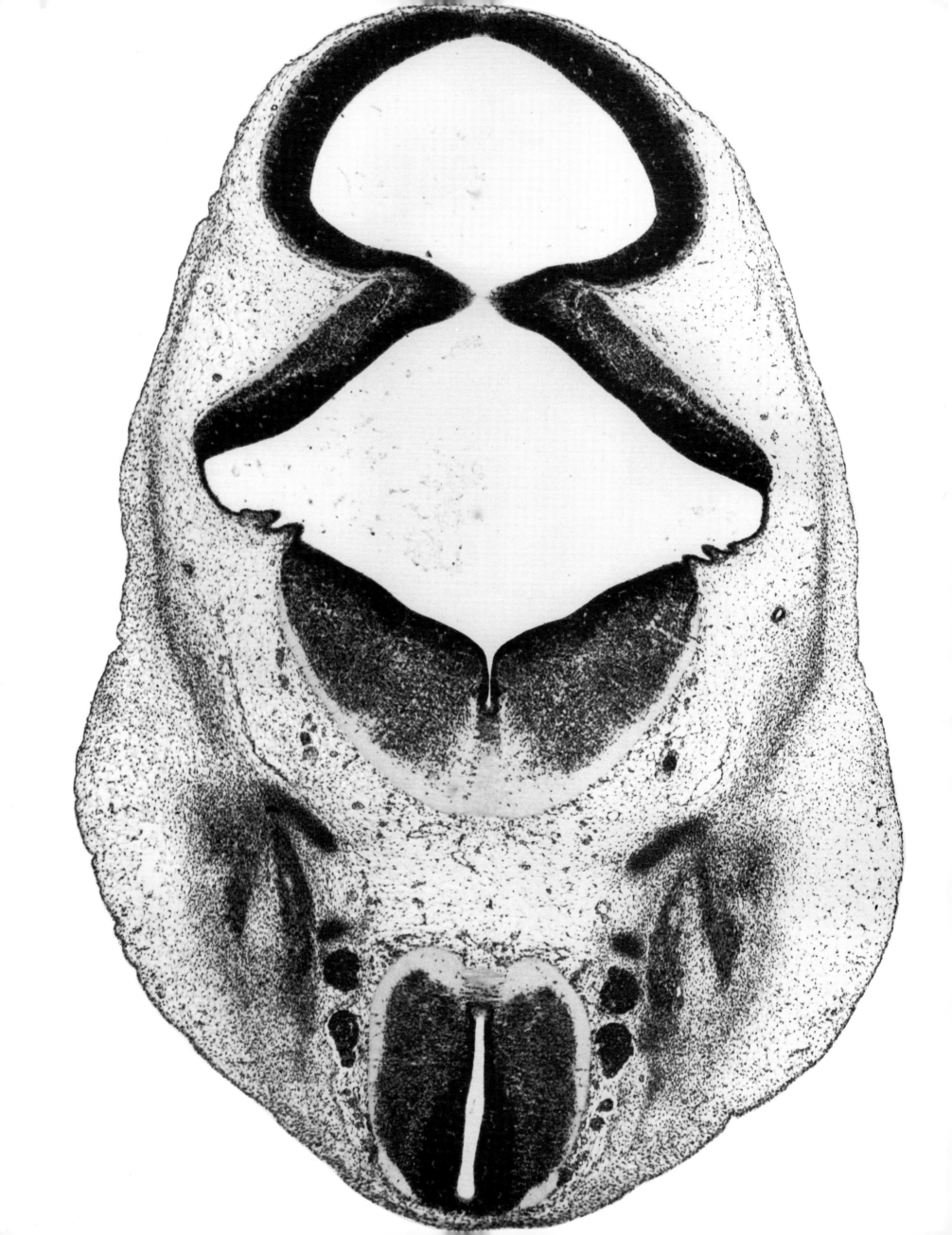

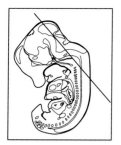

Figure 18
E14 Coronal 18

4   trochlear nucleus
4n   trochlear nerve or its root
4V   fourth ventricle
12   hypoglossal nucleus
Aq   aqueduct (Sylvius)
cb   cerebellar neuroepi
chp   choroid plexus premordium
DH   dorsal horns of spinal cord
dr   dorsal root
DRG   dorsal root ganglion
eld   endolymphatic duct
FovIs   fovea of the isthmus
fp   floor plate of spinal cord neuroepi
IntZ   intermediate zone of spinal gray
Is   isthmus region
LR4V   lateral recess of the 4th ventr
ma   migration a of Rüdeberg
ma1   migration a1 of Rüdeberg
ma2b1   migration a2b1 of Rüdeberg
Mesen   mesencephalon
Meten   metencephalon
mlf   medial longitudinal fasciculus
Myelen   myelencephalon
ne   neuroepithelium
ntz   nuclear transitory zone
rfp   roof plate of spinal cord neuroepi
sc   superior colliculus neuroepi
sl   sulcus limitans
Sp5   spinal trigeminal nucleus
trg   germinal trigone
trs   transverse sinus
vc   ventral commissure of spinal cord
vert   vertebral artery
VH   ventral horn

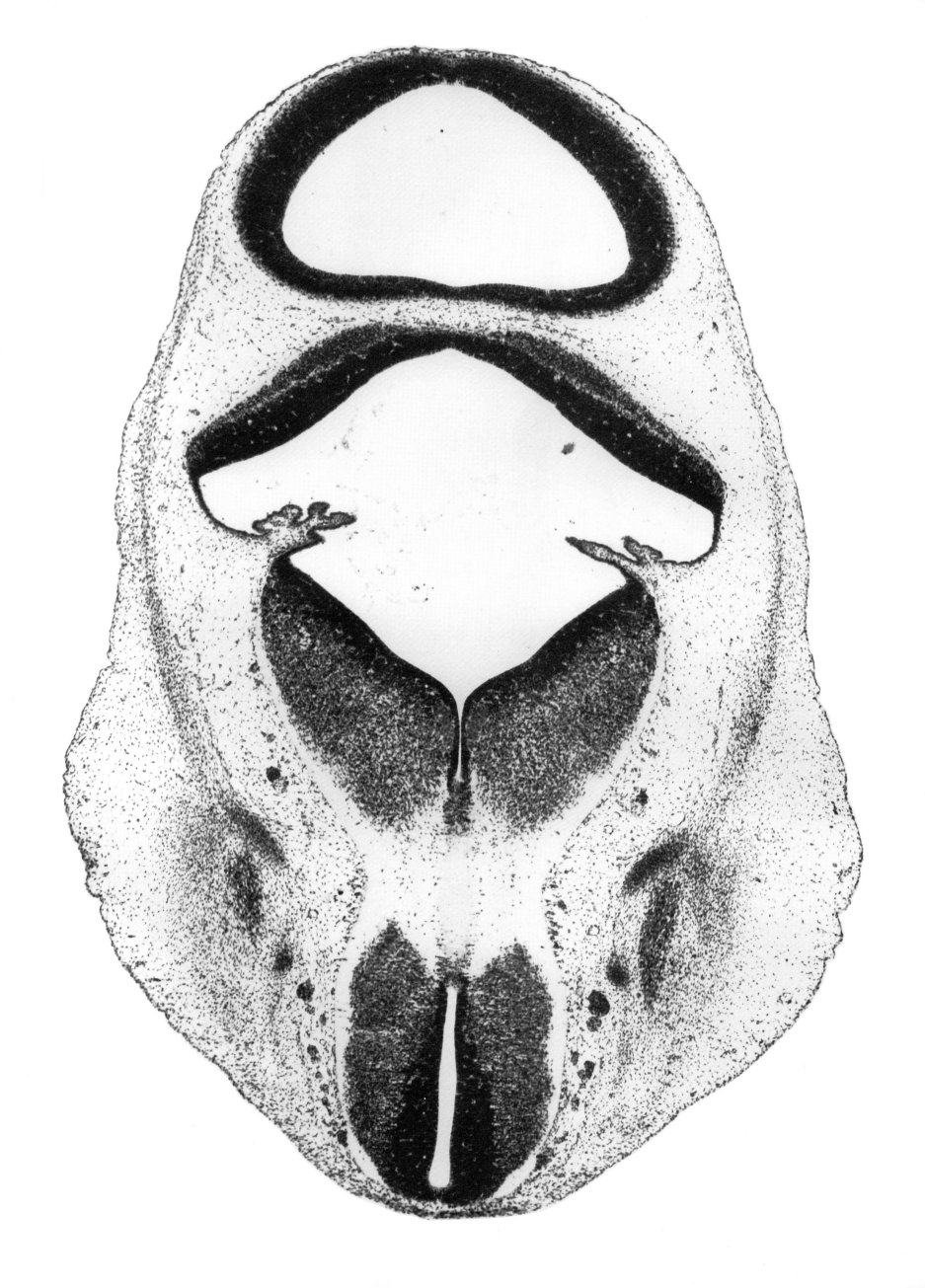

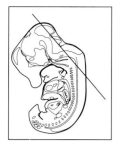

Figure 19
E14 Coronal 19

4V   fourth ventricle
12   hypoglossal nucleus
Aq   aqueduct (Sylvius)
cb   cerebellar neuroepi
CbMeF   cerebellomesencephalic flexure
chp   choroid plexus premordium
DH   dorsal horns of spinal cord
DRG   dorsal root ganglion
fp   floor plate of spinal cord neuroepi
ic   inferior colliculus neuroepi
IntZ   intermediate zone of spinal gray
LR4V   lateral recess of the 4th ventr
Mesen   mesencephalon
Meten   metencephalon
mlf   medial longitudinal fasciculus
Myelen   myelencephalon
ne   neuroepithelium
ntz   nuclear transitory zone
rfp   roof plate of spinal cord neuroepi
sc   superior colliculus neuroepi
sl   sulcus limitans
Sp5   spinal trigeminal nucleus
trg   germinal trigone
VH   ventral horn

sc

Mesen

Aq

ic
CbMeF
ntz

cb

Meten

LR4V

trg
4V
chp

Myelen
sl

Sp5   12

mlf

DRG   VH   fp   DRG

IntZ

sl

DH   ne

rfp

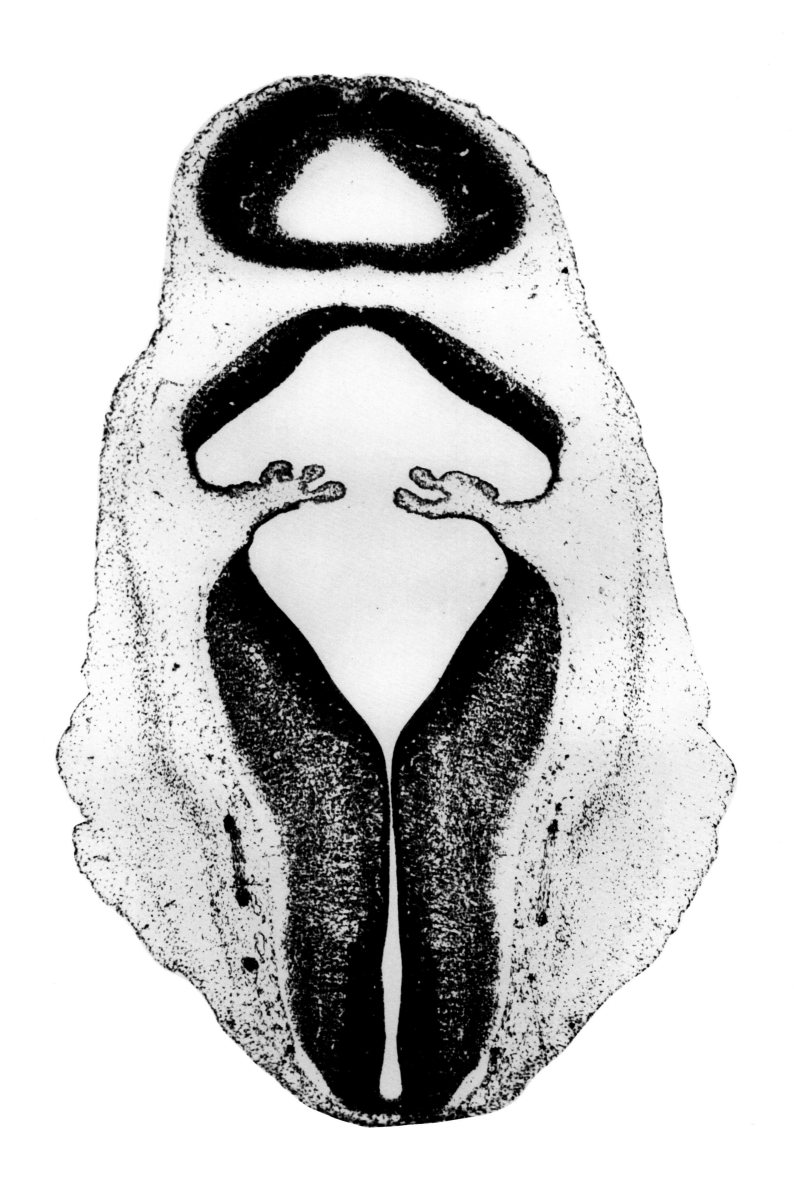

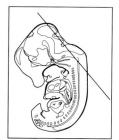

Figure 20
E14 Coronal 20

4V   fourth ventricle
12   hypoglossal nucleus
Aq   aqueduct (Sylvius)
cb   cerebellar neuroepi
chp   choroid plexus premordium
dr   dorsal root
ic   inferior colliculus neuroepi
LR4V   lateral recess of the 4th ventr
Md   medulla
Mesen   mesencephalon
Meten   metencephalon
Myelen   myelencephalon
ntz   nuclear transitory zone
sc   superior colliculus neuroepi
Spinal   spinal cord
trg   germinal trigone

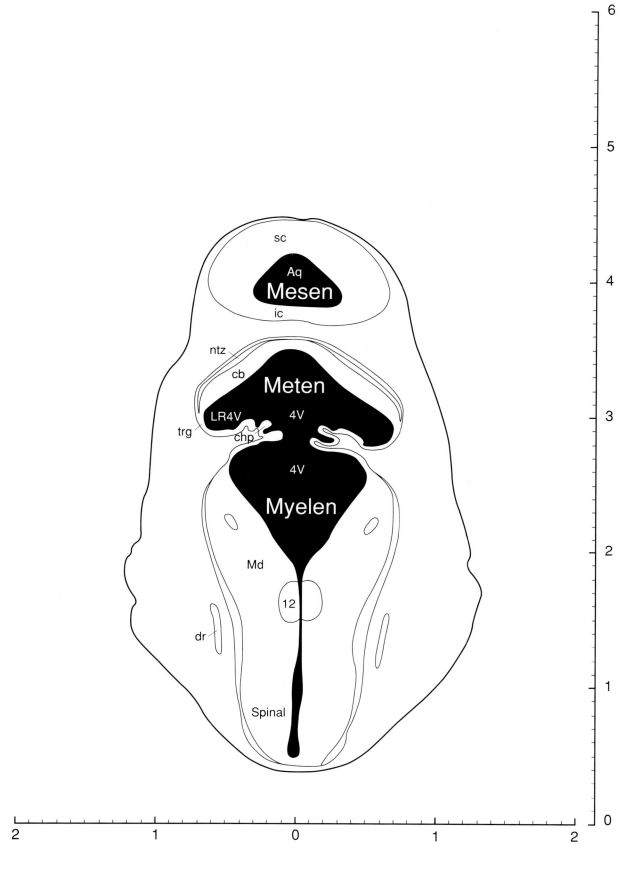

E14 Sagittal Section Plan

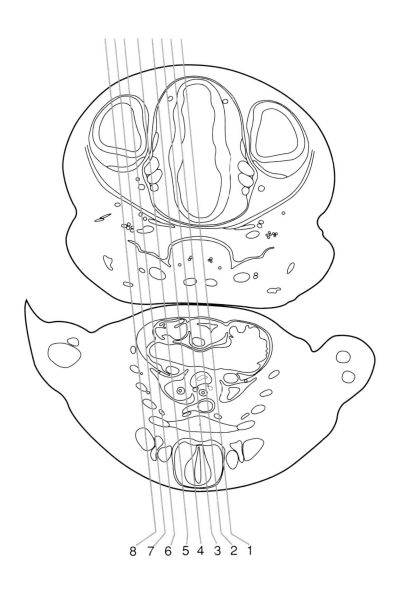

8 7 6 5 4 3 2 1

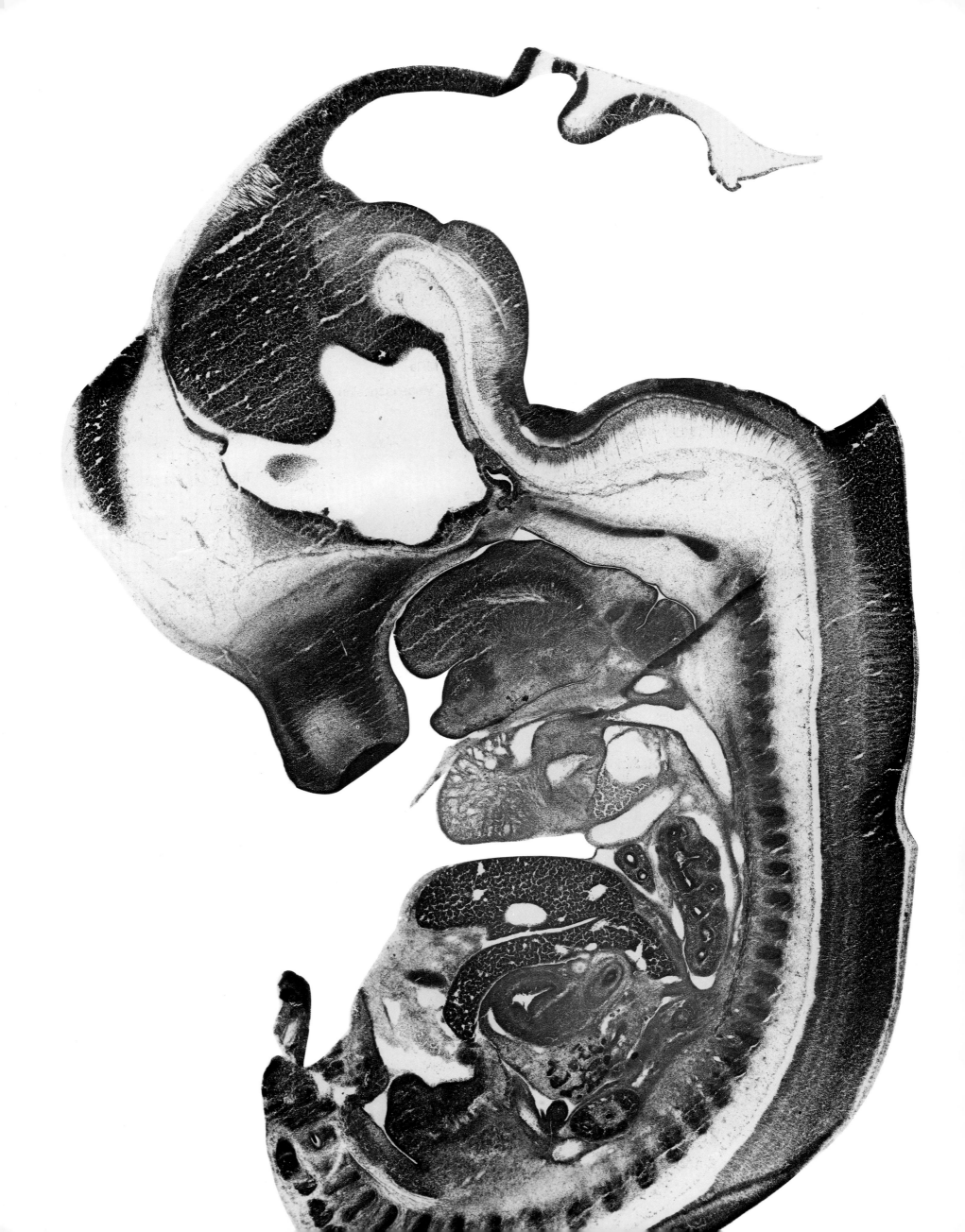

Figure 21
E14 Sagittal 1

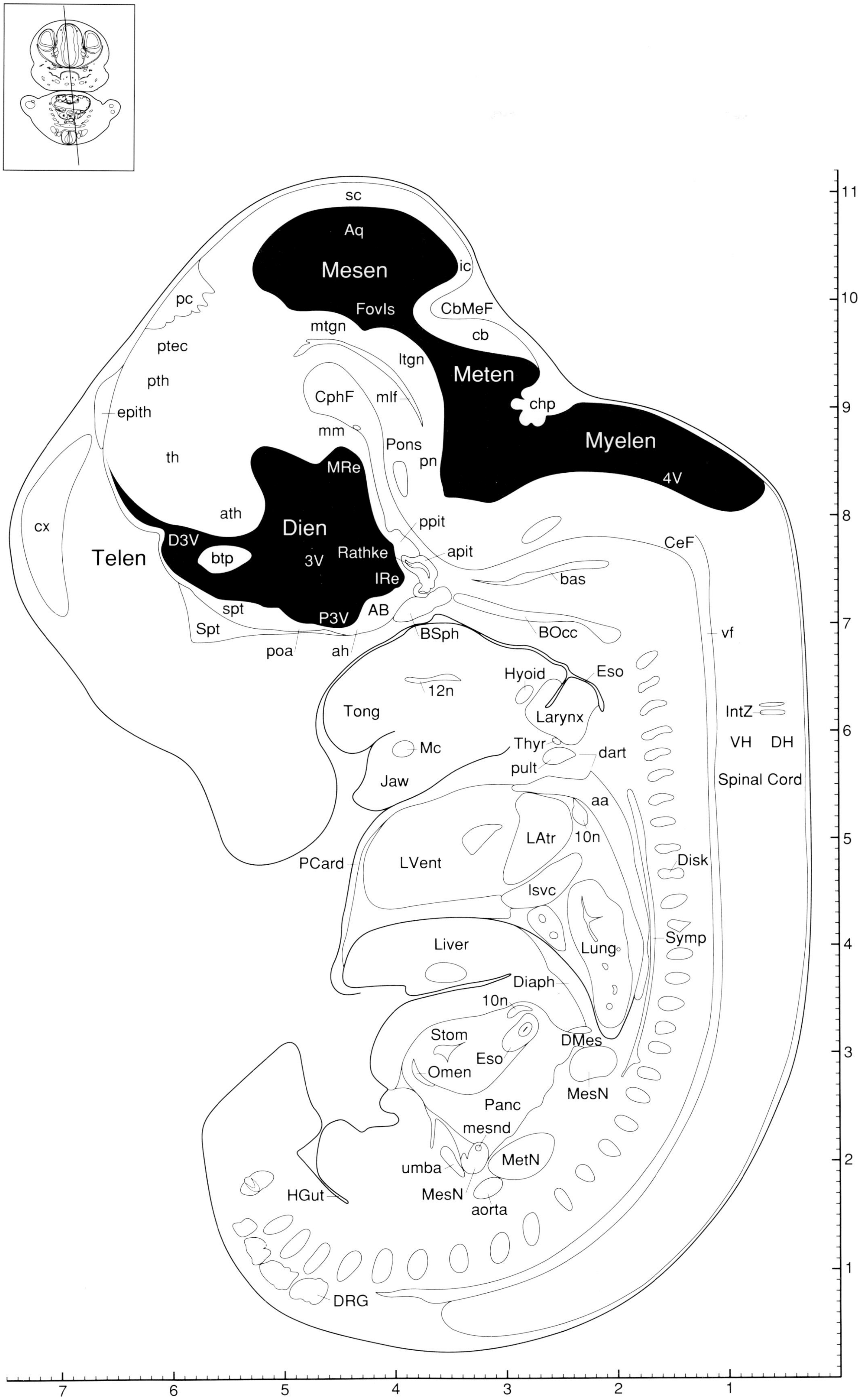

3V   3rd ventricle
4V   4th ventricle
10n   vagus nerve
12n   hypoglossal nerve or its root
aa   aortic arch
AB   anterobasal nucleus
ah   anterior hypothalamic neuroepithel
aorta   aorta
apit   anterior pituitary anlage
Aq   aqueduct
ath   anterior thalamic neuroepithelium
bas   basilar artery
BOcc   basioccipital bone
BSph   basisphenoid bone
btp   basal telecephalic plate, posterior p
cb   cerebellar neuroepithelium
CbMeF   cerebellomesencephalic flex
CeF   cervical flexure
chp   choroid plexus primordium
CphF   cephalic flexure
cx   cortical neuroepithelium
D3V   dorsal third ventricle
dart   ductus arteriosus
DH   dorsal horns of spinal cord
Diaph   diaphragm
Dien   diencephalon
Disk   intervertebral disk
DMes   dorsal mesentery
DRG   dorsal root ganglion
epith   epithalamic neuroepithelium
Eso   esophagus
FovIs   fovea of the isthmus
HGut   hindgut
Hyoid   hyoid bone
ic   inferior colliculus neuroepithelium
IntZ   intermediate zone of spinal gray
IRe   infundibular recess 3rd ventricle
Jaw   jaw
Larynx   larynx
LAtr   left atrium
Liver   liver
lsvc   left superior vena cava
ltgn   lower tegmental neuroepithelium
Lung   lung
LVent   left ventricle
Mc   Meckel's cartilage
Mesen   mesencephalon
MesN   mesonephros
mesnd   mesonephric duct
Meten   metencephalon
MetN   metanephros
mlf   medial longitudinal fasciculus
mm   mammillary body neuroepithel
MRe   mammillary recess 3rd ventricle
mtgn   medial tegmental neuroepithel
Myelen   myelencephalon
Omen   omental bursa
P3V   preoptic recess of the 3rd vent
Panc   pancreas
pc   posterior commissure
PCard   pericardium
pn   pontine nuclei neuroepithelium
poa   preoptic area neuroepithelium
Pons   pons
ppit   posterior pituitary
ptec   pretectum neuroepithelium
pth   posterior thalamic neuroepithelium
pult   pulmonary trunk
Rathke   Rathke's pouch
sc   superior colliculus neuroepithelium
Spinal Cord   spinal cord
Spt   septum
spt   septal neuroepithelium
Stom   stomach
Symp   sympathetic trunk
Telen   telencephalon
th   thalamic neuroepithelium
Thyr   thyroid gland
Tong   tongue
umba   umbilical artery
vf   ventral funiculus
VH   ventral horn

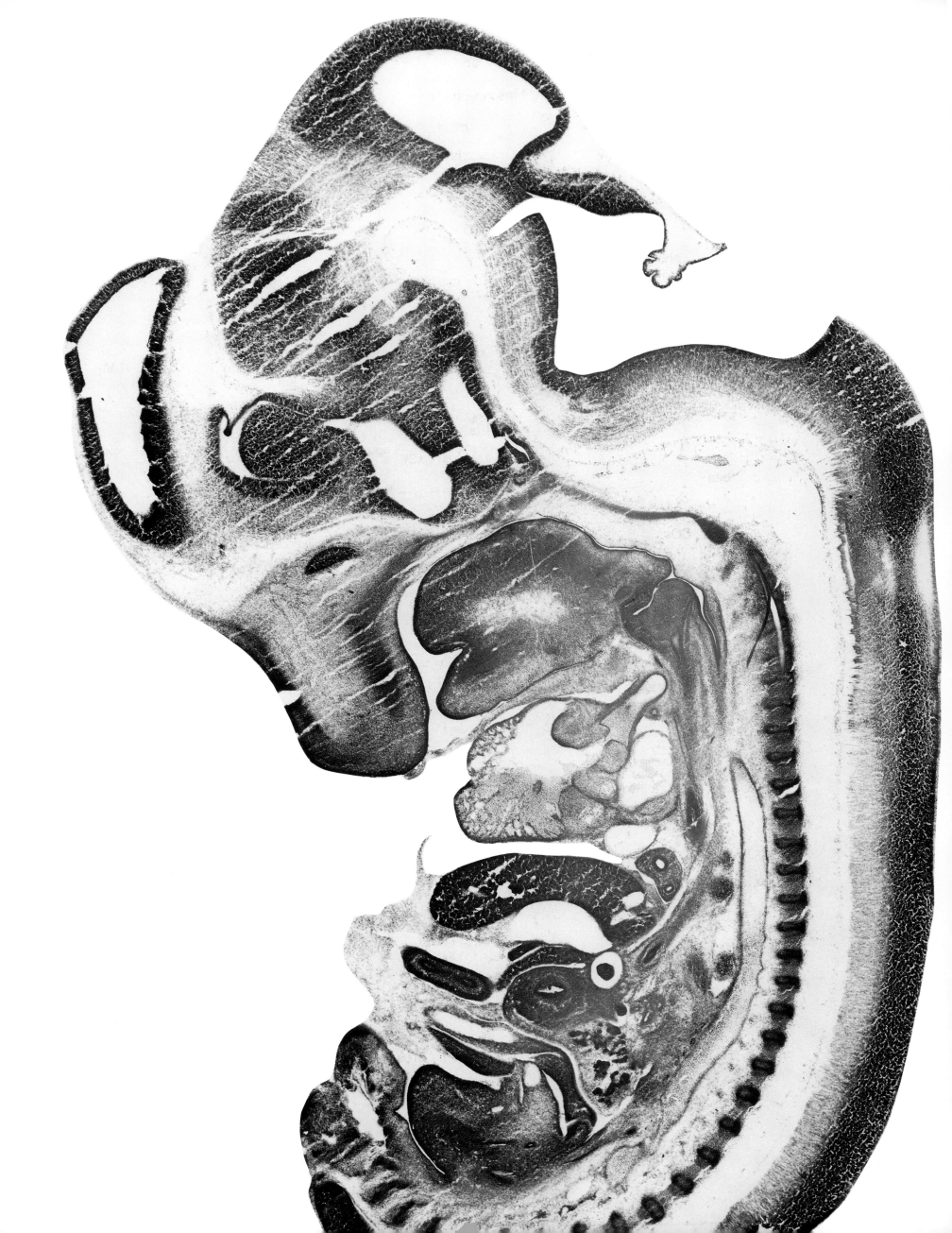

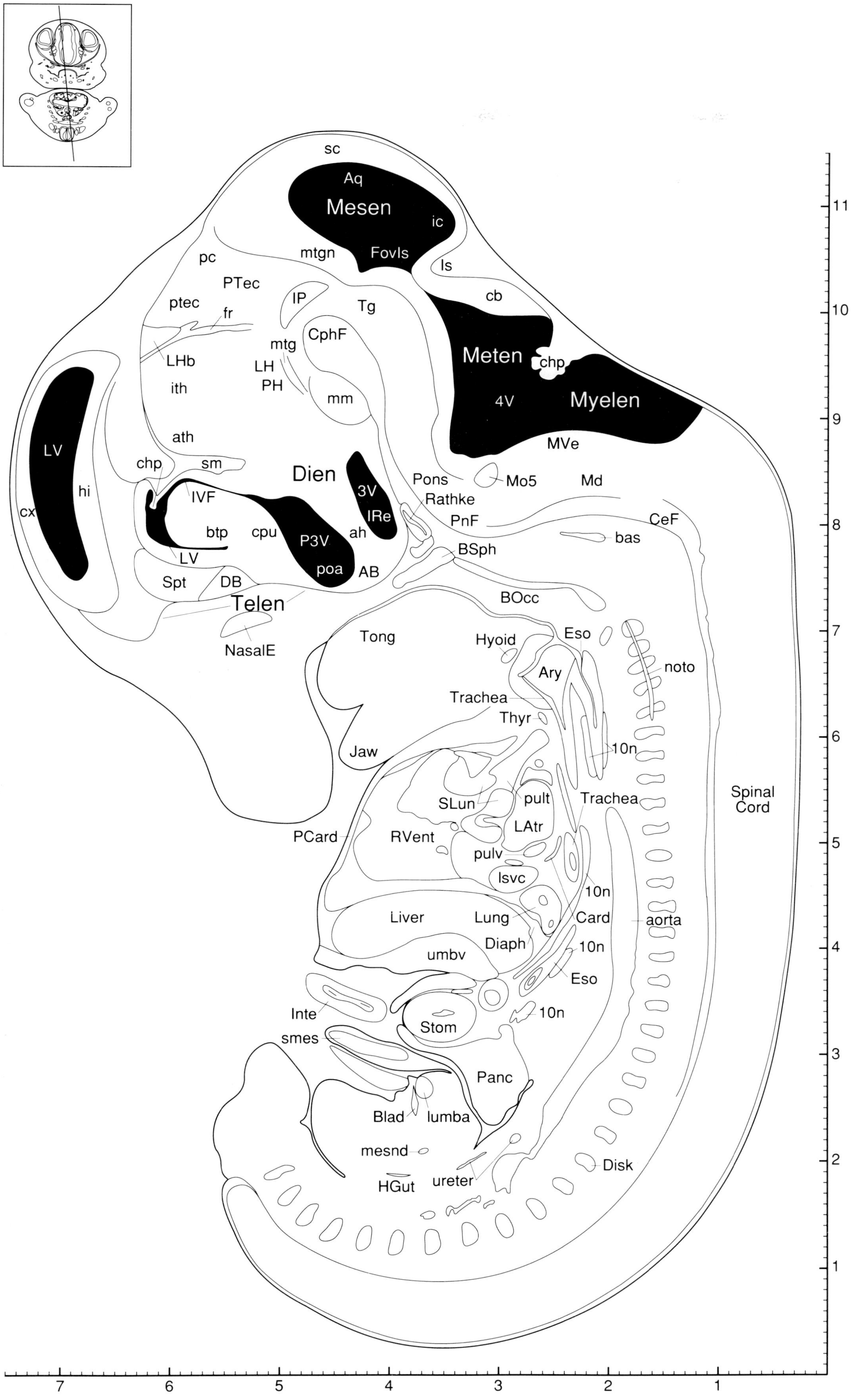

Figure 22
E14 Sagittal 2

3V 3rd ventricle
4V 4th ventricle
10n vagus nerve
AB anterobasal nucleus
ah anterior hypothalamic neuroepithel
aorta aorta
Aq aqueduct
Ary arytenoid swelling of the larynx
ath anterior thalamic neuroepithelium
bas basilar artery
Blad urinary bladder
BOcc basioccipital bone
BSph basisphenoid bone
btp basal telecephalic plate, posteror p
Card cardiac ganglion
cb cerebellar neuroepithelium
chp choroid plexus primordium
CphF cephalic flexure
cpu caudate putamen neuroepithelium
cx cortical neuroepithelium
DB nuclei of the diagonal band
Diaph diaphragm
Dien diencephalon
Disk intervetebral disk
Eso esophagus
FovIs fovea of the isthmus
fr fasciculus retroflexus
HGut hindgut
hi hippocampal formation neuroepi
Hyoid hyoid bone
ic inferior colliculus neuroepithelium
Inte intestine
IP interpeduncular nu
IRe infundibular recess 3rd ventricle
Is isthmus region
ith intermediate thalamic ne
IVF interventricular foramen
Jaw jaw
LAtr left atrium
LH lateral hypothalamic area
LHb lateral habenular nu
Liver liver
lsvc left superior vena cava
lumba left umbilical artery
Lung lung
LV lateral ventricle
Md medulla
Mesen mesencephalon
mesnd mesonephric duct
Meten metencephalon
mm mammillary body neuroepithel
Mo5 motor trigeminal nu
mtg mammillotegmental tract
mtgn medial tegmental neuroepithel
MVe medial vestibular nu
Myelen myelencephalon
NasalE nasal epithelium
noto notochord
P3V preoptic recess of the 3rd vent
Panc pancreas
pc posterior commissure
PCard pericardium
PH posterior hypoth area
PnF pontine flexure
poa preoptic area neuroepithelium
Pons pons
PTec pretectum
ptec pretectum neuroepithelium
pult pulmonary trunk
pulv pulmonary vein
Rathke Rathke's pouch
RVent right ventricle
sc superior colliculus neuroepithelium
SLun semilunar valves
sm stria medullaris thalami
smes superior mesenteric artery
Spinal Cord spinal cord
Spt septum
Stom stomach
Telen telencephalon
Tg tegmentum
Thyr thyroid gland
Tong tongue
Trachea trachea
umbv umbilical vein
ureter ureter

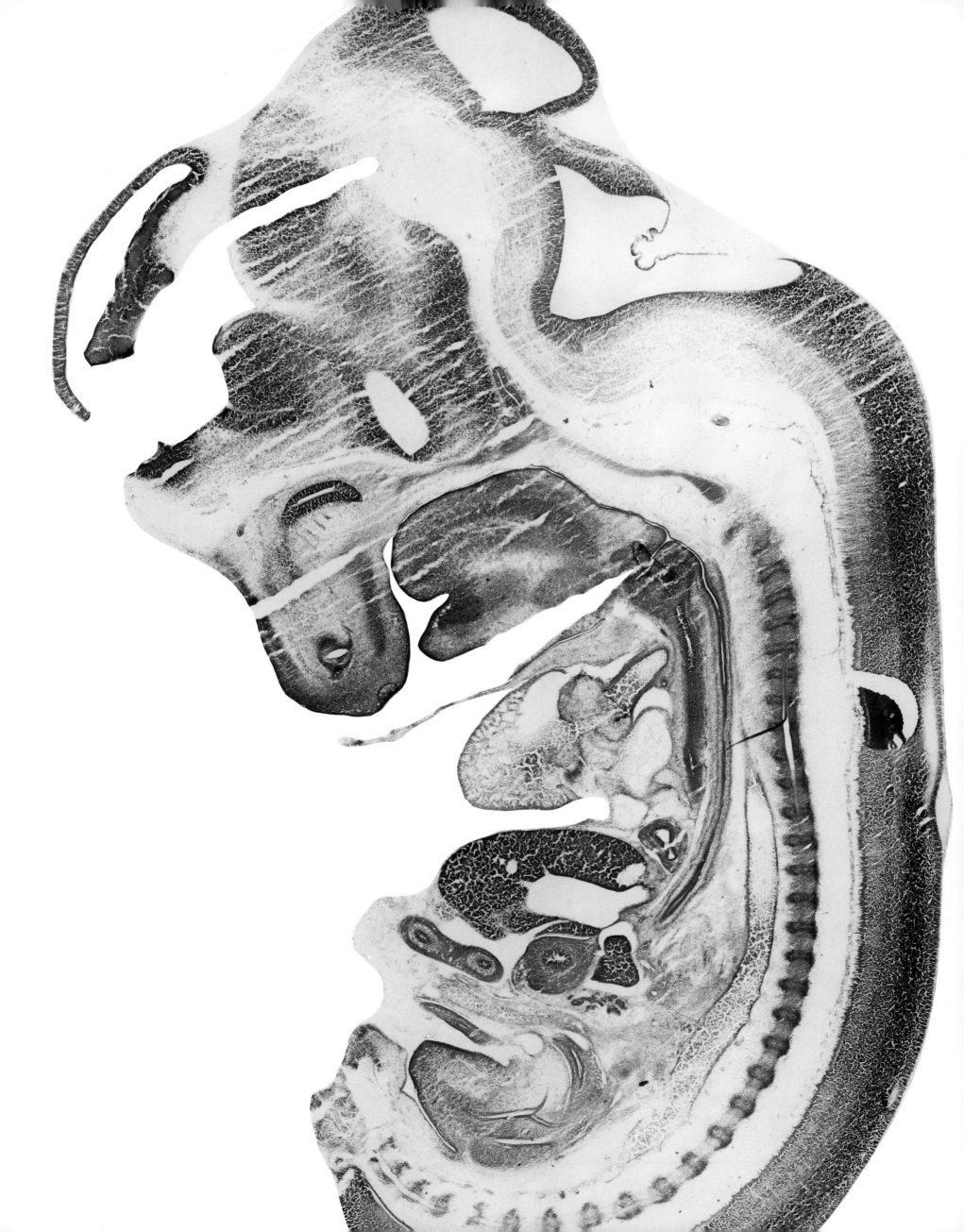

Figure 23
E14 Sagittal 3

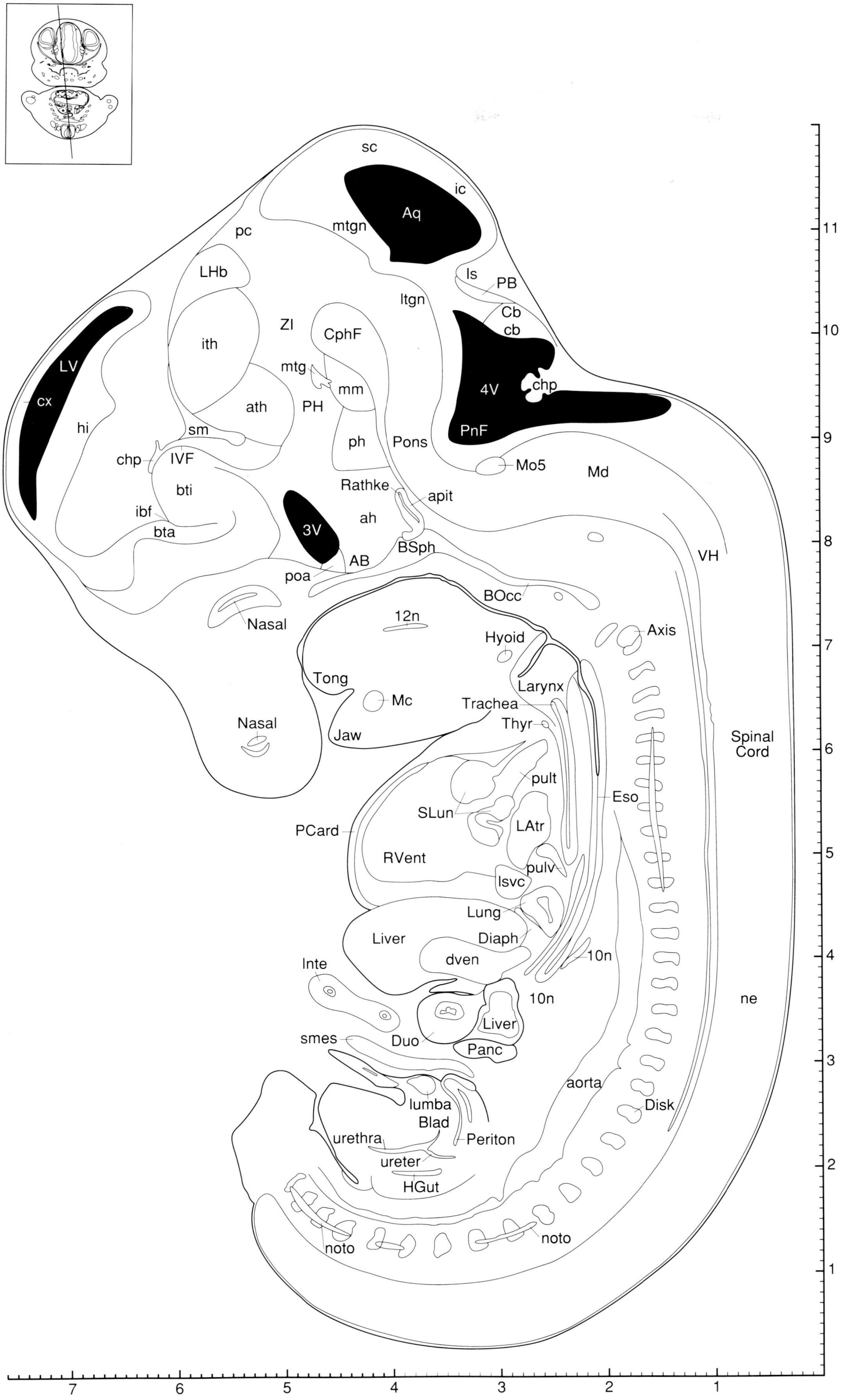

3V   3rd ventricle
4V   4th ventricle
10n   vagus nerve
12n   hypoglossal nerve or its root
AB   anterobasal nucleus
ah   anterior hypothalamic neuroepithel
aorta   aorta
apit   anterior pituitary anlage
Aq   aqueduct
ath   anterior thalamic neuroepithelium
Axis   axis (C2 vertebra)
Blad   urinary bladder
BOcc   basioccipital bone
BSph   basisphenoid bone
bta   basal telencephalic plate, ant part
bti   basal telencephalic plate, interm
Cb   cerebellum
cb   cerebellar neuroepithelium
chp   choroid plexus primordium
CphF   cephalic flexure
cx   cortical neuroepithelium
Diaph   diaphragm
Disk   intervertebral disk
Duo   duodenum
dven   ductus venosus
Eso   esophagus
HGut   hindgut
hi   hippocampal formation neuroepi
Hyoid   hyoid bone
ibf   interbasal plate fissure
ic   inferior colliculus neuroepithelium
Inte   intestine
Is   isthmus region
ith   intermediate thalamic ne
IVF   interventricular foramen
Jaw   jaw
Larynx   larynx
LAtr   left atrium
LHb   lateral habenular nu
Liver   liver
lsvc   left superior vena cava
ltgn   lower tegmental neuroepithelium
lumba   left umbilical artery
Lung   lung
LV   lateral ventricle
Mc   Meckel's cartilage
Md   medulla
mm   mammillary body neuroepithel
Mo5   motor trigeminal nu
mtg   mammillotegmental tract
mtgn   medial tegmental neuroepithel
Nasal   nasal cavity
ne   neuroepithelium
noto   notochord
Panc   pancreas
PB   parabrachial nuclei
pc   posterior commissure
PCard   pericardium
Periton   peritoneal cavity
PH   posterior hypoth area
ph   posterior hypothalamic ne
PnF   pontine flexure
poa   preoptic area neuroepithelium
Pons   pons
pult   pulmonary trunk
pulv   pulmonary vein
Rathke   Rathke's pouch
RVent   right ventricle
sc   superior colliculus neuroepithelium
SLun   semilunar valves
sm   stria medullaris thalami
smes   superior mesenteric artery
Spinal   spinal cord
Thyr   thyroid gland
Tong   tongue
Trachea   trachea
ureter   ureter
urethra   urethra
VH   ventral horn
ZI   zona incerta

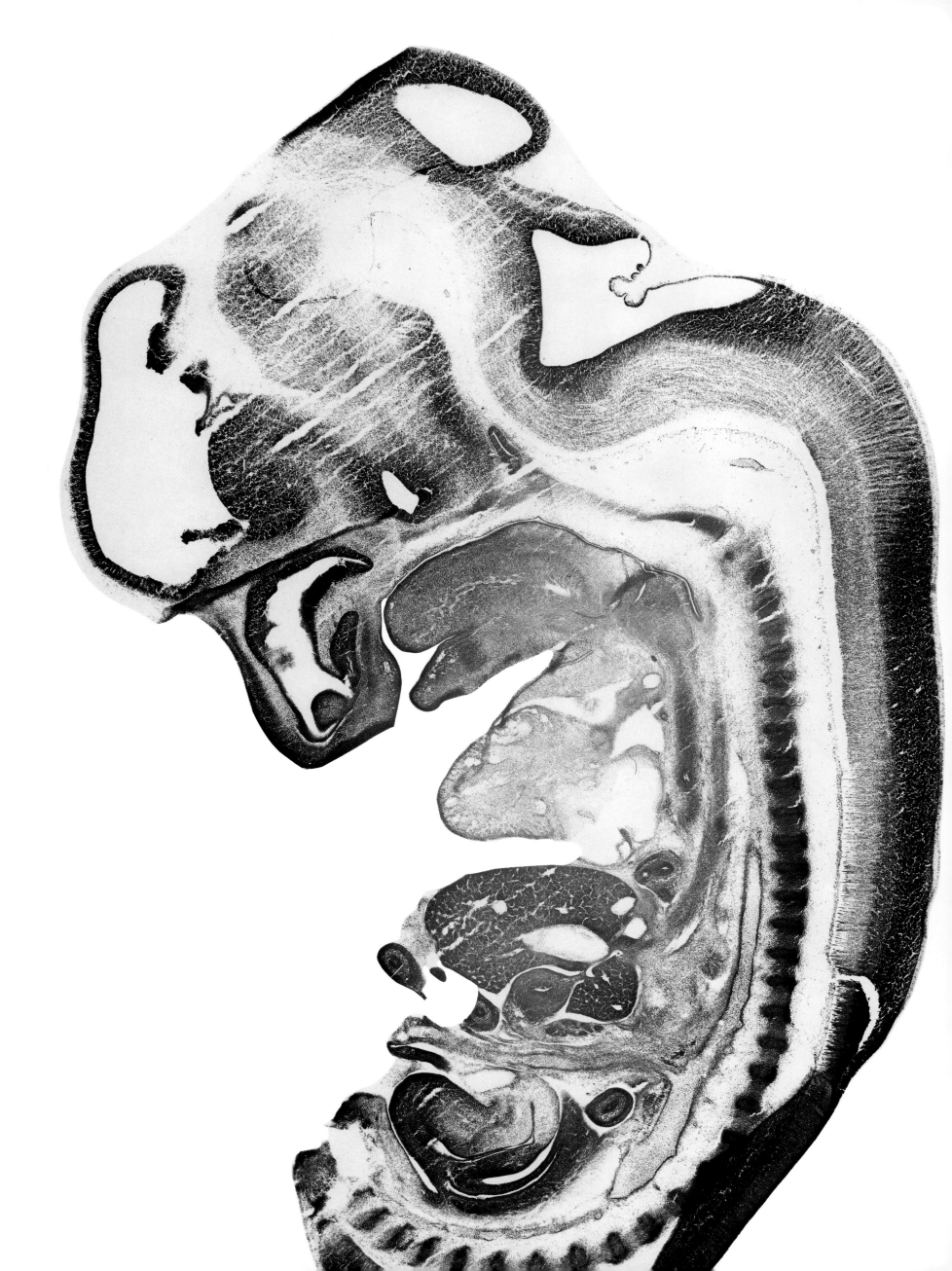

Figure 24
E14 Sagittal 4

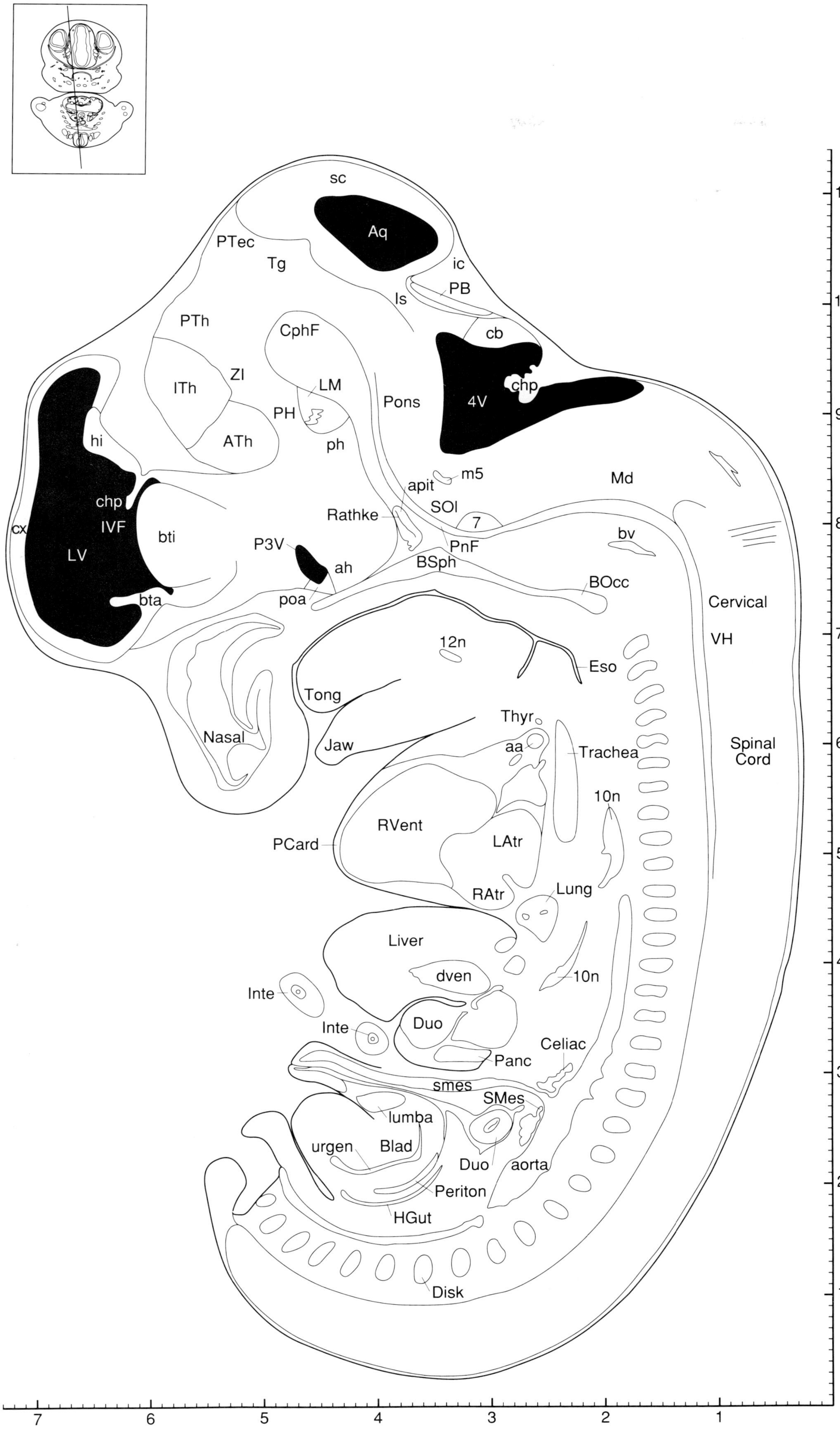

4V   4th ventricle
7   facial nu
10n   vagus nerve
12n   hypoglossal nerve or its root
aa   aortic arch
ah   anterior hypothalamic neuroepithel
aorta   aorta
apit   anterior pituitary anlage
Aq   aqueduct
ATh   anterior thalamus
Blad   urinary bladder
BOcc   basioccipital bone
BSph   basisphenoid bone
bta   basal telencephalic plate, ant part
bti   basal telencephalic plate, interm
bv   blood vessel
cb   cerebellar neuroepithelium
Celiac   celiac ganglion
Cervical   cervical spinal cord
chp   choroid plexus primordium
CphF   cephalic flexure
cx   cortical neuroepithelium
Disk   intervertebral disk
Duo   duodenum
dven   ductus venosus
Eso   esophagus
HGut   hindgut
hi   hippocampal formation neuroepi
ic   inferior colliculus neuroepithelium
Inte   intestine
Is   isthmus region
ITh   intermediate thalamus
IVF   interventricular foramen
Jaw   jaw
LAtr   left atrium
Liver   liver
LM   lateral mammillary nu
lumba   left umbilical artery
Lung   lung
LV   lateral ventricle
m5   motor root trigeminal nerve
Md   medulla
Nasal   nasal cavity
P3V   preoptic recess of the 3rd vent
Panc   pancreas
PB   parabrachial nuclei
PCard   pericardium
Periton   peritoneal cavity
PH   posterior hypoth area
ph   posterior hypothalamic ne
PnF   pontine flexure
poa   preoptic area neuroepithelium
Pons   pons
PTec   pretectum
PTh   posterior thalamus
Rathke   Rathke's pouch
RAtr   right atrium
RVent   right ventricle
sc   superior colliculus neuroepithelium
SMes   superior mesenteric ganglion
smes   superior mesenteric artery
SOl   superior olive
Spinal   spinal cord
Tg   tegmentum
Thyr   thyroid gland
Tong   tongue
Trachea   trachea
urgen   urogenital tract
VH   ventral horn
ZI   zona incerta

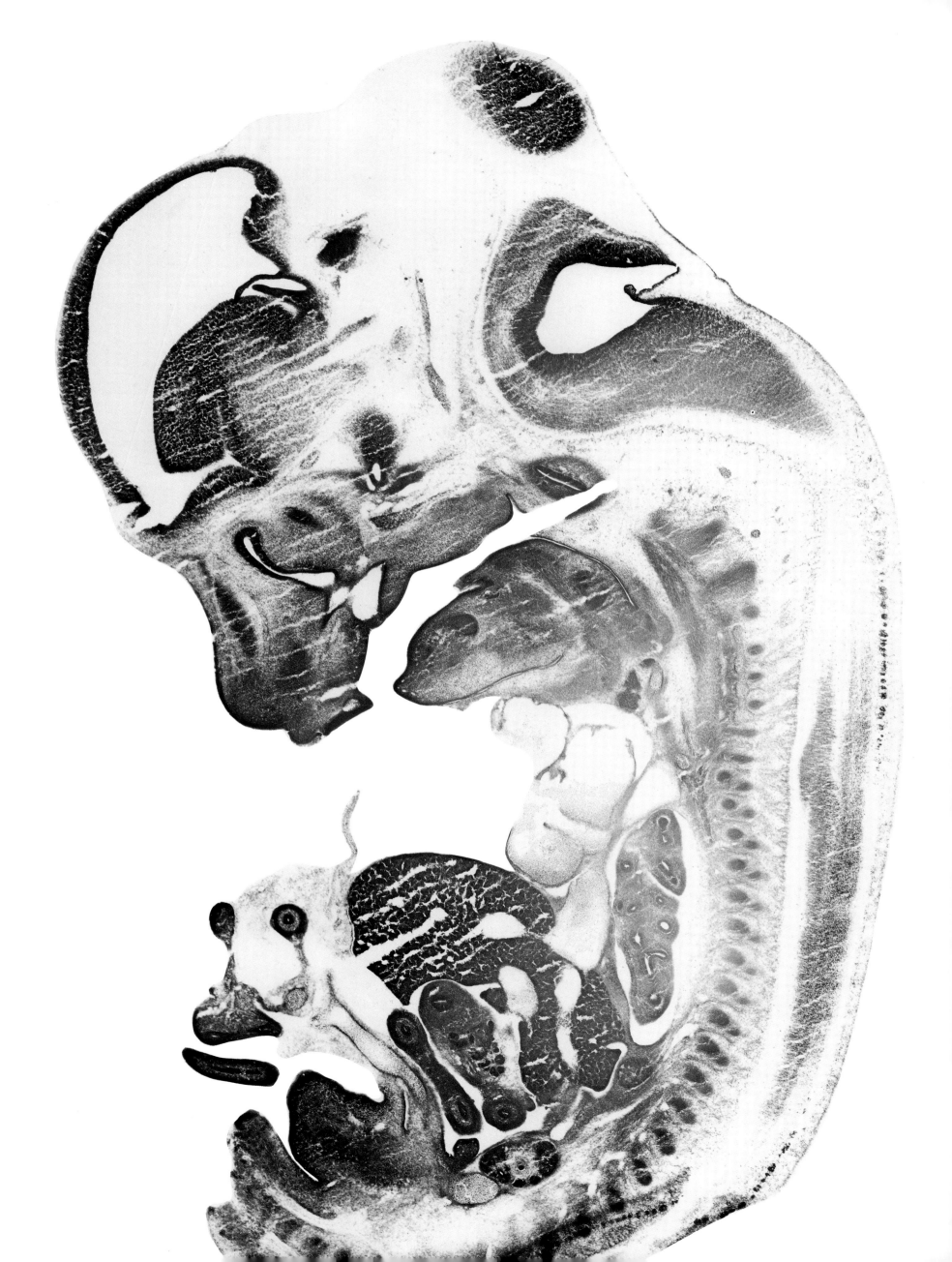

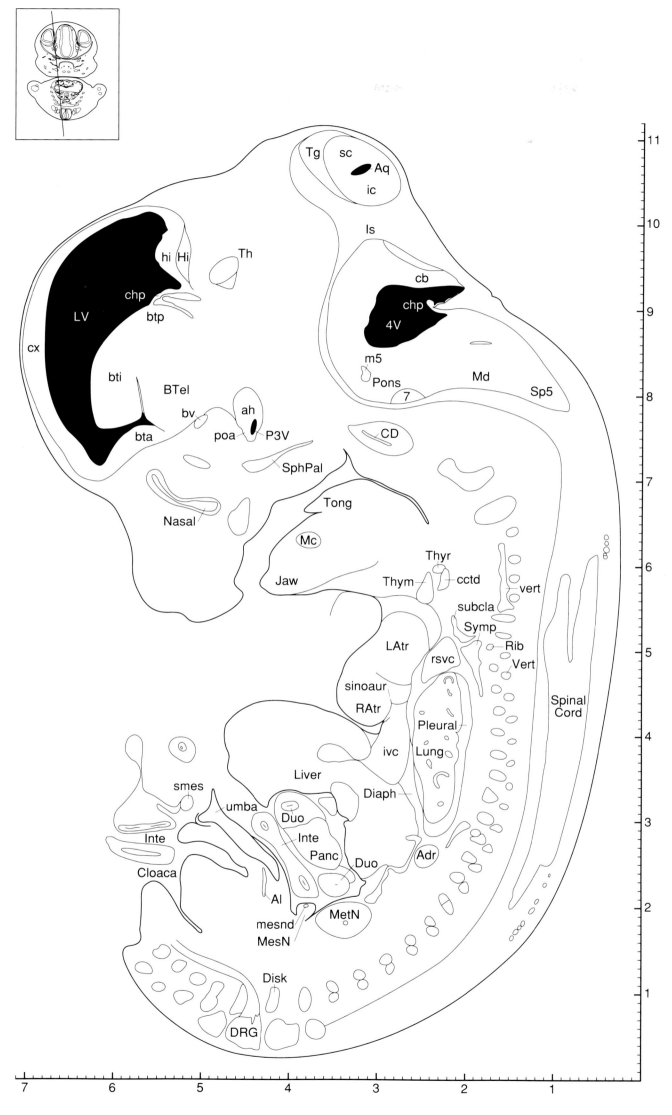

Figure 25
E14 Sagittal 5

4V   4th ventricle
7   facial nu
Adr   adrenal gland
ah   anterior hypothalamic neuroepithel
Al   allantois
Aq   aqueduct
bta   basal telencephalic plate, ant part
BTel   basal telencephalon
bti   basal telencephalic plate, interm
btp   basal telecephalic plate, posteror p
bv   blood vessel
cb   cerebellar neuroepithelium
cctd   common carotid artery
CD   cochlear duct
chp   choroid plexus primordium
Cloaca   cloaca
cx   cortical neuroepithelium
Diaph   diaphragm
Disk   intervertebral disk
DRG   dorsal root ganglion
Duo   duodenum
Hi   hippocampal formation
hi   hippocampal formation neuroepi
ic   inferior colliculus neuroepithelium
Inte   intestine
Is   isthmus region
ivc   inferior vena cava
Jaw   jaw
LAtr   left atrium
Liver   liver
Lung   lung
LV   lateral ventricle
m5   motor root trigeminal nerve
Mc   Meckel's cartilage
Md   medulla
MesN   mesonephros
mesnd   mesonephric duct
MetN   metanephros
Nasal   nasal cavity
P3V   preoptic recess of the 3rd vent
Panc   pancreas
Pleural   pleural cavity
poa   preoptic area neuroepithelium
Pons   pons
RAtr   right atrium
Rib   rib
rsvc   right superior vena cava
sc   superior colliculus neuroepithelium
sinoaur   sinoauricular valve
smes   superior mesenteric artery
Sp5   spinal trigeminal nucleus
SphPal   sphenopalatine ganglion
Spinal   spinal cord
subcla   subclavian artery
Symp   sympathetic trunk
Tg   tegmentum
Th   thalamus
Thym   thymus gland
Thyr   thyroid gland
Tong   tongue
umba   umbilical artery
Vert   vertebra
vert   vertebral artery

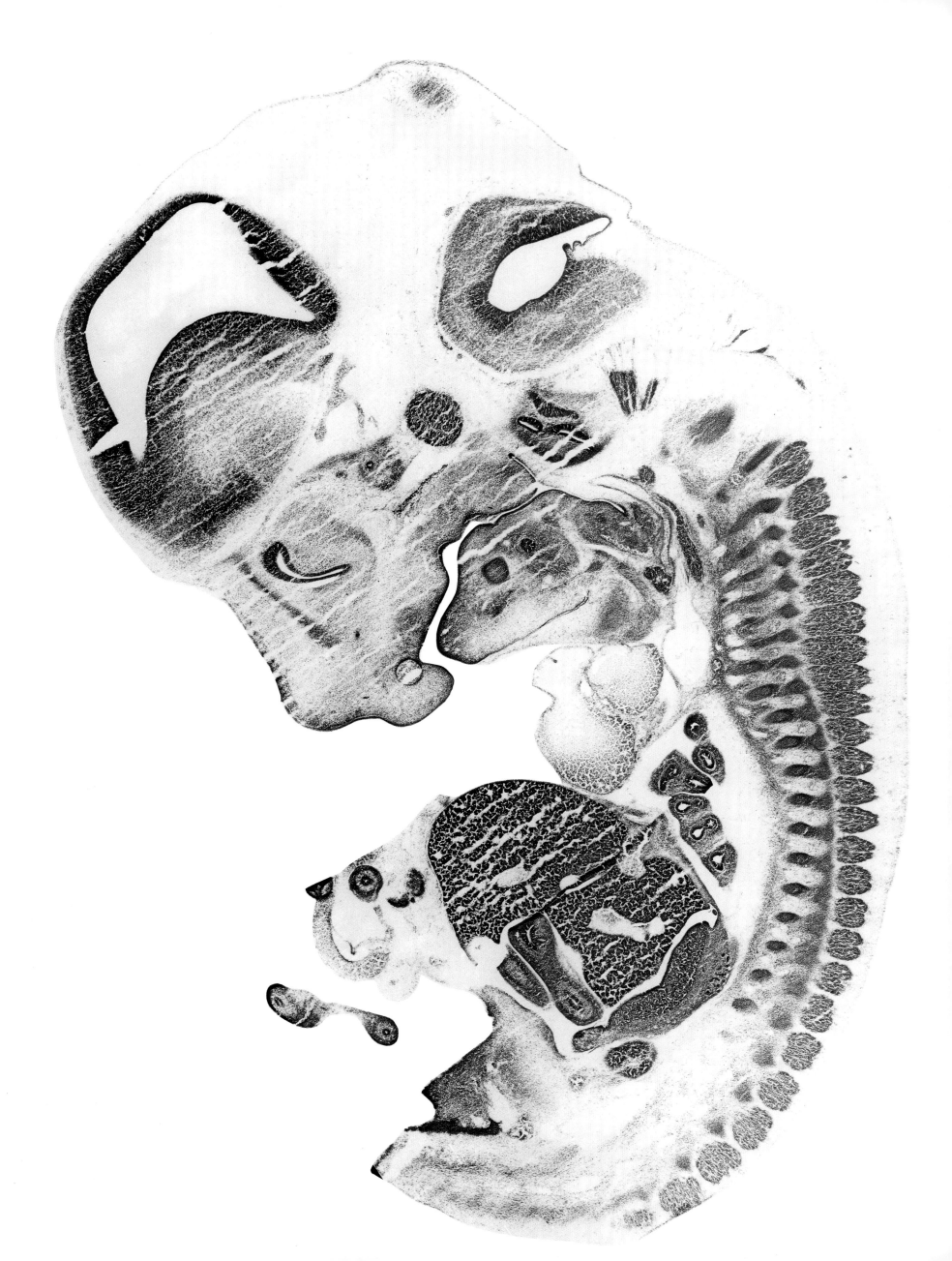

Figure 26
E14 Sagittal 6

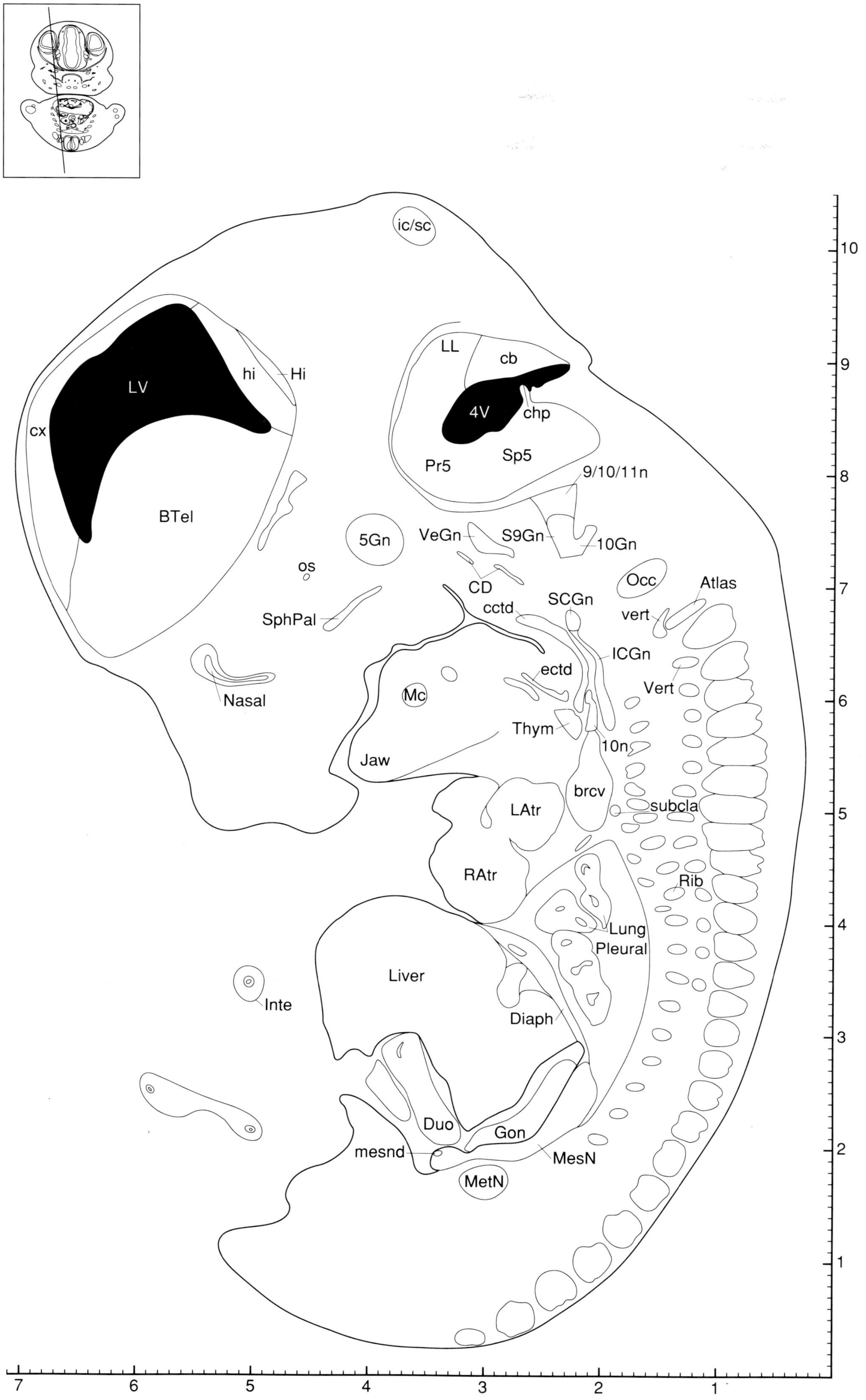

4V   4th ventricle
5Gn   trigeminal ganglion
9n   glossopharyngeal nerve
10Gn vagal ganglion
10n   vagus nerve
11n   spinal accessory nerve
Atlas   atlas (C1 vertebra)
brcv brachiocephalic vein
BTel basal telencephalon
cb   cerebellar neuroepithelium
cctd common carotid artery
CD   cochlear duct
chp   choroid plexus primordium
cx   cortical neuroepithelium
Diaph   diaphragm
Duo   duodenum
ectd external carotid a
Gon   gonad
Hi   hippocampal formation
hi   hippocampal formation neuroepi
ic   inferior colliculus neuroepithel
ICGn inferior cervical ganglion
Inte intestine
Jaw   jaw
LAtr left atrium
Liver   liver
LL   nuclei of the lateral lemniscus
Lung lung
LV   lateral ventricle
Mc   Meckel's cartilage
MesN mesonephros
mesnd   mesonephric duct
MetN metanephros
Nasal   nasal cavity
Occ   occipital bone
os   optic stalk
Pleural   pleural cavity
Pr5   principal sensory trigeminal nu
RAtr right atrium
Rib   rib
S9Gn superior glossopharyngeal gang
sc   superior colliculus neuroepithel
SCGn superior cervical ganglion
Sp5   spinal trigeminal nucleus
SphPal   sphenopalatine ganglion
subcla   subclavian artery
Thym thymus gland
VeGn vestibular ganglion
Vert vertebra
vert vertebral artery

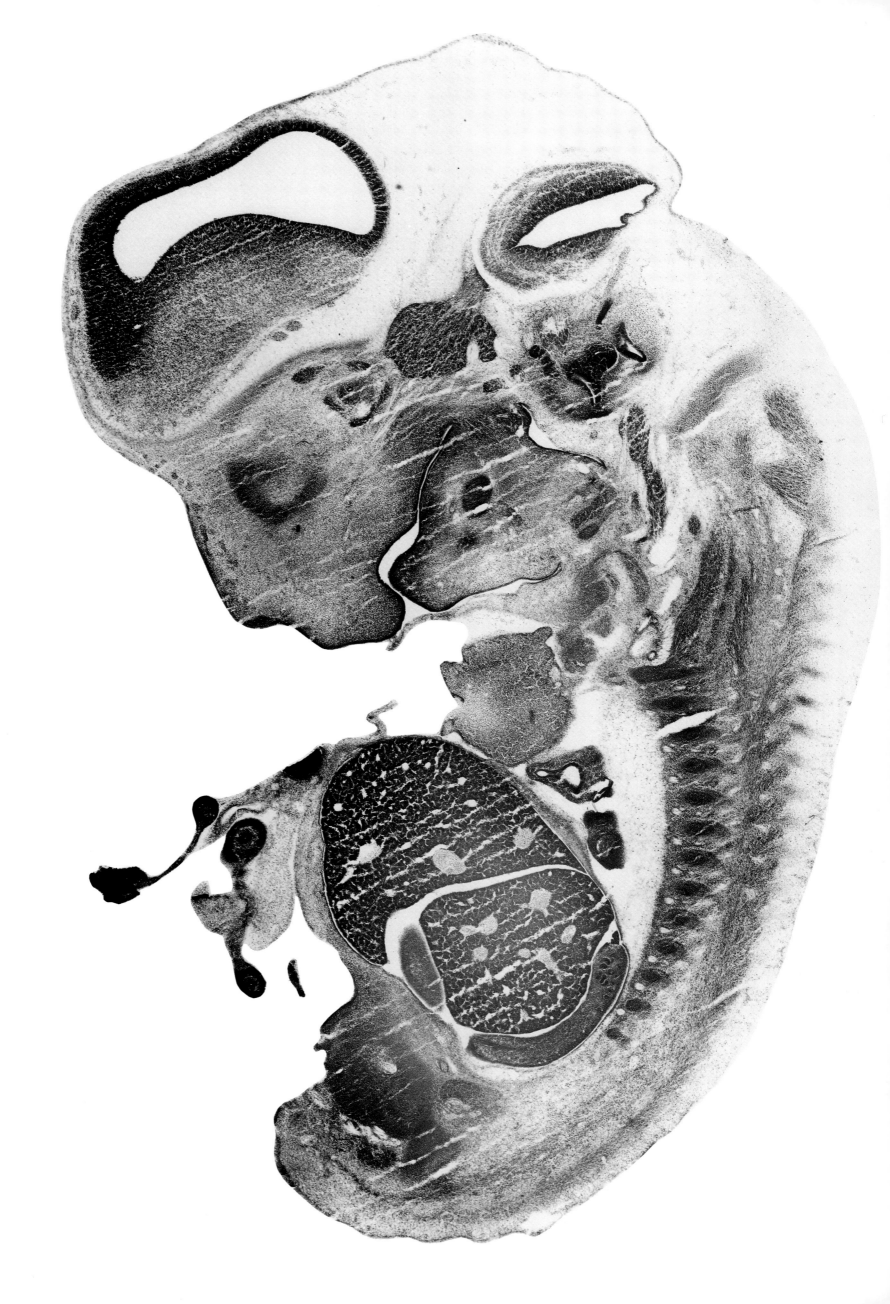

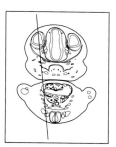

Figure 27
E14 Sagittal 7

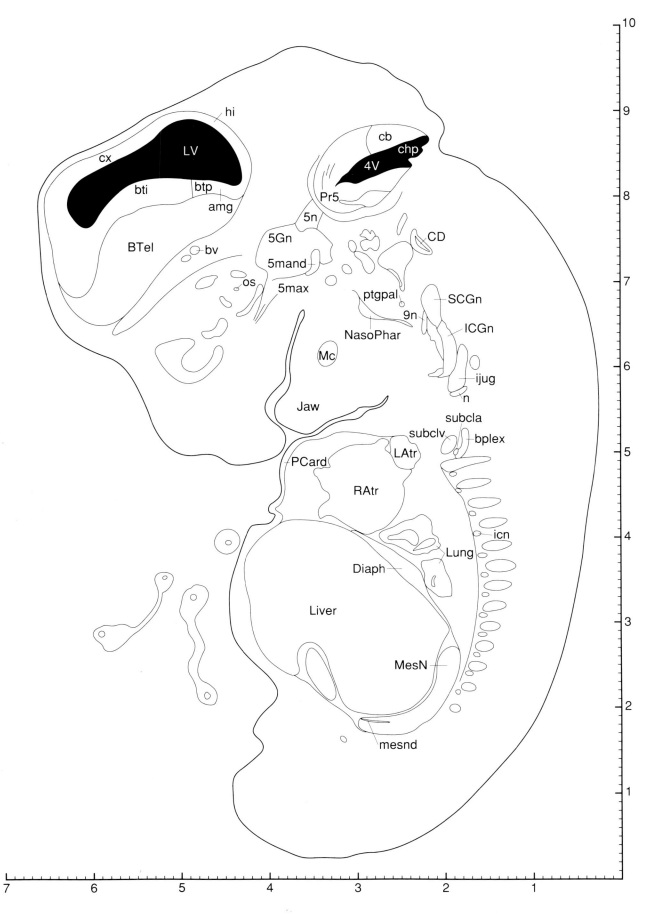

4V   4th ventricle
5Gn   trigeminal ganglion
5mand   mandibular nerve
5max   maxillary nerve
5n   trigeminal nerve
9n   glossopharyngeal nerve
amg   amygdaloid neuroepithelium
bplex   brachial plexus
BTel   basal telencephalon
bti   basal telencephalic plate, interm
btp   basal telecephalic plate, posteror p
bv   blood vessel
cb   cerebellar neuroepithelium
CD   cochlear duct
chp   choroid plexus primordium
cx   cortical neuroepithelium
Diaph   diaphragm
hi   hippocampal formation neuroepi
ICGn   inferior cervical ganglion
icn   intercostal nerve
ijug   internal jugular vein
Jaw   jaw
LAtr   left atrium
Liver   liver
Lung   lung
LV   lateral ventricle
Mc   Meckel's cartilage
MesN   mesonephros
mesnd   mesonephric duct
n   nerve
NasoPhar   nasopharyngeal cavity
os   optic stalk
PCard   pericardium
Pr5   principal sensory trigeminal nu
ptgpal   pterygopalatine artery
RAtr   right atrium
SCGn   superior cervical ganglion
subcla   subclavian artery
subclv   subclavian vein

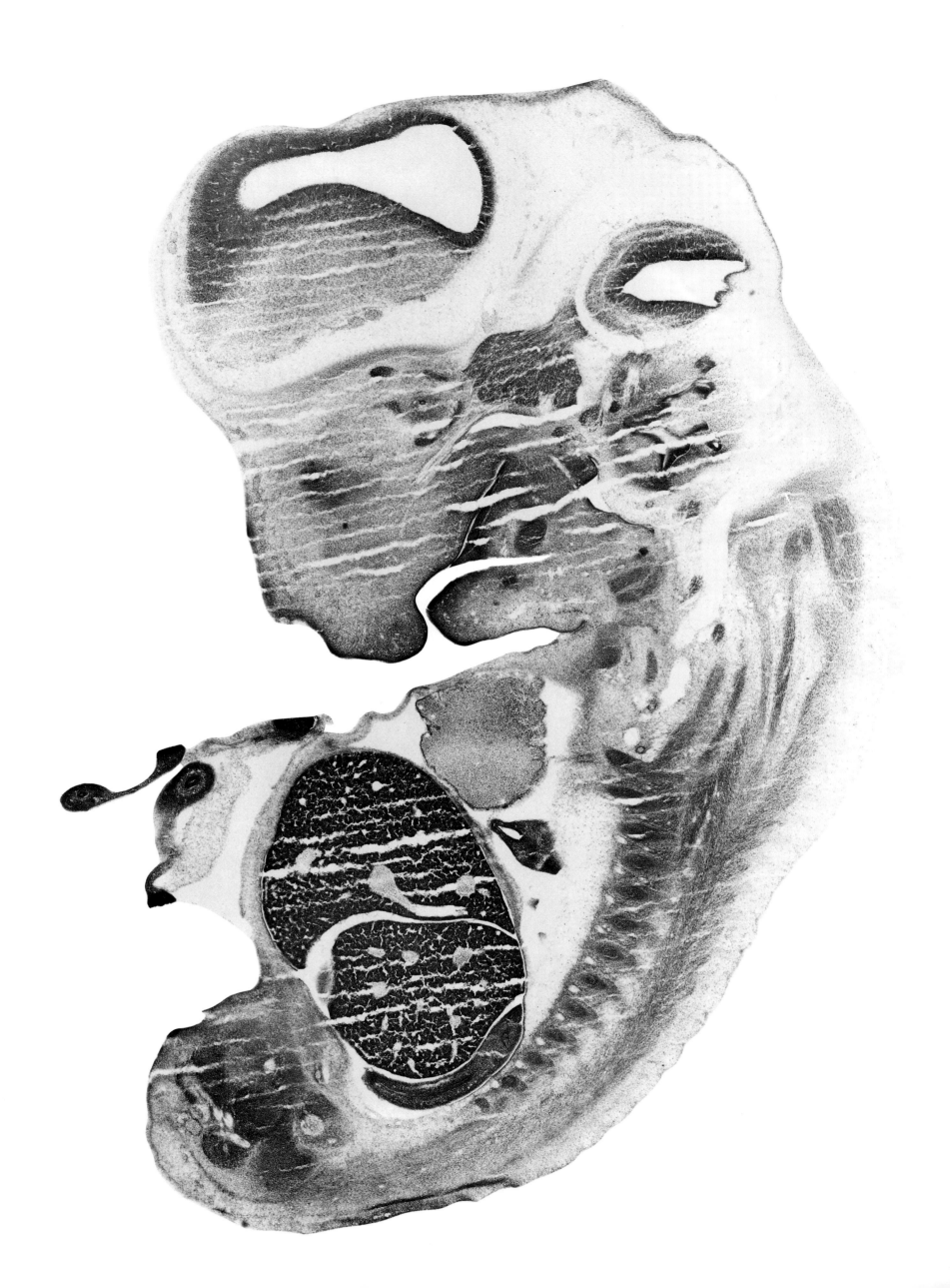

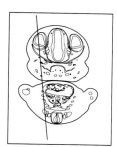

Figure 28
E14 Sagittal 8

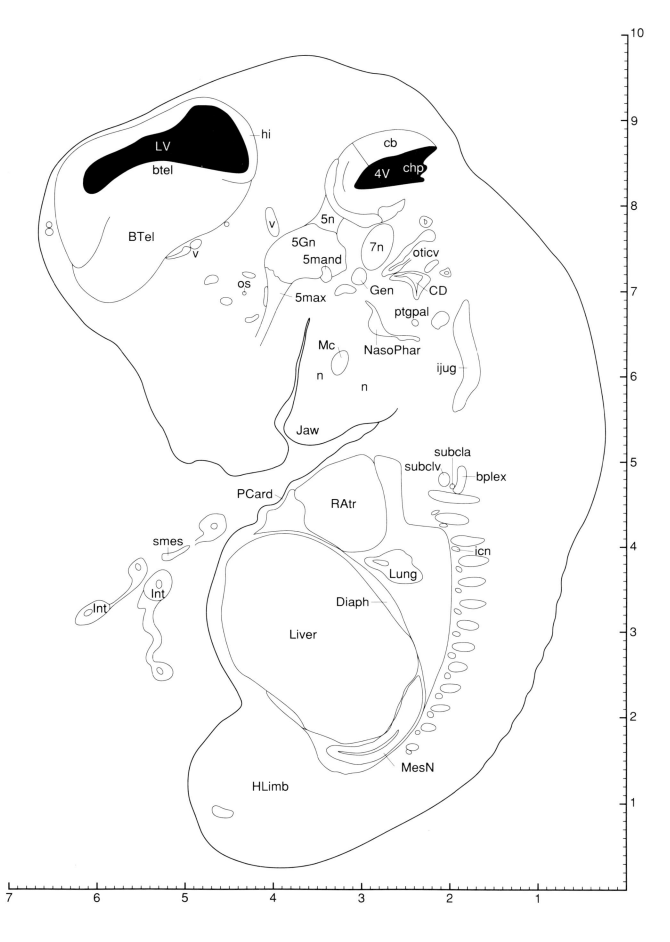

4V   4th ventricle
5Gn   trigeminal ganglion
5mand   mandibular nerve
5max   maxillary nerve
5n   trigeminal nerve
7n   facial nerve or its root
bplex   brachial plexus
BTel   basal telencephalon
btel   basal telencephalic neuroepithel
cb   cerebellar neuroepithelium
CD   cochlear duct
chp   choroid plexus primordium
Diaph   diaphragm
Gen   geniculate ganglion
hi   hippocampal formation neuroepi
HLimb   hind limb
icn   intercostal nerve
ijug   internal jugular vein
Int   interposed cerebellar nu
Jaw   jaw
Liver   liver
Lung   lung
LV   lateral ventricle
Mc   Meckel's cartilage
MesN   mesonephros
n   nerve
NasoPhar   nasopharyngeal cavity
os   optic stalk
oticv   otic vesicle
PCard   pericardium
ptgpal   pterygopalatine artery
RAtr   right atrium
smes   superior mesenteric artery
subcla   subclavian artery
subclv   subclavian vein
v   vein

E16 Coronal Section Plan

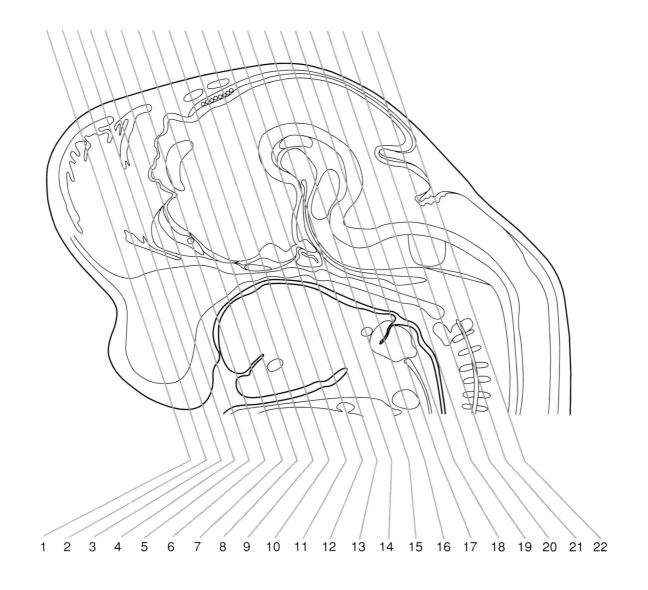

1 2 3 4 5 6 7 8 9 10 11 12 13 14 15 16 17 18 19 20 21 22

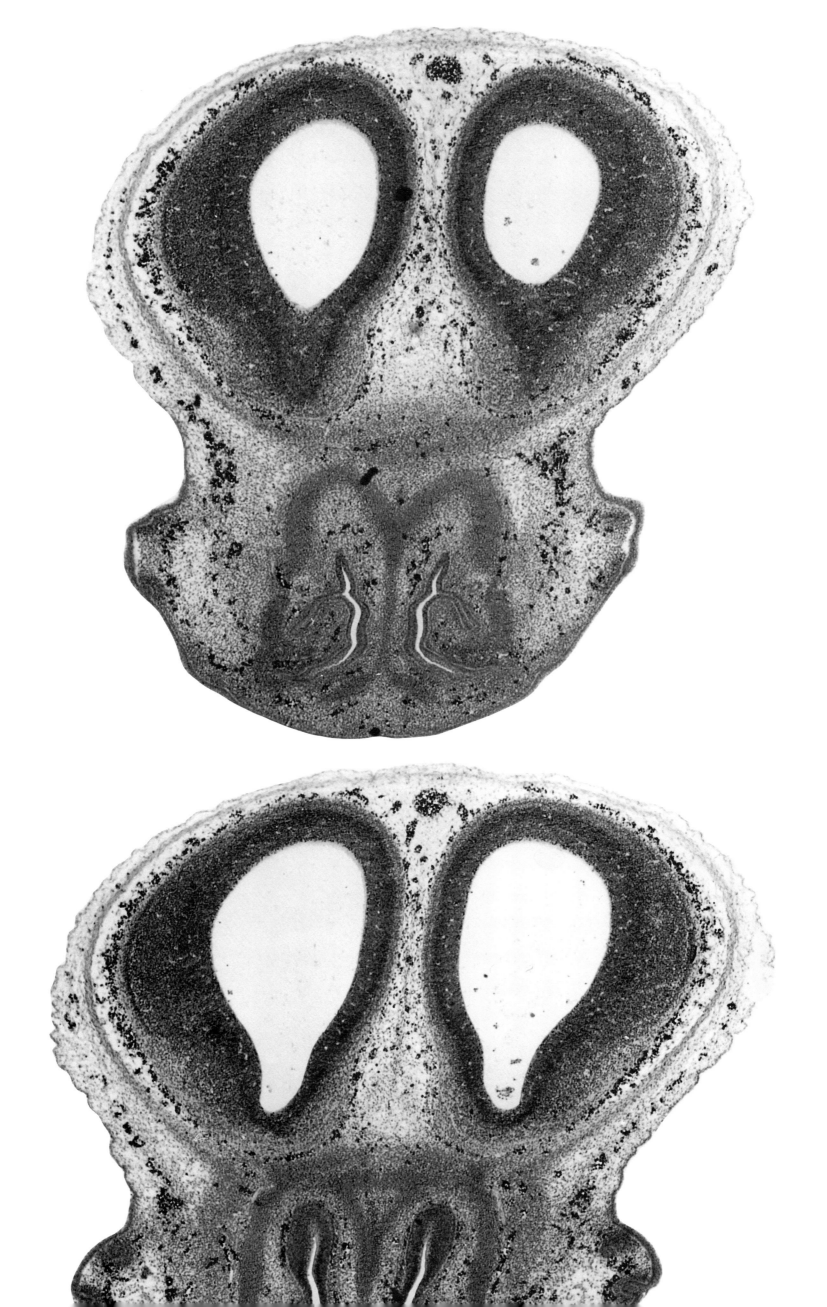

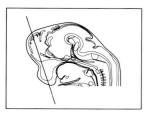

Figure 29
E16 Coronal 1

1   cortical layer 1
bta   basal telencephalic plate, anterior
cx   cortical neuroepi
CxP   cortical plate
EPl   external plexiform layer of the olf bulb
Ethmoid   ethmoid bone
Gl   glomerular layer of the olf bulb
ICx   intermediate cortical layer
LV   lateral ventricle
MaxT   maxilloturbinate
Mi   mitral cell layer of the olf bulb
NasT   nasal turbinate
obn   olfactory bulb neuroepi
olf   olfactory nerve
sss   superior sagittal sinus
SubV   subventricular cortical layer
Vomer   vomer

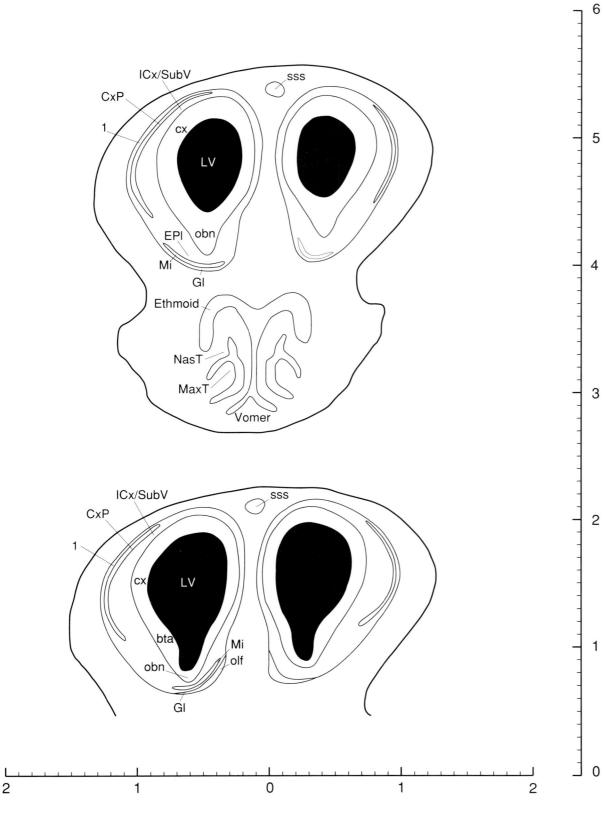

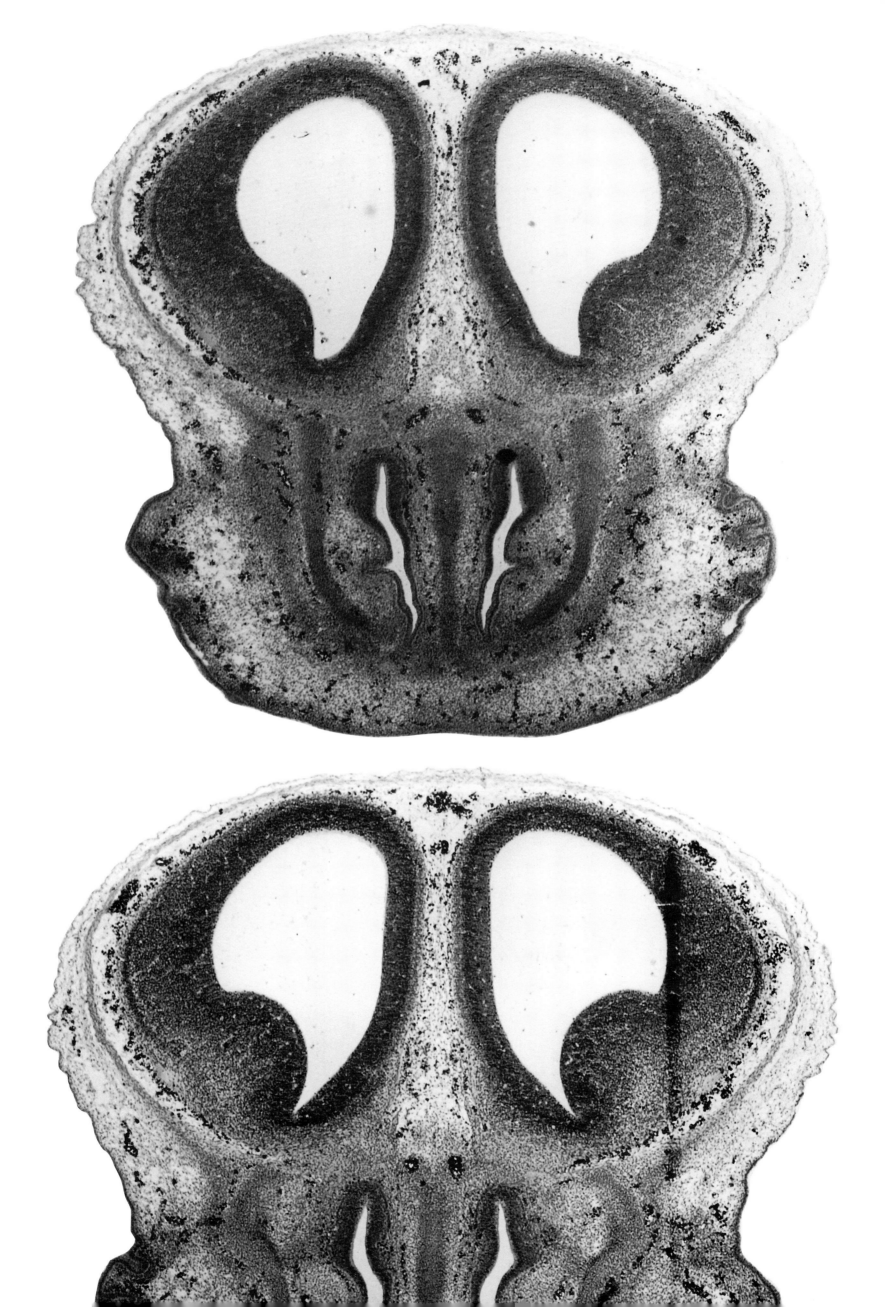

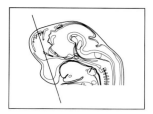

Figure 30
E16 Coronal 2

1   cortical layer 1
acer   anterior cerebral artery
CrP   cribiform plate
cx   cortical neuroepi
CxP   cortical plate
Ecto1   ectoturbinate 1
Ethmoid   ethmoid bone
ICx   intermediate cortical layer
lo   lateral olfactory tract
LV   lateral ventricle
Maxilla   maxilla
Mi   mitral cell layer of the olf bulb
NasT   nasal turbinate
obn   olfactory bulb neuroepi
olf   olfactory nerve
olfa   olfactory artery
Pir   piriform cortex
respepith   respiratory epithelium
Spt   septum
sss   superior sagittal sinus
SubV   subventricular cortical layer
term   terminal nerve
Tu   olfactory tubercle
Vomer   vomer

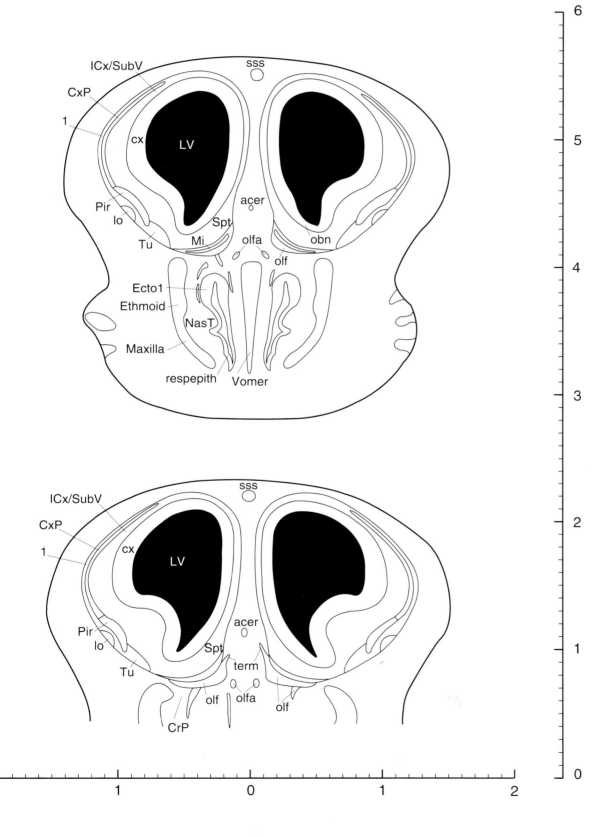

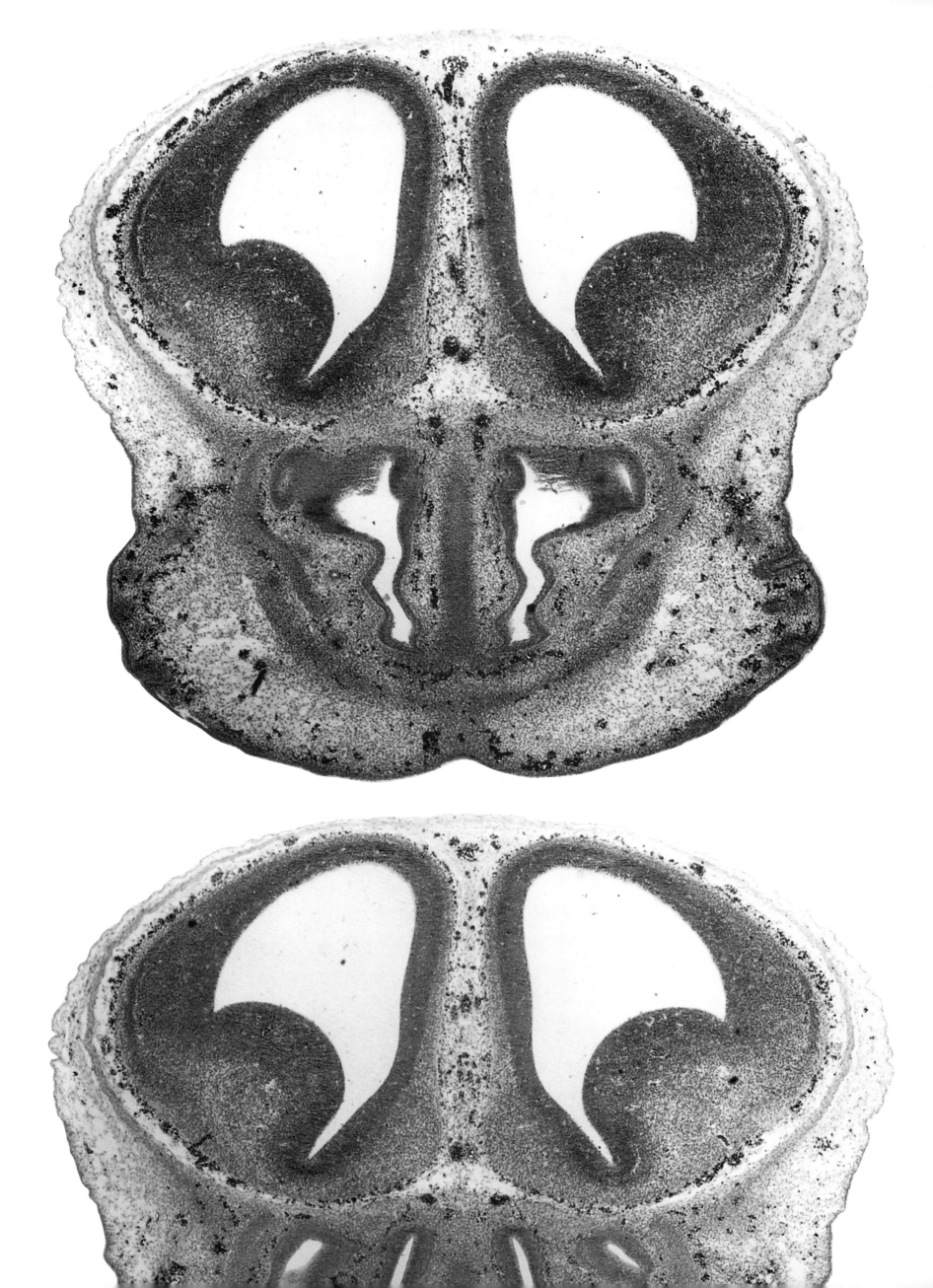

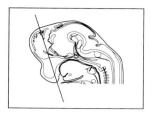

Figure 31
E16 Coronal 3

1   cortical layer 1
acer   anterior cerebral artery
bta   basal telencephalic plate, anterior
BTel   basal telencephalon
CrP   cribiform plate
cx   cortical neuroepi
CxP   cortical plate
Ecto1   ectoturbinate 1
Ecto2,2'   ectoturbinates 2, 2'
EndoII   endoturbinate II
Ethmoid   ethmoid bone
Frontal   frontal bone
hi   hippocampal formation neuroepi
ICx   intermediate cortical layer
lo   lateral olfactory tract
LV   lateral ventricle
Maxilla   maxilla
NasT   nasal turbinate
olf   olfactory nerve
olfa   olfactory artery
PDR   posterior dorsal recess of nasal cavity
Pir   piriform cortex
Spt   septum
spt   septal neuroepi
sss   superior sagittal sinus
SubV   subventricular cortical layer
Tu   olfactory tubercle
Vomer   vomer

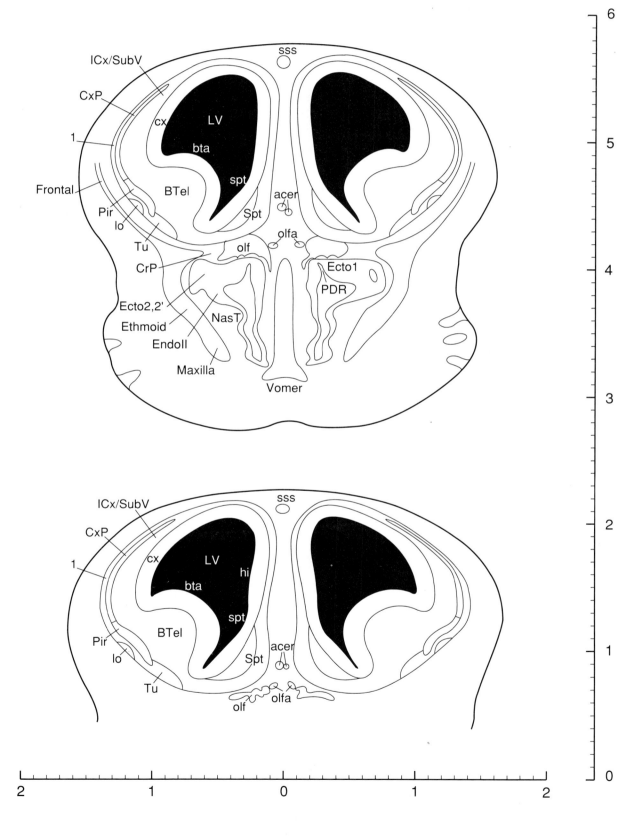

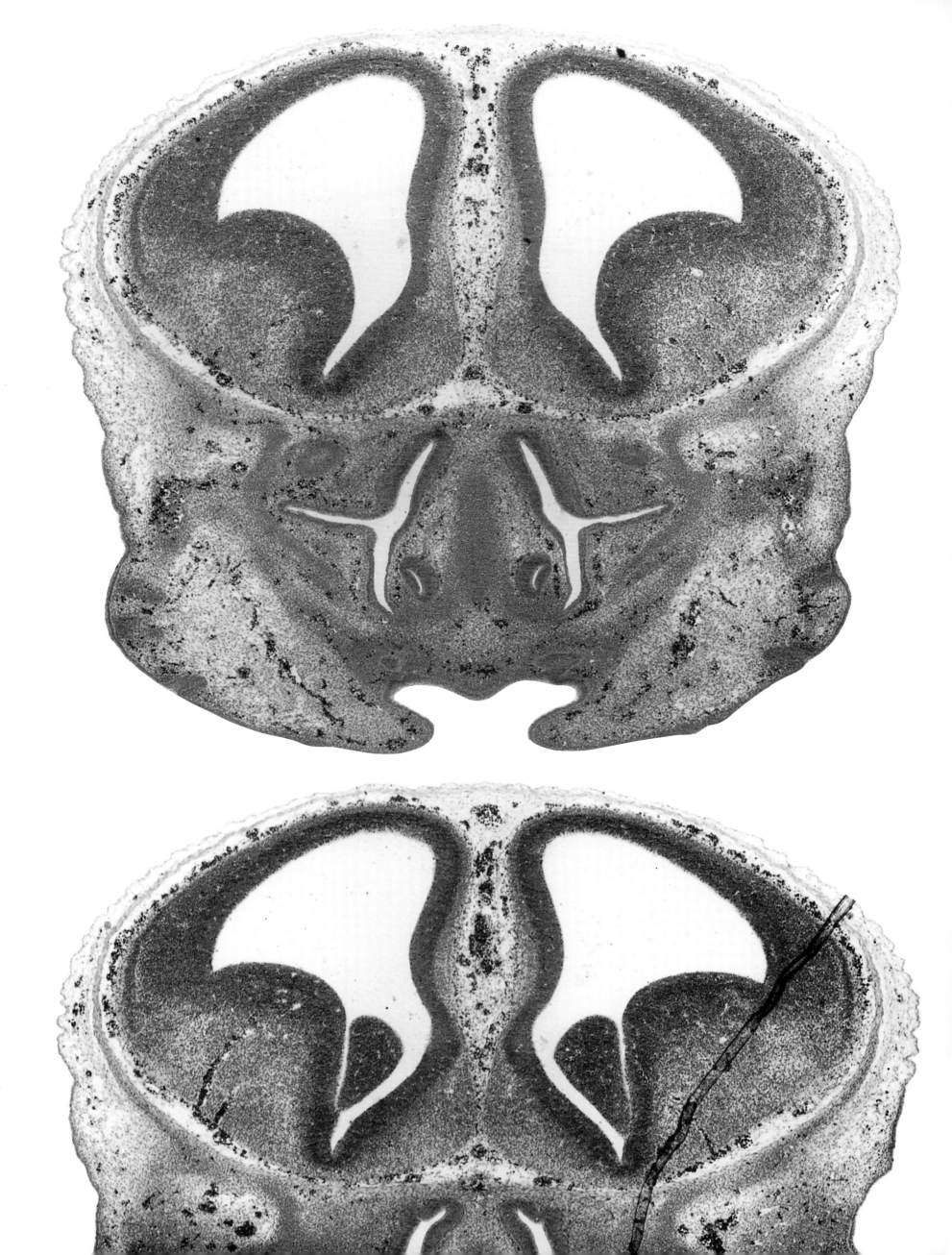

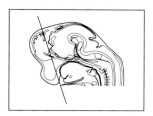

Figure 32
E16 Coronal 4

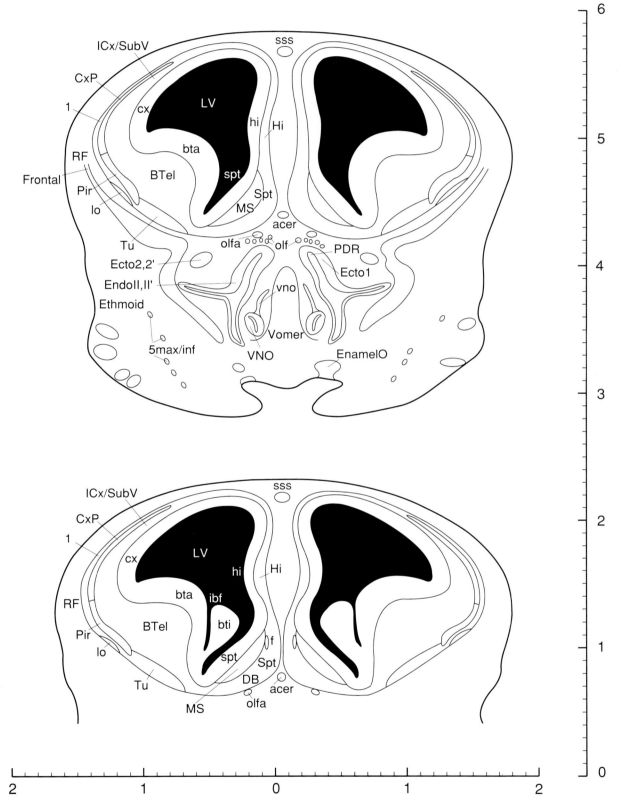

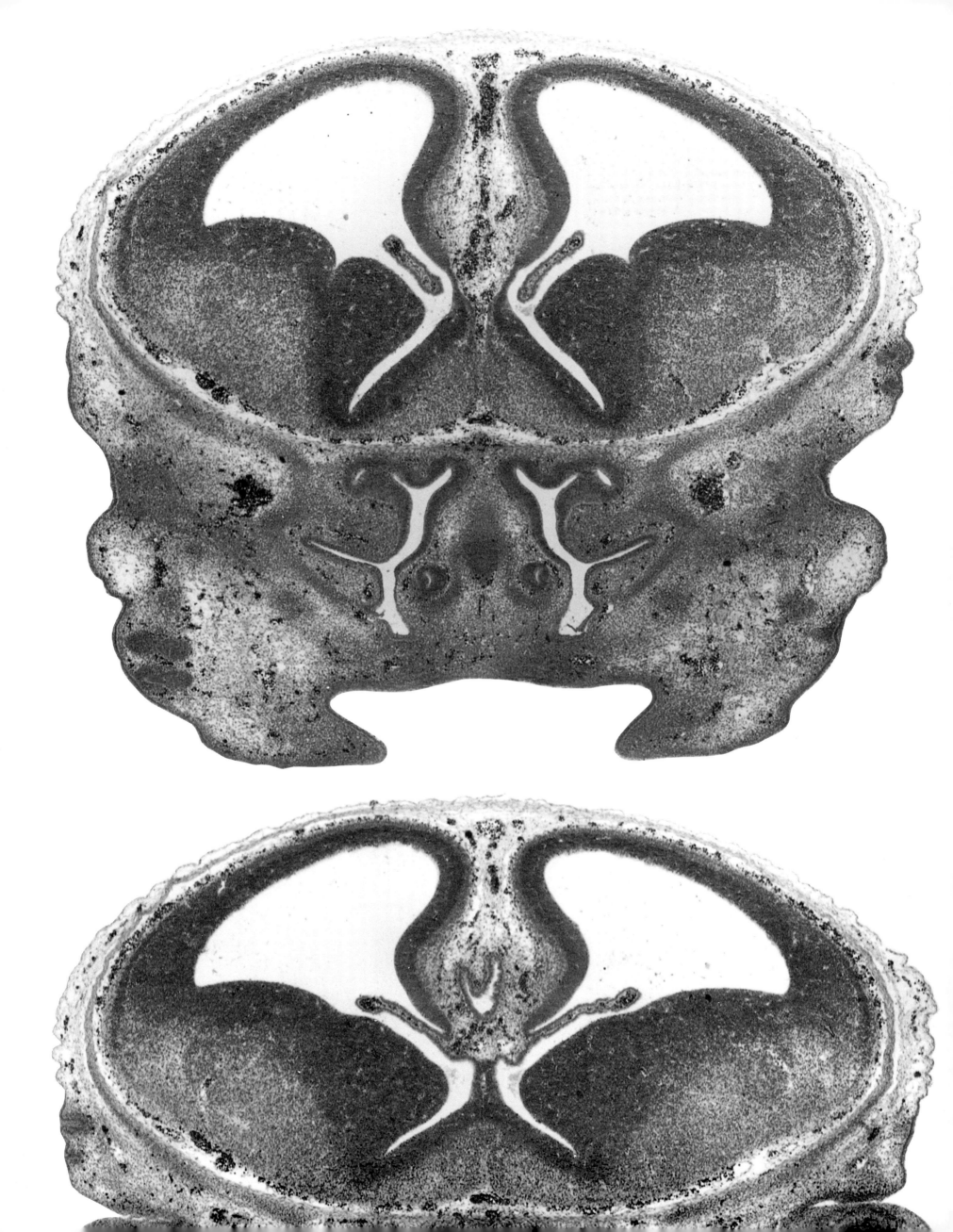

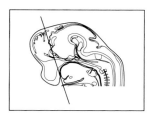

Figure 33
E16 Coronal 5

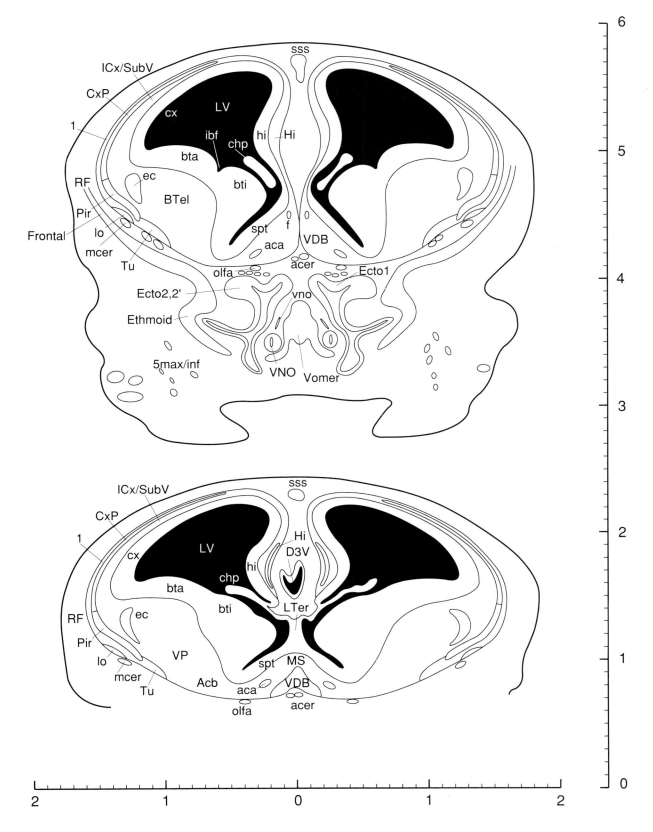

1   cortical layer 1
5max   maxillary nerve
aca   anterior commissure, anterior part
Acb   accumbens nucleus
acer   anterior cerebral artery
bta   basal telencephalic plate, anterior
BTel   basal telencephalon
bti   basal telen plate, intermediate
chp   choroid plexus premordium
cx   cortical neuroepi
CxP   cortical plate
D3V   dorsal third ventricle
ec   external capsule
Ecto1   ectoturbinate 1
Ecto2,2'   ectoturbinates 2, 2'
Ethmoid   ethmoid bone
f   fornix
Frontal   frontal bone
Hi   hippocampal formation
hi   hippocampal formation neuroepi
ibf   interbasal plate fissure
ICx   intermediate cortical layer
inf   infraorbital branch
lo   lateral olfactory tract
LTer   lamina terminalis
LV   lateral ventricle
mcer   middle cerebral artery
MS   medial septal nucleus
olfa   olfactory artery
Pir   piriform cortex
RF   rhinal fissure
spt   septal neuroepi
sss   superior sagittal sinus
SubV   subventricular cortical layer
Tu   olfactory tubercle
VDB   nu vertical limb diagonal band
VNO   vomeronasal organ
vno   nerve of vomeronasal organ
Vomer   vomer
VP   ventral pallidum

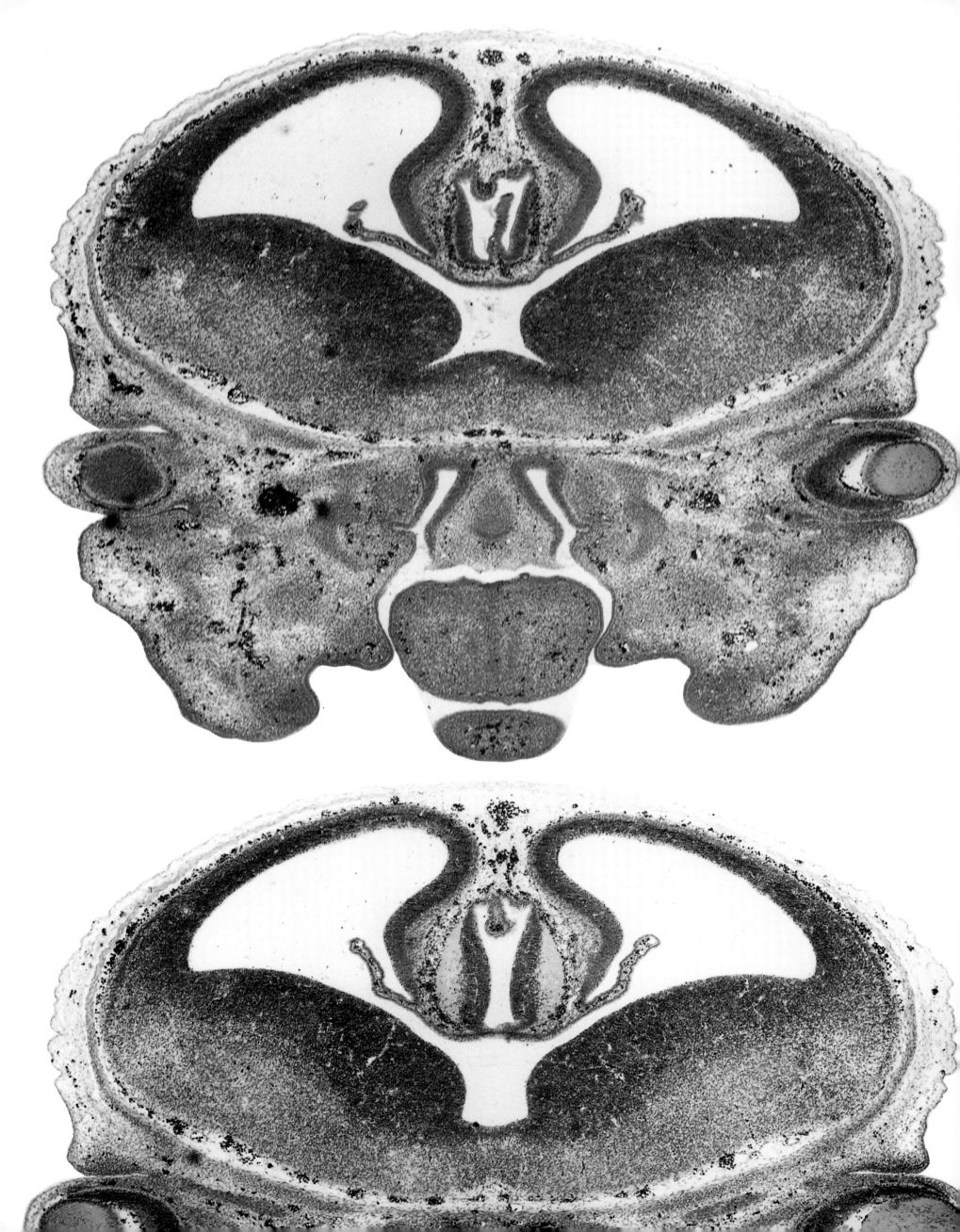

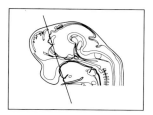

Figure 34
E16 Coronal 6

1   cortical layer 1
5max   maxillary nerve
ac   anterior commissure
aca   anterior commissure, anterior
Acb   accumbens nucleus
acer   anterior cerebral artery
btp   basal telencephalic plate, posterior
chp   choroid plexus premordium
Cornea   cornea
CPu   caudate putamen (striatum)
cx   cortical neuroepi
CxP   cortical plate
ec   external capsule
Ethmoid   ethmoid bone
HDB   nu horizon limb diagonal band
Hi   hippocampal formation
hi   hippocampal formation neuroepi
ICx   intermediate cortical layer
inf   infraorbital branch
inl   inferior neuroepith lobule of dien
IOb   inferior oblique muscle
IVF   interventricular foramen
Jaw   jaw
Lens   lens
lo   lateral olfactory tract
LV   lateral ventricle
mcer   middle cerebral artery
MnPO   median preoptic nucleus
MS   medial septal nucleus
olfa   olfactory artery
ophv   ophthalmic vein
Orbit   orbital cavity
PDR   posterior dorsal recess of nasal cavity
Pi   pineal gland
Pir   piriform cortex
PSphW   presphenoid wing
Retina   retina
RF   rhinal fissure
snl   superior neuroepith lobule of dien
sss   superior sagittal sinus
SubV   subventricular cortical layer
Tong   tongue
Tu   olfactory tubercle
V3V   ventral third ventricle
VDB   nu vertical limb diagonal band
vnl   ventral neuroepith lobule of dien
Vomer   vomer
VP   ventral pallidum

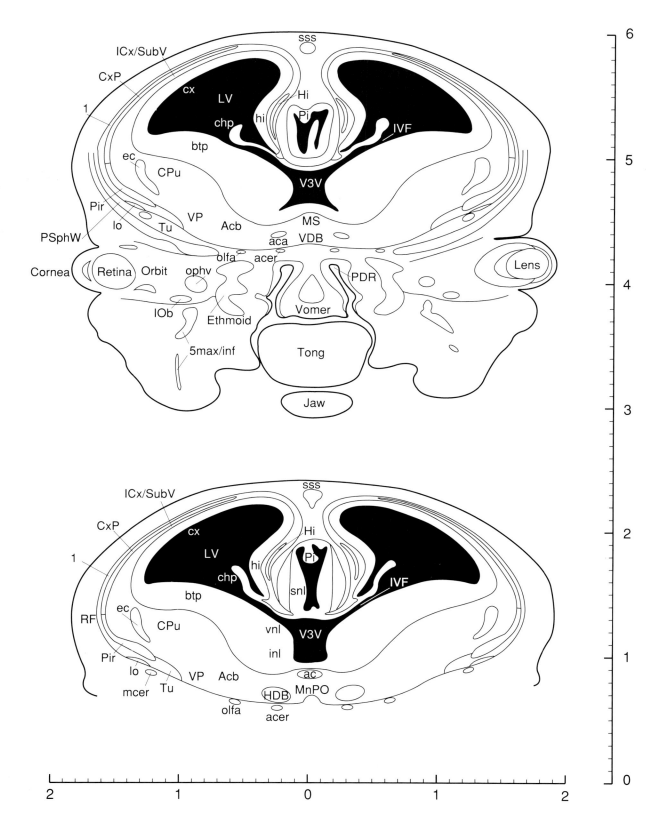

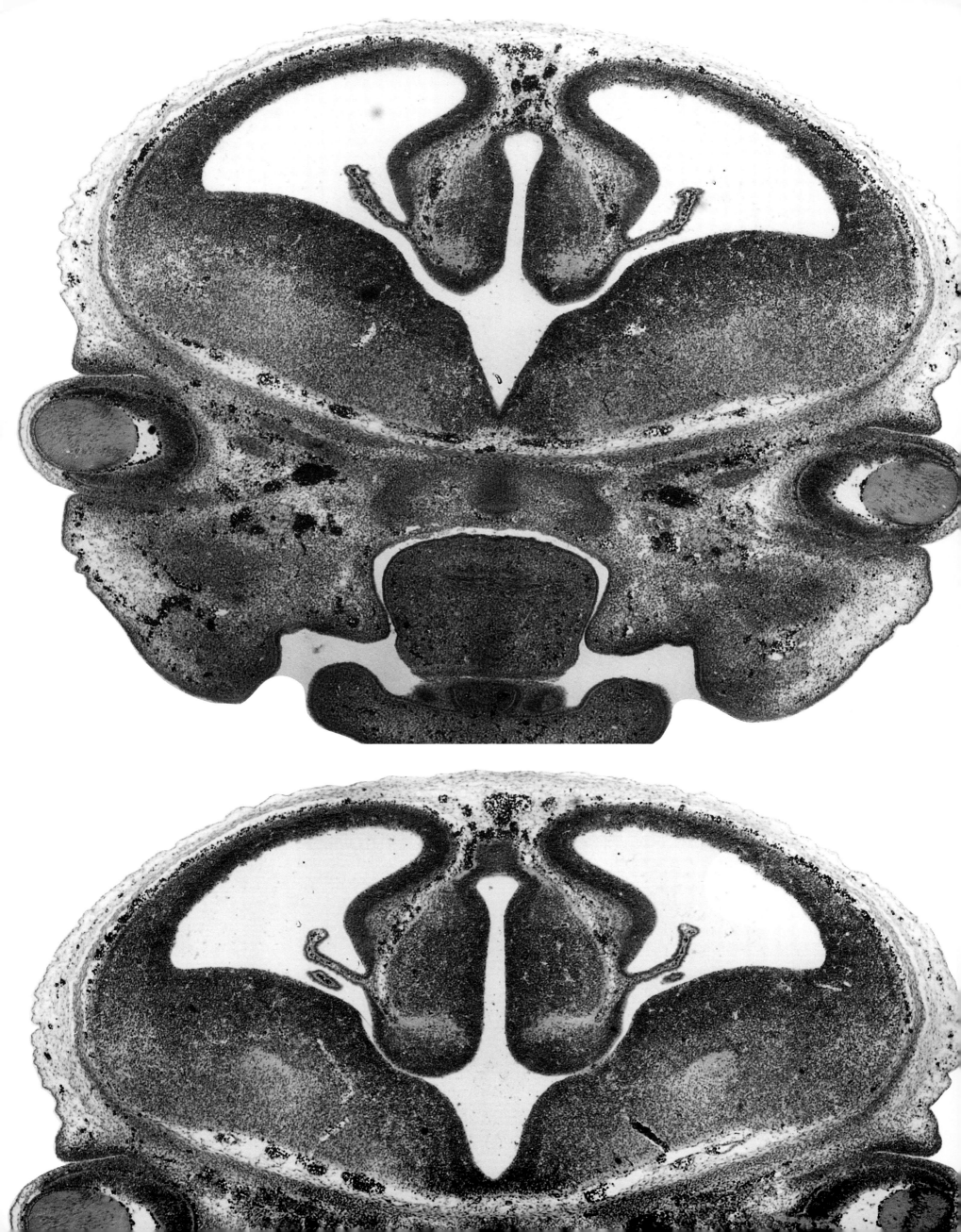

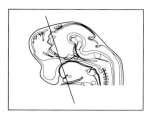

Figure 35
E16 Coronal 7

1   cortical layer 1
3V   third ventricle
5max   maxillary nerve
acer   anterior cerebral artery
ATh   anterior thalamus
BSph   basisphenoid bone
BST   bed nucleus of the stria terminalis
btp   basal telencephalic plate, posterior
chp   choroid plexus premordium
Cornea   cornea
CPu   caudate putamen (striatum)
cx   cortical neuroepi
CxP   cortical plate
D3V   dorsal third ventricle
dnl   dorsal neuroepith lobule of dien
ec   external capsule
EnamelO   enamel organ (of tooth)
HDB   nu horizon limb diagonal band
Hi   hippocampal formation
hi   hippocampal formation neuroepi
ialvn   inferior alveolar nerve
icap   internal capsule
ICx   intermediate cortical layer
inf   infraorbital branch
inl   inferior neuroepi lobule of dien
IOb   inferior oblique muscle
IVF   interventricular foramen
Jaw   jaw
Lens   lens
LHb   lateral habenular nucleus
lo   lateral olfactory tract
LPO   lateral preoptic area
LV   lateral ventricle
Maxilla   maxilla
mcer   middle cerebral artery
mfb   medial forebrain bundle
MHb   medial habenular nucleus
mnl   middle neuroepi lobule of dien
MPO   medial preoptic nucleus
MRec   medial rectus muscle
n   nerve
ophv   ophthalmic vein
Pi   pineal gland
Pig   pigment epithelium of retina
Pir   piriform cortex
PiRe   pineal recess of the third ventricle
poa   preoptic area neuroepi
PSphW   presphenoid wing
Retina   retina
RF   rhinal fissure
sm   stria medullaris of the thalamus
snl   superior neuroepith lobule of dien
SOb   superior oblique muscle
sss   superior sagittal sinus
SubV   subventricular cortical layer
Tong   tongue
trfv   transverse temporal vein
V3V   ventral third ventricle
vnl   ventral neuroepi lobule of dien
VP   ventral pallidum

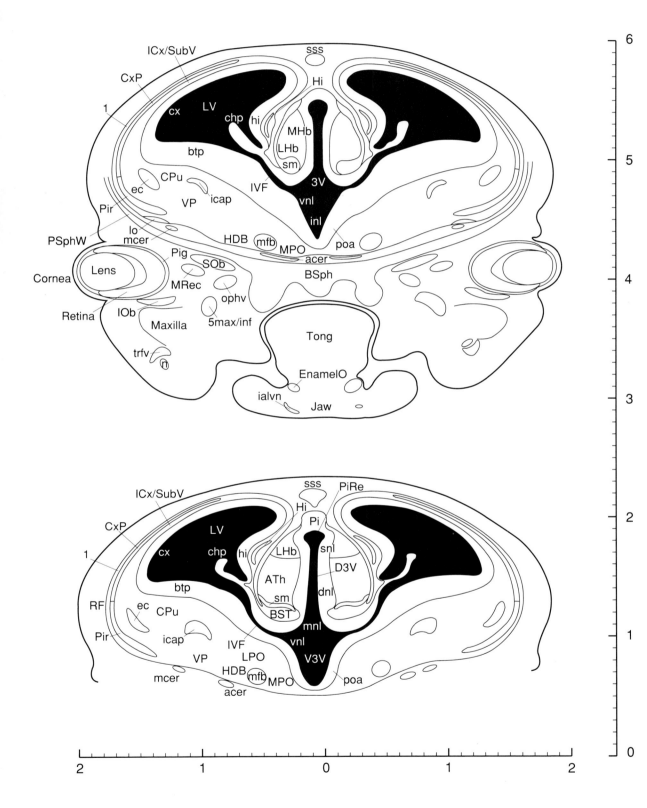

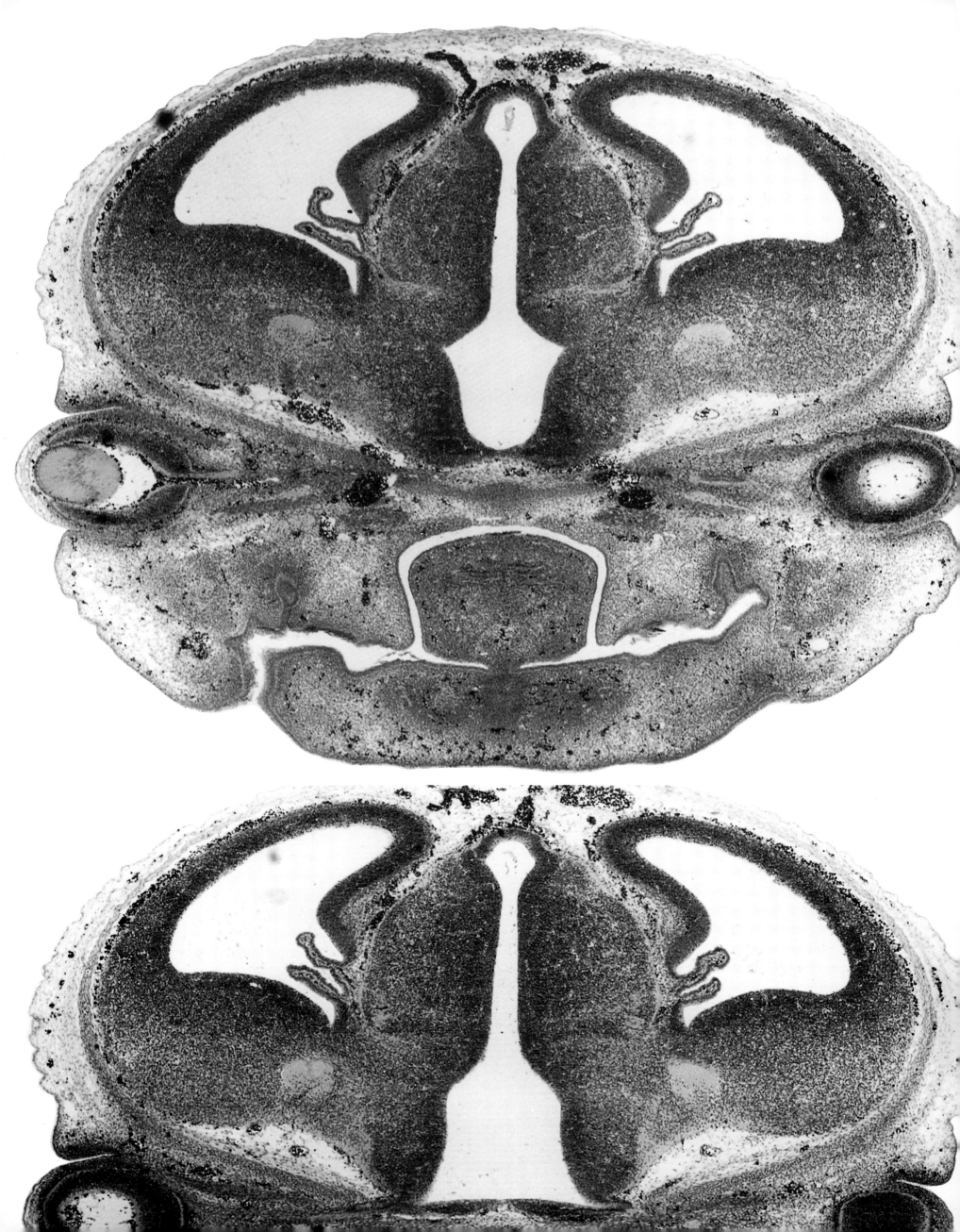

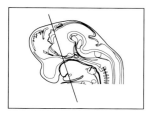

Figure 36
E16 Coronal 8

1 cortical layer 1
2n optic nerve
3n oculomotor nerve or its root
5max maxillary nerve
acer anterior cerebral artery
Amg amygdala
BSph basisphenoid bone
BST bed nu of the stria terminalis
btp basal telencephalic plate, posterior
chp choroid plexus premordium
Cil ciliary ganglion
Cornea cornea
CPu caudate putamen (striatum)
cx cortical neuroepi
CxP cortical plate
D3V dorsal third ventricle
dnl dorsal neuroepith lobule of dien
EnamelO enamel organ (of tooth)
EP entopeduncular nucleus
GP globus pallidus
gpal greater palatine nerve
Hi hippocampal formation
hi hippocampal formation neuroepi
icap internal capsule
ictd internal carotid artery
ICx intermediate cortical layer
inl inferior neuroepith lobule of dien
IRec inferior rectus muscle
Jaw jaw
LD laterodorsal thalamic nucleus
Lens lens
LHb lateral habenular nucleus
LPO lateral preoptic area
LV lateral ventricle
Maxilla maxilla
Mc Meckel's cartilage
mcer middle cerebral artery
mfb medial forebrain bundle
MHb medial habenula nucleus
mnl middle neuroepith lobule of dien
MPO medial preoptic nucleus
myhy mylohyoid nerve
n nerve
OptRe optic recess of third ventricle
Pi pineal gland
Pir piriform cortex
PiRe pineal recess of the third ventricle
poa preoptic area neuroepi
PSphW presphenoid wing
Retina retina
RF rhinal fissure
rp reticular protuberance
Rt reticular thalamic nucleus
sm stria medullaris of the thalamus
snl superior neuroepith lobule of dien
SphPal sphenopalatine ganglion
SRec superior rectus muscle
SubV subventricular cortical layer
Th thalamus
Tong tongue
trfv transverse temporal vein
trs transverse sinus
v vein (unidentified)
V3V ventral third ventricle
vnl ventral neuroepith lobule of dien

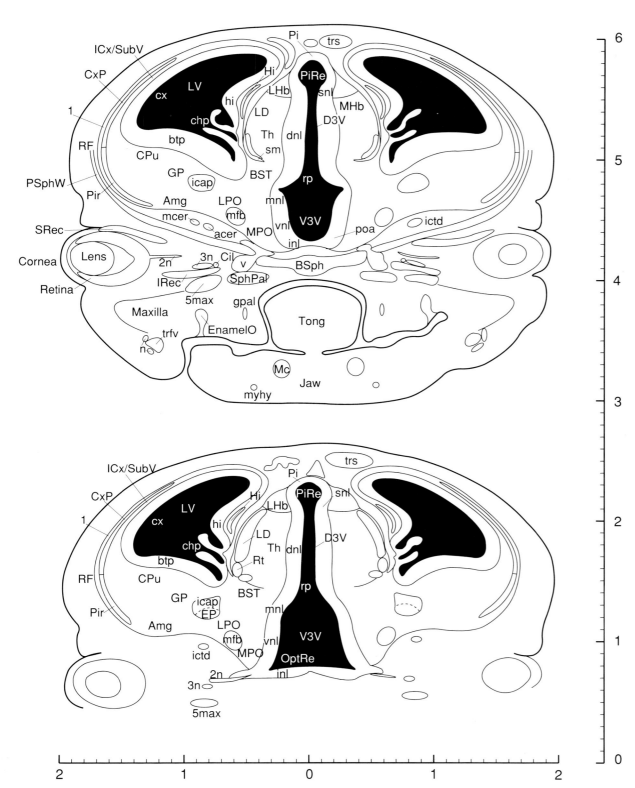

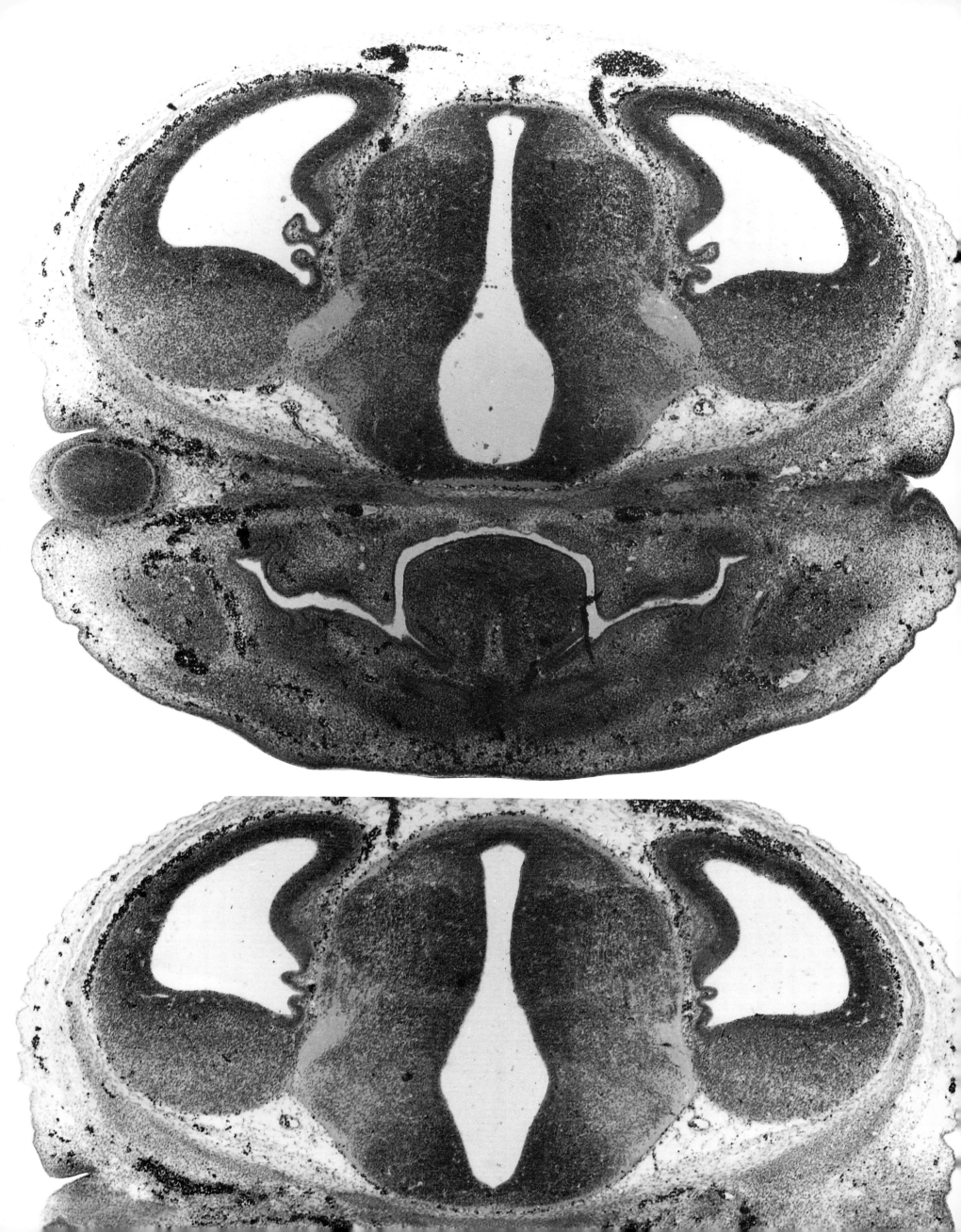

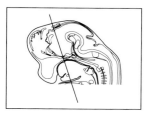

Figure 37
E16 Coronal 9

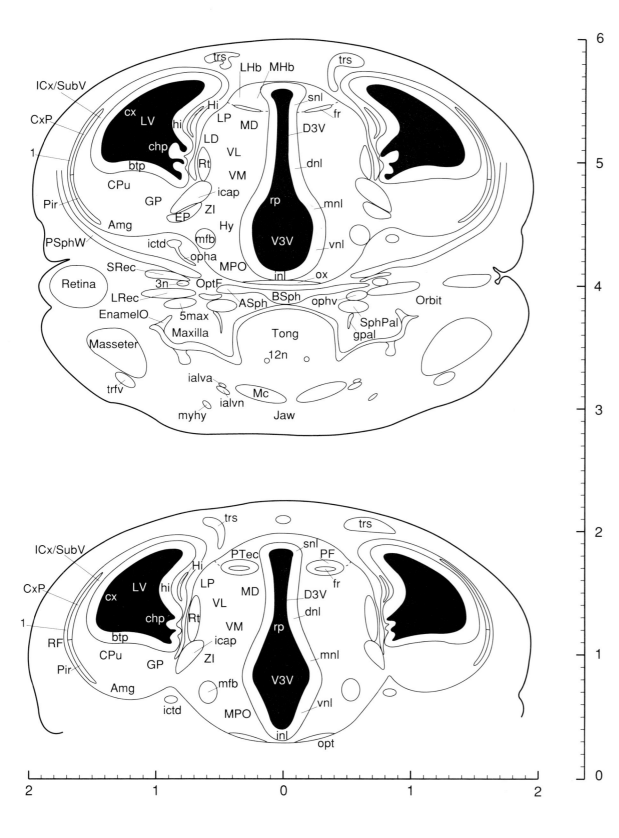

1   cortical layer 1
3n   oculomotor nerve or its root
5max   maxillary nerve
12n   hypoglossal nerve or its root
Amg   amygdala
ASph   alisphenoid bone
BSph   basisphenoid bone
btp   basal telencephalic plate, posterior
chp   choroid plexus premordium
CPu   caudate putamen (striatum)
cx   cortical neuroepi
CxP   cortical plate
D3V   dorsal third ventricle
dnl   dorsal neuroepith lobule of dien
EnamelO   enamel organ (of tooth)
EP   entopeduncular nucleus
fr   fasciculus retroflexus
GP   globus pallidus
gpal   greater palatine nerve
Hi   hippocampal formation
hi   hippocampal formation neuroepi
Hy   hypothalamus
ialva   inferior alveolar artery
ialvn   inferior alveolar nerve
icap   internal capsule
ictd   internal carotid artery
ICx   intermediate cortical layer
inl   inferior neuroepith lobule of dien
Jaw   jaw
LD   laterodorsal thalamic nucleus
LHb   lateral habenular nucleus
LP   lateral posterior thalamic nu
LRec   lateral rectus muscle
LV   lateral ventricle
Masseter   masseter muscle
Maxilla   maxilla
Mc   Meckel's cartilage
MD   mediodorsal thalamic nucleus
mfb   medial forebrain bundle
MHb   medial habenular nucleus
mnl   middle neuroepith lobule of dien
MPO   medial preoptic nucleus
myhy   mylohyoid nerve
opha   ophthalmic artery
ophv   ophthalmic vein
opt   optic tract
OptF   optic foramen
Orbit   orbital cavity
ox   optic chiasm
PF   parafascicular thalamic nucleus
Pir   piriform cortex
PSphW   presphenoid wing
PTec   pretectum
Retina   retina
RF   rhinal fissure
rp   reticular protuberance
Rt   reticular thalamic nucleus
snl   superior neuroepith lobule of dien
SphPal   sphenopalatine ganglion
SRec   superior rectus muscle
SubV   subventricular cortical layer
Tong   tongue
trfv   transverse temporal vein
trs   transverse sinus
V3V   ventral third ventricle
VL   ventrolateral thalamic nucleus
VM   ventromedial thalamic nucleus
vnl   ventral neuroepith lobule of dien
ZI   zona incerta

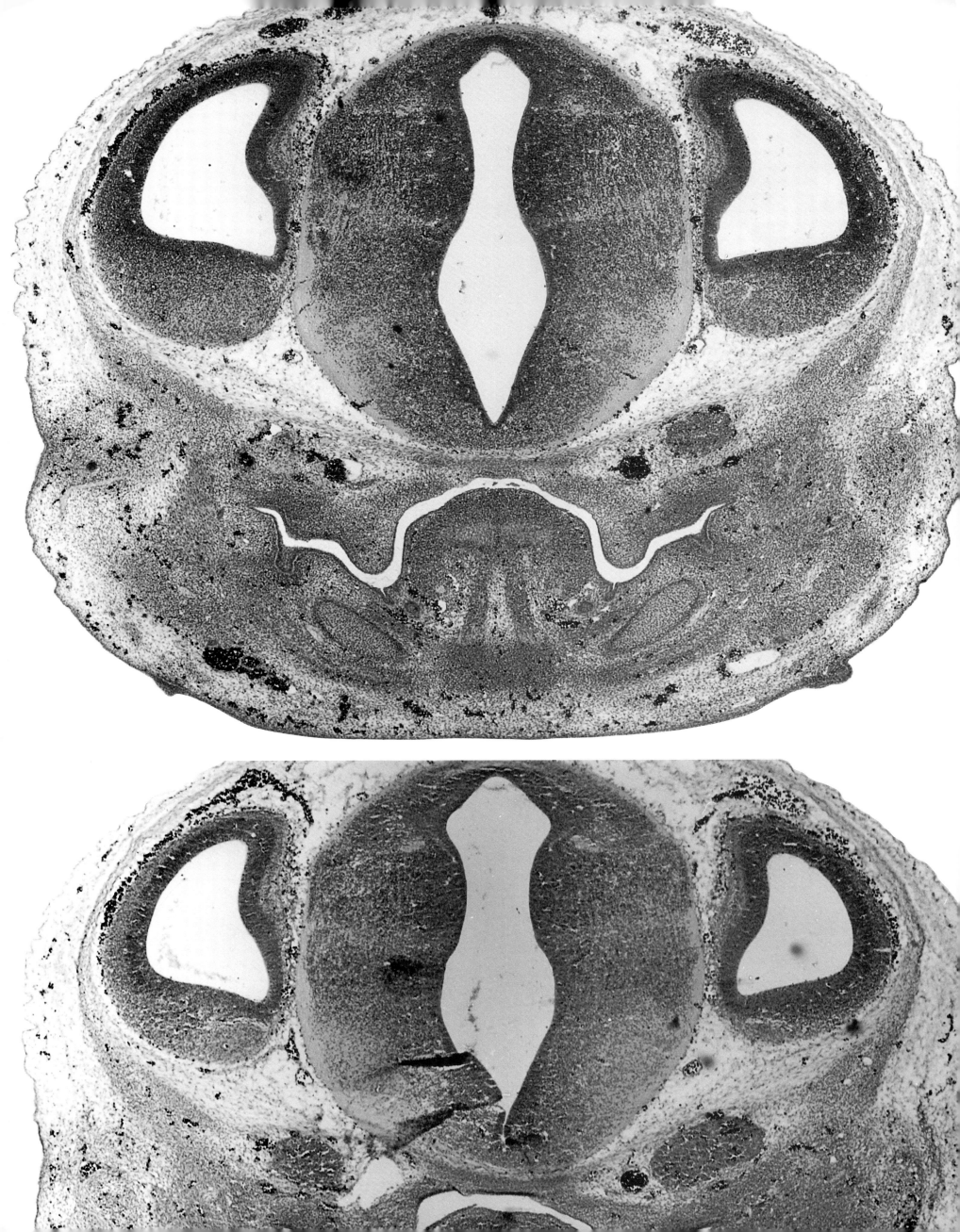

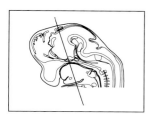

Figure 38
E16 Coronal 10

1   cortical layer 1
3n   oculomotor nerve or its root
5Gn   trigeminal ganglion
5mand   mandibular nerve
5max   maxillary nerve
6n   abducens nerve or its root
12n   hypoglossal nerve or its root
Amg   amygdala
BSph   basisphenoid bone
btp   basal telencephalic plate, posterior
cav   cavernous sinus
CPu   caudate putamen (striatum)
cx   cortical neuroepi
D3V   dorsal third ventricle
DM   dorsomedial hypothalamic nu
dnl   dorsal neuroepith lobule of dien
ectring   ectodermal ring
EnamelO   enamel organ (of tooth)
fr   fasciculus retroflexus
genioglossus   genioglossus muscle
Hi   hippocampal formation
hi   hippocampal formation neuroepi
ialva   inferior alveolar artery
ialvn   inferior alveolar nerve
icap   internal capsule
ictd   internal carotid artery
inl   inferior neuroepith lobule of dien
Jaw   jaw
LG   lateral geniculate nucleus
LH   lateral hypothalamic area
LP   lateral posterior thalamic nu
LV   lateral ventricle
Masseter   masseter muscle
Maxilla   maxilla
Mc   Meckel's cartilage
MD   mediodorsal thalamic nucleus
Me   medial amygdaloid nucleus
mfb   medial forebrain bundle
mnl   middle neuroepith lobule of dien
opt   optic tract
Pa   paraventricular hypothalamic nu
Pal   palatine bone
pc   posterior commissure
PF   parafascicular thalamic nucleus
PSphW   presphenoid wing
PTec   pretectum
rp   reticular protuberance
Rt   reticular thalamic nucleus
snl   superior neuroepith lobule of dien
SubV   subventricular cortical layer
Temp   temporal muscle
Tong   tongue
trs   transverse sinus
v   vein (unidentified)
V3V   ventral third ventricle
VMH   ventromedial hypothalamic nu
vnl   ventral neuroepith lobule of dien
VPL   ventral posterolateral thalamic nucleu
VPM   ventral posteromedial thalamic nucle
ZI   zona incerta

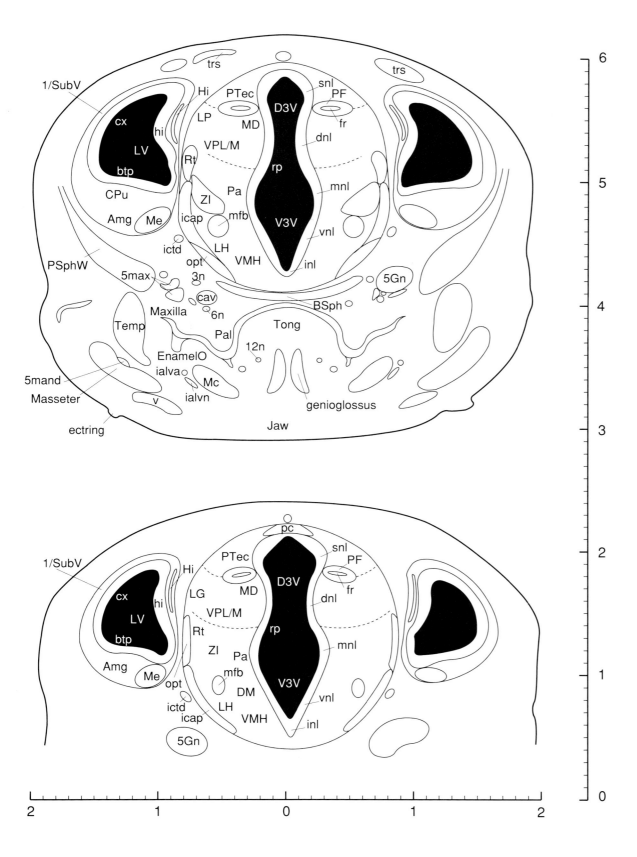

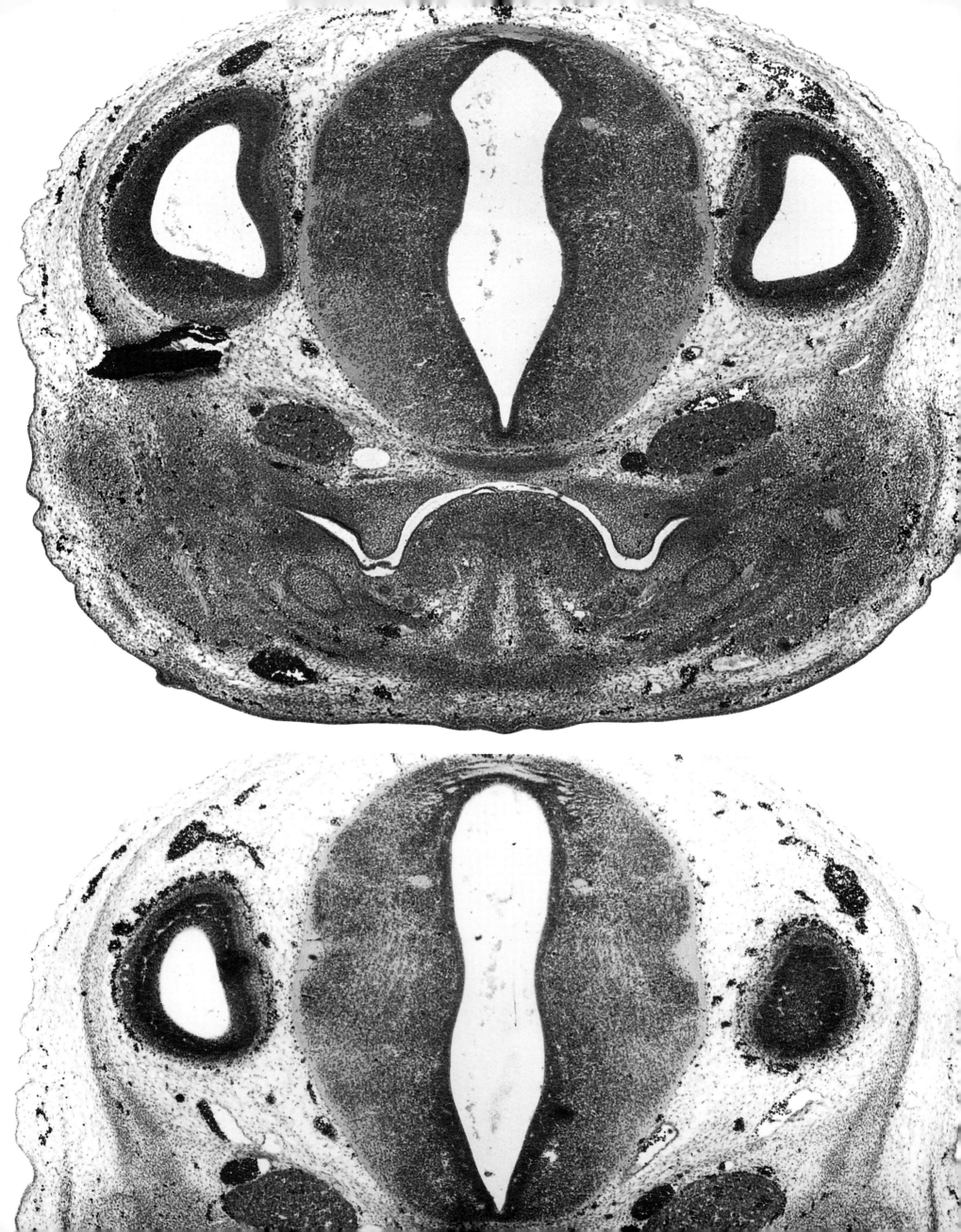

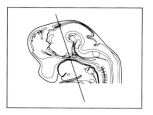

Figure 39
E16 Coronal 11

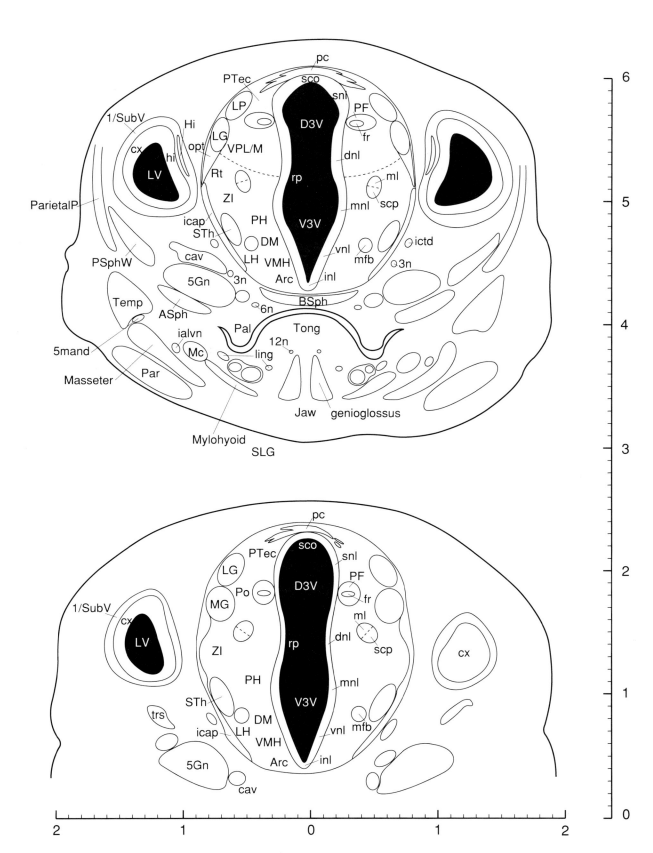

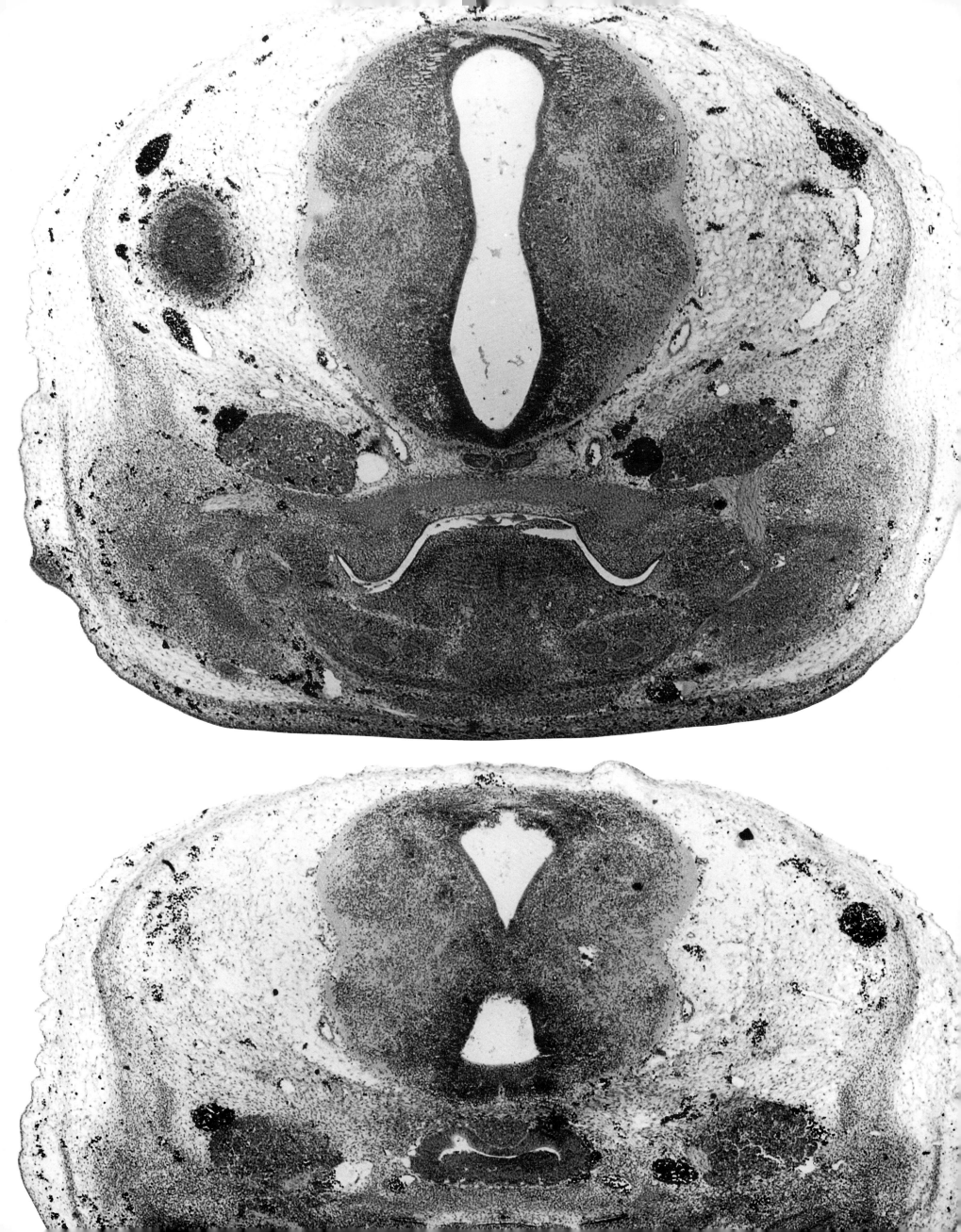

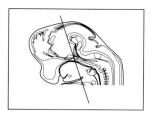

Figure 40
E16 Coronal 12

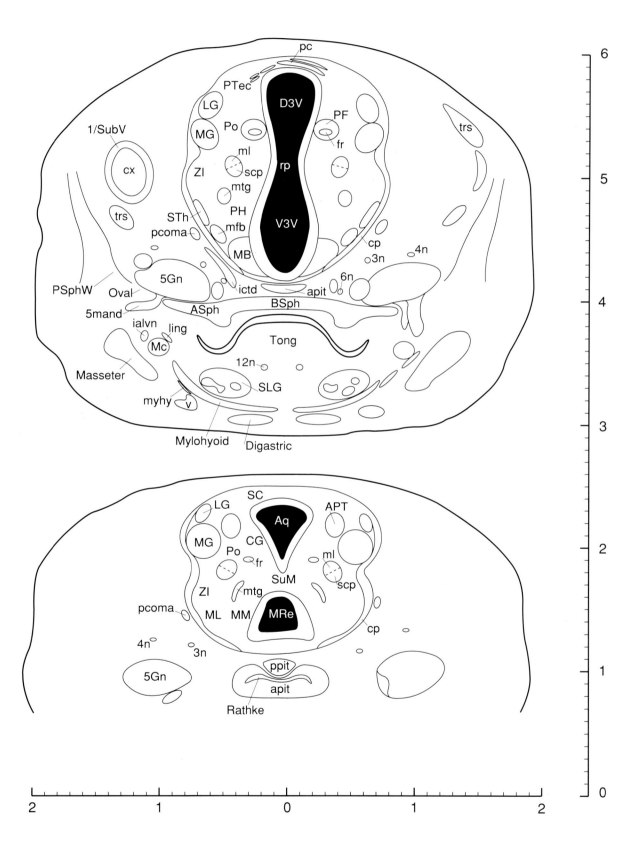

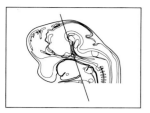

Figure 41
E16 Coronal 13

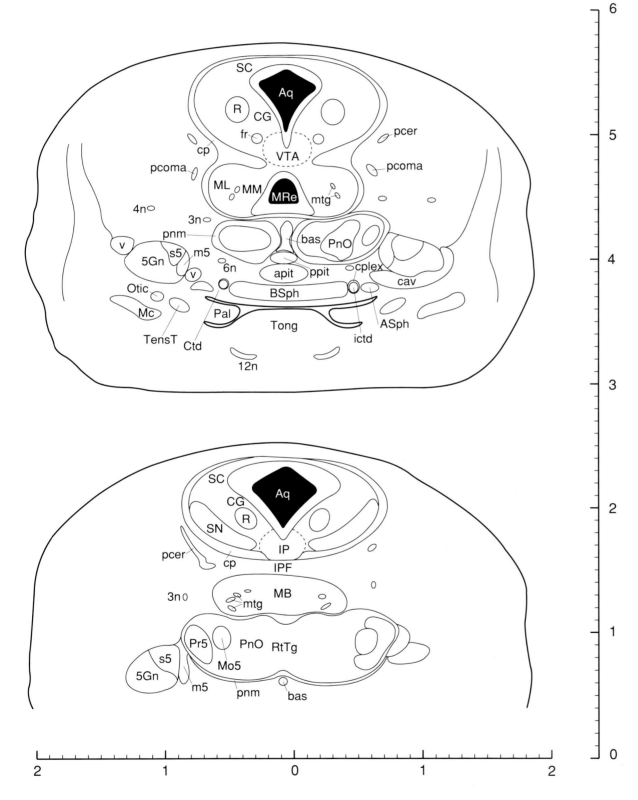

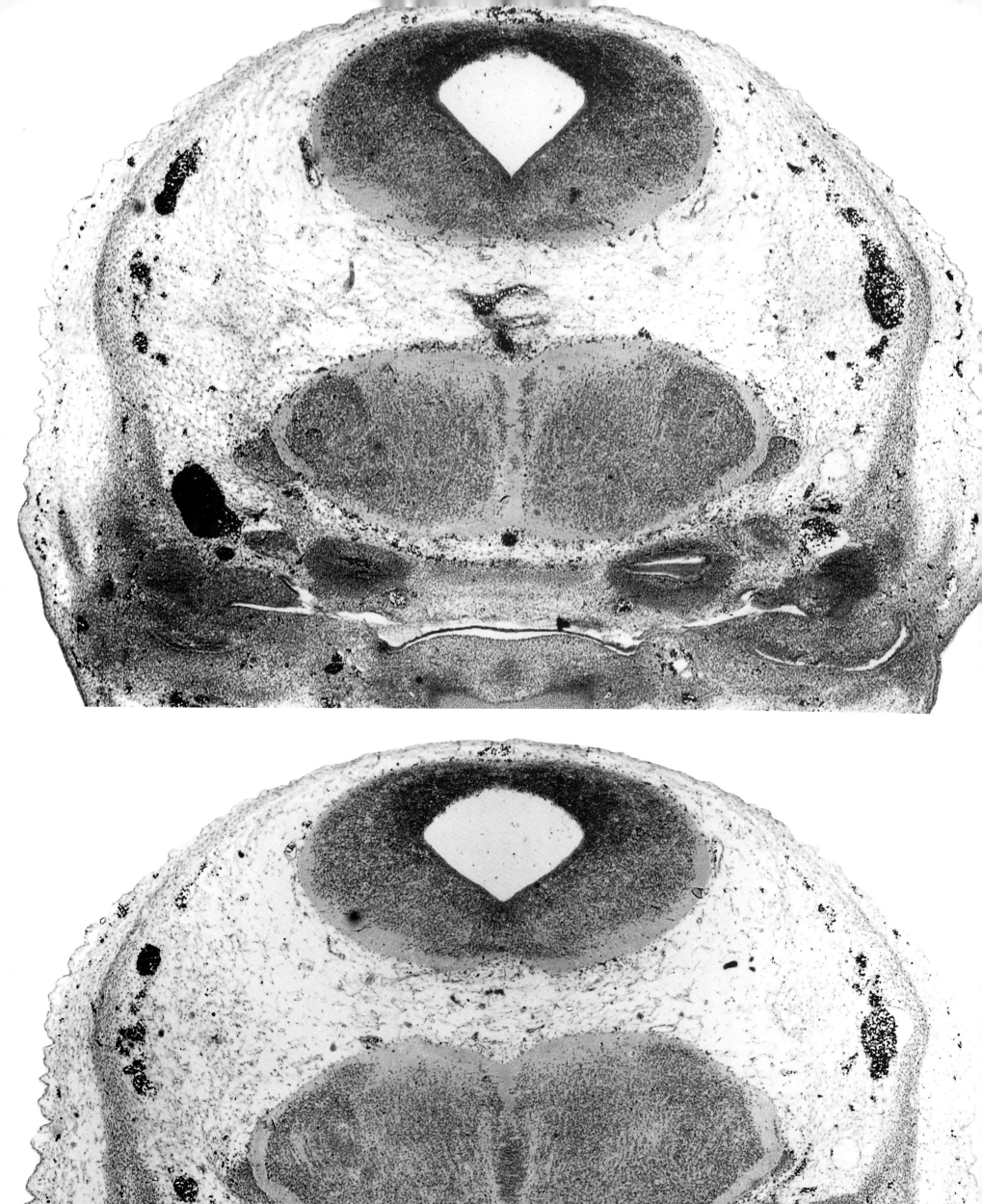

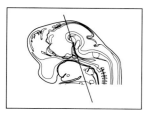

Figure 42
E16 Coronal 14

3n oculomotor nerve or its root
7n facial nerve or its root
10n vagus nerve
12n hypoglossal nerve or its root
Aq aqueduct (Sylvius)
bas basilar artery
BOcc basioccipital bone
CD cochlear duct
CG central (periaqueductal) gray
cp cerebral peduncle, basal part
cplex carotid plexus
DR dorsal raphe nucleus
EAM external auditory meatus
Gen geniculate ganglion
Hyoid hyoid bone
ictd internal carotid artery
IPF interpeduncular fossa
m5 motor root of the trigeminal nerve
Malleus malleus (ossicle)
Mesen mesencephalon
mlf medial longitudinal fasciculus
MnR median raphe nucleus
Mo5 motor trigeminal nucleus
pcer posterior cerebral artery
PnC pontine reticular nu, caudal part
pnm pontine migration
PnO pontine reticular nu, oral part
Pr5 principal sensory trigeminal nu
R red nucleus
RtTg reticulotegmental nu of the pons
s5 sensory root of the trigeminal nerve
SC superior colliculus
scba superior cerebellar artery
SN substantia nigra
SOl superior olive
sp5 spinal trigeminal tract
Sp5O spinal trigeminal nu, oral part
TensT tensor tympani muscle
trs transverse sinus
VeGn vestibular ganglion
VLL ventral nu of the lateral lemniscus

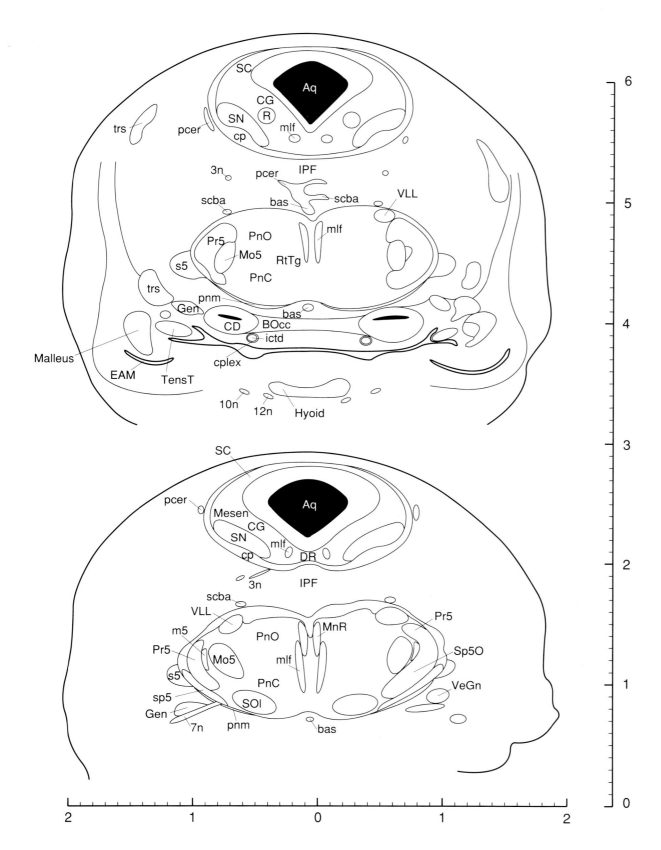

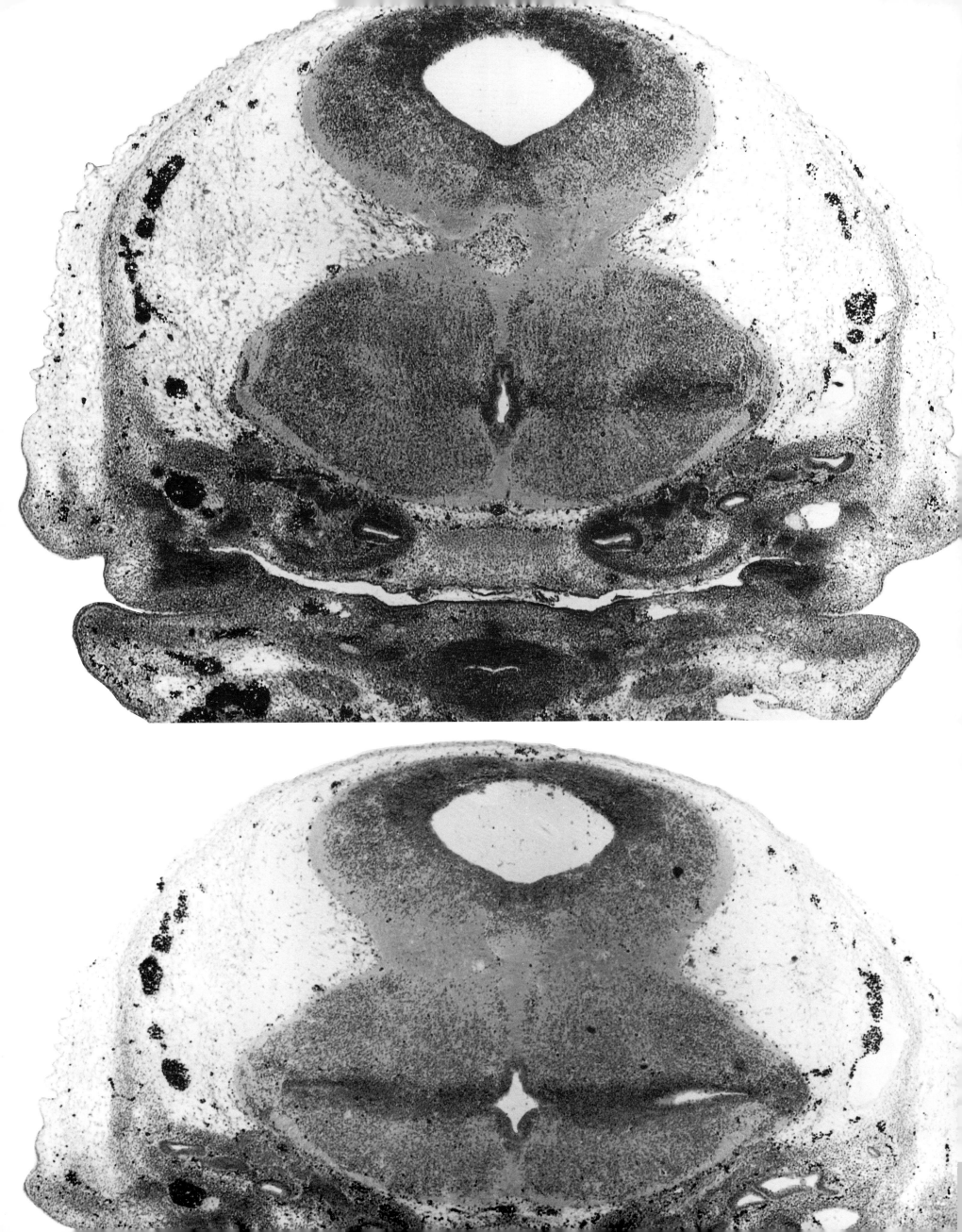

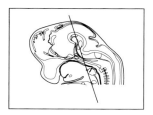

Figure 43
E16 Coronal 15

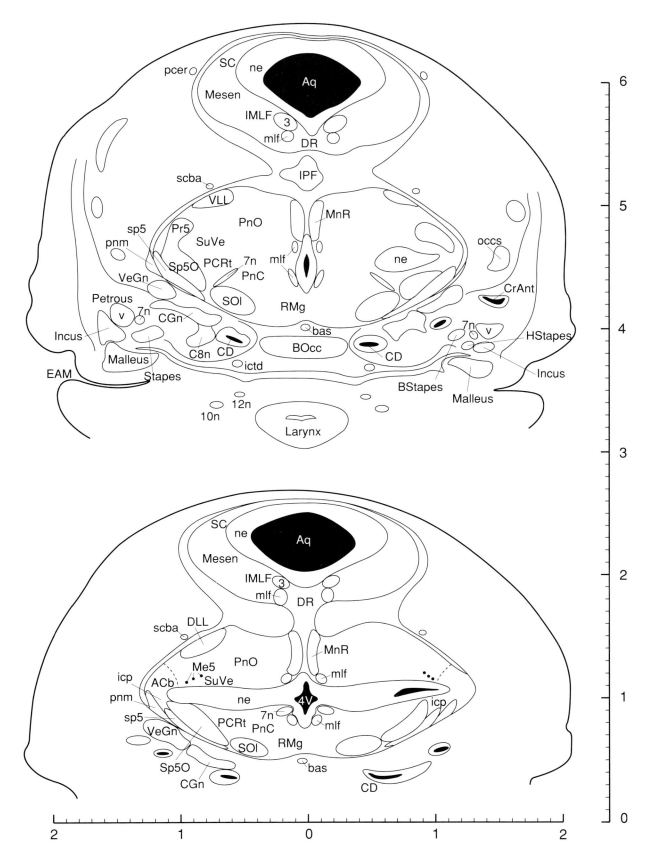

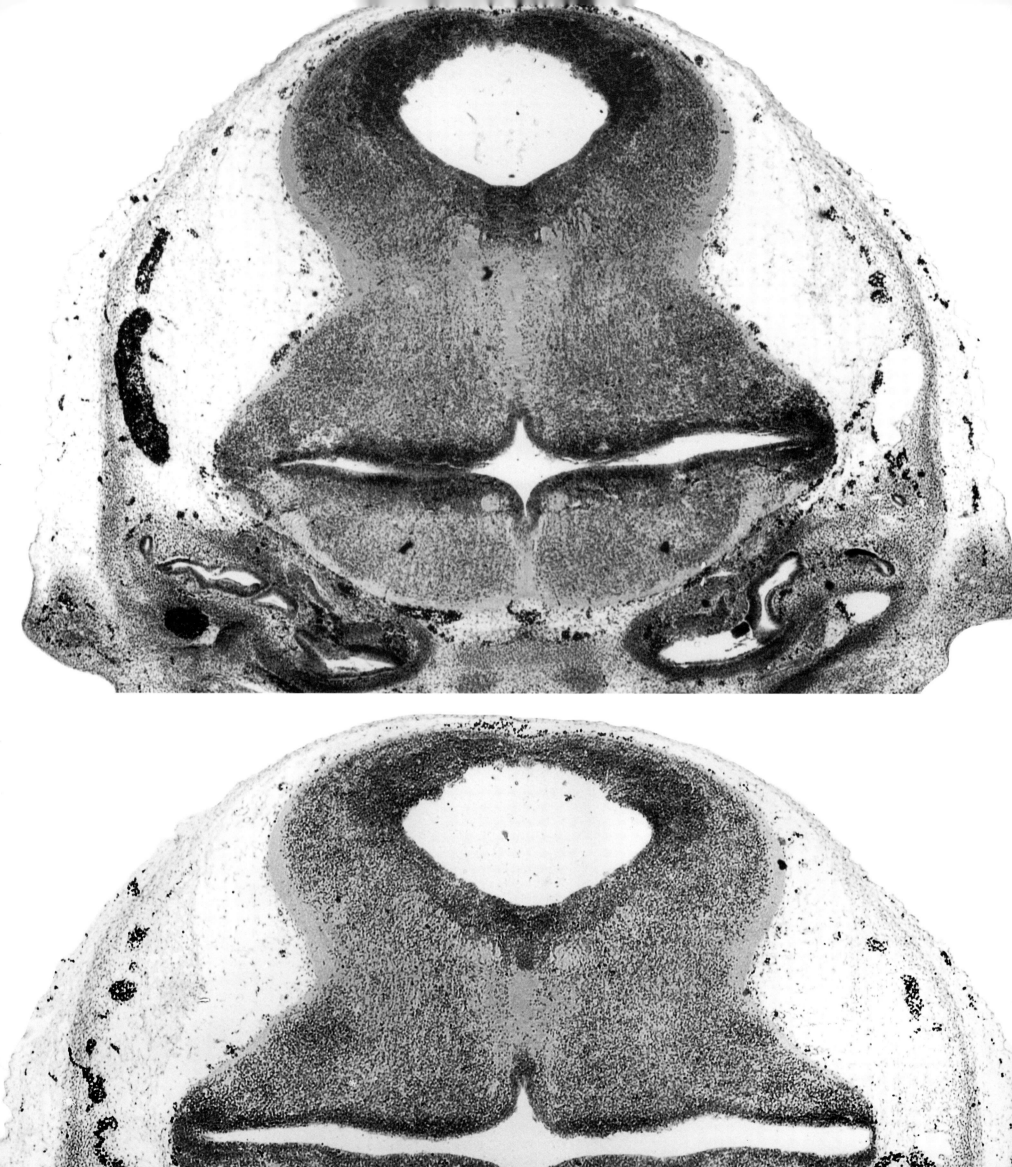

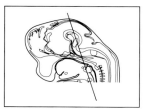

Figure 44
E16 Coronal 16

4   trochlear nucleus
4V   fourth ventricle
6   abducens nucleus
7   facial nucleus
7n   facial nerve or its root
8cn   cochlear root of the vestibulocochlear nerve
8n   vestibulocochlear nerve
8vn   vestibular root of the 8th nerve
ACb   anterior cerebellum
aica   anterior inferior cerebellar artery
Ant   anterior semicircular duct
Aq   aqueduct (Sylvius)
aud   auditory neuroepi
bas   basilar artery
BOcc   basioccipital bone
bp   basal plate neuroepi
CD   cochlear duct
CGn   cochlear (spiral) ganglion
DLL   dorsal nu of the lateral lemniscus
DR   dorsal raphe nucleus
EGL   external germinal layer of Cb
Gi   gigantocellular reticular nucleus
Hor   horizontal semicircular duct
icp   inferior cerebellar peduncle
ictd   internal carotid artery
IRt   intermediate reticular zone
lf   lateral fissure
LPGi   lateral paragigantocellular nu
LVe   lateral vestibular nucleus
Me5   mesencephalic trigeminal nu
mlf   medial longitudinal fasciculus
MnR   median raphe nucleus
MVe   medial vestibular nucleus
PCRt   parvocellular reticular nu
Petrous   petrous part of the temp bone
PnC   pontine reticular nu, caudal part
PnO   pontine reticular nu, oral part
PrH   prepositus hypoglossal nucleus
RMg   raphe magnus nucleus
Sacc   saccule
SC   superior colliculus
scba   superior cerebellar artery
scp   superior cerebellar peduncle
sl   sulcus limitans
SOl   superior olive
sol   solitary tract
sp5   spinal trigeminal tract
Sp5O   spinal trigeminal nu, oral part
Stapedius   stapedius muscle
SuVe   superior vestibular nucleus
trg   germinal trigone
trs   transverse sinus
Utr   utricle
ve   vestibular area neuroepi
VeGn   vestibular ganglion
VTA   ventral tegmental area (Tsai)

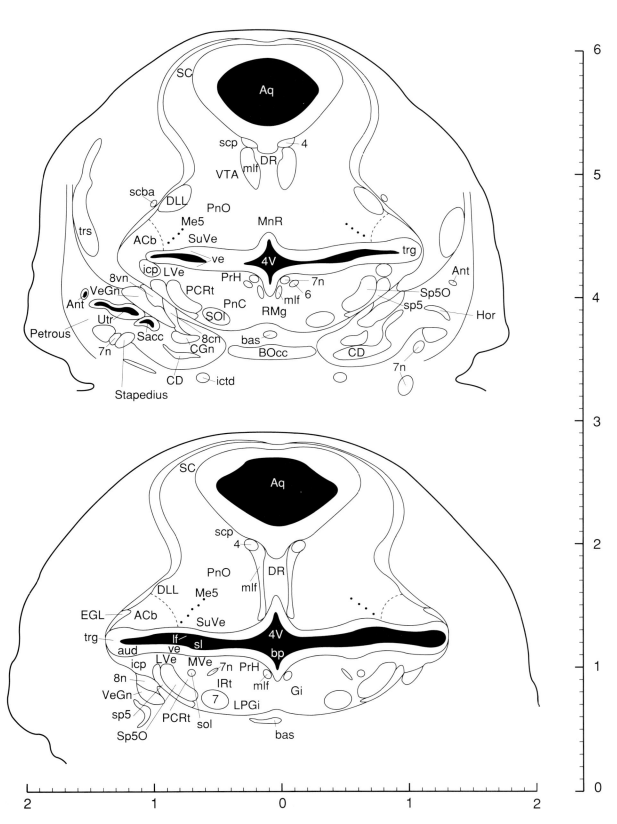

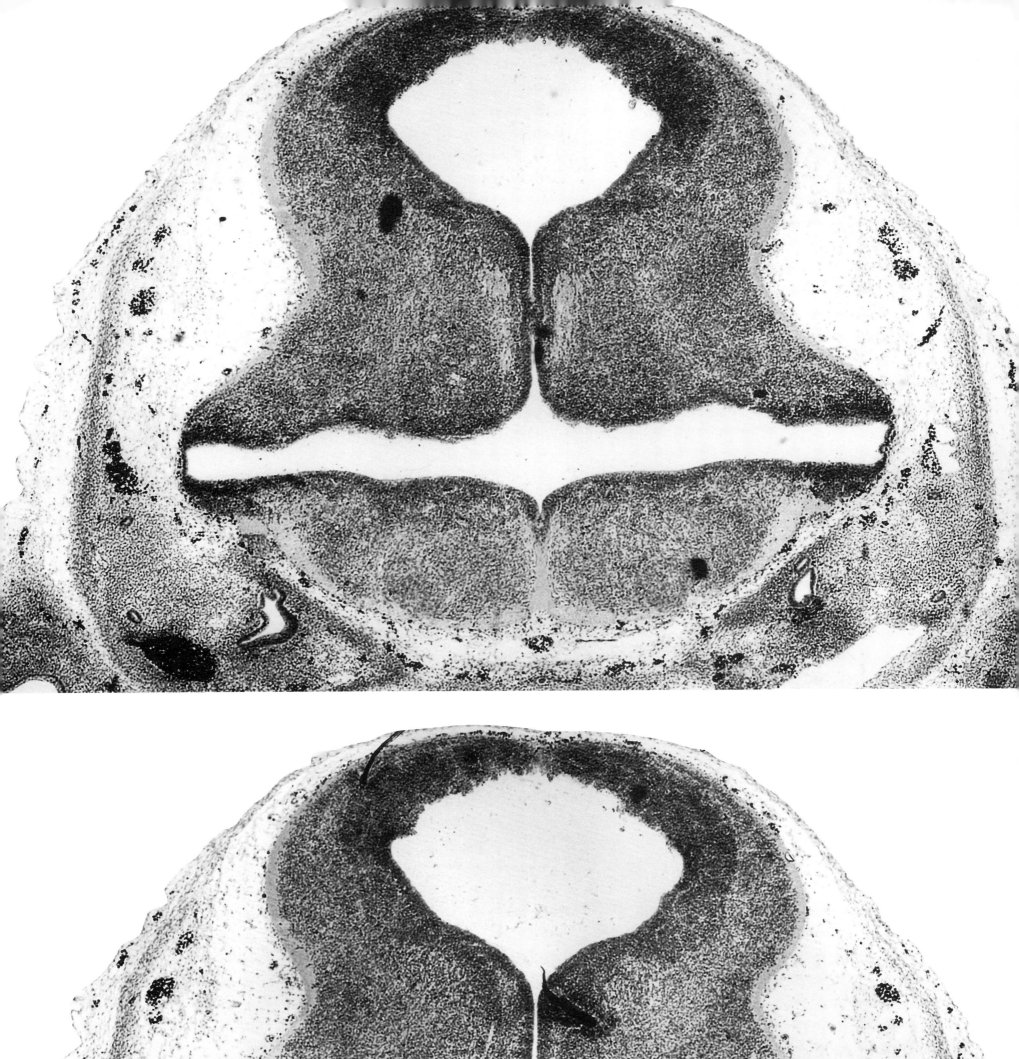

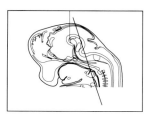

Figure 45
E16 Coronal 17

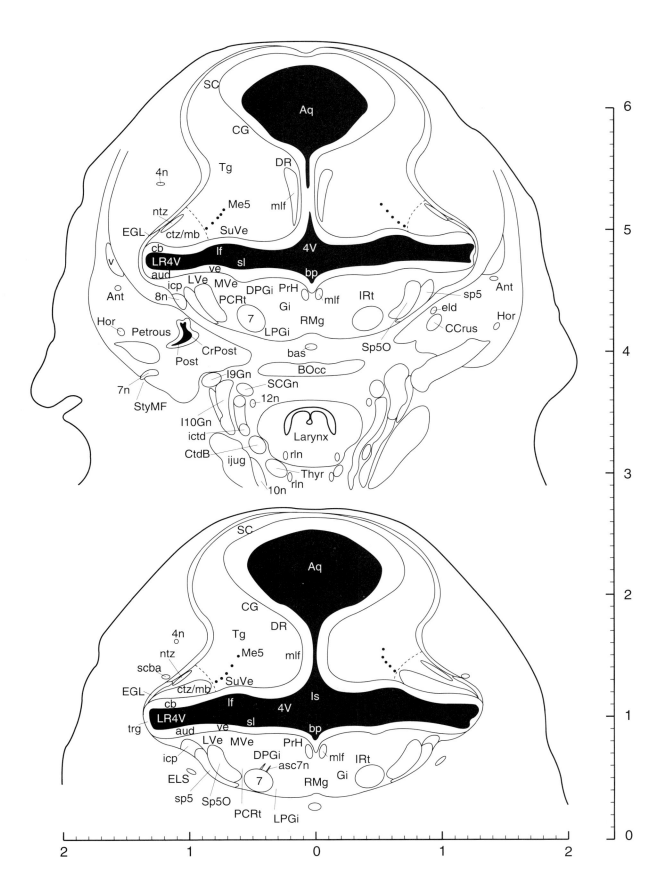

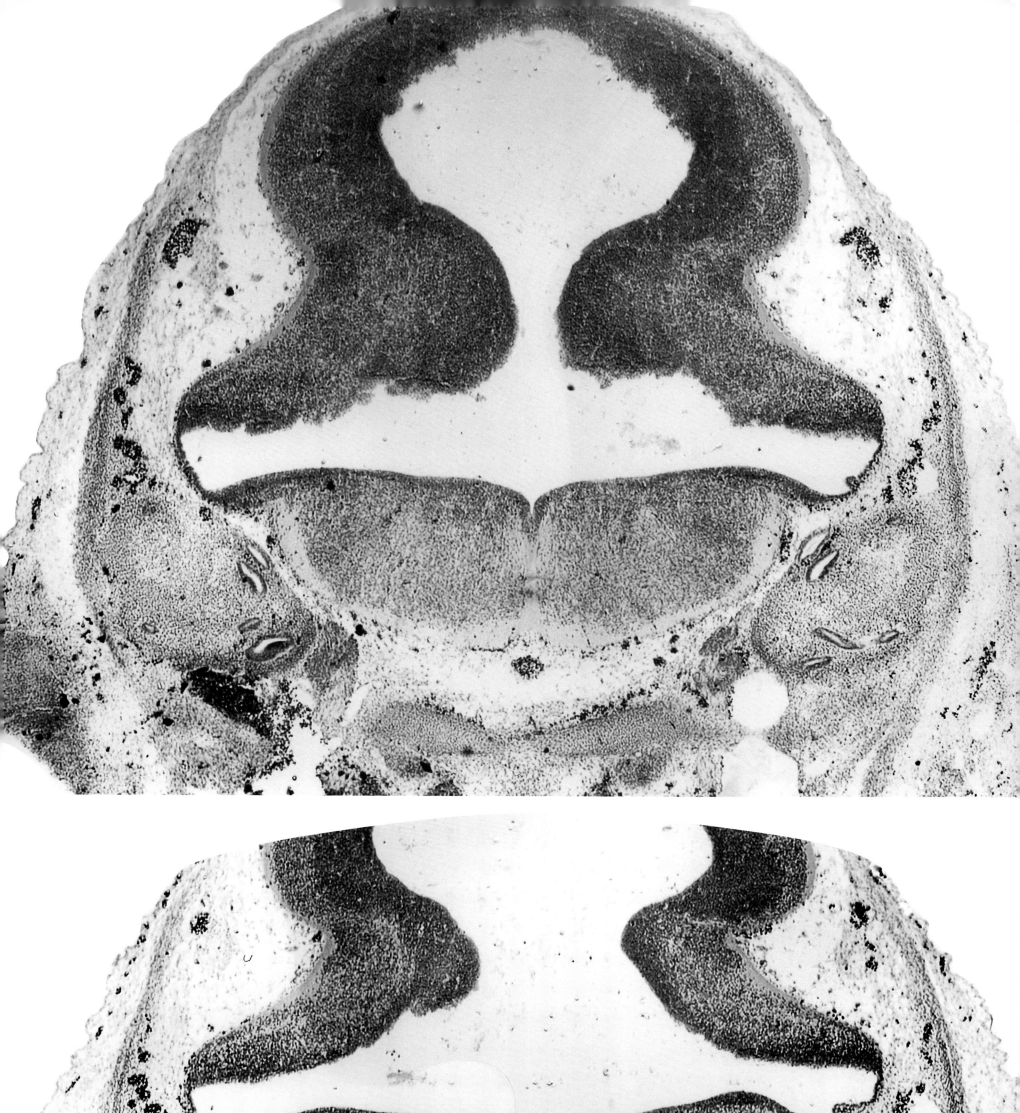

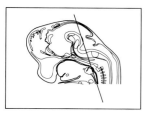

Figure 46
E16 Coronal 18

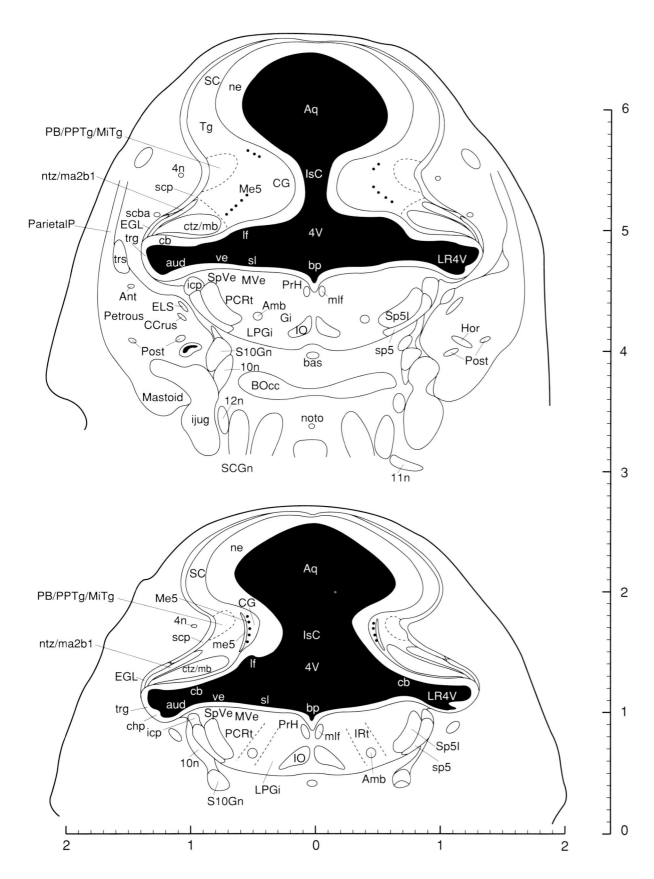

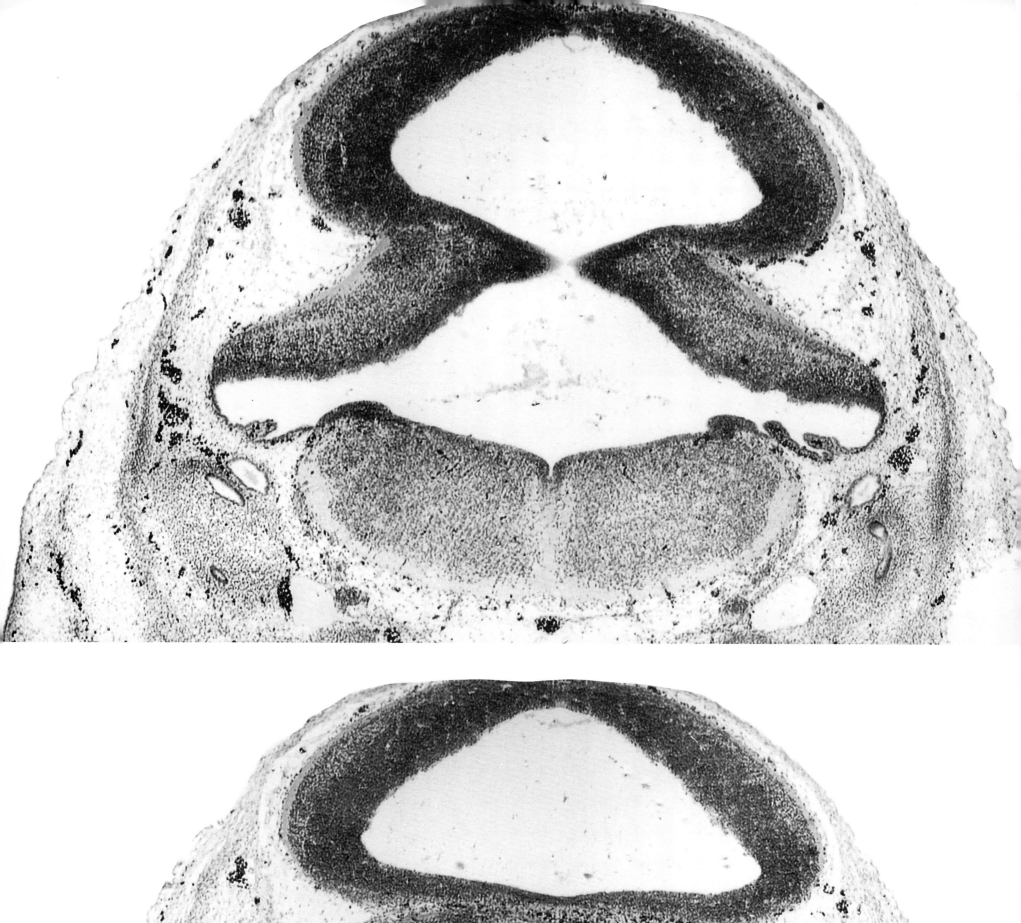
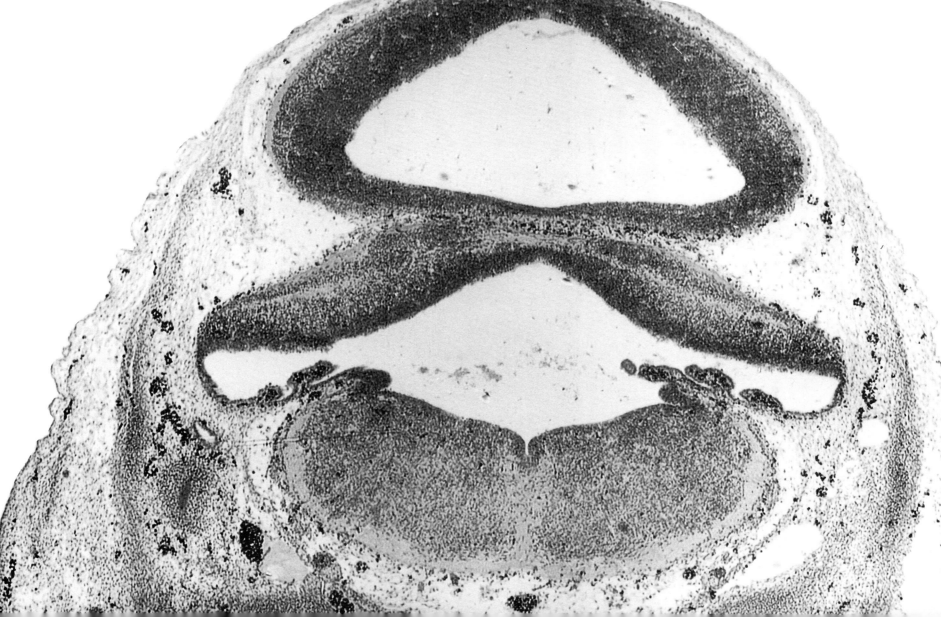

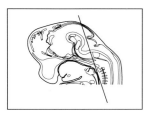

Figure 47
E16 Coronal 19

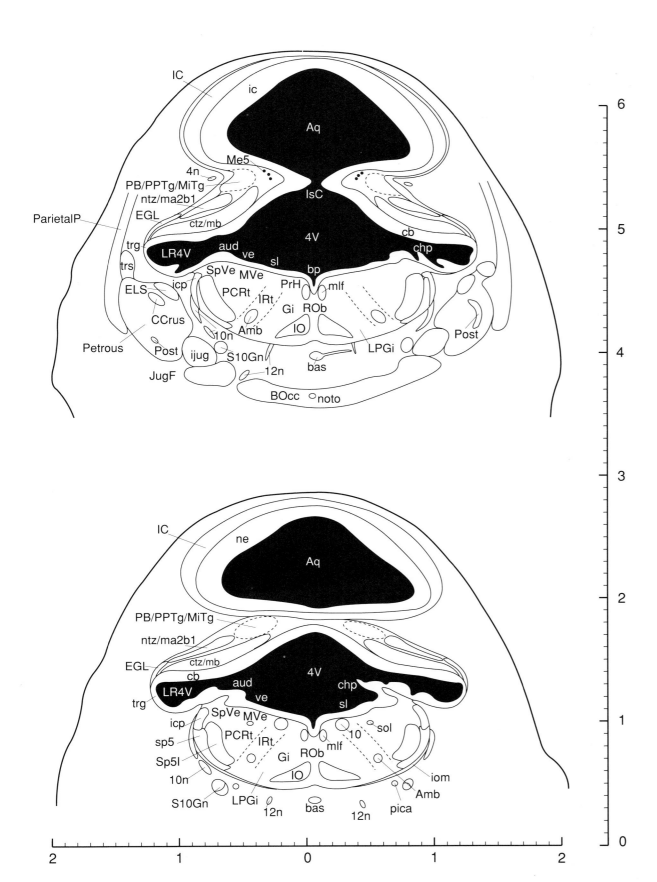

4n   trochlear nerve or its root
4V   fourth ventricle
10   dorsal motor nucleus of vagus
10n   vagus nerve
12n   hypoglossal nerve or its root
Amb   ambiguus nucleus
Aq   aqueduct (Sylvius)
aud   auditory neuroepi
bas   basilar artery
BOcc basioccipital bone
bp   basal plate neuroepi
cb   cerebellar neuroepi
CCrus common crus of anterior and posterior semicircular ducts
chp   choroid plexus premordium
ctz   cortical transitory zone
EGL   external germinal layer of Cb
ELS   endolymphatic sac
Gi   gigantocellular reticular nucleus
IC   inferior colliculus
ic   inferior colliculus neuroepi
icp   inferior cerebellar peduncle
ijug internal jugular vein
IO   inferior olive
iom   inferior olivary migration
IRt   intermediate reticular zone
IsC   isthmal canal
JugF jugular foramen
LPGi lateral paragigantocellular nu
LR4V lateral recess of the 4th ventr
ma2b1 migration a2b1 of Rüdeberg
mb   migration b of Rüdeberg
Me5   mesencephalic trigeminal nu
MiTg microcellular tegmental nu
mlf   medial longitudinal fasciculus
MVe   medial vestibular nucleus
ne   neuroepithelium
noto notochord
ntz   nuclear transitory zone
ParietalP parietal plate
PB   parabrachial nuclei
PCRt parvocellular reticular nu
Petrous petrous part of the temp bone
pica posterior inferior cerebellar artery
Post posterior semicircular duct
PPTg pedunculopontine tegmental nu
PrH   prepositus hypoglossal nucleus
ROb   raphe obscurus nucleus
S10Gn superior vagal (jugular) ganglion
sl   sulcus limitans
sol   solitary tract
sp5   spinal trigeminal tract
Sp5I spinal trigeminal nu, interpolar
SpVe spinal vestibular nucleus
trg   germinal trigone
trs   transverse sinus
ve   vestibular area neuroepi

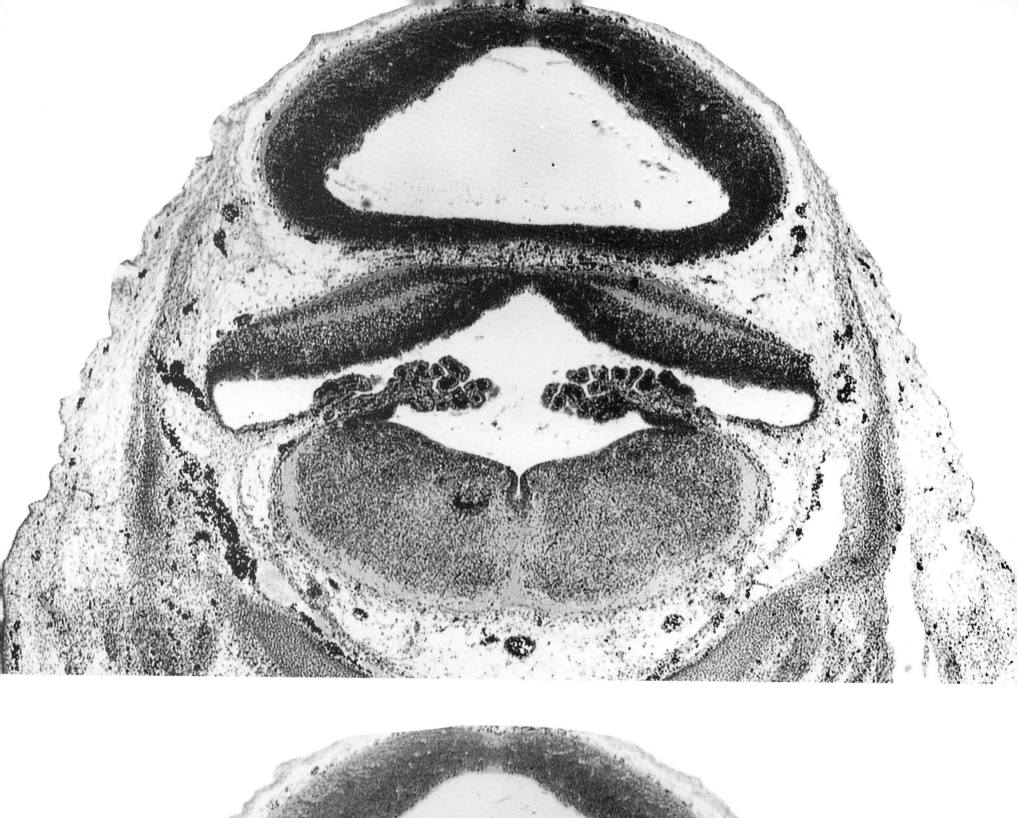
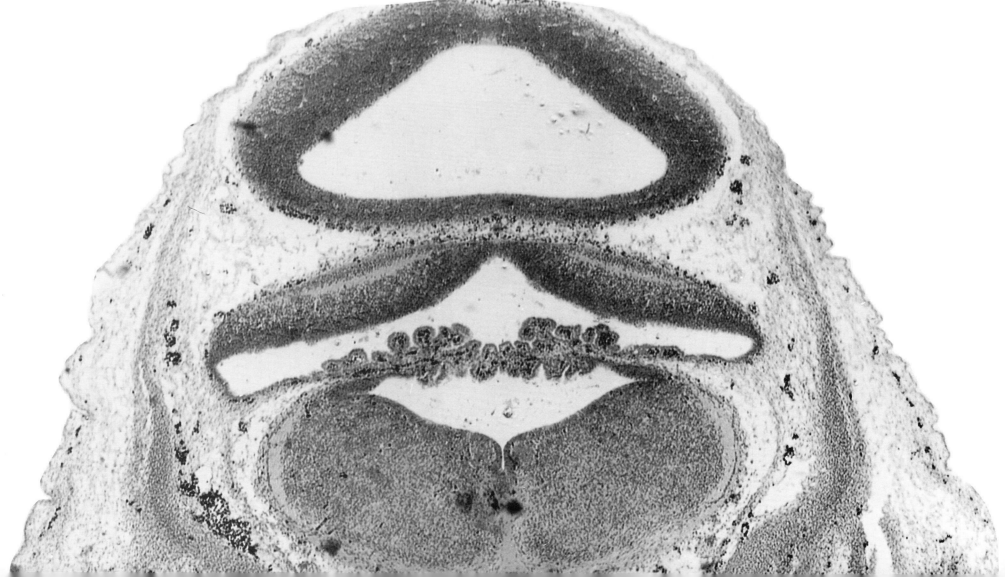

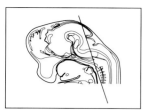

Figure 48
E16 Coronal 20

4n   trochlear nerve or its root
4V   fourth ventricle
10   dorsal motor nucleus of vagus
10n   vagus nerve
12   hypoglossal nucleus
12n   hypoglossal nerve or its root
Amb   ambiguus nucleus
Aq   aqueduct (Sylvius)
Atlas   atlas (C1 vertebra)
bas   basilar artery
chp   choroid plexus premordium
ctz   cortical transitory zone
EGL   external germinal layer of Cb
Gi   gigantocellular reticular nucleus
IC   inferior colliculus
ic   inferior colliculus neuroepi
icp   inferior cerebellar peduncle
IO   inferior olive
iom   inferior olivary migration
IRt   intermediate reticular zone
LPGi   lateral paragigantocellular nucleus
LR4V   lateral recess of the 4th ventr
ma2b1   migration a2b1 of Rüdeberg
mb   migration b of Rüdeberg
mlf   medial longitudinal fasciculus
MVe   medial vestibular nucleus
noto   notochord
ntz   nuclear transitory zone
Occ   occipital bone
occs   occipital sinus
ParietalP   parietal plate
PCRt   parvocellular reticular nu
pica   posterior inferior cerebellar artery
S10Gn   superior vagal (jugular) ganglion
sl   sulcus limitans
Sol   nucleus of the solitary tract
sp5   spinal trigeminal tract
Sp5I   spinal trigeminal nu, interpolar
SpVe   spinal vestibular nucleus
trg   germinal trigone
vert   vertebral artery

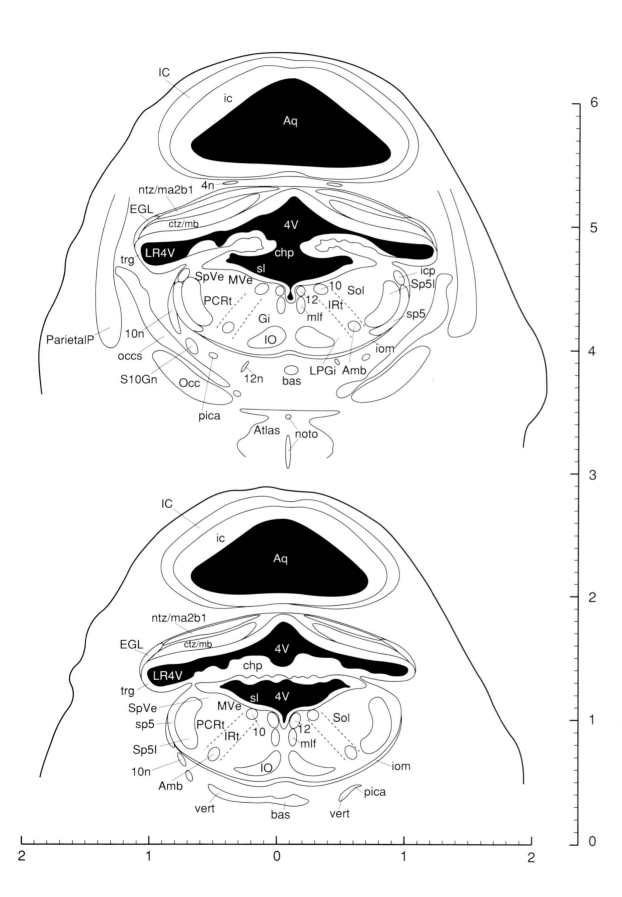

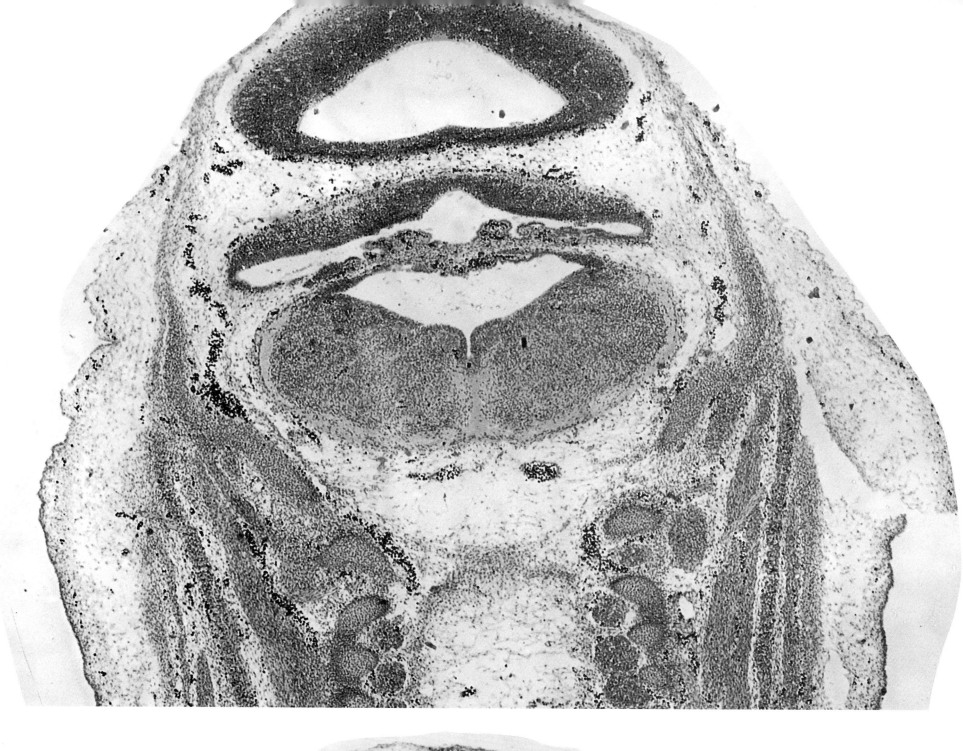
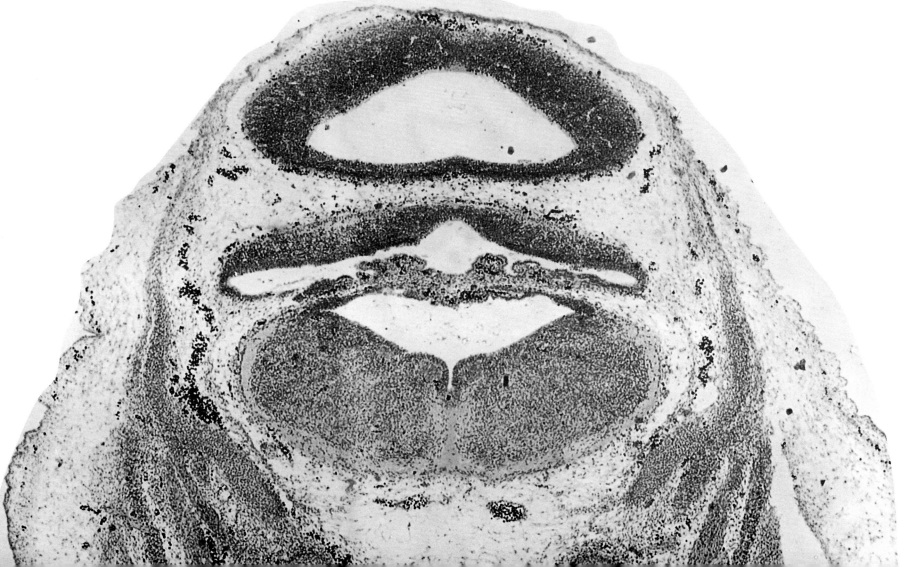

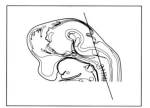

Figure 49
E16 Coronal 21

4V   fourth ventricle
10   dorsal motor nucleus of vagus
10n   vagus nerve
12   hypoglossal nucleus
12n   hypoglossal nerve or its root
Aq   aqueduct (Sylvius)
chp   choroid plexus premordium
ctz   cortical transitory zone of Cb
EGL   external germinal layer of Cb
Gi   gigantocellular reticular nucleus
IC   inferior colliculus
icp   inferior cerebellar peduncle
IO   inferior olive
iom   inferior olivary migration
IRt   intermediate reticular zone
LR4V   lateral recess of the 4th ventr
LRt   lateral reticular nucleus
mlf   medial longitudinal fasciculus
ntz   nuclear transitory zone
Occ   occipital bone
occs   occipital sinus
PCb   posterior cerebellum
PCRt   parvocellular reticular nu
sl   sulcus limitans
Sol   nucleus of the solitary tract
sp5   spinal trigeminal tract
Sp5I   spinal trigeminal nu, interpolar
SpVe   spinal vestibular nucleus
ve   vestibular area neuroepi
vert   vertebral artery

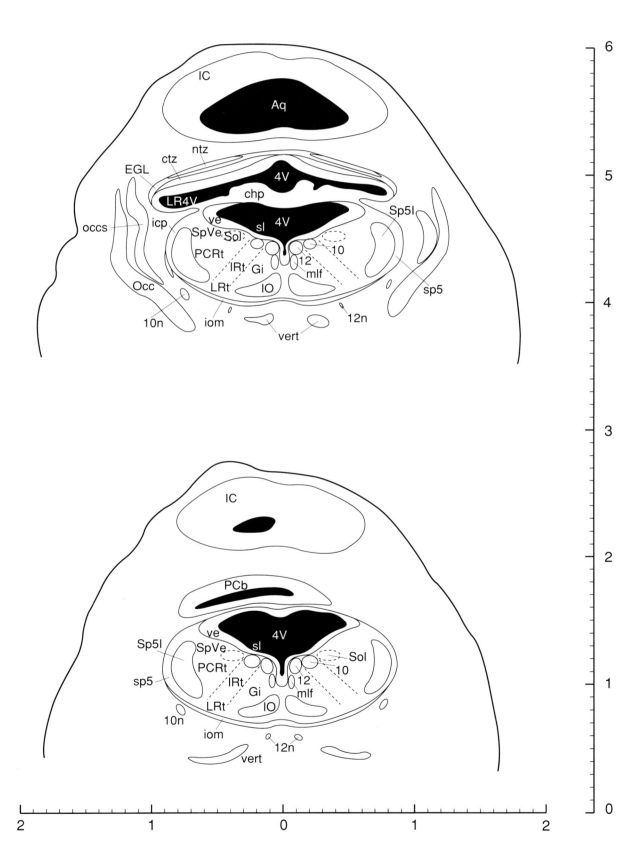

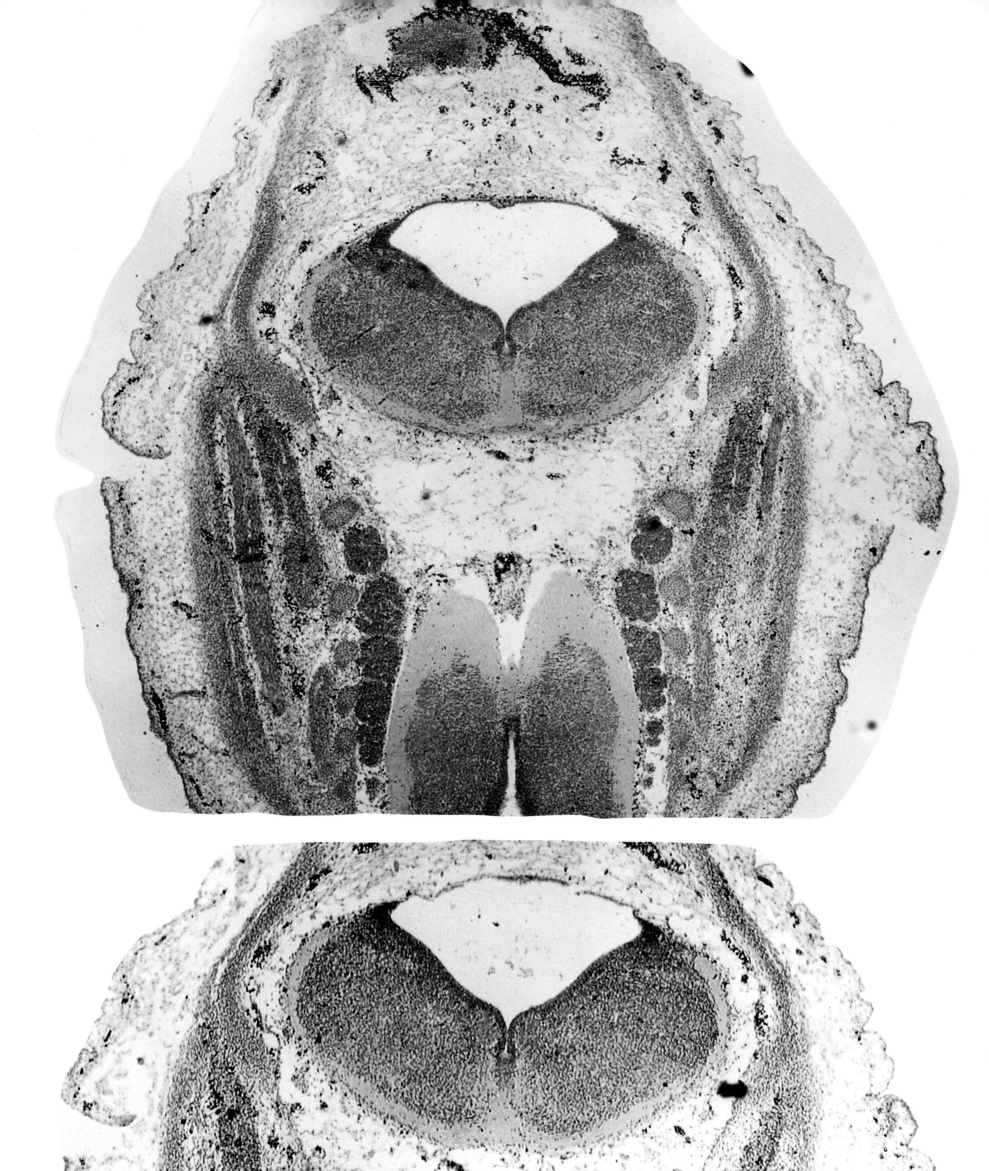

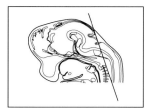

Figure 50
E16 Coronal 22

4V   fourth ventricle
10   dorsal motor nucleus of vagus
12   hypoglossal nucleus
aspina   anterior spinal artery
bp   basal plate neuroepi
IC   inferior colliculus
IO   inferior olive
iom   inferior olivary migration
IRt   intermediate reticular zone
LRt   lateral reticular nucleus
MdD   medullary reticular nu, dors
MdV   medullary reticular nu, vental
mlf   medial longitudinal fasciculus
Occ   occipital bone
sl   sulcus limitans
Sol   nucleus of the solitary tract
sp5   spinal trigeminal tract
Sp5C   spinal trigeminal nu, caudal part
SpVe   spinal vestibular nucleus
ve   vestibular area neuroepi
vert   vertebral artery

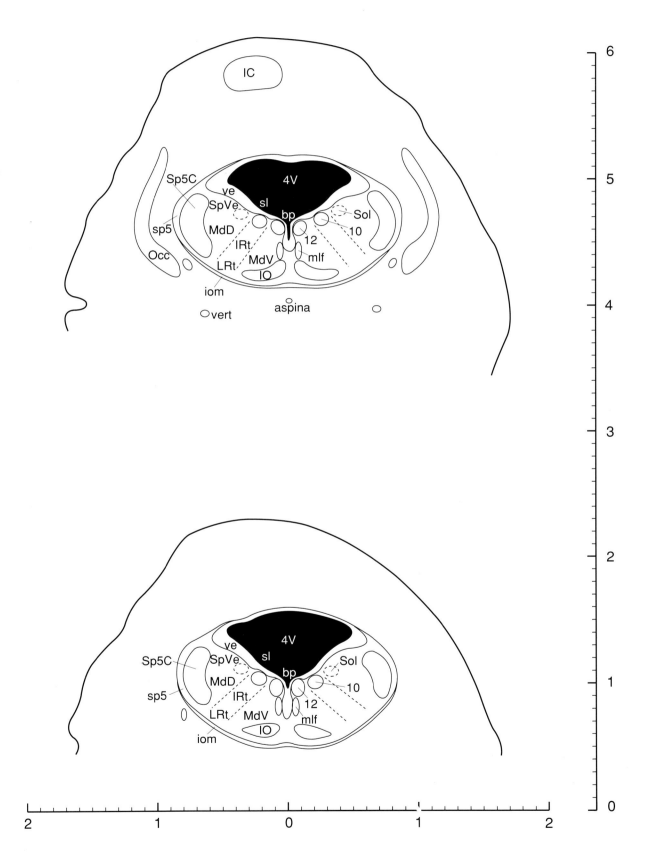

E16 Sagittal Section Plan

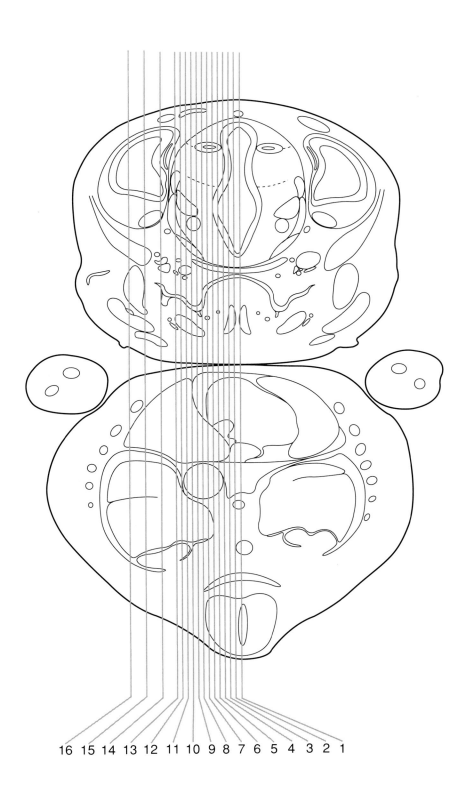

16 15 14 13 12 11 10 9 8 7 6 5 4 3 2 1

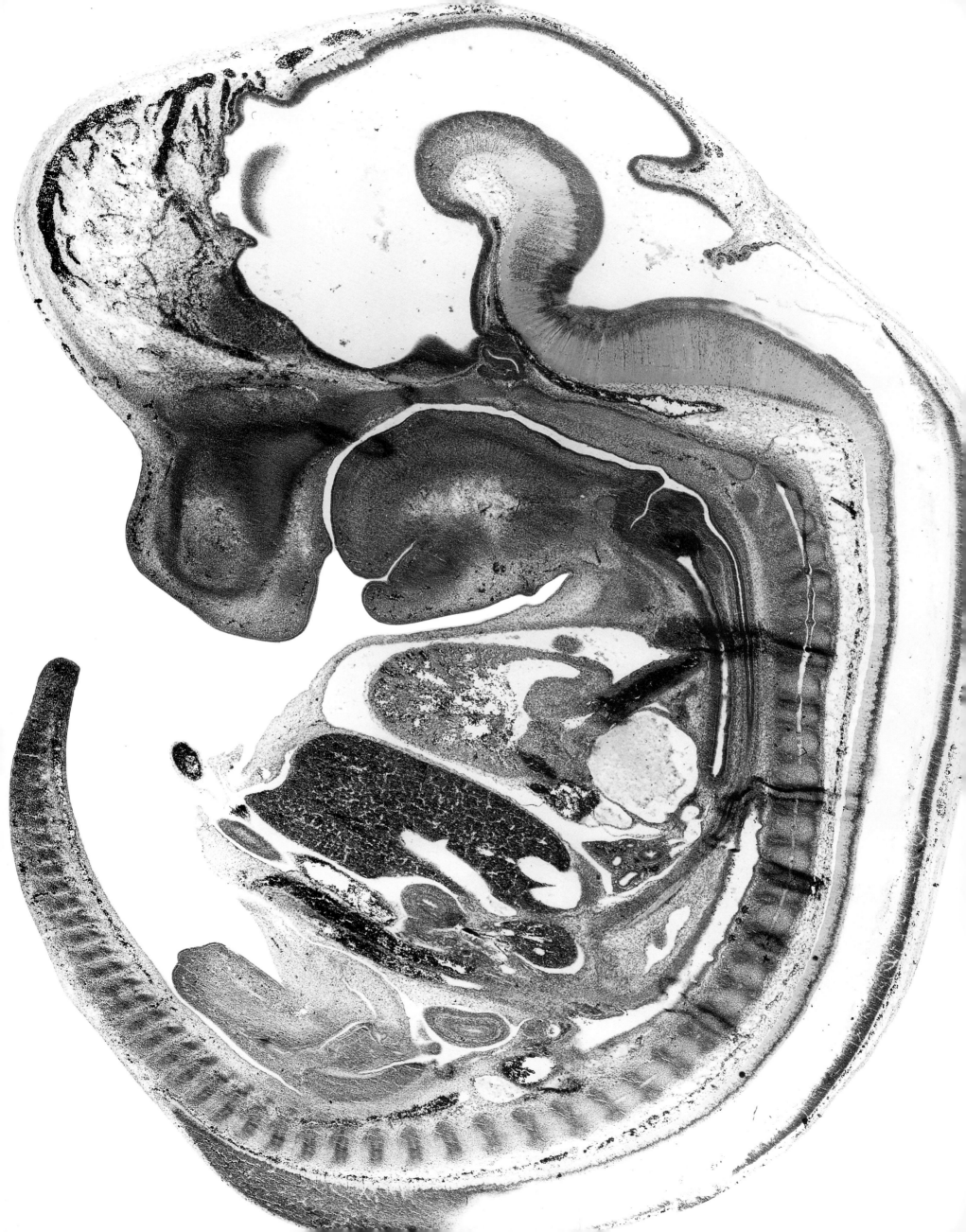

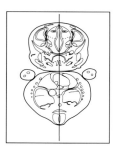

Figure 51
E16 Sagittal 1

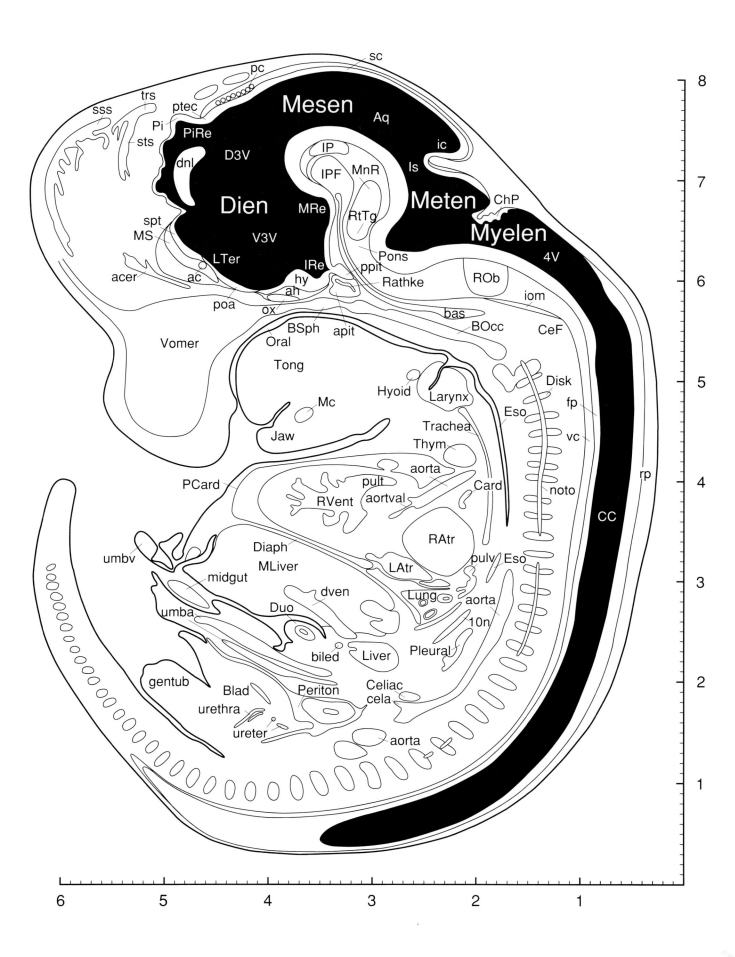

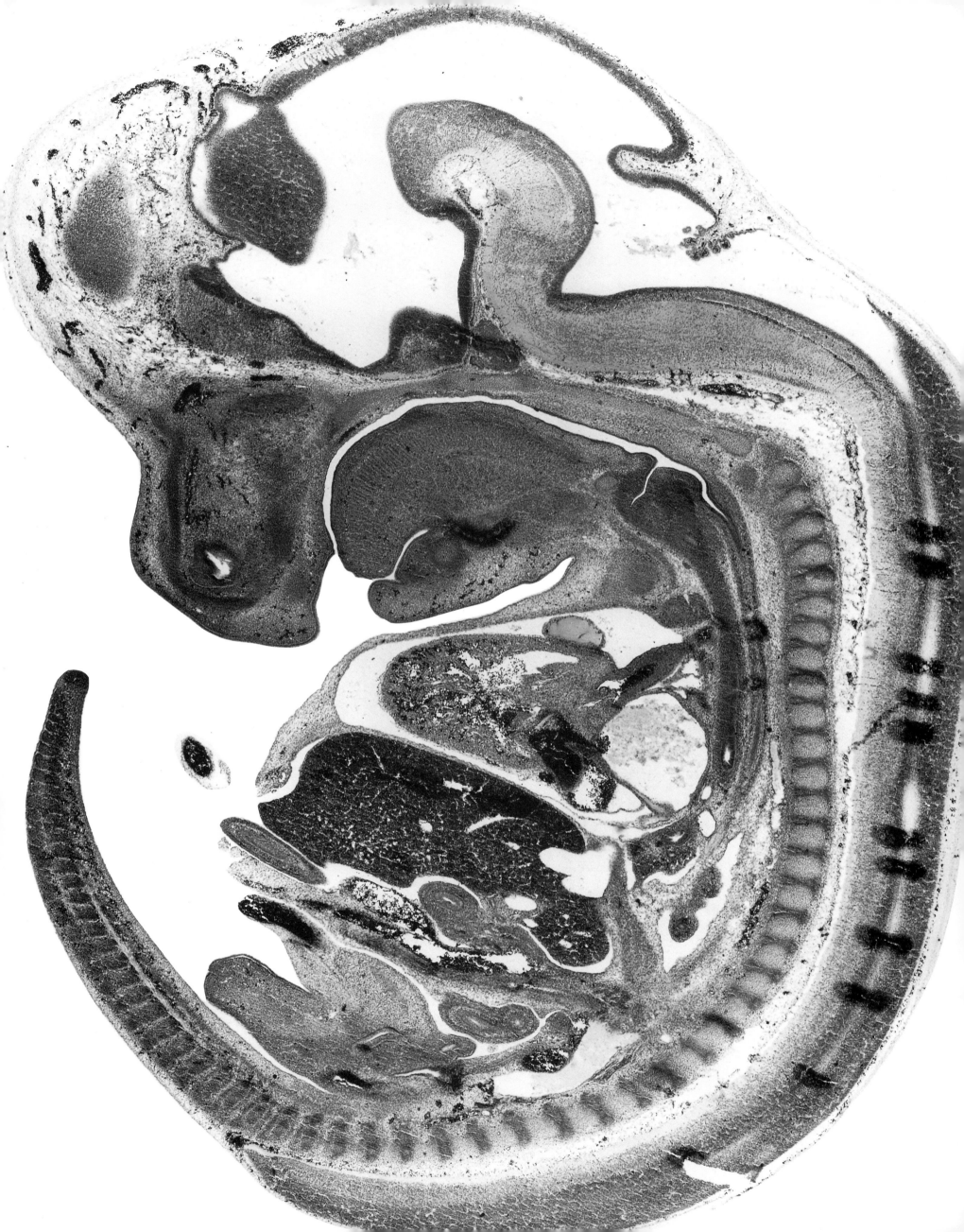

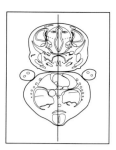

Figure 52
E16 Sagittal 2

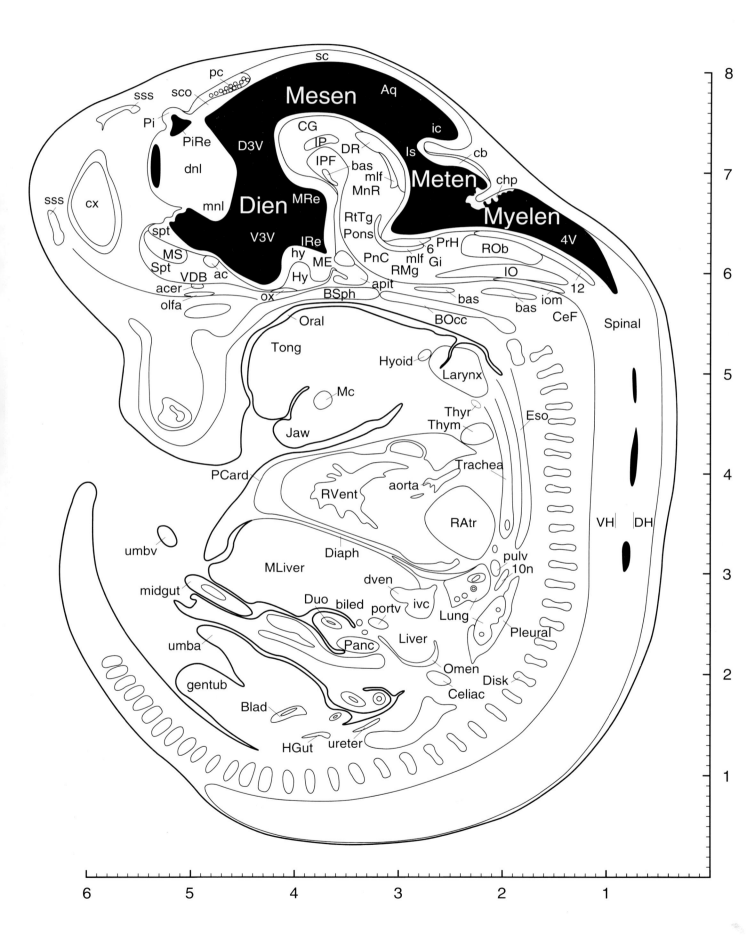

4V   fourth ventricle
6   abducens nucleus
10n   vagus nerve
12   hypoglossal nucleus
A8   A8 dopamine cells
ac   anterior commissure
acer   anterior cerebral artery
aorta   aorta
apit   anterior pituitary anlage
bas   basilar artery
biled   bile duct
Blad   urinary bladder
BOcc   basioccipital bone
BSph   basisphenoid bone
cb   cerebellar neuroepi
CeF   cervical flexure
Celiac celiac ganglion
CG   central (periaqueductal) gray
chp   choroid plexus premordium
cx   cortical neuroepi
D3V   dorsal third ventricle
DH   dorsal horns of spinal cord
Diaph   diaphragm
Dien   diencephalon
Disk   intervertebral disk
dnl   dorsal neuroepith lobule of dien
DR   dorsal raphe nucleus
Duo   duodenum
dven   ductus venosus
Eso   esophagus
gentub genital tubercle
Gi   gigantocellular reticular nu
HGut   hindgut
Hy   hypothalamus
hy   hypothalamic neuroepi
Hyoid   hyoid bone
ic   inferior colliculus neuroepi
IO   inferior olive
iom   inferior olivary migration
IP   interpeduncular nucleus
IPF   interpeduncular fossa
IRe   infundibular recess of the 3rd ventr
Is   isthmus region
ivc   inferior vena cava
Jaw   jaw
Larynx larynx
Liver   liver
Lung   lung
Mc   Meckel's cartilage
ME   median eminence
Mesen   mesencephalon
Meten   metencephalon
midgut midgut in physiological herniation
mlf   medial longitudinal fasciculus
MLiver middle lobe of liver
mnl   middle neuroepith lobule of dien
MnR   median raphe nucleus
MRe   mammillary recess of the 3rd ventr
MS   medial septal nucleus
Myelen myelencephalon
olfa   olfactory artery
Omen   omental bursa
Oral   oral cavity
ox   optic chiasm
Panc   pancreas
pc   posterior commissure
PCard   pericardium
Pi   pineal gland
PiRe   pineal recess of the third ventricle
Pleural pleural cavity
PnC   pontine reticular nu, caudal part
Pons   pons
portv   portal vein
PrH   prepositus hypoglossal nucleus
pulv   pulmonary vein
RAtr   right atrium
RMg   raphe magnus nucleus
ROb   raphe obscurus nucleus
RtTg   reticulotegmental nu of the pons
RVent   right ventricle
sc   superior colliculus neuroepi
sco   subcommissural organ neuroepi
Spinal spinal cord
Spt   septum
spt   septal neuroepi
sss   superior sagittal sinus
Thym   thymus gland
Thyr   thyroid gland
Tong   tongue
Trachea trachea
umba   umbilical artery
umbv   umbilical vein
ureter ureter
V3V   ventral third ventricle
VDB   nu vertical limb diagonal band
VH   ventral horn

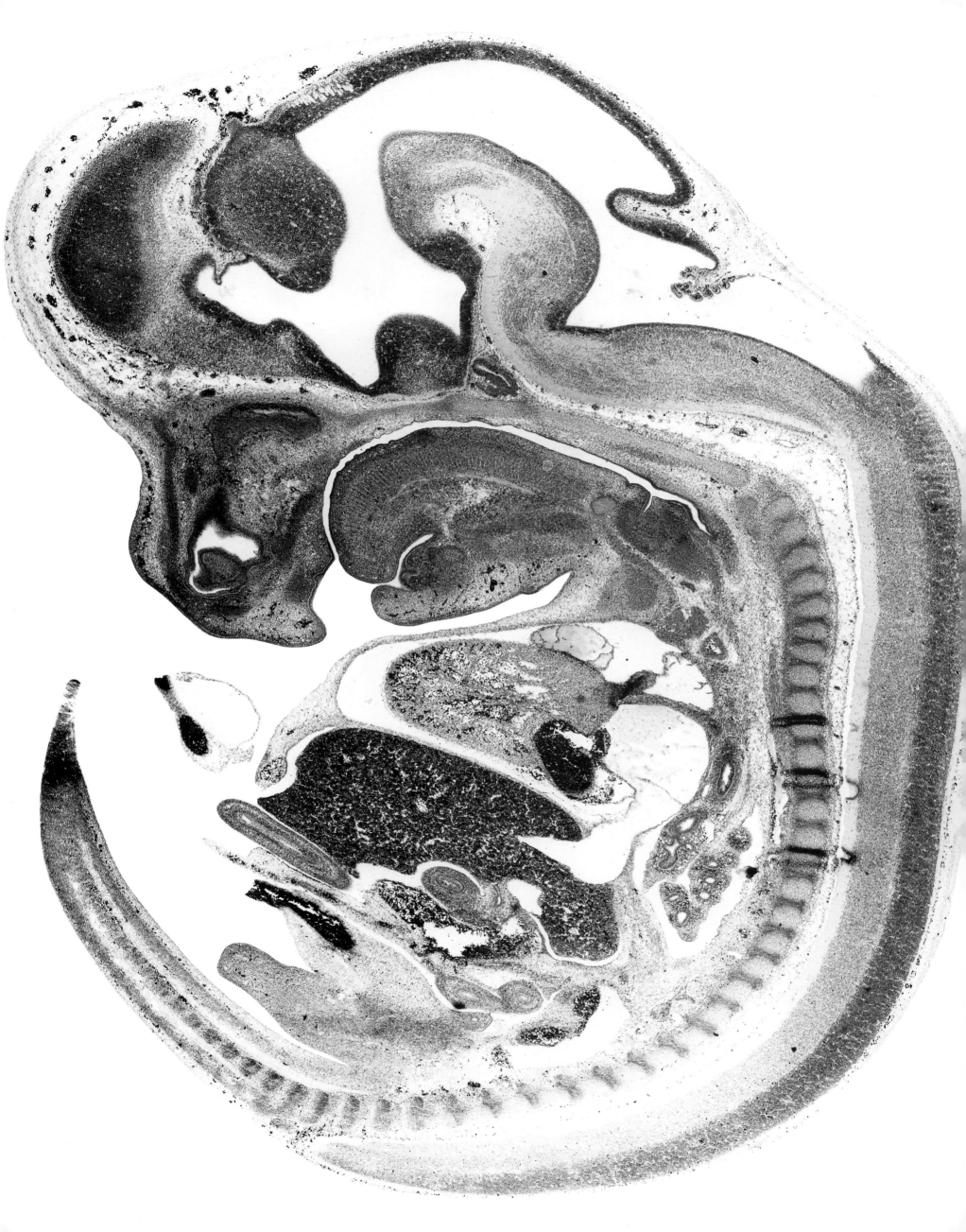

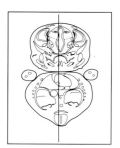

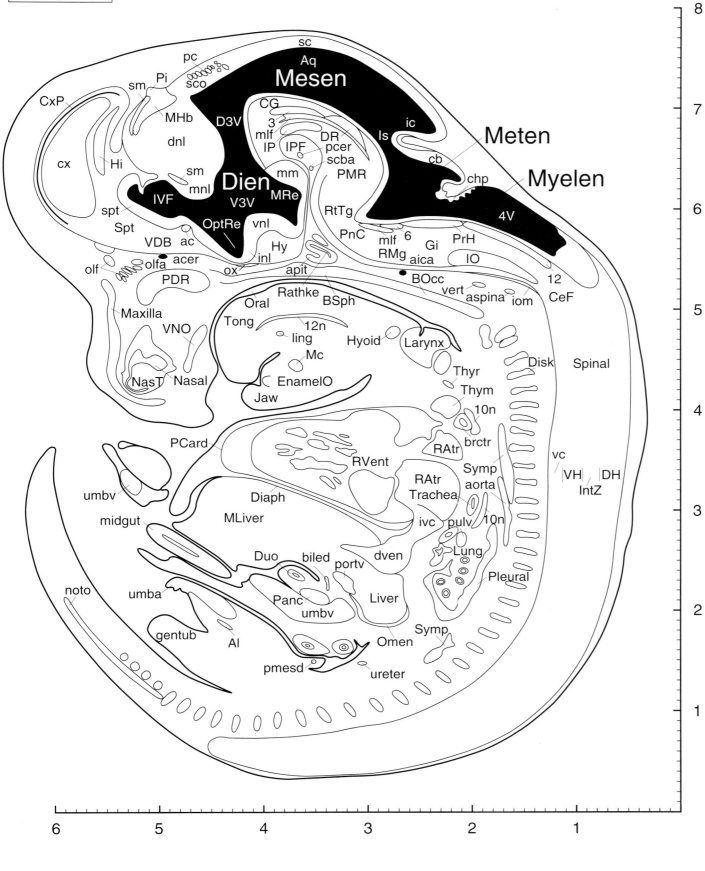

Figure 53

E16 Sagittal 3

3   oculomotor nucleus
4V   fourth ventricle
6   abducens nucleus
10n   vagus nerve
12   hypoglossal nucleus
12n   hypoglossal nerve or its root
ac   anterior commissure
acer   anterior cerebral artery
aica   anterior inferior cerebellar artery

Al   allantois
aorta   aorta
apit   anterior pituitary anlage
Aq   aqueduct (Sylvius)
aspina anterior spinal artery
biled   bile duct
BOcc   basioccipital bone
brctr   brachiocephalic trunk
BSph   basisphenoid bone
cb   cerebellar neuroepi
CeF   cervical flexure
CG   central (periaqueductal) gray
chp   choroid plexus premordium
cx   cortical neuroepi
CxP   cortical plate

D3V   dorsal third ventricle
DH   dorsal horns of spinal cord
Diaph   diaphragm
Dien   diencephalon
Disk   intervertebral disk
dnl   dorsal neuroepith lobule of dien
DR   dorsal raphe nucleus
Duo   duodenum
dven   ductus venosus
EnamelO enamel organ (of tooth)
gentub genital tubercle
Gi   gigantocellular reticular nucleus
Hi   hippocampal formation
Hy   hypothalamus
Hyoid   hyoid bone
ic   inferior colliculus neuroepi
inl   inferior neuroepith lobule of dien
IntZ   intermediate zone of spinal gray
IO   inferior olive
iom   inferior olivary migration
IP   interpeduncular nucleus
IPF   interpeduncular fossa
Is   isthmus region
ivc   inferior vena cava
IVF   interventricular foramen
Jaw   jaw
Larynx larynx
ling   lingual nerve
Liver   liver
Lung   lung
Maxilla maxilla
Mc   Meckel's cartilage
Mesen   mesencephalon
Meten   metencephalon
MHb   medial habenular nucleus
midgut midgut in physiological herniation
mlf   medial longitudinal fasciculus
MLiver middle lobe of liver
mm   mammillary body neuroepi
mnl   middle neuroepith lobule of dien
MRe   mammillary recess of the 3rd ventr
Myelen myelencephalon
Nasal   nasal cavity
NasT   nasal turbinate
noto   notochord
olf   olfactory nerve
olfa   olfactory artery
Omen   omental bursa
OptRe   optic recess of third ventricle
Oral   oral cavity
ox   optic chiasm
Panc   pancreas
pc   posterior commissure
PCard   pericardium
pcer   posterior cerebral artery
PDR   posterior dorsal recess of nasal cavity
Pi   pineal gland
Pleural pleural cavity
pmesd   paramesonephric duct
PMR   paramedian raphe nucleus
PnC   pontine reticular nu, caudal part
portv   portal vein
PrH   prepositus hypoglossal nucleus
pulv   pulmonary vein
Rathke Rathke's pouch
RAtr   right atrium
RMg   raphe magnus nucleus
RtTg   reticulotegmental nu of the pons
RVent   right ventricle
sc   superior colliculus neuroepi
scba   superior cerebellar artery
sco   subcommissural organ neuroepi
sm   stria medullaris of the thalamus
Spinal spinal cord
Spt   septum
spt   septal neuroepi
Symp   sympathetic trunk
Thym   thymus gland
Thyr   thyroid gland
Tong   tongue
Trachea trachea
umba   umbilical artery
umbv   umbilical vein
ureter ureter
V3V   ventral third ventricle
vc   ventral commissure of spinal cord
VDB   nu vertical limb diagonal band
vert   vertebral artery
VH   ventral horn
vnl   ventral neuroepith lobule of dien
VNO   vomeronasal organ

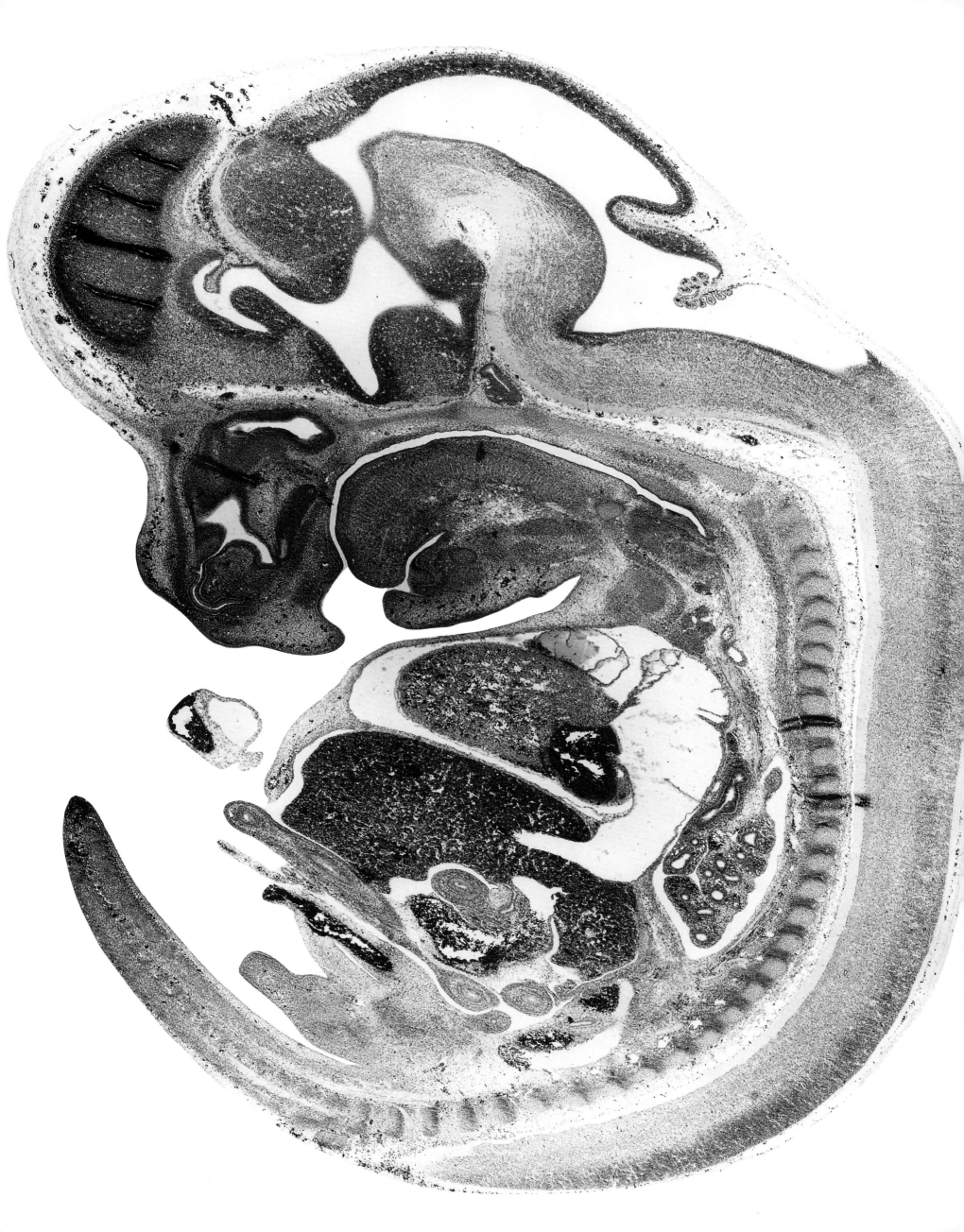

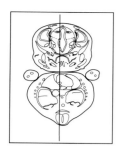

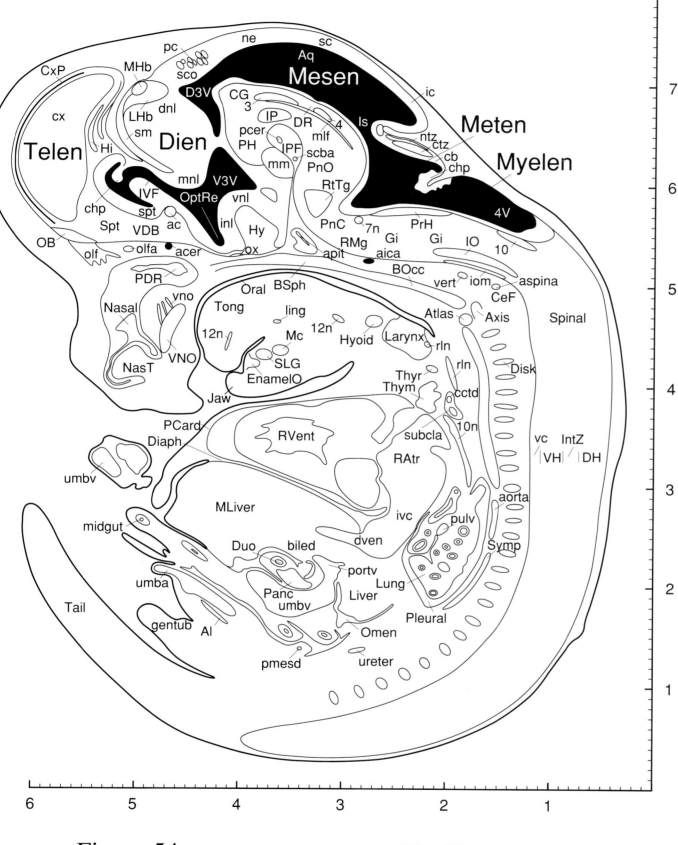

Figure 54

E16 Sagittal 4

3   oculomotor nucleus
4   trochlear nucleus
4V   fourth ventricle
7n   facial nerve or its root
10   dorsal motor nucleus of vagus
10n   vagus nerve
12n   hypoglossal nerve or its root
ac   anterior commissure
acer   anterior cerebral artery
aica   anterior inferior cerebellar artery
Al   allantois

aorta   aorta
apit   anterior pituitary anlage
Aq   aqueduct (Sylvius)
aspina anterior spinal artery
Atlas   atlas (C1 vertebra)
Axis   axis (C2 vertebra)
biled   bile duct
BOcc   basioccipital bone
BSph   basisphenoid bone
cb   cerebellar neuroepi
cctd   common carotid artery
CeF   cervical flexure
CG   central (periaqueductal) gray
chp   choroid plexus premordium
ctz   cortical transitory zone of Cb
cx   cortical neuroepi
CxP   cortical plate

D3V   dorsal third ventricle
DH   dorsal horns of spinal cord
Diaph   diaphragm
Dien   diencephalon
Disk   intervertebral disk
dnl   dorsal neuroepith lobule of dien
DR   dorsal raphe nucleus
Duo   duodenum
dven   ductus venosus
EnamelO enamel organ (of tooth)
gentub genital tubercle
Gi   gigantocellular reticular nucleus
Hi   hippocampal formation
Hy   hypothalamus
Hyoid   hyoid bone
ic   inferior colliculus neuroepi
inl   inferior neuroepith lobule of dien
IntZ   intermediate zone of spinal gray
IO   inferior olive
iom   inferior olivary migration
IP   interpeduncular nucleus
IPF   interpeduncular fossa
Is   isthmus region
ivc   inferior vena cava
IVF   interventricular foramen
Jaw   jaw
Larynx larynx
LHb   lateral habenular nucleus
ling   lingual nerve
Liver   liver
Lung   lung
Mc   Meckel's cartilage
Mesen   mesencephalon
Meten   metencephalon
MHb   medial habenular nucleus
midgut midgut in physiological herniation
mlf   medial longitudinal fasciculus
MLiver middle lobe of liver
mm   mammillary body neuroepi
mnl   middle neuroepith lobule of dien
Myelen myelencephalon
Nasal   nasal cavity
NasT   nasal turbinate
ne   neuroepithelium
ntz   nuclear transitory zone
OB   olfactory bulb
olf   olfactory nerve
olfa   olfactory artery
Omen   omental bursa
OptRe   optic recess of third ventricle
Oral   oral cavity
ox   optic chiasm
Panc   pancreas
pc   posterior commissure
PCard   pericardium
pcer   posterior cerebral artery
PDR   posterior dorsal recess of nasal cavity
PH   posterior hypothalamic area
Pleural pleural cavity
pmesd   paramesonephric duct
PnC   pontine reticular nu, caudal part
PnO   pontine reticular nu, oral part
portv   portal vein
PrH   prepositus hypoglossal nucleus
pulv   pulmonary vein
RAtr   right atrium
rln   recurrent laryngeal nerve
RMg   raphe magnus nucleus
RtTg   reticulotegmental nu of the pons
RVent   right ventricle
sc   superior colliculus neuroepi
scba   superior cerebellar artery
sco   subcommissural organ neuroepi
SLG   sublingual gland
sm   stria medullaris of the thalamus
Spinal spinal cord
Spt   septum
spt   septal neuroepi
subcla subclavian artery
Symp   sympathetic trunk
Tail   tail
Telen   telencephalon
Thym   thymus gland
Thyr   thyroid gland
Tong   tongue
umba   umbilical artery
umbv   umbilical vein
ureter ureter
V3V   ventral third ventricle
vc   ventral commissure of spinal cord
VDB   nu vertical limb diagonal band
vert   vertebral artery
VH   ventral horn
vnl   ventral neuroepith lobule of dien
VNO   vomeronasal organ
vno   nerve of vomeronasal organ

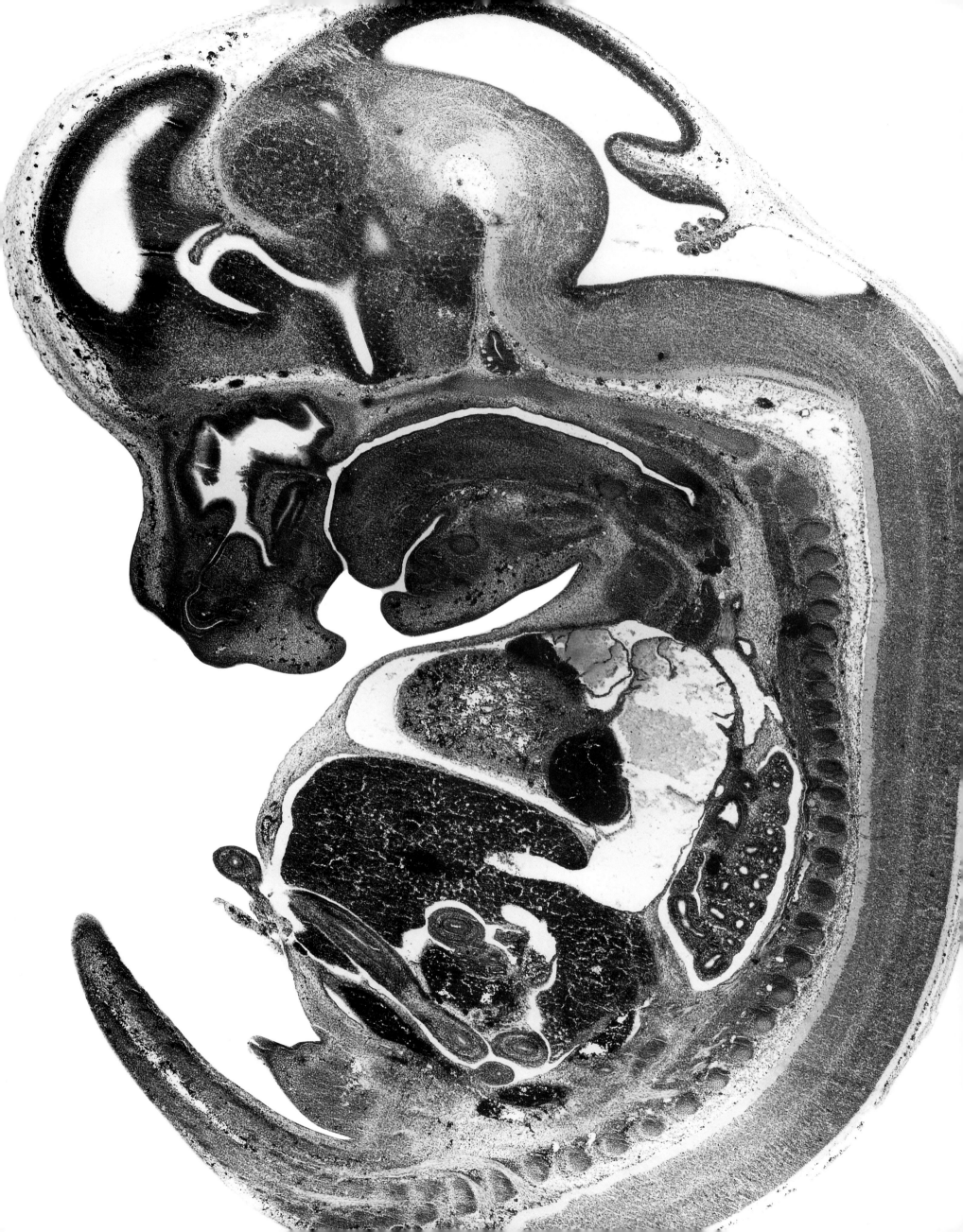

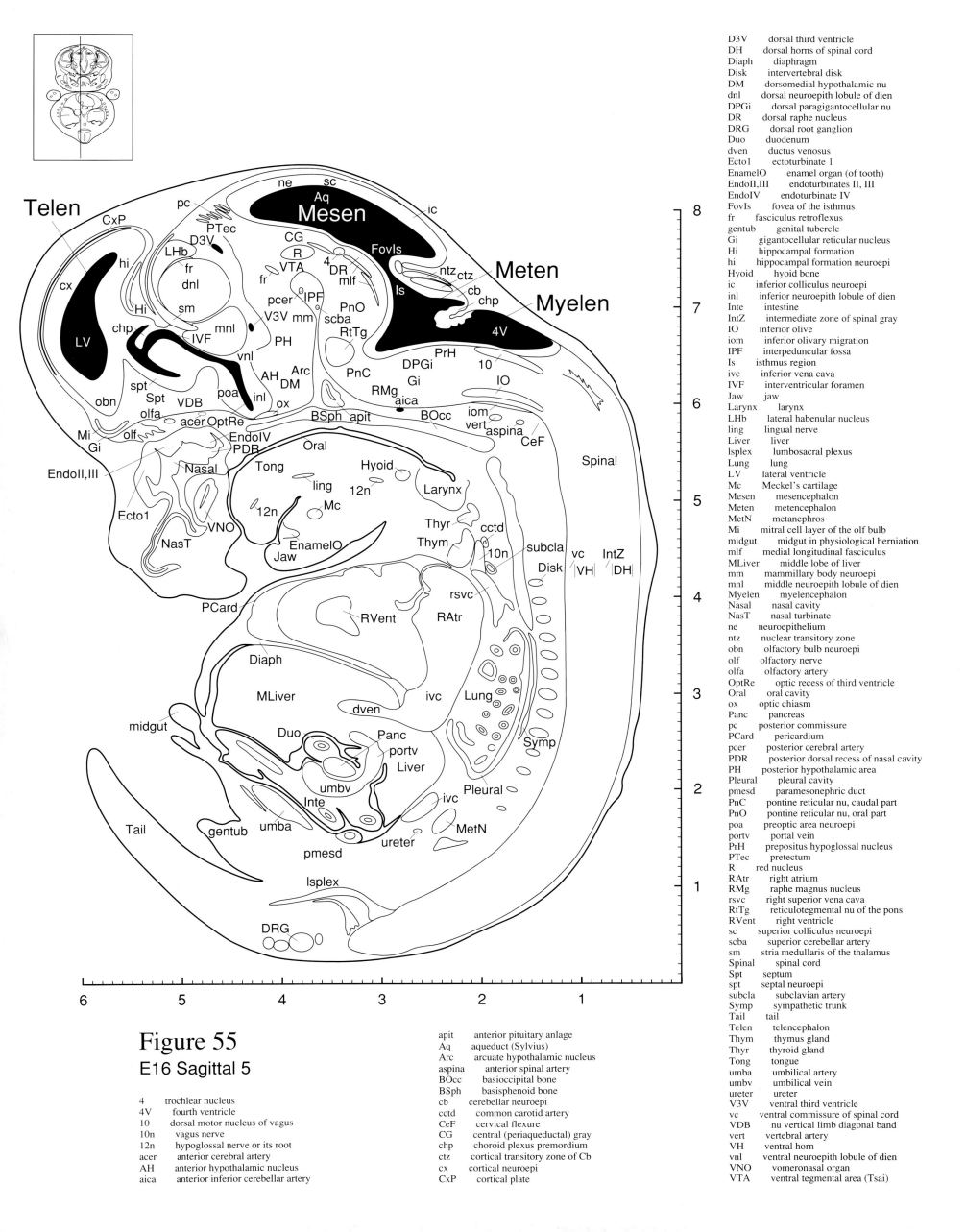

Figure 55
E16 Sagittal 5

4 trochlear nucleus
4V fourth ventricle
10 dorsal motor nucleus of vagus
10n vagus nerve
12n hypoglossal nerve or its root
acer anterior cerebral artery
AH anterior hypothalamic nucleus
aica anterior inferior cerebellar artery

apit anterior pituitary anlage
Aq aqueduct (Sylvius)
Arc arcuate hypothalamic nucleus
aspina anterior spinal artery
BOcc basioccipital bone
BSph basisphenoid bone
cb cerebellar neuroepi
cctd common carotid artery
CeF cervical flexure
CG central (periaqueductal) gray
chp choroid plexus premordium
ctz cortical transitory zone of Cb
cx cortical neuroepi
CxP cortical plate

D3V dorsal third ventricle
DH dorsal horns of spinal cord
Diaph diaphragm
Disk intervertebral disk
DM dorsomedial hypothalamic nu
dnl dorsal neuroepith lobule of dien
DPGi dorsal paragigantocellular nu
DR dorsal raphe nucleus
DRG dorsal root ganglion
Duo duodenum
dven ductus venosus
Ecto1 ectoturbinate 1
EnamelO enamel organ (of tooth)
EndoII,III endoturbinates II, III
EndoIV endoturbinate IV
FovIs fovea of the isthmus
fr fasciculus retroflexus
gentub genital tubercle
Gi gigantocellular reticular nucleus
Hi hippocampal formation
hi hippocampal formation neuroepi
Hyoid hyoid bone
ic inferior colliculus neuroepi
inl inferior neuroepith lobule of dien
Inte intestine
IntZ intermediate zone of spinal gray
IO inferior olive
iom inferior olivary migration
IPF interpeduncular fossa
Is isthmus region
ivc inferior vena cava
IVF interventricular foramen
Jaw jaw
Larynx larynx
LHb lateral habenular nucleus
ling lingual nerve
Liver liver
lsplex lumbosacral plexus
Lung lung
LV lateral ventricle
Mc Meckel's cartilage
Mesen mesencephalon
Meten metencephalon
MetN metanephros
Mi mitral cell layer of the olf bulb
midgut midgut in physiological herniation
mlf medial longitudinal fasciculus
MLiver middle lobe of liver
mm mammillary body neuroepi
mnl middle neuroepith lobule of dien
Myelen myelencephalon
Nasal nasal cavity
NasT nasal turbinate
ne neuroepithelium
ntz nuclear transitory zone
obn olfactory bulb neuroepi
olf olfactory nerve
olfa olfactory artery
OptRe optic recess of third ventricle
Oral oral cavity
ox optic chiasm
Panc pancreas
pc posterior commissure
PCard pericardium
pcer posterior cerebral artery
PDR posterior dorsal recess of nasal cavity
PH posterior hypothalamic area
Pleural pleural cavity
pmesd paramesonephric duct
PnC pontine reticular nu, caudal part
PnO pontine reticular nu, oral part
poa preoptic area neuroepi
portv portal vein
PrH prepositus hypoglossal nucleus
PTec pretectum
R red nucleus
RAtr right atrium
RMg raphe magnus nucleus
rsvc right superior vena cava
RtTg reticulotegmental nu of the pons
RVent right ventricle
sc superior colliculus neuroepi
scba superior cerebellar artery
sm stria medullaris of the thalamus
Spinal spinal cord
Spt septum
spt septal neuroepi
subcla subclavian artery
Symp sympathetic trunk
Tail tail
Telen telencephalon
Thym thymus gland
Thyr thyroid gland
Tong tongue
umba umbilical artery
umbv umbilical vein
ureter ureter
V3V ventral third ventricle
vc ventral commissure of spinal cord
VDB nu vertical limb diagonal band
vert vertebral artery
VH ventral horn
vnl ventral neuroepith lobule of dien
VNO vomeronasal organ
VTA ventral tegmental area (Tsai)

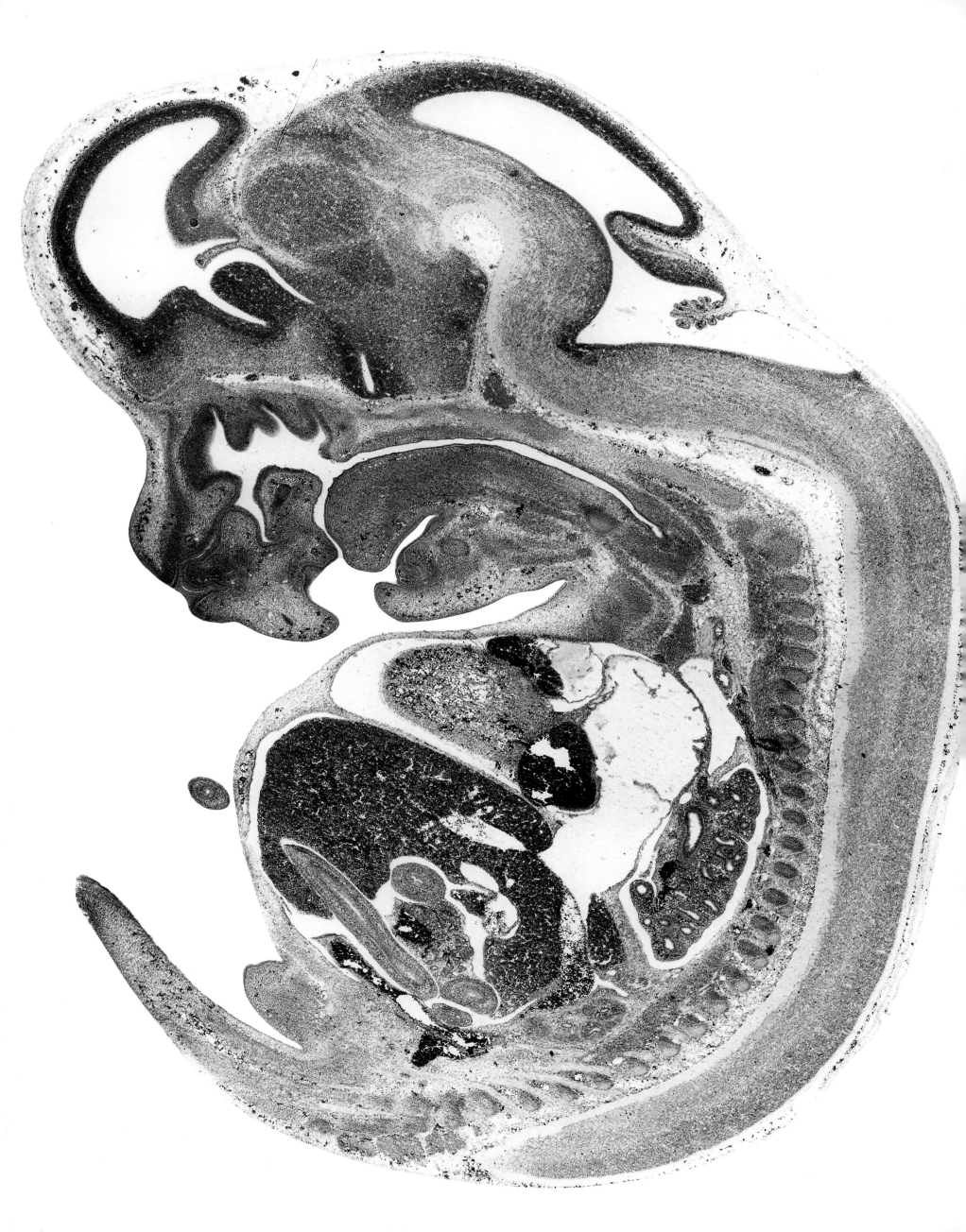

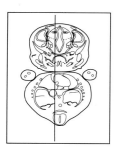

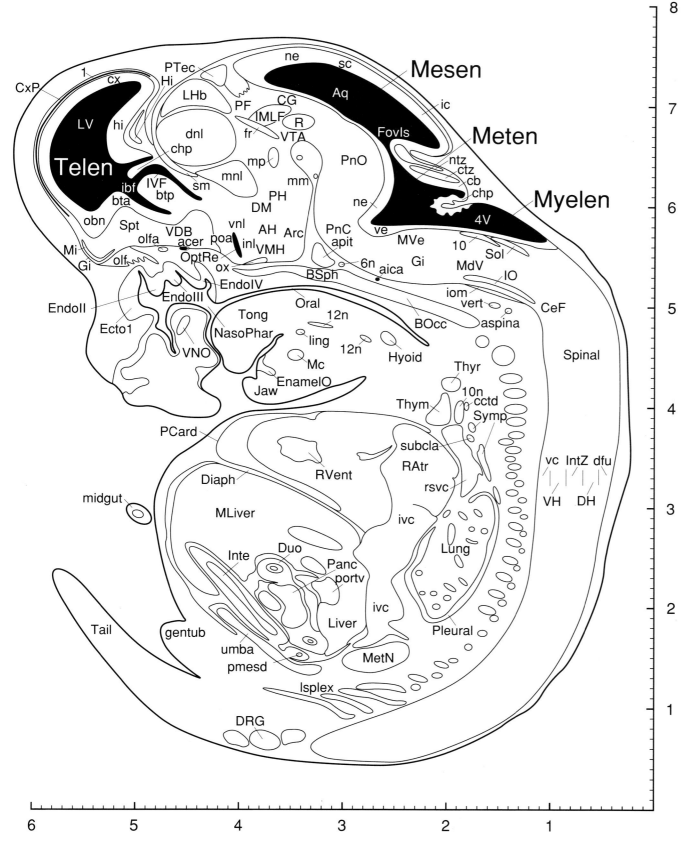

Figure 56

E16 Sagittal 6

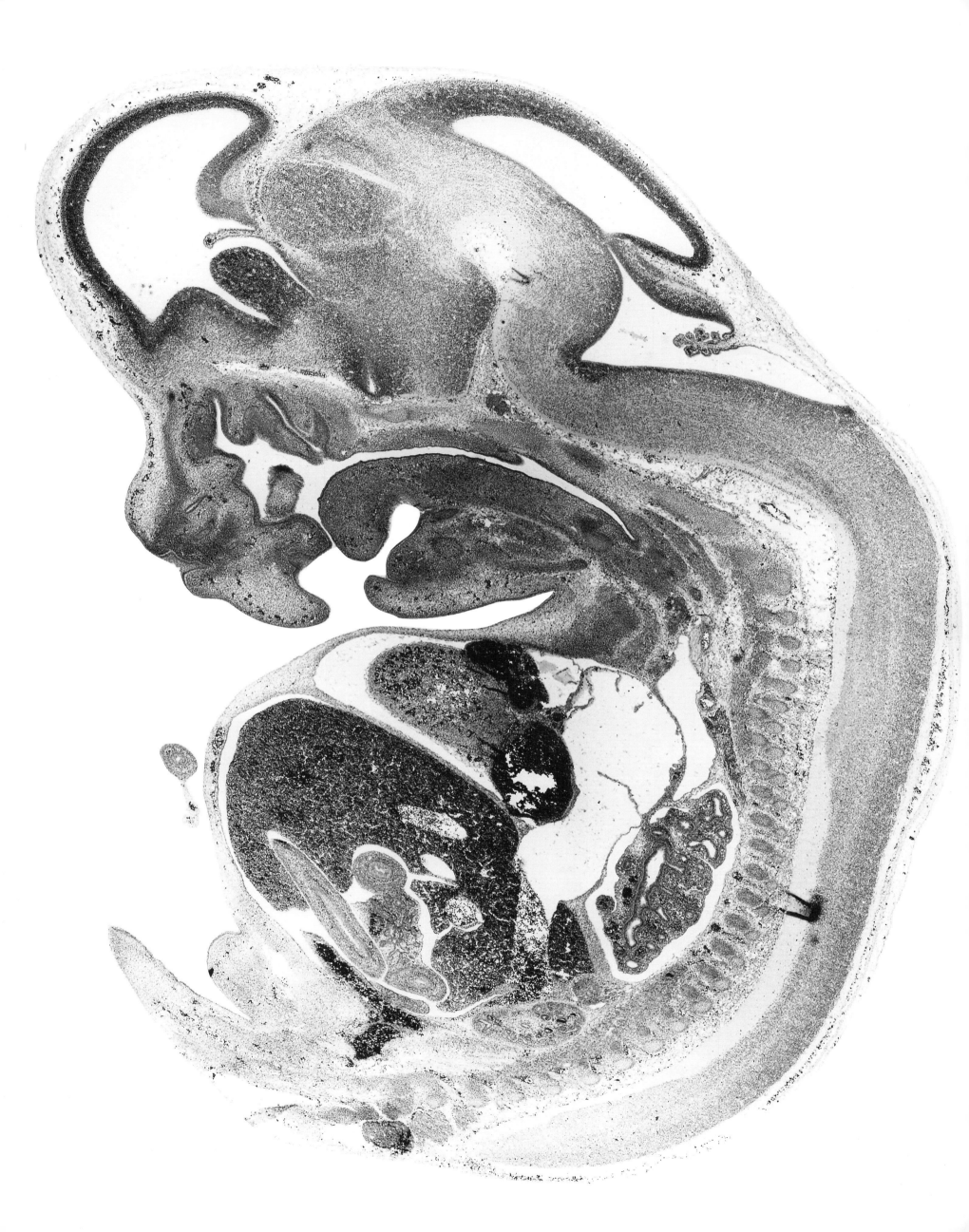

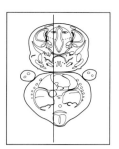

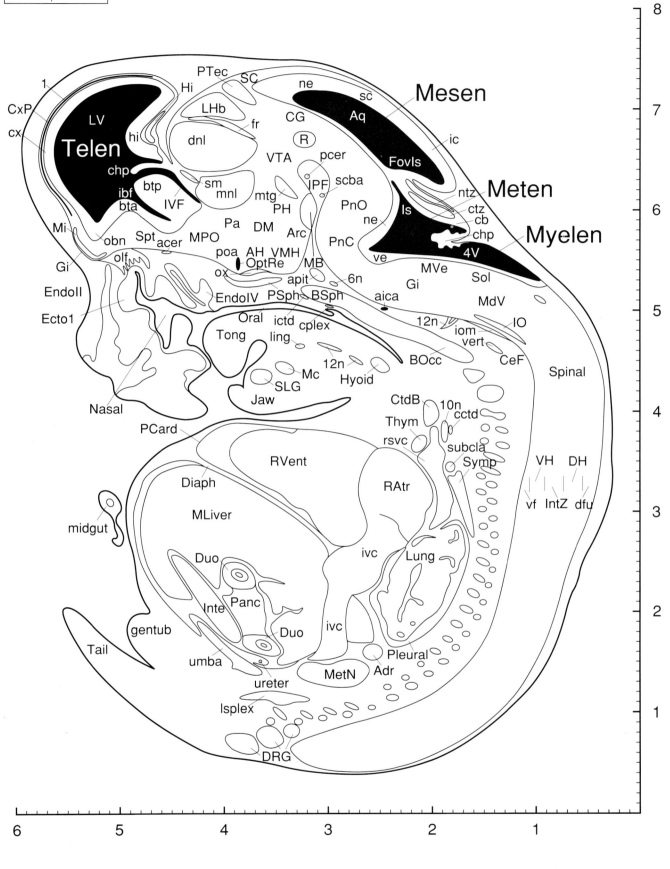

Figure 57
E16 Sagittal 7

1	cortical layer 1
4V	fourth ventricle
6n	abducens nerve or its root
10n	vagus nerve
12n	hypoglossal nerve or its root
acer	anterior cerebral artery
aica	anterior inferior cerebellar artery
Adr	adrenal gland

AH	anterior hypothalamic nucleus
apit	anterior pituitary anlage
Aq	aqueduct (Sylvius)
Arc	arcuate hypothalamic nucleus
BOcc	basioccipital bone
BSph	basisphenoid bone
bta	basal telencephalic plate, anterior
btp	basal telencephalic plate, posterior
cb	cerebellar neuroepi
cctd	common carotid artery
CeF	cervical flexure
CG	central (periaqueductal) gray
chp	choroid plexus premordium
cplex	carotid plexus

CtdB	carotid body
ctz	cortical transitory zone of Cb
cx	cortical neuroepi
CxP	cortical plate
dfu	dorsal funiculus of spinal cord
DH	dorsal horns of spinal cord
Diaph	diaphragm
DM	dorsomedial hypothalamic nu
dnl	dorsal neuroepith lobule of dien
DRG	dorsal root ganglion
Duo	duodenum
Ecto1	ectoturbinate 1
EndoII	endoturbinate II
EndoIV	endoturbinate IV
FovIs	fovea of the isthmus
fr	fasciculus retroflexus
gentub	genital tubercle
Gi	gigantocellular reticular nucleus
Hi	hippocampal formation
hi	hippocampal formation neuroepi
Hyoid	hyoid bone
ibf	interbasal plate fissure
ic	inferior colliculus neuroepi
ictd	internal carotid artery
Inte	intestine
IntZ	intermediate zone of spinal gray
IO	inferior olive
iom	inferior olivary migration
IPF	interpeduncular fossa
Is	isthmus region
ivc	inferior vena cava
IVF	interventricular foramen
Jaw	jaw
LHb	lateral habenular nucleus
ling	lingual nerve
lsplex	lumbosacral plexus
Lung	lung
LV	lateral ventricle
MB	mammillary body
Mc	Meckel's cartilage
MdV	medullary reticular nu, ventral
Mesen	mesencephalon
Meten	metencephalon
MetN	metanephros
Mi	mitral cell layer of the olf bulb
midgut	midgut in physiological herniation
MLiver	middle lobe of liver
mnl	middle neuroepith lobule of dien
MPO	medial preoptic nucleus
mtg	mammillotegmental tract
MVe	medial vestibular nucleus
Myelen	myelencephalon
Nasal	nasal cavity
ne	neuroepithelium
ntz	nuclear transitory zone
obn	olfactory bulb neuroepi
olf	olfactory nerve
OptRe	optic recess of third ventricle
Oral	oral cavity
ox	optic chiasm
Pa	paraventricular hypothalamic nu
Panc	pancreas
PCard	pericardium
pcer	posterior cerebral artery
PH	posterior hypothalamic area
Pleural	pleural cavity
PnC	pontine reticular nucleus, caudal part
PnO	pontine reticular nu, oral part
poa	preoptic area neuroepi
PSph	presphenoid bone
PTec	pretectum
R	red nucleus
RAtr	right atrium
rsvc	right superior vena cava
RVent	right ventricle
SC	superior colliculus
sc	superior colliculus neuroepi
scba	superior cerebellar artery
SLG	sublingual gland
sm	stria medullaris of the thalamus
Sol	nucleus of the solitary tract
Spinal	spinal cord
Spt	septum
subcla	subclavian artery
Symp	sympathetic trunk
Tail	tail
Telen	telencephalon
Thym	thymus gland
Tong	tongue
umba	umbilical artery
ureter	ureter
ve	vestibular area neuroepi
vert	vertebral artery
vf	ventral funiculus
VH	ventral horn
VMH	ventromedial hypothalamic nu
VTA	ventral tegmental area (Tsai)

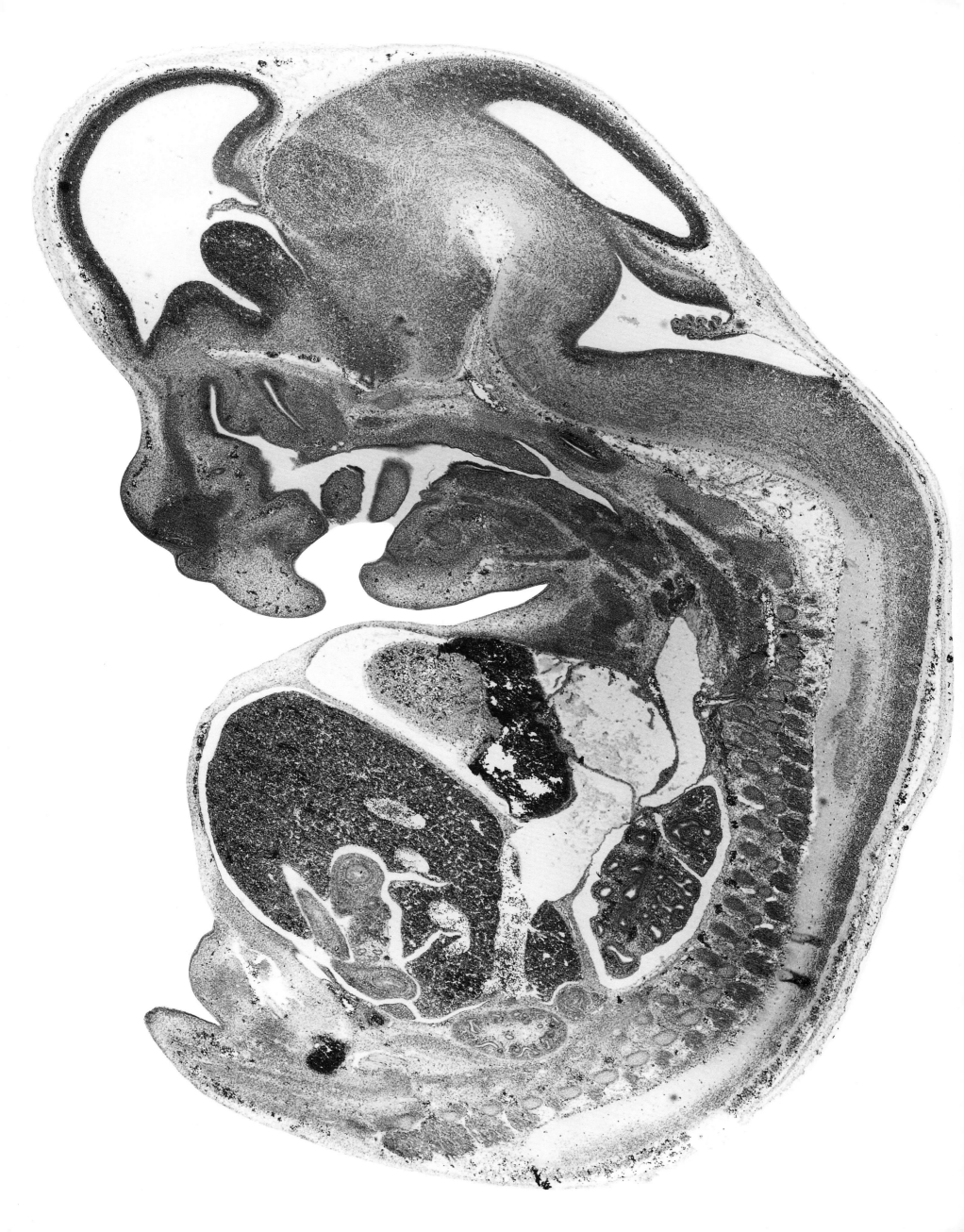

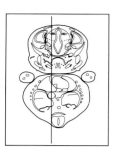

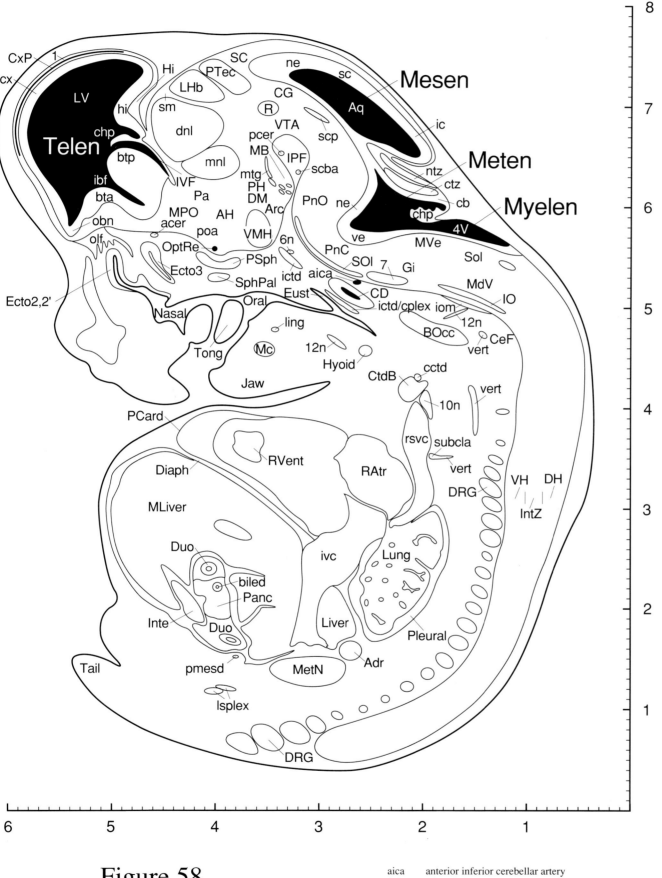

Figure 58

E16 Sagittal 8

1	cortical layer 1
4V	fourth ventricle
6n	abducens nerve or its root
7	facial nucleus
10n	vagus nerve
12n	hypoglossal nerve or its root
acer	anterior cerebral artery

aica	anterior inferior cerebellar artery
Adr	adrenal gland
AH	anterior hypothalamic nucleus
Aq	aqueduct (Sylvius)
Arc	arcuate hypothalamic nucleus
biled	bile duct
BOcc	basiooccipital bone
bta	basal telencephalic plate, anterior
btp	basal telencephalic plate, posterior
cb	cerebellar neuroepi
cctd	common carotid artery
CD	cochlear duct
CeF	cervical flexure

CG	central (periaqueductal) gray
chp	choroid plexus premordium
cplex	carotid plexus
CtdB	carotid body
ctz	cortical transitory zone of Cb
cx	cortical neuroepi
CxP	cortical plate
DH	dorsal horns of spinal cord
Diaph	diaphragm
DM	dorsomedial hypothalamic nu
dnl	dorsal neuroepith lobule of dien
DRG	dorsal root ganglion
Duo	duodenum
Ecto2,2'	ectoturbinates 2, 2'
Ecto3	ectoturbinate 3
Eust	Eustachian tube
Gi	gigantocellular reticular nu
Hi	hippocampal formation
hi	hippocampal formation neuroepi
Hyoid	hyoid bone
ibf	interbasal plate fissure
ic	inferior colliculus neuroepi
ictd	internal carotid artery
Inte	intestine
IntZ	intermediate zone of spinal gray
IO	inferior olive
iom	inferior olivary migration
IPF	interpeduncular fossa
ivc	inferior vena cava
IVF	interventricular foramen
Jaw	jaw
LHb	lateral habenular nucleus
ling	lingual nerve
Liver	liver
lsplex	lumbosacral plexus
Lung	lung
LV	lateral ventricle
MB	mammillary body
Mc	Meckel's cartilage
MdV	medullary reticular nu, ventral
Mesen	mesencephalon
Meten	metencephalon
MetN	metanephros
MLiver	middle lobe of liver
mnl	middle neuroepith lobule of dien
MPO	medial preoptic nucleus
mtg	mammillotegmental tract
MVe	medial vestibular nucleus
Myelen	myelencephalon
Nasal	nasal cavity
ne	neuroepithelium
ntz	nuclear transitory zone
obn	olfactory bulb neuroepi
olf	olfactory nerve
OptRe	optic recess of third ventricle
Oral	oral cavity
Pa	paraventricular hypothalamic nu
Panc	pancreas
PCard	pericardium
pcer	posterior cerebral artery
PH	posterior hypothalamic area
Pleural	pleural cavity
pmesd	paramesonephric duct
PnC	pontine reticular nu, caudal part
PnO	pontine reticular nu, oral part
poa	preoptic area neuroepi
PSph	presphenoid bone
PTec	pretectum
R	red nucleus
RAtr	right atrium
rsvc	right superior vena cava
RVent	right ventricle
SC	superior colliculus
sc	superior colliculus neuroepi
scba	superior cerebellar artery
scp	superior cerebellar peduncle
sm	stria medullaris of the thalamus
SOl	superior olive
Sol	nucleus of the solitary tract
SphPal	sphenopalatine ganglion
subcla	subclavian artery
Tail	tail
Telen	telencephalon
Tong	tongue
ve	vestibular area neuroepi
vert	vertebral artery
VH	ventral horn
VMH	ventromedial hypothalamic nu
VTA	ventral tegmental area (Tsai)

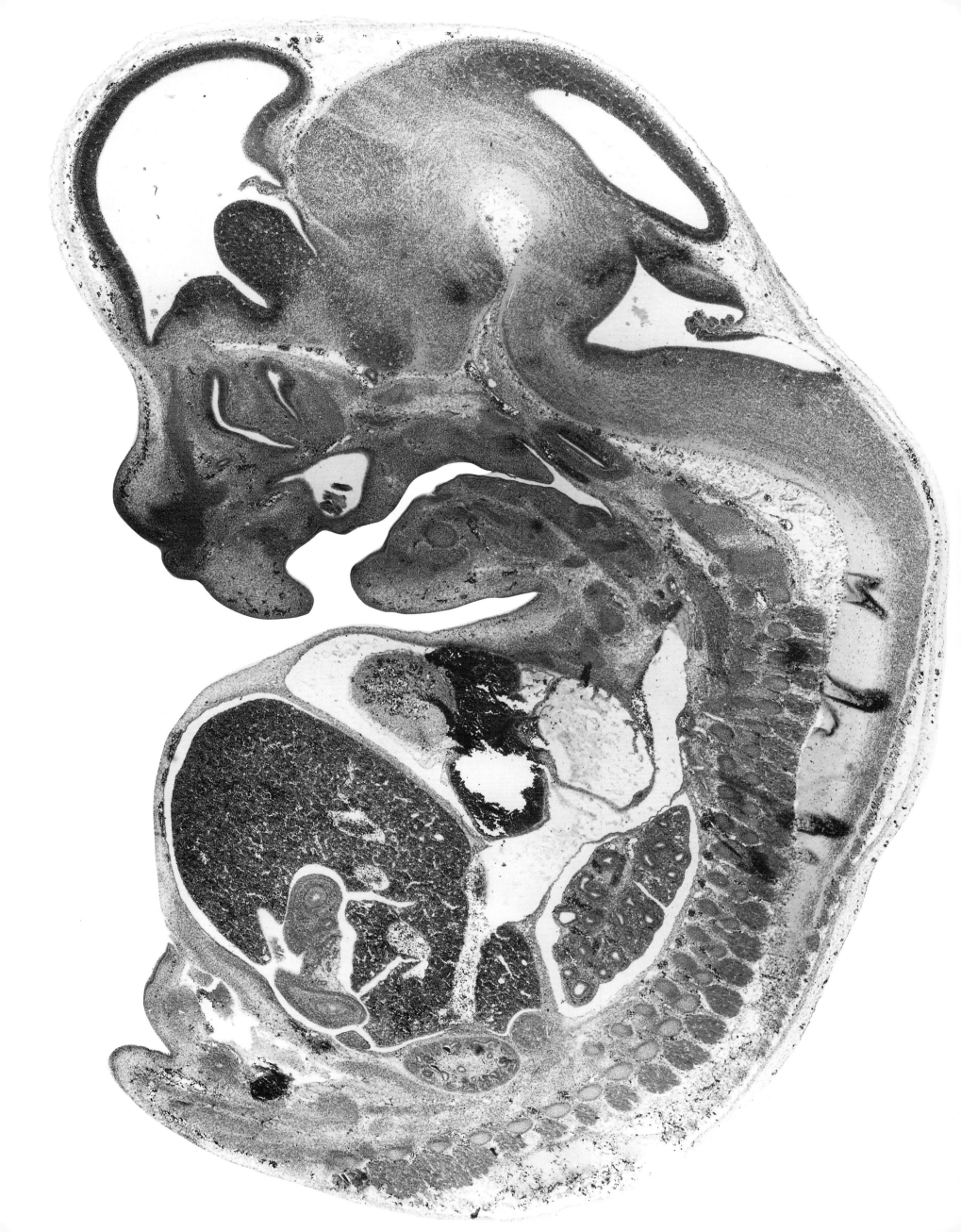

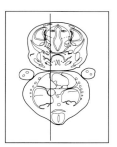

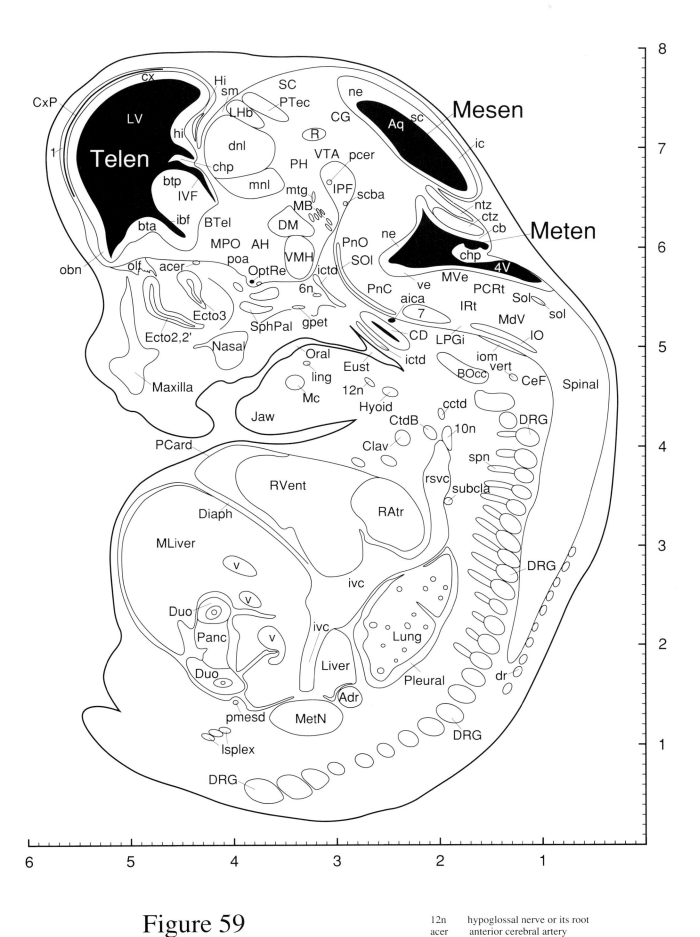

Figure 59

E16 Sagittal 9

1   cortical layer 1
4V   fourth ventricle
6n   abducens nerve or its root
7   facial nucleus
10n   vagus nerve

12n   hypoglossal nerve or its root
acer   anterior cerebral artery
Adr   adrenal gland
AH   anterior hypothalamic nucleus
aica   anterior inferior cerebellar artery
Aq   aqueduct (Sylvius)
BOcc   basioccipital bone
bta   basal telencephalic plate, anterior
BTel   basal telencephalon
btp   basal telencephalic plate, posterior

cb   cerebellar neuroepi
cctd   common carotid artery
CD   cochlear duct
CeF   cervical flexure
CG   central (periaqueductal) gray
chp   choroid plexus premordium
Clav   clavicle
CtdB   carotid body
ctz   cortical transitory zone of Cb
cx   cortical neuroepi
CxP   cortical plate
Diaph   diaphragm
DM   dorsomedial hypothalamic nu
dnl   dorsal neuroepith lobule of dien
dr   dorsal root
DRG   dorsal root ganglion
Duo   duodenum
Ecto2,2'   ectoturbinates 2, 2'
Ecto3   ectoturbinate 3
Eust   Eustachian tube
gpet   greater petrosal nerve
Hi   hippocampal formation
hi   hippocampal formation neuroepi
Hyoid   hyoid bone
ibf   interbasal plate fissure
ic   inferior colliculus neuroepi
ictd   internal carotid artery
IO   inferior olive
iom   inferior olivary migration
IPF   interpeduncular fossa
IRt   intermediate reticular zone
ivc   inferior vena cava
IVF   interventricular foramen
Jaw   jaw
LHb   lateral habenular nucleus
ling   lingual nerve
Liver   liver
LPGi   lateral paragigantocellular nu
lsplex   lumbosacral plexus
Lung   lung
LV   lateral ventricle
Maxilla   maxilla
MB   mammillary body
Mc   Meckel's cartilage
MdV   medullary reticular nu, ventral
Mesen   mesencephalon
Meten   metencephalon
MetN   metanephros
MLiver   middle lobe of liver
mnl   middle neuroepith lobule of dien
MPO   medial preoptic nucleus
mtg   mammillotegmental tract
MVe   medial vestibular nucleus
Nasal   nasal cavity
ne   neuroepithelium
ntz   nuclear transitory zone
obn   olfactory bulb neuroepi
olf   olfactory nerve
OptRe   optic recess of third ventricle
Oral   oral cavity
Panc   pancreas
PCard   pericardium
pcer   posterior cerebral artery
PCRt   parvocellular reticular nu
PH   posterior hypothalamic area
Pleural   pleural cavity
pmesd   paramesonephric duct
PnC   pontine reticular nu, caudal part
PnO   pontine reticular nu, oral part
poa   preoptic area neuroepi
PTec   pretectum
R   red nucleus
RAtr   right atrium
rsvc   right superior vena cava
RVent   right ventricle
SC   superior colliculus
sc   superior colliculus neuroepi
scba   superior cerebellar artery
sm   stria medullaris of the thalamus
SOl   superior olive
Sol   nucleus of the solitary tract
sol   solitary tract
SphPal   sphenopalatine ganglion
Spinal   spinal cord
spn   spinal nerve
subcla   subclavian artery
Telen   telencephalon
v   vein (unidentified)
ve   vestibular area neuroepi
vert   vertebral artery
VMH   ventromedial hypothalamic nu
VTA   ventral tegmental area (Tsai)

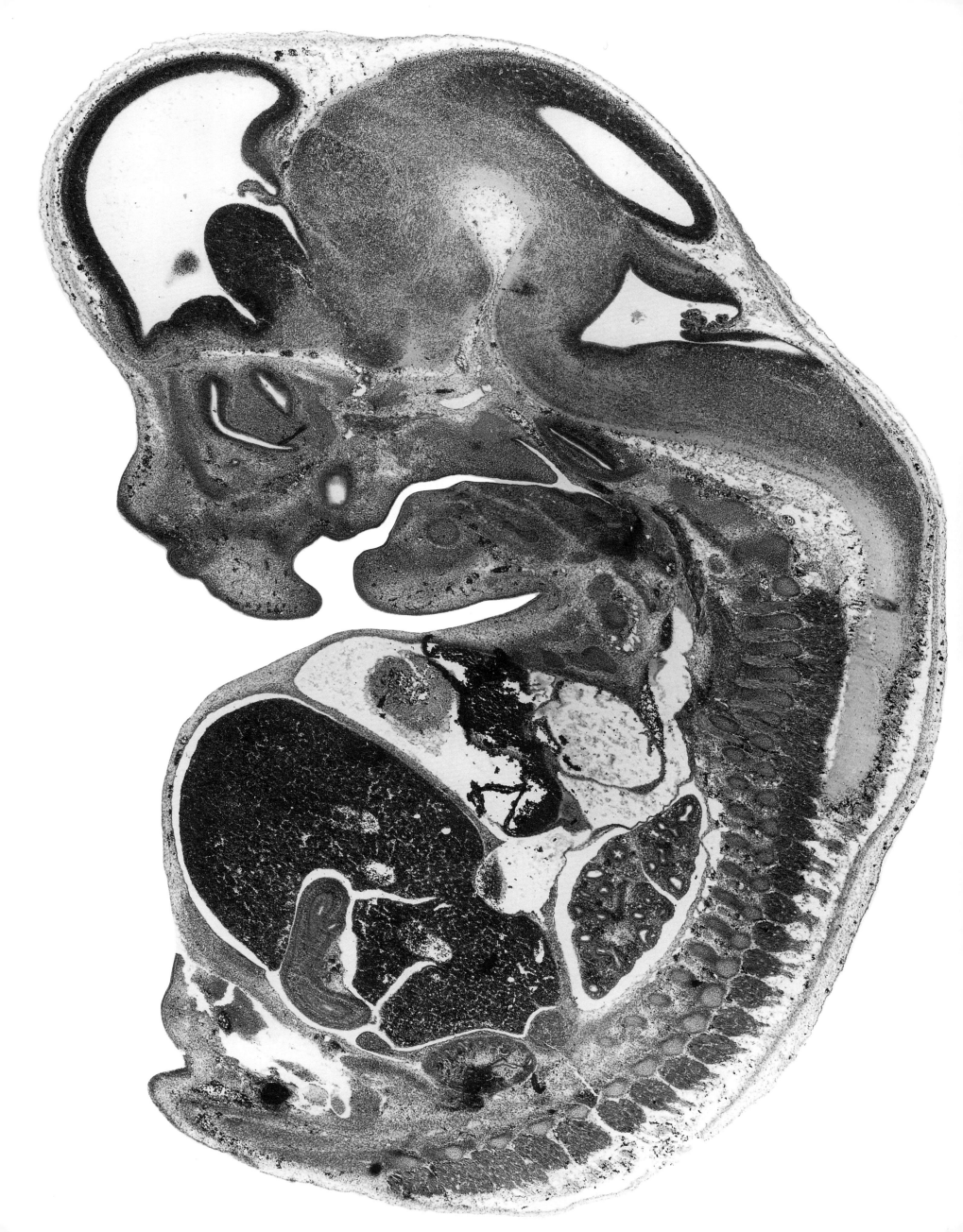

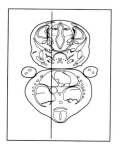

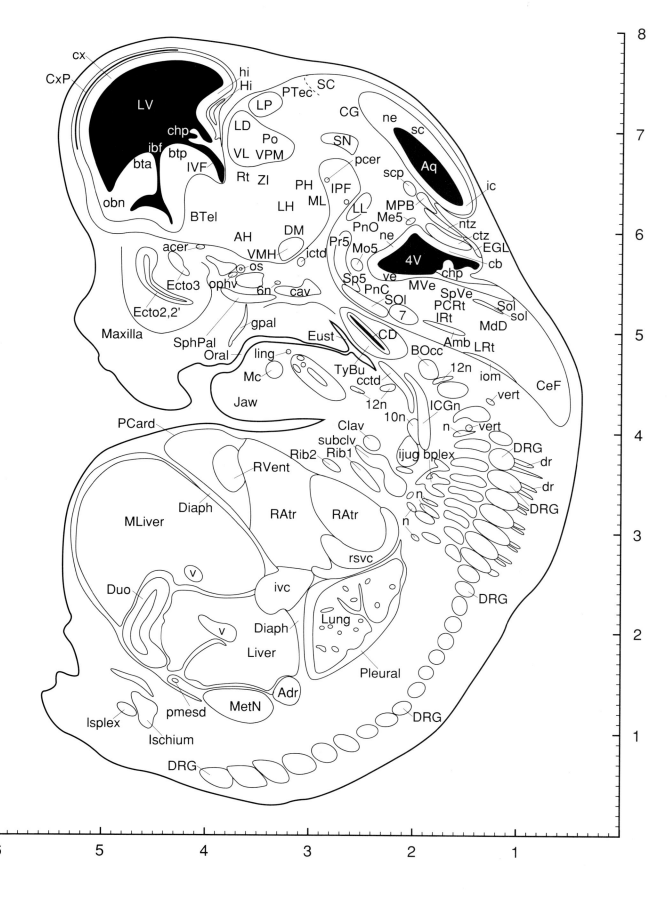

Figure 60

E16 Sagittal 10

4V   fourth ventricle
7   facial nucleus
7n   facial nerve or its root
10n   vagus nerve
12n   hypoglossal nerve or its root

acer   anterior cerebral artery
Adr   adrenal gland
AH   anterior hypothalamic nucleus
Aq   aqueduct (Sylvius)
BOcc   basioccipital bone
bta   basal telencephalic plate, anterior
BTel   basal telencephalon
btp   basal telencephalic plate, posterior
cav   cavernous sinus
cb   cerebellar neuroepi
cctd   common carotid artery

CD   cochlear duct
CeF   cervical flexure
CG   central (periaqueductal) gray
chp   choroid plexus premordium
Clav   clavicle
ctz   cortical transitory zone of Cb
cx   cortical neuroepi
CxP   cortical plate
Diaph   diaphragm
DM   dorsomedial hypothalamic nu
dr   dorsal root
DRG   dorsal root ganglion
Duo   duodenum
Ecto2,2'   ectoturbinates 2, 2'
Ecto3   ectoturbinate 3
EGL   external germinal layer of Cb
Eust   Eustachian tube
gpet   greater petrosal nerve
Hi   hippocampal formation
hi   hippocampal formation neuroepi
ibf   interbasal plate fissure
ic   inferior colliculus neuroepi
ICGn   inferior cervical ganglion
ictd   internal carotid artery
ijug   internal jugular vein
IO   inferior olive
iom   inferior olivary migration
IPF   interpeduncular fossa
IRt   intermediate reticular zone
ivc   inferior vena cava
IVF   interventricular foramen
Jaw   jaw
LD   laterodorsal thalamic nucleus
LH   lateral hypothalamic area
ling   lingual nerve
Liver   liver
LP   lateral posterior thalamic nu
LPGi   lateral paragigantocellular nu
lsplex   lumbosacral plexus
Lung   lung
LV   lateral ventricle
Maxilla   maxilla
MB   mammillary body
Mc   Meckel's cartilage
MetN   metanephros
MLiver   middle lobe of liver
MPO   medial preoptic nucleus
MVe   medial vestibular nucleus
n   nerve
ne   neuroepithelium
ntz   nuclear transitory zone
obn   olfactory bulb neuroepi
olf   olfactory nerve
ophv   ophthalmic vein
Oral   oral cavity
os   optic stalk
PCard   pericardium
pcer   posterior cerebral artery
PCRt   parvocellular reticular nu
PH   posterior hypothalamic area
Pleural   pleural cavity
pmesd   paramesonephric duct
PnC   pontine reticular nu, caudal part
PnO   pontine reticular nu, oral part
Po   posterior thalamic nuclear group
poa   preoptic area neuroepi
PTec   pretectum
RAtr   right atrium
Rib1   rib 1
RR   retrorubral nucleus
rsvc   right superior vena cava
Rt   reticular thalamic nucleus
RVent   right ventricle
SC   superior colliculus
sc   superior colliculus neuroepi
scba   superior cerebellar artery
SCGn   superior cervical ganglion
SLG   sublingual gland
SOl   superior olive
Sol   nucleus of the solitary tract
sol   solitary tract
SphPal   sphenopalatine ganglion
Spinal   spinal cord
spn   spinal nerve
subcla   subclavian artery
SuVe   superior vestibular nucleus
v   vein (unidentified)
ve   vestibular area neuroepi
vert   vertebral artery
VL   ventrolateral thalamic nucleus
VMH   ventromedial hypothalamic nu
VPM   ventral posteromedial thalamic nu
VTA   ventral tegmental area (Tsai)
ZI   zona incerta

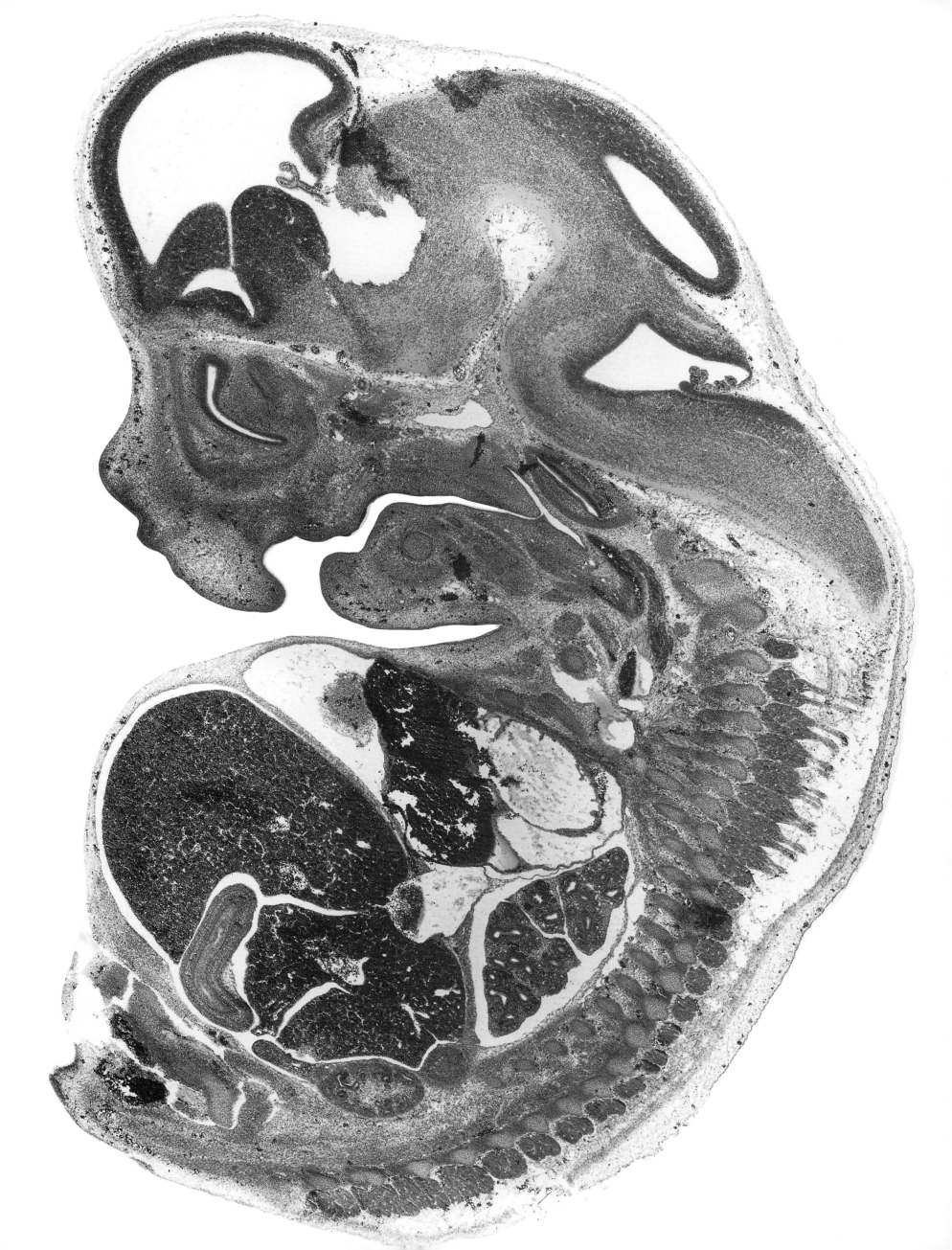

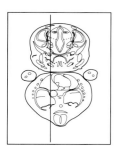

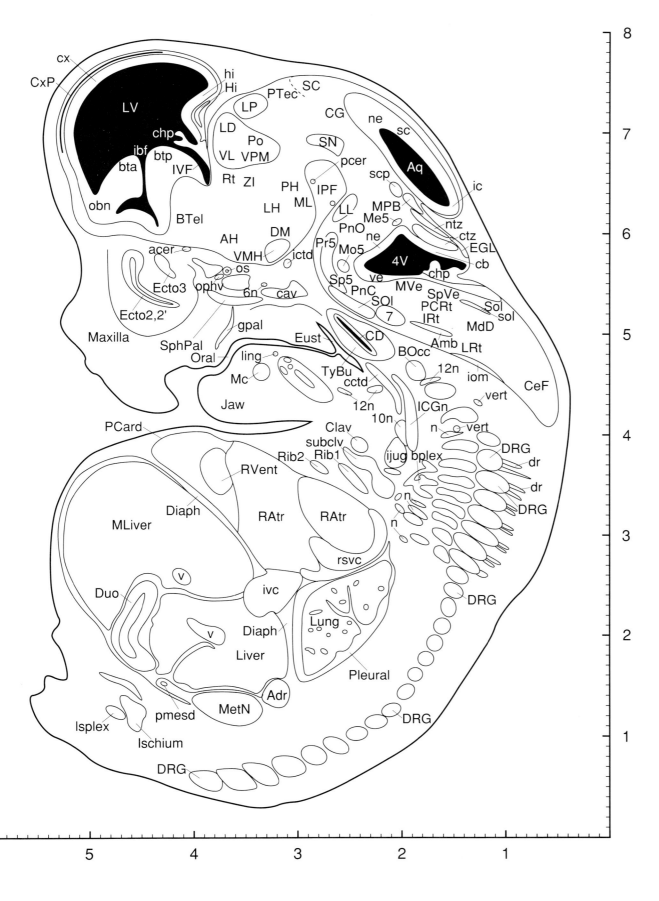

Figure 61
E16 Sagittal 11

4V   fourth ventricle
6n   abducens nerve or its root
7   facial nucleus
10n   vagus nerve
12n   hypoglossal nerve or its root

acer   anterior cerebral artery
Adr   adrenal gland
AH   anterior hypothalamic nucleus
Amb   ambiguus nucleus Aq   aqueduct (Sylvius)
BOcc   basioccipital bone
bplex   brachial plexus
BSph   basisphenoid bone
bta   basal telencephalic plate, anterior part
BTel   basal telencephalon
btp   basal telencephalic plate, posterior

cav   cavernous sinus
cb   cerebellar neuroepi
cctd   common carotid artery
CD   cochlear duct
CeF   cervical flexure
CG   central (periaqueductal) gray
chp   choroid plexus premordium
Clav   clavicle
ctz   cortical transitory zone of Cb
cx   cortical neuroepi
CxP   cortical plate
Diaph   diaphragm
DM   dorsomedial hypothalamic nu
dr   dorsal root
DRG   dorsal root ganglion
Duo   duodenum
Ecto2,2'   ectoturbinates 2, 2'
Ecto3   ectoturbinate 3
EGL   external germinal layer of Cb
Eust   Eustachian tube
gpal   greater palatine nerve
Hi   hippocampal formation
hi   hippocampal formation neuroepi
ibf   interbasal plate fissure
ic   inferior colliculus neuroepi
ICGn   inferior cervical ganglion
ictd   internal carotid artery
ijug   internal jugular vein
iom   inferior olivary migration
IPF   interpeduncular fossa
IRt   intermediate reticular zone
Ischium   ischium
ivc   inferior vena cava
IVF   interventricular foramen
Jaw   jaw
LD   laterodorsal thalamic nucleus
LH   lateral hypothalamic area
ling   lingual nerve
Liver   liver
LL   nuclei of the lateral lemniscus
LP   lateral posterior thalamic nucleus
LRt   lateral reticular nucleus
lsplex   lumbosacral plexus
Lung   lung
LV   lateral ventricle
Maxilla   maxilla
Mc   Meckel's cartilage
MdD   medullary reticular nu, dorsal part
Me5   mesencephalic trigeminal nu
MetN   metanephros
ML   medial mammillary nu, lateral part
MLiver   middle lobe of liver
Mo5   motor trigeminal nucleus
MPB   medial parabrachial nucleus
MVe   medial vestibular nucleus
n   nerve
ne   neuroepithelium
ntz   nuclear transitory zone
obn   olfactory bulb neuroepi
ophv   ophthalmic vein
Oral   oral cavity
os   optic stalk
PCard   pericardium
pcer   posterior cerebral artery
PCRt   parvocellular reticular nu
PH   posterior hypothalamic area
Pleural   pleural cavity
pmesd   paramesonephric duct
PnC   pontine reticular nu, caudal part
PnO   pontine reticular nu, oral part
Po   posterior thalamic nuclear group
Pr5   principal sensory trigeminal nu
PTec   pretectum
RAtr   right atrium
Rib1   rib 1
Rib2   rib 2
rsvc   right superior vena cava
Rt   reticular thalamic nucleus
RVent   right ventricle
SC   superior colliculus
sc   superior colliculus neuroepi
scp   superior cerebellar peduncle
SN   substantia nigra
SOl   superior olive
Sol   nucleus of the solitary tract
sol   solitary tract
Sp5   spinal trigeminal nucleus
SphPal   sphenopalatine ganglion
SpVe   spinal vestibular nucleus
subclv   subclavian vein
svc   superior vena cava
TyBu   tympanic bulla
v   vein (unidentified)
ve   vestibular area neuroepi
vert   vertebral artery
VL   ventrolateral thalamic nucleus
VMH   ventromedial hypothalamic nu
VPM   ventral posteromedial thalamic nu
ZI   zona incerta

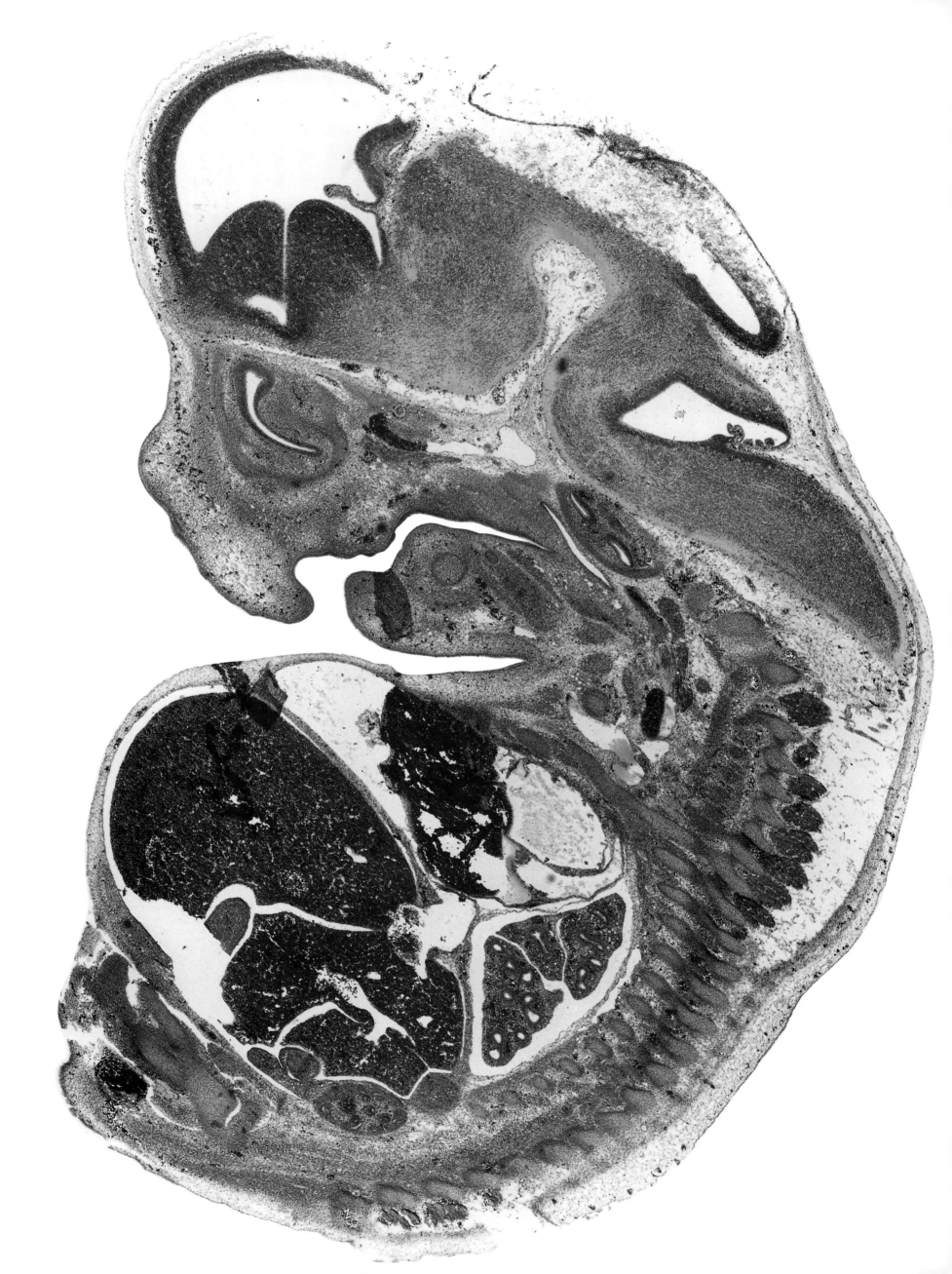

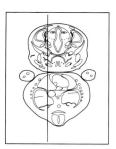

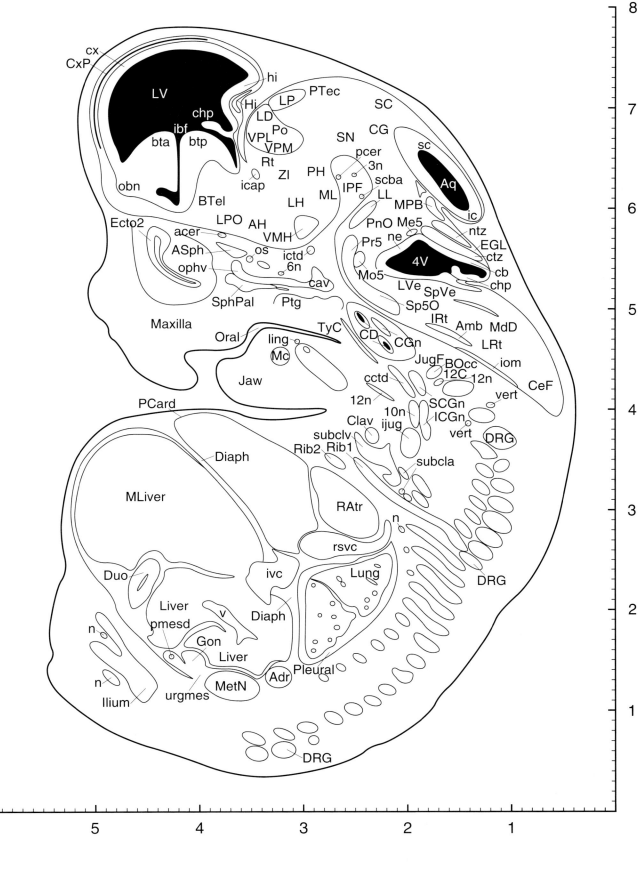

Figure 62
E16 Sagittal 12

3n   oculomotor nerve or its root
4V   fourth ventricle
6n   abducens nerve or its root
10n   vagus nerve
12C   hypoglossal canal
12n   hypoglossal nerve or its root

acer   anterior cerebral artery
Adr   adrenal gland
AH   anterior hypothalamic nucleus
Amb   ambiguus nucleus
Aq   aqueduct (Sylvius)
ASph   alisphenoid bone
BOcc   basioccipital bone
bta   basal telencephalic plate, anterior
BTel   basal telencephalon
btp   basal telencephalic plate, posterior
cav   cavernous sinus

cb   cerebellar neuroepi
cctd   common carotid artery
CD   cochlear duct
CeF   cervical flexure
CG   central (periaqueductal) gray
CGn   cochlear (spiral) ganglion
chp   choroid plexus premordium
Clav   clavicle
ctz   cortical transitory zone of Cb
cx   cortical neuroepi
CxP   cortical plate
Diaph   diaphragm
DRG   dorsal root ganglion
Duo   duodenum
Ecto2   ectoturbinate 2
EGL   external germinal layer of cerebellum
Gon   gonad
Hi   hippocampal formation
hi   hippocampal formation neuroepi
ibf   interbasal plate fissure
ic   inferior colliculus neuroepi
icap   internal capsule
ICGn   inferior cervical ganglion
ictd   internal carotid artery
ijug   internal jugular vein
Ilium   ilium
iom   inferior olivary migration
IPF   interpeduncular fossa
IRt   intermediate reticular zone
ivc   inferior vena cava
Jaw   jaw
JugF   jugular foramen
LD   laterodorsal thalamic nucleus
LH   lateral hypothalamic area
ling   lingual nerve
Liver   liver
LL   nuclei of the lateral lemniscus
LP   lateral posterior thalamic nucleus
LPO   lateral preoptic area
LRt   lateral reticular nucleus
Lung   lung
LV   lateral ventricle
LVe   lateral vestibular nucleus
Maxilla   maxilla
Mc   Meckel's cartilage
MdD   medullary reticular nu, dorsal part
Me5   mesencephalic trigeminal nu
MetN   metanephros
ML   medial mammillary nu, lateral part
MLiver   middle lobe of liver
Mo5   motor trigeminal nucleus
MPB   medial parabrachial nucleus
n   nerve
ne   neuroepithelium
ntz   nuclear transitory zone
obn   olfactory bulb neuroepi
ophv   ophthalmic vein
Oral   oral cavity
os   optic stalk
PCard   pericardium
pcer   posterior cerebral artery
PH   posterior hypothalamic area
Pleural   pleural cavity
pmesd   paramesonephric duct
PnO   pontine reticular nu, oral part
Po   posterior thalamic nuclear group
Pr5   principal sensory trigeminal nu
PTec   pretectum
Ptg   pterygoid bone
RAtr   right atrium
Rib1   rib 1
Rib2   rib 2
rsvc   right superior vena cava
Rt   reticular thalamic nucleus
SC   superior colliculus
sc   superior colliculus neuroepi
scba   superior cerebellar artery
SCGn   superior cervical ganglion
SN   substantia nigra
Sp5O   spinal trigeminal nu, oral part
SphPal   sphenopalatine ganglion
SpVe   spinal vestibular nucleus
subcla   subclavian artery
subclv   subclavian vein
TyC   tympanic cavity
urgmes   urogenital mesentery
v   vein (unidentified)
vert   vertebral artery
VMH   ventromedial hypothalamic nu
VPL   ventral posterolateral thalamic nu
VPM   ventral posteromedial thalamic nu
ZI   zona incerta

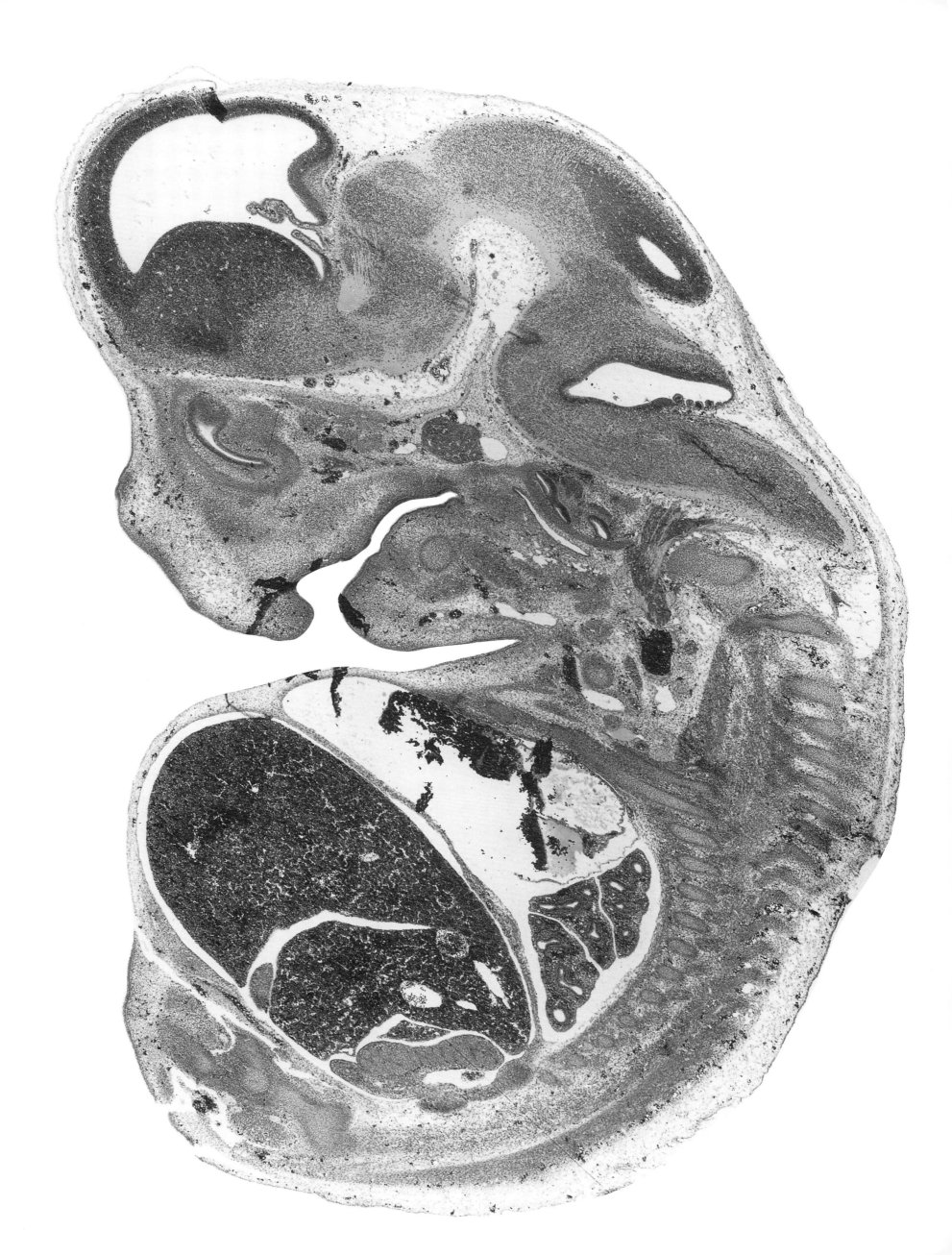

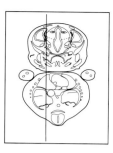

Figure 63
E16 Sagittal 13

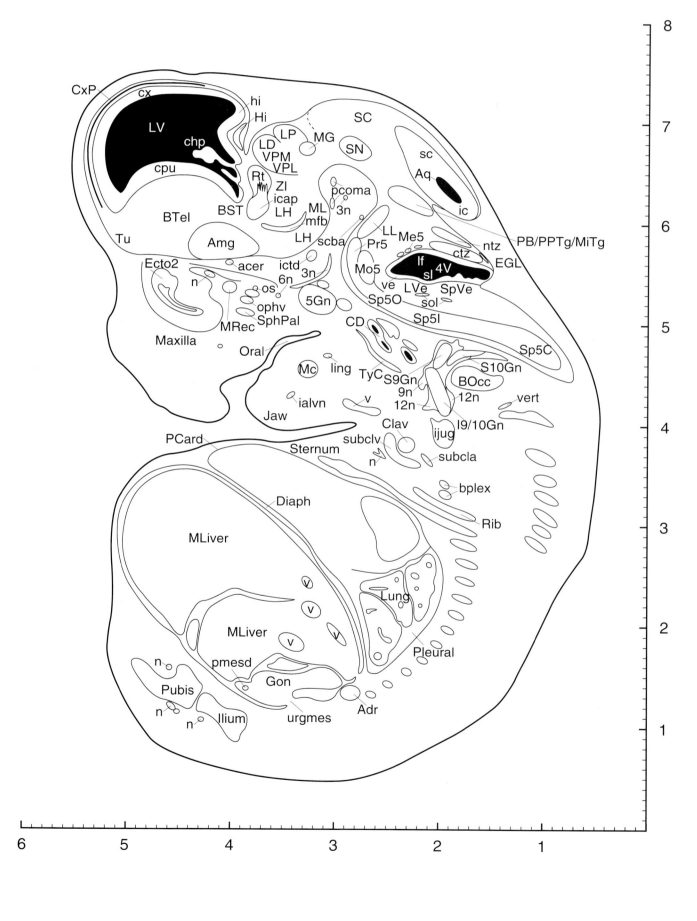

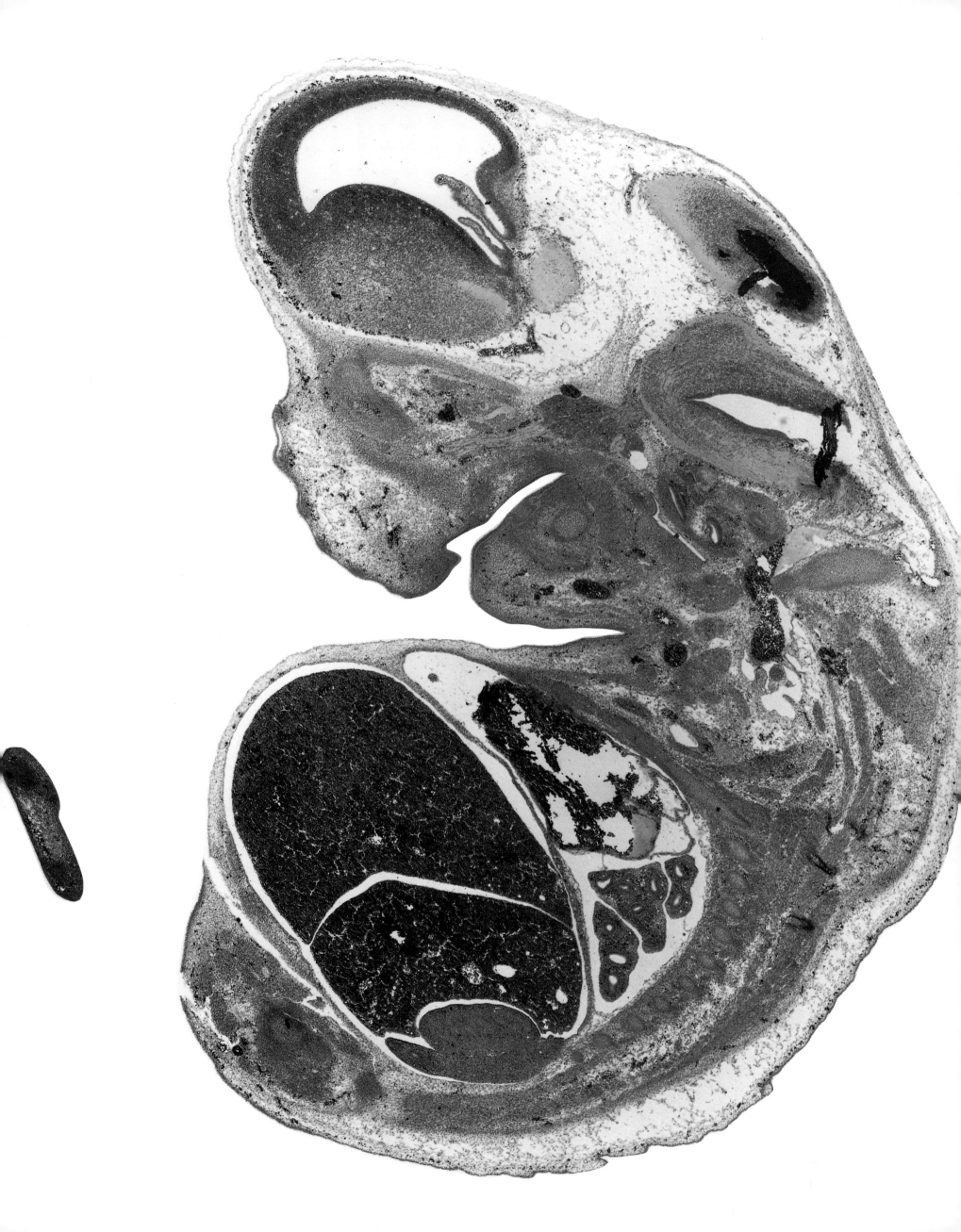

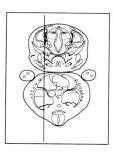

Figure 64
E16 Sagittal 14

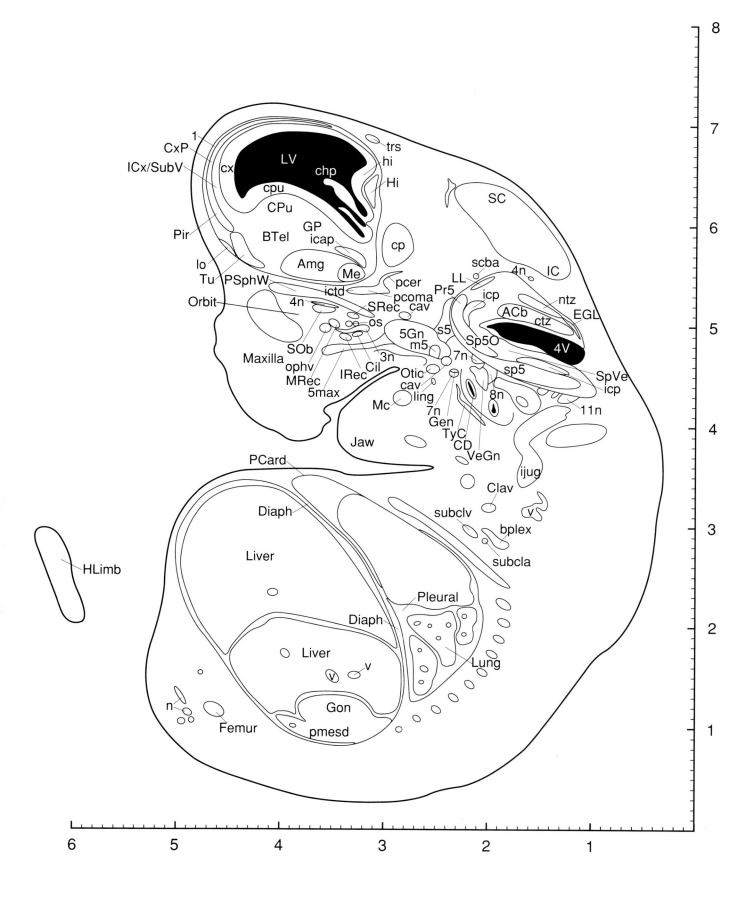

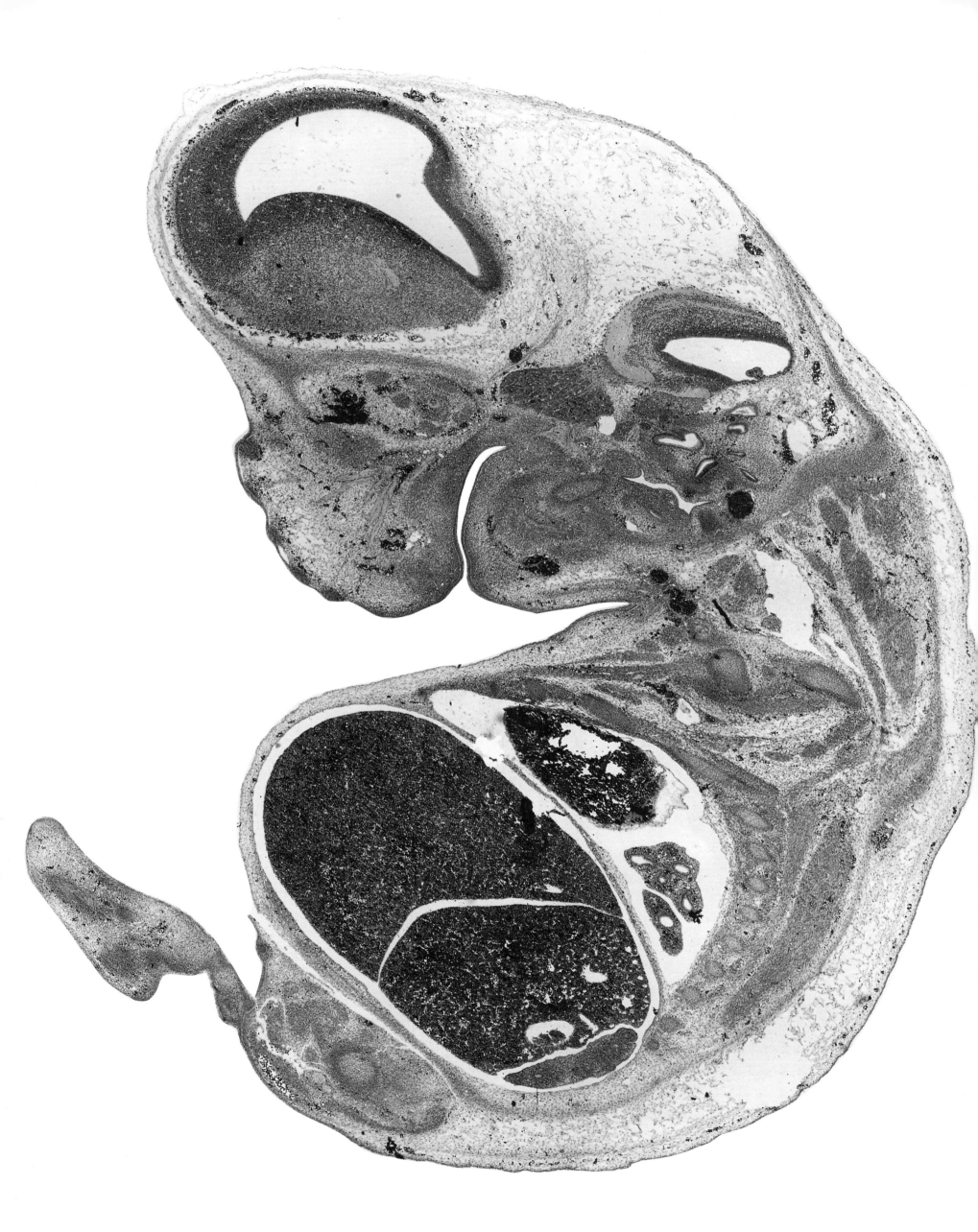

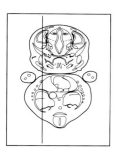

Figure 65
E16 Sagittal 15

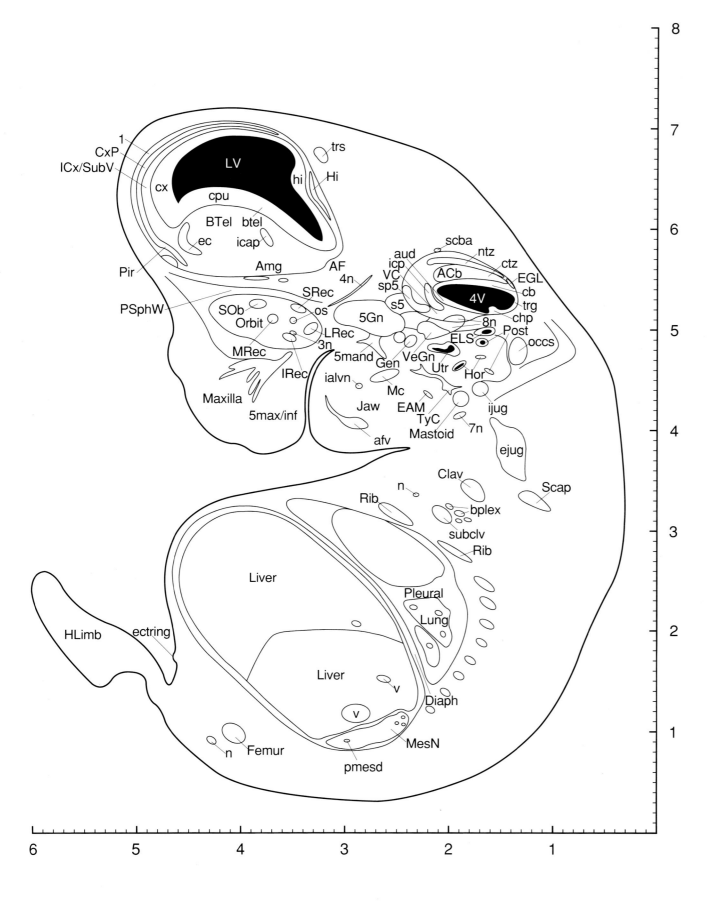

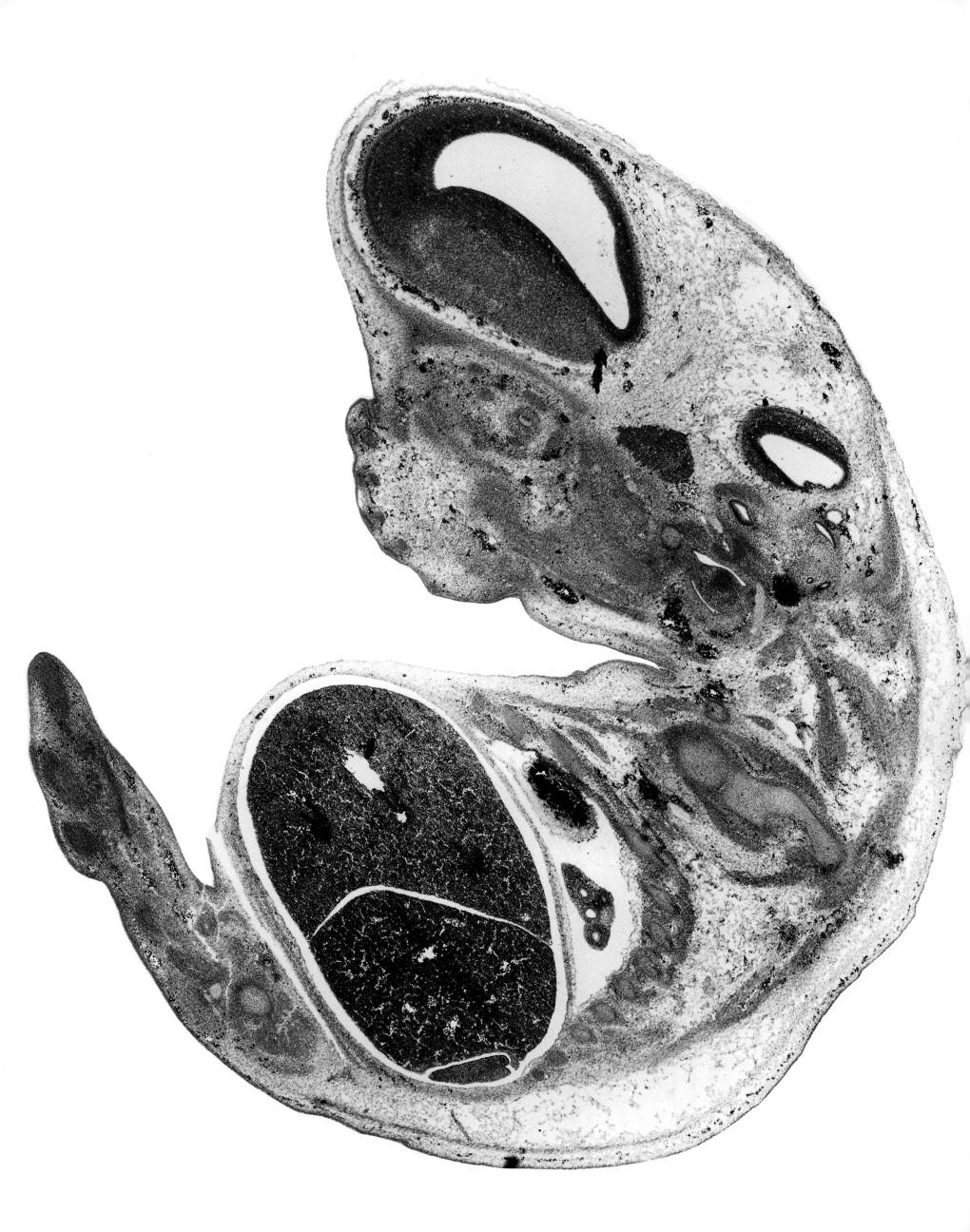

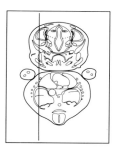

Figure 66
E16 Sagittal 16

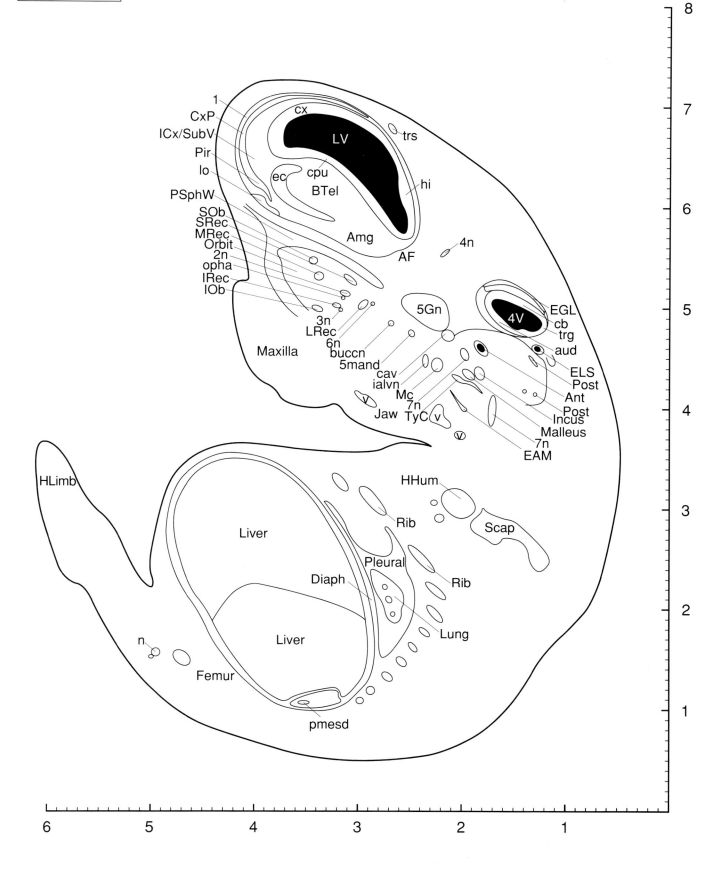

1 cortical layer 1
2n optic nerve
3n oculomotor nerve or its root
4n trochlear nerve or its root
4V fourth ventricle
5Gn trigeminal ganglion
5mand mandibular nerve
6n abducens nerve or its root
7n facial nerve or its root
AF amygdaloid fissure
Amg amygdala
Ant anterior semicircular duct
aud auditory neuroepi
BTel basal telencephalon
buccn buccal nerve
cav cavernous sinus
cb cerebellar neuroepi
cpu caudate putamen neuroepi
cx cortical neuroepi
CxP cortical plate
Diaph diaphragm
EAM external auditory meatus
ec external capsule
EGL external germinal layer of Cb
ELS endolymphatic sac
Femur femur
HHum head of humerus
hi hippocampal formation neuroepi
HLimb hind limb
ialvn inferior alveolar nerve
ICx intermediate cortical layer
Incus incus (ossicle)
IOb inferior oblique muscle
IRec inferior rectus muscle
Jaw jaw
Liver liver
lo lateral olfactory tract
LRec lateral rectus muscle
Lung lung
LV lateral ventricle
Malleus malleus (ossicle)
Maxilla maxilla
Mc Meckel's cartilage
MRec medial rectus muscle
n nerve
opha ophthalmic artery
Orbit orbital cavity
Pir piriform cortex
Pleural pleural cavity
pmesd paramesonephric duct
Post posterior semicircular duct
PSphW presphenoid wing
Rib rib
Scap scapula
SOb superior oblique muscle
SRec superior rectus muscle
SubV subventricular cortical layer
trg germinal trigone
trs transverse sinus
TyC tympanic cavity
v vein (unidentified)

E17 Coronal Section

Figure 67
E17 Coronal 1

Gl   glomerular layer olfactory bulb
Mi   mitral cell layer olfactory bulb
NSpt   nasal septum
obn   olfactory bulb neuroepithelium
OV   olfactory ventricle
VNC   vomeronasal cavity
VNO   vomeronasal organ

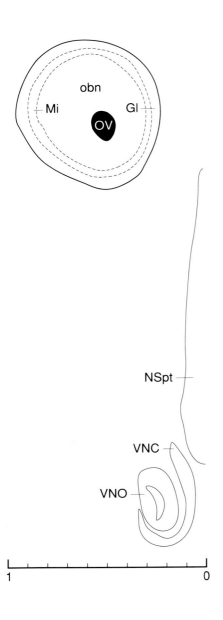

Figure 68
E17 Coronal 2

1   cortical layer 1
CxP   cortical plate
ICx   intermediate cortical layer
lo   lateral olfactory tract
LV   lateral ventricle
ne   neuroepithelium
Pir   piriform cortex
SubV   subventricular cortical layer

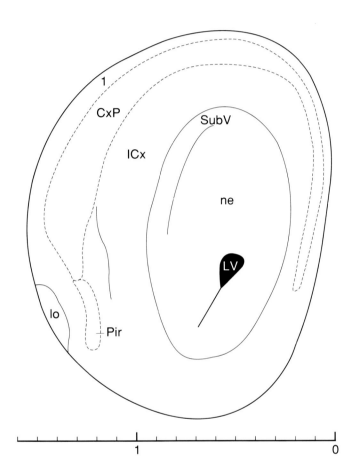

1

0

Figure 69
E17 Coronal 3

1   cortical layer 1
ac   anterior commissure
acer   anterior cerebral artery
btel   basal telencephalic neuroepithel
cx   cortical neuroepithelium
CxP   cortical plate
CxS   cortical subplate
ICx   intermediate cortical layer
Lens   lens
lo   lateral olfactory tract
LV   lateral ventricle
ne   neuroepithelium
olfa   olfactory artery
Pig   pigment epithelium of retina
Pir   piriform cortex
Pu   putamen
RF   rhinal fissure
Spt   septum
spt   septal neuroepithelium
SubV   subventricular cortical layer
Tu   olfactory tubercule
Vent   ventricular zone of the retina
Vitr   vitreous of the eye

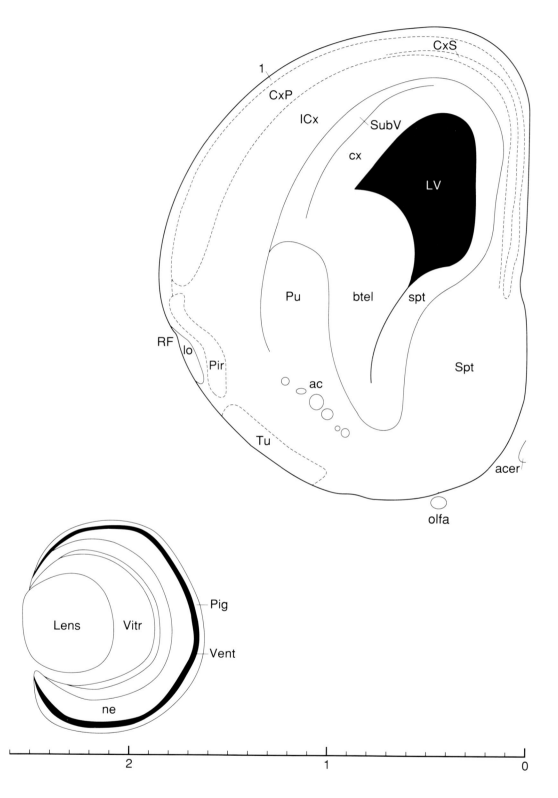

Figure 70
E17 Coronal 4

1 cortical layer 1
2n optic nerve
acer anterior cerebral artery
BTel basal telencephalon
btel basal telencephalic neuroepithel
chp choroid plexus primordium
cpu caudate putamen neuroepithelium
cx cortical neuroepithelium
CxP cortical plate
f fornix
ICx intermediate cortical layer
lo lateral olfactory tract
LV lateral ventricle
mcer middle cerebral artery
olfa olfactory artery
Pir piriform cortex
Pu putamen
RF rhinal fissure
Spt septum
spt septal neuroepithelium
SubV subventricular cortical layer

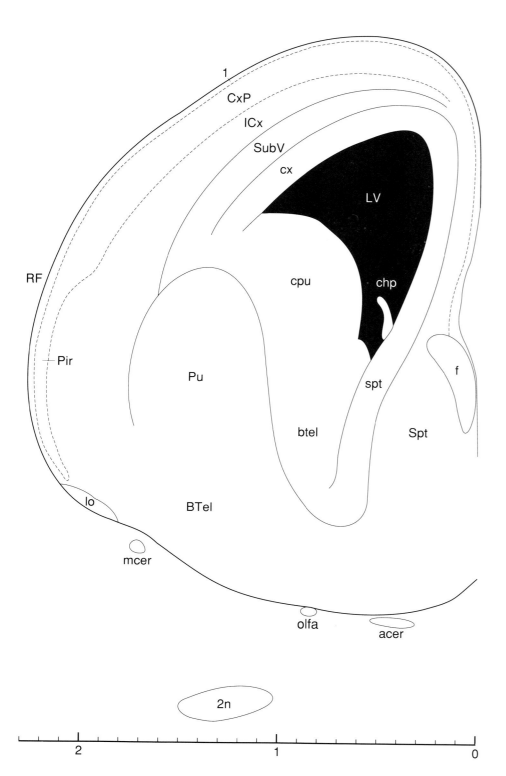

Figure 71
E17 Coronal 5

1   cortical layer 1
2n   optic nerve
3V   3rd ventricle
acer   anterior cerebral artery
BTel   basal telencephalon
btel   basal telencephalic neuroepithel
chp   choroid plexus primordium
cpu   caudate putamen neuroepithelium
cx   cortical neuroepithelium
CxP   cortical plate
fi   fimbria hippocampus
glia   glia
GP   globus pallidus
hi   hippocampal formation neuroepi
ic   inferior colliculus neuroepithelium
ICx   intermediate cortical layer
IVF   interventricular foramen
lo   lateral olfactory tract
LPO   lateral preoptic area
LV   lateral ventricle
mcer   middle cerebral artery
mfb   medial forebrain bundle
MPO   medial preoptic nu
OptF   optic foramen
Pir   piriform cortex
Pu   putamen
RF   rhinal fissure
SCh   suprachiasmatic nu
SFO   subfornical organ
sm   stria medullaris thalami
SubV   subventricular cortical layer

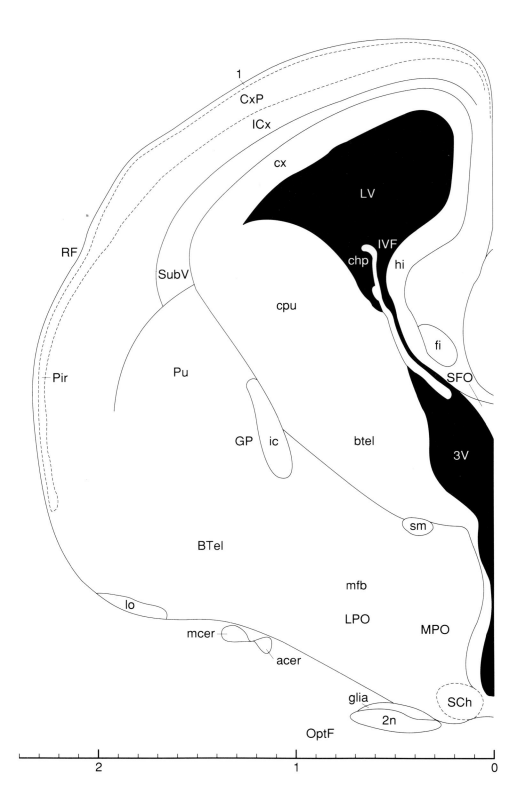

Figure 72
E17 Coronal 6

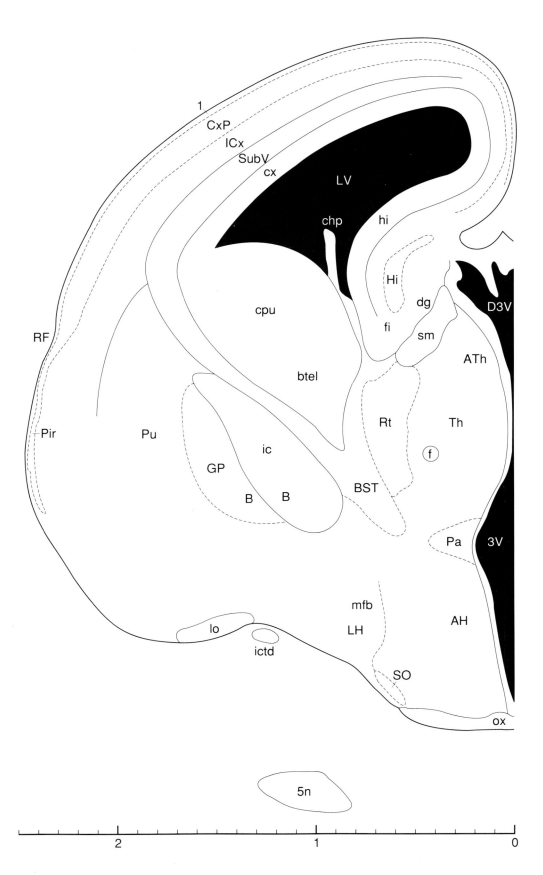

1   cortical layer 1
3V   3rd ventricle
5n   trigeminal nerve
AH   anterior hypoth nu
ATh   anterior thalamus
B   basal nu Meynert
BST   bed nu stria terminalis
btel   basal telencephalic neuroepithel
chp   choroid plexus primordium
cpu   caudate putamen neuroepithelium
cx   cortical neuroepithelium
CxP   cortical plate
D3V   dorsal third ventricle
dg   dentate gyrus neuroepithelium
f   fornix
fi   fimbria hippocampus
GP   globus pallidus
Hi   hippocampal formation
hi   hippocampal formation neuroepi
ic   inferior colliculus neuroepithelium
ictd   internal carotid artery
ICx   intermediate cortical layer
LH   lateral hypothalamic area
lo   lateral olfactory tract
LV   lateral ventricle
mfb   medial forebrain bundle
ox   optic chiasm
Pa   paraventricular hypoth nu
Pir   piriform cortex
Pu   putamen
RF   rhinal fissure
Rt   reticular thal nu
sm   stria medullaris thalami
SO   supraoptic nu
SubV   subventricular cortical layer
Th   thalamus

Figure 73
E17 Coronal 7

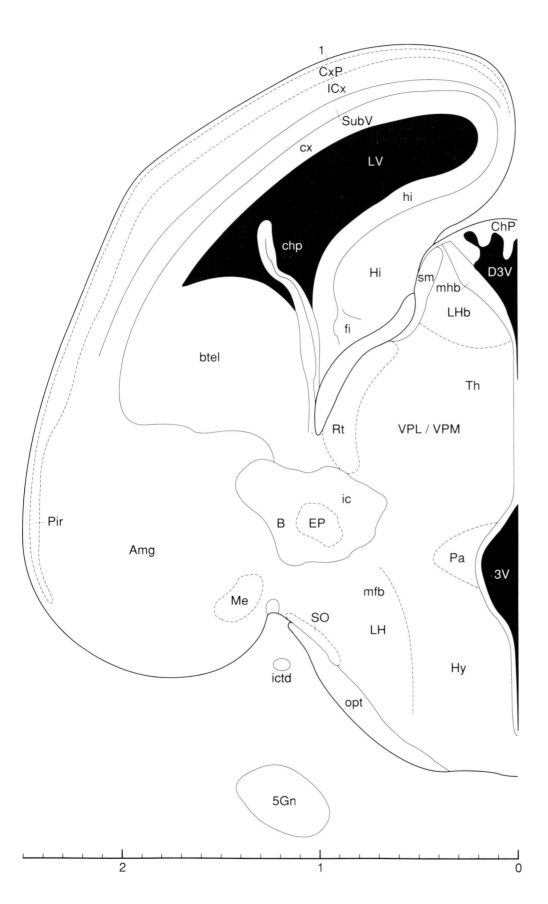

1   cortical layer 1
3V   3rd ventricle
5Gn   trigeminal ganglion
Amg   amygdala
B   basal nu Meynert
btel   basal telencephalic neuroepithel
ChP   choroid plexus
chp   choroid plexus primordium
cx   cortical neuroepithelium
CxP   cortical plate
D3V   dorsal third ventricle
EP   entopeduncular nu
fi   fimbria hippocampus
Hi   hippocampal formation
hi   hippocampal formation neuroepi
Hy   hypothalamus
ic   inferior colliculus neuroepithelium
ictd   internal carotid artery
ICx   intermediate cortical layer
LH   lateral hypothalamic area
LHb   lateral habenular nu
LV   lateral ventricle
Me   medial amygdaloid nu
mfb   medial forebrain bundle
mhb   medial habenula neuroepithel
opt   optic tract
Pa   paraventricular hypoth nu
Pir   piriform cortex
Rt   reticular thal nu
sm   stria medullaris thalami
SO   supraoptic nu
SubV   subventricular cortical layer
Th   thalamus
VPL   ventral posterolateral thal nu
VPM   ventral posteromedial thal nu

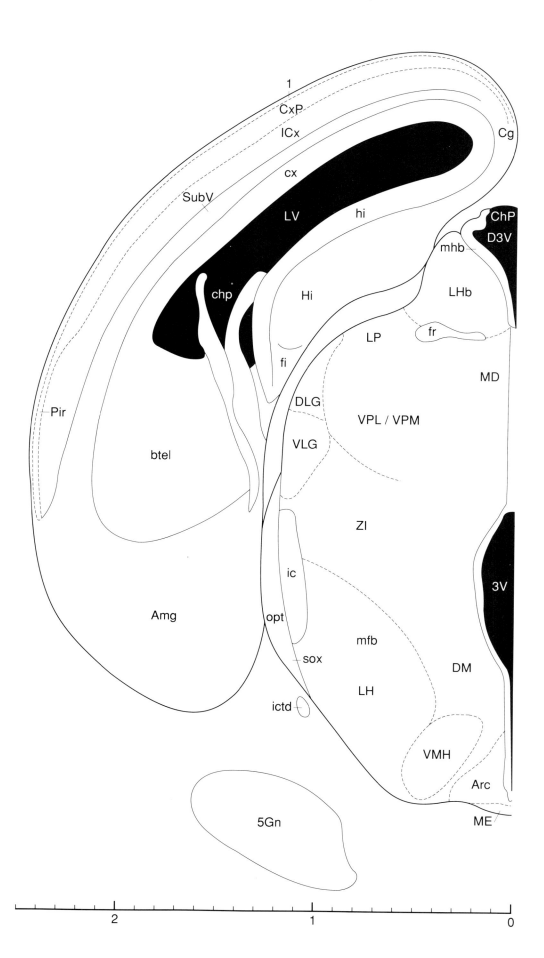

Figure 74
E17 Coronal 8

1   cortical layer 1
3V   3rd ventricle
5Gn   trigeminal ganglion
Amg   amygdala
Arc   arcuate hypoth nu
btel   basal telencephalic neuroepithel
Cg   cingulate cortex
ChP   choroid plexus
chp   choroid plexus primordium
cx   cortical neuroepithelium
CxP   cortical plate
D3V   dorsal third ventricle
DLG   dorsal lateral geniculate nu
DM   dorsomedial hypothalamic nu
fi   fimbria hippocampus
fr   fasciculus retroflexus
Hi   hippocampal formation
hi   hippocampal formation neuroepi
ic   inferior colliculus neuroepithelium
ictd   internal carotid artery
ICx   intermediate cortical layer
LH   lateral hypothalamic area
LHb   lateral habenular nu
LP   lateral posterior thal nu
LV   lateral ventricle
MD   mediodorsal thal nu
ME   median eminence
mfb   medial forebrain bundle
mhb   medial habenula neuroepithel
opt   optic tract
Pir   piriform cortex
sox   supraoptic decussation
SubV   subventricular cortical layer
VLG   ventral lateral geniculate nu
VMH   ventromedial hypoth nu
VPL   ventral posterolateral thal nu
VPM   ventral posteromedial thal nu
ZI   zona incerta

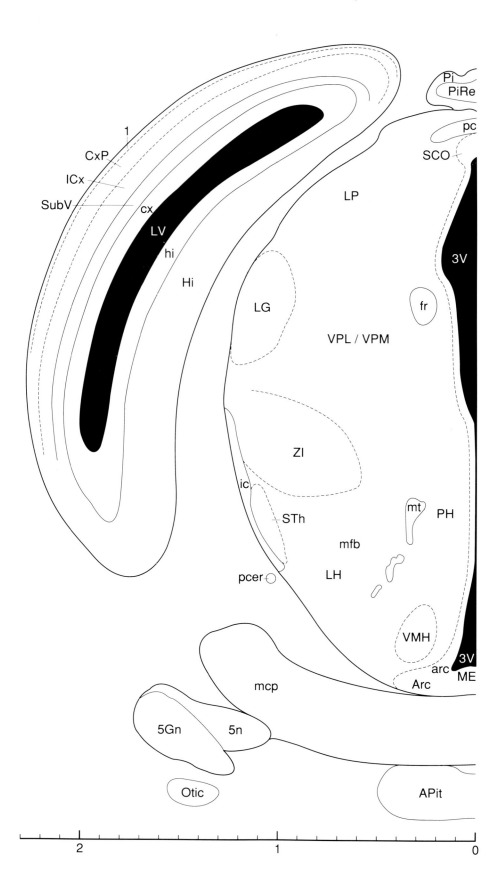

Figure 75
E17 Coronal 9

1   cortical layer 1
3V   3rd ventricle
5Gn   trigeminal ganglion
5n   trigeminal nerve
APit   anterior lobe pituitary
Arc   arcuate hypoth nu
arc   arcuate nu neuroepithelium
cx   cortical neuroepithelium
CxP   cortical plate
fr   fasciculus retroflexus
Hi   hippocampal formation
hi   hippocampal formation neuroepi
ic   inferior colliculus neuroepithelium
ICx   intermediate cortical layer
LG   lateral geniculate nucleus
LH   lateral hypothalamic area
LP   lateral posterior thal nu
LV   lateral ventricle
mcp   middle cerebellar peduncle
ME   median eminence
mfb   medial forebrain bundle
mt   mammillothalamic tract
Otic   otic ganglion
pc   posterior commissure
pcer   posterior cerebral artery
PH   posterior hypoth area
Pi   pineal gland
PiRe   pineal recess of 3V
SCO   subcommissural organ
STh   subthalamic nu
SubV   subventricular cortical layer
VMH   ventromedial hypoth nu
VPL   ventral posterolateral thal nu
VPM   ventral posteromedial thal nu
ZI   zona incerta

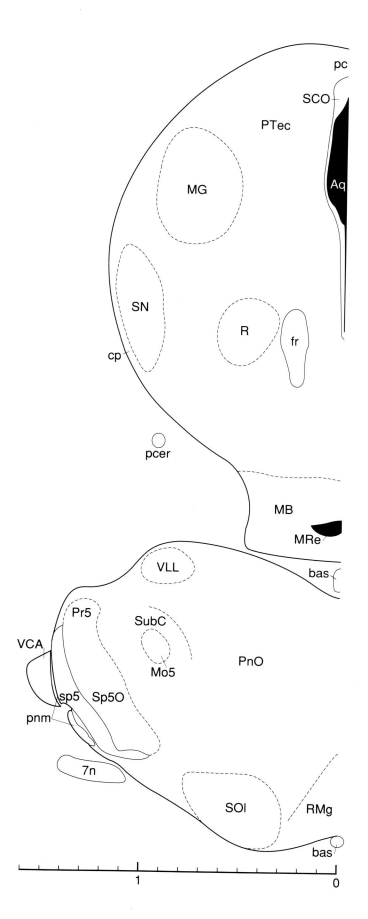

Figure 76
E17 Coronal 10

7n   facial nerve or its root
Aq   aqueduct
bas   basilar artery
cp   cerebral peduncle, basal
fr   fasciculus retroflexus
MB   mammillary body
MG   medial geniculate nu
Mo5   motor trigeminal nu
MRe   mammillary recess 3rd ventricle
pc   posterior commissure
pcer   posterior cerebral artery
pnm   pontine migration
PnO   pontine reticular nu, oral
Pr5   principal sensory trigeminal nu
PTec   pretectum
R   red nu
RMg   raphe magnus nu
SCO   subcommissural organ
SN   substantia nigra
SOl   superior olive
sp5   spinal trigeminal tract
Sp5O   spinal trigeminal nu, oral
SubC   subcoeruleus nu
VCA   ventral cochlear nu, anterior
VLL   ventral nu lateral lemniscus

Figure 77
E17 Coronal 11

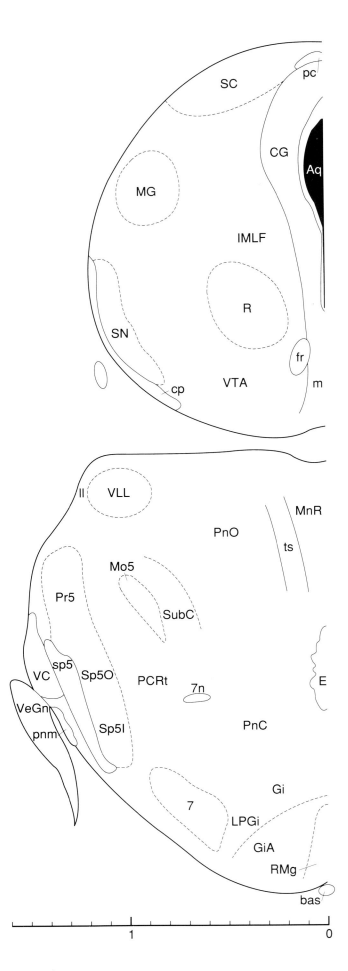

7 facial nu
7n facial nerve or its root
Aq aqueduct
bas basilar artery
CG central gray
cp cerebral peduncle, basal
E ependyma and subependymal layer
fr fasciculus retroflexus
Gi gigantocellular reticular nucleus
GiA gigantocell reticular nu, alpha
IMLF interstitial nu mlf
ll lateral lemniscus
LPGi lateral paragigantocellular nu
m migration of neurons
MG medial geniculate nu
MnR median raphe nu
Mo5 motor trigeminal nu
pc posterior commissure
PCRt parvocellular reticular nu
PnC pontine reticular nu, caudal
pnm pontine migration
PnO pontine reticular nu, oral
Pr5 principal sensory trigeminal nu
R red nu
RMg raphe magnus nu
SC superior colliculus
SN substantia nigra
sp5 spinal trigeminal tract
Sp5I spinal trigem nu, interpolar
Sp5O spinal trigeminal nu, oral
SubC subcoeruleus nu
ts tectospinal tract
VC ventral cochlear nu
VeGn vestibular ganglion
VLL ventral nu lateral lemniscus
VTA ventral tegmental area

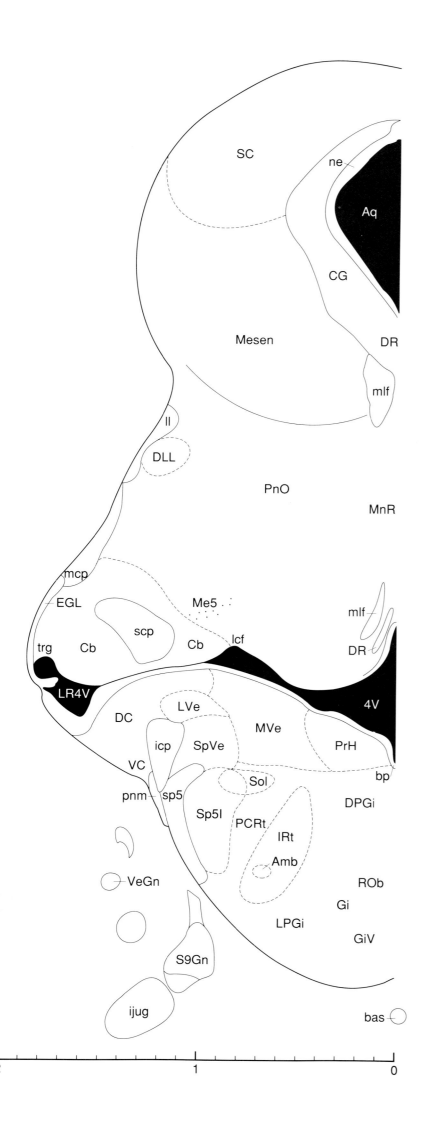

Figure 78
E17 Coronal 12

4V   4th ventricle
Amb   ambiguus nu
Aq   aqueduct
bas   basilar artery
bp   basal plate neuroepithelium
Cb   cerebellum
CG   central gray
DC   dorsal cochlear nu
DLL   dorsal nu lateral lemniscus
DPGi   dorsal paragigantocellular nu
DR   dorsal raphe nu
EGL   external germinal layer of cb
Gi   gigantocellular reticular nucleus
GiV   gigantocell reticular nu, vent
icp   inferior cerebellar peduncle
ijug   internal jugular vein
IRt   intermediate reticular zone
lcf   lateral cerebellar fissure
ll   lateral lemniscus
LPGi   lateral paragigantocellular nu
LR4V   lateral recess 4th ventricle
LVe   lateral vestibular nu
mcp   middle cerebellar peduncle
Me5   mesencephalic trigeminal nu
Mesen   mesencephalon
mlf   medial longitudinal fasciculus
MnR   median raphe nu
MVe   medial vestibular nu
ne   neuroepithelium
PCRt   parvocellular reticular nu
pnm   pontine migration
PnO   pontine reticular nu, oral
PrH   prepositus hypoglossal nu
ROb   raphe obscurus nu
S9Gn   superior glossopharyngeal gang
SC   superior colliculus
scp   superior cerebellar peduncle
Sol   nu solitary tract
sp5   spinal trigeminal tract
Sp5I   spinal trigem nu, interpolar
SpVe   spinal vestibular nu
trg   germinal trigone
VC   ventral cochlear nu
VeGn   vestibular ganglion

Figure 79
E17 Coronal 13

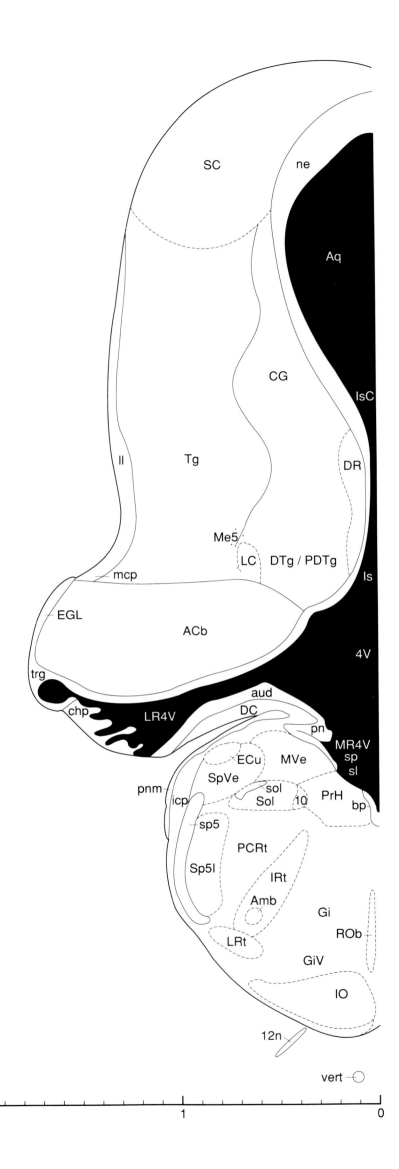

4V   4th ventricle
10   dorsal motor nu vagus
12n   hypoglossal nerve or its root
ACb   anterior cerebellum
Amb   ambiguus nu
Aq   aqueduct
aud   auditory neuroepithelium
bp   basal plate neuroepithelium
CG   central gray
chp   choroid plexus primordium
DC   dorsal cochlear nu
DR   dorsal raphe nu
DTg   dorsal tegmental nu
ECu   external cuneate nu
EGL   external germinal layer of cb
Gi   gigantocellular reticular nucleus
GiV   gigantocell reticular nu, vent
icp   inferior cerebellar peduncle
IO   inferior olive
IRt   intermediate reticular zone
Is   isthmus region
IsC   isthmal canal
LC   locus coeruleus
ll   lateral lemniscus
LR4V lateral recess 4th ventricle
LRt   lateral reticular nu
mcp   middle cerebellar peduncle
Me5   mesencephalic trigeminal nu
MR4V medial recesses of the 4th vent
MVe   medial vestibular nu
ne   neuroepithelium
PCRt parvocellular reticular nu
PDTg Posterodorsal tegmental nu
pn   pontine nuclei neuroepithelium
pnm   pontine migration
PrH   prepositus hypoglossal nu
ROb   raphe obscurus nu
SC   superior colliculus
sl   sulcus limitans
Sol   nu solitary tract
sol   solitary tract
sp   superficial plate neuroepithelium
sp5   spinal trigeminal tract
Sp5I spinal trigem nu, interpolar
SpVe spinal vestibular nu
Tg   tegmentum
trg   germinal trigone
vert vertebral artery

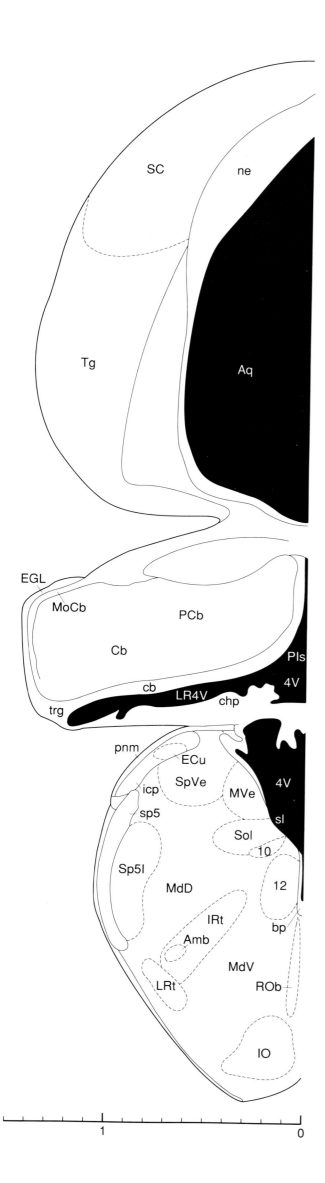

Figure 80
E17 Coronal 14

4V   4th ventricle
10   dorsal motor nu vagus
12   hypoglossal nu
Amb   ambiguus nu
Aq   aqueduct
bp   basal plate neuroepithelium
Cb   cerebellum
cb   cerebellar neuroepithelium
chp   choroid plexus primordium
ECu   external cuneate nu
EGL   external germinal layer of cb
icp   inferior cerebellar peduncle
IO   inferior olive
IRt   intermediate reticular zone
LR4V   lateral recess 4th ventricle
LRt   lateral reticular nu
MdD   medullary reticular nu, dors
MdV   medullary reticular nu, ventral
MoCb   molecular layer of cerebellum
MVe   medial vestibular nu
ne   neuroepithelium
PCb   posterior cerebellum
PIs   posterior isthmal recess
pnm   pontine migration
ROb   raphe obscurus nu
SC   superior colliculus
sl   sulcus limitans
Sol   nu solitary tract
sp5   spinal trigeminal tract
Sp5I   spinal trigem nu, interpolar
SpVe   spinal vestibular nu
Tg   tegmentum
trg   germinal trigone

E19 Coronal Section Plan

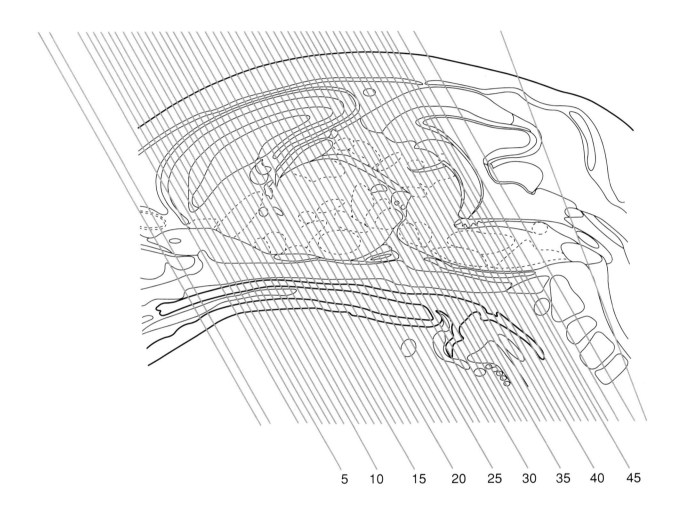

5   10   15   20   25   30   35   40   45

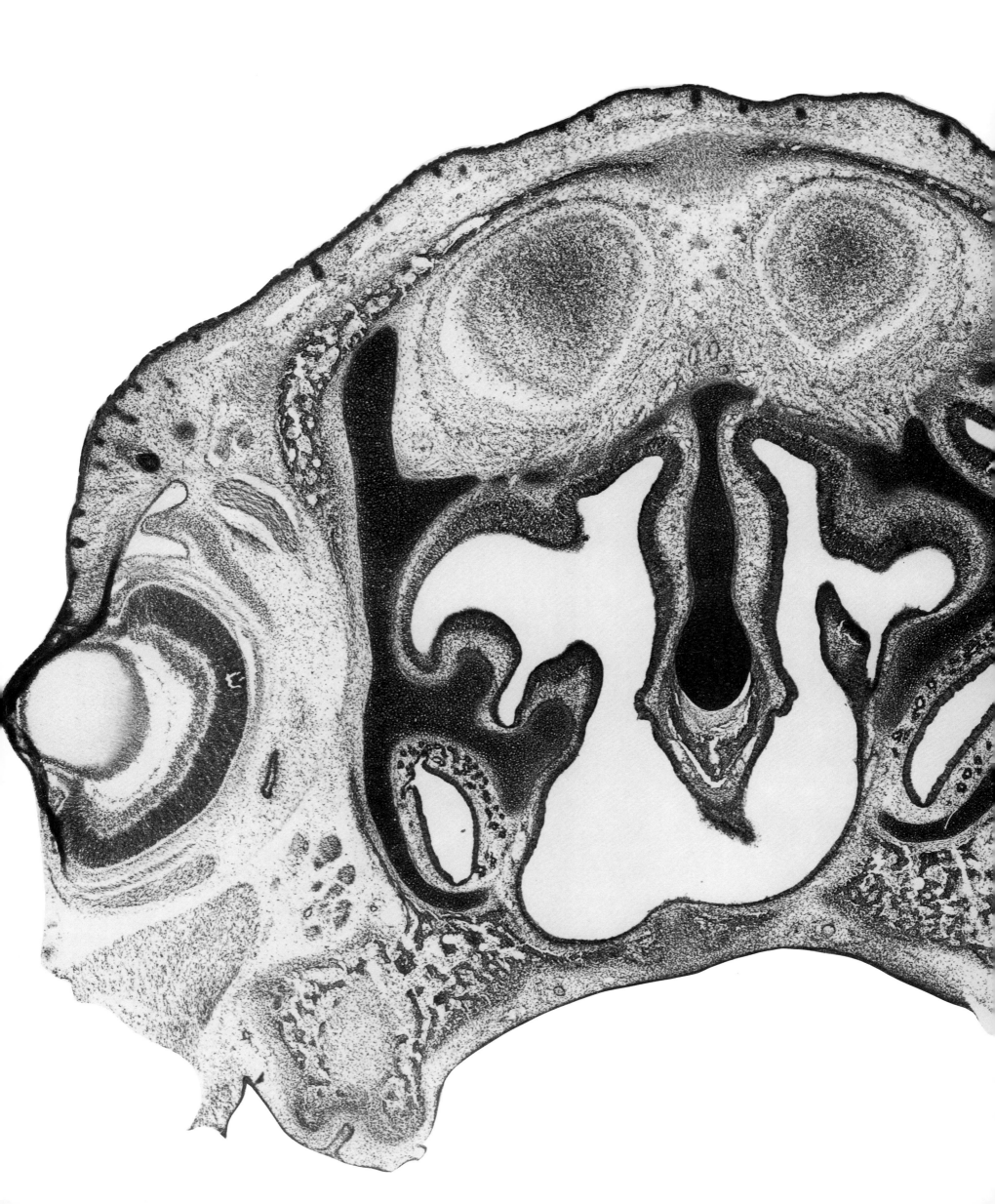

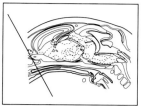

Figure 81
E19 Coronal 1

5max   maxillary nerve
CrP   cribriform plate
DentLam   dental lamina
dpal   descending palatine artery
EPl   external plexiform layer olf bulb
Ethmoid   ethmoid bone
Frontal   frontal bone
Gl   glomerular layer olfactory bulb
Harderian   Harderian gland
InfM   inf meatus nasal cav
InPl   inner plexiform layer of retina
iorb   inferior orbital artery
Lens   lens
Maxilla   maxilla
Mi   mitral cell layer olfactory bulb
MRec   medial rectus muscle
Nasal   nasal cavity
ne   neuroepithelium
NSpt   nasal septum
obn   olfactory bulb neuroepithelium
OF   optic fiber layer
olfa   olfactory artery
ON   olf nerve layer
Oral   oral cavity
Pig   pigment epithelium of retina
psa   post sup alveolar artery
RGn   retinal ganglion cell layer
SOb   superior oblique muscle
Vent   ventricular zone of the retina
Vomer   vomer

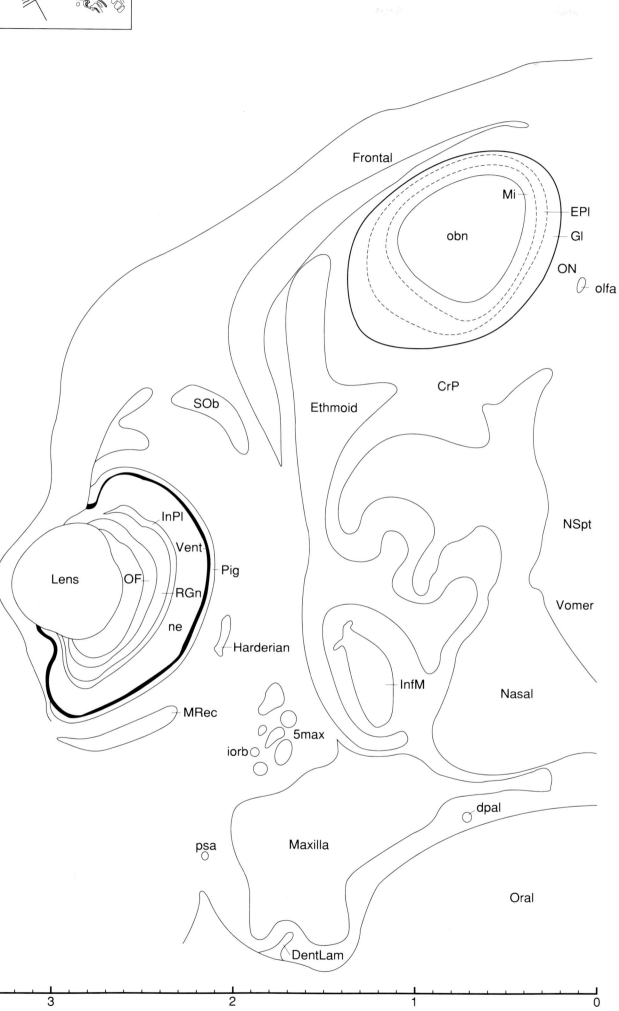

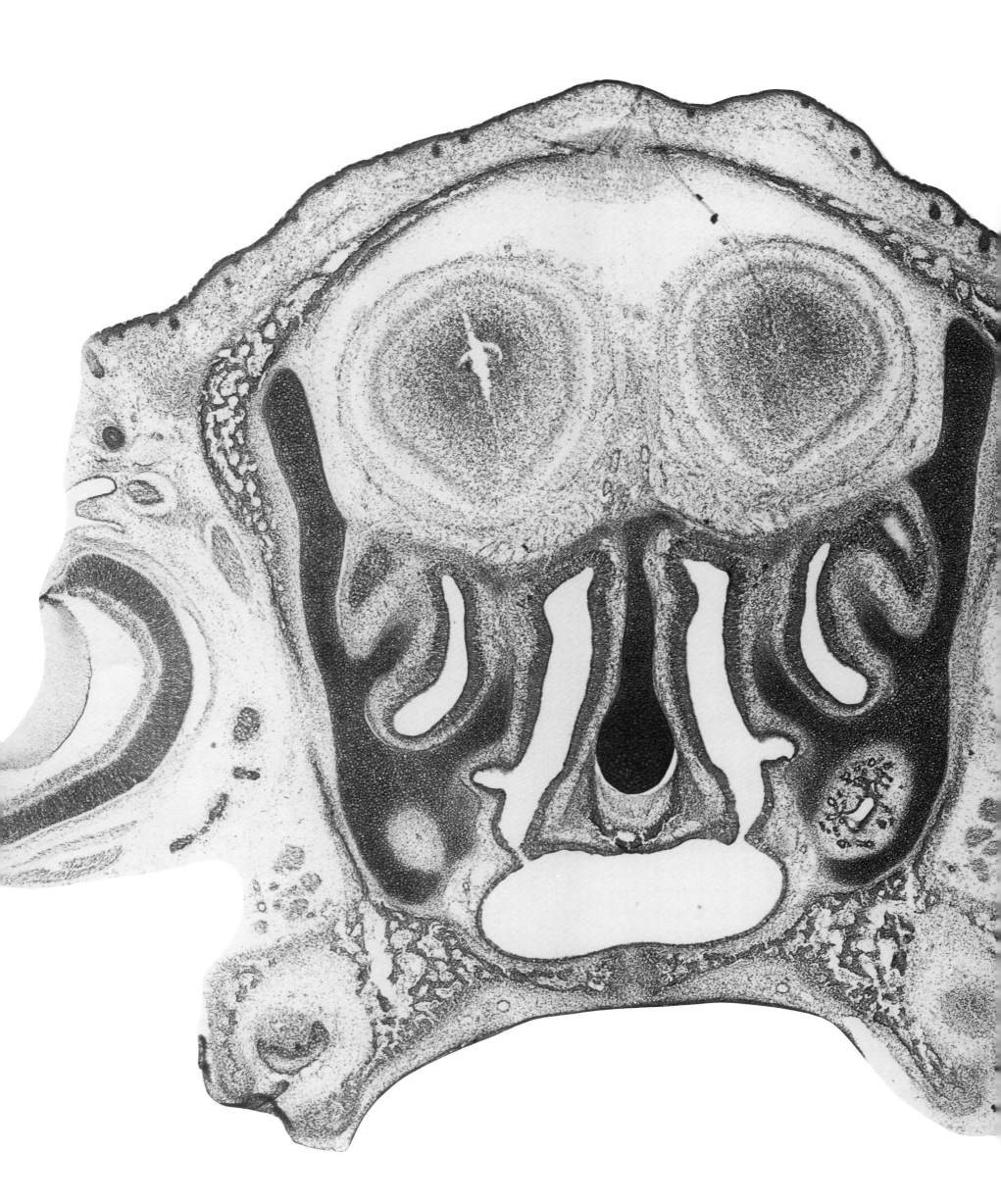

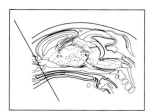

Figure 82
E19 Coronal 2

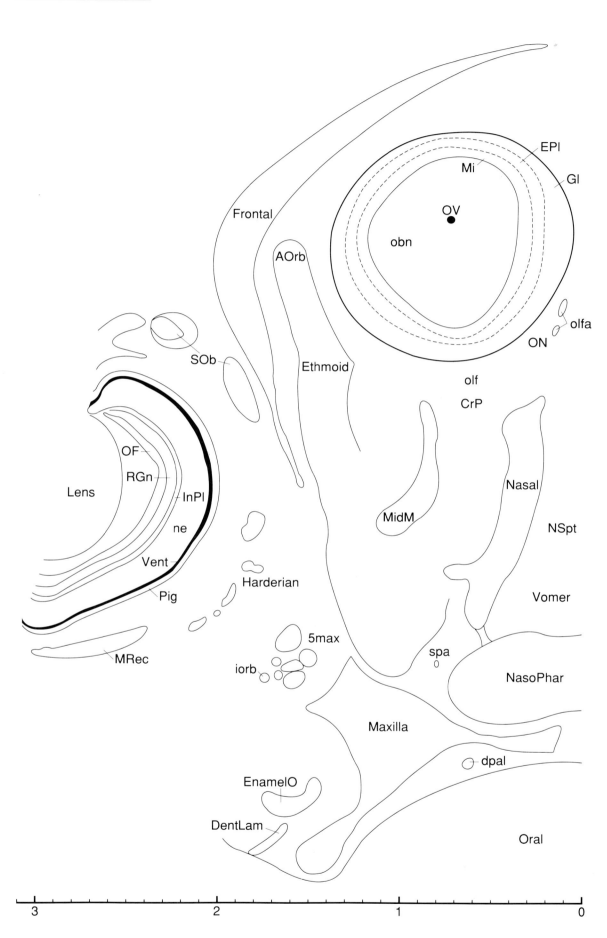

3   2   1   0

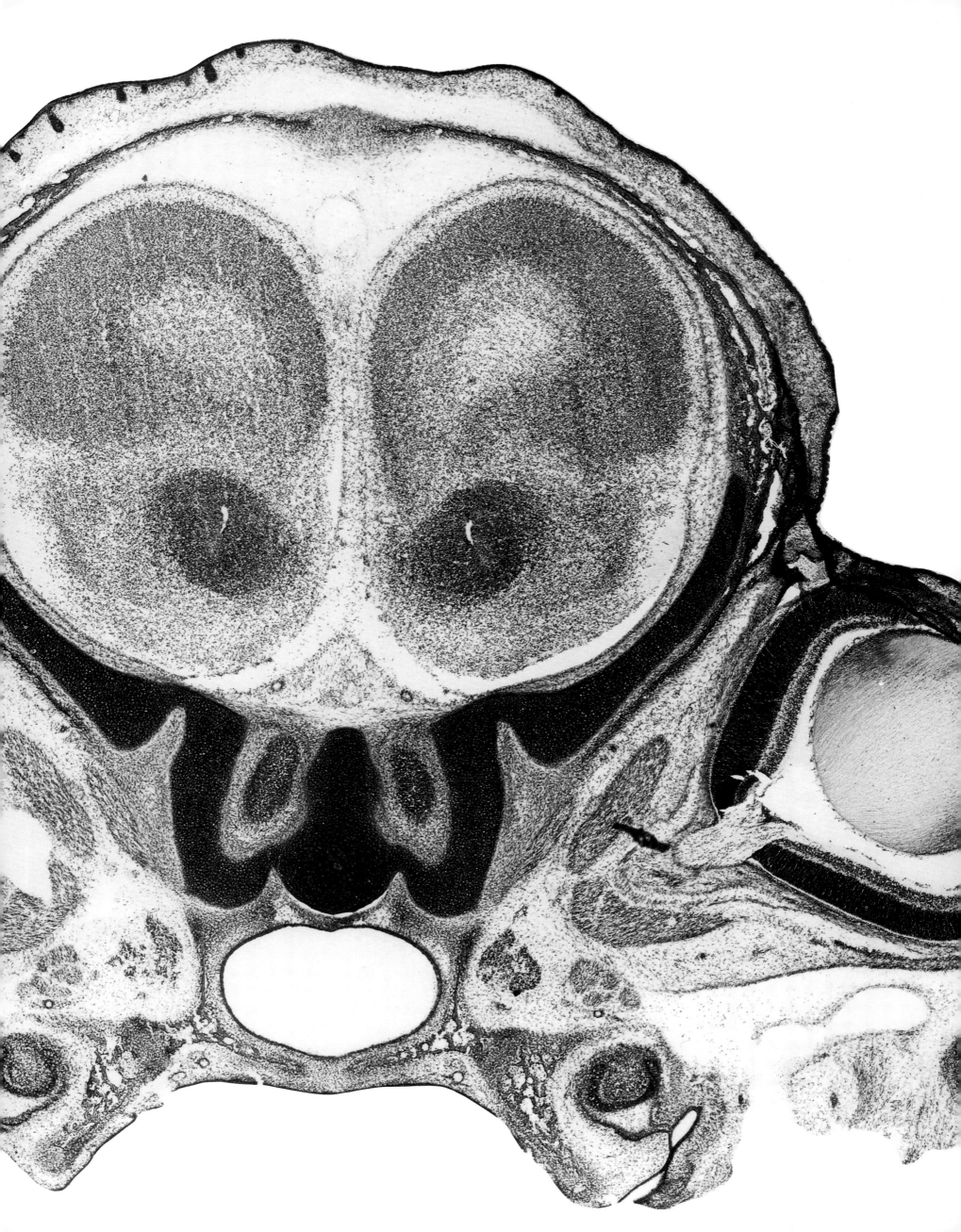

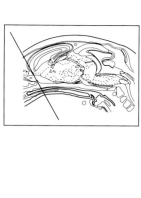

Figure 83
E19 Coronal 3

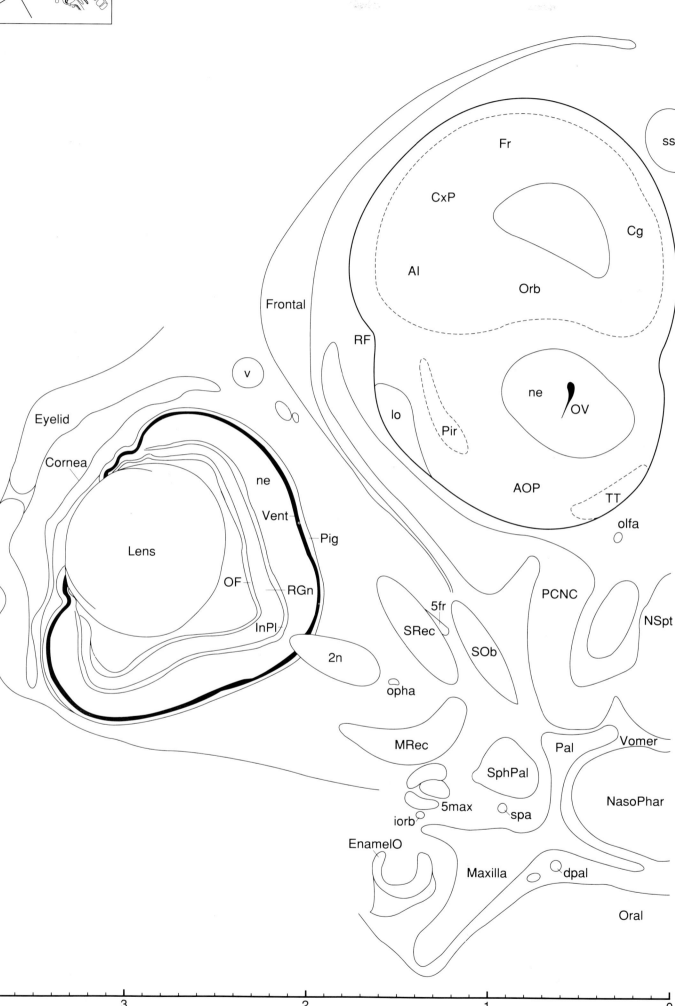

2n   optic nerve
5fr   frontal branch of trigeminal n
5max   maxillary nerve
AI   agranular insular Cx
AOP   anterior olfactory nu, posterior
Cg   cingulate cortex
Cornea   cornea
CxP   cortical plate
dpal   descending palatine artery
EnamelO   enamel organ (of tooth)
Eyelid   eyelid
Fr   frontal cortex
Frontal   frontal bone
InPl   inner plexiform layer of retina
iorb   inferior orbital artery
Lens   lens
lo   lateral olfactory tract
Maxilla   maxilla
MRec   medial rectus muscle
NasoPhar   nasopharyngeal cavity
ne   neuroepithelium
NSpt   nasal septum
OF   optic fiber layer
olfa   olfactory artery
opha   ophthalmic artery
Oral   oral cavity
Orb   orbital cortex
OV   olfactory ventricle
Pal   palatine bone
PCNC   posterior cupula of the nasal
Pig   pigment epithelium of retina
Pir   piriform cortex
RF   rhinal fissure
RGn   retinal ganglion cell layer
SOb   superior oblique muscle
spa   sphenopalatine artery
SphPal   sphenopalatine ganglion
SRec   superior rectus muscle
sss   superior sagittal sinus
TT   tenia tecta
v   vein
Vent   ventricular zone of the retina
Vomer   vomer

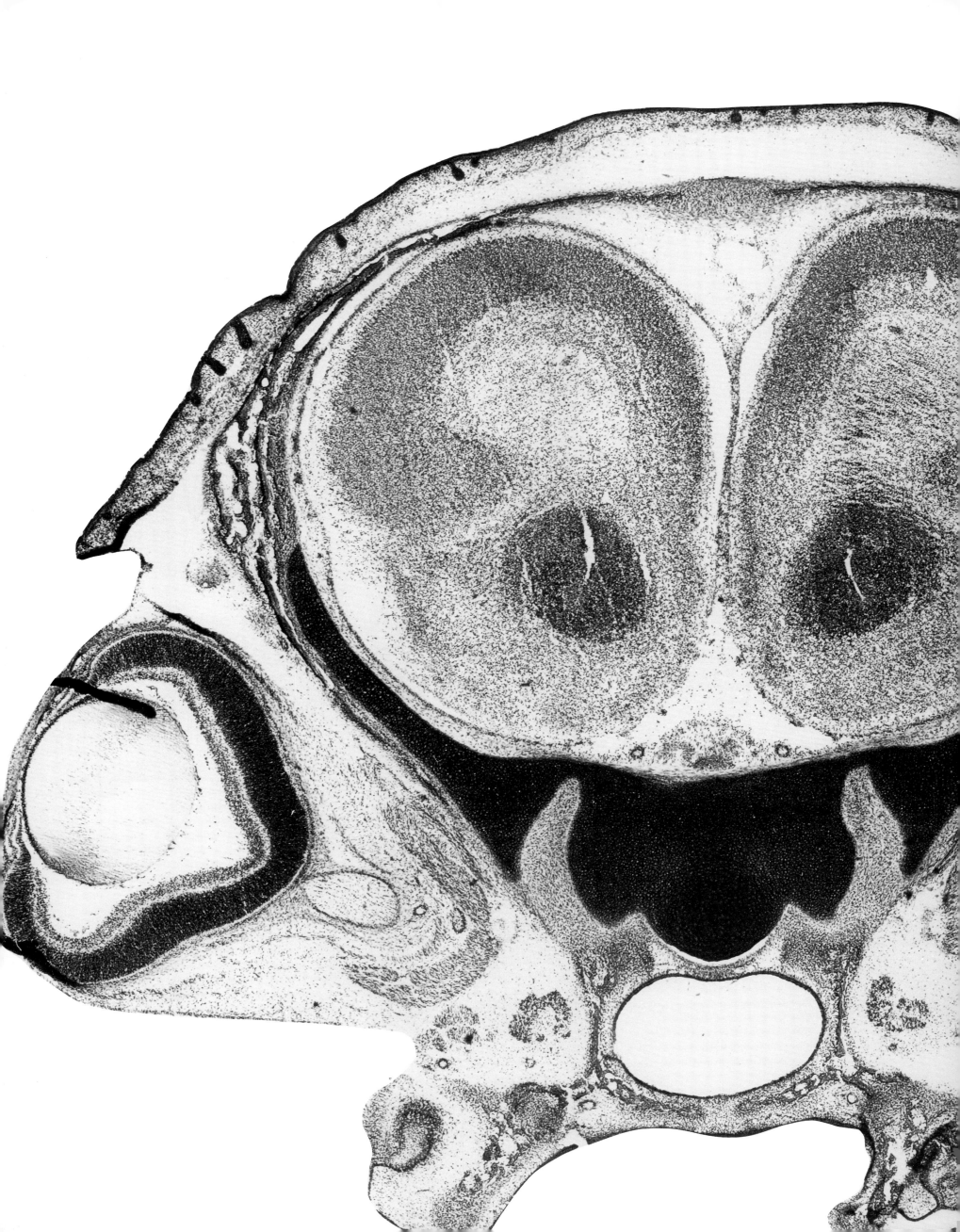

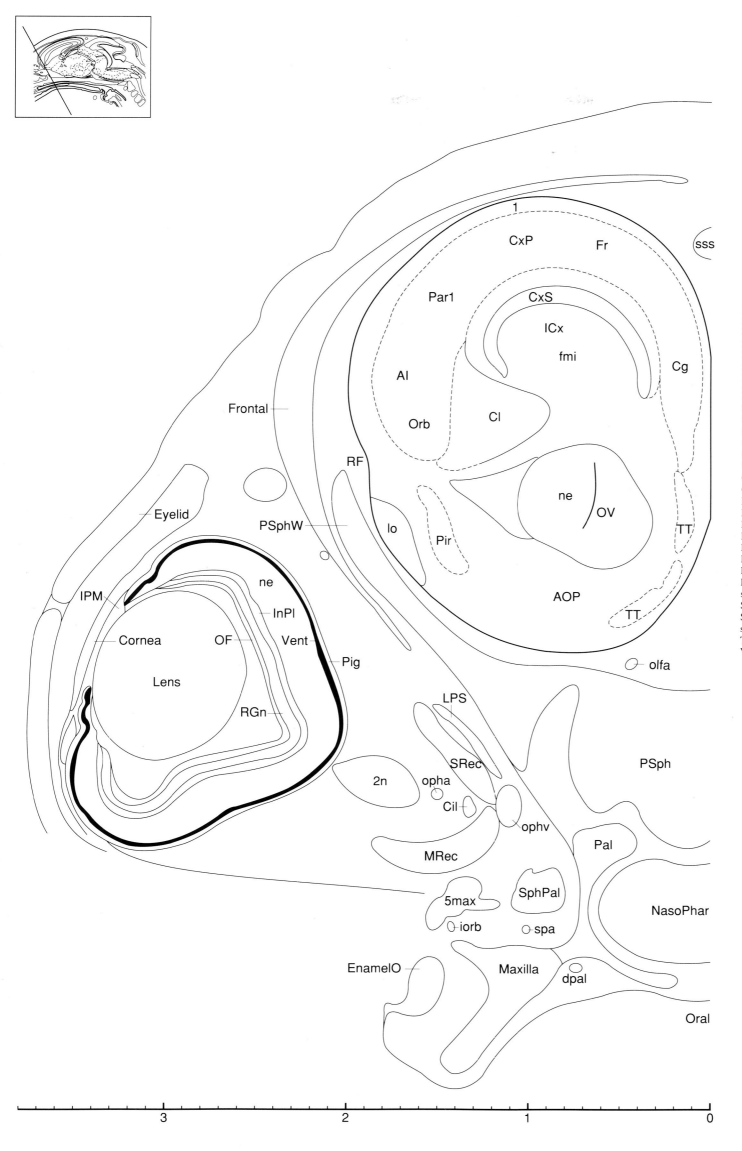

Figure 84
E19 Coronal 4

1 cortical layer 1
2n optic nerve
5max maxillary nerve
AI agranular insular Cx
AOP anterior olfactory nu, posterior
Cg cingulate cortex
Cil ciliary ganglion
Cl claustrum
Cornea cornea
CxP cortical plate
CxS cortical subplate
dpal descending palatine artery
EnamelO enamel organ (of tooth)
Eyelid eyelid
fmi forceps minor corpus callosum
Fr frontal cortex
Frontal frontal bone
ICx intermediate cortical layer
InPl inner plexiform layer of retina
iorb inferior orbital artery
IPM iridopupillary membrane
Lens lens
lo lateral olfactory tract
LPS levator palpebrae superioris mus
Maxilla maxilla
MRec medial rectus muscle
NasoPhar nasopharyngeal cavity
ne neuroepithelium
OF optic fiber layer
olfa olfactory artery
opha ophthalmic artery
ophv ophthalmic vein
Oral oral cavity
Orb orbital cortex
OV olfactory ventricle
Pal palatine bone
Par1 parietal cortex, area 1
Pig pigment epithelium of retina
Pir piriform cortex
PSph presphenoid bone
PSphW presphenoid wing
RF rhinal fissure
RGn retinal ganglion cell layer
spa sphenopalatine artery
SphPal sphenopalatine ganglion
SRec superior rectus muscle
sss superior sagittal sinus
TT tenia tecta
Vent ventricular zone of the retina

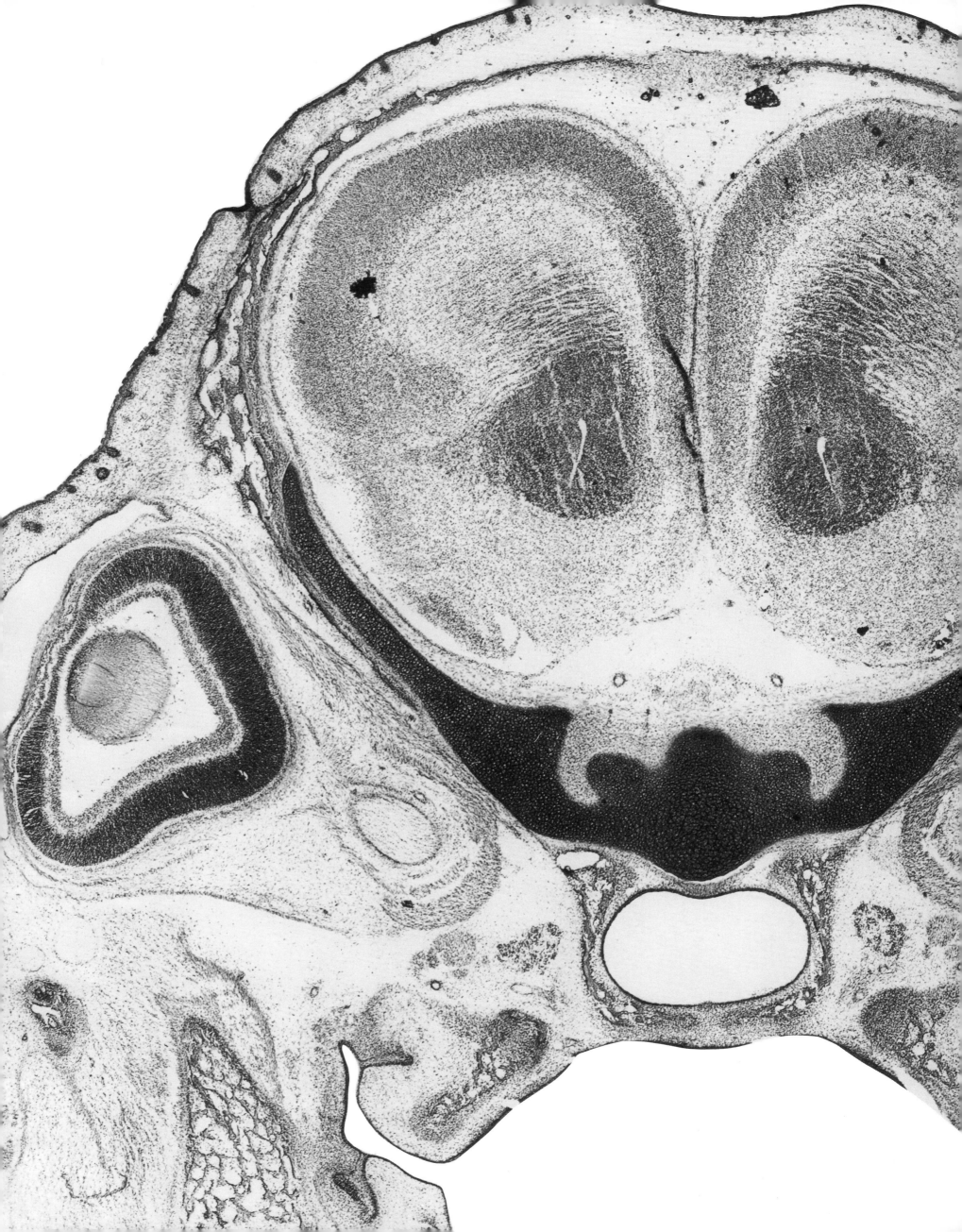

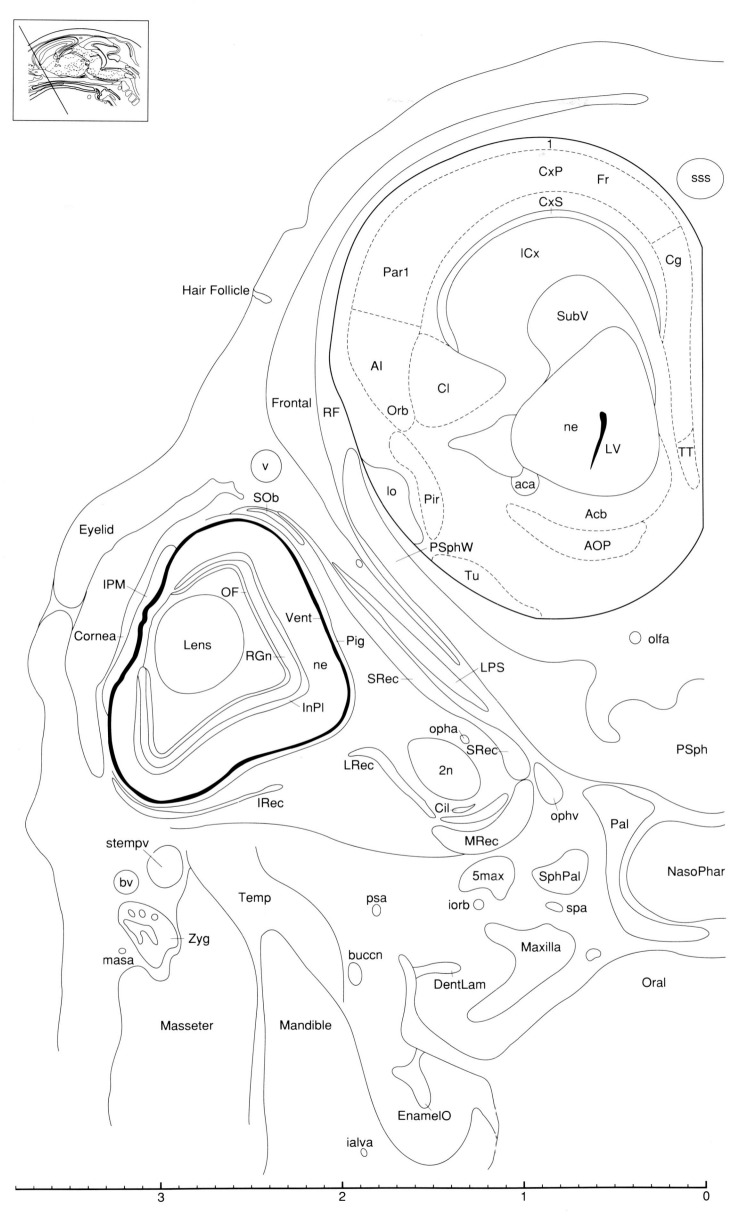

Figure 85
E19 Coronal 5

1 cortical layer 1
2n optic nerve
5max maxillary nerve
aca anterior commissure, anterior
Acb accumbens nu
AI agranular insular Cx
AOP anterior olfactory nu, posterior
buccn buccal nerve
bv blood vessel
Cg cingulate cortex
Cil ciliary ganglion
Cl claustrum
Cornea cornea
CxP cortical plate
CxS cortical subplate
DentLam dental lamina
EnamelO enamel organ (of tooth)
Eyelid eyelid
Fr frontal cortex
Frontal frontal bone
Hair Follicle hair follicle
ialva inferior alveolar artery
ICx intermediate cortical layer
InPl inner plexiform layer of retina
iorb inferior orbital artery
IPM iridopupillary membrane
IRec inferior rectus muscle
Lens lens
lo lateral olfactory tract
LPS levator palpebrae superioris mus
LRec lateral rectus muscle
LV lateral ventricle
Mandible mandible
masa masseteric artery
Masseter masseter muscle
Maxilla maxilla
MRec medial rectus muscle
NasoPhar nasopharyngeal cavity
ne neuroepithelium
OF optic fiber layer
olfa olfactory artery
opha ophthalmic artery
ophv ophthalmic vein
Oral oral cavity
Orb orbital cortex
Pal palatine bone
Par1 parietal cortex, area 1
Pig pigment epithelium of retina
Pir piriform cortex
psa post sup alveolar artery
PSph presphenoid bone
PSphW presphenoid wing
RF rhinal fissure
RGn retinal ganglion cell layer
SOb superior oblique muscle
spa sphenopalatine artery
SphPal sphenopalatine ganglion
SRec superior rectus muscle
sss superior sagittal sinus
stempv superficial temporal vein
SubV subventricular cortical layer
Temp temporal muscle
TT tenia tecta
Tu olfactory tubercule
v vein
Vent ventricular zone of the retina
Zyg zygomatic arch

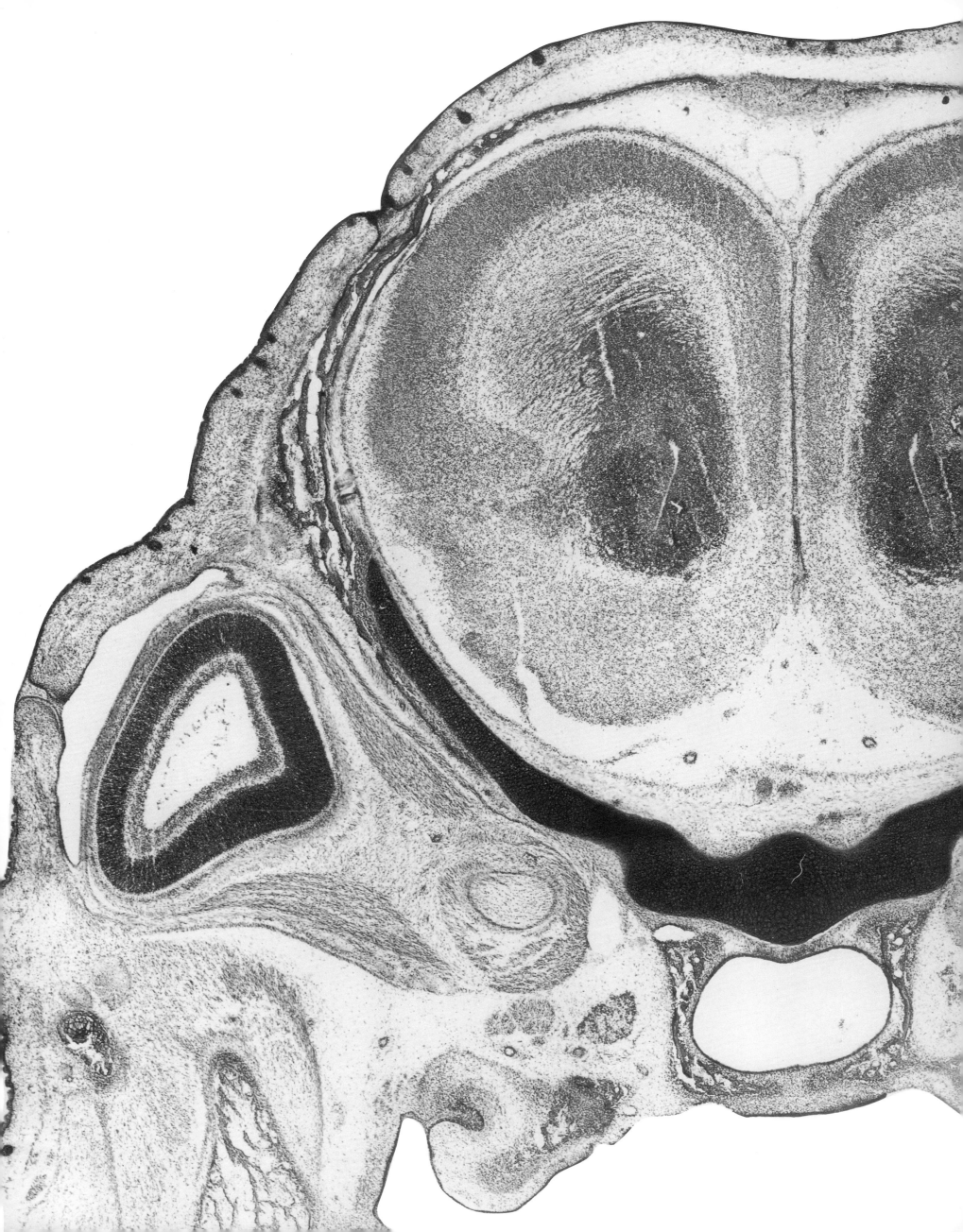

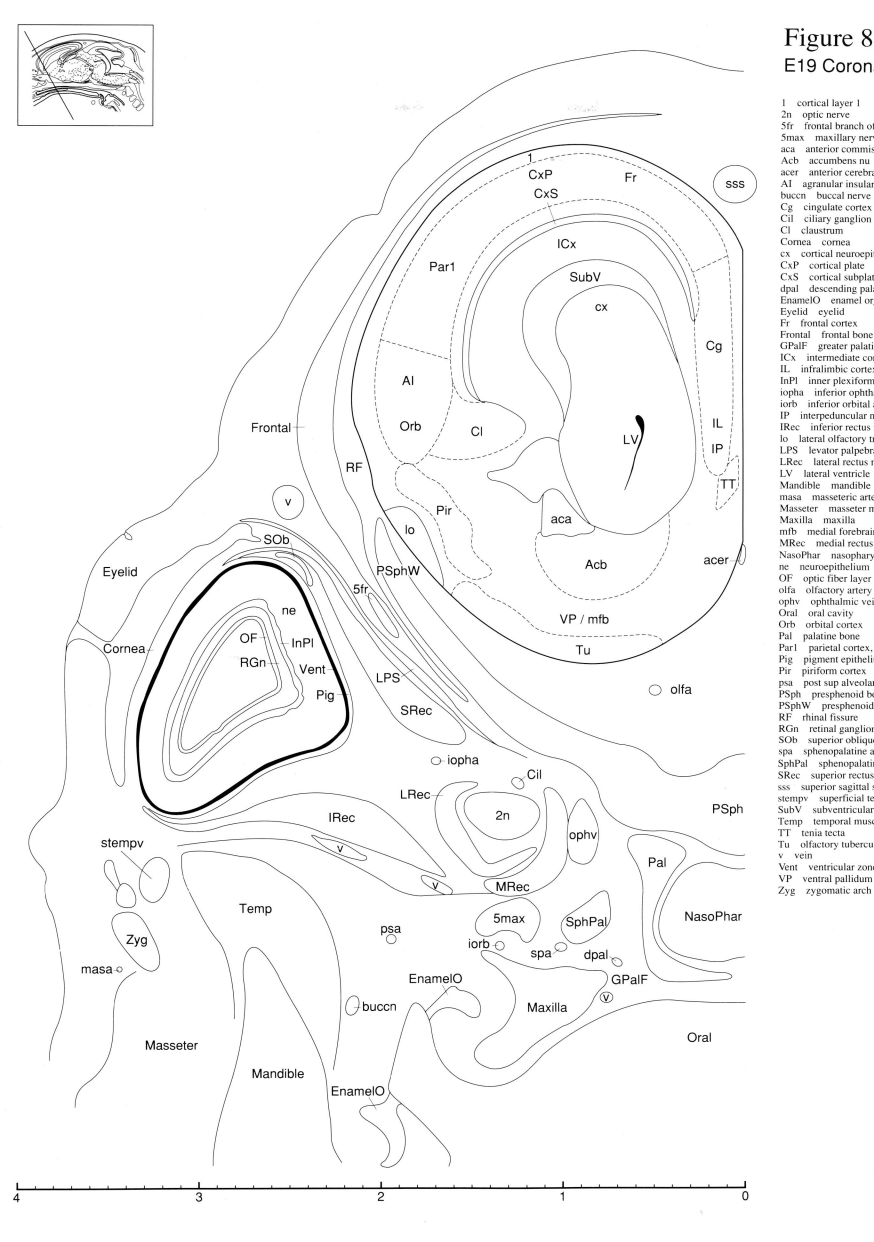

Figure 86
E19 Coronal 6

1   cortical layer 1
2n   optic nerve
5fr   frontal branch of trigeminal n
5max   maxillary nerve
aca   anterior commissure, anterior
Acb   accumbens nu
acer   anterior cerebral artery
AI   agranular insular Cx
buccn   buccal nerve
Cg   cingulate cortex
Cil   ciliary ganglion
Cl   claustrum
Cornea   cornea
cx   cortical neuroepithelium
CxP   cortical plate
CxS   cortical subplate
dpal   descending palatine artery
EnamelO   enamel organ (of tooth)
Eyelid   eyelid
Fr   frontal cortex
Frontal   frontal bone
GPalF   greater palatine foramen
ICx   intermediate cortical layer
IL   infralimbic cortex
InPl   inner plexiform layer of retina
iopha   inferior ophthalmic artery
iorb   inferior orbital artery
IP   interpeduncular nu
IRec   inferior rectus muscle
lo   lateral olfactory tract
LPS   levator palpebrae superioris mus
LRec   lateral rectus muscle
LV   lateral ventricle
Mandible   mandible
masa   masseteric artery
Masseter   masseter muscle
Maxilla   maxilla
mfb   medial forebrain bundle
MRec   medial rectus muscle
NasoPhar   nasopharyngeal cavity
ne   neuroepithelium
OF   optic fiber layer
olfa   olfactory artery
ophv   ophthalmic vein
Oral   oral cavity
Orb   orbital cortex
Pal   palatine bone
Par1   parietal cortex, area 1
Pig   pigment epithelium of retina
Pir   piriform cortex
psa   post sup alveolar artery
PSph   presphenoid bone
PSphW   presphenoid wing
RF   rhinal fissure
RGn   retinal ganglion cell layer
SOb   superior oblique muscle
spa   sphenopalatine artery
SphPal   sphenopalatine ganglion
SRec   superior rectus muscle
sss   superior sagittal sinus
stempv   superficial temporal vein
SubV   subventricular cortical layer
Temp   temporal muscle
TT   tenia tecta
Tu   olfactory tubercule
v   vein
Vent   ventricular zone of the retina
VP   ventral pallidum
Zyg   zygomatic arch

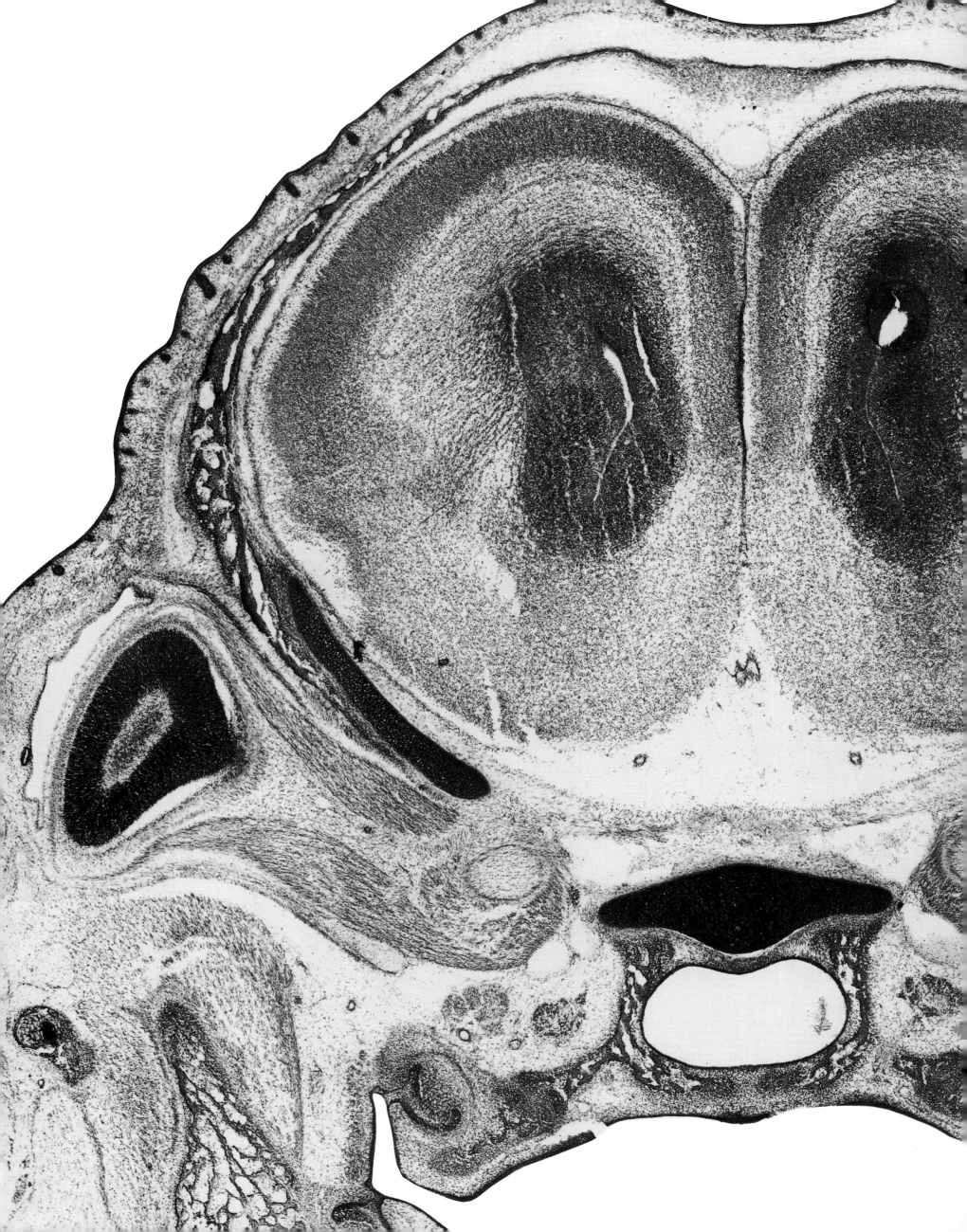

Figure 87
E19 Coronal 7

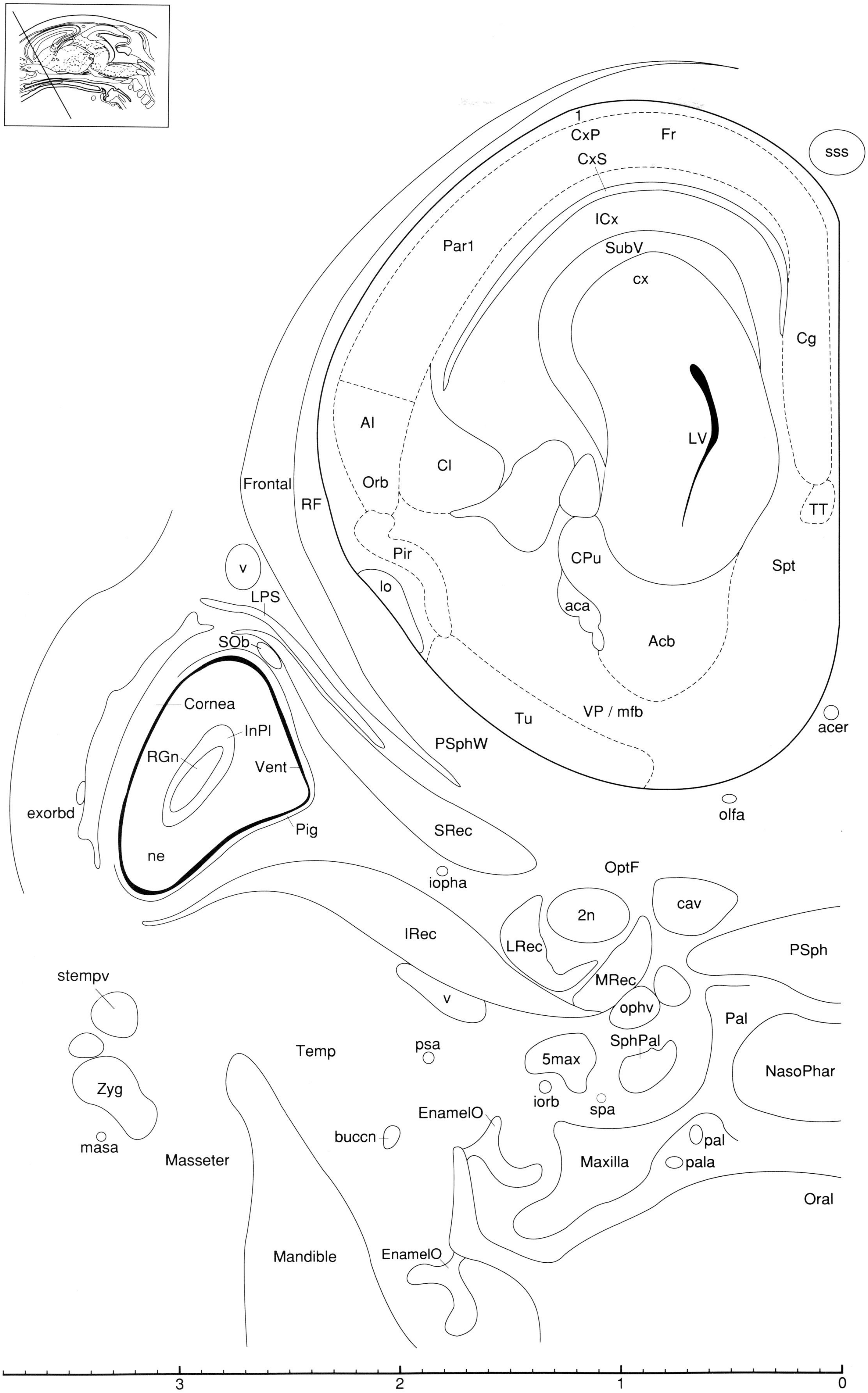

1   cortical layer 1
2n   optic nerve
5max   maxillary nerve
aca   anterior commissure, anterior
Acb   accumbens nu
acer   anterior cerebral artery
AI   agranular insular Cx
buccn   buccal nerve
cav   cavernous sinus
Cg   cingulate cortex
Cl   claustrum
Cornea   cornea
CPu   caudate putamen
cx   cortical neuroepithelium
CxP   cortical plate
CxS   cortical subplate
EnamelO   enamel organ (of tooth)
exorbd   duct of exorbital gland
Fr   frontal cortex
Frontal   frontal bone
ICx   intermediate cortical layer
InPl   inner plexiform layer of retina
iopha   inferior ophthalmic artery
iorb   inferior orbital artery
IRec   inferior rectus muscle
lo   lateral olfactory tract
LPS   levator palpebrae superioris mus
LRec   lateral rectus muscle
LV   lateral ventricle
Mandible   mandible
masa   masseteric artery
Masseter   masseter muscle
Maxilla   maxilla
mfb   medial forebrain bundle
MRec   medial rectus muscle
NasoPhar   nasopharyngeal cavity
ne   neuroepithelium
olfa   olfactory artery
ophv   ophthalmic vein
OptF   optic foramen
Oral   oral cavity
Orb   orbital cortex
Pal   palatine bone
pal   palatine nerve
pala   palatine artery
Par1   parietal cortex, area 1
Pig   pigment epithelium of retina
Pir   piriform cortex
psa   post sup alveolar artery
PSph   presphenoid bone
PSphW   presphenoid wing
RF   rhinal fissure
RGn   retinal ganglion cell layer
SOb   superior oblique muscle
spa   sphenopalatine artery
SphPal   sphenopalatine ganglion
Spt   septum
SRec   superior rectus muscle
sss   superior sagittal sinus
stempv   superficial temporal vein
SubV   subventricular cortical layer
Temp   temporal muscle
TT   tenia tecta
Tu   olfactory tubercule
v   vein
Vent   ventricular zone of the retina
VP   ventral pallidum
Zyg   zygomatic arch

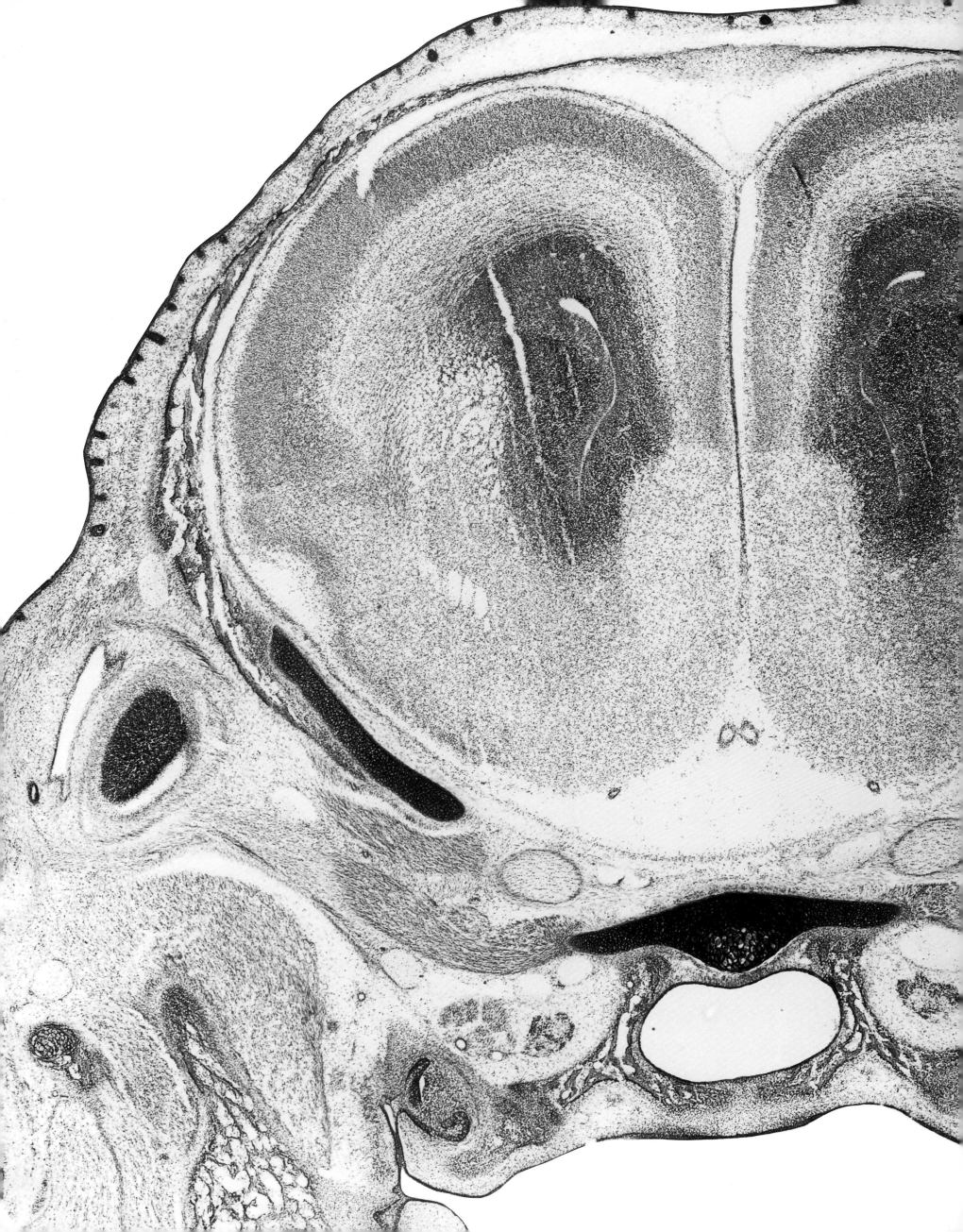

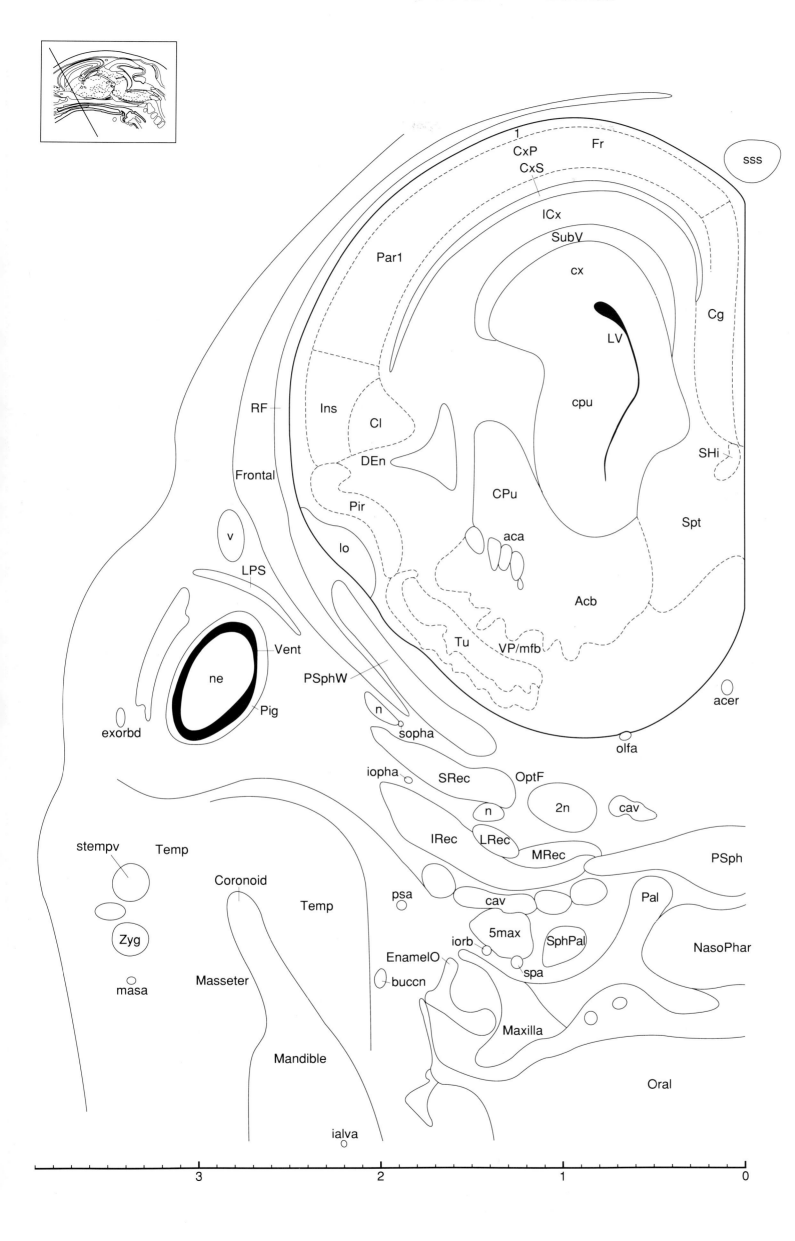

Figure 88
E19 Coronal 8

1 cortical layer 1
2n optic nerve
5max maxillary nerve
aca anterior commissure, anterior
Acb accumbens nu
acer anterior cerebral artery
buccn buccal nerve
cav cavernous sinus
Cg cingulate cortex
Cl claustrum
Coronoid coronoid process, mandible
CPu caudate putamen
cpu caudate putamen neuroepithelium
cx cortical neuroepithelium
CxP cortical plate
CxS cortical subplate
DEn dorsal endopiriform nu
EnamelO enamel organ (of tooth)
exorbd duct of exorbital gland
Fr frontal cortex
Frontal frontal bone
ialva inferior alveolar artery
ICx intermediate cortical layer
Ins insular cortex
iopha inferior ophthalmic artery
iorb inferior orbital artery
IRec inferior rectus muscle
lo lateral olfactory tract
LPS levator palpebrae superioris mus
LRec lateral rectus muscle
LV lateral ventricle
Mandible mandible
masa masseteric artery
Masseter masseter muscle
Maxilla maxilla
mfb medial forebrain bundle
MRec medial rectus muscle
n nerve
NasoPhar nasopharyngeal cavity
ne neuroepithelium
olfa olfactory artery
OptF optic foramen
Oral oral cavity
Pal palatine bone
Par1 parietal cortex, area 1
Pig pigment epithelium of retina
Pir piriform cortex
psa post sup alveolar artery
PSph presphenoid bone
PSphW presphenoid wing
RF rhinal fissure
SHi septohippocampal nu
sopha superior ophthalmic artery
spa sphenopalatine artery
SphPal sphenopalatine ganglion
Spt septum
SRec superior rectus muscle
sss superior sagittal sinus
stempv superficial temporal vein
SubV subventricular cortical layer
Temp temporal muscle
Tu olfactory tubercle
v vein
Vent ventricular zone of the retina
VP ventral pallidum
Zyg zygomatic arch

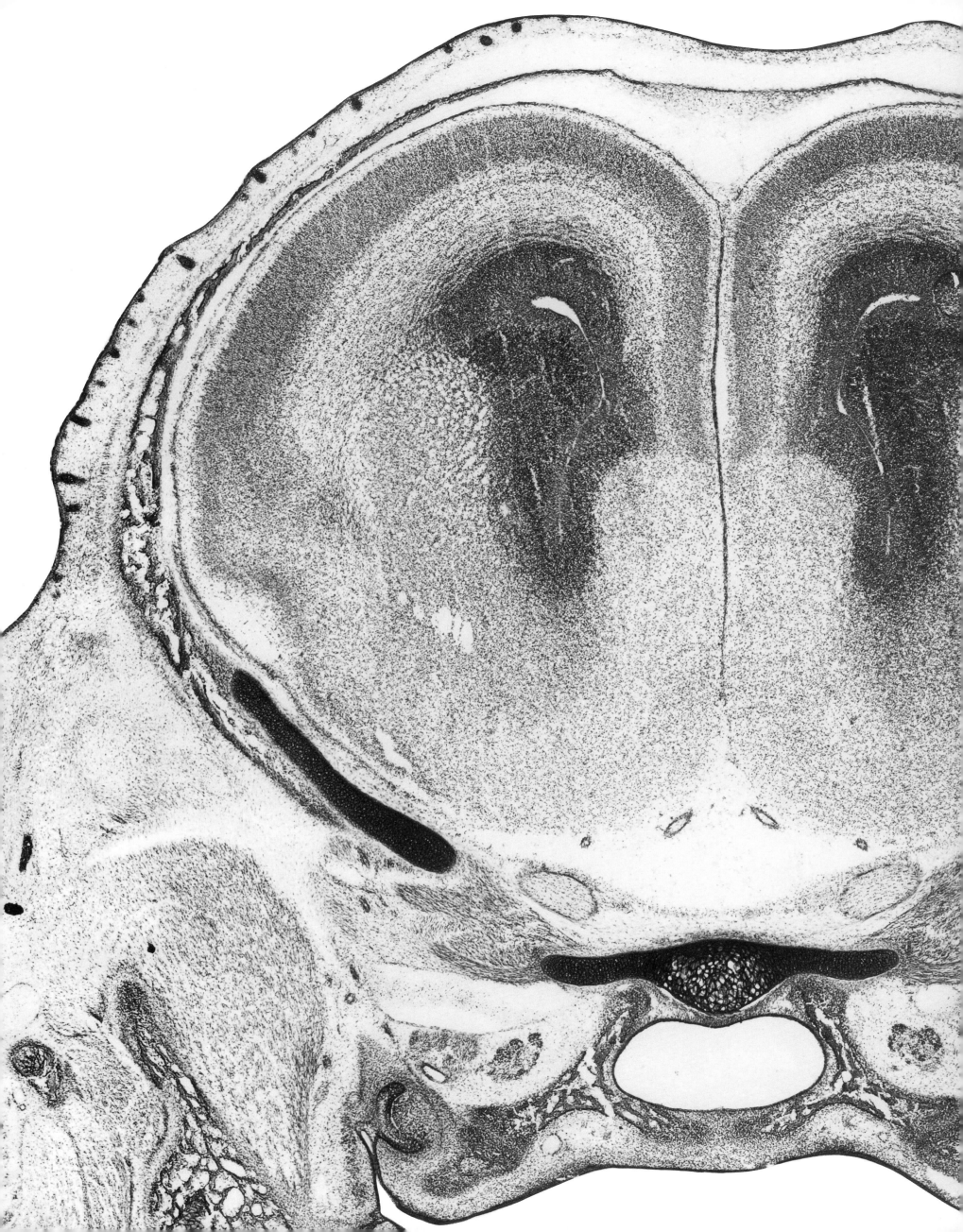

Figure 89
E19 Coronal 9

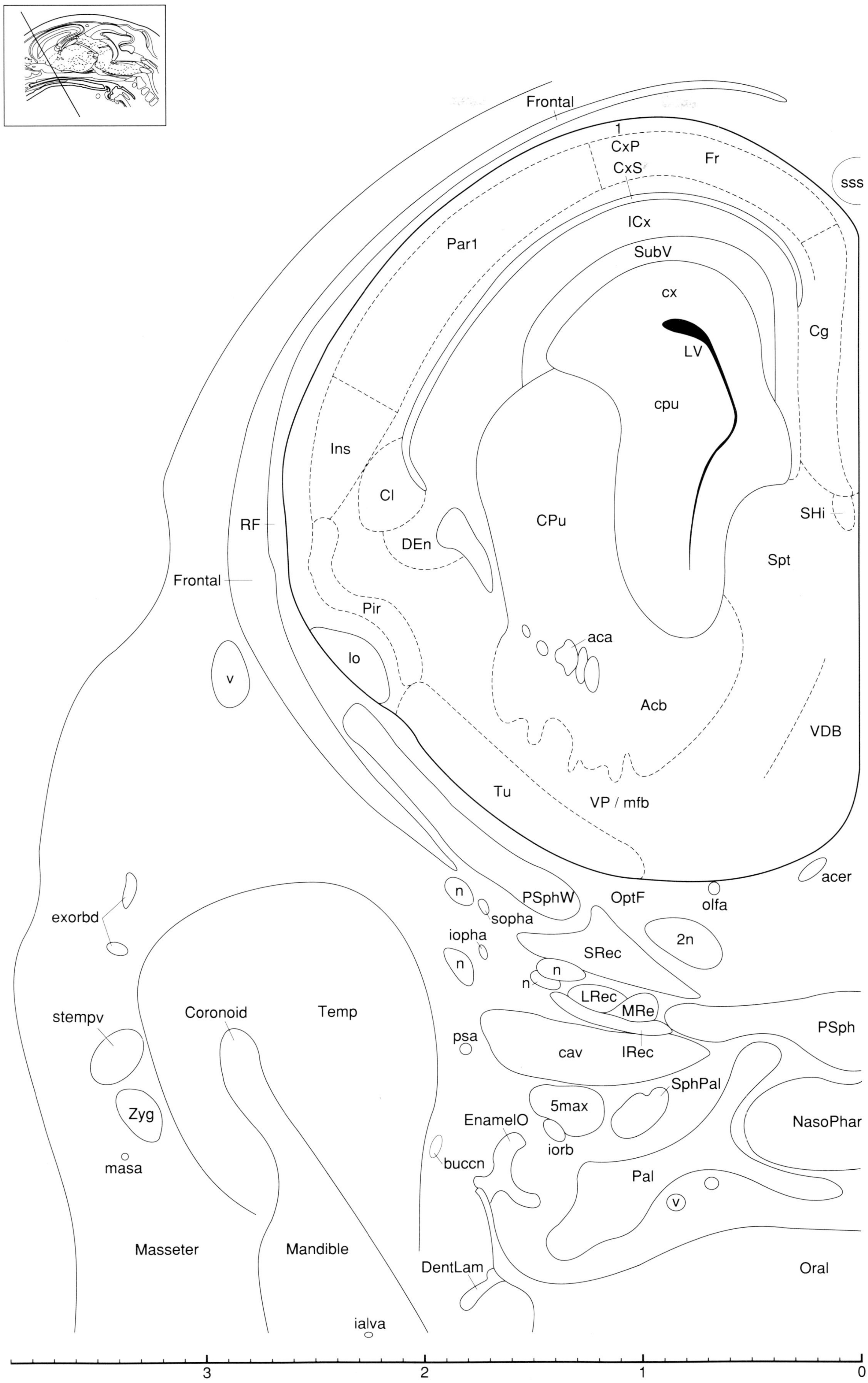

1   cortical layer 1
2n   optic nerve
5max   maxillary nerve
aca   anterior commissure, anterior
Acb   accumbens nu
acer   anterior cerebral artery
buccn   buccal nerve
cav   cavernous sinus
Cg   cingulate cortex
Cl   claustrum
Coronoid   coronoid process, mandible
CPu   caudate putamen
cpu   caudate putamen neuroepithelium
cx   cortical neuroepithelium
CxP   cortical plate
CxS   cortical subplate
DEn   dorsal endopiriform nu
DentLam   dental lamina
EnamelO   enamel organ (of tooth)
exorbd   duct of exorbital gland
Fr   frontal cortex
Frontal   frontal bone
ialva   inferior alveolar artery
ICx   intermediate cortical layer
Ins   insular cortex
iopha   inferior ophthalmic artery
iorb   inferior orbital artery
IRec   inferior rectus muscle
lo   lateral olfactory tract
LRec   lateral rectus muscle
LV   lateral ventricle
Mandible   mandible
masa   masseteric artery
Masseter   masseter muscle
mfb   medial forebrain bundle
MRe   mammillary recess 3rd ventricle
n   nerve
NasoPhar   nasopharyngeal cavity
olfa   olfactory artery
OptF   optic foramen
Oral   oral cavity
Pal   palatine bone
Par1   parietal cortex, area 1
Pir   piriform cortex
psa   post sup alveolar artery
PSph   presphenoid bone
PSphW   presphenoid wing
RF   rhinal fissure
SHi   septohippocampal nu
sopha   superior ophthalmic artery
SphPal   sphenopalatine ganglion
Spt   septum
SRec   superior rectus muscle
sss   superior sagittal sinus
stempv   superficial temporal vein
SubV   subventricular cortical layer
Temp   temporal muscle
Tu   olfactory tubercule
v   vein
VDB   nu vertical limb diagonal band
VP   ventral pallidum
Zyg   zygomatic arch

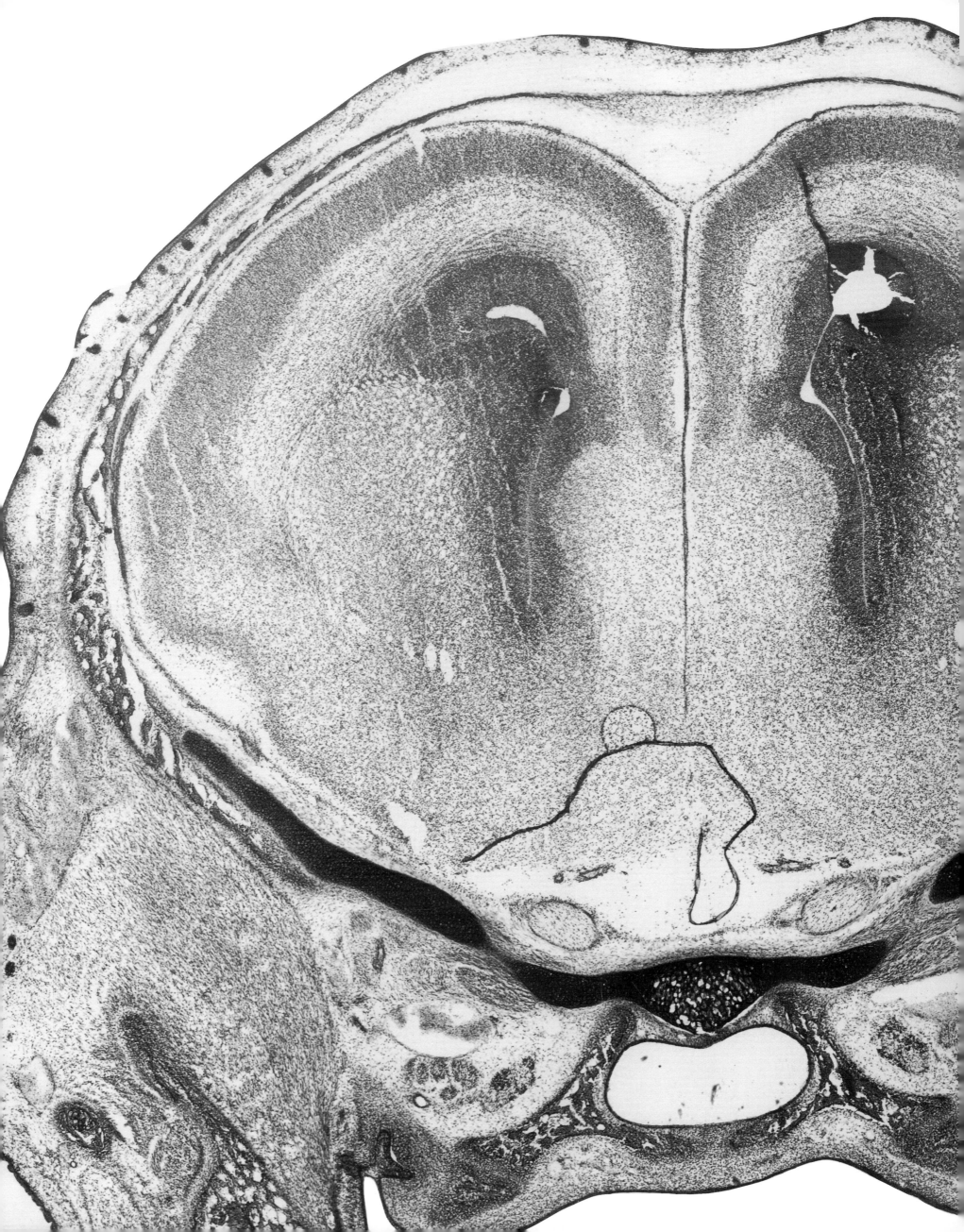

Figure 90
E19 Coronal 10

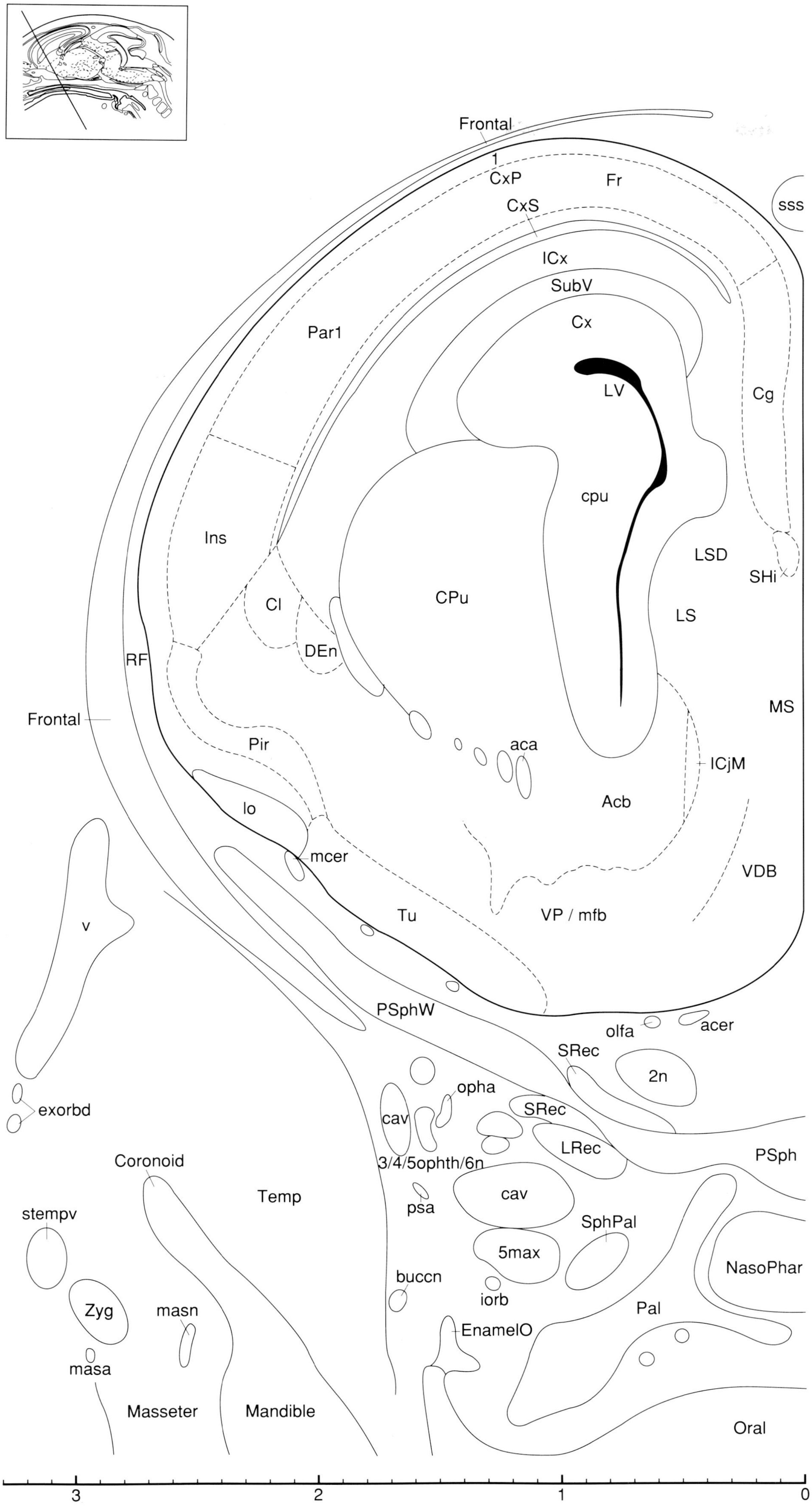

Frontal
1
CxP   Fr
CxS
ICx
SubV
Cx
sss
Par1
LV
Cg
cpu
ICx
SubV
Cx
LSD
SHi
Ins
LS
Cl
CPu
DEn
MS
RF
Frontal
Pir
aca
ICjM
lo
Acb
mcer
VDB
Tu   VP / mfb
PSphW
v
olfa   acer
SRec
exorbd
opha   2n
cav   SRec
Coronoid   LRec
stempv   Temp   psa   cav   PSph
SphPal
5max
buccn
Zyg   masn   iorb   NasoPhar
Pal
masa   EnamelO
Masseter   Mandible   Oral

1   cortical layer 1
2n   optic nerve
3n   oculomotor nerve or its root
4n   trochlear nerve or its root
5max   maxillary nerve
5ophth   ophthalmic n of trigem
6n   abducens nerve or its root
aca   anterior commissure, anterior
Acb   accumbens nu
acer   anterior cerebral artery
buccn   buccal nerve
cav   cavernous sinus
Cg   cingulate cortex
Cl   claustrum
Coronoid   coronoid process, mandible
CPu   caudate putamen
cpu   caudate putamen neuroepithelium
Cx   cerebral cortex
CxP   cortical plate
CxS   cortical subplate
DEn   dorsal endopiriform nu
EnamelO   enamel organ (of tooth)
exorbd   duct of exorbital gland
Fr   frontal cortex
Frontal   frontal bone
ICjM   islands of Calleja, major island
ICx   intermediate cortical layer
Ins   insular cortex
iorb   inferior orbital artery
lo   lateral olfactory tract
LRec   lateral rectus muscle
LS   lateral septal nu
LSD   lateral septal nu, dorsal
LV   lateral ventricle
Mandible   mandible
masa   masseteric artery
masn   masseteric nerve
Masseter   masseter muscle
mcer   middle cerebral artery
mfb   medial forebrain bundle
MS   medial septal nu
NasoPhar   nasopharyngeal cavity
olfa   olfactory artery
opha   ophthalmic artery
Oral   oral cavity
Pal   palatine bone
Par1   parietal cortex, area 1
Pir   piriform cortex
psa   post sup alveolar artery
PSph   presphenoid bone
PSphW   presphenoid wing
RF   rhinal fissure
SHi   septohippocampal nu
SphPal   sphenopalatine ganglion
SRec   superior rectus muscle
sss   superior sagittal sinus
stempv   superficial temporal vein
SubV   subventricular cortical layer
Temp   temporal muscle
Tu   olfactory tubercule
v   vein
VDB   nu vertical limb diagonal band
VP   ventral pallidum
Zyg   zygomatic arch

3   2   1   0

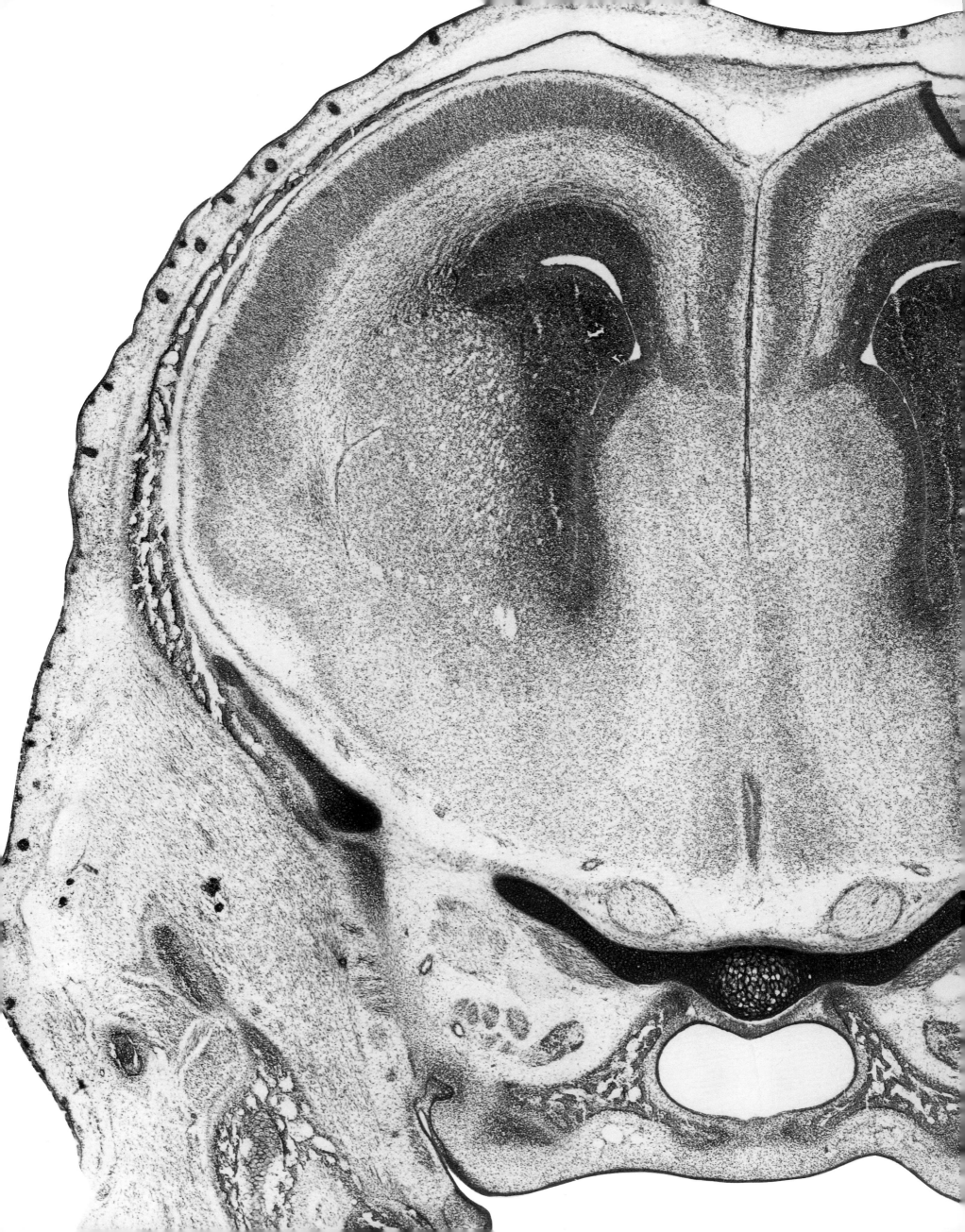

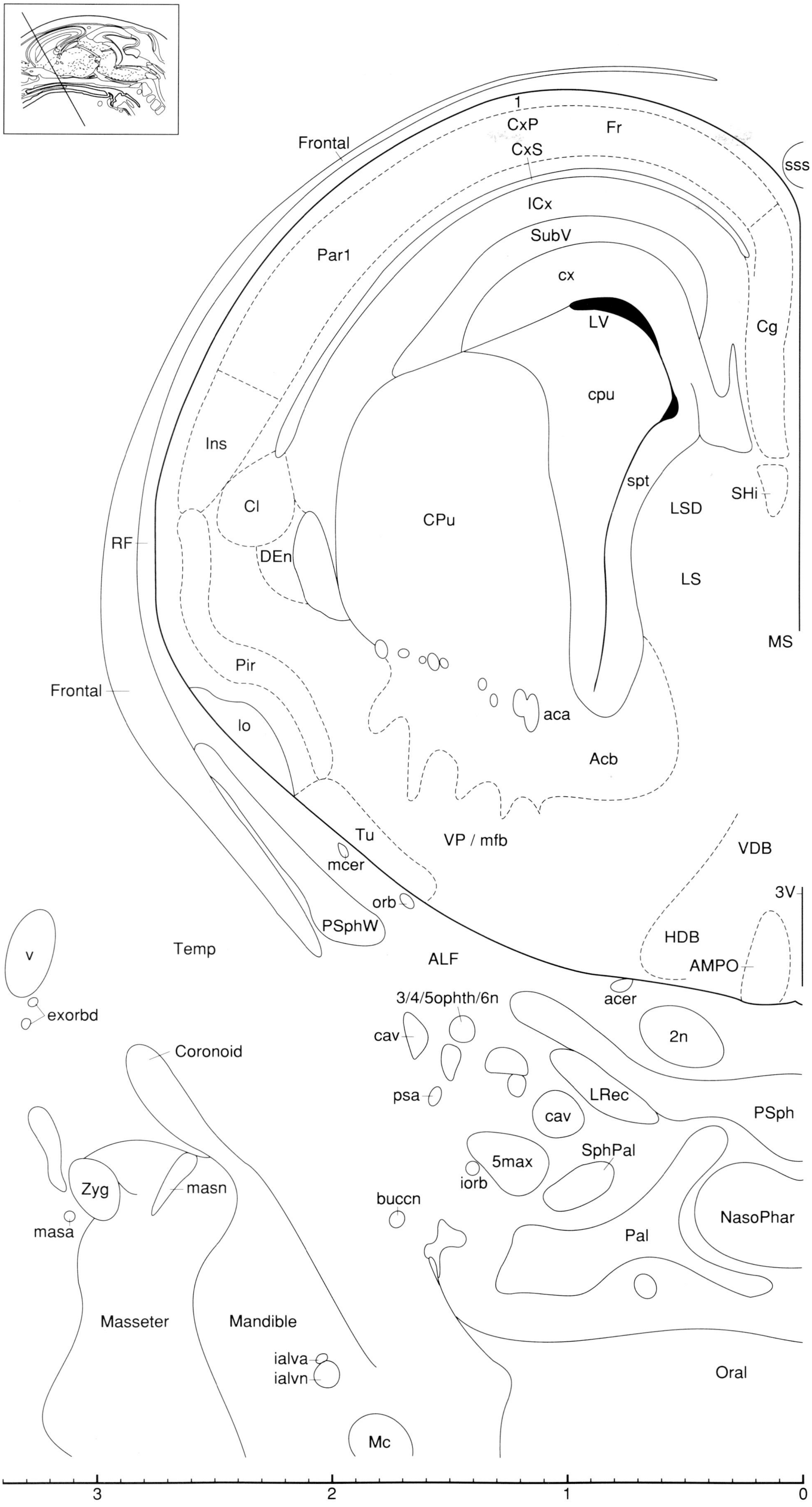

Figure 91
E19 Coronal 11

1 cortical layer 1
2n optic nerve
3n oculomotor nerve or its root
3V 3rd ventricle
5max maxillary nerve
4n trochlear nerve or its root
5ophth ophthalmic n of trigem
6n abducens nerve or its root
aca anterior commissure, anterior
Acb accumbens nu
acer anterior cerebral artery
ALF anterior lacerated foramen
AMPO anterior medial preoptic nu
buccn buccal nerve
cav cavernous sinus
Cg cingulate cortex
Cl claustrum
Coronoid coronoid process, mandible
CPu caudate putamen
cpu caudate putamen neuroepithelium
cx cortical neuroepithelium
CxP cortical plate
CxS cortical subplate
DEn dorsal endopiriform nu
exorbd duct of exorbital gland
Fr frontal cortex
Frontal frontal bone
HDB nu horiz limb diagonal band
ialva inferior alveolar artery
ialvn inferior alveolar nerve
ICx intermediate cortical layer
Ins insular cortex
iorb inferior orbital artery
lo lateral olfactory tract
LRec lateral rectus muscle
LS lateral septal nu
LSD lateral septal nu, dorsal
LV lateral ventricle
Mandible mandible
masa masseteric artery
masn masseteric nerve
Masseter masseter muscle
Mc Meckel's cartilage
mcer middle cerebral artery
mfb medial forebrain bundle
MS medial septal nu
NasoPhar nasopharyngeal cavity
Oral oral cavity
orb orbital artery
Pal palatine bone
Par1 parietal cortex, area 1
Pir piriform cortex
psa post sup alveolar artery
PSph presphenoid bone
PSphW presphenoid wing
RF rhinal fissure
SHi septohippocampal nu
SphPal sphenopalatine ganglion
spt septal neuroepithelium
sss superior sagittal sinus
SubV subventricular cortical layer
Temp temporal muscle
Tu olfactory tubercle
v vein
VDB nu vertical limb diagonal band
VP ventral pallidum
Zyg zygomatic arch

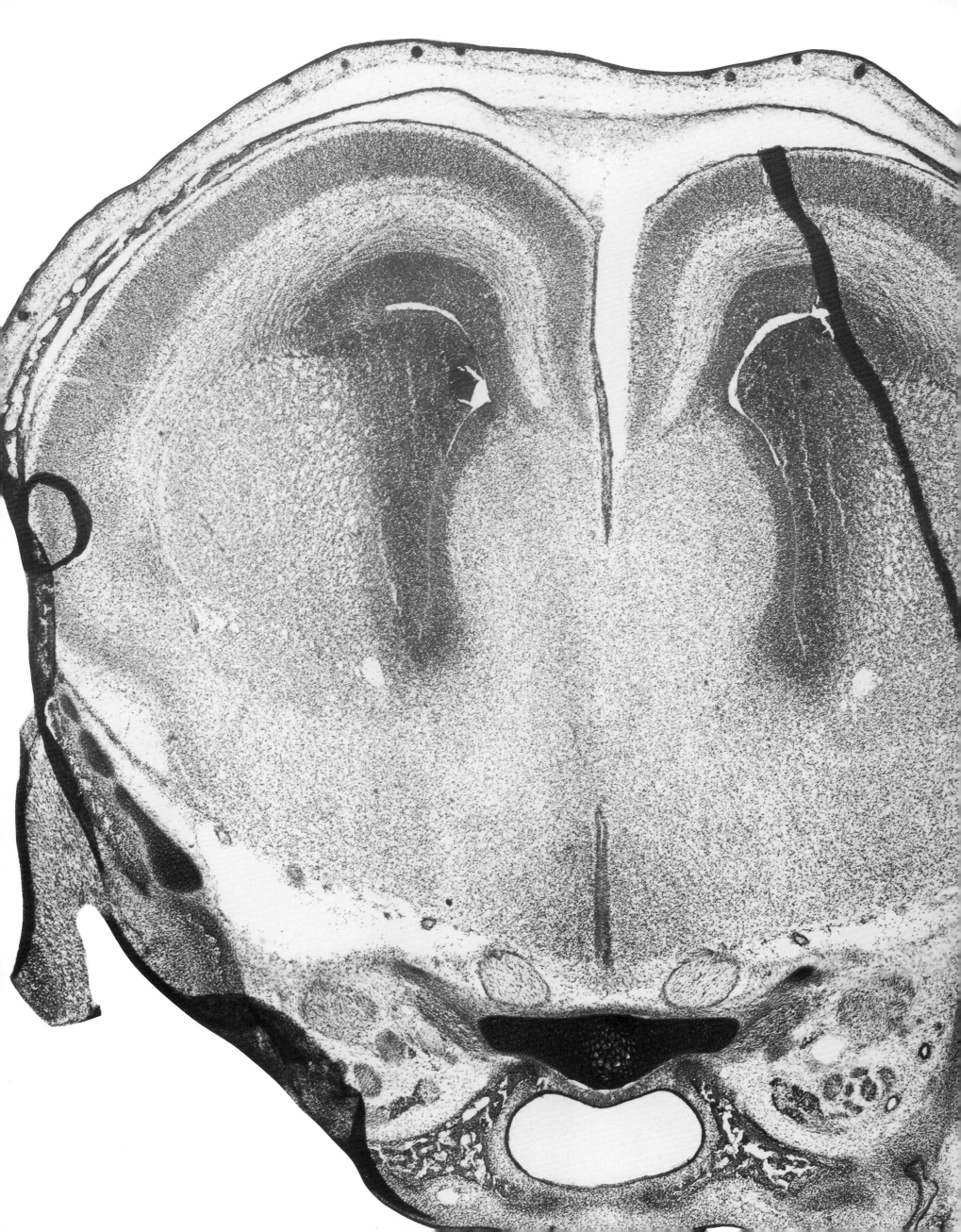

Figure 92
E19 Coronal 12

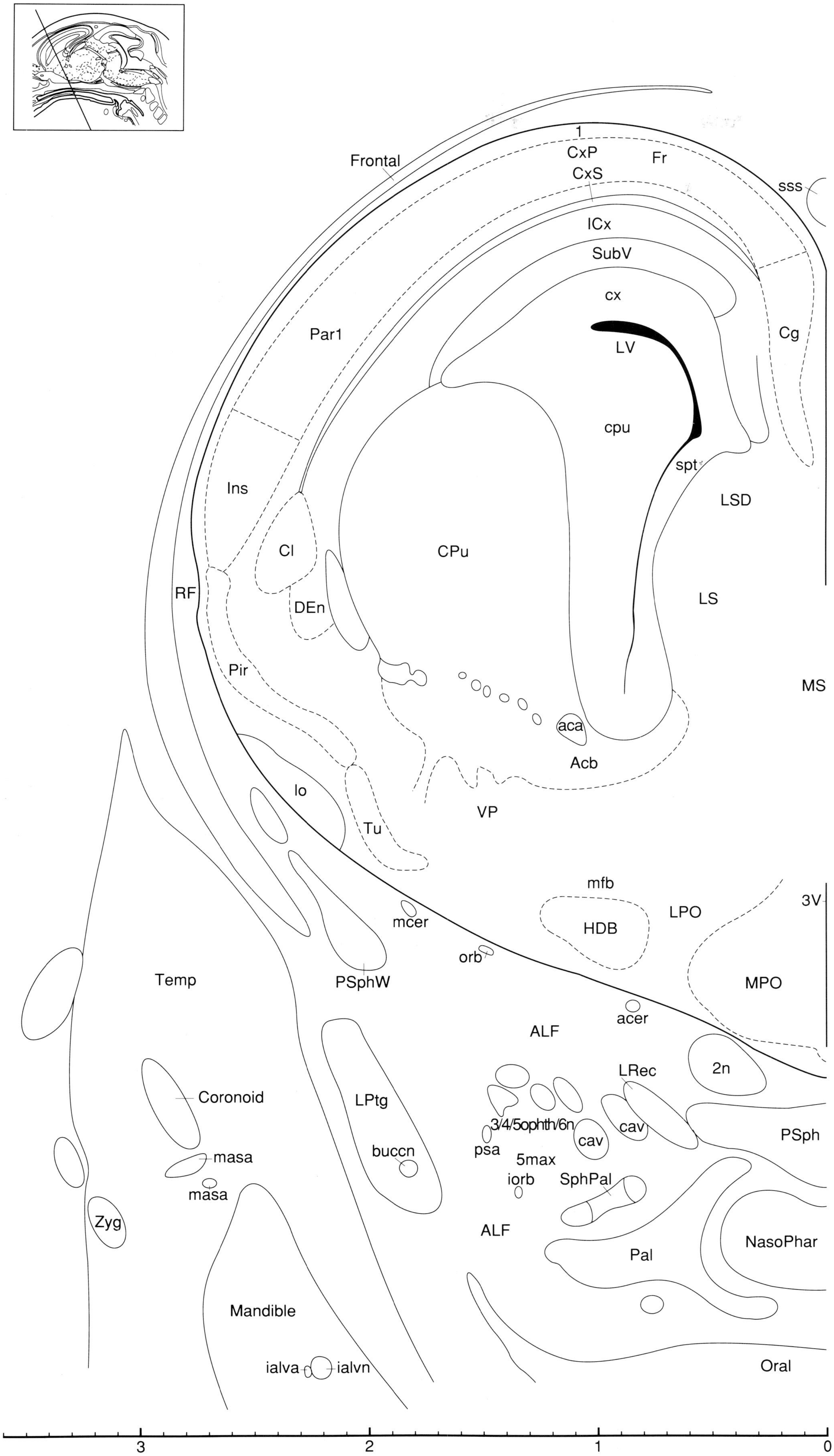

1 cortical layer 1
2n optic nerve
3n oculomotor nerve or its root
3V 3rd ventricle
4n trochlear nerve or its root
5max maxillary nerve
5ophth ophthalmic n of trigem
6n abducens nerve or its root
aca anterior commissure, anterior
Acb accumbens nu
acer anterior cerebral artery
ALF anterior lacerated foramen
buccn buccal nerve
cav cavernous sinus
Cg cingulate cortex
Cl claustrum
Coronoid coronoid process, mandible
CPu caudate putamen
cpu caudate putamen neuroepithelium
cx cortical neuroepithelium
CxP cortical plate
CxS cortical subplate
DEn dorsal endopiriform nu
Fr frontal cortex
Frontal frontal bone
HDB nu horiz limb diagonal band
ialva inferior alveolar artery
ialvn inferior alveolar nerve
ICx intermediate cortical layer
Ins insular cortex
iorb inferior orbital artery
lo lateral olfactory tract
LPO lateral preoptic area
LPtg lateral pterygoid
LRec lateral rectus muscle
LS lateral septal nu
LSD lateral septal nu, dorsal
LV lateral ventricle
Mandible mandible
masa masseteric artery
mcer middle cerebral artery
mfb medial forebrain bundle
MPO medial preoptic nu
MS medial septal nu
NasoPhar nasopharyngeal cavity
Oral oral cavity
orb orbital artery
Pal palatine bone
Par1 parietal cortex, area 1
Pir piriform cortex
psa post sup alveolar artery
PSph presphenoid bone
PSphW presphenoid wing
RF rhinal fissure
SphPal sphenopalatine ganglion
spt septal neuroepithelium
sss superior sagittal sinus
SubV subventricular cortical layer
Temp temporal muscle
Tu olfactory tubercule
VP ventral pallidum
Zyg zygomatic arch

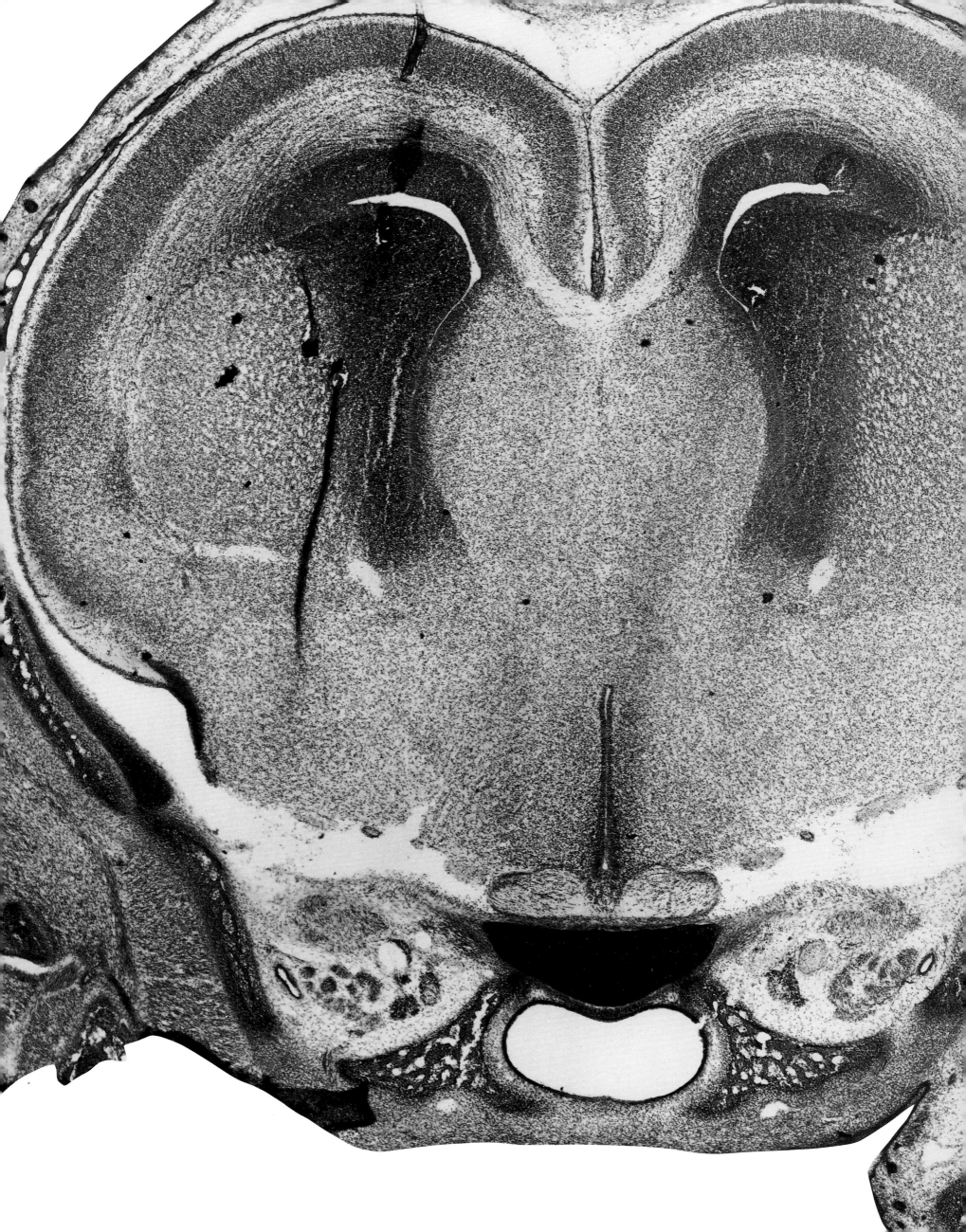

Figure 93
E19 Coronal 13

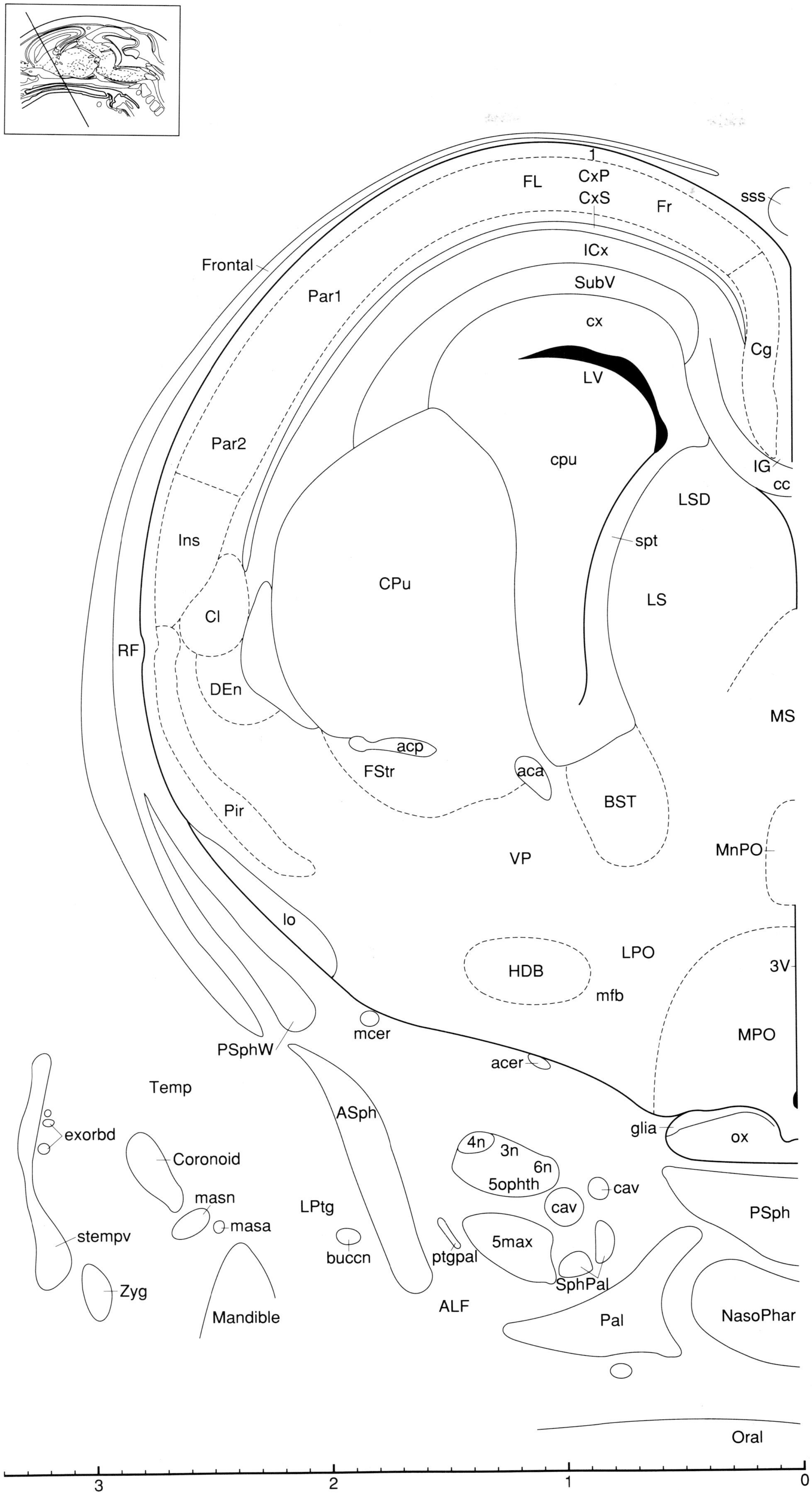

1 cortical layer 1
3n oculomotor nerve or its root
3V 3rd ventricle
4n trochlear nerve or its root
5max maxillary nerve
5ophth ophthalmic n of trigem
6n abducens nerve or its root
aca anterior commissure, anterior
acer anterior cerebral artery
acp anterior commissure, posterior
ALF anterior lacerated foramen
ASph alisphenoid bone
BST bed nu stria terminalis
buccn buccal nerve
cav cavernous sinus
cc corpus callosum
Cg cingulate cortex
Cl claustrum
Coronoid coronoid process, mandible
CPu caudate putamen
cpu caudate putamen neuroepithelium
cx cortical neuroepithelium
CxP cortical plate
CxS cortical subplate
DEn dorsal endopiriform nu
exorbd duct of exorbital gland
FL forelimb area of cortex
Fr frontal cortex
Frontal frontal bone
FStr fundus striati
glia glia
HDB nu horiz limb diagonal band
ICx intermediate cortical layer
IG indusium griseum
Ins insular cortex
lo lateral olfactory tract
LPO lateral preoptic area
LPtg lateral pterygoid
LS lateral septal nu
LSD lateral septal nu, dorsal
LV lateral ventricle
Mandible mandible
masa masseteric artery
masn masseteric nerve
mcer middle cerebral artery
mfb medial forebrain bundle
MnPO median preoptic nu
MPO medial preoptic nu
MS medial septal nu
NasoPhar nasopharyngeal cavity
Oral oral cavity
ox optic chiasm
Pal palatine bone
Par1 parietal cortex, area 1
Par2 parietal cortex, area 2
Pir piriform cortex
PSph presphenoid bone
PSphW presphenoid wing
ptgpal pterygopalatine artery
RF rhinal fissure
SphPal sphenopalatine ganglion
spt septal neuroepithelium
sss superior sagittal sinus
stempv superficial temporal vein
SubV subventricular cortical layer
Temp temporal muscle
VP ventral pallidum
Zyg zygomatic arch

3   2   1   0

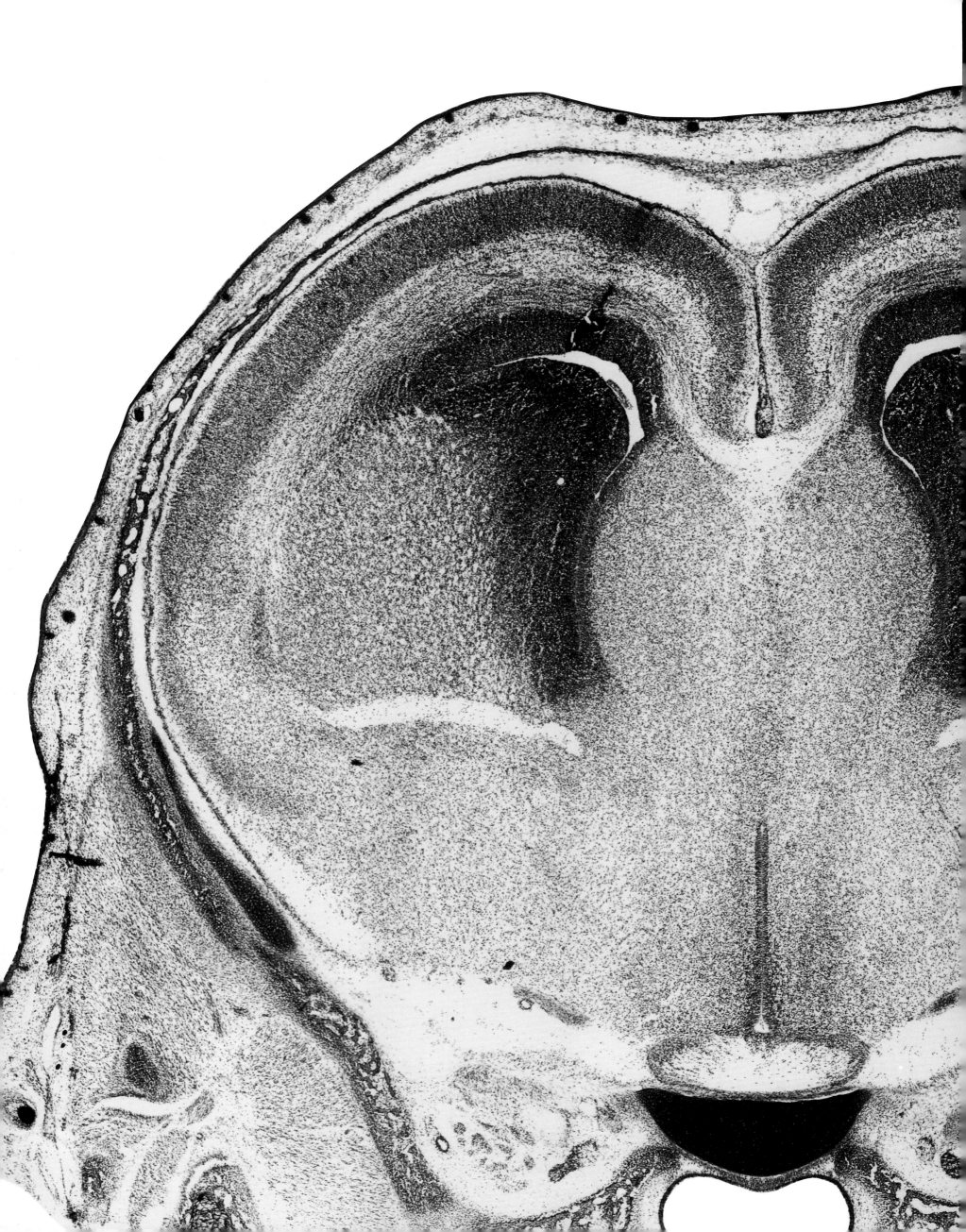

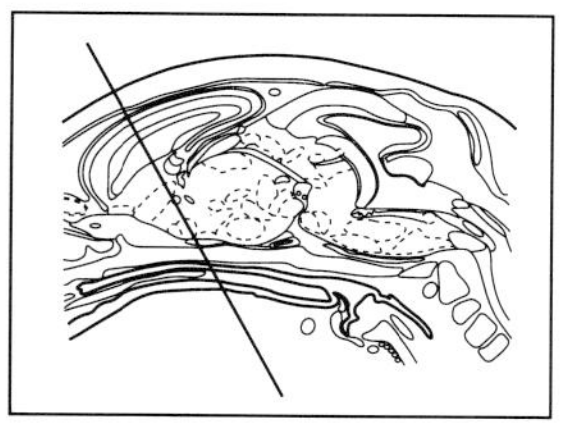

Figure 94
E19 Coronal 14

1   cortical layer 1
3n   oculomotor nerve or its root
3V   3rd ventricle
4n   trochlear nerve or its root
5max   maxillary nerve
5ophth   ophthalmic n of trigem
6n   abducens nerve or its root
AA   anterior amygdaloid area
aca   anterior commissure, anterior
acer   anterior cerebral artery
acp   anterior commissure, posterior
ALF   anterior lacerated foramen
ASph   alisphenoid bone
B   basal nu Meynert
BST   bed nu stria terminalis
buccn   buccal nerve
cav   cavernous sinus
cc   corpus callosum
Cg   cingulate cortex
Cl   claustrum
Coronoid   coronoid process, mandible
CPu   caudate putamen
cpu   caudate putamen neuroepithelium
cx   cortical neuroepithelium
CxP   cortical plate
CxS   cortical subplate
DEn   dorsal endopiriform nu
dtn   deep temporal nerve
exorbd   duct of exorbital gland
FL   forelimb area of cortex
Fr   frontal cortex
Frontal   frontal bone
FStr   fundus striati
glia   glia
HDB   nu horiz limb diagonal band
ICx   intermediate cortical layer
IG   indusium griseum
Ins   insular cortex
lo   lateral olfactory tract
LPO   lateral preoptic area
LPtg   lateral pterygoid
LS   lateral septal nu
LSD   lateral septal nu, dorsal
LV   lateral ventricle
Mandible   mandible
masa   masseteric artery
masn   masseteric nerve
Masseter   masseter muscle
mcer   middle cerebral artery
mfb   medial forebrain bundle
MnPO   median preoptic nu
MPO   medial preoptic nu
MS   medial septal nu
NasoPhar   nasopharyngeal cavity
ox   optic chiasm
Par1   parietal cortex, area 1
Par2   parietal cortex, area 2
Pir   piriform cortex
pof   postero-orbital follicle
PSph   presphenoid bone
PSphW   presphenoid wing
ptgcn   nerve of the pterygoid canal
ptgpal   pterygopalatine artery
RF   rhinal fissure
SphPal   sphenopalatine ganglion
spt   septal neuroepithelium
sss   superior sagittal sinus
SubV   subventricular cortical layer
Temp   temporal muscle
v   vein
VP   ventral pallidum
Zyg   zygomatic arch

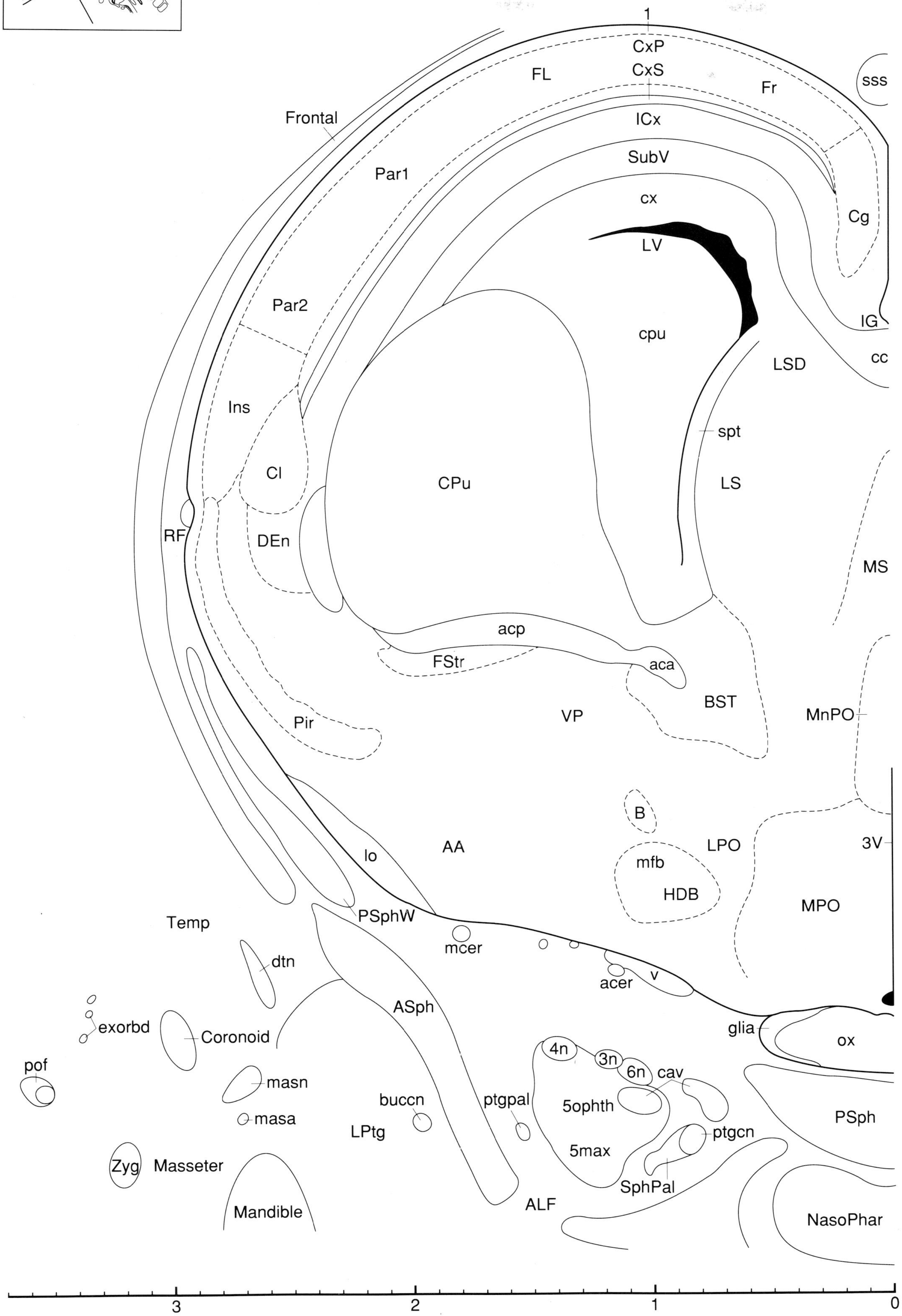

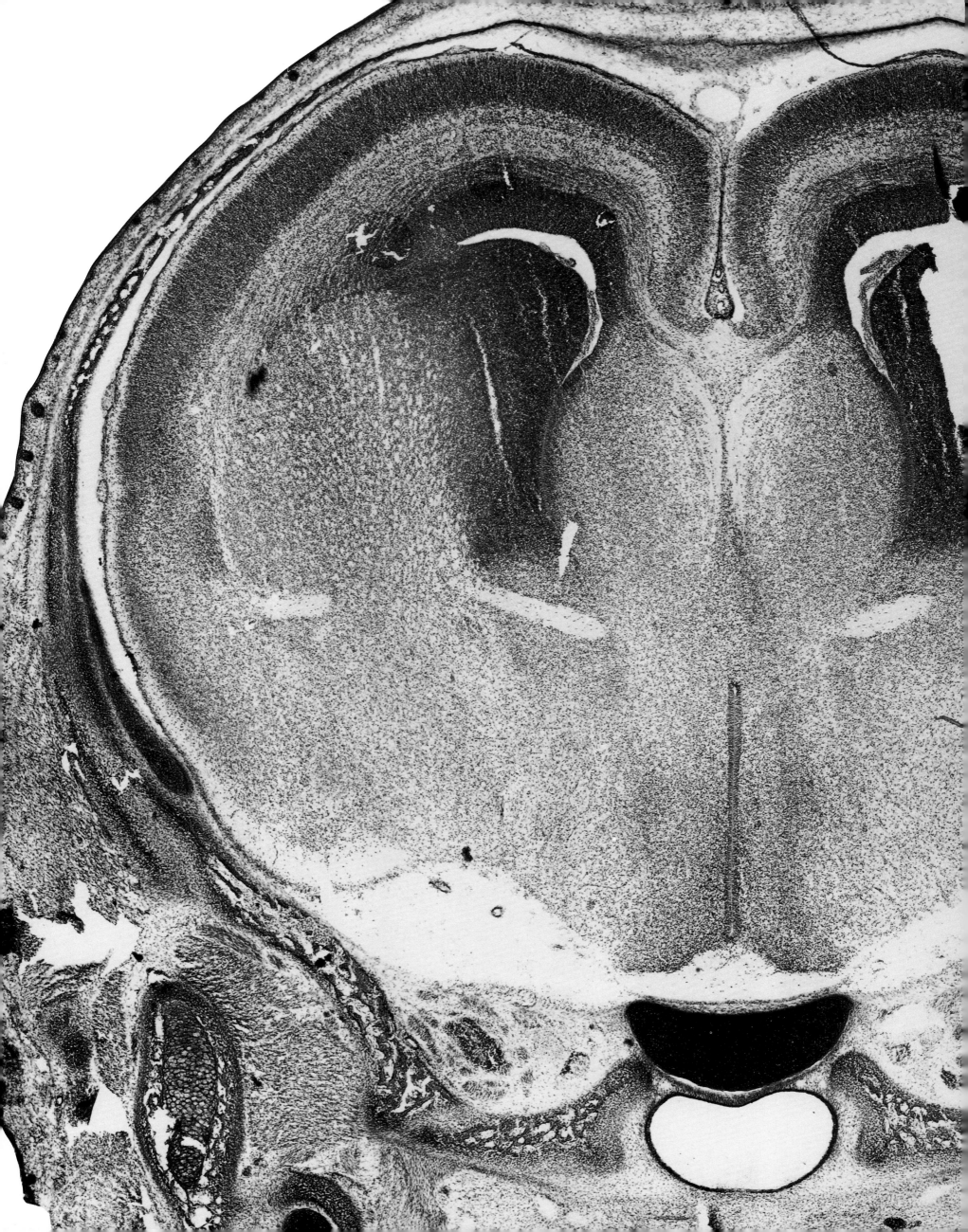

Figure 95
E19 Coronal 15

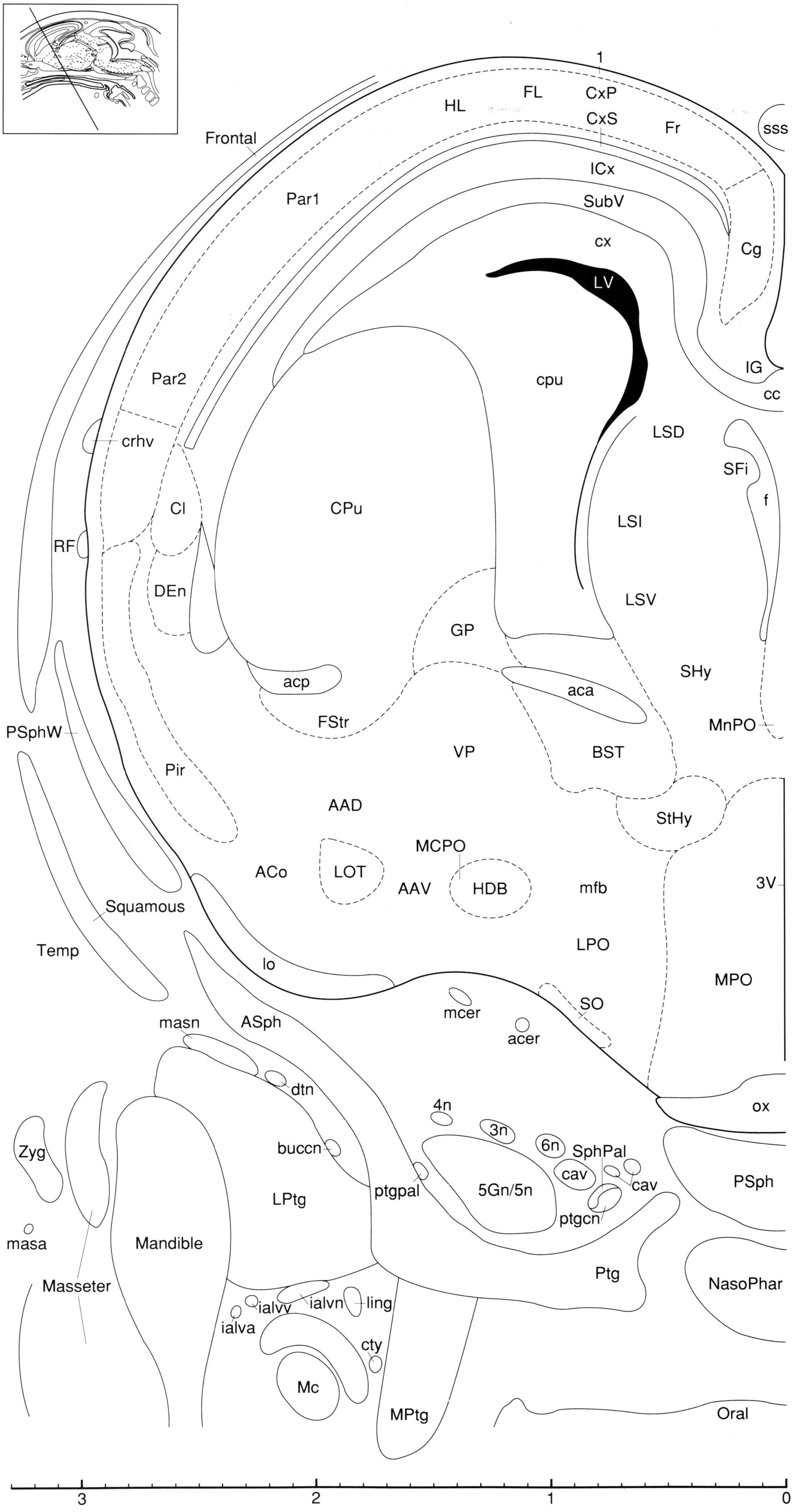

1 cortical layer 1
3n oculomotor nerve or its root
3V 3rd ventricle
4n trochlear nerve or its root
5Gn trigeminal ganlion
5n trigeminal nerve
6n abducens nerve or its root
AAD anterior amygdaloid area, dorsal
AAV anterior amygdaloid area, ventral
aca anterior commissure, anterior
acer anterior cerebral artery
ACo anterior cortical amygdaloid nu
acp anterior commissure, posterior
ASph alisphenoid bone
BST bed nu stria terminalis
buccn buccal nerve
cav cavernous sinus
cc corpus callosum
Cg cingulate cortex
Cl claustrum
CPu caudate putamen
cpu caudate putamen neuroepithelium
crhv caudal rhinal vein
cty chorda tympani nerve
cx cortical neuroepithelium
CxP cortical plate
CxS cortical subplate
DEn dorsal endopiriform nu
dtn deep temporal nerve
f fornix
FL forelimb area of cortex
Fr frontal cortex
Frontal frontal bone
FStr fundus striati
GP globus pallidus
HDB nu horiz limb diagonal band
HL hindlimb area of cortex
ialva inferior alveolar artery
ialvn inferior alveolar nerve
ialvv inferior alveolar vein
ICx intermediate cortical layer
IG indusium griseum
ling lingual nerve
lo lateral olfactory tract
LOT nu lateral olfactory tract
LPO lateral preoptic area
LPtg lateral pterygoid
LSD lateral septal nu, dorsal
LSI lateral septal nu, intermediate
LSV lateral septal nu, ventral
LV lateral ventricle
Mandible mandible
masa masseteric artery
masn masseteric nerve
Masseter masseter muscle
Mc Meckel's cartilage
mcer middle cerebral artery
MCPO magnocellular preoptic nu
mfb medial forebrain bundle
MnPO median preoptic nu
MPO medial preoptic nu
MPtg medial pterygoid muscle
NasoPhar nasopharyngeal cavity
Oral oral cavity
ox optic chiasm
Par1 parietal cortex, area 1
Par2 parietal cortex, area 2
Pir piriform cortex
PSph presphenoid bone
PSphW presphenoid wing
Ptg pterygoid bone
ptgcn nerve of the pterygoid canal
ptgpal pterygopalatine artery
RF rhinal fissure
SFi septofimbrial nu
SHy septohypothalamic nu
SO supraoptic nu
SphPal sphenopalatine ganglion
Squamous squamous part of temp
sss superior sagittal sinus
StHy striohypothalamic nu
SubV subventricular cortical layer
Temp temporal muscle
VP ventral pallidum
Zyg zygomatic arch

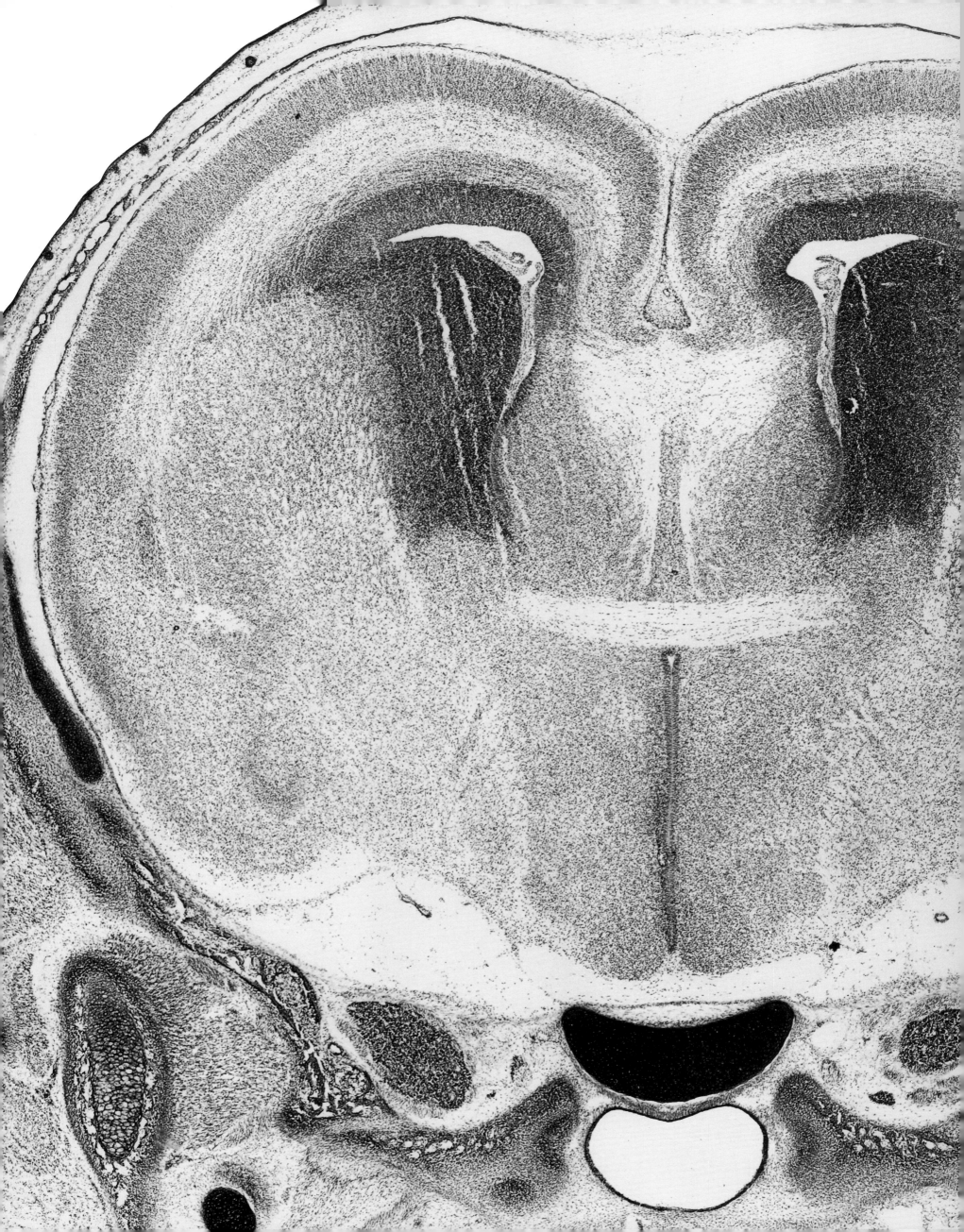

Figure 96
E19 Coronal 16

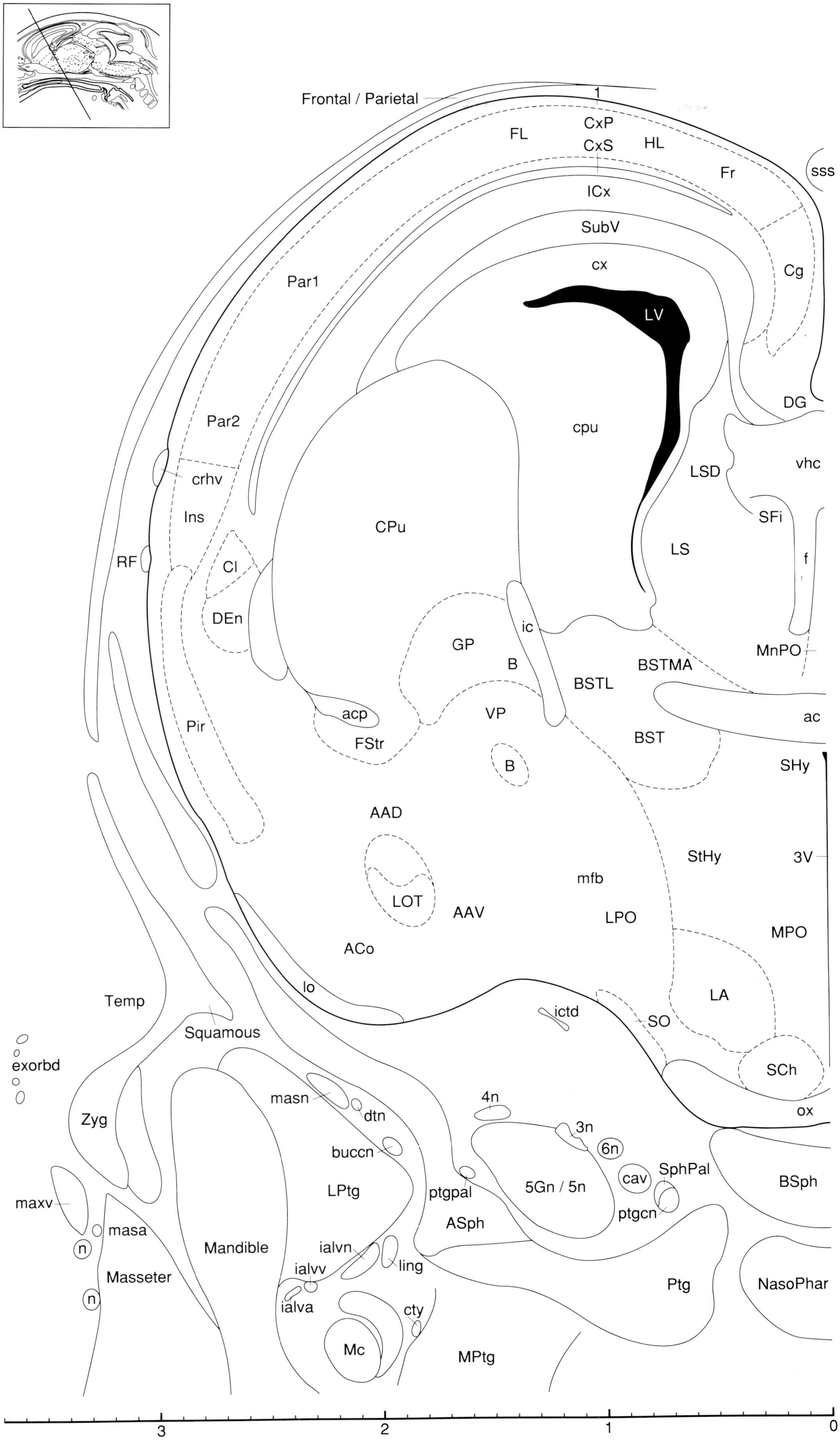

Frontal / Parietal

1

FL CxP HL
CxS
Fr
sss
ICx
SubV
Cg
cx
Par1
LV
DG
Par2
cpu
vhc
crhv
LSD
SFi
Ins
CPu
LS
f
RF
Cl
DEn
ic
MnPO
GP
BSTMA
B
BSTL
Pir
acp
VP
ac
FStr
BST
B
SHy
LA
AAD
StHy
3V
mfb
LOT
AAV
LPO
MPO
ACo
LA
lo
ictd
SO
Temp
SCh
Squamous
ox
exorbd
4n
masn
dtn
3n
Zyg
buccn
6n
SphPal
maxv
LPtg
ptgpal
5Gn / 5n
cav
BSph
masa
ptgcn
n
ASph
Mandible
ialvn
Masseter
ialvv
ling
n
ialva
cty
Ptg
NasoPhar
Mc
MPtg

1   cortical layer 1
3n   oculomotor nerve or its root
3V   3rd ventricle
4n   trochlear nerve or its root
5Gn   trigeminal ganlion
5n   trigeminal nerve
6n   abducens nerve or its root
AAD   anterior amygdaloid area, dorsal
AAV   anterior amygdaloid area, ventral
ac   anterior commissure
ACo   anterior cortical amygdaloid nu
acp   anterior commissure, posterior
ASph   alisphenoid bone
B   basal nu Meynert
BSph   basisphenoid bone
BST   bed nu stria terminalis
BSTL   bed nu st, lateral div
BSTMA   bed nu st, med div, ant
buccn   buccal nerve
cav   cavernous sinus
Cg   cingulate cortex
Cl   claustrum
CPu   caudate putamen
cpu   caudate putamen neuroepithelium
crhv   caudal rhinal vein
cty   chorda tympani nerve
cx   cortical neuroepithelium
CxP   cortical plate
CxS   cortical subplate
DEn   dorsal endopiriform nu
DG   dentate gyrus
dtn   deep temporal nerve
exorbd   duct of exorbital gland
f   fornix
FL   forelimb area of cortex
Fr   frontal cortex
Frontal   frontal bone
FStr   fundus striati
GP   globus pallidus
HL   hindlimb area of cortex
ialva   inferior alveolar artery
ialvn   inferior alveolar nerve
ialvv   inferior alveolar vein
ic   inferior colliculus neuroepithelium
ictd   internal carotid artery
ICx   intermediate cortical layer
Ins   insular cortex
LA   lateroanterior hypoth nu
ling   lingual nerve
lo   lateral olfactory tract
LOT   nu lateral olfactory tract
LPO   lateral preoptic area
LPtg   lateral pterygoid
LS   lateral septal nu
LSD   lateral septal nu, dorsal
LV   lateral ventricle
Mandible   mandible
masa   masseteric artery
masn   masseteric nerve
Masseter   masseter muscle
maxv   maxillary vein
Mc   Meckel's cartilage
mfb   medial forebrain bundle
MnPO   median preoptic nu
MPO   medial preoptic nu
MPtg   medial pterygoid muscle
n   nerve
NasoPhar   nasopharyngeal cavity
ox   optic chiasm
Par1   parietal cortex, area 1
Par2   parietal cortex, area 2
Parietal   parietal bone
Pir   piriform cortex
Ptg   pterygoid bone
ptgcn   nerve of the pterygoid canal
ptgpal   pterygopalatine artery
RF   rhinal fissure
SCh   suprachiasmatic nu
SFi   septofimbrial nu
SHy   septohypothalamic nu
SO   supraoptic nu
SphPal   sphenopalatine ganglion
Squamous   squamous part of temp
sss   superior sagittal sinus
StHy   striohypothalamic nu
SubV   subventricular cortical layer
Temp   temporal muscle
vhc   ventral hip commissure
VP   ventral pallidum
Zyg   zygomatic arch

3   2   1   0

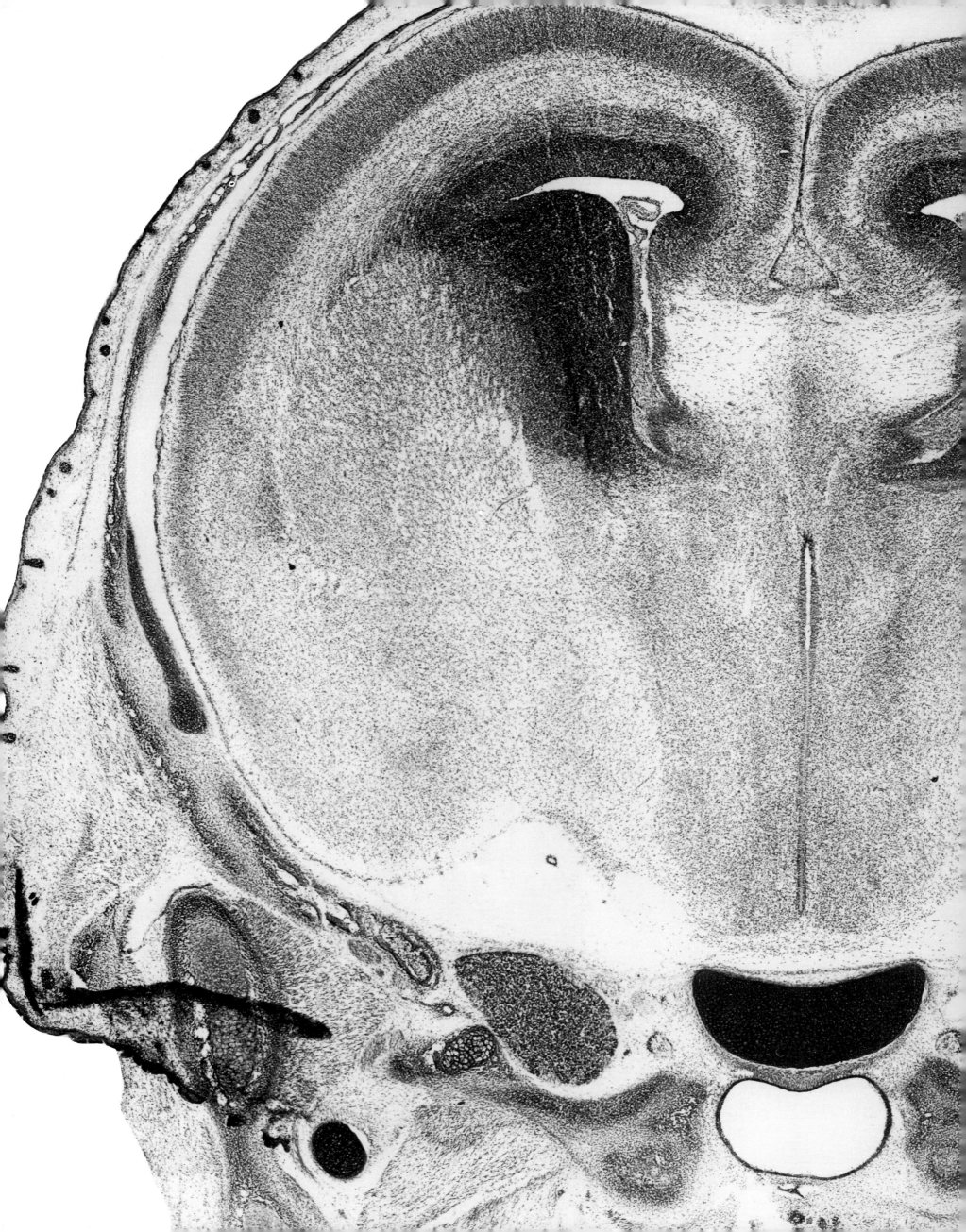

Figure 97
E19 Coronal 17

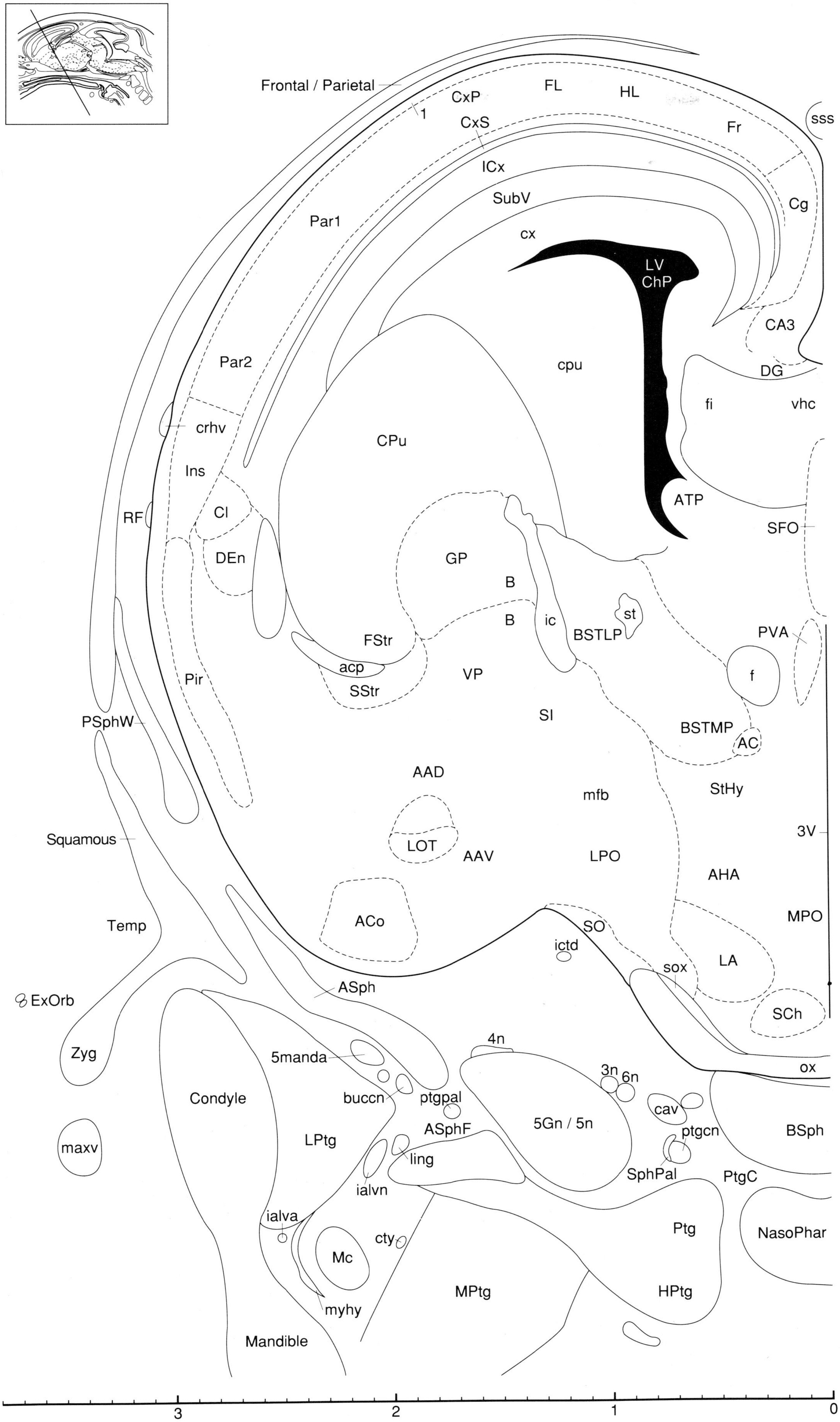

Frontal / Parietal

CxP FL HL
1 CxS Fr sss
ICx Cg
Par1 SubV
cx CA3
LV
ChP
Par2 cpu DG
fi vhc
crhv
Ins
CPu ATP
RF Cl
DEn SFO
GP B
B ic st
FStr BSTLP PVA
acp VP f
Pir SStr
SI BSTMP
PSphW AC
AAD mfb StHy
Squamous 3V
LOT AAV LPO AHA
Temp MPO
ACo SO LA
ictd
sox SCh
ASph
ExOrb ox
Zyg 4n
5manda 3n 6n
Condyle buccn ptgpal cav
ASphF 5Gn / 5n ptgcn BSph
maxv LPtg ling SphPal PtgC
ialvn
ialva cty Ptg NasoPhar
Mc
myhy MPtg HPtg
Mandible

3 2 1 0

1 cortical layer 1
3n oculomotor nerve or its root
3V 3rd ventricle
4n trochlear nerve or its root
5Gn trigeminal ganlion
5manda mandibular nerve, ant trunk
5n trigeminal nerve
6n abducens nerve or its root
AAD anterior amygdaloid area, dorsal
AAV anterior amygdaloid area, ventral
AC anterior commissural nu
ACo anterior cortical amygdaloid nu
acp anterior commissure, posterior
AHA anterior hypoth area, anterior
ASph alisphenoid bone
ASphF alisphenoid foramen
ATP anterior transitional promontory
B basal nu Meynert
BSph basisphenoid bone
BSTLP bed nu st, lat div, post
BSTMP bed nu st, med div, posterior
buccn buccal nerve
CA3 CA3 field of the hippocampus
cav cavernous sinus
Cg cingulate cortex
ChP choroid plexus
Cl claustrum
Condyle condyloid process of mandib
CPu caudate putamen
cpu caudate putamen neuroepithelium
crhv caudal rhinal vein
cty chorda tympani nerve
cx cortical neuroepithelium
CxP cortical plate
CxS cortical subplate
DEn dorsal endopiriform nu
DG dentate gyrus
ExOrb exorbital lacrimal gland
f fornix
fi fimbria hippocampus
FL forelimb area of cortex
Fr frontal cortex
Frontal frontal bone
FStr fundus striati
GP globus pallidus
HL hindlimb area of cortex
HPtg hamulus of the pterygoid bone
ialva inferior alveolar artery
ialvn inferior alveolar nerve
ic inferior colliculus neuroepithelium
ictd internal carotid artery
ICx intermediate cortical layer
Ins insular cortex
LA lateroanterior hypoth nu
ling lingual nerve
LOT nu lateral olfactory tract
LPO lateral preoptic area
LPtg lateral pterygoid
LV lateral ventricle
Mandible mandible
maxv maxillary vein
Mc Meckel's cartilage
mfb medial forebrain bundle
MPO medial preoptic nu
MPtg medial pterygoid muscle
myhy mylohyoid nerve
NasoPhar nasopharyngeal cavity
ox optic chiasm
Par1 parietal cortex, area 1
Par2 parietal cortex, area 2
Parietal parietal bone
Pir piriform cortex
PSphW presphenoid wing
Ptg pterygoid bone
PtgC pterygoid canal
ptgcn nerve of the pterygoid canal
ptgpal pterygopalatine artery
PVA paraventricular thal nu, anterior
RF rhinal fissure
SCh suprachiasmatic nu
SFO subfornical organ
SI substantia innominata
SO supraoptic nu
sox supraoptic decussation
SphPal sphenopalatine ganglion
Squamous squamous part of temp
sss superior sagittal sinus
SStr substriatal area
st stria terminalis
StHy striohypothalamic nu
SubV subventricular cortical layer
Temp temporal muscle
vhc ventral hip commissure
VP ventral pallidum
Zyg zygomatic arch

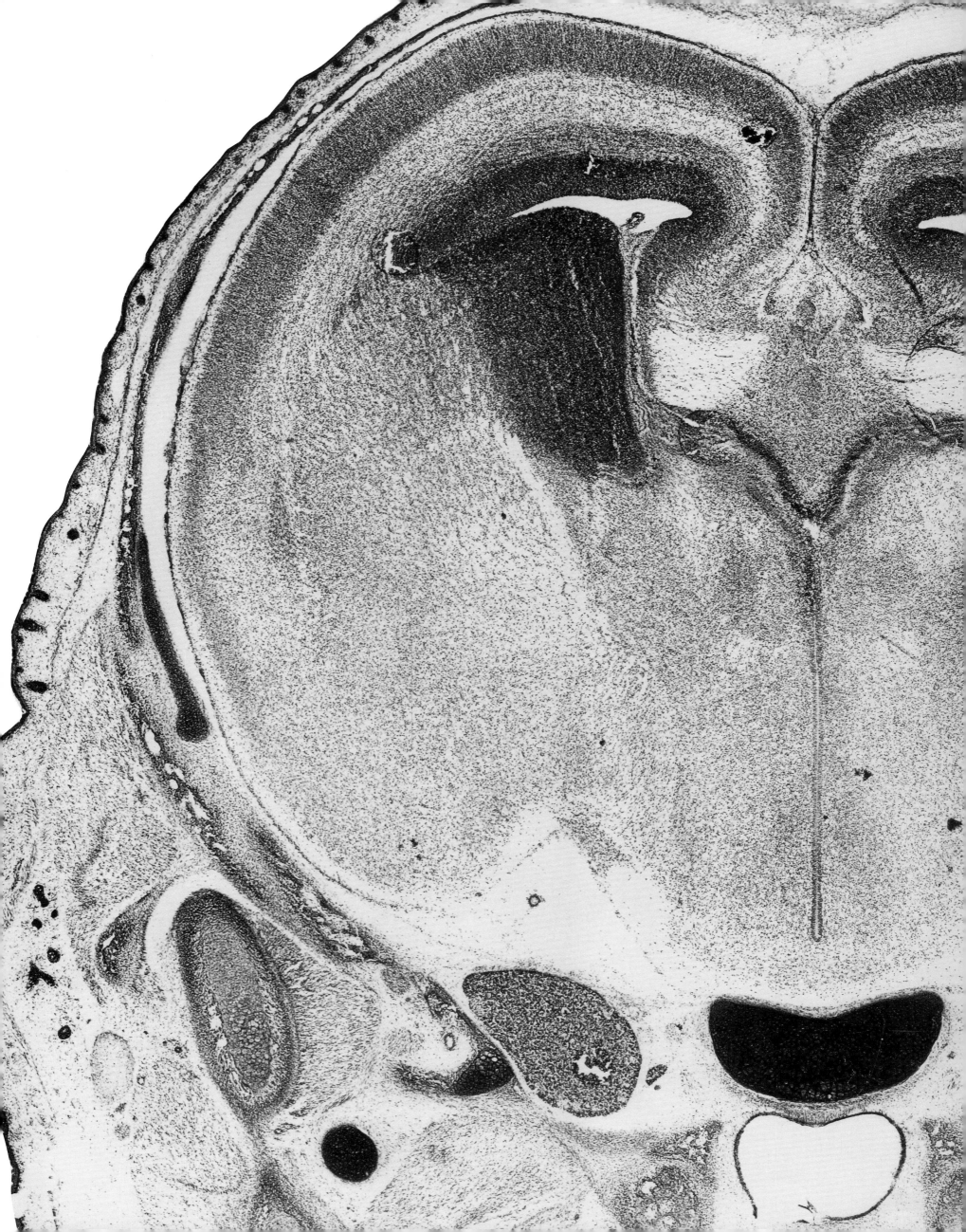

Figure 98
E19 Coronal 18

Frontal / Parietal

1
CxP
CxS
HL
FL
Fr
sss
ICx
Par1
SubV
Cg
cx
LV
CA3
Par2
cpu
DG
crhv
fi
PRh
IVF
SFO
RF
GP
st
sm
DEn
B
PT
PVA
Pir
ic
BST
SI
f
AStr
SStr
sm
PSphW
SM
BSTMP
Squamous
Me
3V
Temp
mfb
ACo
MeAV
LH
BAOT
SO
AHA
Syn
ictd
opt
LA
ExOrb
Zyg
4n
sox
AB
5manda
3n
6n
Condyle
5mand
cav
BSph
maxv
LPtg
5Gn / 5n
masa
ptgpal
ASph
aute
ptgcn
PtgC
n
ialvv
cty
ialva
Mc
Ptg
NasoPhar
MPtg
HPtg

1   cortical layer 1
3n   oculomotor nerve or its root
3V   3rd ventricle
4n   trochlear nerve or its root
5Gn   trigeminal ganlion
5mand   mandibular nerve
5manda   mandibular nerve, ant trunk
5n   trigeminal nerve
6n   abducens nerve or its root
AB   anterobasal nucleus
ACo   anterior cortical amygdaloid nu
AHA   anterior hypoth area, anterior
ASph   alisphenoid bone
AStr   amygdalostriatal transition area
aute   auriculotemporal nerve
B   basal nu Meynert
BAOT   bed nu accessory olfactory tr
BSph   basisphenoid bone
BST   bed nu stria terminalis
BSTMP   bed nu st, med div, posterior
CA3   CA3 field of the hippocampus
cav   cavernous sinus
Cg   cingulate cortex
Condyle   condyloid process of mandib
CPu   caudate putamen
cpu   caudate putamen neuroepithelium
crhv   caudal rhinal vein
cty   chorda tympani nerve
cx   cortical neuroepithelium
CxP   cortical plate
CxS   cortical subplate
DEn   dorsal endopiriform nu
DG   dentate gyrus
ExOrb   exorbital lacrimal gland
f   fornix
fi   fimbria hippocampus
FL   forelimb area of cortex
Fr   frontal cortex
Frontal   frontal bone
GP   globus pallidus
HL   hindlimb area of cortex
HPtg   hamulus of the pterygoid bone
ialva   inferior alveolar artery
ialvv   inferior alveolar vein
ic   inferior colliculus neuroepithelium
ictd   internal carotid artery
ICx   intermediate cortical layer
IVF   interventricular foramen
LA   lateroanterior hypoth nu
LH   lateral hypothalamic area
LPtg   lateral pterygoid
LV   lateral ventricle
masa   masseteric artery
maxv   maxillary vein
Mc   Meckel's cartilage
Me   medial amygdaloid nu
MeAV   medial amyg nu, anterventral
mfb   medial forebrain bundle
MPtg   medial pterygoid muscle
n   nerve
NasoPhar   nasopharyngeal cavity
opt   optic tract
Par1   parietal cortex, area 1
Par2   parietal cortex, area 2
Parietal   parietal bone
Pir   piriform cortex
PRh   perirhinal cortex
PSphW   presphenoid wing
PT   paratenial thal nu
Ptg   pterygoid bone
PtgC   pterygoid canal
ptgcn   nerve of the pterygoid canal
ptgpal   pterygopalatine artery
PVA   paraventricular thal nu, anterior
RF   rhinal fissure
SFO   subfornical organ
SI   substantia innominata
SM   nu stria medullaris
sm   stria medullaris thalami
SO   supraoptic nu
sox   supraoptic decussation
Squamous   squamous part of temp
sss   superior sagittal sinus
SStr   substriatal area
st   stria terminalis
SubV   subventricular cortical layer
Syn   synovial cavity
Temp   temporal muscle
Zyg   zygomatic arch

3   2   1   0

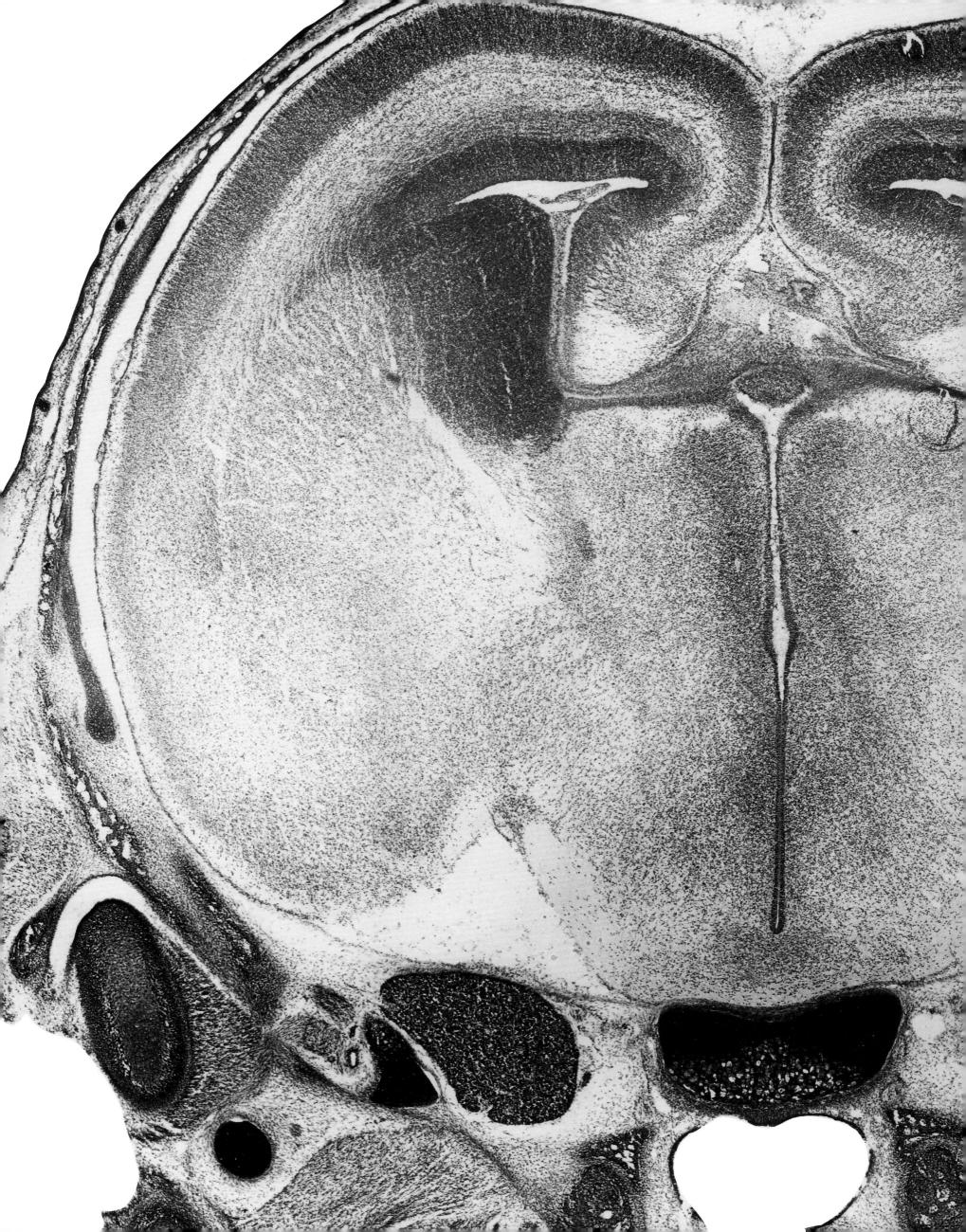

Figure 99
E19 Coronal 19

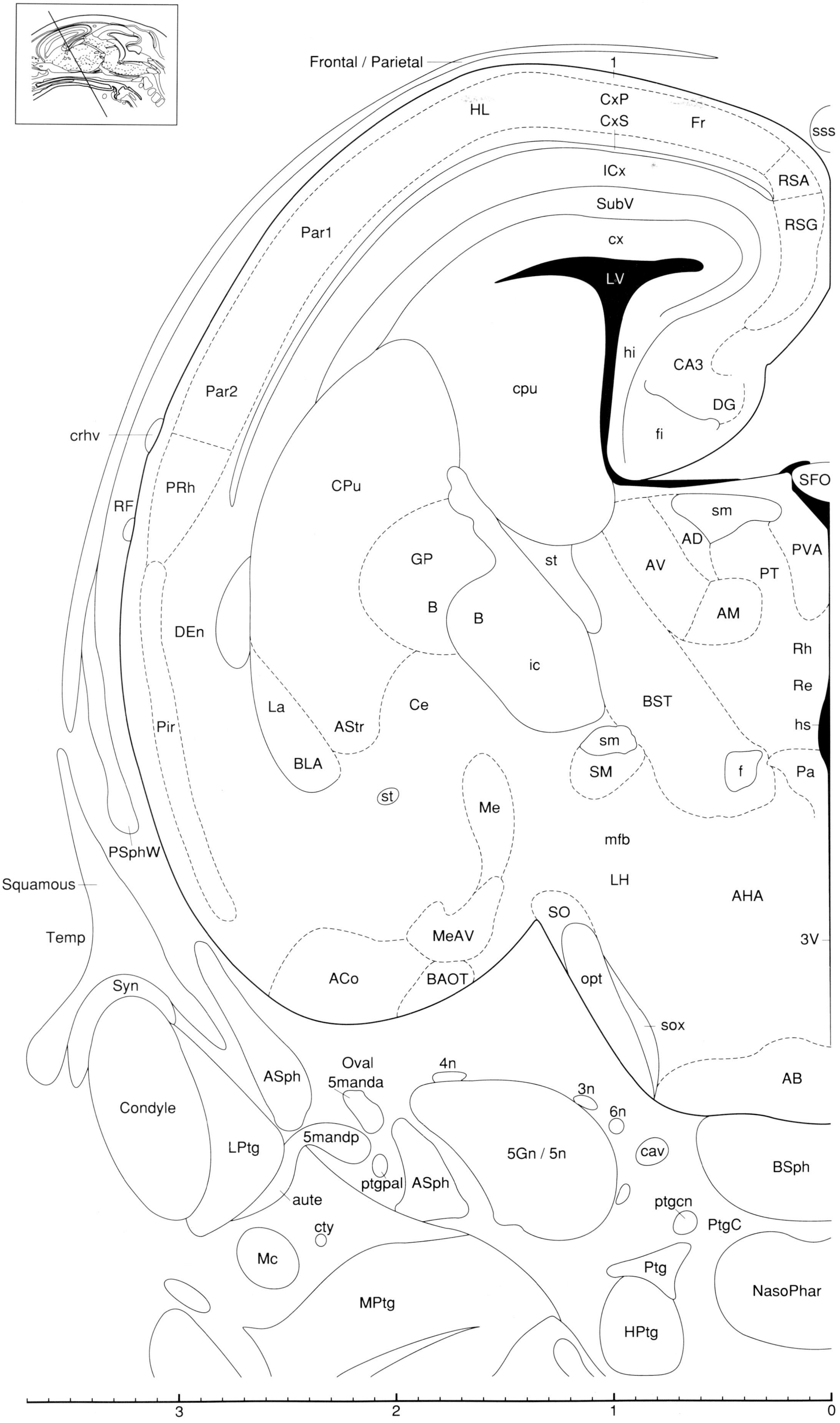

1 cortical layer 1
3n oculomotor nerve or its root
3V 3rd ventricle
4n trochlear nerve or its root
5Gn trigeminal ganglion
5manda mandibular nerve, ant trunk
5mandp mandibular nerve, post trunk
5n trigeminal nerve
6n abducens nerve or its root
AB anterobasal nucleus
ACo anterior cortical amygdaloid nu
AD anterodorsal thal nu
AHA anterior hypoth area, anterior
AM anteromedial thal nu
ASph alisphenoid bone
AStr amygdalostriatal transition area
aute auriculotemporal nerve
AV anteroventral thai nu
B basal nu Meynert
BAOT bed nu accessory olfactory tr
BLA basolateral amygdaloid nu, ant
BSph basisphenoid bone
BST bed nu stria terminalis
CA3 CA3 field of the hippocampus
cav cavernous sinus
Ce central amygdaloid nu
Condyle condyloid process of mandib
CPu caudate putamen
cpu caudate putamen neuroepithelium
crhv caudal rhinal vein
cty chorda tympani nerve
cx cortical neuroepithelium
CxP cortical plate
CxS cortical subplate
DEn dorsal endopiriform nu
DG dentate gyrus
f fornix
fi fimbria hippocampus
Fr frontal cortex
Frontal frontal bone
GP globus pallidus
hi hippocampal formation neuroepi
HL hindlimb area of cortex
HPtg hamulus of the pterygoid bone
hs hypothalamic sulcus
ic inferior colliculus neuroepithelium
ICx intermediate cortical layer
La lateral amygdaloid nu
LH lateral hypothalamic area
LPtg lateral pterygoid
LV lateral ventricle
Mc Meckel's cartilage
Me medial amygdaloid nu
MeAV medial amyg nu, anterventral
mfb medial forebrain bundle
MPtg medial pterygoid muscle
NasoPhar nasopharyngeal cavity
opt optic tract
Oval oval foramen
Pa paraventricular hypoth nu
Par1 parietal cortex, area 1
Par2 parietal cortex, area 2
Parietal parietal bone
Pir piriform cortex
PRh perirhinal cortex
PSphW presphenoid wing
PT paratenial thal nu
Ptg pterygoid bone
PtgC pterygoid canal
ptgcn nerve of the pterygoid canal
ptgpal pterygopalatine artery
PVA paraventricular thal nu, anterior
Re reuniens thal nu
RF rhinal fissure
Rh rhomboid thal nu
RSA retrosplenial agranular Cx
RSG retrosplenial granular Cx
SFO subfornical organ
SM nu stria medullaris
sm stria medullaris thalami
SO supraoptic nu
sox supraoptic decussation
Squamous squamous part of temp
sss superior sagittal sinus
st stria terminalis
SubV subventricular cortical layer
Syn synovial cavity
Temp temporal muscle

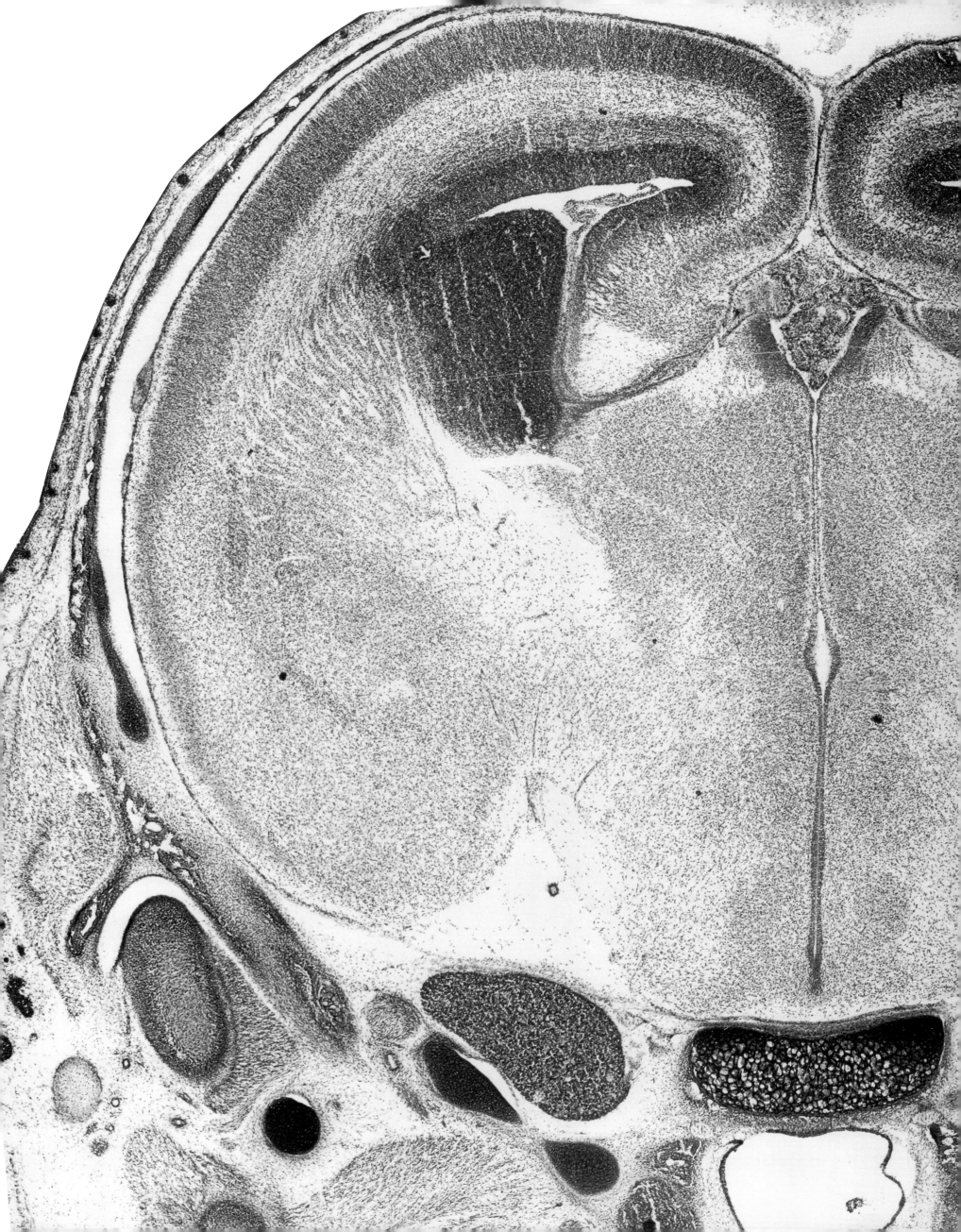

Figure 100
E19 Coronal 20

1   cortical layer 1
3n   oculomotor nerve or its root
3V   3rd ventricle
4n   trochlear nerve or its root
5Gn   trigeminal ganlion
5mand   mandibular nerve
5n   trigeminal nerve
6n   abducens nerve or its root
AB   anterobasal nucleus
ACo   anterior cortical amygdaloid nu
AH   anterior hypoth nu
AM   anteromedial thal nu
Arc   arcuate hypoth nu
ASph   alisphenoid bone
AStr   amygdalostriatal transition area
aute   auriculotemporal nerve
AV   anteroventral thal nu
B   basal nu Meynert
BL   basolateral amygdaloid nu
BM   basomedial amygdaloid nu
BSph   basisphenoid bone
BST   bed nu stria terminalis
CA1   CA1 field of the hippocampus
cav   cavernous sinus
Ce   central amygdaloid nu
Condyle   condyloid process of mandib
CPu   caudate putamen
cpu   caudate putamen neuroepithelium
crhv   caudal rhinal vein
cty   chorda tympani nerve
cx   cortical neuroepithelium
CxP   cortical plate
CxS   cortical subplate
D3V   dorsal third ventricle
DEn   dorsal endopiriform nu
DG   dentate gyrus
eml   external medullary lamina
ExOrb   exorbital lacrimal gland
f   fornix
fi   fimbria hippocampus
Fr   frontal cortex
Frontal   frontal bone
GP   globus pallidus
hi   hippocampal formation neuroepi
HL   hindlimb area of cortex
HPtg   hamulus of the pterygoid bone
hs   hypothalamic sulcus
ic   inferior colliculus neuroepithelium
ictd   internal carotid artery
ICx   intermediate cortical layer
IM   intercalated amygdaloid nu, main
La   lateral amygdaloid nu
LD   laterodorsal thal nu
LH   lateral hypothalamic area
LHb   lateral habenular nu
LPtg   lateral pterygoid
LV   lateral ventricle
Mandible   mandible
masa   masseteric artery
maxa   maxillary artery
maxv   maxillary vein
Mc   Meckel's cartilage
ME   median eminence
Me   medial amygdaloid nu
MeAV   medial amyg nu, anterventral
mfb   medial forebrain bundle
MHb   medial habenular nu
MPtg   medial pterygoid muscle
mptg   medial pterygoid nerve
n   nerve
NasoPhar   nasopharyngeal cavity
opt   optic tract
Oval   oval foramen
Pa   paraventricular hypoth nu
Par1   parietal cortex, area 1
Par2   parietal cortex, area 2
Parietal   parietal bone
Pir   piriform cortex
PLCo   posterolateral cortical amyg nu
PRh   perirhinal cortex
PSphW   presphenoid wing
Ptg   pterygoid bone
PtgC   pterygoid canal
ptgcn   nerve of the pterygoid canal
ptgpal   pterygopalatine artery
PVA   paraventricular thal nu, anterior
Re   reuniens thal nu
RF   rhinal fissure
Rh   rhomboid thal nu
RSA   retrosplenial agranular Cx
RSG   retrosplenial granular Cx
Rt   reticular thal nu
sm   stria medullaris thalami
SO   supraoptic nu
sox   supraoptic decussation
Squamous   squamous part of temp
sss   superior sagittal sinus
st   stria terminalis
SubV   subventricular cortical layer
Syn   synovial cavity
Temp   temporal muscle
vaf   ventral amygdalofugal pathway
VMH   ventromedial hypoth nu

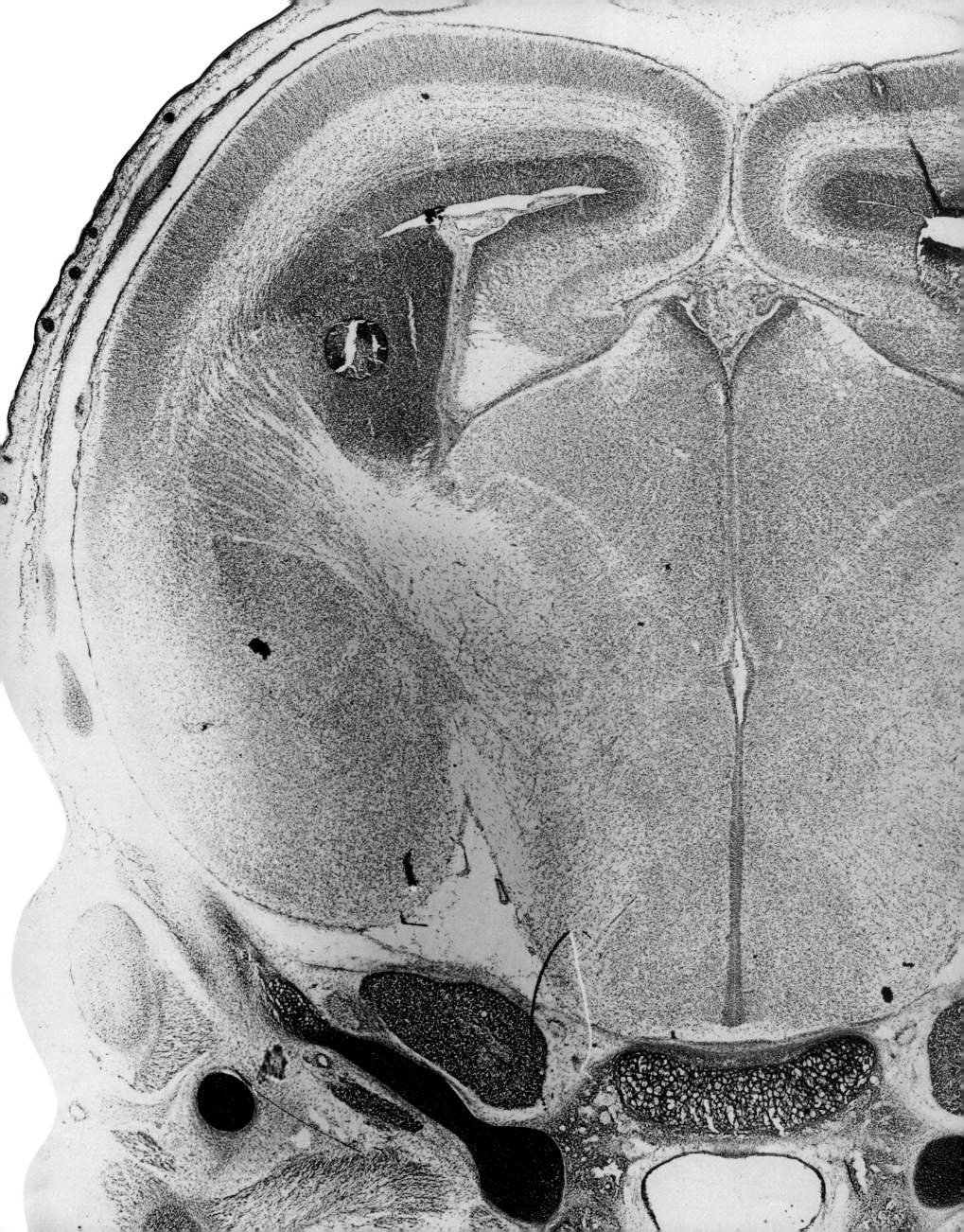

Figure 101
E19 Coronal 21

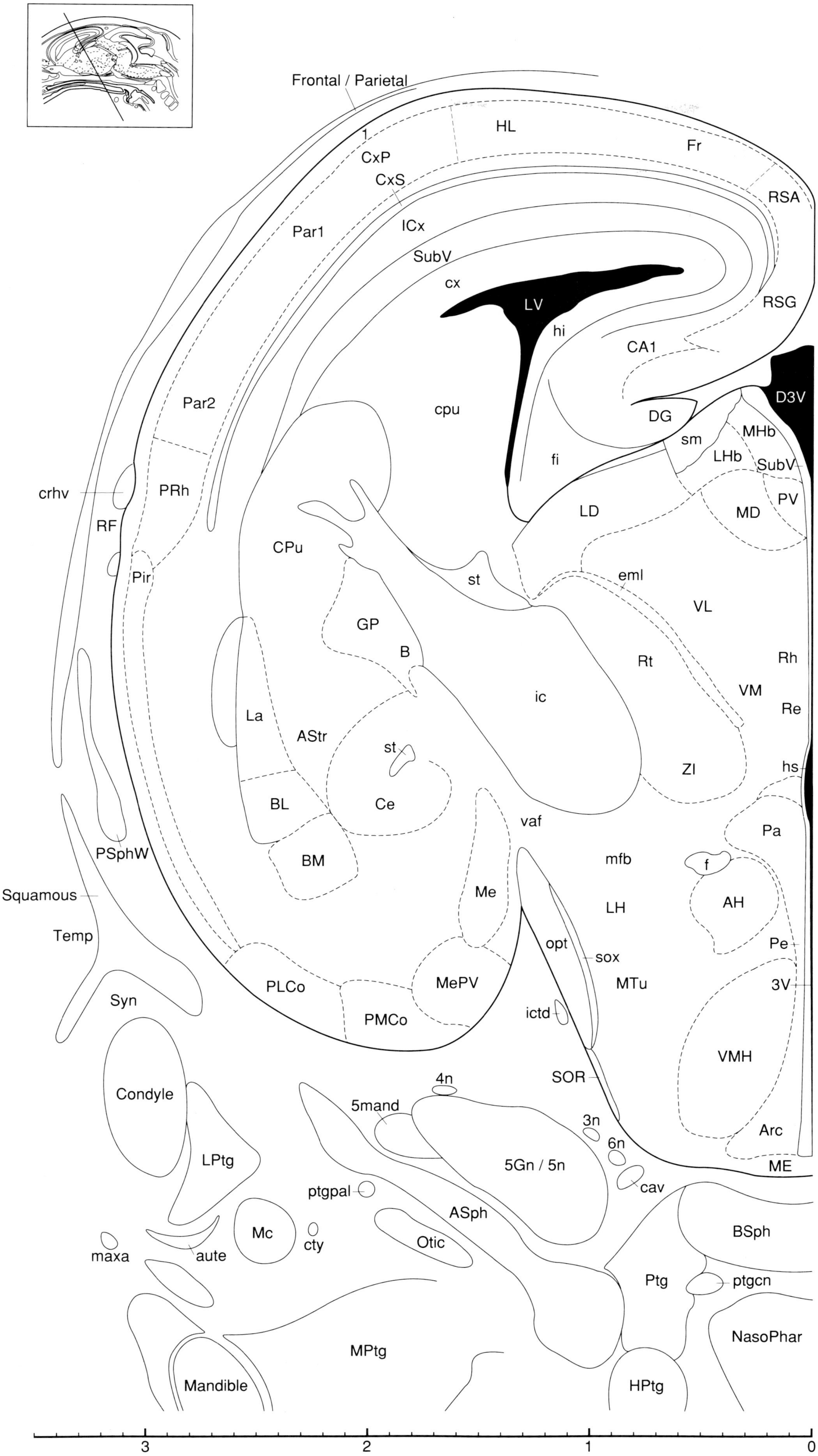

1   cortical layer 1
3n   oculomotor nerve or its root
3V   3rd ventricle
4n   trochlear nerve or its root
5Gn trigeminal ganlion
5mand mandibular nerve
5n   trigeminal nerve
6n   abducens nerve or its root
AH   anterior hypoth nu
Arc arcuate hypoth nu
ASph alisphenoid bone
AStr amygdalostriatal transition area
aute auriculotemporal nerve
B   basal nu Meynert
BL   basolateral amygdaloid nu
BM   basomedial amygdaloid nu
BSph basisphenoid bone
CA1 CA1 field of the hippocampus
cav cavernous sinus
Ce   central amygdaloid nu
Condyle condyloid process of mandib
CPu caudate putamen
cpu caudate putamen neuroepithelium
crhv caudal rhinal vein
cty chorda tympani nerve
cx   cortical neuroepithelium
CxP cortical plate
CxS cortical subplate
D3V dorsal third ventricle
DG   dentate gyrus
eml external medullary lamina
f   fornix
fi   fimbria hippocampus
Fr   frontal cortex
Frontal frontal bone
GP   globus pallidus
hi   hippocampal formation neuroepi
HL   hindlimb area of cortex
HPtg hamulus of the pterygoid bone
hs   hypothalamic sulcus
ic   inferior colliculus neuroepithelium
ictd internal carotid artery
ICx intermediate cortical layer
La   lateral amygdaloid nu
LD   laterodorsal thal nu
LH   lateral hypothalamic area
LHb lateral habenular nu
LPtg lateral pterygoid
LV   lateral ventricle
Mandible mandible
maxa maxillary artery
Mc   Meckel's cartilage
MD   mediodorsal thal nu
ME   median eminence
Me   medial amygdaloid nu
MePV medial amyg nu, posteroventr
mfb medial forebrain bundle
MHb medial habenular nu
MPtg medial pterygoid muscle
MTu medial tuberal nu
NasoPhar nasopharyngeal cavity
opt optic tract
Otic otic ganglion
Pa   paraventricular hypoth nu
Par1 parietal cortex, area 1
Par2 parietal cortex, area 2
Parietal parietal bone
Pe   periventricular hypoth nu
Pir piriform cortex
PLCo posterolateral cortical amyg nu
PMCo posteromed cortical amyg nu
PRh perirhinal cortex
PSphW presphenoid wing
Ptg pterygoid bone
ptgcn nerve of the pterygoid canal
ptgpal pterygopalatine artery
PV   paraventricular thal nu
Re   reuniens thal nu
RF   rhinal fissure
Rh   rhomboid thal nu
RSA retrosplenial agranular Cx
RSG retrosplenial granular Cx
Rt   reticular thal nu
sm   stria medullaris thalami
SOR supraoptic nu, retrochiasmatic
sox supraoptic decussation
Squamous squamous part of temp
st   stria terminalis
SubV subventricular cortical layer
Syn synovial cavity
Temp temporal muscle
vaf ventral amygdalofugal pathway
VL   ventrolateral thal nu
VM   ventromedial thal
VMH ventromedial hypoth nu
ZI   zona incerta

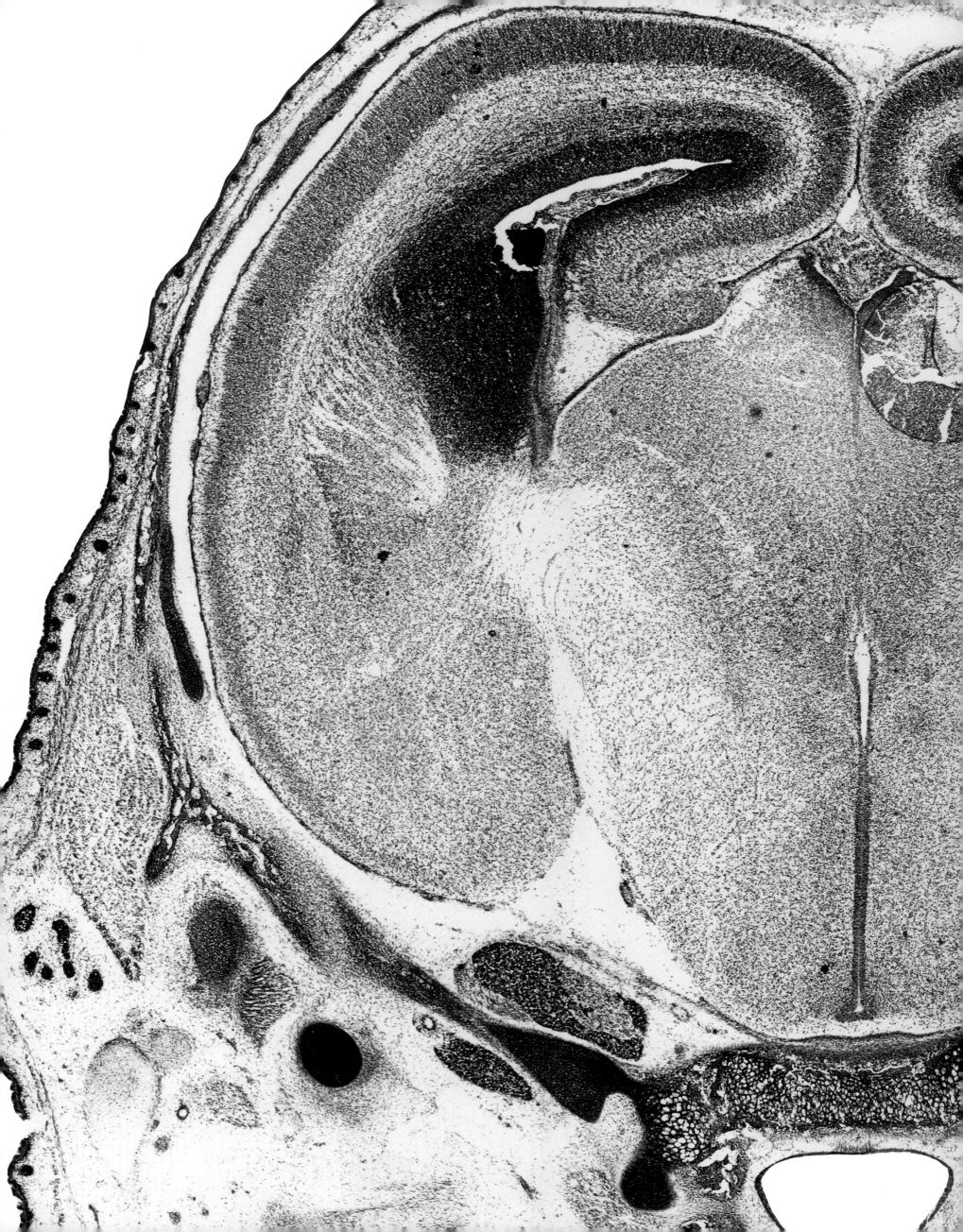

Figure 102
E19 Coronal 22

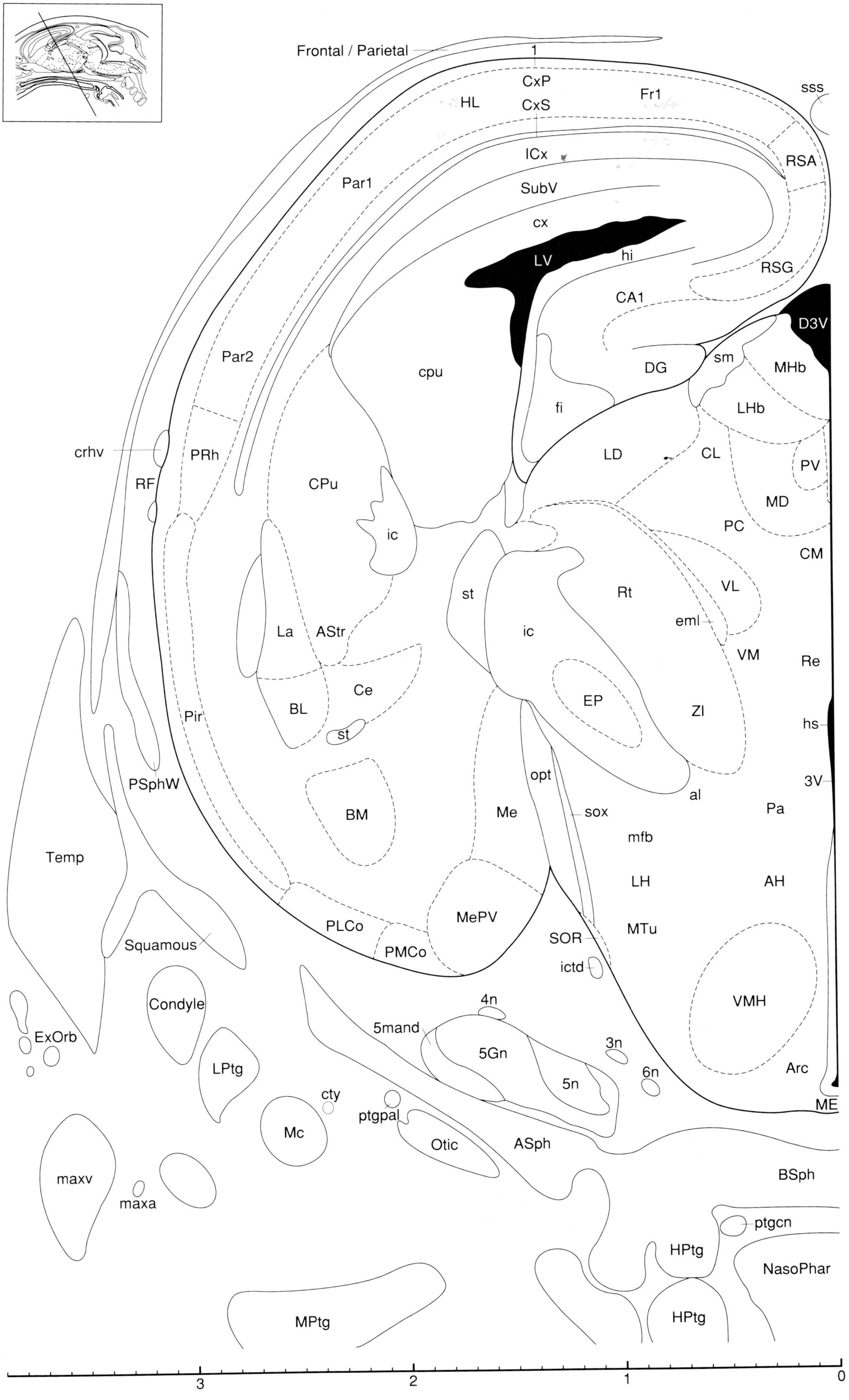

Frontal / Parietal

1

CxP
CxS
Fr1
HL
sss
ICx
RSA
SubV
Par1
cx
LV
hi
RSG
CA1
D3V
Par2
DG
sm
MHb
cpu
LHb
fi
LD
CL
PV
crhv
PRh
MD
RF
CPu
PC
CM
ic
VL
st
Rt
eml
ic
VM
Re
La
AStr
Ce
ZI
hs
BL
EP
Pir
3V
st
opt
al
Pa
PSphW
Me
sox
mfb
BM
Temp
LH
AH
MePV
MTu
PLCo
SOR
PMCo
ictd
VMH
Condyle
4n
5mand
3n
ExOrb
5Gn
6n
LPtg
5n
Arc
cty
ME
ptgpal
Mc
Otic
ASph
BSph
maxv
maxa
ptgcn
HPtg
NasoPhar
MPtg
HPtg

3   2   1   0

1   cortical layer 1
3n   oculomotor nerve or its root
3V   3rd ventricle
4n   trochlear nerve or its root
5Gn   trigeminal ganglion
5mand   mandibular nerve
5n   trigeminal nerve
6n   abducens nerve or its root
AH   anterior hypoth nu
al   ansa lenticularis
Arc   arcuate hypoth nu
ASph   alisphenoid bone
AStr   amygdalostriatal transition area
BL   basolateral amygdaloid nu
BM   basomedial amygdaloid nu
BSph   basisphenoid bone
CA1   CA1 field of the hippocampus
Ce   central amygdaloid nu
CL   centrolateral thal nu
CM   central medial thal nu
Condyle   condyloid process of mandib
CPu   caudate putamen
cpu   caudate putamen neuroepithelium
crhv   caudal rhinal vein
cty   chorda tympani nerve
cx   cortical neuroepithelium
CxP   cortical plate
CxS   cortical subplate
D3V   dorsal third ventricle
DG   dentate gyrus
eml   external medullary lamina
EP   entopeduncular nu
ExOrb   exorbital lacrimal gland
fi   fimbria hippocampus
Fr1   frontal cortex, area 1
Frontal   frontal bone
hi   hippocampal formation neuroepi
HL   hindlimb area of cortex
HPtg   hamulus of the pterygoid bone
hs   hypothalamic sulcus
ic   inferior colliculus neuroepithelium
ictd   internal carotid artery
ICx   intermediate cortical layer
La   lateral amygdaloid nu
LD   laterodorsal thal nu
LH   lateral hypothalamic area
LHb   lateral habenular nu
LPtg   lateral pterygoid
LV   lateral ventricle
maxa   maxillary artery
maxv   maxillary vein
Mc   Meckel's cartilage
MD   mediodorsal thal nu
ME   median eminence
Me   medial amygdaloid nu
MePV   medial amyg nu, posteroventr
mfb   medial forebrain bundle
MHb   medial habenular nu
MPtg   medial pterygoid muscle
MTu   medial tuberal nu
NasoPhar   nasopharyngeal cavity
opt   optic tract
Otic   otic ganglion
Pa   paraventricular hypoth nu
Par1   parietal cortex, area 1
Par2   parietal cortex, area 2
Parietal   parietal bone
PC   paracentral thal nu
Pir   piriform cortex
PLCo   posterolateral cortical amyg nu
PMCo   posteromed cortical amyg nu
PRh   perirhinal cortex
PSphW   presphenoid wing
ptgcn   nerve of the pterygoid canal
ptgpal   pterygopalatine artery
PV   paraventricular thal nu
Re   reuniens thal nu
RF   rhinal fissure
RSA   retrosplenial agranular Cx
RSG   retrosplenial granular Cx
Rt   reticular thal nu
sm   stria medullaris thalami
SOR   supraoptic nu, retrochiasmatic
sox   supraoptic decussation
Squamous   squamous part of temp
sss   superior sagittal sinus
st   stria terminalis
SubV   subventricular cortical layer
Temp   temporal muscle
VL   ventrolateral thal nu
VM   ventromedial thal
VMH   ventromedial hypoth nu
ZI   zona incerta

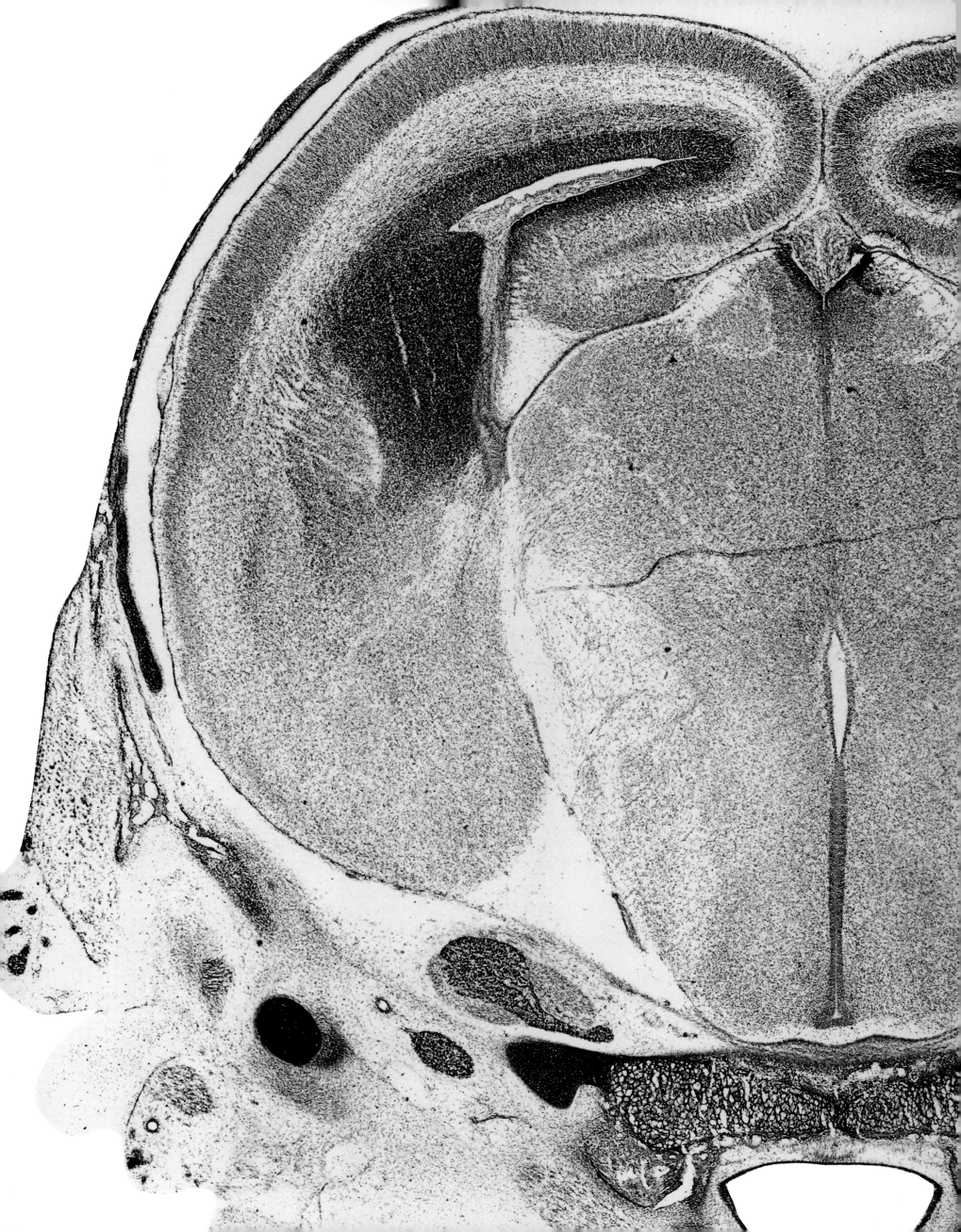

Figure 103
E19 Coronal 23

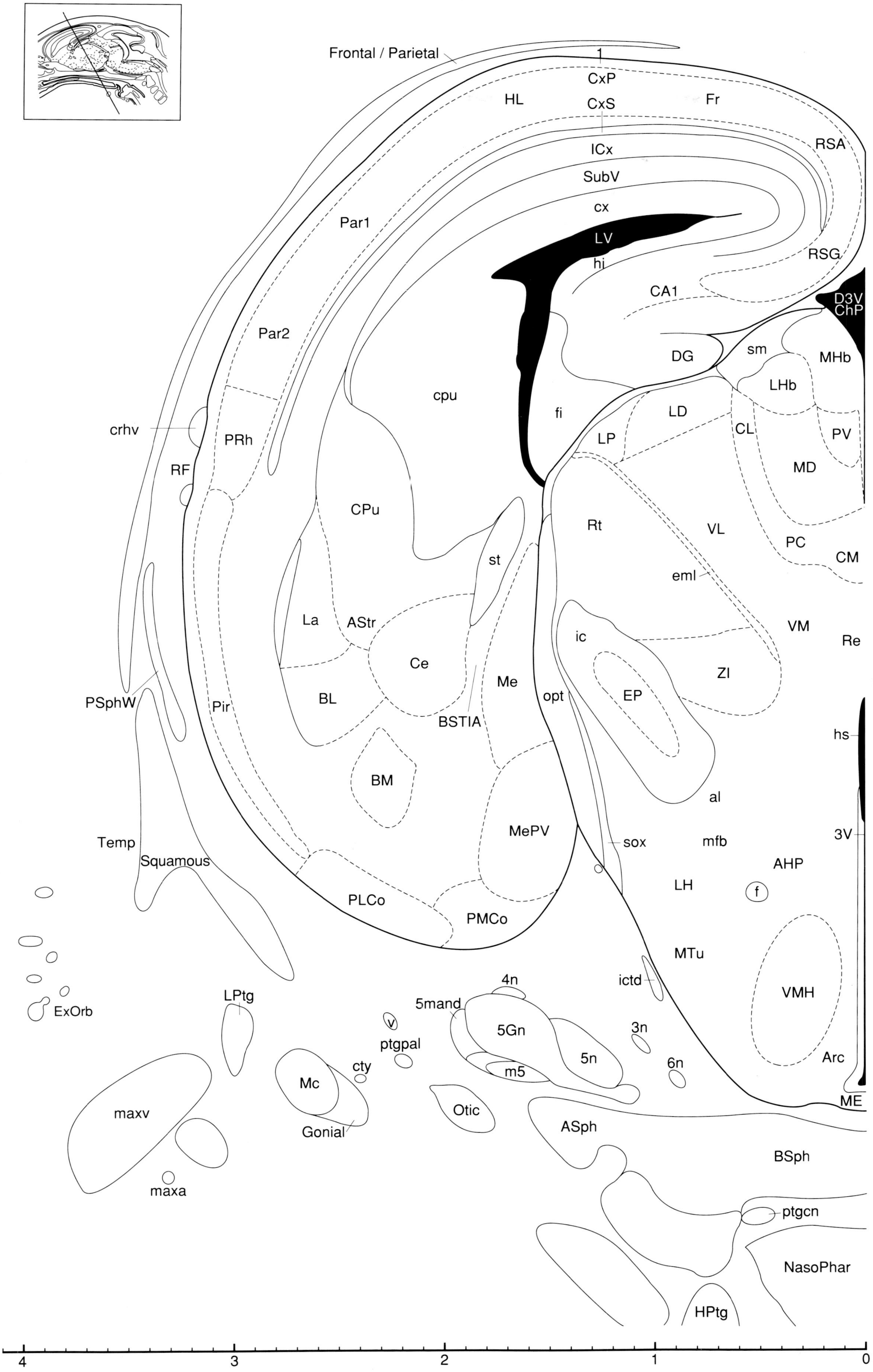

1   cortical layer 1
3n   oculomotor nerve or its root
3V   3rd ventricle
4n   trochlear nerve or its root
5Gn trigeminal ganglion
5mand mandibular nerve
5n   trigeminal nerve
6n   abducens nerve or its root
AHP anterior hypoth area, posterior
al   ansa lenticularis
Arc arcuate hypoth nu
ASph alisphenoid bone
AStr amygdalostriatal transition area
BL   basolateral amygdaloid nu
BM   basomedial amygdaloid nu
BSph basisphenoid bone
BSTIA bed nu st, intraamyg div
CA1 CA1 field of the hippocampus
Ce   central amygdaloid nu
ChP choroid plexus
CL   centrolateral thal nu
CM   central medial thal nu
CPu caudate putamen
cpu caudate putamen neuroepithelium
crhv caudal rhinal vein
cty chorda tympani nerve
cx   cortical neuroepithelium
CxP cortical plate
CxS cortical subplate
D3V dorsal third ventricle
DG   dentate gyrus
eml external medullary lamina
EP   entopeduncular nu
ExOrb exorbital lacrimal gland
f   fornix
fi   fimbria hippocampus
Fr   frontal cortex
Frontal frontal bone
Gonial gonial
hi   hippocampal formation neuroepi
HL   hindlimb area of cortex
HPtg hamulus of the pterygoid bone
hs   hypothalamic sulcus
ic   inferior colliculus neuroepithelium
ictd internal carotid artery
ICx intermediate cortical layer
La   lateral amygdaloid nu
LD   laterodorsal thal nu
LH   lateral hypothalamic area
LHb lateral habenular nu
LP   lateral posterior thal nu
LPtg lateral pterygoid
LV   lateral ventricle
m5   motor root of trigeminal nerve
maxa maxillary artery
maxv maxillary vein
Mc   Meckel's cartilage
MD   mediodorsal thal nu
ME   median eminence
Me   medial amygdaloid nu
MePV medial amyg nu, posteroventr
mfb medial forebrain bundle
MHb medial habenular nu
MTu medial tuberal nu
NasoPhar nasopharyngeal cavity
opt optic tract
Otic otic ganglion
Par1 parietal cortex, area 1
Par2 parietal cortex, area 2
Parietal parietal bone
PC   paracentral thal nu
Pir piriform cortex
PLCo posterolateral cortical amyg nu
PMCo posteromed cortical amyg nu
PRh perirhinal cortex
PSphW presphenoid wing
ptgcn nerve of the pterygoid canal
ptgpal pterygopalatine artery
PV   paraventricular thal nu
Re   reuniens thal nu
RF   rhinal fissure
RSA retrosplenial agranular Cx
RSG retrosplenial granular Cx
Rt   reticular thal nu
sm   stria medullaris thalami
sox supraoptic decussation
Squamous squamous part of temp
st   stria terminalis
SubV subventricular cortical layer
Temp temporal muscle
v   vein
VL   ventrolateral thal nu
VM   ventromedial thal
VMH ventromedial hypoth nu
ZI   zona incerta

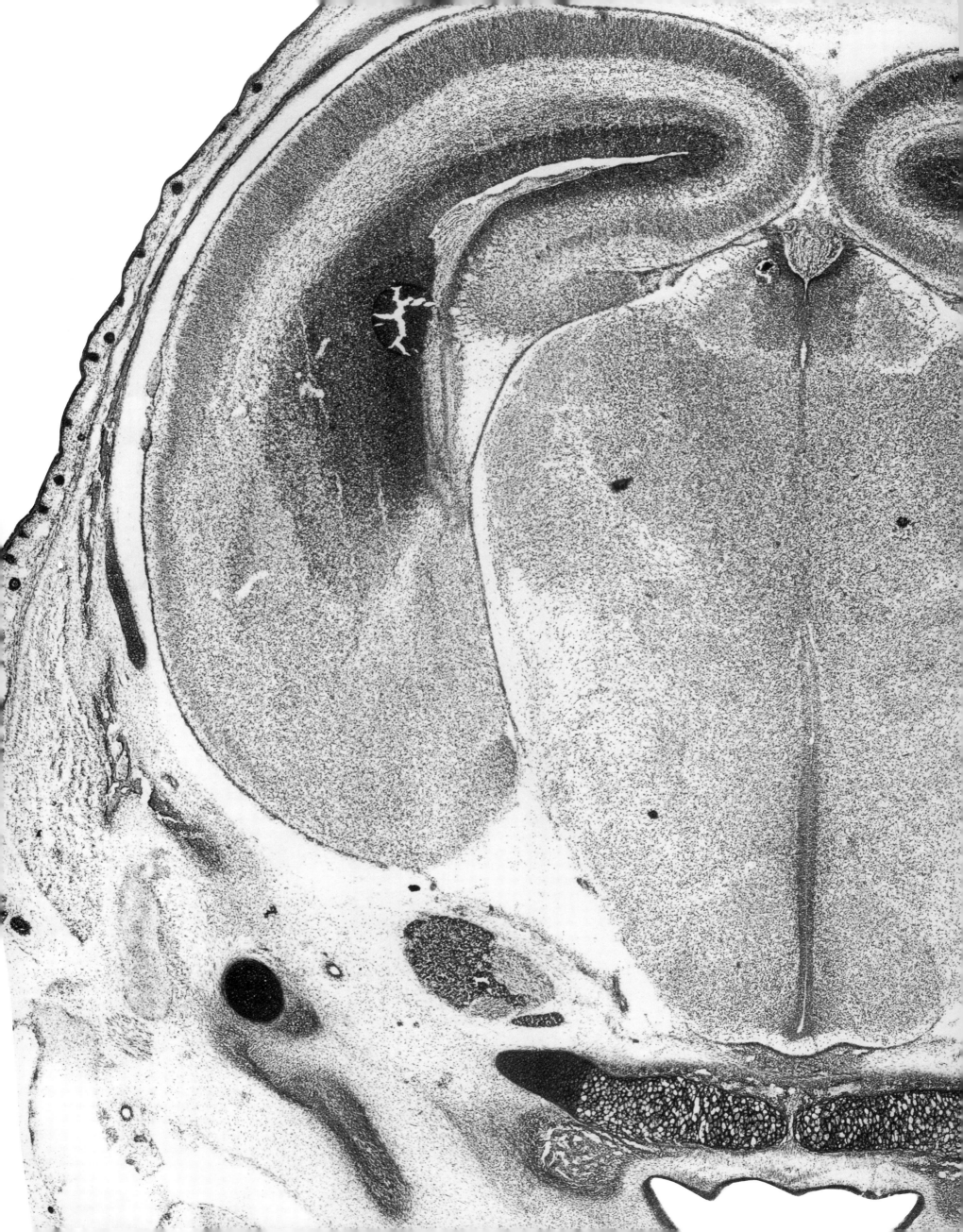

Figure 104

E19 Coronal 24

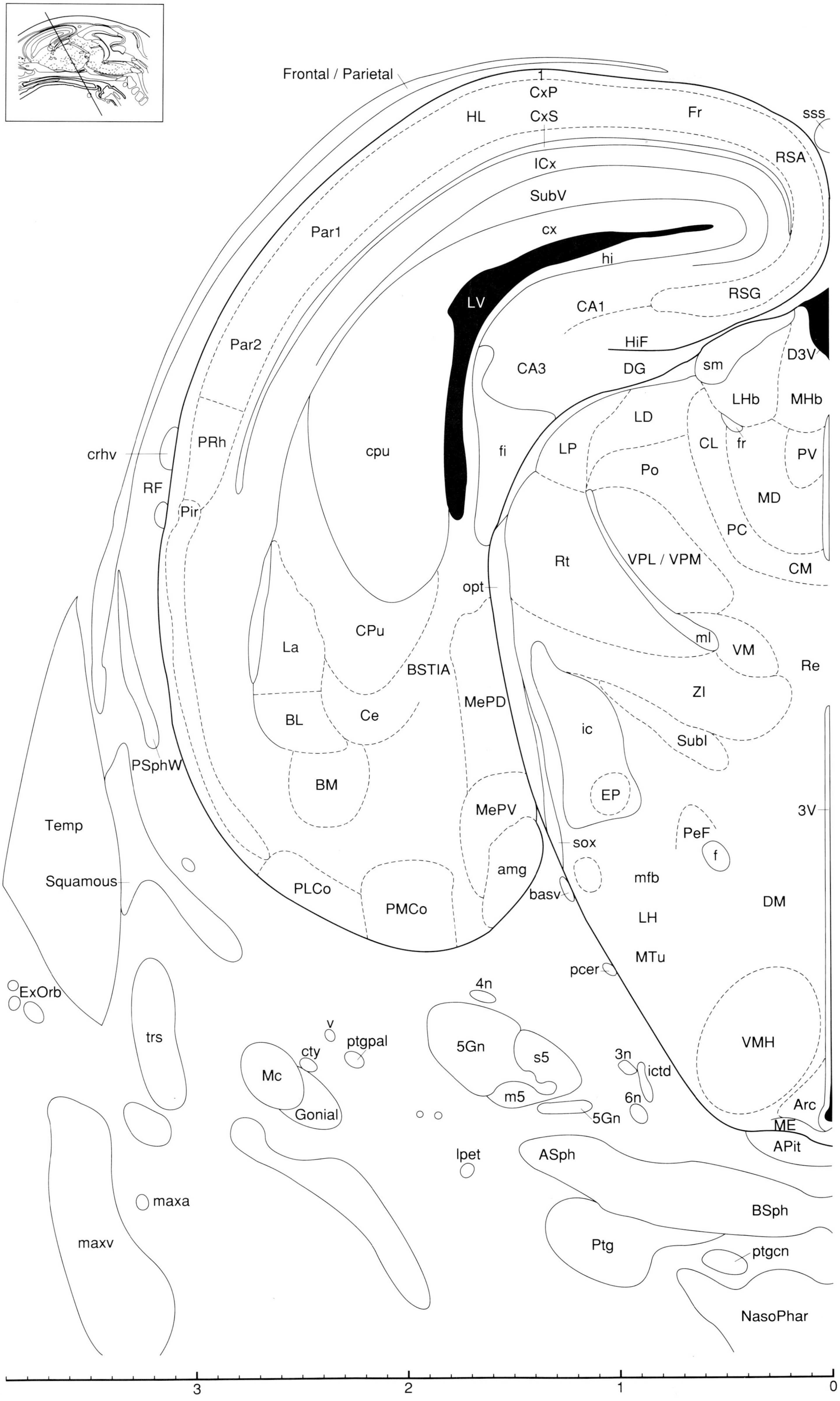

1 cortical layer 1
3n oculomotor nerve or its root
3V 3rd ventricle
4n trochlear nerve or its root
5Gn trigeminal ganglion
6n abducens nerve or its root
amg amygdaloid neuroepithelium
APit anterior lobe pituitary
Arc arcuate hypoth nu
ASph alisphenoid bone
basv basal vein
BL basolateral amygdaloid nu
BM basomedial amygdaloid nu
BSph basisphenoid bone
BSTIA bed nu st, intraamyg div
CA1 CA1 field of the hippocampus
CA3 CA3 field of the hippocampus
Ce central amygdaloid nu
CL centrolateral thal nu
CM central medial thal nu
CPu caudate putamen
cpu caudate putamen neuroepithelium
crhv caudal rhinal vein
cty chorda tympani nerve
cx cortical neuroepithelium
CxP cortical plate
CxS cortical subplate
D3V dorsal third ventricle
DG dentate gyrus
DM dorsomedial hypothalamic nu
EP entopeduncular nu
ExOrb exorbital lacrimal gland
f fornix
fi fimbria hippocampus
Fr frontal cortex
fr fasciculus retroflexus
Frontal frontal bone
Gonial gonial
hi hippocampal formation neuroepi
HiF hippocampal fissure
HL hindlimb area of cortex
ic inferior colliculus neuroepithelium
ictd internal carotid artery
ICx intermediate cortical layer
La lateral amygdaloid nu
LD laterodorsal thal nu
LH lateral hypothalamic area
LHb lateral habenular nu
LP lateral posterior thal nu
lpet lesser petrosal nerve
LV lateral ventricle
m5 motor root of trigeminal nerve
maxa maxillary artery
maxv maxillary vein
Mc Meckel's cartilage
MD mediodorsal thal nu
ME median eminence
MePD medial amyg nu, posterodorsal
MePV medial amyg nu, posteroventr
mfb medial forebrain bundle
MHb medial habenular nu
ml medial lemniscus
MTu medial tuberal nu
NasoPhar nasopharyngeal cavity
opt optic tract
Par1 parietal cortex, area 1
Par2 parietal cortex, area 2
Parietal parietal bone
PC paracentral thal nu
pcer posterior cerebral artery
PeF perifornical nu
Pir piriform cortex
PLCo posterolateral cortical amyg nu
PMCo posteromed cortical amyg nu
Po posterior thal nuclear group
PRh perirhinal cortex
PSphW presphenoid wing
Ptg pterygoid bone
ptgcn nerve of the pterygoid canal
ptgpal pterygopalatine artery
PV paraventricular thal nu
Re reuniens thal nu
RF rhinal fissure
RSA retrosplenial agranular Cx
RSG retrosplenial granular Cx
Rt reticular thal nu
s5 sensory root of trigeminal nerve
sm stria medullaris thalami
sox supraoptic decussation
Squamous squamous part of temp
sss superior sagittal sinus
SubI subincertal nu
SubV subventricular cortical layer
Temp temporal muscle
trs transverse sinus
v vein
VM ventromedial thal
VMH ventromedial hypoth nu
VPL ventral posterolateral thal nu
VPM ventral posteromedial thal nu
ZI zona incerta

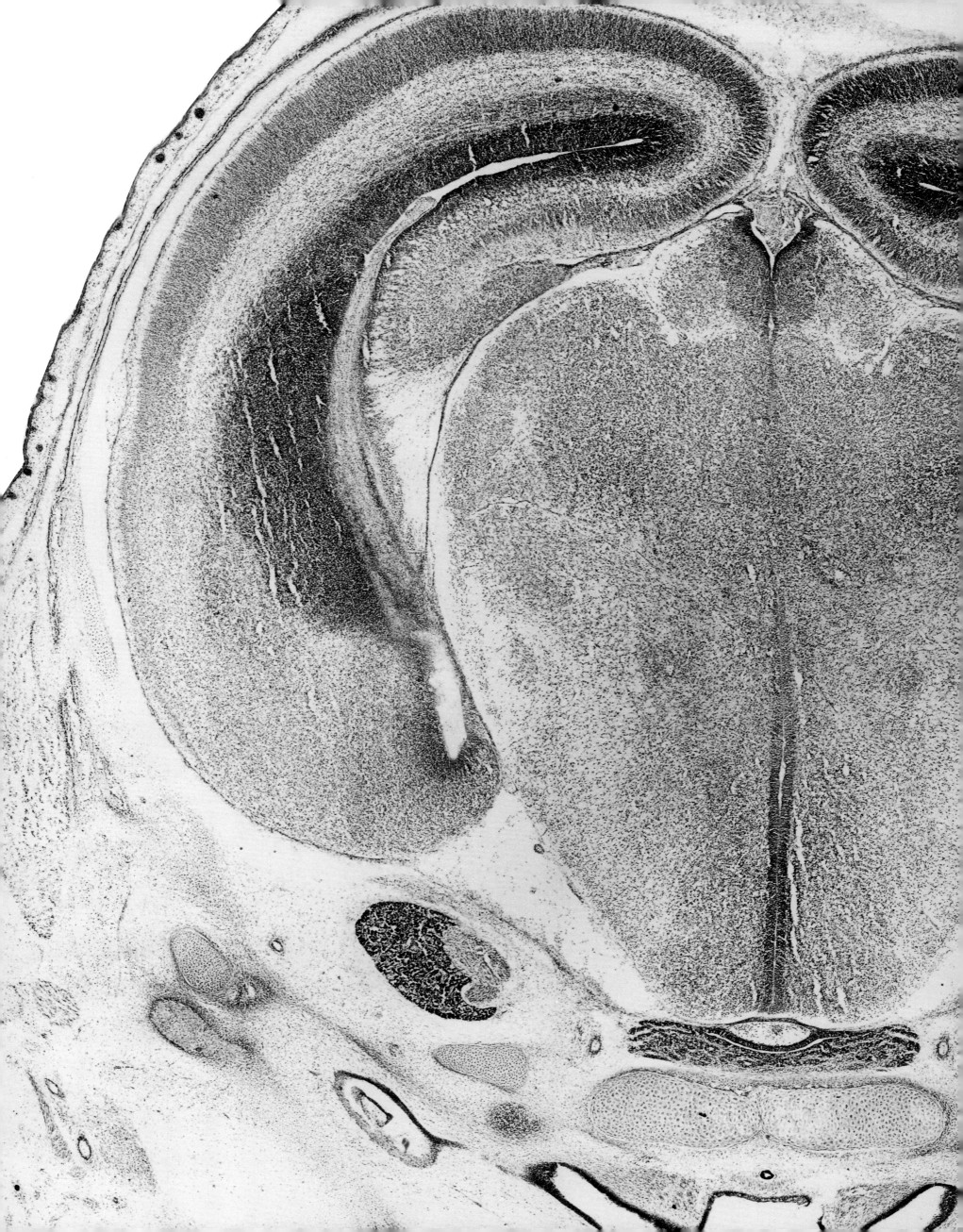

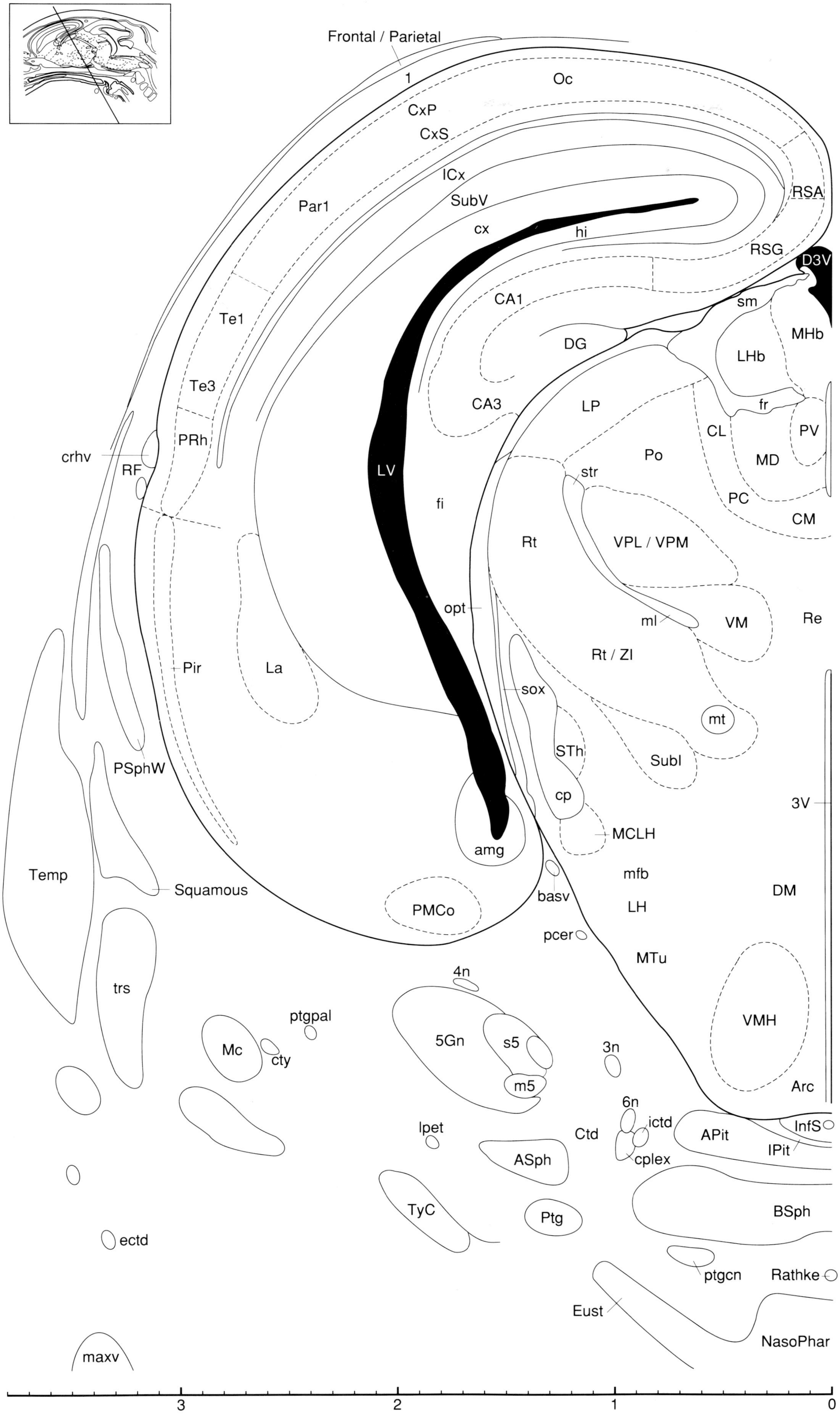

Figure 105
E19 Coronal 25

1 cortical layer 1
3n oculomotor nerve or its root
3V 3rd ventricle
4n trochlear nerve or its root
5Gn trigeminal ganglion
6n abducens nerve or its root
amg amygdaloid neuroepithelium
APit anterior lobe pituitary
Arc arcuate hypoth nu
ASph alisphenoid bone
basv basal vein
BSph basisphenoid bone
CA1 CA1 field of the hippocampus
CA3 CA3 field of the hippocampus
CL centrolateral thal nu
CM central medial thal nu
cp cerebral peduncle, basal
cplex carotid plexus
crhv caudal rhinal vein
Ctd carotid canal
cty chorda tympani nerve
cx cortical neuroepithelium
CxP cortical plate
CxS cortical subplate
D3V dorsal third ventricle
DG dentate gyrus
DM dorsomedial hypothalamic nu
ectd external carotid a
Eust Eustachian tube
fi fimbria hippocampus
fr fasciculus retroflexus
Frontal frontal bone
hi hippocampal formation neuroepi
ictd internal carotid artery
ICx intermediate cortical layer
InfS infundibular stem
IPit intermediate lobe pituitary
La lateral amygdaloid nu
LH lateral hypothalamic area
LHb lateral habenular nu
LP lateral posterior thal nu
lpet lesser petrosal nerve
LV lateral ventricle
m5 motor root of trigeminal nerve
maxv maxillary vein
Mc Meckel's cartilage
MCLH magnocellular nu lat hypoth
MD mediodorsal thal nu
mfb medial forebrain bundle
MHb medial habenular nu
ml medial lemniscus
mt mammillothalamic tract
MTu medial tuberal nu
NasoPhar nasopharyngeal cavity
Oc occipital cortex
opt optic tract
Par1 parietal cortex, area 1
Parietal parietal bone
PC paracentral thal nu
pcer posterior cerebral artery
Pir piriform cortex
PMCo posteromed cortical amyg nu
Po posterior thal nuclear group
PRh perirhinal cortex
PSphW presphenoid wing
Ptg pterygoid bone
ptgcn nerve of the pterygoid canal
ptgpal pterygopalatine artery
PV paraventricular thal nu
Rathke Rathke's pouch
Re reuniens thal nu
RF rhinal fissure
RSA retrosplenial agranular Cx
RSG retrosplenial granular Cx
Rt reticular thal nu
s5 sensory root of trigeminal nerve
sm stria medullaris thalami
sox supraoptic decussation
Squamous squamous part of temp
STh subthalamic nu
str superior thal radiation
SubI subincertal nu
SubV subventricular cortical layer
Te1 temporal cortex, area 1
Te3 temporal cortex, area 3
Temp temporal muscle
trs transverse sinus
TyC tympanic cavity
VM ventromedial thal
VMH ventromedial hypoth nu
VPL ventral posterolateral thal nu
VPM ventral posteromedial thal nu
ZI zona incerta

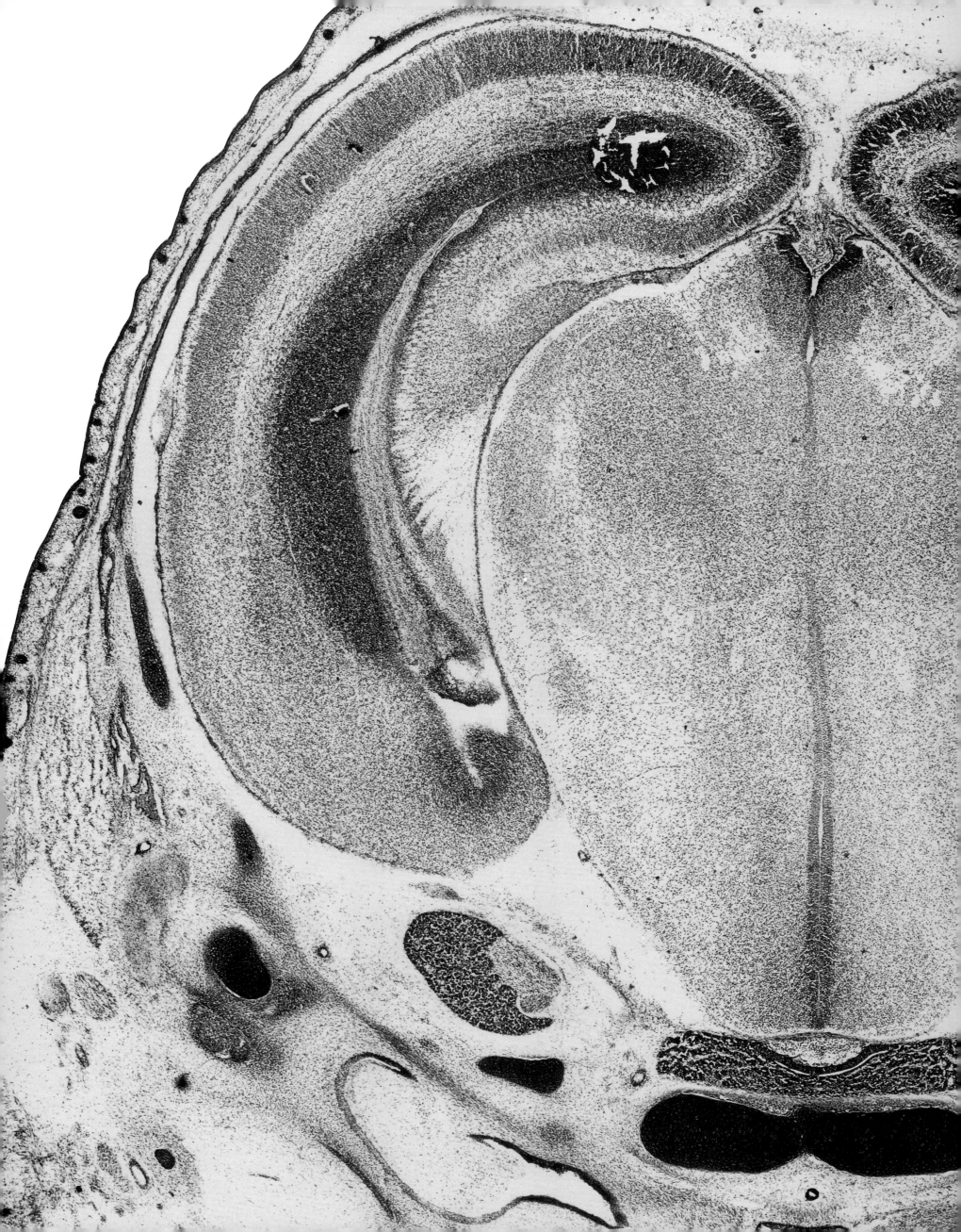

Figure 106
E19 Coronal 26

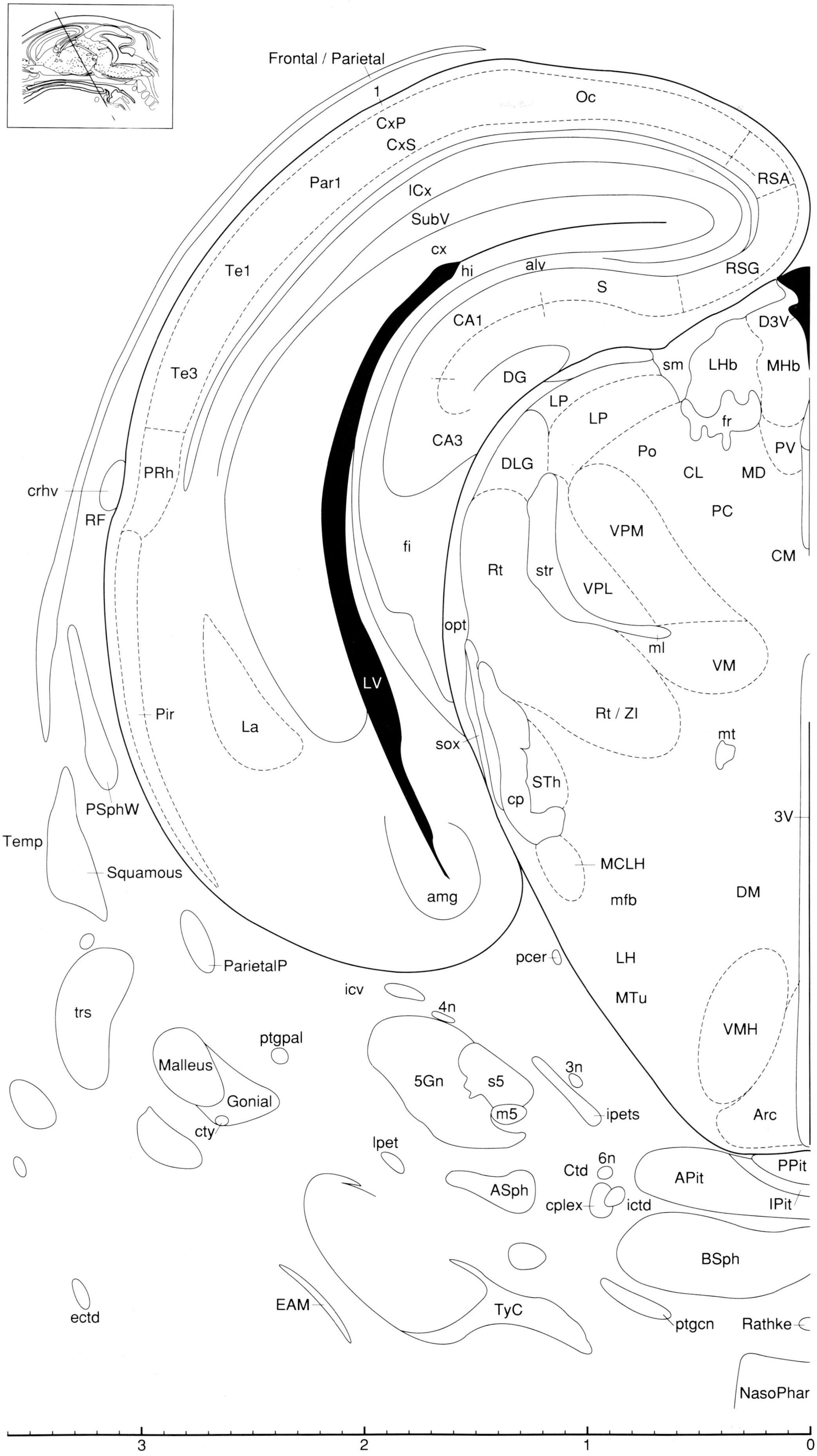

1 cortical layer 1
3n oculomotor nerve or its root
3V 3rd ventricle
4n trochlear nerve or its root
5Gn trigeminal ganglion
6n abducens nerve or its root
alv alveus of hippocampus
amg amygdaloid neuroepithelium
APit anterior lobe pituitary
Arc arcuate hypoth nu
ASph alisphenoid bone
BSph basisphenoid bone
CA1 CA1 field of the hippocampus
CA3 CA3 field of the hippocampus
CL centrolateral thal nu
CM central medial thal nu
cp cerebral peduncle, basal
cplex carotid plexus
crhv caudal rhinal vein
Ctd carotid canal
cty chorda tympani nerve
cx cortical neuroepithelium
CxP cortical plate
CxS cortical subplate
D3V dorsal third ventricle
DG dentate gyrus
DLG dorsal lateral geniculate nu
DM dorsomedial hypothalamic nu
EAM external auditory meatus
ectd external carotid a
fi fimbria hippocampus
fr fasciculus retroflexus
Frontal frontal bone
Gonial gonial
hi hippocampal formation neuroepi
ictd internal carotid artery
icv inferior cerebral vein
ICx intermediate cortical layer
ipets inferior petrosal sinus
IPit intermediate lobe pituitary
La lateral amygdaloid nu
LH lateral hypothalamic area
LHb lateral habenular nu
LP lateral posterior thal nu
lpet lesser petrosal nerve
LV lateral ventricle
m5 motor root of trigeminal nerve
Malleus malleus (ossicle)
MCLH magnocellular nu lat hypoth
MD mediodorsal thal nu
mfb medial forebrain bundle
MHb medial habenular nu
ml medial lemniscus
mt mammillothalamic tract
MTu medial tuberal nu
NasoPhar nasopharyngeal cavity
Oc occipital cortex
opt optic tract
Par1 parietal cortex, area 1
Parietal parietal bone
ParietalP parietal plate
PC paracentral thal nu
pcer posterior cerebral artery
Pir piriform cortex
Po posterior thal nuclear group
PPit posterior lobe pituitary
PRh perirhinal cortex
PSphW presphenoid wing
ptgcn nerve of the pterygoid canal
ptgpal pterygopalatine artery
PV paraventricular thal nu
Rathke Rathke's pouch
RF rhinal fissure
RSA retrosplenial agranular Cx
RSG retrosplenial granular Cx
Rt reticular thal nu
S subiculum
s5 sensory root of trigeminal nerve
sm stria medullaris thalami
sox supraoptic decussation
Squamous squamous part of temp
STh subthalamic nu
str superior thal radiation
SubV subventricular cortical layer
Te1 temporal cortex, area 1
Te3 temporal cortex, area 3
Temp temporal muscle
trs transverse sinus
TyC tympanic cavity
VM ventromedial thal nu
VMH ventromedial hypoth nu
VPL ventral posterolateral thal nu
VPM ventral posteromedial thal nu
ZI zona incerta

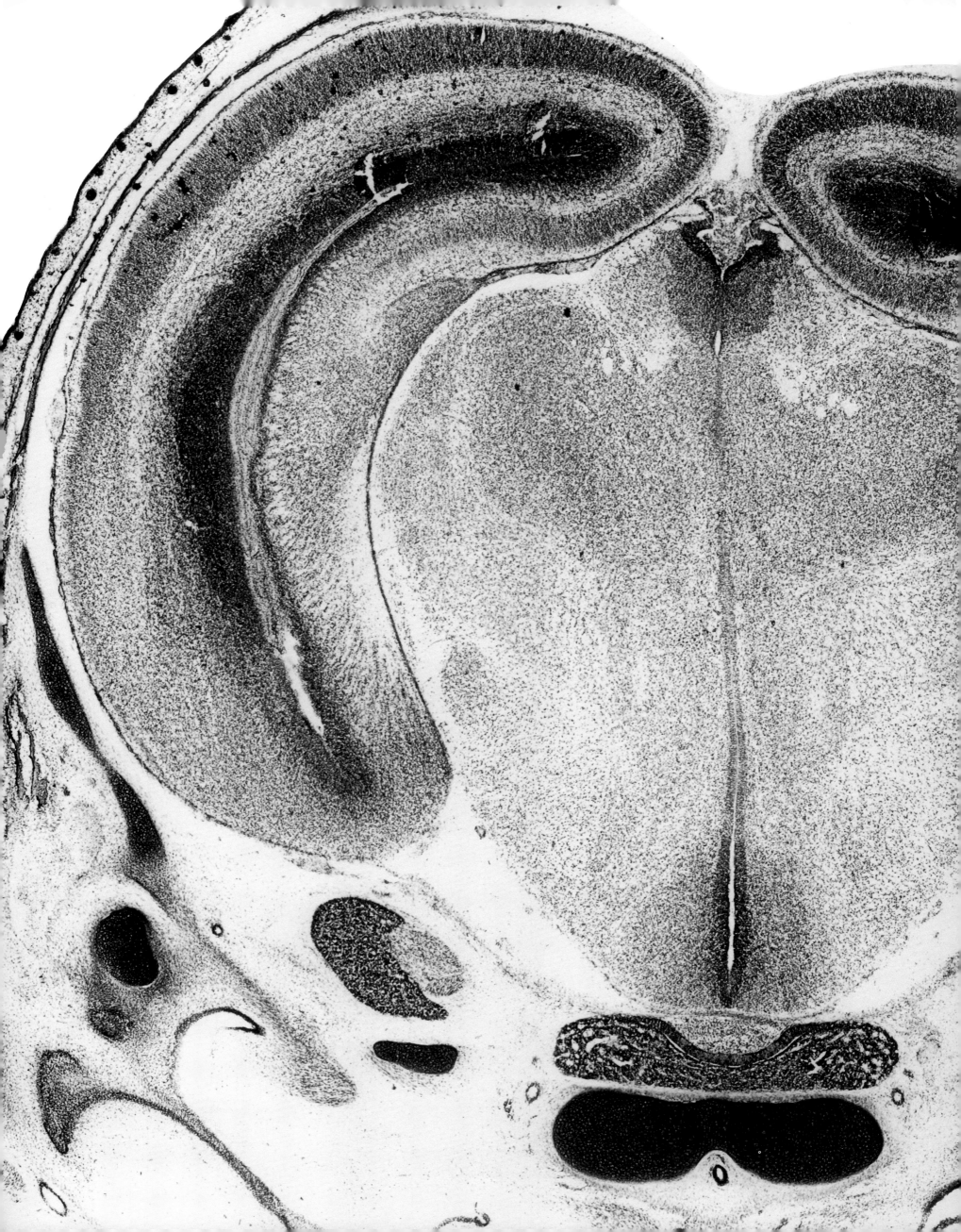

Figure 107
E19 Coronal 27

1 cortical layer 1
3n oculomotor nerve or its root
3V 3rd ventricle
4n trochlear nerve or its root
5Gn trigeminal ganglion
6n abducens nerve or its root
amg amygdaloid neuroepithelium
APit anterior lobe pituitary
Arc arcuate hypoth nu
arc arcuate nu neuroepithelium
ASph alisphenoid bone
basv basal vein
BSph basisphenoid bone
CA1 CA1 field of the hippocampus
CA3 CA3 field of the hippocampus
CM central medial thal nu
cp cerebral peduncle, basal
cplex carotid plexus
crhv caudal rhinal vein
Ctd carotid canal
cty chorda tympani nerve
cx cortical neuroepithelium
CxP cortical plate
CxS cortical subplate
D3V dorsal third ventricle
DG dentate gyrus
DLG dorsal lateral geniculate nu
DM dorsomedial hypothalamic nu
dpet deep petrosal nerve
Dura dura membrane
EAM external auditory meatus
ectd external carotid a
f fornix
fi fimbria hippocampus
fr fasciculus retroflexus
Frontal frontal bone
gpet greater petrosal nerve
hi hippocampal formation neuroepi
ictd internal carotid artery
icv inferior cerebral vein
ICx intermediate cortical layer
ipets inferior petrosal sinus
IPit intermediate lobe pituitary
LH lateral hypothalamic area
LHb lateral habenular nu
LP lateral posterior thal nu
lpet lesser petrosal nerve
LV lateral ventricle
m5 sensory root of trigeminal nerve
Malleus malleus (ossicle)
MCLH magnocellular nu lat hypoth
MD mediodorsal thal nu
mfb medial forebrain bundle
MHb medial habenular nu
ml medial lemniscus
mt mammillothalamic tract
MTu medial tuberal nu
NasoPhar nasopharyngeal cavity
Oc occipital cortex
opt optic tract
Par1 parietal cortex, area 1
Parietal parietal bone
ParietalP parietal plate
pcer posterior cerebral artery
PH posterior hypoth area
Pir piriform cortex
Po posterior thal nuclear group
PPit posterior lobe pituitary
PRh perirhinal cortex
PrS presubiculum
PSphW presphenoid wing
ptgpal pterygopalatine artery
PV paraventricular thal nu
Rathke Rathke's pouch
RF rhinal fissure
RSA retrosplenial agranular Cx
RSG retrosplenial granular Cx
Rt reticular thal nu
S subiculum
s5 sensory root of trigeminal nerve
sm stria medullaris thalami
sox supraoptic decussation
Squamous squamous part of temp
STh subthalamic nu
str superior thal radiation
SubG subgeniculate nu
SubV subventricular cortical layer
Te1 temporal cortex, area 1
Te3 temporal cortex, area 3
Temp temporal muscle
TensT tensor tympani muscle
trs transverse sinus
TyC tympanic cavity
VM ventromedial thal
VMH ventromedial hypoth nu
VPL ventral posterolat thal nu
VPM ventral posteromedial thal nu
ZI zona incerta

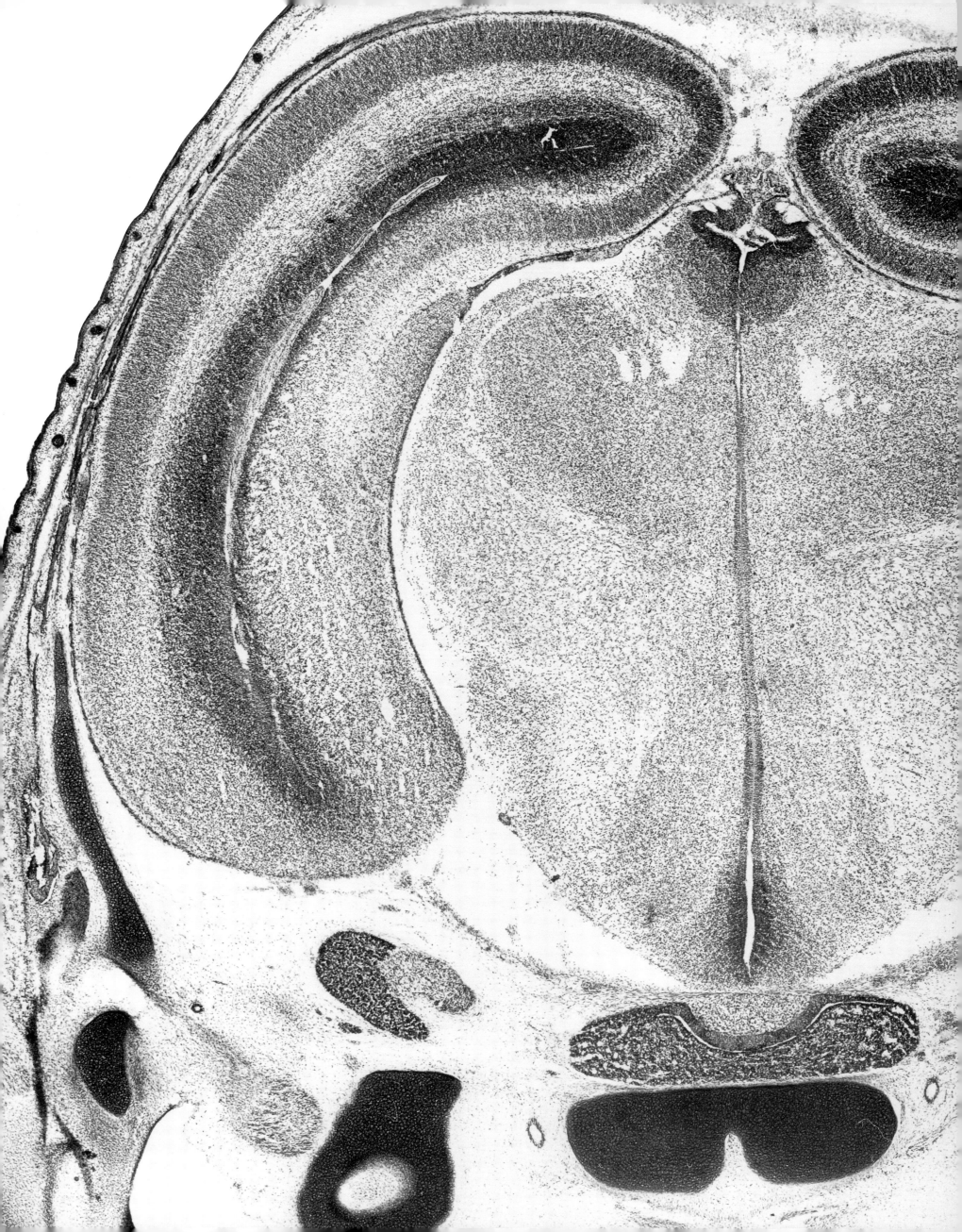

Figure 108
E19 Coronal 28

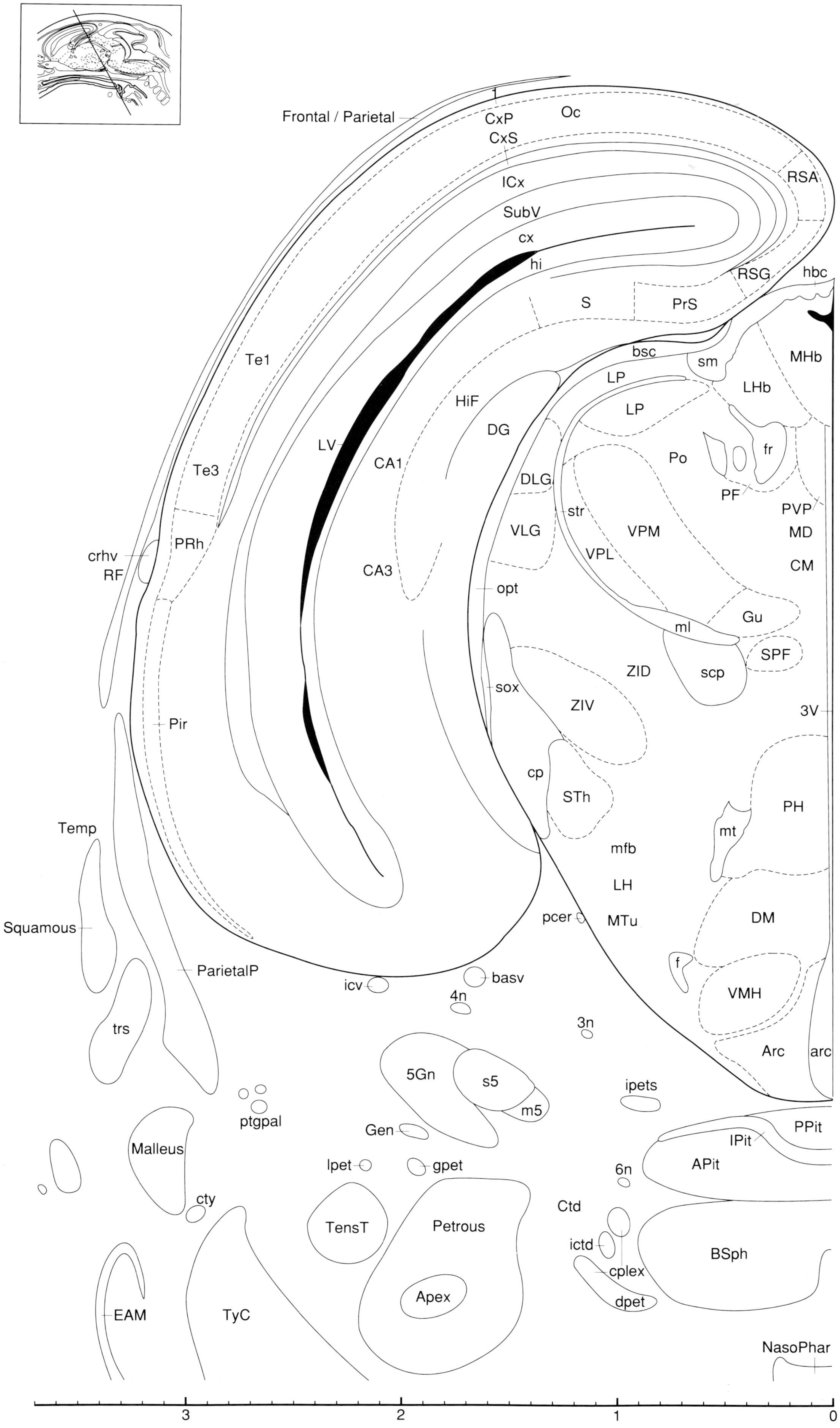

1   cortical layer 1
3n   oculomotor nerve or its root
3V   3rd ventricle
4n   trochlear nerve or its root
5Gn trigeminal ganglion
6n   abducens nerve or its root
Apex   apex of the cochlea
APit   anterior lobe pituitary
Arc   arcuate hypoth nu
arc   arcuate nu neuroepithelium
basv   basal vein
bsc   brachium superior colliculus
BSph   basisphenoid bone
CA1   CA1 field of the hippocampus
CA3   CA3 field of the hippocampus
CM   central medial thal nu
cp   cerebral peduncle, basal
cplex   carotid plexus
crhv   caudal rhinal vein
Ctd   carotid canal
cty   chorda tympani nerve
cx   cortical neuroepithelium
CxP   cortical plate
CxS   cortical subplate
DG   dentate gyrus
DLG   dorsal lateral geniculate nu
DM   dorsomedial hypothalamic nu
dpet   deep petrosal nerve
EAM   external auditory meatus
f   fornix
fr   fasciculus retroflexus
Frontal   frontal bone
Gen   geniculate ganglion
gpet   greater petrosal nerve
Gu   gustatory thalamic nu
hbc   habenular commissure
hi   hippocampal formation neuroepi
HiF   hippocampal fissure
ictd   internal carotid artery
icv   inferior cerebral vein
ICx   intermediate cortical layer
ipets   inferior petrosal sinus
IPit   intermediate lobe pituitary
LH   lateral hypothalamic area
LHb   lateral habenular nu
LP   lateral posterior thal nu
lpet   lesser petrosal nerve
LV   lateral ventricle
m5   motor root of trigeminal nerve
Malleus   malleus (ossicle)
MD   mediodorsal thal nu
mfb   medial forebrain bundle
MHb   medial habenular nu
ml   medial lemniscus
mt   mammillothalamic tract
MTu   medial tuberal nu
NasoPhar   nasopharyngeal cavity
Oc   occipital cortex
opt   optic tract
Parietal   parietal bone
ParietalP   parietal plate
pcer   posterior cerebral artery
Petrous   petrous part, temporal bone
PF   parafascicular thal nu
PH   posterior hypoth area
Pir   piriform cortex
Po   posterior thal nuclear group
PPit   posterior lobe pituitary
PRh   perirhinal cortex
PrS   presubiculum
ptgpal   pterygopalatine artery
PVP   paraventricular thal nu, post
RF   rhinal fissure
RSA   retrosplenial agranular Cx
RSG   retrosplenial granular Cx
S   subiculum
s5   sensory root of trigeminal nerve
scp   superior cerebellar peduncle
sm   stria medullaris thalami
sox   supraoptic decussation
SPF   subparafascicular thal nu
Squamous   squamous part of temp
STh   subthalamic nu
str   superior thal radiation
SubV   subventricular cortical layer
Te1   temporal cortex, area 1
Te3   temporal cortex, area 3
Temp   temporal muscle
TensT   tensor tympani muscle
trs   transverse sinus
TyC   tympanic cavity
VLG   ventral lateral geniculate nu
VMH   ventromedial hypoth nu
VPL   ventral posterolat thal nu
VPM   ventral posteromedial thal nu
ZID   zona incerta, dorsal
ZIV   zona incerta, ventral

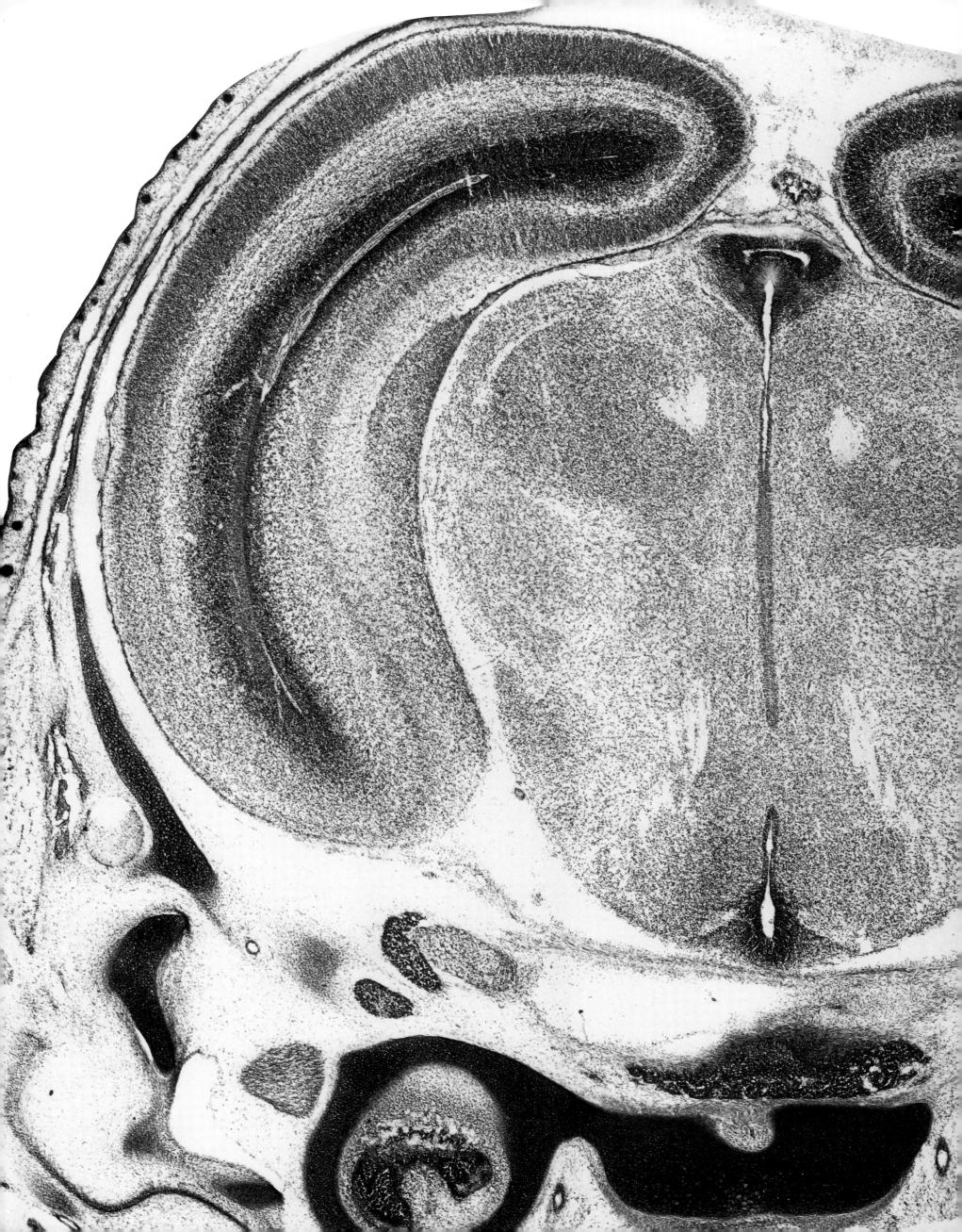

Figure 109
E19 Coronal 29

1 cortical layer 1
3n oculomotor nerve or its root
3V 3rd ventricle
4n trochlear nerve or its root
5Gn trigeminal ganglion
6n abducens nerve or its root
APit anterior lobe pituitary
Arc arcuate hypoth nu
arc arcuate nu neuroepithelium
basv basal vein
bsc brachium superior colliculus
BSph basisphenoid bone
CA1 CA1 field of the hippocampus
CGn cochlear (spiral) ganglion
cp cerebral peduncle, basal
cplex carotid plexus
crhv caudal rhinal vein
cty chorda tympani nerve
cx cortical neuroepithelium
CxP cortical plate
CxS cortical subplate
DG dentate gyrus
DLG dorsal lateral geniculate nu
EAM external auditory meatus
ectd external carotid a
f fornix
fr fasciculus retroflexus
Frontal frontal bone
Gen geniculate ganglion
Gu gustatory thalamic nu
HandleM handle of Malleus
hi hippocampal formation neuroepi
ictd internal carotid artery
icv inferior cerebral vein
ICx intermediate cortical layer
IPit intermediate lobe pituitary
LH lateral hypothalamic area
LHb lateral habenular nu
LM lateral mammillary nu
LP lateral posterior thal nu
LV lateral ventricle
m5 motor root of trigeminal nerve
Malleus malleus (ossicle)
mfb medial forebrain bundle
MHb medial habenular nu
ML medial mammillary nu, lateral
ml medial lemniscus
MLF middle lacerated foramen
MM medial mammillary nu, medial
mp mammillary peduncle
mt mammillothalamic tract
Oc occipital cortex
opt optic tract
Parietal parietal bone
ParietalP parietal plate
pcer posterior cerebral artery
Petrous petrous part, temporal bone
PF parafascicular thal nu
PH posterior hypoth area
Pir piriform cortex
PiRe pineal recess of 3V
PMV premammillary nu, ventral
Po posterior thal nuclear group
PPit posterior lobe pituitary
PrC precommissural nu
PRh perirhinal cortex
PrS presubiculum
ptgpal pterygopalatine artery
PVP paraventricular thal nu, post
RF rhinal fissure
RSA retrosplenial agranular Cx
RSG retrosplenial granular Cx
S subiculum
s5 sensory root of trigeminal nerve
SCO subcommissural organ
scp superior cerebellar peduncle
SMT submammillothalamic nu
SPF subparafascicular thal nu
Squamous squamous part of temp
STh subthalamic nu
str superior thal radiation
SubV subventricular cortical layer
SuM supramammillary nu
Te1 temporal cortex, area 1
Te3 temporal cortex, area 3
Temp temporal muscle
TensT tensor tympani muscle
TM tuberomammillary nu
trs transverse sinus
TyC tympanic cavity
TyM tympanic membrane
VLG ventral lateral geniculate nu
VPL ventral posterolat thal nu
VPM ventral posteromedial thal nu
ZID zona incerta, dorsal
ZIV zona incerta, ventral

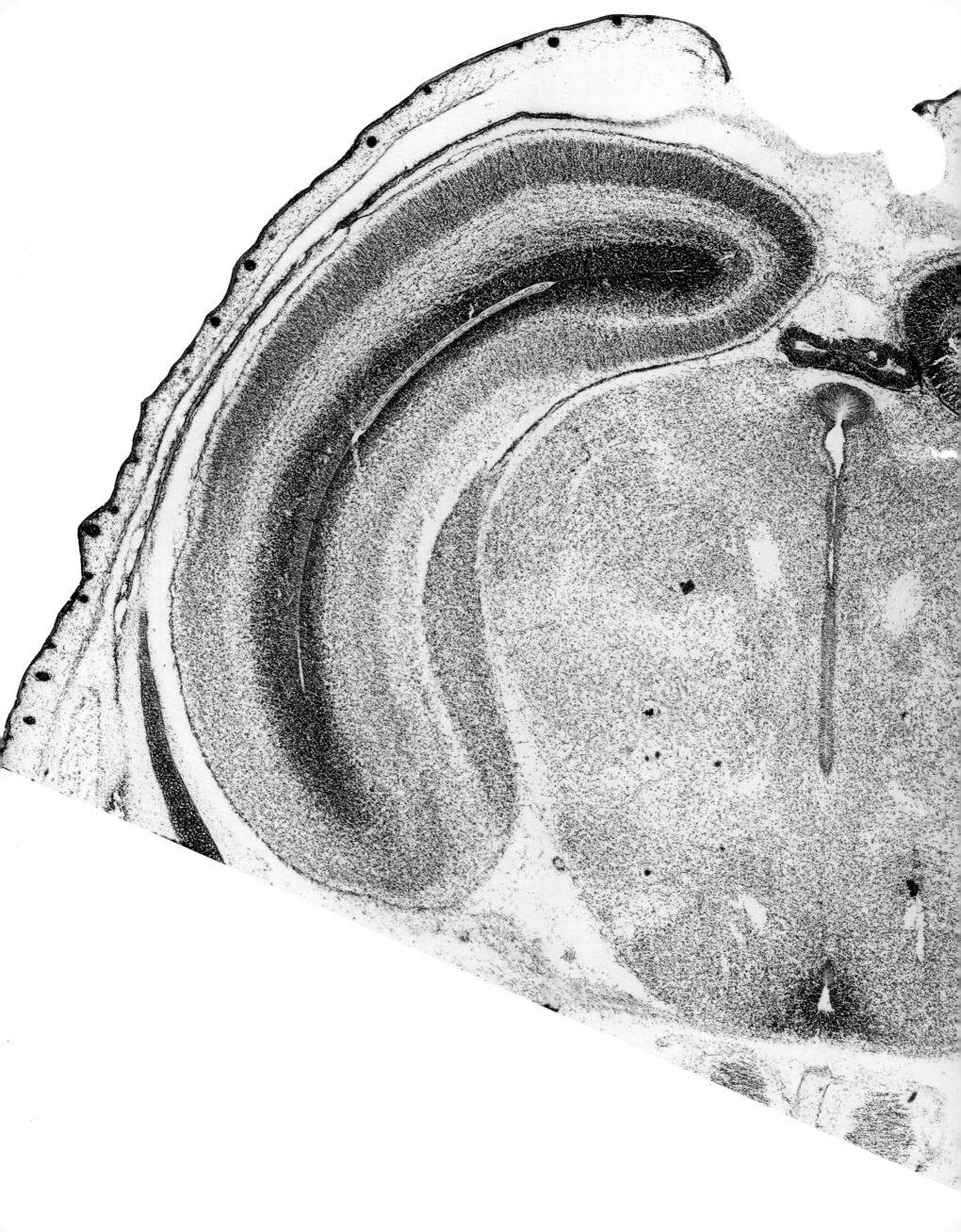

Figure 110
E19 Coronal 30

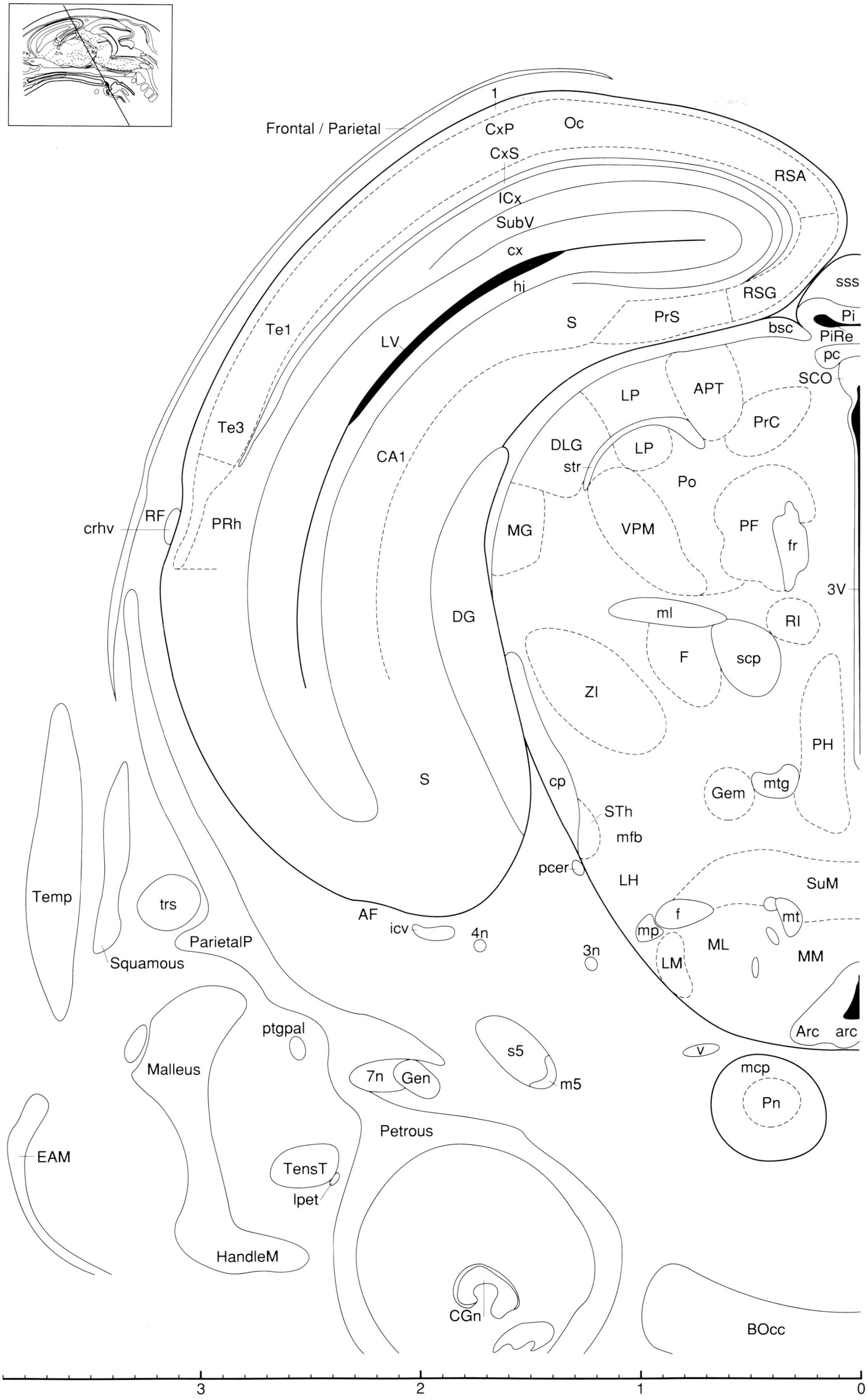

1 cortical layer 1
3n oculomotor nerve or its root
3V 3rd ventricle
4n trochlear nerve or its root
7n facial nerve or its root
AF amygdaloid fissure
APT anterior pretectal nu
Arc arcuate hypoth nu
arc arcuate nu neuroepithelium
BOcc basioccipital bone
bsc brachium superior colliculus
CA1 CA1 field of the hippocampus
CGn cochlear (spiral) ganglion
cp cerebral peduncle, basal
crhv caudal rhinal vein
cx cortical neuroepithelium
CxP cortical plate
CxS cortical subplate
DG dentate gyrus
DLG dorsal lateral geniculate nu
EAM external auditory meatus
F nu fields of Forel
f fornix
fr fasciculus retroflexus
Frontal frontal bone
Gem gemini hypoth nu
Gen geniculate ganglion
HandleM handle of Malleus
hi hippocampal formation neuroepi
icv inferior cerebral vein
ICx intermediate cortical layer
LH lateral hypothalamic area
LM lateral mammillary nu
LP lateral posterior thal nu
lpet lesser petrosal nerve
LV lateral ventricle
m5 motor root of trigeminal nerve
Malleus malleus (ossicle)
mcp middle cerebellar peduncle
mfb medial forebrain bundle
MG medial geniculate nu
ML medial mammillary nu, lateral
ml medial lemniscus
MM medial mammillary nu, medial
mp mammillary peduncle
mt mammillothalamic tract
mtg mammillotegmental tract
Oc occipital cortex
Parietal parietal bone
ParietalP parietal plate
pc posterior commissure
pcer posterior cerebral artery
Petrous petrous part, temporal bone
PF parafascicular thal nu
PH posterior hypoth area
Pi pineal gland
PiRe pineal recess of 3V
Pn pontine nuclei
Po posterior thal nuclear group
PrC precommissural nu
PRh perirhinal cortex
PrS presubiculum
ptgpal pterygopalatine artery
RF rhinal fissure
RI rostral interstitial nu mlf
RSA retrosplenial agranular Cx
RSG retrosplenial granular Cx
S subiculum
s5 sensory root of trigeminal nerve
SCO subcommissural organ
scp superior cerebellar peduncle
Squamous squamous part of temp
sss superior sagittal sinus
STh subthalamic nu
str superior thal radiation
SubV subventricular cortical layer
SuM supramammillary nu
Te1 temporal cortex, area 1
Te3 temporal cortex, area 3
Temp temporal muscle
TensT tensor tympani muscle
trs transverse sinus
v vein
VPM ventral posteromedial thal nu
ZI zona incerta

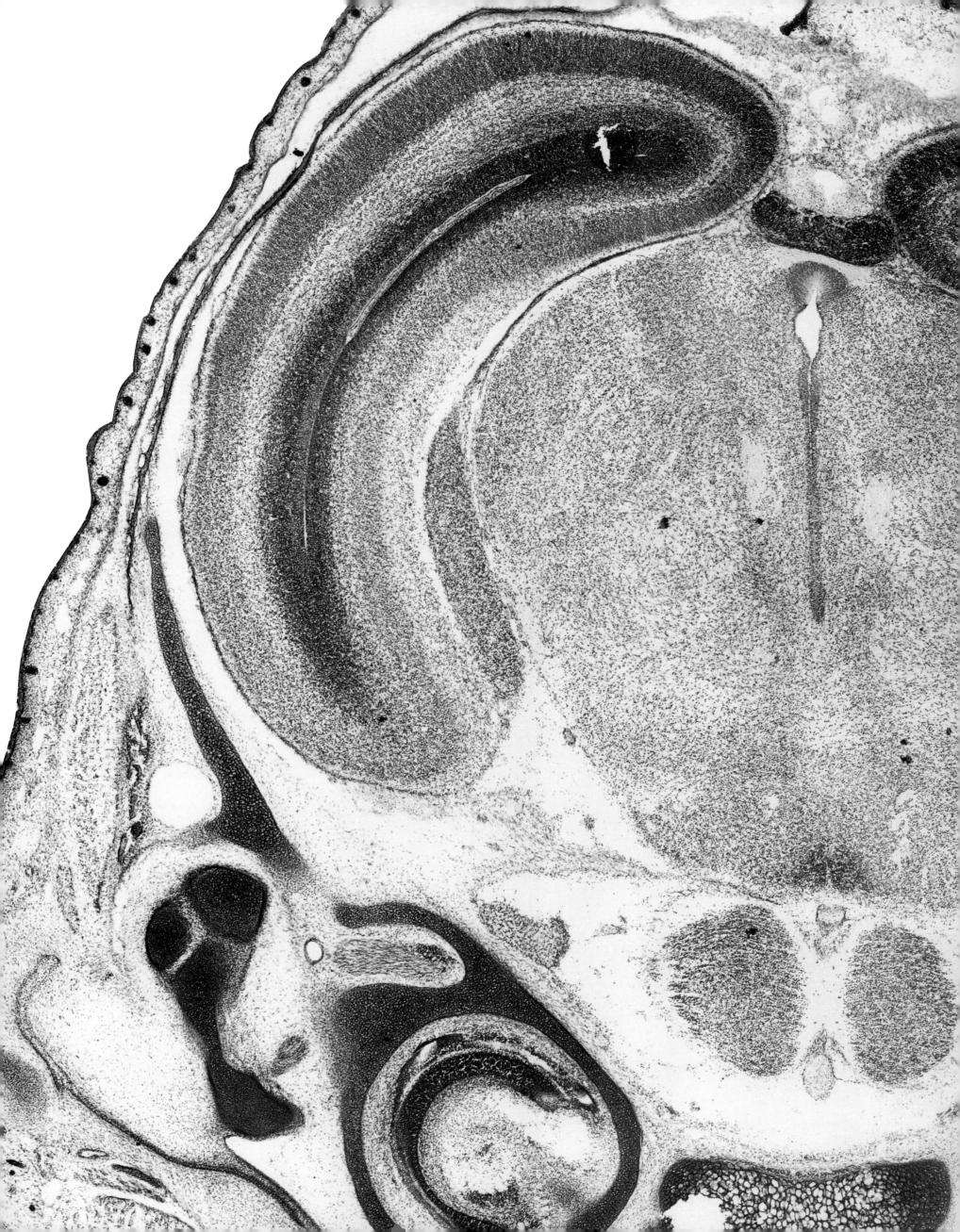

Figure 111
E19 Coronal 31

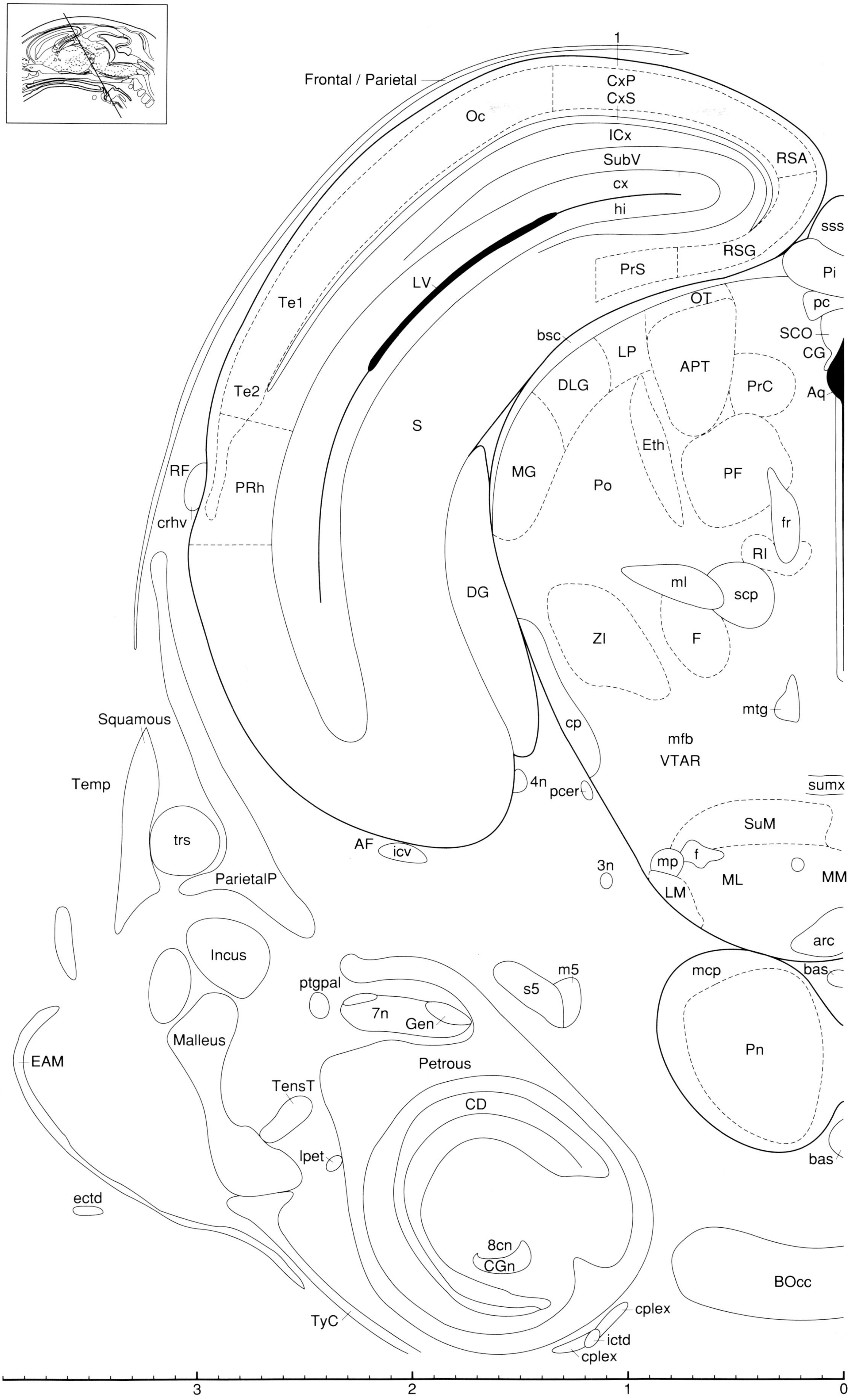

1   cortical layer 1
3n   oculomotor nerve or its root
4n   trochlear nerve or its root
7n   facial nerve or its root
8cn   cochlear root vestibulococh nerve
AF   amygdaloid fissure
APT   anterior pretectal nu
Aq   aqueduct
arc   arcuate nu neuroepithelium
bas   basilar artery
BOcc   basioccipital bone
bsc   brachium superior colliculus
CD   cochlear duct
CG   central gray
CGn   cochlear (spiral) ganglion
cp   cerebral peduncle, basal
cplex   carotid plexus
crhv   caudal rhinal vein
cx   cortical neuroepithelium
CxP   cortical plate
CxS   cortical subplate
DG   dentate gyrus
DLG   dorsal lateral geniculate nu
EAM   external auditory meatus
ectd   external carotid a
Eth   ethmoid thal nu
F   nu fields of Forel
f   fornix
fr   fasciculus retroflexus
Frontal   frontal bone
Gen   geniculate ganglion
hi   hippocampal formation neuroepi
ictd   internal carotid artery
icv   inferior cerebral vein
ICx   intermediate cortical layer
Incus   incus (ossicle)
LM   lateral mammillary nu
LP   lateral posterior thal nu
lpet   lesser petrosal nerve
LV   lateral ventricle
m5   motor root of trigeminal nerve
Malleus   malleus (ossicle)
mcp   middle cerebellar peduncle
mfb   medial forebrain bundle
MG   medial geniculate nu
ML   medial mammillary nu, lateral
ml   medial lemniscus
MM   medial mammillary nu, medial
mp   mammillary peduncle
mtg   mammillotegmental tract
Oc   occipital cortex
OT   nu optic tract
Parietal   parietal bone
ParietalP   parietal plate
pc   posterior commissure
pcer   posterior cerebral artery
Petrous   petrous part, temporal bone
PF   parafascicular thal nu
Pi   pineal gland
Pn   pontine nuclei
Po   posterior thal nuclear group
PrC   precommissural nu
PRh   perirhinal cortex
PrS   presubiculum
ptgpal   pterygopalatine artery
RF   rhinal fissure
RI   rostral interstitial nu mlf
RSA   retrosplenial agranular Cx
RSG   retrosplenial granular Cx
S   subiculum
s5   sensory root of trigeminal nerve
SCO   subcommissural organ
scp   superior cerebellar peduncle
Squamous   squamous part of temp
sss   superior sagittal sinus
SubV   subventricular cortical layer
SuM   supramammillary nu
sumx   supramammillary decussation
Te1   temporal cortex, area 1
Te2   temporal cortex, area 2
Temp   temporal muscle
TensT   tensor tympani muscle
trs   transverse sinus
TyC   tympanic cavity
VTAR   vent tegmental area, rostral
ZI   zona incerta

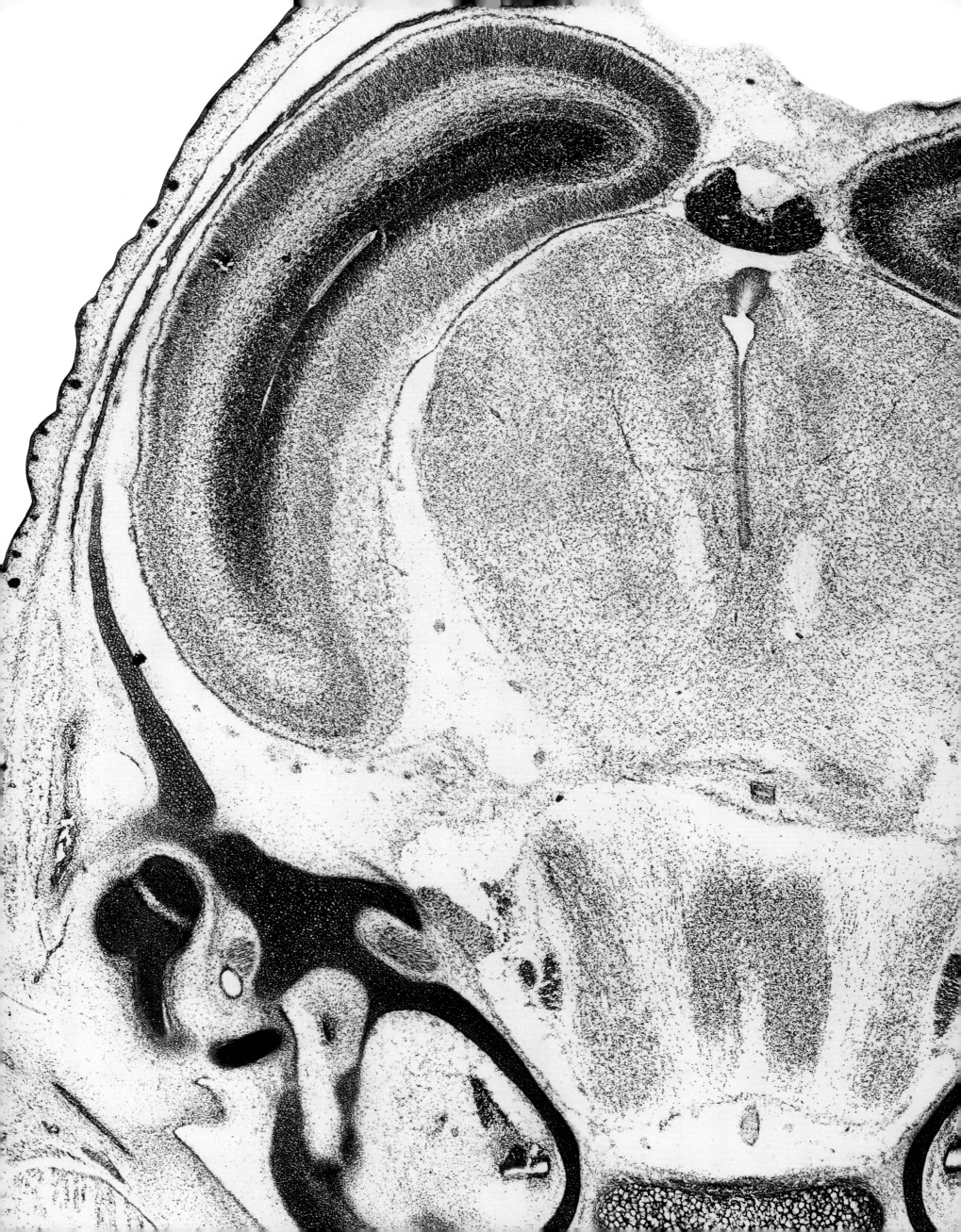

Figure 112
E19 Coronal 32

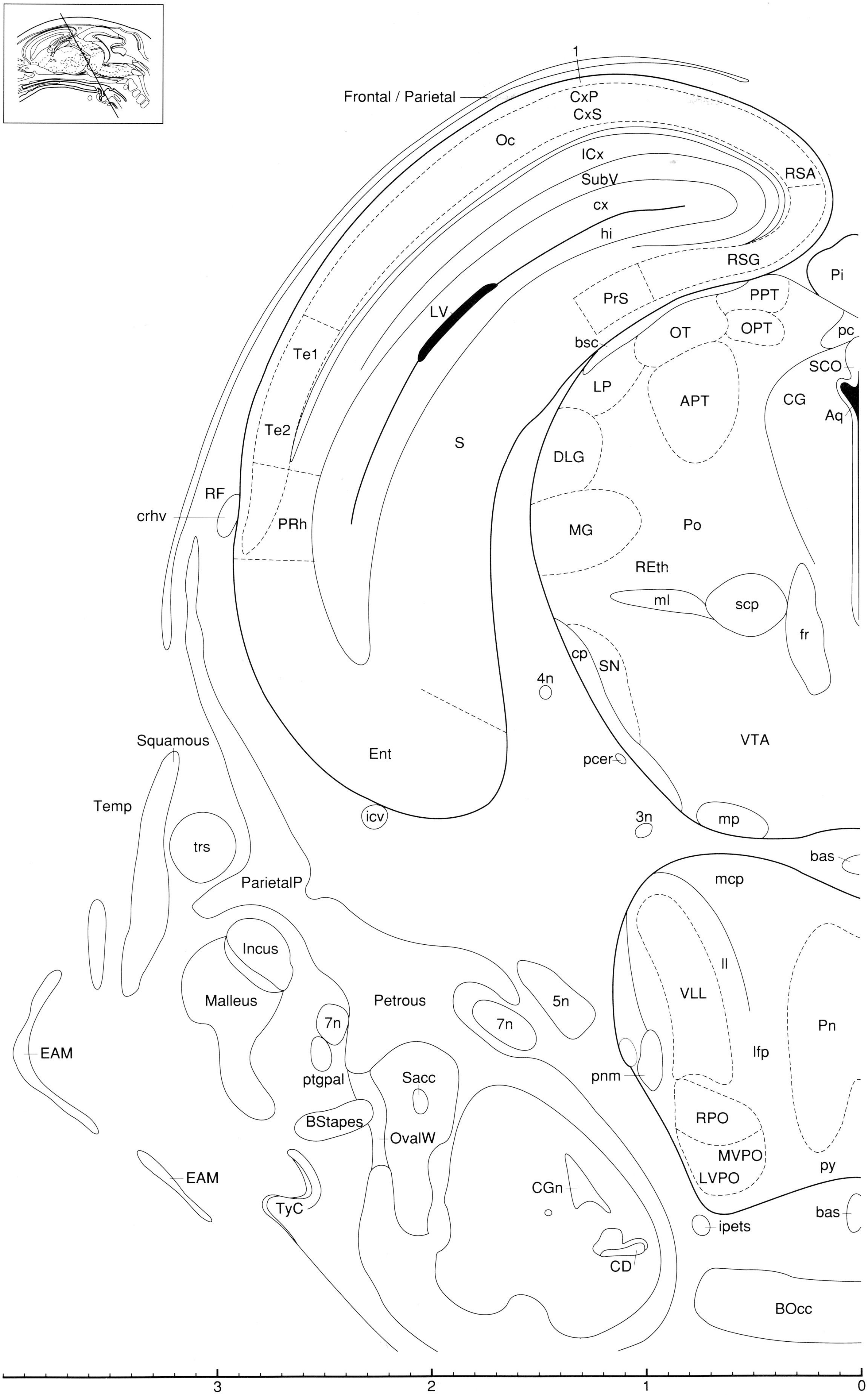

Frontal / Parietal

1   cortical layer 1
3n   oculomotor nerve or its root
4n   trochlear nerve or its root
5n   trigeminal nerve
7n   facial nerve or its root
APT   anterior pretectal nu
Aq   aqueduct
bas   basilar artery
BOcc   basioccipital bone
bsc   brachium superior colliculus
BStapes   base of stapes
CD   cochlear duct
CG   central gray
CGn   cochlear (spiral) ganglion
cp   cerebral peduncle, basal
crhv   caudal rhinal vein
cx   cortical neuroepithelium
CxP   cortical plate
CxS   cortical subplate
DLG   dorsal lateral geniculate nu
EAM   external auditory meatus
Ent   entorhinal cortex
fr   fasciculus retroflexus
Frontal   frontal bone
hi   hippocampal formation neuroepi
icv   inferior cerebral vein
ICx   intermediate cortical layer
Incus   incus (ossicle)
ipets   inferior petrosal sinus
lfp   longitudinal fasciculus pons
ll   lateral lemniscus
LP   lateral posterior thal nu
LV   lateral ventricle
LVPO   lateroventral periolivary nu
Malleus   malleus (ossicle)
mcp   middle cerebellar peduncle
MG   medial geniculate nu
ml   medial lemniscus
mp   mammillary peduncle
MVPO   medioventral periolivary nu
Oc   occipital cortex
OPT   olivary pretectal nu
OT   nu optic tract
OvalW   oval window
Parietal   parietal bone
ParietalP   parietal plate
pc   posterior commissure
pcer   posterior cerebral artery
Petrous   petrous part, temporal bone
Pi   pineal gland
Pn   pontine nuclei
pnm   pontine migration
Po   posterior thal nuclear group
PPT   posterior pretectal nu
PRh   perirhinal cortex
PrS   presubiculum
ptgpal   pterygopalatine artery
py   pyramidal tract
REth   retroethmoid nu
RF   rhinal fissure
RPO   rostral periolivary region
RSA   retrosplenial agranular Cx
RSG   retrosplenial granular Cx
S   subiculum
Sacc   saccule
SCO   subcommissural organ
scp   superior cerebellar peduncle
SN   substantia nigra
Squamous   squamous part of temp
SubV   subventricular cortical layer
Te1   temporal cortex, area 1
Te2   temporal cortex, area 2
Temp   temporal muscle
trs   transverse sinus
TyC   tympanic cavity
VLL   ventral nu lateral lemniscus
VTA   ventral tegmental area

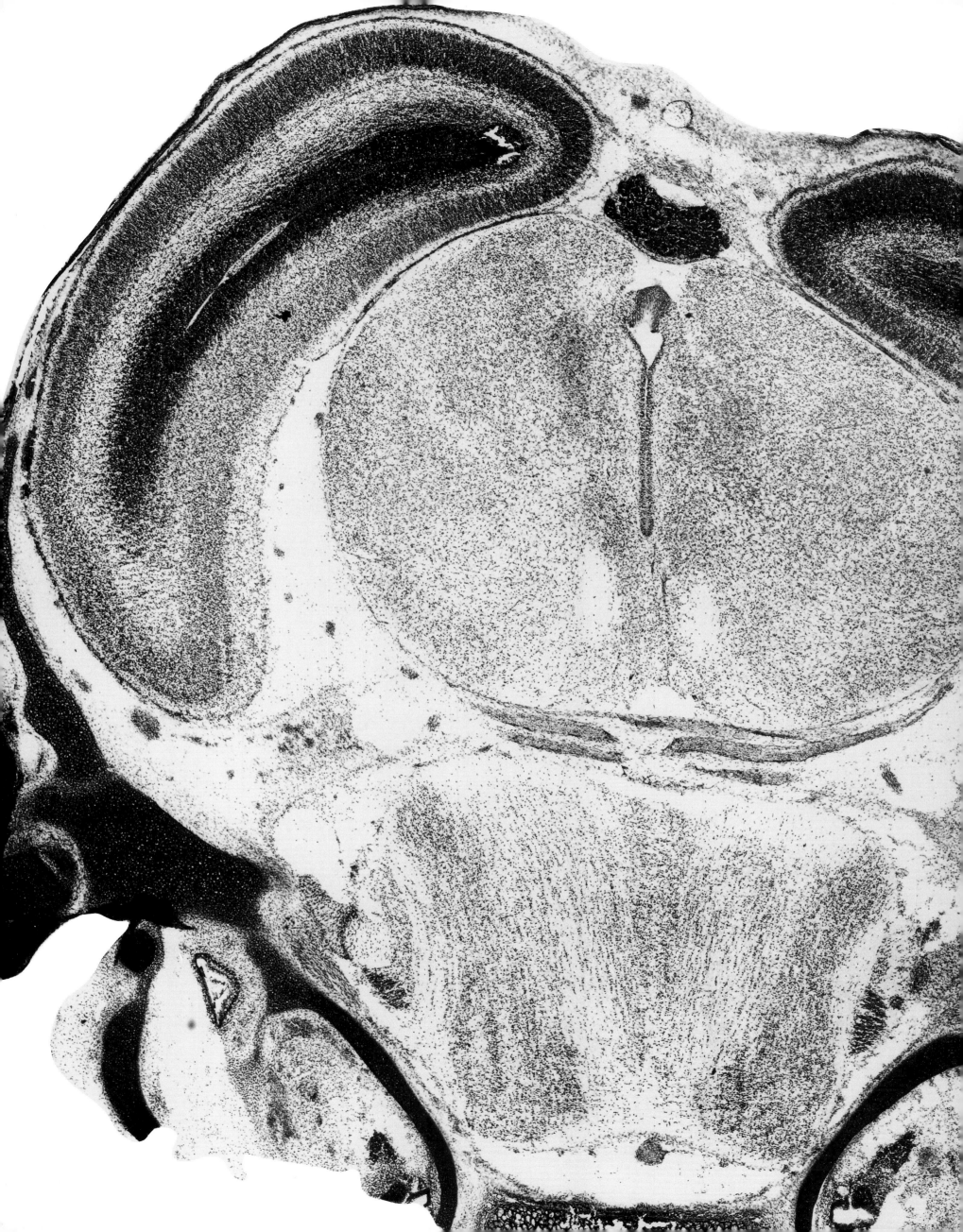

Figure 113
E19 Coronal 33

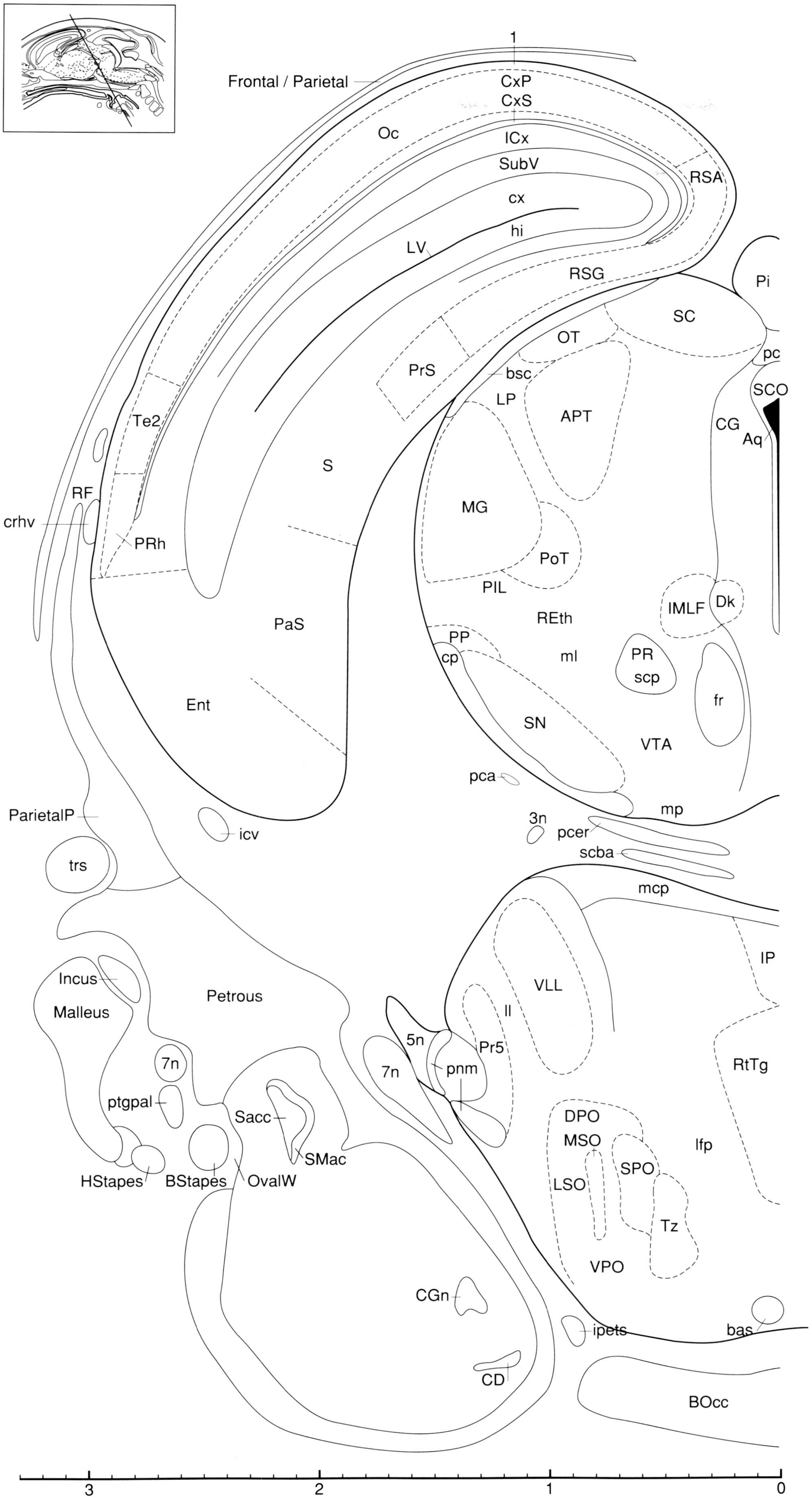

Frontal / Parietal

1
CxP
CxS
Oc
ICx
SubV
cx
hi
LV
RSA
RSG
Pi
SC
OT
PrS
bsc
LP
APT
pc
SCO
CG
Aq
Te2
S
MG
RF
crhv
PRh
PoT
PIL
REth
IMLF
Dk
PaS
PP
ml
PR
scp
cp
fr
Ent
SN
VTA
pca
ParietalP
icv
mp
3n
pcer
trs
scba
mcp
Petrous
IP
Incus
VLL
Malleus
ll
5n
Pr5
7n
pnm
RtTg
ptgpal
DPO
Sacc
MSO
lfp
SMac
LSO
SPO
HStapes BStapes OvalW
Tz
VPO
CGn
ipets
bas
CD
BOcc

3   2   1   0

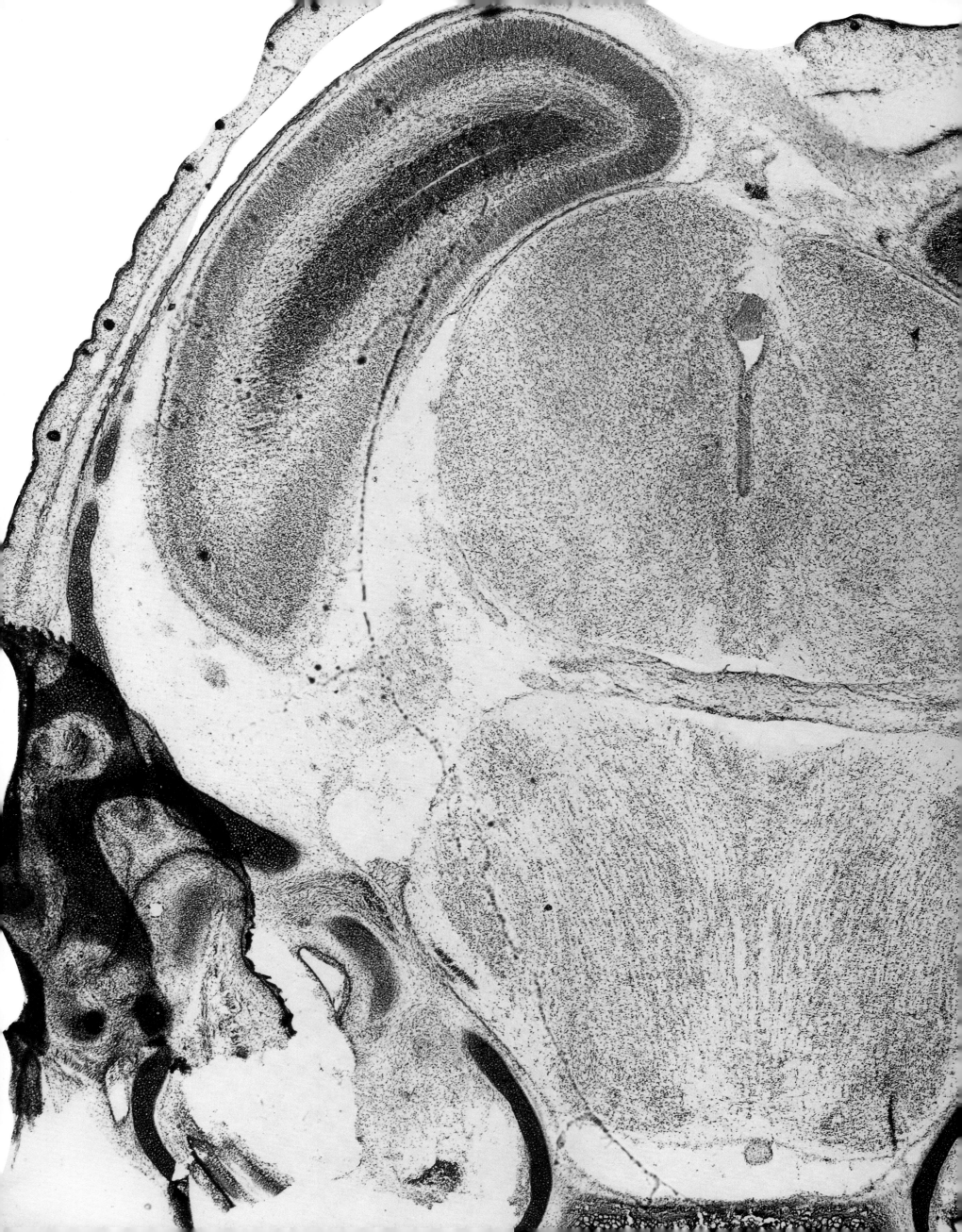

Figure 114
E19 Coronal 34

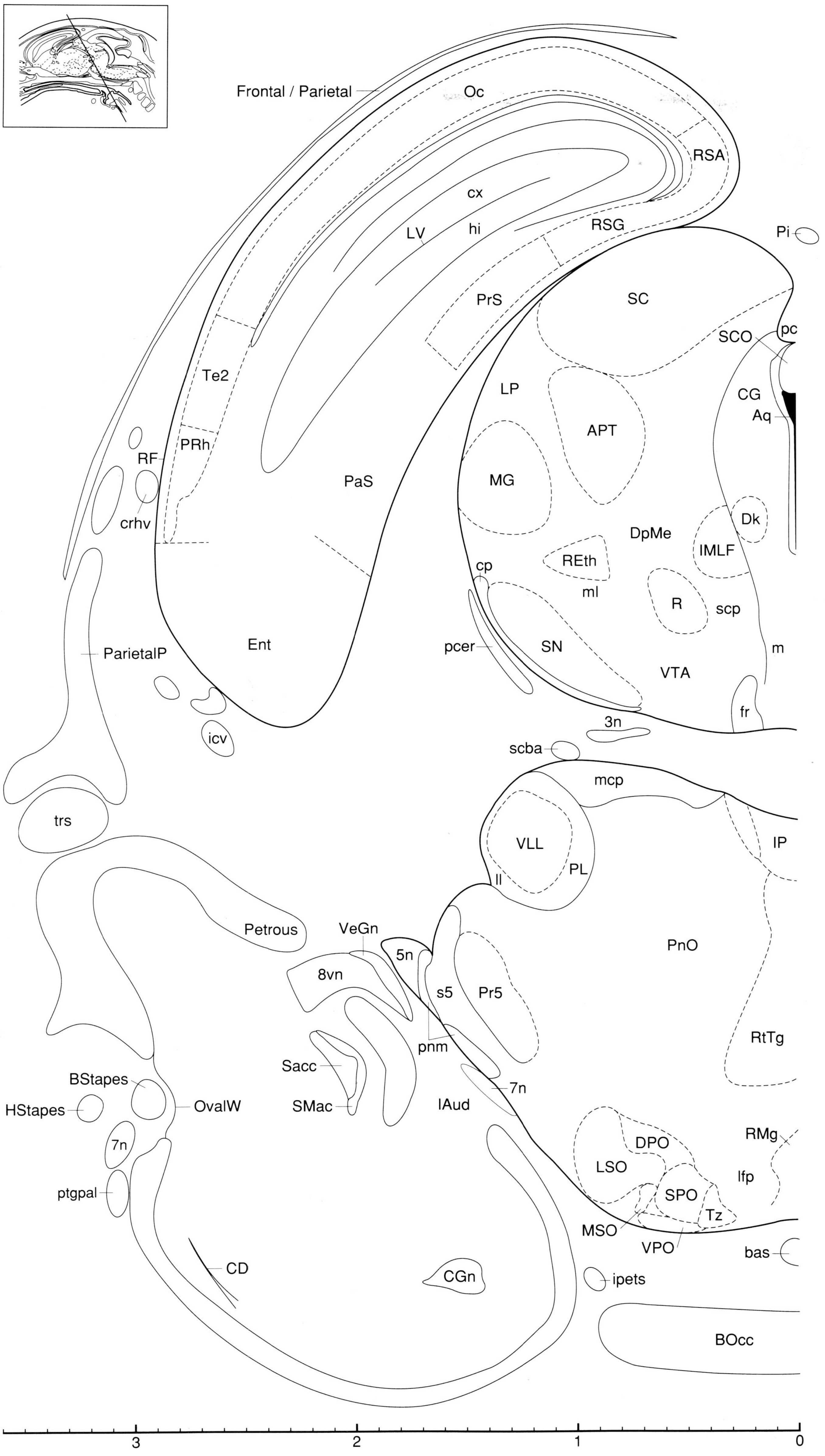

3n   oculomotor nerve or its root
5n   trigeminal nerve
7n   facial nerve or its root
8vn   vestibular root vestibulococh n
APT   anterior pretectal nu
Aq   aqueduct
bas   basilar artery
BOcc basioccipital bone
BStapes   base of stapes
CD   cochlear duct
CG   central gray
CGn   cochlear (spiral) ganglion
cp   cerebral peduncle, basal
crhv caudal rhinal vein
cx   cortical neuroepithelium
Dk   nucleus Darkschewitsch
DpMe deep mesencephalic nu
DPO   dorsal periolivary region
Ent   entorhinal cortex
fr   fasciculus retroflexus
Frontal   frontal bone
hi   hippocampal formation neuroepi
HStapes   head of stapes
IAud interanteromedial thal nu
icv   inferior cerebral vein
IMLF interstitial nu mlf
IP   interpeduncular nu
ipets inferior petrosal sinus
lfp   longitudinal fasciculus pons
ll   lateral lemniscus
LP   lateral posterior thal nu
LSO   lateral superior olive
LV   lateral ventricle
m   migration of neurons
mcp   middle cerebellar peduncle
MG   medial geniculate nu
ml   medial lemniscus
MSO   medial superior olive
Oc   occipital cortex
OvalW oval window
Parietal   parietal bone
ParietalP parietal plate
PaS   parasubiculum
pc   posterior commissure
pcer posterior cerebral artery
Petrous   petrous part, temporal bone
Pi   pineal gland
PL   paralemniscal nu
pnm   pontine migration
PnO   pontine reticular nu, oral
Pr5   principal sensory trigeminal nu
PRh   perirhinal cortex
PrS   presubiculum
ptgpal pterygopalatine artery
R   red nu
REth retroethmoid nu
RF   rhinal fissure
RMg   raphe magnus nu
RSA   retrosplenial agranular Cx
RSG   retrosplenial granular Cx
RtTg reticulotegmental nu pons
s5   sensory root trigeminal nerve
Sacc saccule
SC   superior colliculus
scba superior cerebellar artery
SCO   subcommissural organ
scp   superior cerebellar peduncle
SMac saccular macula
SN   substantia nigra
SPO   superior paraolivary nu
Te2   temporal cortex, area 2
trs   transverse sinus
Tz   nu trapezoid body
VeGn vestibular ganglion
VLL   ventral nu lateral lemniscus
VPO   ventral periolivary nuclei
VTA   ventral tegmental area

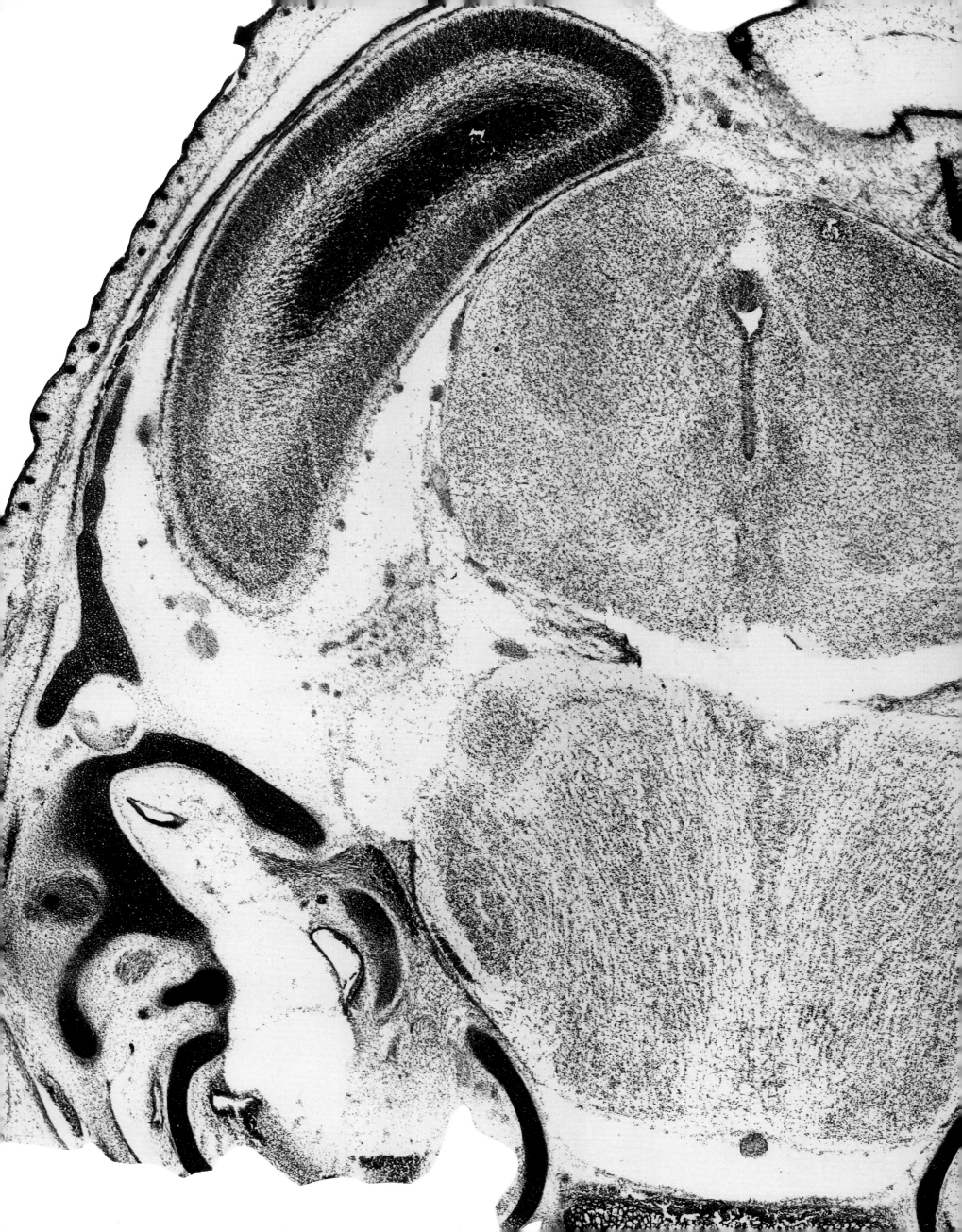

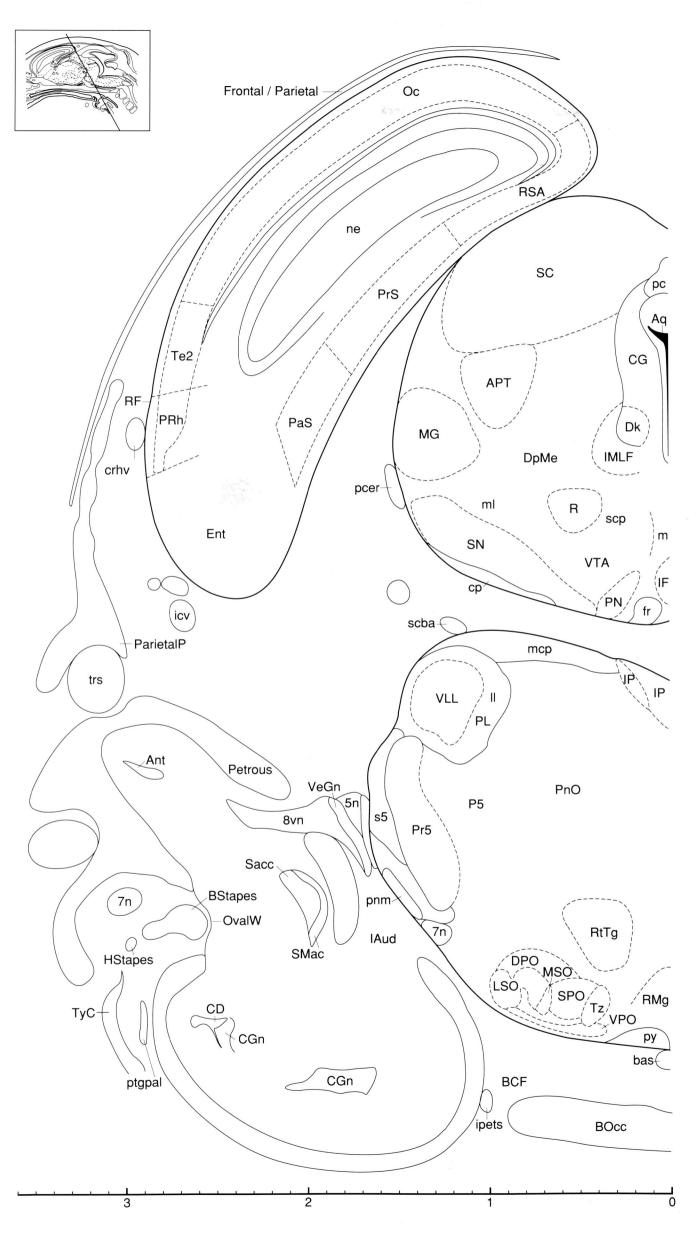

Figure 115
E19 Coronal 35

5n   trigeminal nerve
7n   facial nerve or its root
8vn   vestibular root vestibulococh n
Ant   anterior semicircular duct
APT   anterior pretectal nu
Aq   aqueduct
bas   basilar artery
BCF   basocochlear fissure
BOcc   basioccipital bone
BStapes   base of stapes
CD   cochlear duct
CG   central gray
CGn   cochlear (spiral) ganglion
cp   cerebral peduncle, basal
crhv   caudal rhinal vein
Dk   nucleus Darkschewitsch
DpMe   deep mesencephalic nu
DPO   dorsal periolivary region
Ent   entorhinal cortex
fr   fasciculus retroflexus
Frontal   frontal bone
HStapes   head of stapes
IAud   interanteromedial thal nu
icv   inferior cerebral vein
IF   interfascicular nucleus
IMLF   interstitial nu mlf
IP   interpeduncular nu
ipets   inferior petrosal sinus
ll   lateral lemniscus
LSO   lateral superior olive
m   migration of neurons
mcp   middle cerebellar peduncle
MG   medial geniculate nu
ml   medial lemniscus
MSO   medial superior olive
ne   neuroepithelium
Oc   occipital cortex
OvalW   oval window
P5   peritrigeminal zone
Parietal   parietal bone
ParietalP   parietal plate
PaS   parasubiculum
pc   posterior commissure
pcer   posterior cerebral artery
Petrous   petrous part, temporal bone
PL   paralemniscal nu
PN   paranigral nu
pnm   pontine migration
PnO   pontine reticular nu, oral
Pr5   principal sensory trigeminal nu
PRh   perirhinal cortex
PrS   presubiculum
ptgpal   pterygopalatine artery
py   pyramidal tract
R   red nu
RF   rhinal fissure
RMg   raphe magnus nu
RSA   retrosplenial agranular Cx
RtTg   reticulotegmental nu pons
s5   sensory root trigeminal nerve
Sacc   saccule
SC   superior colliculus
scba   superior cerebellar artery
scp   superior cerebellar peduncle
SMac   saccular macula
SN   substantia nigra
SPO   superior paraolivary nu
Te2   temporal cortex, area 2
trs   transverse sinus
TyC   tympanic cavity
Tz   nu trapezoid body
VeGn   vestibular ganglion
VLL   ventral nu lateral lemniscus
VPO   ventral periolivary nuclei
VTA   ventral tegmental area

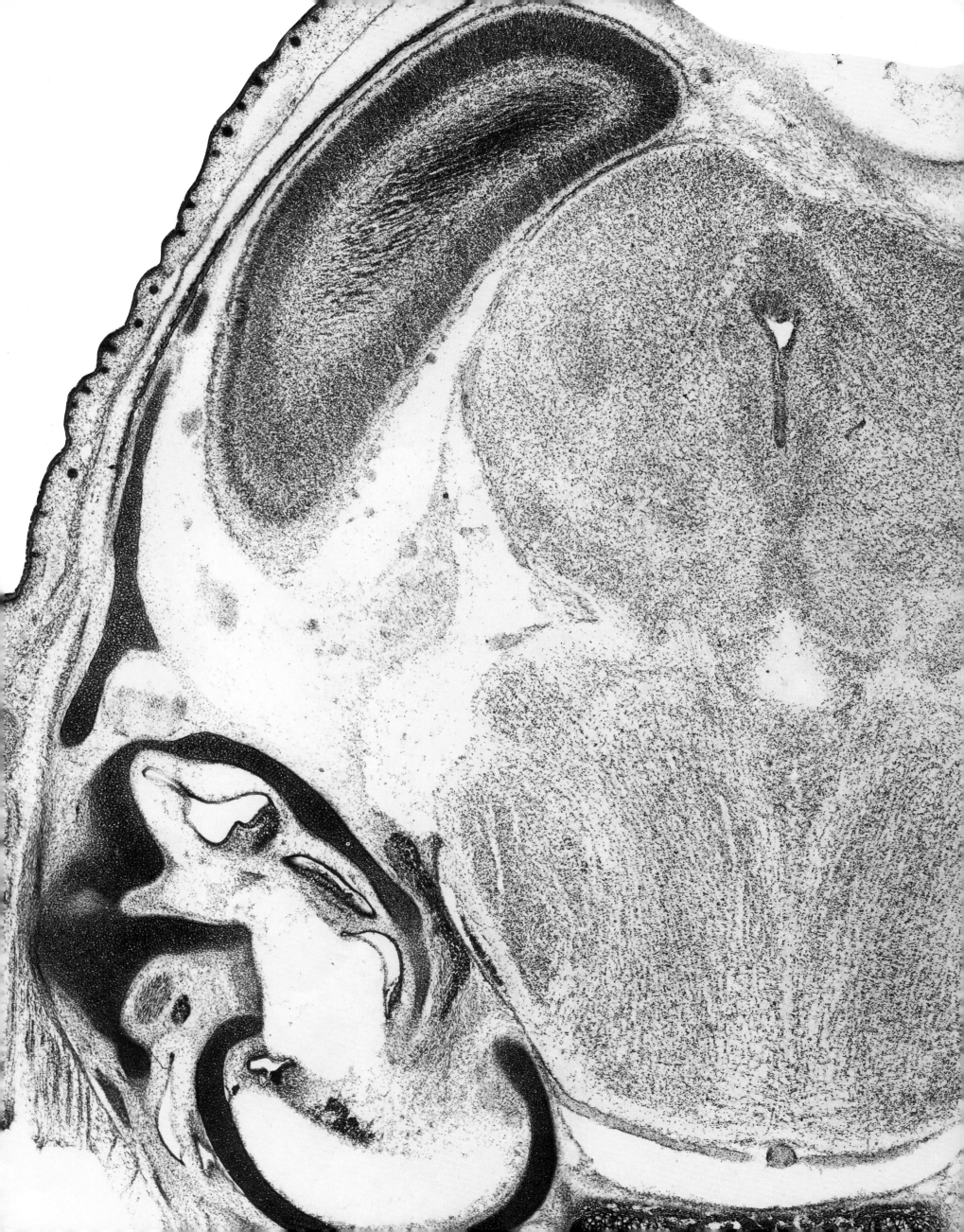

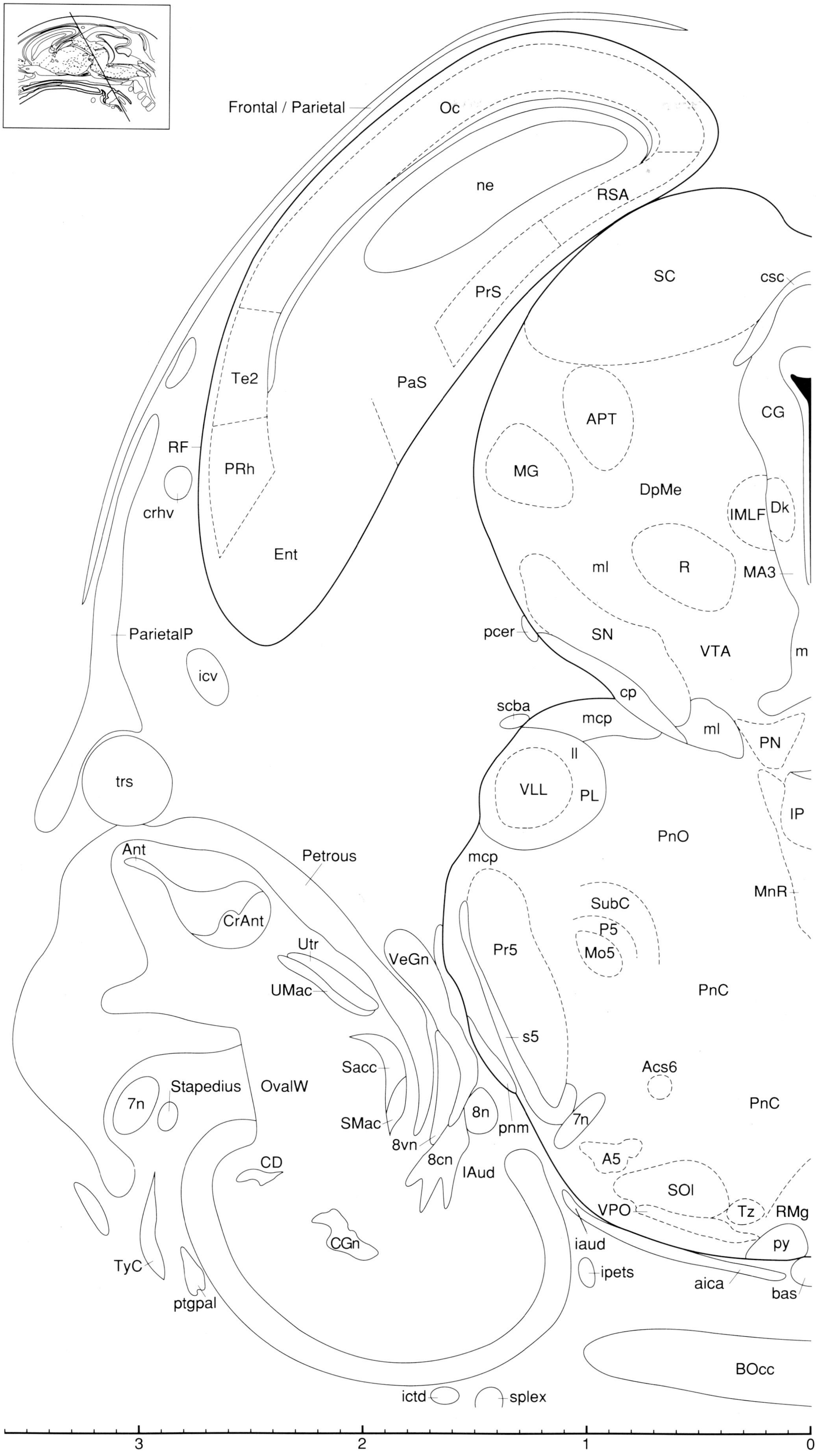

Figure 116
E19 Coronal 36

Frontal / Parietal

Oc

ne

RSA

SC

csc

PrS

Te2

PaS

APT

CG

RF

MG

DpMe

IMLF

Dk

PRh

crhv

Ent

ml

R

MA3

ParietalP

pcer

SN

VTA

m

icv

scba

cp

mcp

ml

PN

ll

trs

VLL

PL

IP

Ant

Petrous

mcp

PnO

MnR

CrAnt

SubC

Utr

P5

Pr5

Mo5

PnC

VeGn

UMac

s5

Acs6

PnC

Sacc

Stapedius

OvalW

7n

SMac

8n

7n

8vn

pnm

CD

8cn

IAud

A5

SOl

TyC

VPO

Tz

RMg

CGn

iaud

py

ptgpal

ipets

aica

bas

BOcc

ictd

splex

3 2 1 0

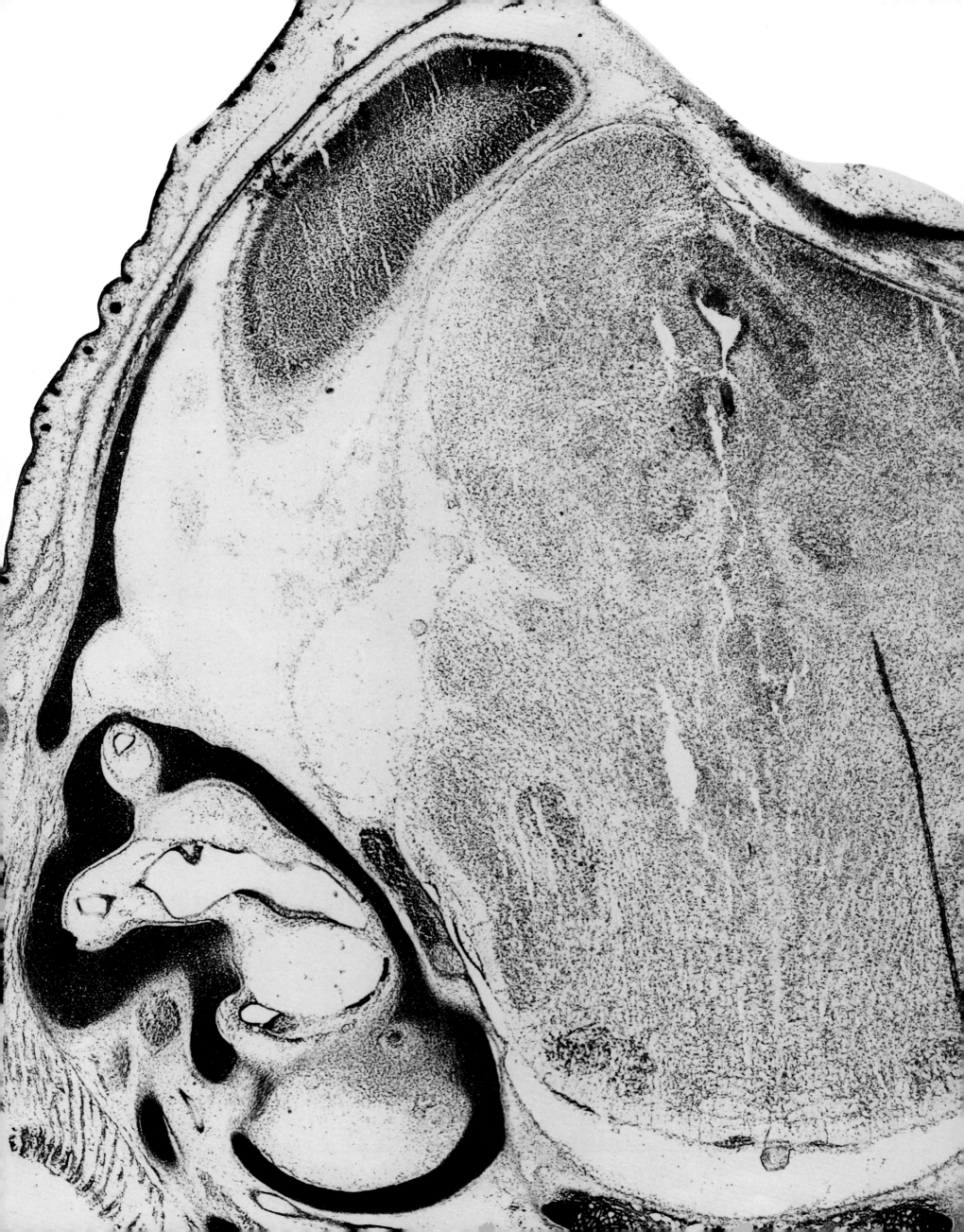

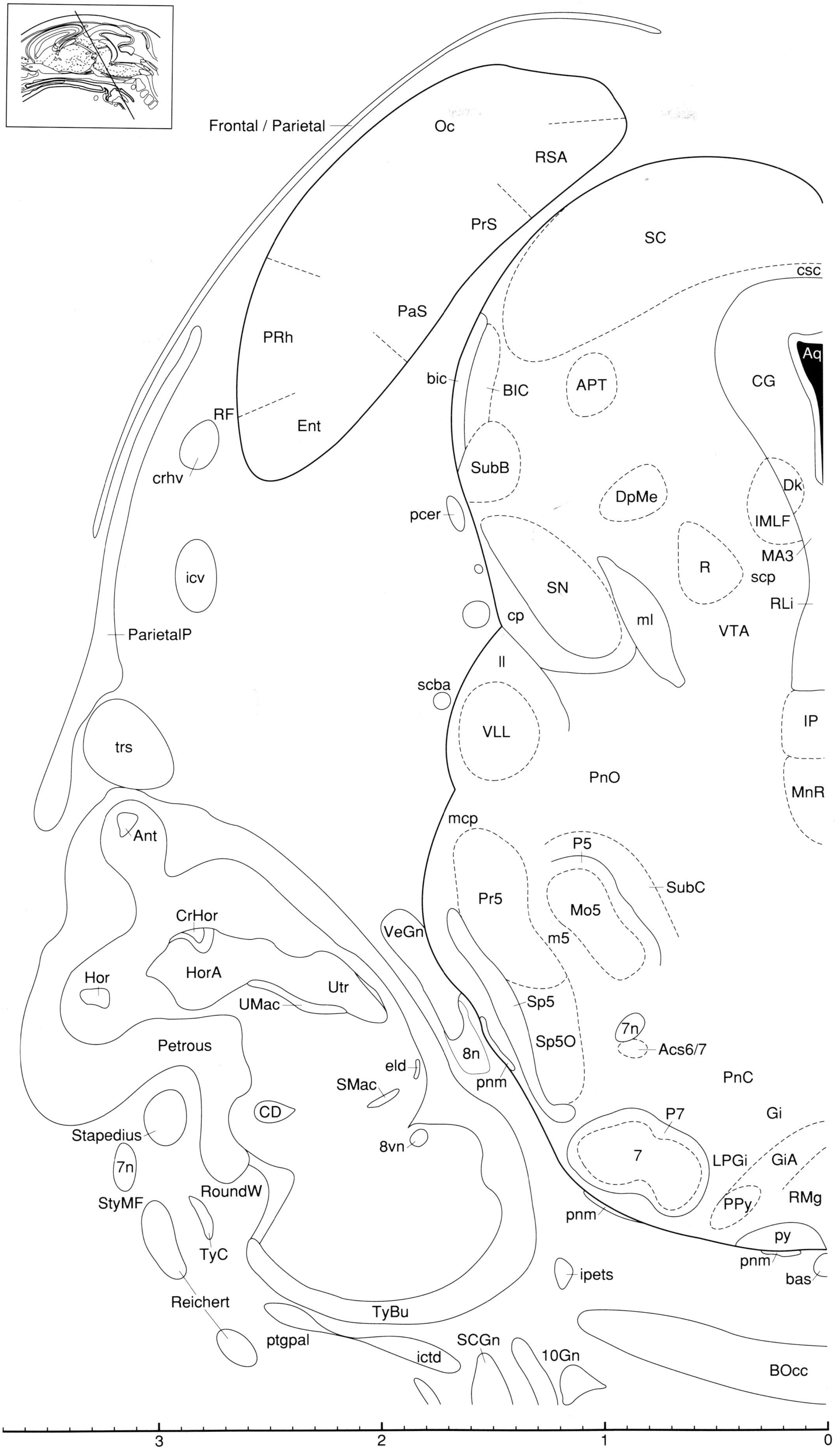

Figure 117
E19 Coronal 37

Frontal / Parietal
Oc
RSA
PrS
SC
csc
PaS
PRh
bic
BIC
APT
CG
Aq
RF
Ent
SubB
DpMe
Dk
crhv
pcer
IMLF
MA3
icv
SN
R
scp
RLi
cp
ml
VTA
ParietalP
ll
scba
VLL
IP
trs
PnO
MnR
mcp
P5
Ant
Pr5
SubC
CrHor
Mo5
VeGn
m5
Hor
HorA
Utr
Sp5
UMac
Sp5O
7n
Petrous
8n
Acs6/7
eld
pnm
PnC
SMac
8vn
P7
Gi
CD
7
LPGi
GiA
Stapedius
RoundW
PPy
RMg
7n
StyMF
pnm
TyC
py
pnm
Reichert
bas
ipets
ptgpal
TyBu
ictd
SCGn
10Gn
BOcc

7   facial nu
7n   facial nerve or its root
8n   vestibulocochlear nerve
8vn   vestibular root vestibulococh n
10Gn   vagal ganglion
Acs6   accessory abducens nu
Acs7   accessory facial nu
Ant   anterior semicircular duct
APT   anterior pretectal nu
Aq   aqueduct
bas   basilar artery
BIC   nu brachium inferior colliculus
bic   brachium inferior colliculus
BOcc   basioccipital bone
CD   cochlear duct
CG   central gray
cp   cerebral peduncle, basal
CrHor   crista of horizontal ampulla
crhv   caudal rhinal vein
csc   commissure superior colliculus
Dk   nucleus Darkschewitsch
DpMe   deep mesencephalic nu
eld   endolymphatic duct
Ent   entorhinal cortex
Frontal   frontal bone
Gi   gigantocellular reticular nucleus
GiA   gigantocell reticular nu, alpha
Hor   horizontal semicircular duct
HorA   ampulla of hor semicircular duct
ictd   internal carotid artery
icv   inferior cerebral vein
IMLF   interstitial nu mlf
IP   interpeduncular nu
ipets   inferior petrosal sinus
ll   lateral lemniscus
LPGi   lateral paragigantocellular nu
m5   motor root trigeminal nerve
MA3   medial acs oculomotor nu
mcp   middle cerebellar peduncle
ml   medial lemniscus
MnR   median raphe nu
Mo5   motor trigeminal nu
Oc   occipital cortex
P5   peritrigeminal zone
P7   perifacial zone
Parietal   parietal bone
ParietalP   parietal plate
PaS   parasubiculum
pcer   posterior cerebral artery
Petrous   petrous part, temporal bone
PnC   pontine reticular nu, caudal
pnm   pontine migration
PnO   pontine reticular nu, oral
PPy   parapyramidal reticular nu
Pr5   principal sensory trigeminal nu
PRh   perirhinal cortex
PrS   presubiculum
ptgpal   pterygopalatine artery
py   pyramidal tract
R   red nu
Reichert   Reichert's cartilage
RF   rhinal fissure
RLi   rostral linear nu raphe
RMg   raphe magnus nu
RoundW   round window
RSA   retrosplenial agranular Cx
SC   superior colliculus
scba   superior cerebellar artery
SCGn   superior cervical ganglion
scp   superior cerebellar peduncle
SMac   saccular macula
SN   substantia nigra
Sp5   spinal trigeminal nucleus
Sp5O   spinal trigeminal nu, oral
Stapedius   stapedius muscle
StyMF   stylomastoid foramen
SubB   subbrachial nu
SubC   subcoeruleus nu
trs   transverse sinus
TyBu   tympanic bulla
TyC   tympanic cavity
UMac   utricular macula
Utr   utricle
VeGn   vestibular ganglion
VLL   ventral nu lateral lemniscus
VTA   ventral tegmental area

3
2
1
0

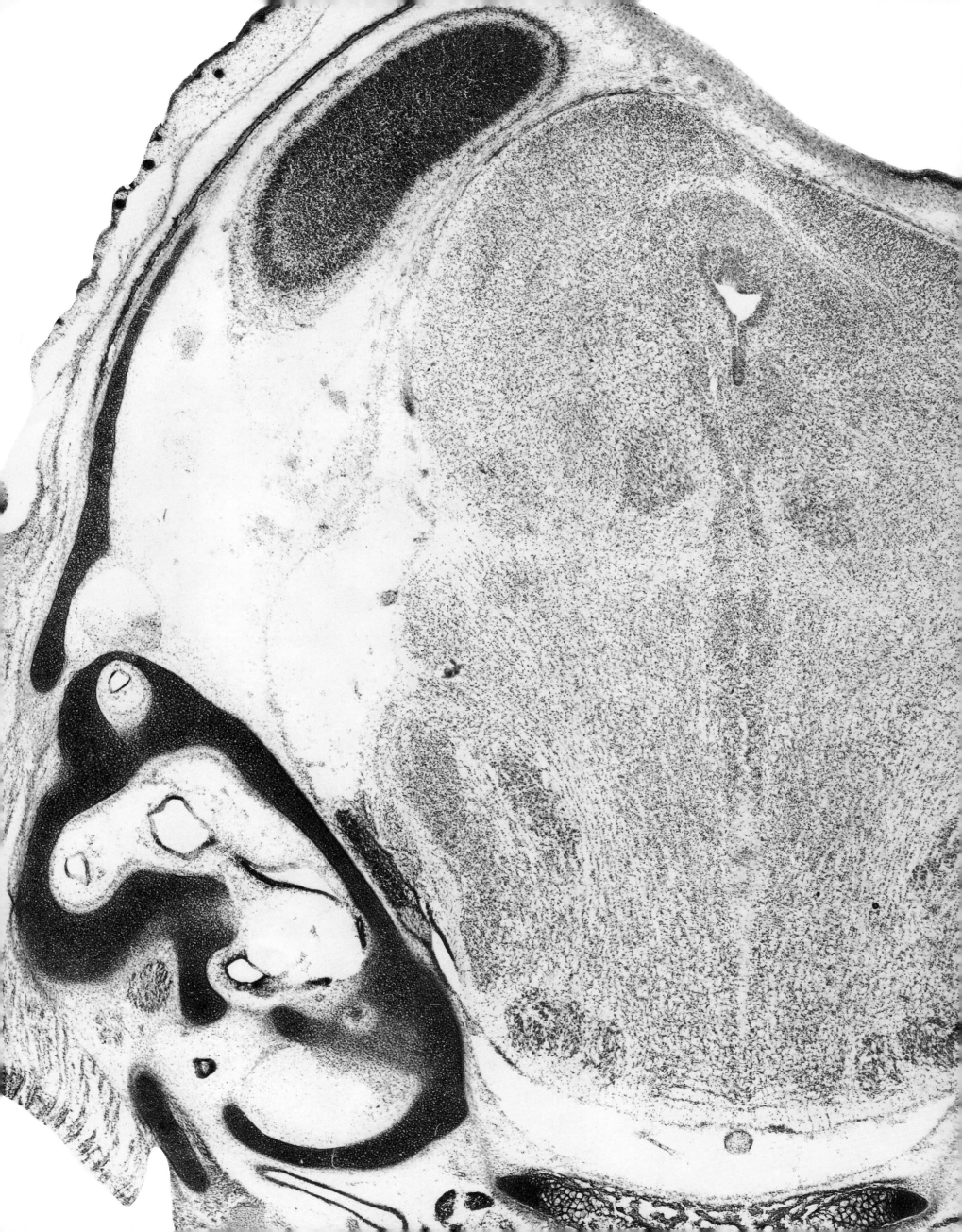

Figure 118
E19 Coronal 38

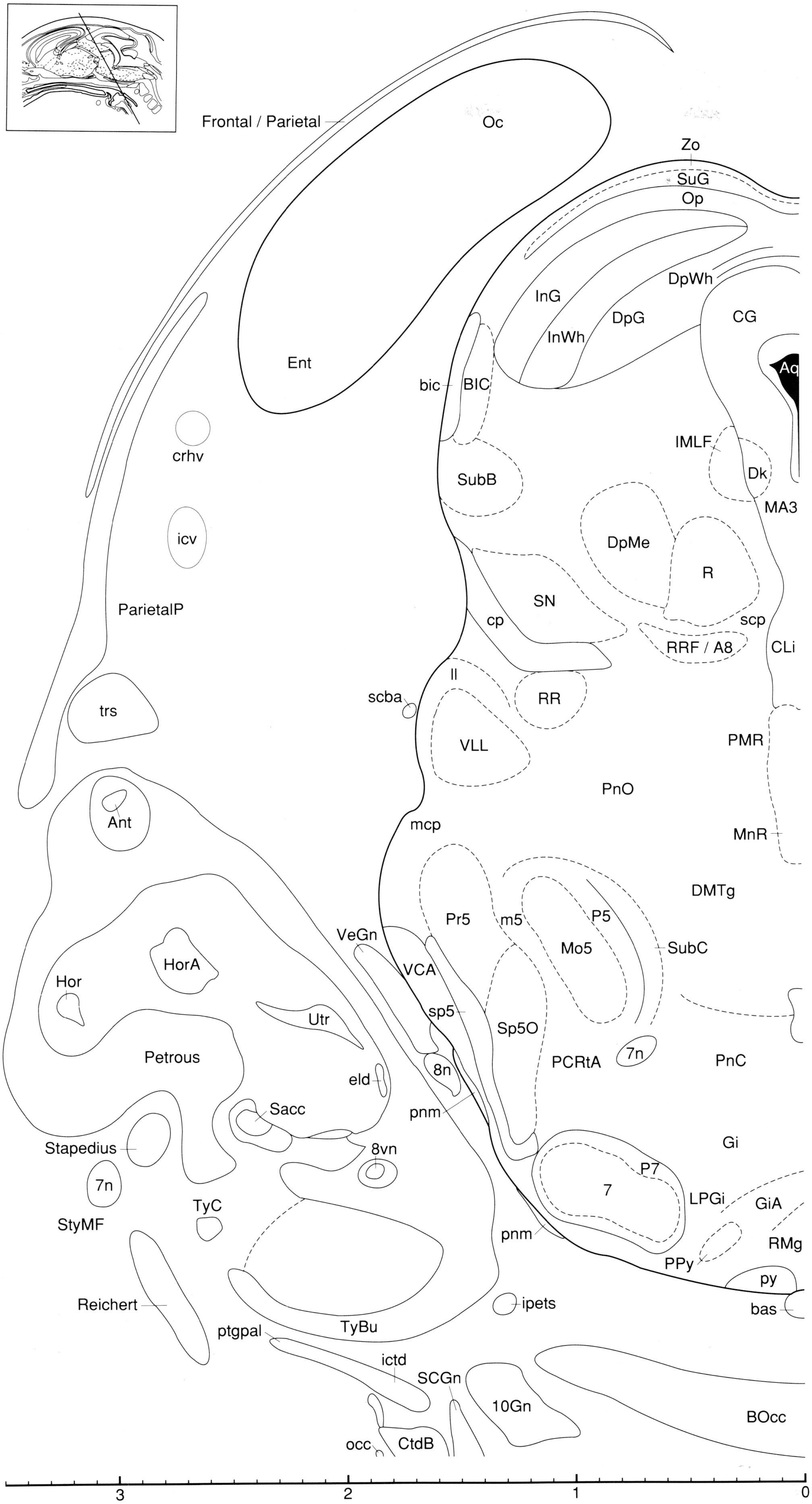

7   facial nu
7n   facial nerve or its root
8n   vestibulocochlear nerve
8vn   vestibular root vestibulococh n
10Gn   vagal ganglion
A8   A8 dopamine cells
Ant   anterior semicircular duct
Aq   aqueduct
bas   basilar artery
BIC   nu brachium inferior colliculus
bic   brachium inferior colliculus
BOcc   basioccipital bone
CG   central gray
CLi   caudal linear nu raphe
cp   cerebral peduncle, basal
crhv   caudal rhinal vein
CtdB   carotid body
Dk   nucleus Darkschewitsch
DMTg   dorsomedial tegmental area
DpG   deep gray layer of superior coll
DpMe   deep mesencephalic nu
DpWh   deep white layer of super coll
eld   endolymphatic duct
Ent   entorhinal cortex
Frontal   frontal bone
Gi   gigantocellular reticular nucleus
GiA   gigantocell reticular nu, alpha
Hor   horizontal semicircular duct
HorA   ampulla of hor semicircular duct
ictd   internal carotid artery
icv   inferior cerebral vein
IMLF   interstitial nu mlf
InG   intermediate gray layer sup col
InWh   intermediate white layer sup col
ipets inferior petrosal sinus
ll   lateral lemniscus
LPGi   lateral paragigantocellular nu
m5   motor root trigeminal nerve
MA3   medial acs oculomotor nu
mcp   middle cerebellar peduncle
MnR   median raphe nu
Mo5   motor trigeminal nu
Oc   occipital cortex
occ   occipital artery
Op   optic nerve layer superior collicul
P5   peritrigeminal zone
P7   perifacial zone
Parietal   parietal bone
ParietalP parietal plate
PCRtA parvocellular ret nu, alpha
Petrous   petrous part, temporal bone
PMR   paramedian raphe
PnC   pontine reticular nu, caudal
pnm   pontine migration
PnO   pontine reticular nu, oral
PPy   parapyramidal reticular nu
Pr5   principal sensory trigeminal nu
ptgpal pterygopalatine artery
py   pyramidal tract
R   red nu
Reichert   Reichert's cartilage
RMg   raphe magnus nu
RR   retrorubral nu
RRF   retrorubral field
Sacc   saccule
scba   superior cerebellar artery
SCGn   superior cervical ganglion
scp   superior cerebellar peduncle
SN   substantia nigra
sp5   spinal trigeminal tract
Sp5O   spinal trigeminal nu, oral
Stapedius   stapedius muscle
StyMF stylomastoid foramen
SubB   subbrachial nu
SubC   subcoeruleus nu
SuG   superficial gray layer sup coll
trs   transverse sinus
TyBu   tympanic bulla
TyC   tympanic cavity
Utr   utricle
VCA   ventral cochlear nu, anterior
VeGn   vestibular ganglion
VLL   ventral nu lateral lemniscus
Zo   zonal layer of the superior coll

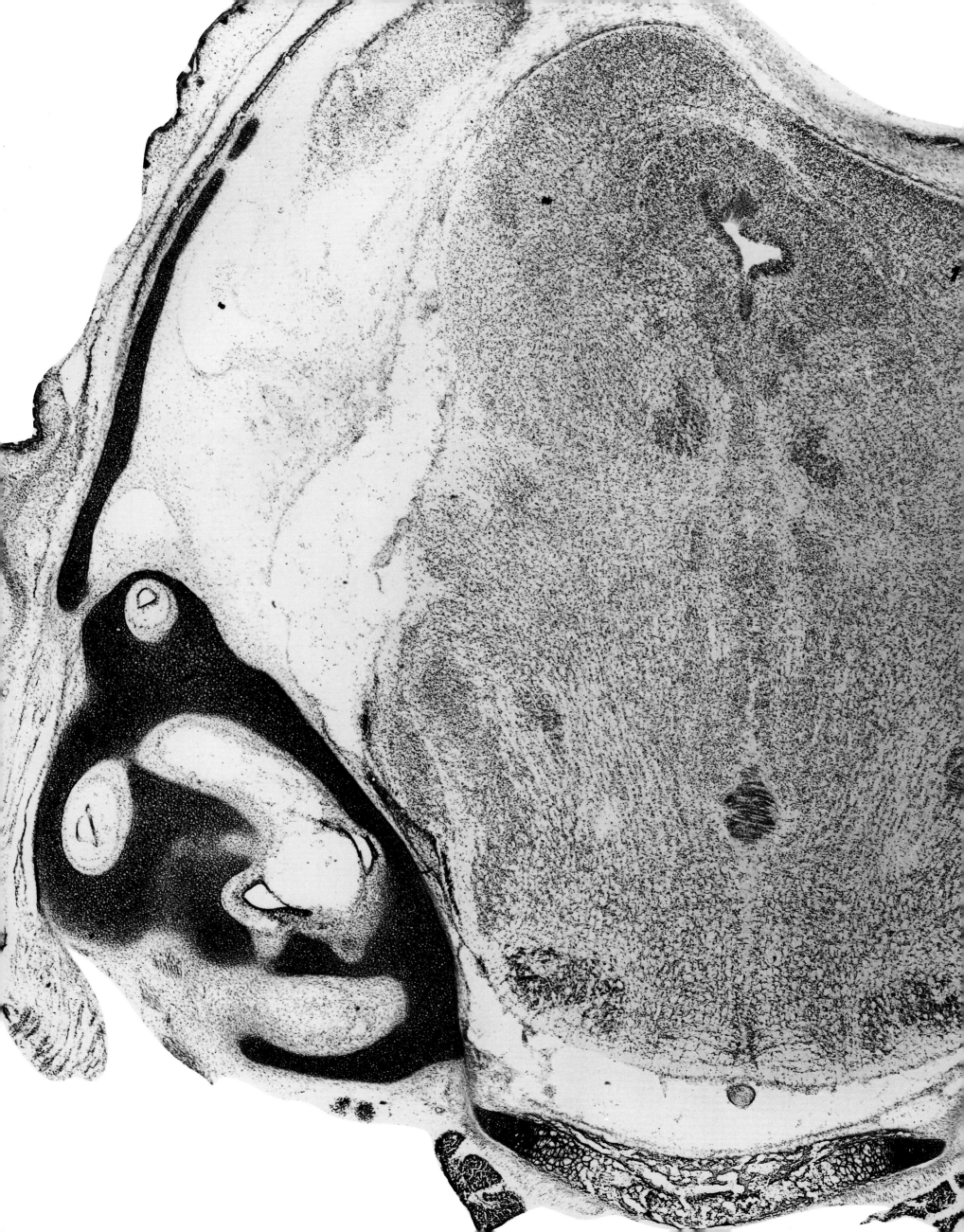

Figure 119
E19 Coronal 39

Frontal / Parietal

Oc

SC

bic   BIC

CG   Aq

SubB

IMLF   Dk

trs

DpMe

3

cp

SN   scp

R

RRF / A8

ParietalP

ll

RR

CLi

trs

scba   VLL

VTg   MnR

PnO

Endolymph

PMR

Ant

mcp

Perilymph

DMTg

Mo5   SubC

VeGn   Pr5   m5   P5

Hor

VCA

Petrous   Utr   8n   sp5   PCRtA

7n

ne

Sacc   eld   Sp5O

SMac   pnm

PnC

8vn   Acs6/7

StyMF   Stapedius   P7   Gi

7n   7   LPGi

GiA

v   RMg

ty9   TyBu   PPy   py

bas

I9Gn

Mastoid   occ   ictd

StM   11n   10Gn   BOcc

ectd

3   2   1   0

3   oculomotor nu
7   facial nu
7n   facial nerve or its root
8n   vestibulocochlear nerve
8vn vestibular root vestibulococh n
10Gn vagal ganglion
11n spinal accessory nerve
A8   A8 dopamine cells
Acs6 accessory abducens nu
Acs7 accessory facial nu
Ant anterior semicircular duct
Aq   aqueduct
bas basilar artery
BIC nu brachium inferior colliculus
bic brachium inferior colliculus
BOcc basioccipital bone
CG   central gray
CLi caudal linear nu raphe
cp   cerebral peduncle, basal
Dk   nucleus Darkschewitsch
DMTg dorsomedial tegmental area
DpMe deep mesencephalic nu
ectd external carotid a
eld endolymphatic duct
Endolymph endolymph
Frontal frontal bone
Gi   gigantocellular reticular nucleus
GiA gigantocell reticular nu, alpha
Hor horizontal semicircular duct
I9Gn inferior glossopharyngeal gang
ictd internal carotid artery
IMLF interstitial nu mlf
ll   lateral lemniscus
LPGi lateral paragigantocellular nu
m5   motor root trigeminal nerve
Mastoid mastoid process, temp
mcp middle cerebellar peduncle
MnR median raphe nu
Mo5 motor trigeminal nu
ne   neuroepithelium
Oc   occipital cortex
occ occipital artery
P5   peritrigeminal zone
P7   perifacial zone
Parietal parietal bone
ParietalP parietal plate
PCRtA parvocellular ret nu, alpha
Perilymph perilymph
Petrous petrous part, temporal bone
PMR paramedian raphe
PnC pontine reticular nu, caudal
pnm pontine migration
PnO pontine reticular nu, oral
PPy parapyramidal reticular nu
Pr5 principal sensory trigeminal nu
py   pyramidal tract
R   red nu
RMg raphe magnus nu
RR   retrorubral nu
RRF retrorubral field
Sacc saccule
SC   superior colliculus
scba superior cerebellar artery
scp superior cerebellar peduncle
SMac saccular macula
SN   substantia nigra
sp5 spinal trigeminal tract
Sp5O spinal trigeminal nu, oral
Stapedius stapedius muscle
StM sternomastoid muscle
StyMF stylomastoid foramen
SubB subbrachial nu
SubC subcoeruleus nu
trs transverse sinus
ty9 tympanic branch of 9n
TyBu tympanic bulla
Utr utricle
v   vein
VCA ventral cochlear nu, anterior
VeGn vestibular ganglion
VLL ventral nu lateral lemniscus
VTg ventral tegmental nu

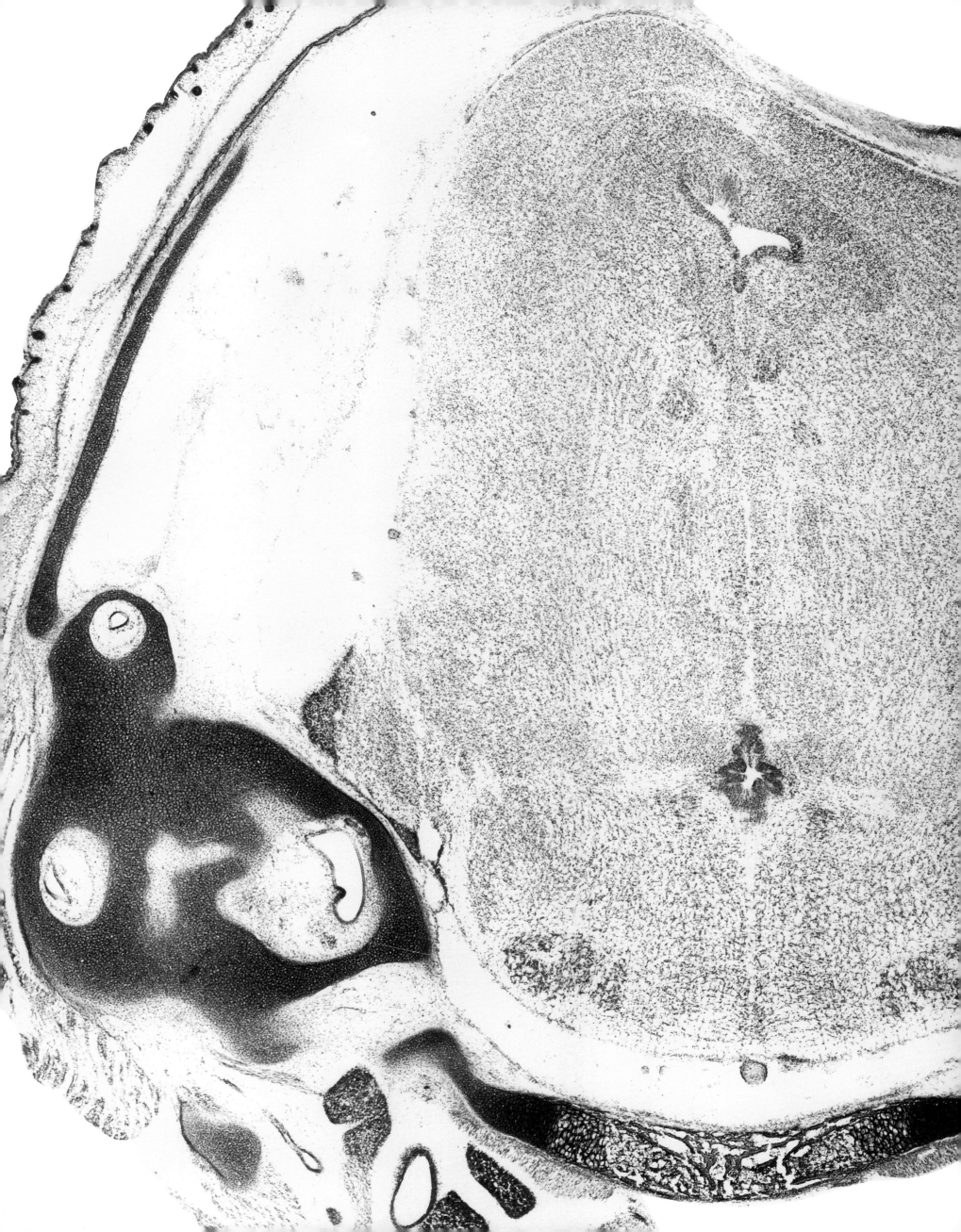

Figure 120
E19 Coronal 40

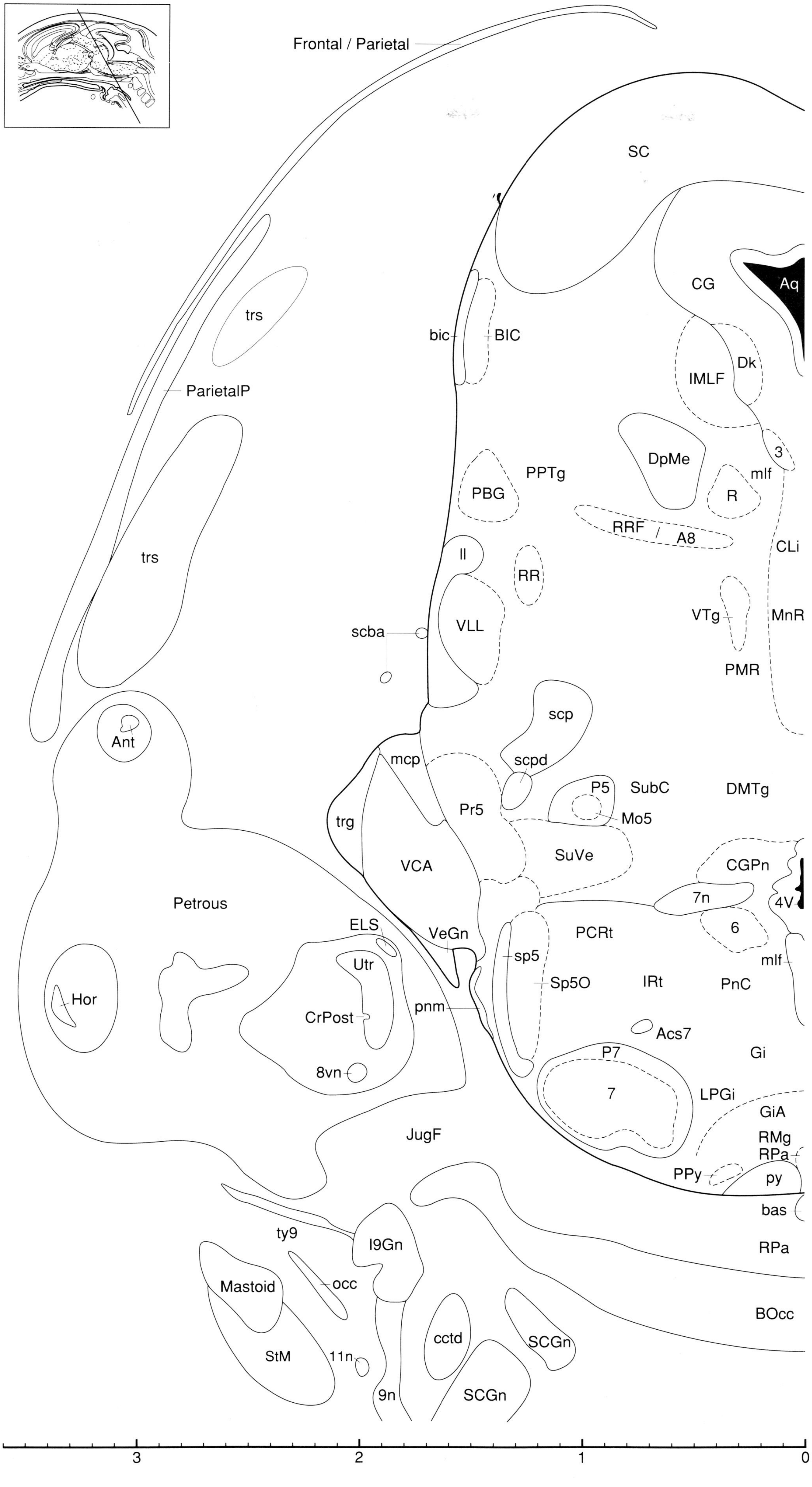

3 oculomotor nu
4V 4th ventricle
6 abducens nu
7 facial nu
7n facial nerve or its root
8vn vestibular root vestibulococh n
9n glossopharyngeal nerve
11n spinal accessory nerve
A8 A8 dopamine cells
Acs7 accessory facial nu
Ant anterior semicircular duct
Aq aqueduct
bas basilar artery
BIC nu brachium inferior colliculus
bic brachium inferior colliculus
BOcc basioccipital bone
cctd common carotid artery
CG central gray
CGPn central gray, pons
CLi caudal linear nu raphe
CrPost crista of posterior ampulla
Dk nucleus Darkschewitsch
DMTg dorsomedial tegmental area
DpMe deep mesencephalic nu
ELS endolymphatic sac
Frontal frontal bone
Gi gigantocellular reticular nucleus
GiA gigantocell reticular nu, alpha
Hor horizontal semicircular duct
I9Gn inferior glossopharyngeal gang
IMLF interstitial nu mlf
IRt intermediate reticular zone
JugF jugular foramen
ll lateral lemniscus
LPGi lateral paragigantocellular nu
Mastoid mastoid process, temp
mcp middle cerebellar peduncle
mlf medial longitudinal fasciculus
MnR median raphe nu
Mo5 motor trigeminal nu
occ occipital artery
P5 peritrigeminal zone
P7 perifacial zone
Parietal parietal bone
ParietalP parietal plate
PBG parabigeminal nu
PCRt parvocellular reticular nu
Petrous petrous part, temporal bone
PMR paramedian raphe
PnC pontine reticular nu, caudal
pnm pontine migration
PPTg pedunculopontine tegmental nu
PPy parapyramidal reticular nu
Pr5 principal sensory trigeminal nu
py pyramidal tract
R red nu
RMg raphe magnus nu
RPa raphe pallidus nu
RR retrorubral nu
RRF retrorubral field
SC superior colliculus
scba superior cerebellar artery
SCGn superior cervical ganglion
scp superior cerebellar peduncle
scpd sup cerebellar ped, descending
sp5 spinal trigeminal tract
Sp5O spinal trigeminal nu, oral
StM sternomastoid muscle
SubC subcoeruleus nu
SuVe superior vestibular nu
trg germinal trigone
trs transverse sinus
ty9 tympanic branch of 9n
Utr utricle
VCA ventral cochlear nu, anterior
VeGn vestibular ganglion
VLL ventral nu lateral lemniscus
VTg ventral tegmental nu

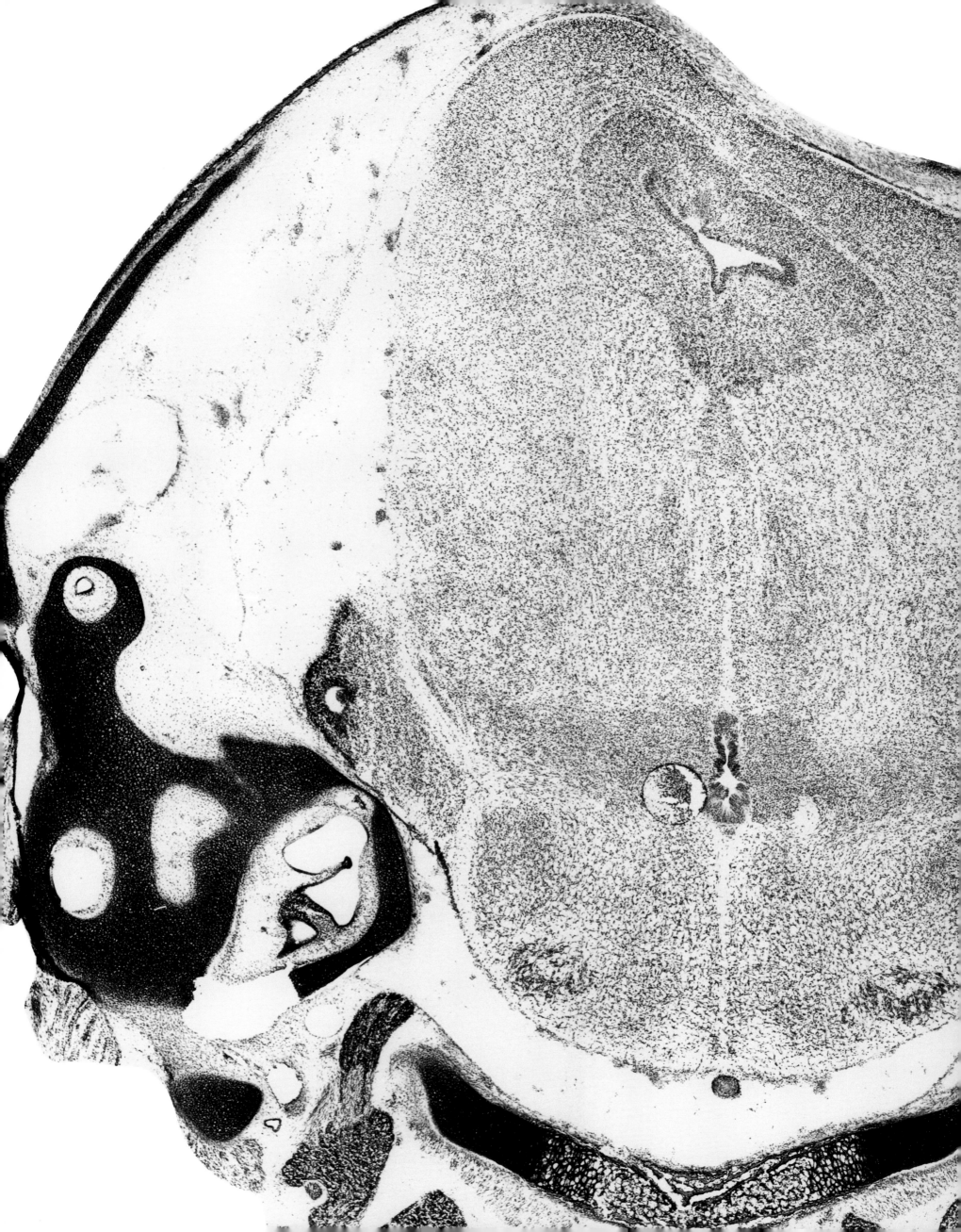

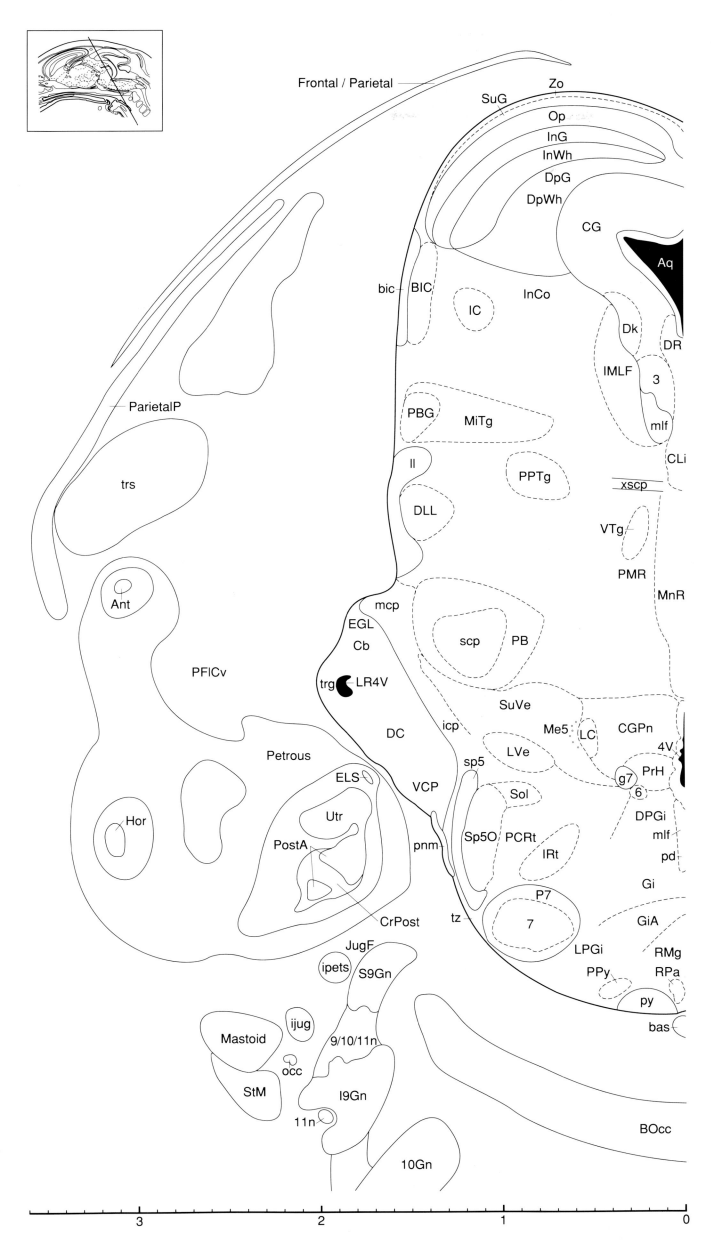

Figure 121
E19 Coronal 41

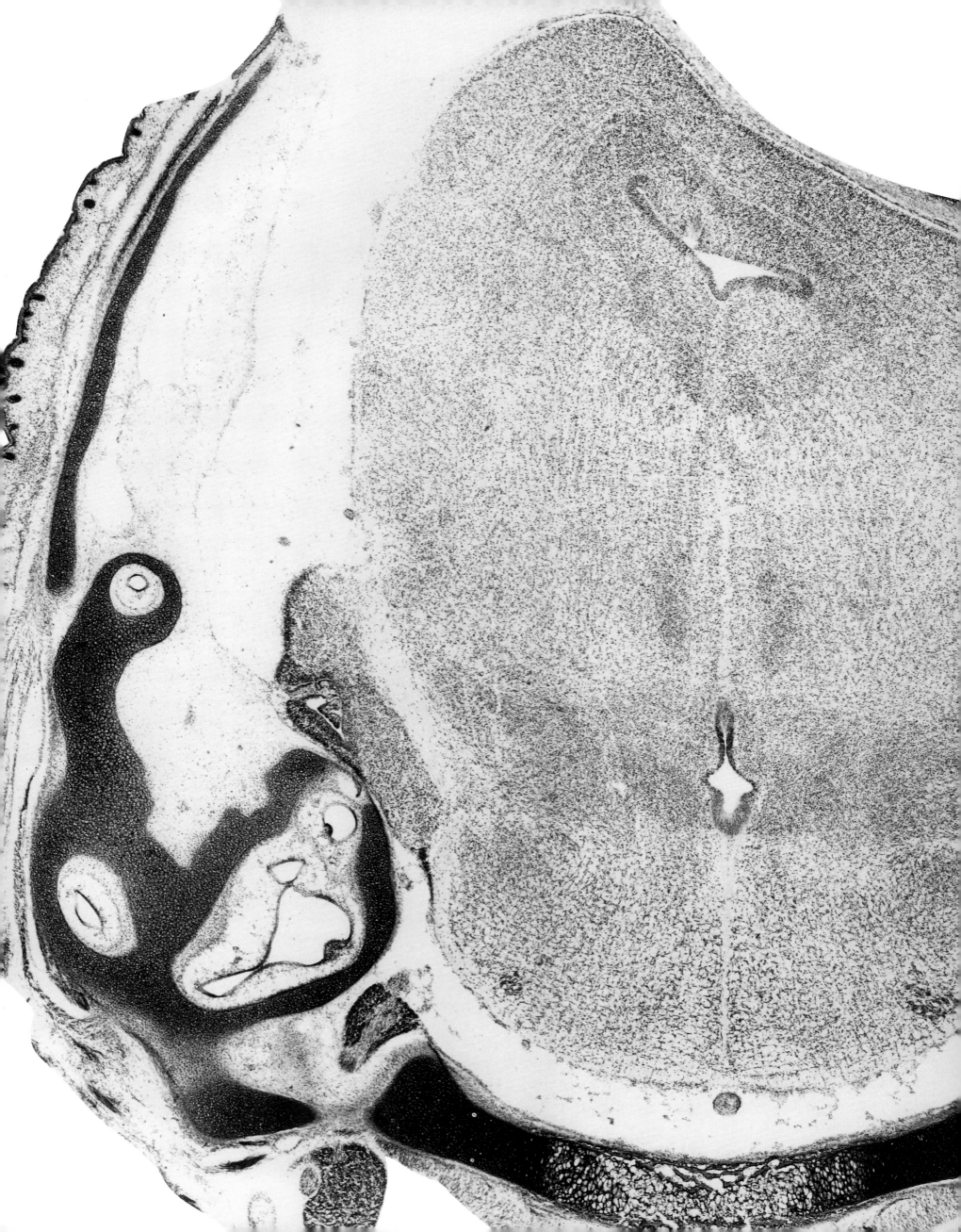

Figure 122
E19 Coronal 42

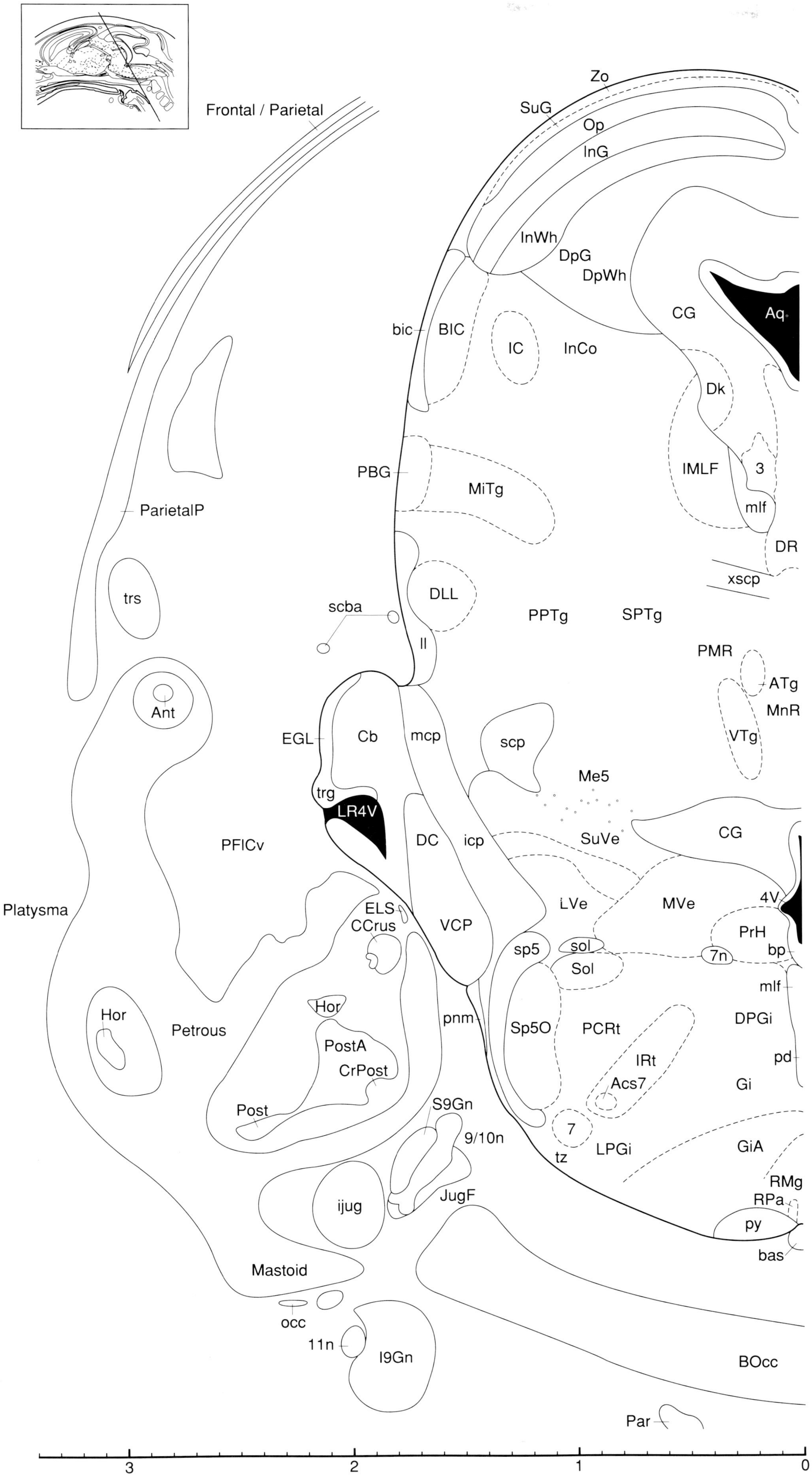

Frontal / Parietal

Zo
SuG
Op
InG
InWh
DpG
DpWh
bic
BIC
IC
InCo
CG
Aq
Dk
PBG
MiTg
IMLF
3
mlf
ParietalP
DR
xscp
trs
DLL
scba
ll
PPTg
SPTg
PMR
ATg
MnR
VTg
EGL
Cb
mcp
scp
Ant
Me5
trg
LR4V
DC
icp
SuVe
CG
PFlCv
ELS
CCrus
VCP
LVe
MVe
4V
PrH
bp
sp5
sol
7n
Sol
Hor
Hor
pnm
Sp5O
PCRt
mlf
Petrous
PostA
CrPost
IRt
Acs7
DPGi
pd
Post
S9Gn
7
LPGi
Gi
9/10n
tz
GiA
ijug
JugF
RMg
RPa
Mastoid
py
occ
bas
11n
I9Gn
BOcc
Par
Platysma

3   2   1   0

3   oculomotor nu
4V   4th ventricle
7   facial nu
7n   facial nerve or its root
9n   glossopharyngeal nerve
10n   vagus nerve
11n   spinal accessory nerve
Acs7   accessory facial nu
Ant   anterior semicircular duct
Aq   aqueduct
ATg   anterior tegmental nu
bas   basilar artery
BIC   nu brachium inferior colliculus
bic   brachium inferior colliculus
BOcc   basioccipital bone
bp   basal plate neuroepithelium
Cb   cerebellum
CCrus   common crus of ant and post s
CG   central gray
CrPost   crista of posterior ampulla
DC   dorsal cochlear nu
Dk   nucleus Darkschewitsch
DLL   dorsal nu lateral lemniscus
DpG   deep gray layer of superior coll
DPGi   dorsal paragigantocellular nu
DpWh   deep white layer of super coll
DR   dorsal raphe nu
EGL   external germinal layer of cb
ELS   endolymphatic sac
Frontal   frontal bone
Gi   gigantocellular reticular nucleus
GiA   gigantocell reticular nu, alpha
Hor   horizontal semicircular duct
I9Gn   inferior glossopharyngeal gang
IC   inferior colliculus
icp   inferior cerebellar peduncle
ijug   internal jugular vein
IMLF   interstitial nu mlf
InCo   intercollicular nu
InG   intermediate gray layer sup col
InWh   intermediate white layer sup col
IRt   intermediate reticular zone
JugF   jugular foramen
ll   lateral lemniscus
LPGi   lateral paragigantocellular nu
LR4V   lateral recess 4th ventricle
LVe   lateral vestibular nu
Mastoid   mastoid process, temp
mcp   middle cerebellar peduncle
Me5   mesencephalic trigeminal nu
MiTg   microcellular tegmental nu
mlf   medial longitudinal fasciculus
MnR   median raphe nu
MVe   medial vestibular nu
occ   occipital artery
Op   optic nerve layer superior collicul
Par   parotid gland
Parietal   parietal bone
ParietalP   parietal plate
PBG   parabigeminal nu
PCRt   parvocellular reticular nu
pd   predorsal bundle
Petrous   petrous part, temporal bone
PFlCv   parafloccular cavity
Platysma   platysma muscle
PMR   paramedian raphe nu
pnm   pontine migration
Post   posterior semicircular duct
PostA   ampulla of post semicirc duct
PPTg   pedunculopontine tegmental nu
PrH   prepositus hypoglossal nu
py   pyramidal tract
RMg   raphe magnus nu
RPa   raphe pallidus nu
S9Gn   superior glossopharyngeal gang
scba   superior cerebellar artery
scp   superior cerebellar peduncle
Sol   nu solitary tract
sol   solitary tract
sp5   spinal trigeminal tract
Sp5O   spinal trigeminal nu, oral
SPTg   subpeduncular tegmental nu
SuG   superficial gray layer sup col
SuVe   superior vestibular nu
trg   germinal trigone
trs   transverse sinus
tz   trapezoid body
VCP   ventral cochlear nu, posterior
VTg   ventral tegmental nu
xscp   decussation sup Cb peduncle
Zo   zonal layer of the superior coll

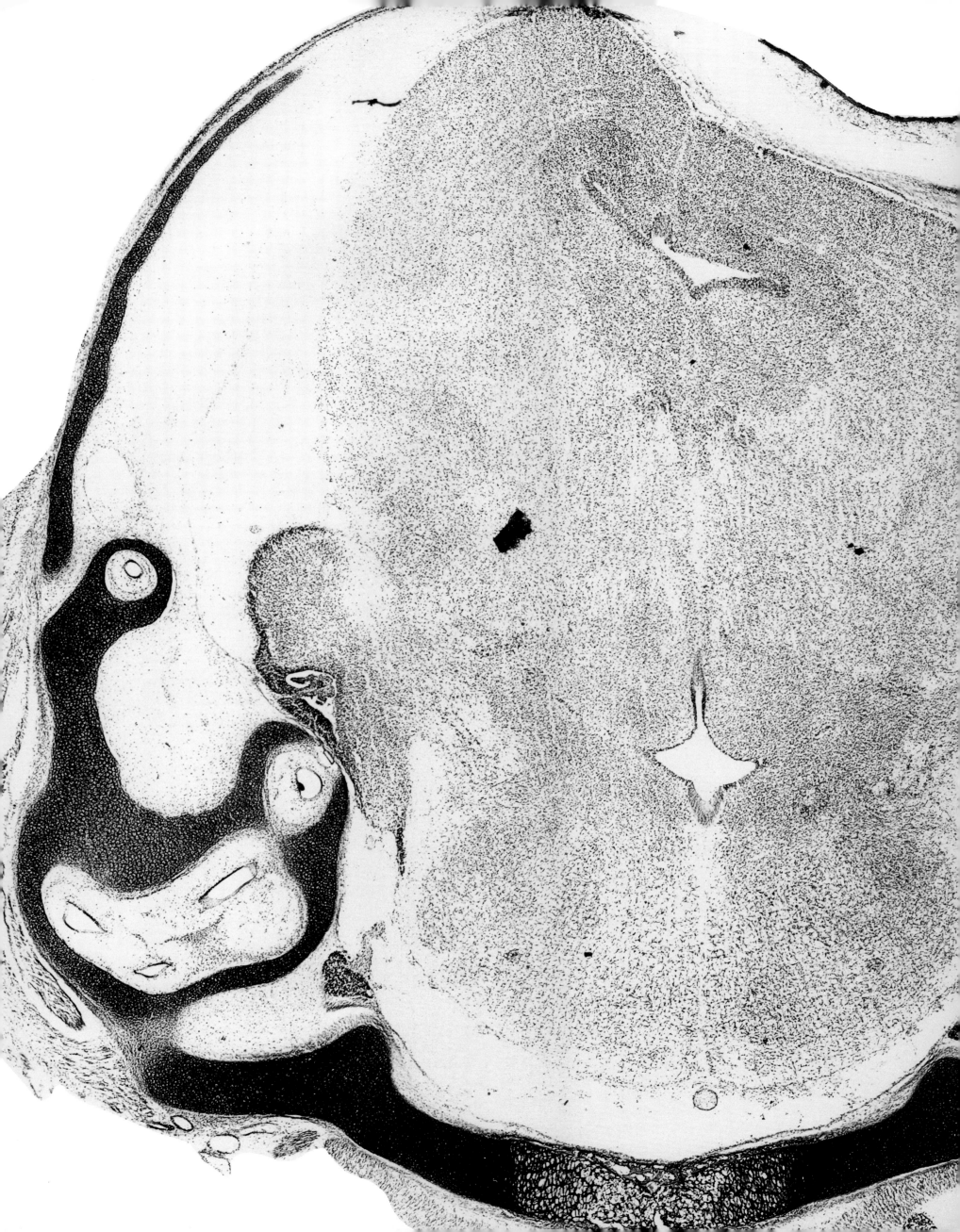

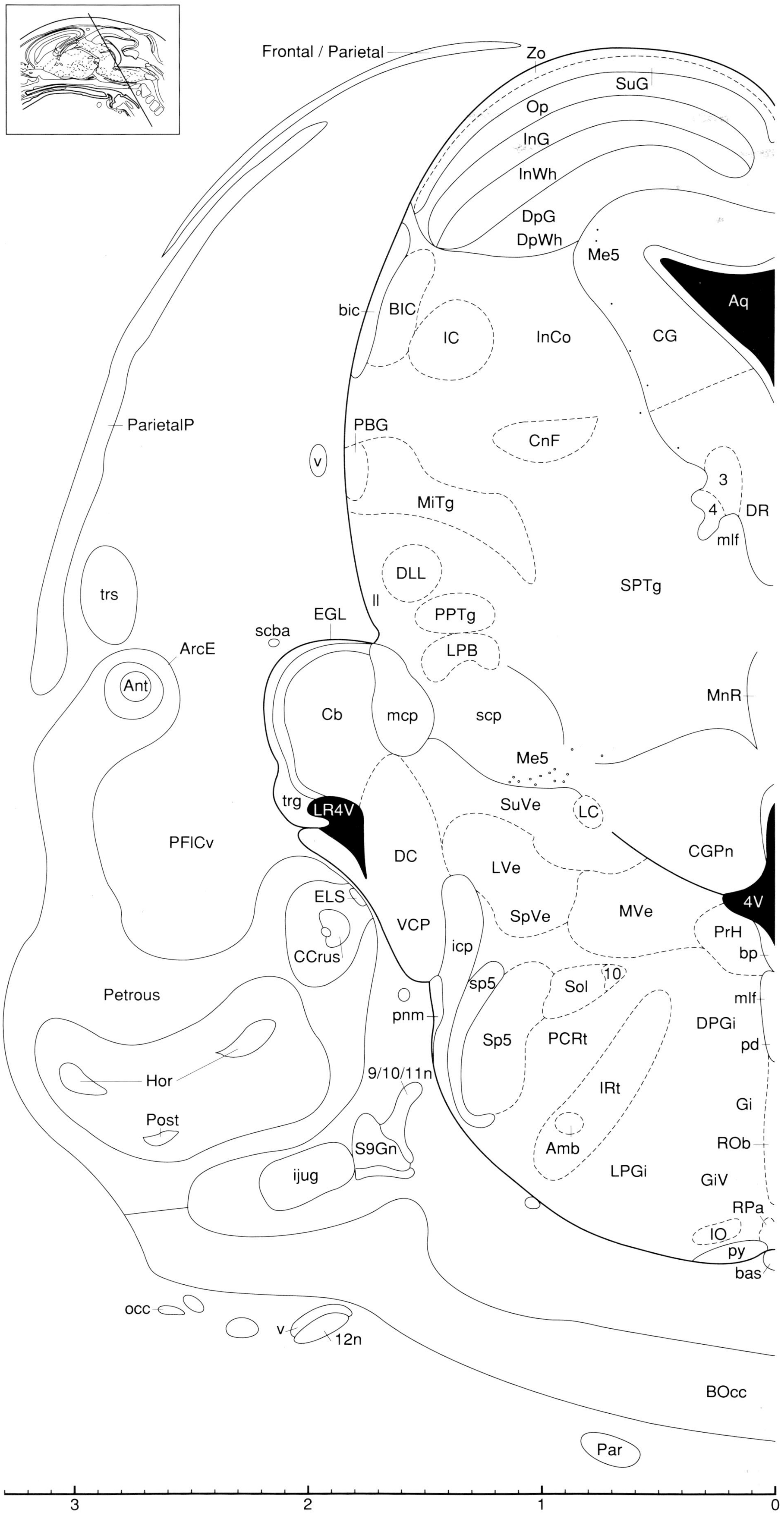

Figure 123
E19 Coronal 43

3   oculomotor nu
4   trochlear nu
4V   4th ventricle
9n   glossopharyngeal nerve
10   dorsal motor nu vagus
10n   vagus nerve
11n   spinal accessory nerve
12n   hypoglossal nerve or its root
Amb   ambiguus nu
Ant   anterior semicircular duct
Aq   aqueduct
ArcE   arcuate eminence
bas   basilar artery
BIC   nu brachium inferior colliculus
bic   brachium inferior colliculus
BOcc   basioccipital bone
bp   basal plate neuroepithelium
Cb   cerebellum
CCrus   common crus of ant and post s
CG   central gray
CGPn   central gray, pons
CnF   cuneiform nu
DC   dorsal cochlear nu
DLL   dorsal nu lateral lemniscus
DpG   deep gray layer of superior coll
DPGi   dorsal paragigantocellular nu
DpWh   deep white layer of super coll
DR   dorsal raphe nu
EGL   external germinal layer of cb
ELS   endolymphatic sac
Frontal   frontal bone
Gi   gigantocellular reticular nucleus
GiV   gigantocell reticular nu, vent
Hor   horizontal semicircular duct
IC   inferior colliculus
icp   inferior cerebellar peduncle
ijug   internal jugular vein
InCo   intercollicular nu
InG   intermediate gray layer sup col
InWh   intermediate white layer sup col
IO   inferior olive
IRt   intermediate reticular zone
LC   locus coeruleus
ll   lateral lemniscus
LPB   lateral parabrachial nu
LPGi   lateral paragigantocellular nu
LR4V   lateral recess 4th ventricle
LVe   lateral vestibular nu
mcp   middle cerebellar peduncle
Me5   mesencephalic trigeminal nu
MiTg   microcellular tegmental nu
mlf   medial longitudinal fasciculus
MnR   median raphe nu
MVe   medial vestibular nu
occ   occipital artery
Op   optic nerve layer superior collicul
Par   parotid gland
Parietal   parietal bone
ParietalP   parietal plate
PBG   parabigeminal nu
PCRt   parvocellular reticular nu
pd   predorsal bundle
Petrous   petrous part, temporal bone
PFlCv   paraflocccular cavity
pnm   pontine migration
Post   posterior semicircular duct
PPTg   pedunculopontine tegmental nu
PrH   prepositus hypoglossal nu
py   pyramidal tract
ROb   raphe obscurus nu
RPa   raphe pallidus nu
S9Gn   superior glossopharyngeal gang
scba   superior cerebellar artery
scp   superior cerebellar peduncle
Sol   nu solitary tract
Sp5   spinal trigeminal nucleus
sp5   spinal trigeminal tract
SPTg   subpeduncular tegmental nu
SpVe   spinal vestibular nu
SuG   superficial gray layer sup col
SuVe   superior vestibular nu
trg   germinal trigone
trs   transverse sinus
v   vein
VCP   ventral cochlear nu, posterior
Zo   zonal layer of the superior coll

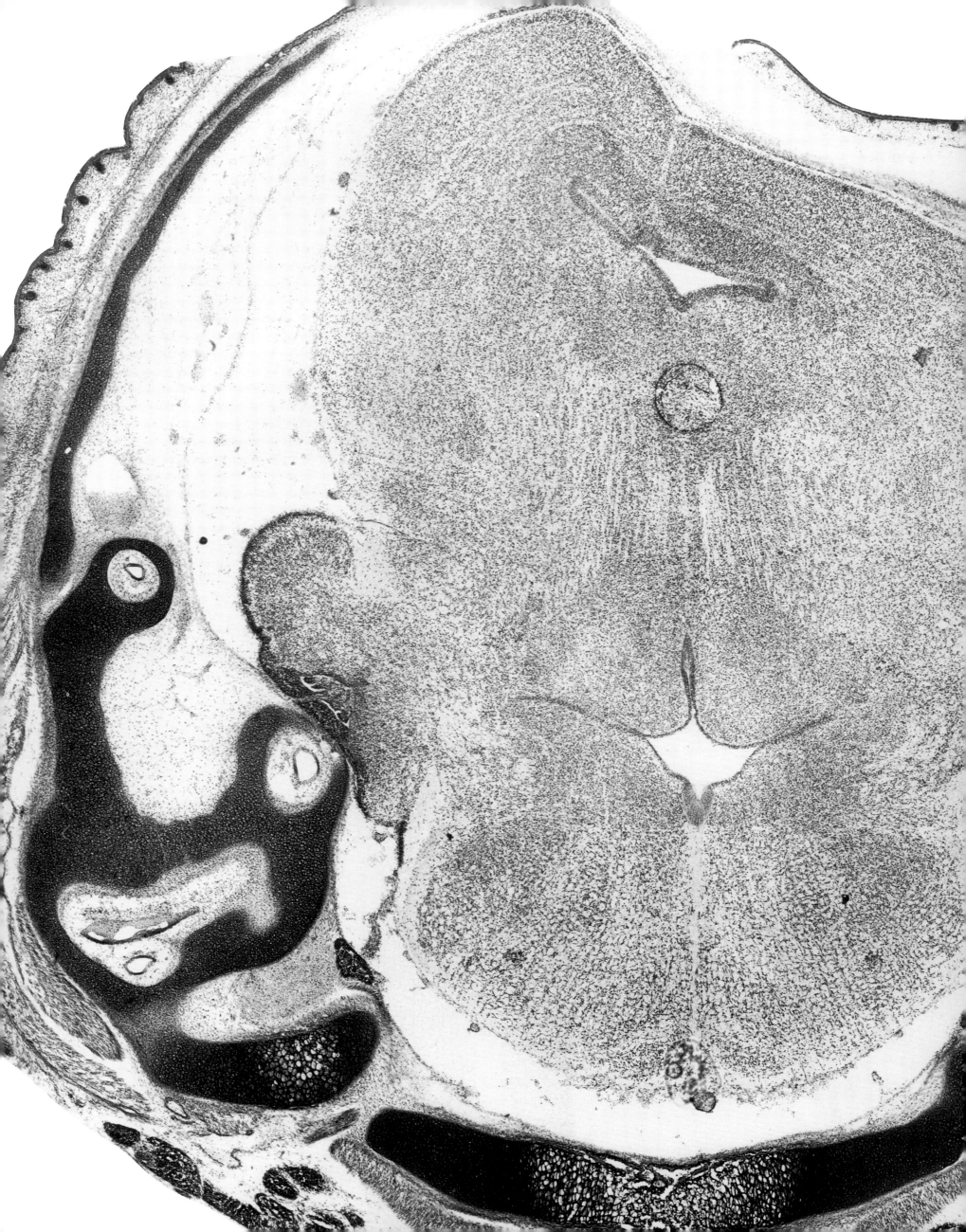

Figure 124
E19 Coronal 44

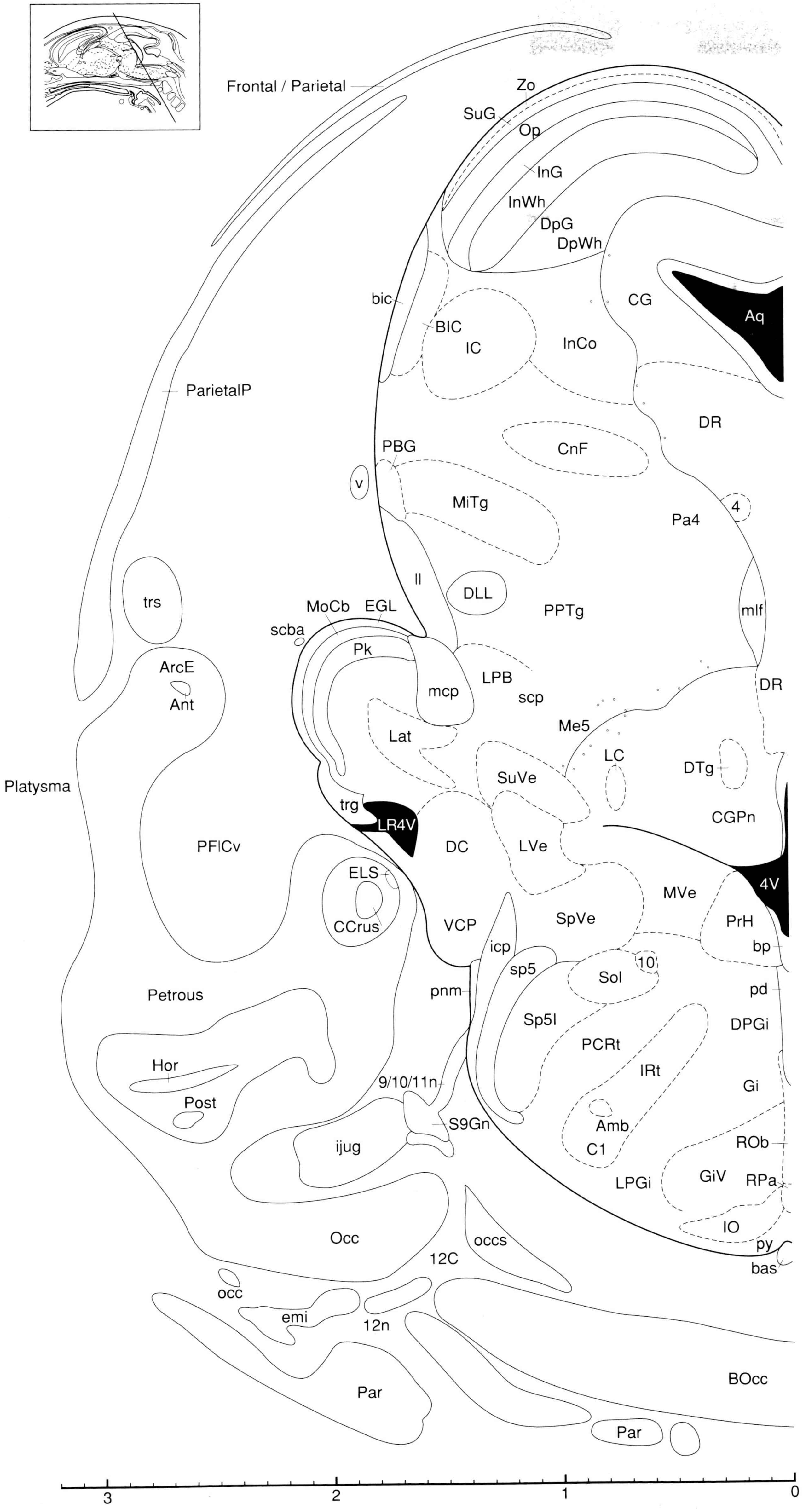

4   trochlear nu
4V   4th ventricle
9n   glossopharyngeal nerve
10   dorsal motor nu vagus
10n   vagus nerve
11n   spinal accessory nerve
12C   hypoglossal canal
12n   hypoglossal nerve or its root
Amb   ambiguus nu
Ant   anterior semicircular duct
Aq   aqueduct
ArcE arcuate eminence
bas   basilar artery
BIC   nu brachium inferior colliculus
bic   brachium inferior colliculus
BOcc basioccipital bone
bp   basal plate neuroepithelium
C1   C1 adrenaline cells
CCrus common crus of ant and post s
CG   central gray
CGPn central gray, pons
CnF   Cuneiform nu
DC   dorsal cochlear nu
DLL   dorsal nu lateral lemniscus
DpG   deep gray layer of superior coll
DPGi dorsal paragigantocellular nu
DpWh deep white layer of super coll
DR   dorsal raphe nu
DTg   dorsal tegmental nu
EGL   external germinal layer of cb
ELS   endolymphatic sac
emi   emissary vein, hypogl canal
Frontal frontal bone
Gi   gigantocellular reticular nucleus
GiV   gigantocell reticular nu, vent
Hor   horizontal semicircular duct
IC   inferior colliculus
icp   inferior cerebellar peduncle
ijug internal jugular vein
InCo intercollicular nu
InG   intermediate gray layer sup col
InWh intermediate white layer sup col
IO   inferior olive
IRt   intermediate reticular zone
Lat   lateral cerebellar nu
LC   locus coeruleus
ll   lateral lemniscus
LPB   lateral parabrachial nu
LPGi lateral paragigantocellular nu
LR4V lateral recess 4th ventricle
LVe   lateral vestibular nu
mcp   middle cerebellar peduncle
Me5   mesencephalic trigeminal nu
MiTg microcellular tegmental nu
mlf   medial longitudinal fasciculus
MoCb molecular layer of cerebellum
MVe   medial vestibular nu
Occ   occipital bone
occ   occipital artery
occs occipital sinus
Op   optic nerve layer superior collicul
Pa4   paratrochlear nu
Par   parotid gland
Parietal parietal bone
ParietalP parietal plate
PBG   parabigeminal nu
PCRt parvocellular reticular nu
pd   predorsal bundle
Petrous petrous part, temporal bone
PFlCv paraflocular cavity
Pk   Purkinje cell layer (cerebellum)
Platysma platysma muscle
pnm   pontine migration
Post posterior semicircular duct
PPTg pedunculopontine tegmental nu
PrH   prepositus hypoglossal nu
py   pyramidal tract
ROb   raphe obscurus nu
RPa   raphe pallidus nu
S9Gn superior glossopharyngeal gang
scba superior cerebellar artery
scp   superior cerebellar peduncle
Sol   nu solitary tract
sp5   spinal trigeminal tract
Sp5I spinal trigem nu, interpolar
SpVe spinal vestibular nu
SuG   superficial gray layer sup col
SuVe superior vestibular nu
trg   germinal trigone
trs   transverse sinus
v   vein
VCP   ventral cochlear nu, posterior
Zo   zonal layer of the superior coll

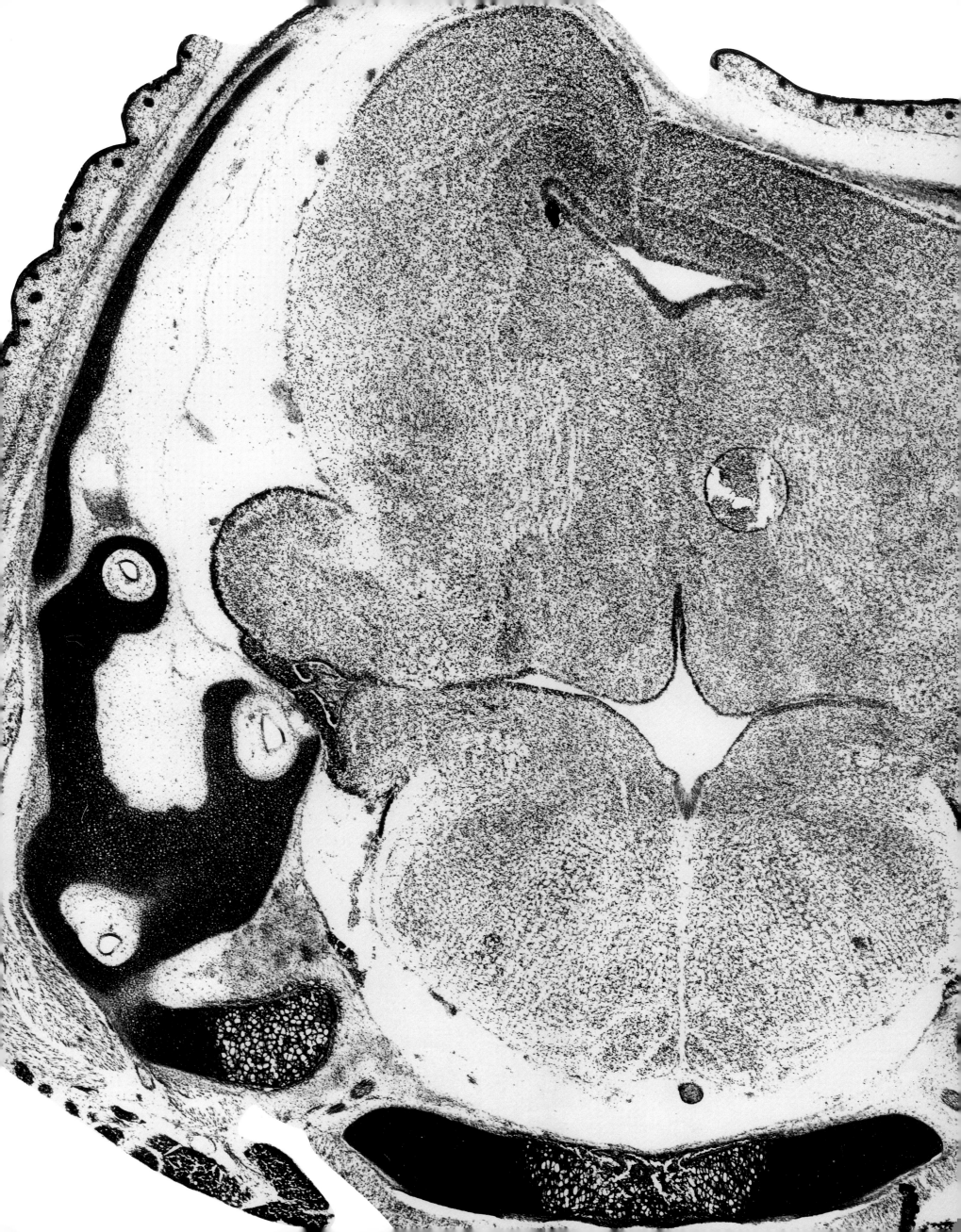

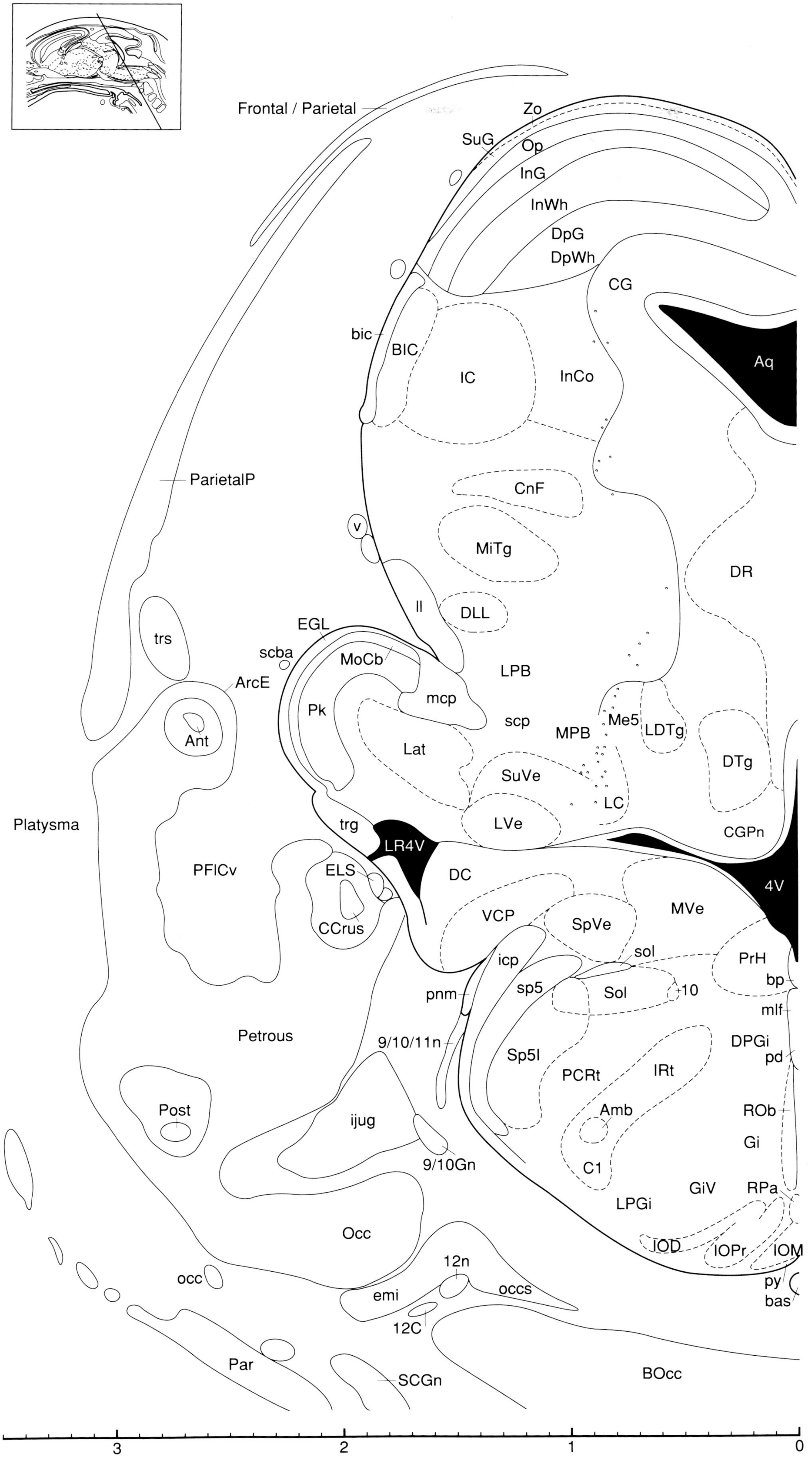

Figure 125
E19 Coronal 45

4V   4th ventricle
9Gn   glossopharyngeal ganglion
9n   glossopharyngeal nerve
10   dorsal motor nu vagus
10Gn vagal ganlion
10n   vagus nerve
11n   spinal accessory nerve
12C   hypoglossal canal
12n   hypoglossal nerve or its root
Amb   ambiguus nu
Ant   anterior semicircular duct
Aq   aqueduct
ArcE arcuate eminence
bas   basilar artery
BIC   nu brachium inferior colliculus
bic   brachium inferior colliculus
BOcc basioccipital bone
bp   basal plate neuroepithelium
C1   C1 adrenaline cells
CCrus common crus of ant and post s
CG   central gray
CGPn central gray, pons
CnF   cuneiform nu
DC   dorsal cochlear nu
DLL   dorsal nu lateral lemniscus
DpG   deep gray layer of superior coll
DPGi dorsal paragigantocellular nu
DpWh deep white layer of super coll
DR   dorsal raphe nu
DTg   dorsal tegmental nu
EGL   external germinal layer of cb
ELS   endolymphatic sac
emi   emissary vein, hypogl canal
Frontal frontal bone
Gi   gigantocellular reticular nucleus
GiV   gigantocell reticular nu, vent
IC   inferior colliculus
icp   inferior cerebellar peduncle
ijug internal jugular vein
InCo intercollicular nu
InG   intermediate gray layer sup col
InWh intermediate white layer sup col
IOD   inferior olive, dorsal nu
IOM   inferior olive, medial nu
IOPr inferior olive, principal nu
IRt   intermediate reticular zone
Lat   lateral cerebellar nu
LC   locus coeruleus
LDTg laterodorsal tegmental nu
ll   lateral lemniscus
LPB   lateral parabrachial nu
LPGi lateral paragigantocellular nu
LR4V lateral recess 4th ventricle
LVe   lateral vestibular nu
mcp   middle cerebellar peduncle
Me5   mesencephalic trigeminal nu
MiTg microcellular tegmental nu
mlf   medial longitudinal fasciculus
MoCb molecular layer of cerebellum
MPB   medial parabrachial nu
MVe   medial vestibular nu
Occ   occipital bone
occ   occipital artery
occs occipital sinus
Op   optic nerve layer superior collicul
Par   parotid gland
Parietal parietal bone
ParietalP parietal plate
PCRt parvocellular reticular nu
pd   predorsal bundle
Petrous petrous part, temporal bone
PFlCv paraflocular cavity
Pk   Purkinje cell layer (cerebellum)
Platysma platysma muscle
pnm   pontine migration
Post posterior semicircular duct
PrH   prepositus hypoglossal nu
py   pyramidal tract
ROb   raphe obscurus nu
RPa   raphe pallidus nu
scba superior cerebellar artery
SCGn superior cervical ganglion
scp   superior cerebellar peduncle
Sol   nu solitary tract
sol   solitary tract
sp5   spinal trigeminal tract
Sp5I spinal trigem nu, interpolar
SpVe spinal vestibular nu
SuG   superficial gray layer sup col
SuVe superior vestibular nu
trg   germinal trigone
trs   transverse sinus
v   vein
VCP   ventral cochlear nu, posterior
Zo   zonal layer of the superior coll

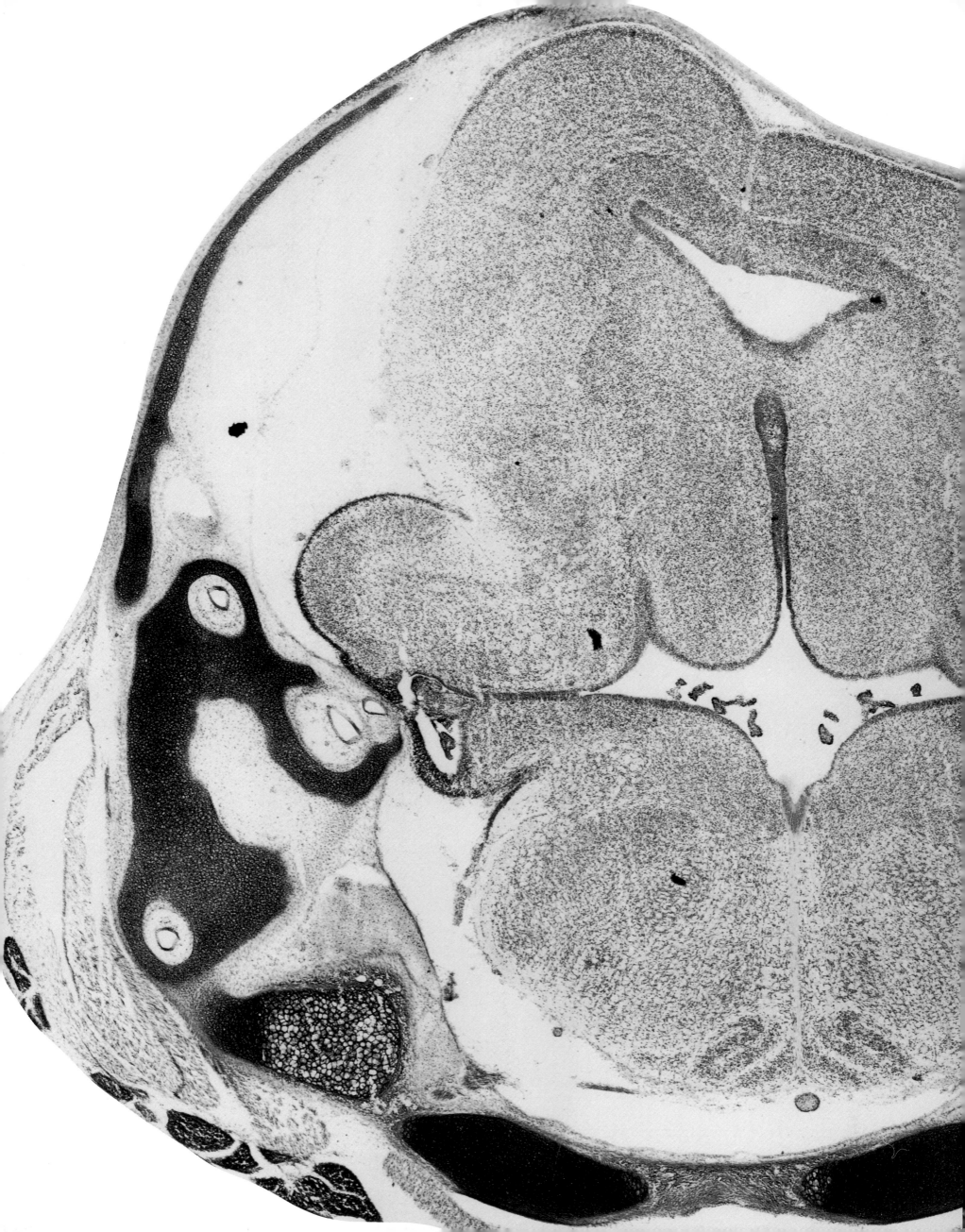

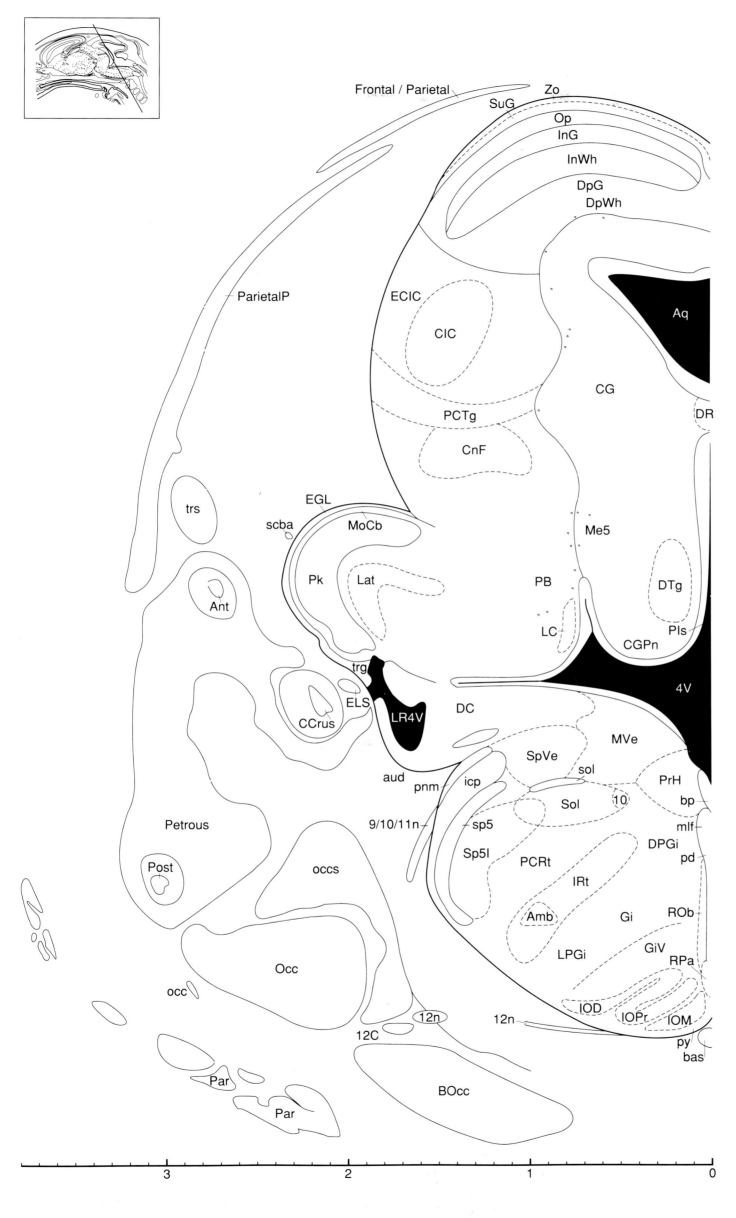

Figure 126
E19 Coronal 46

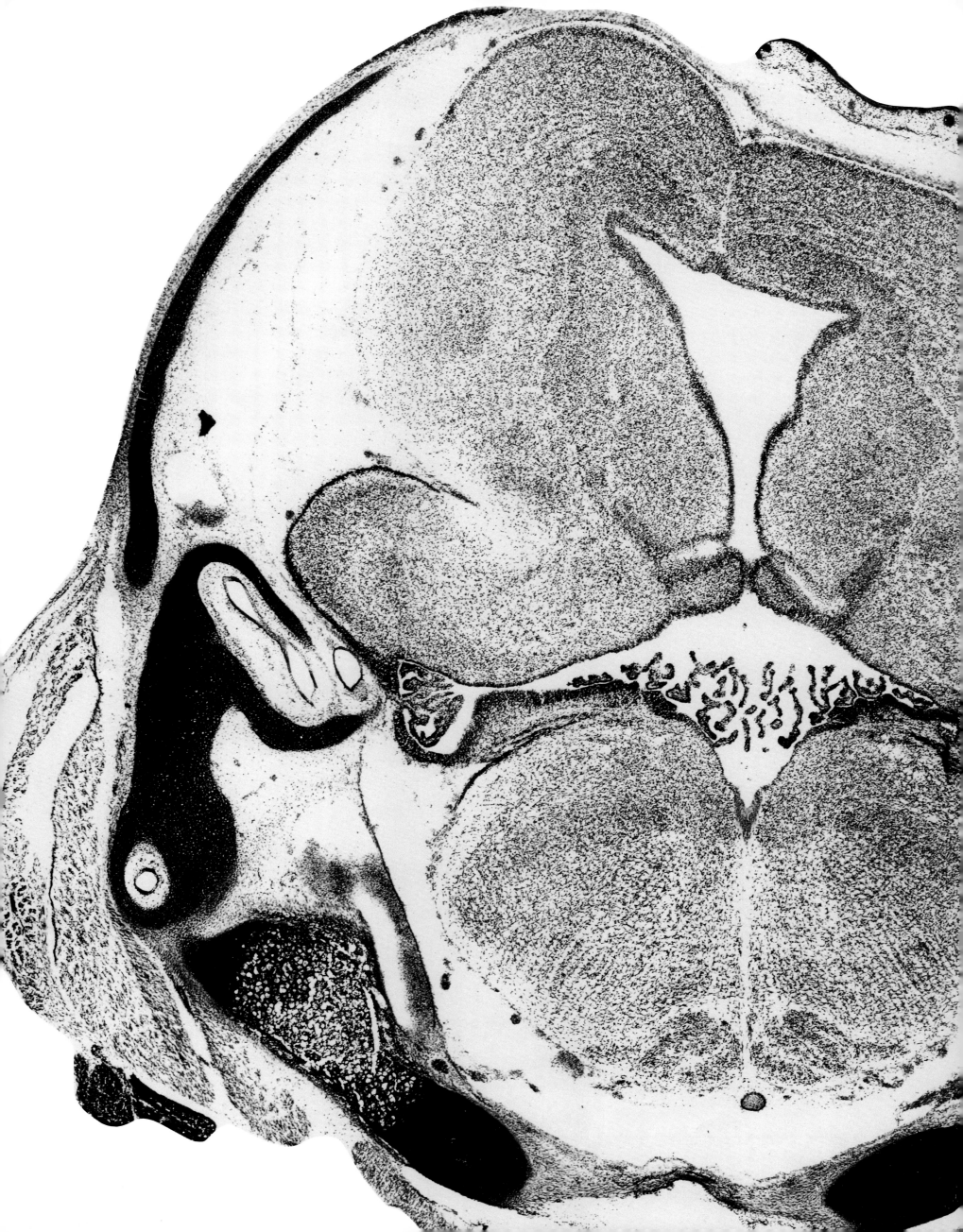

Figure 127
E19 Coronal 47

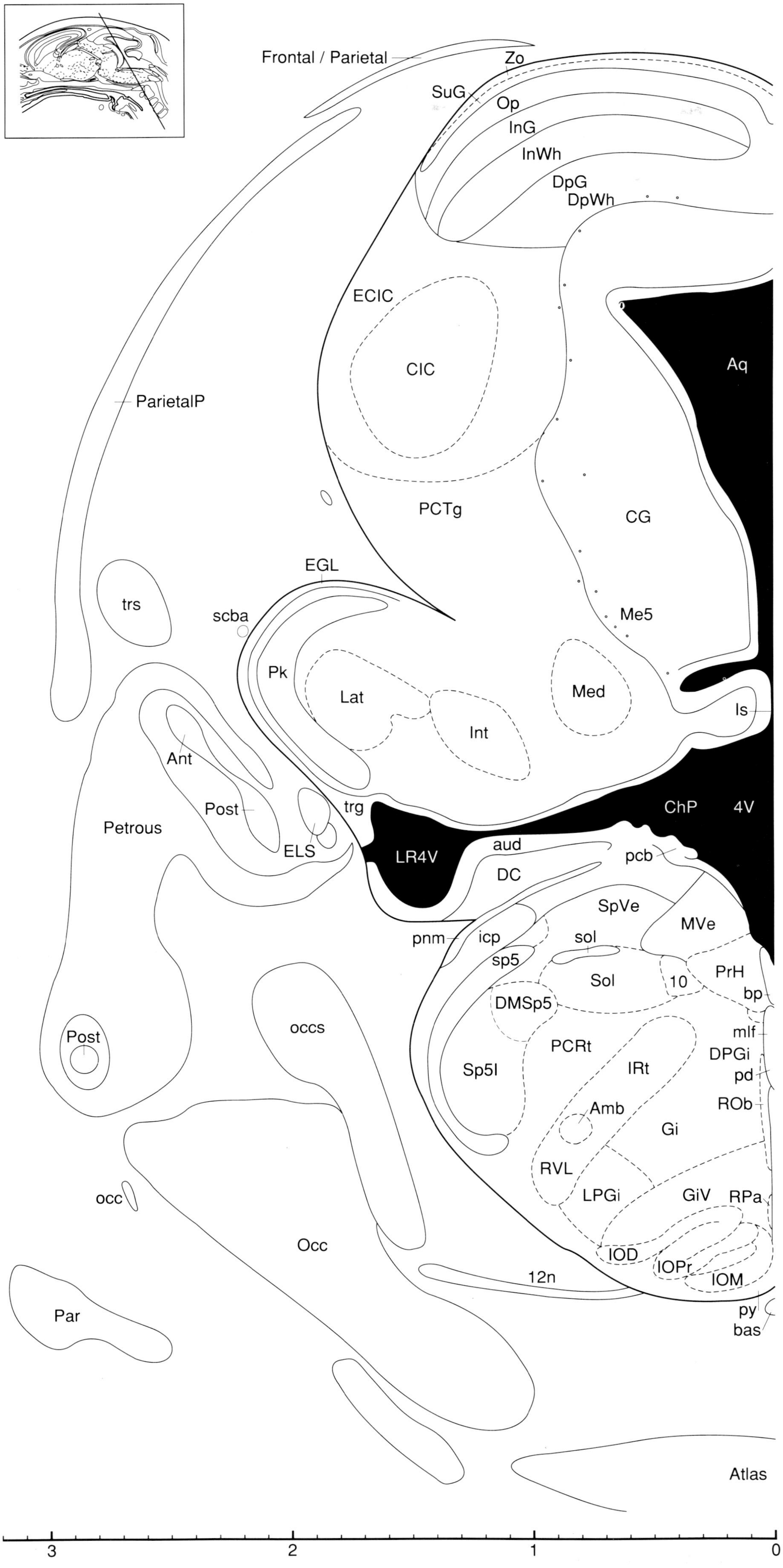

Frontal / Parietal

Zo

SuG

Op

InG

InWh

DpG

DpWh

ECIC

CIC

Aq

ParietalP

PCTg

CG

trs

EGL

scba

Me5

Pk

Lat

Med

Int

Is

Ant

Post

trg

ChP

4V

Petrous

ELS

LR4V

aud

pcb

DC

SpVe

MVe

pnm

icp

sol

sp5

Sol

10

PrH

bp

DMSp5

occs

PCRt

IRt

mlf

DPGi

Sp5I

pd

Amb

Gi

ROb

RVL

LPGi

GiV

RPa

occ

IOD

IOPr

IOM

Occ

12n

py

bas

Par

Atlas

4V   4th ventricle
10   dorsal motor nu vagus
12n   hypoglossal nerve or its root
Amb   ambiguus nu
Ant   anterior semicircular duct
Aq   aqueduct
Atlas atlas (C1 vertebra)
aud   auditory neuroepithelium
bas   basilar artery
bp   basal plate neuroepithelium
CG   central gray
ChP   choroid plexus
CIC   central nu inferior colliculus
DC   dorsal cochlear nu
DMSp5 dorsomedial spinal trig nu
DpG   deep gray layer of superior coll
DPGi   dorsal paragigantocellular nu
DpWh   deep white layer of super coll
ECIC   external cortex inf colliculus
EGL   external germinal layer of cb
ELS   endolymphatic sac
Frontal   frontal bone
Gi   gigantocellular reticular nucleus
GiV   gigantocell reticular nu, vent
icp   inferior cerebellar peduncle
InG   intermediate gray layer sup col
Int   interposed cerebellar nu
InWh   intermediate white layer sup col
IOD   inferior olive, dorsal nu
IOM   inferior olive, medial nu
IOPr   inferior olive, principal nu
IRt   intermediate reticular zone
Is   isthmus region
Lat   lateral cerebellar nu
LPGi   lateral paragigantocellular nu
LR4V   lateral recess 4th ventricle
Me5   mesencephalic trigeminal nu
Med   medial cerebellar nu
mlf   medial longitudinal fasciculus
MVe   medial vestibular nu
Occ   occipital bone
occ   occipital artery
occs   occipital sinus
Op   optic nerve layer superior collicul
Par   parotid gland
Parietal   parietal bone
ParietalP   parietal plate
pcb   precerebellar nuclei ne
PCRt   parvocellular reticular nu
PCTg   paracollicular tegmentum
pd   predorsal bundle
Petrous   petrous part, temporal bone
Pk   Purkinje cell layer (cerebellum)
pnm   pontine migration
Post   posterior semicircular duct
PrH   prepositus hypoglossal nu
py   pyramidal tract
ROb   raphe obscurus nu
RPa   raphe pallidus nu
RVL   rostroventrolateral retic nu
scba   superior cerebellar artery
Sol   nu solitary tract
sol   solitary tract
sp5   spinal trigeminal tract
Sp5I   spinal trigem nu, interpolar
SpVe   spinal vestibular nu
SuG   superficial gray layer sup col
trg   germinal trigone
trs   transverse sinus
Zo   zonal layer of the superior coll

3   2   1   0

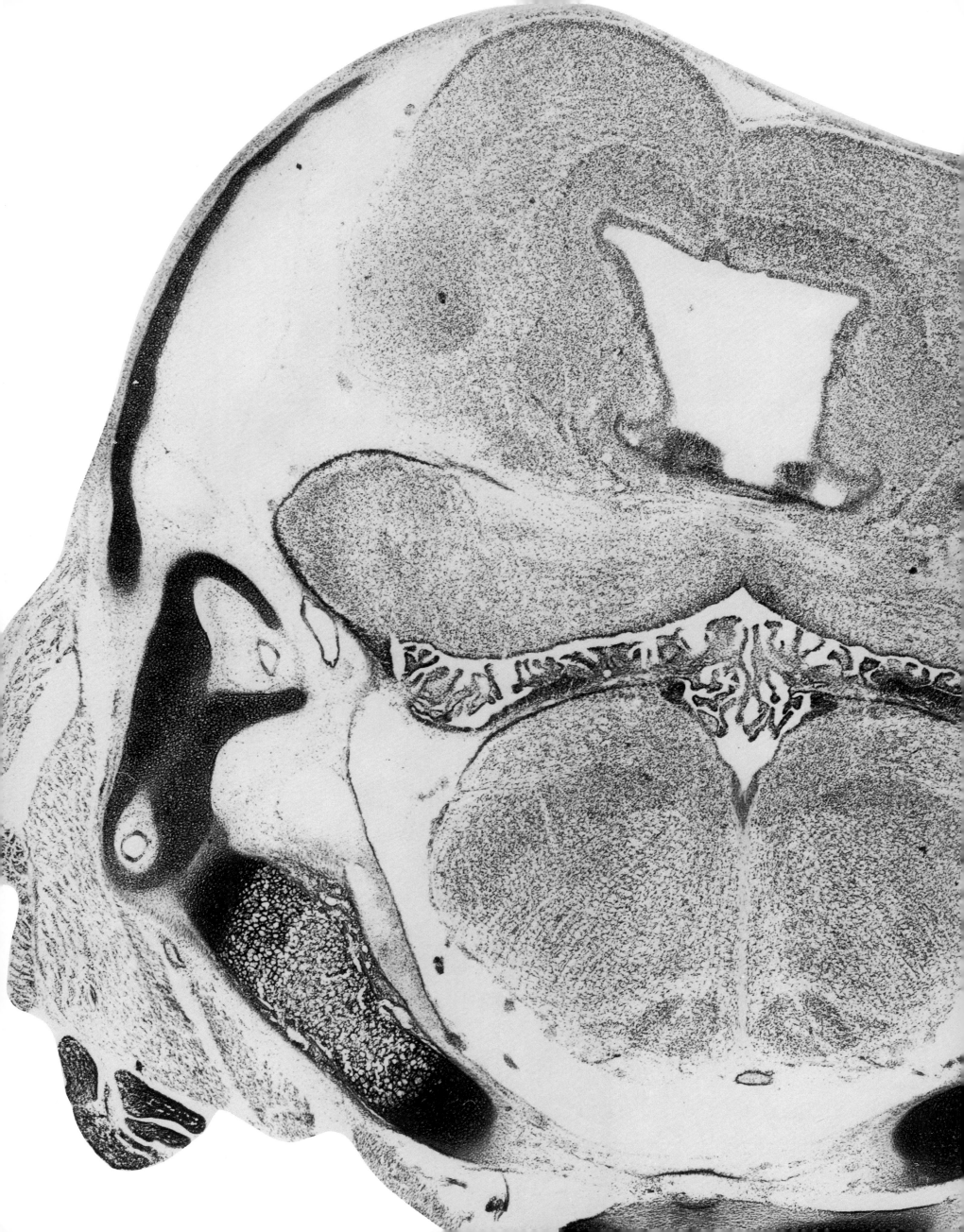

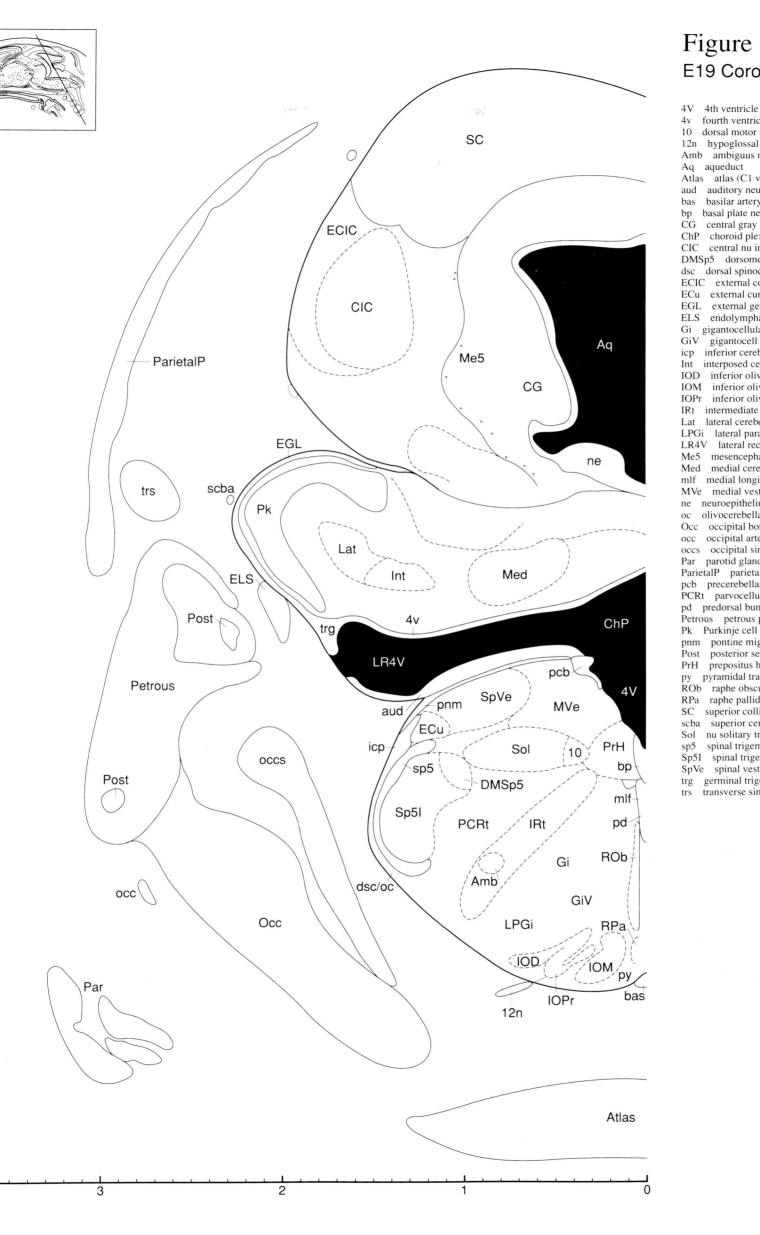

Figure 128
E19 Coronal 48

4V   4th ventricle
4v   fourth ventricle neuroepithelium
10   dorsal motor nu vagus
12n   hypoglossal nerve or its root
Amb   ambiguus nu
Aq   aqueduct
Atlas   atlas (C1 vertebra)
aud   auditory neuroepithelium
bas   basilar artery
bp   basal plate neuroepithelium
CG   central gray
ChP   choroid plexus
CIC   central nu inferior colliculus
DMSp5   dorsomedial spinal trig nu
dsc   dorsal spinocerebellar tract
ECIC   external cortex inf colliculus
ECu   external cuneate nu
EGL   external germinal layer of cb
ELS   endolymphatic sac
Gi   gigantocellular reticular nucleus
GiV   gigantocell reticular nu, vent
icp   inferior cerebellar peduncle
Int   interposed cerebellar nu
IOD   inferior olive, dorsal nu
IOM   inferior olive, medial nu
IOPr   inferior olive, principal nu
IRt   intermediate reticular zone
Lat   lateral cerebellar nu
LPGi   lateral paragigantocellular nu
LR4V   lateral recess 4th ventricle
Me5   mesencephalic trigeminal nu
Med   medial cerebellar nu
mlf   medial longitudinal fasciculus
MVe   medial vestibular nu
ne   neuroepithelium
oc   olivocerebellar tract
Occ   occipital bone
occ   occipital artery
occs   occipital sinus
Par   parotid gland
ParietalP   parietal plate
pcb   precerebellar nuclei ne
PCRt   parvocellular reticular nu
pd   predorsal bundle
Petrous   petrous part, temporal bone
Pk   Purkinje cell layer (cerebellum)
pnm   pontine migration
Post   posterior semicircular duct
PrH   prepositus hypoglossal nu
py   pyramidal tract
ROb   raphe obscurus nu
RPa   raphe pallidus nu
SC   superior colliculus
scba   superior cerebellar artery
Sol   nu solitary tract
sp5   spinal trigeminal tract
Sp5I   spinal trigem nu, interpolar
SpVe   spinal vestibular nu
trg   germinal trigone
trs   transverse sinus

3   2   1   0

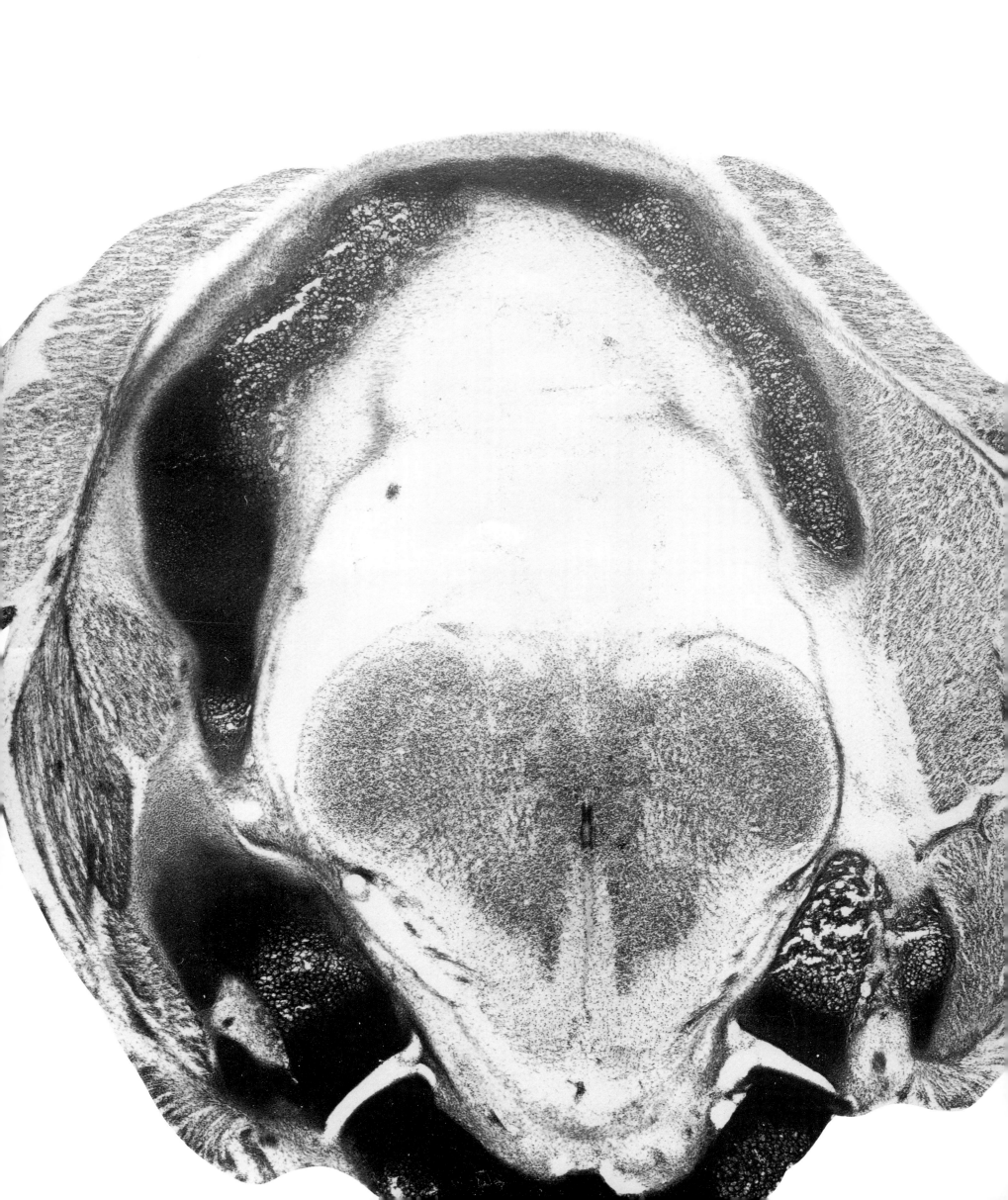

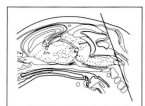

Figure 129
E19 Coronal 49

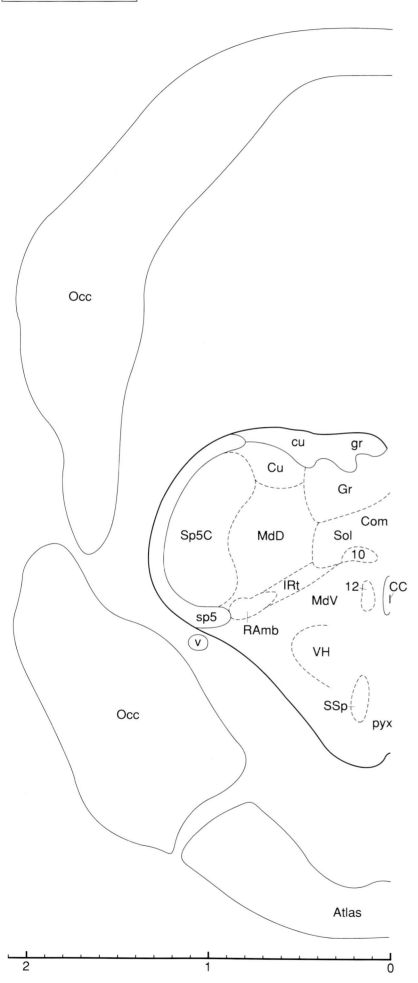

E19 Sagittal Section Plan

15 14 13 12 11 10 9 8 7 6 5 4 3 2 1

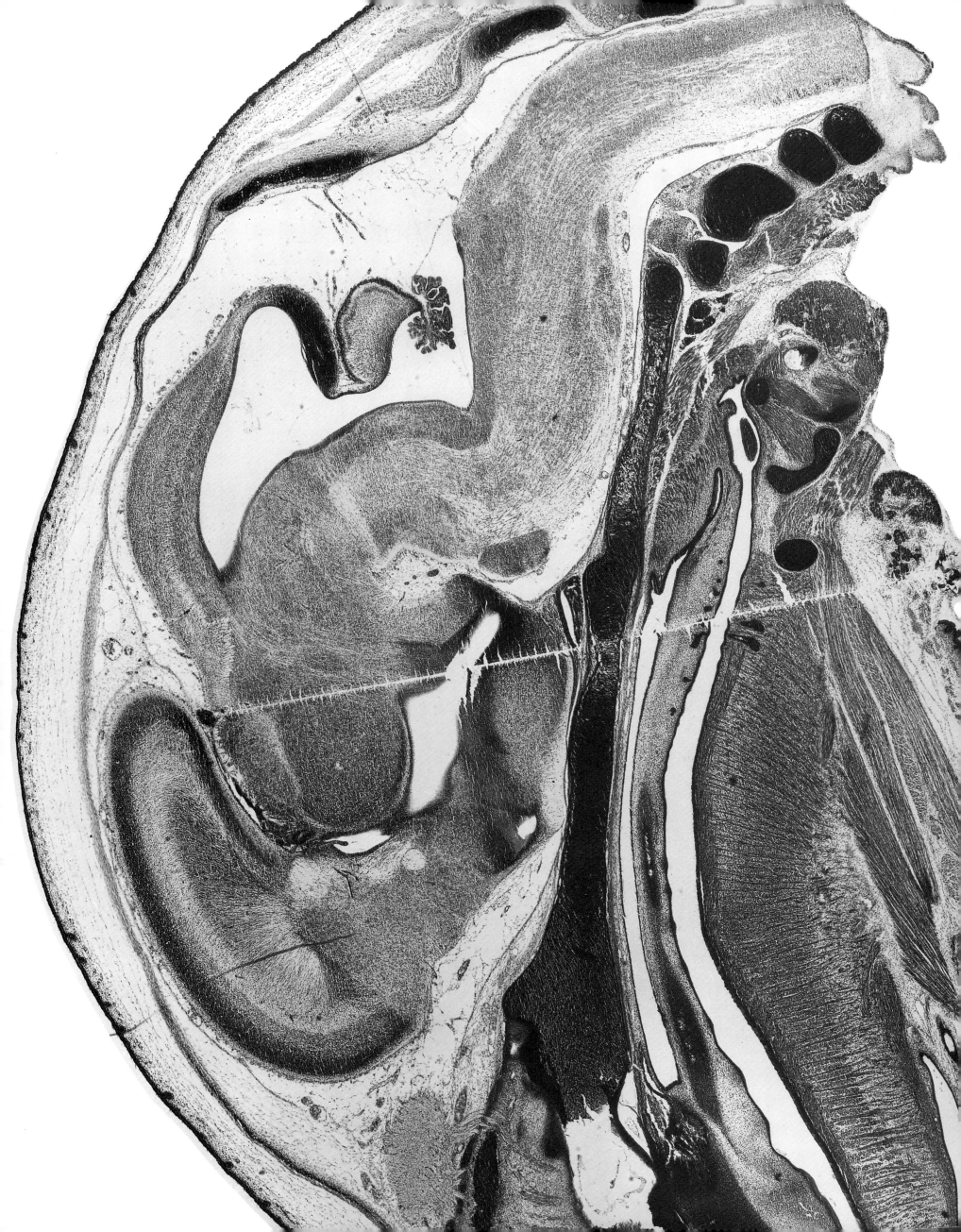

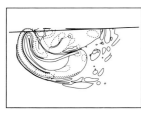

Figure 130
E19 Sagittal 1

1 cortical layer 1
3 oculomotor nucleus
3V third ventricle
4 trochlear nucleus
4V fourth ventricle
10 dorsal motor nucleus of vagus
ac anterior commissure
acer anterior cerebral artery
AH anterior hypothalamic nucleus
aica anterior inferior cerebellar artery
AP area postrema
APit anterior lobe of the pituitary
Aq aqueduct (Sylvius)
Arc arcuate hypothalamic nucleus
Atlas atlas (C1 vertebra)
Axis axis (C2 vertebra)
bas basilar artery
BOcc basioccipital bone
CA1 CA1 field of the hippocampus
CA3 CA3 field of the hippocampus
cc corpus callosum
CeF cervical flexure
CG central (periaqueductal) gray
ChP choroid plexus
CLi caudal linear nu of the raphe
CM central medial thalamic nucleus
CrP cribiform plate
csc commissure of the superior colliculus
CxP cortical plate
CxS cortical subplate
DH dorsal horns of spinal cord
Dk nucleus of Darkschewitsch
DPGi dorsal paragigantocellular nu
DR dorsal raphe nucleus
DTg dorsal tegmental nucleus
EGL external germinal layer of Cb
f fornix
Fr frontal cortex
Frontal frontal bone
Gi gigantocellular reticular nu
GiA gigantocellular reticular nu, alpha
GiV gigantocellular reticular nu, ventral
Gl glomerular layer of the olf bulb
Gr gracile nucleus
gr gracile fasciculus
hbc habenular commissure
hs hypothalamic sulcus
Hyoid hyoid bone
IAM interanteromedial thalamic nu
IC inferior colliculus
ICx intermediate cortical layer
IF interfascicular nucleus
IMLF interstitial nu of the mlf
IO inferior olive
IP interpeduncular nucleus
IPF interpeduncular fossa
IRt intermediate reticular zone

IVF interventricular foramen
LHb lateral habenular nucleus
LSD lateral septal nu, dorsal part
MD mediodorsal thalamic nucleus
ME median eminence
MHb medial habenular nucleus
ml medial lemniscus
mlf medial longitudinal fasciculus
MM medial mammillary nu, medial
MnPO median preoptic nucleus
MnR median raphe nucleus
MoCb molecular layer of cerebellum
MPO medial preoptic nucleus
Nasal nasal cavity
ne neuroepithelium
NSpt nasal septum
oc olivocerebellar tract
Occ occipital bone
olf olfactory nerve
olfa olfactory layer of the olf bulb
OptRe optic recess of third ventricle
Oral oral cavity
ox optic chiasm
Pa paraventricular hypothalamic nu

Pal palatine bone
Palate palate
Parietal parietal bone
pc posterior commissure
PDTg posterodorsal tegmental nu
PF parafascicular thalamic nucleus
PH posterior hypothalamic area
Pi pineal gland
Pk Purkinje cell layer (cerebellum)
Pn pontine nuclei
PnC pontine reticular nu, caudal part
pmm pontine migration
PnO pontine reticular nu, oral part
PPit posterior lobe of the pituitary
PrH prepositus hypoglossal nucleus
PrS presubiculum
PSph presphenoid bone
PTec pretectum
PV paraventricular thalamic nu
py pyramidal tract
Rathke Rathke's pouch
Re reuniens thalamic nucleus
Rh rhomboid thalamic nucleus
RI rostral interstitial nu of mlf

RLi rostral linear nu of the raphe
RMg raphe magnus nucleus
RSA retrosplenial agranular cortex
RSG retrosplenial granular cortex
RtTg reticulotegmental nu of the pons
S subiculum
SC superior colliculus
SFO subfornical organ
sm stria medullaris of the thalamus
Sol nucleus of the solitary tract
sol solitary tract
sox supraoptic decussation
SPF subparafascicular thalamic nu
SubV subventricular cortical layer
SuM supramammillary nucleus
trs transverse sinus
Tu olfactory tubercle
tz trapezoid body
v vein (unidentified)
VDB nu vertical limb diagonal band
vert vertebral artery
vfu ventral funiculus of the spinal cord
VH ventral horn
vhc ventral hippocampal commissure

VMH ventromedial hypothalamic nu
Vomer vomer
xscp decussation of the superior cerebellar peduncle
zo stratum zonale of superior colliculus

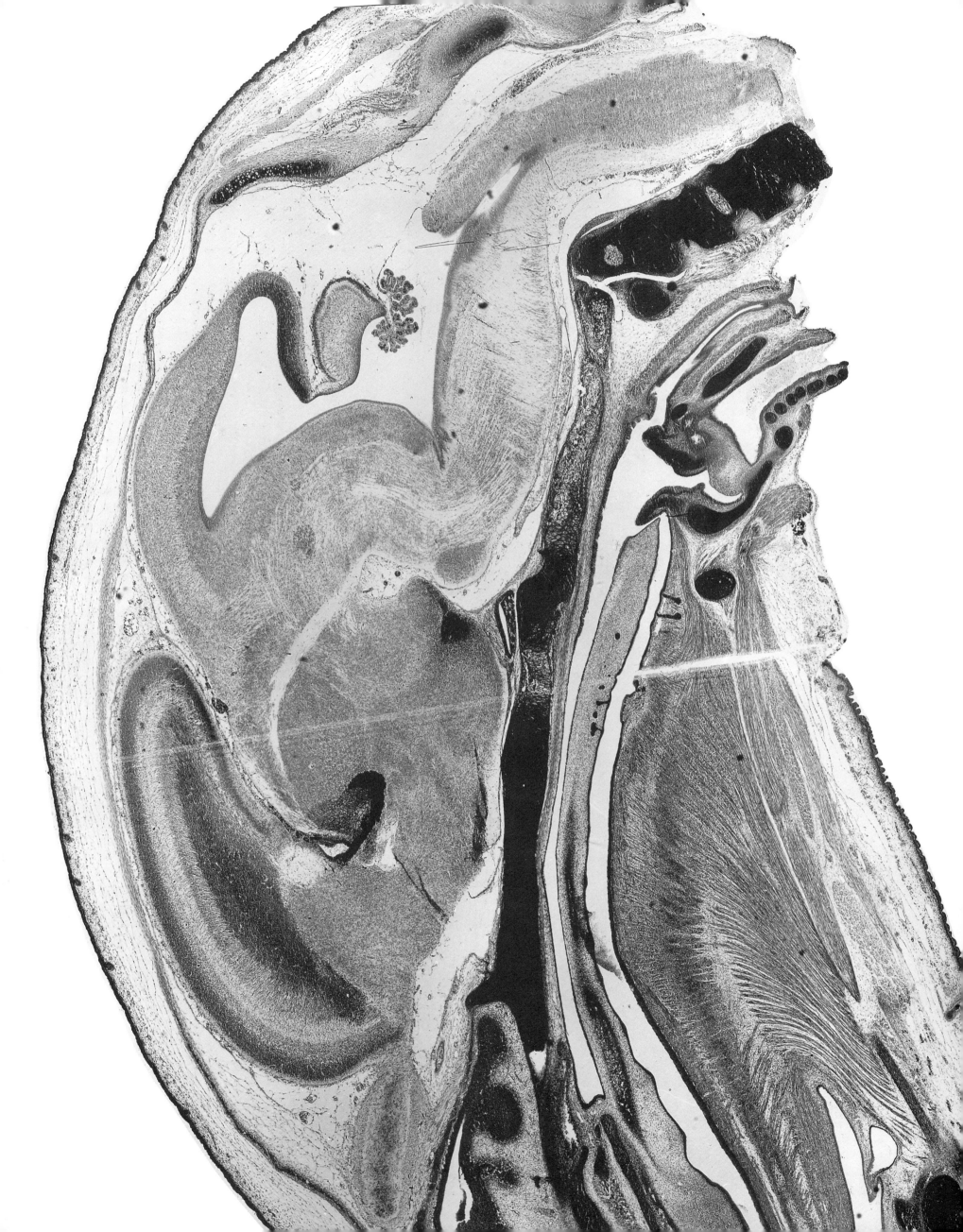

Figure 131
E19 Sagittal 2

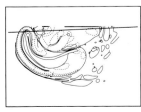

1   cortical layer 1
4   trochlear nucleus
4V   fourth ventricle
12   hypoglossal nucleus
AB   anterobasal nucleus
ac   anterior commissure
Acb   accumbens nucleus
AH   anterior hypothalamic nucleus
AM   anteromedial thalamic nucleus
AMPO   anterior medial preoptic nu
AP   area postrema
APit   anterior lobe of the pituitary
Aq   aqueduct (Sylvius)
Arc   arcuate hypothalamic nucleus
aspina   anterior spinal artery
Atlas   atlas (C1 vertebra)
Axis   axis (C2 vertebra)
bas   basilar artery
BOcc   basioccipital bone
CA1   CA1 field of the hippocampus
CA3   CA3 field of the hippocampus
CC   central canal
cc   corpus callosum
CeF   cervical flexure
CG   central (periaqueductal) gray
CGPn   central gray of the pons
ChP   choroid plexus
Cricoid   cricoid
cx   cortical neuroepi
CxP   cortical plate
CxS   cortical subplate
DA   dorsal hypothalamic area
dfu   dorsal funiculus of spinal cord
DM   dorsomedial hypothalamic nu
DR   dorsal raphe nucleus
DTg   dorsal tegmental nucleus
EGL   external germinal layer of Cb
eml   external medullary lamina
Epiglottis   epiglottis
EPI   external plexiform layer of the olf bulb
Eso   esophagus
f   fornix
Fr   frontal cortex
fr   fasciculus retroflexus
Frontal   frontal bone
Gi   gigantocellular reticular nucleus
GiA   gigantocellular reticular nu, alpha
Gl   glomerular layer of the olf bulb
Gr   gracile nucleus
gr   gracile fasciculus
Gu   gustatory thalamic nucleus
hbc   habenular commissure
Hyoid   hyoid bone
IC   inferior colliculus
ICx   intermediate cortical layer
IMLF   interstitial nu of the mlf
IP   interpeduncular nucleus

IPF   interpeduncular fossa
IVF   interventricular foramen
LA   lateroanterior hypothalamic nu
lfp   longitudinal fasciculus of the pons
LHb   lateral habenular nucleus
LS   lateral septal nucleus
LSD   lateral septal nu, dorsal part
MCPC   magnocellular nu of the posterior commissure
MD   mediodorsal thalamic nucleus
ME   median eminence
Mi   mitral cell layer of the olf bulb
mlf   medial longitudinal fasciculus
MM   medial mammillary nu, medial
MoCb   molecular layer of cerebellum
MPO   medial preoptic nucleus
MS   medial septal nucleus
mtg   mammillotegmental tract
Nasal   nasal cavity
NasoPhar   nasopharyngeal cavity
ne   neuroepithelium
Oc   occipital cortex
Occ   occipital bone
olf   olfactory nerve
olfa   olfactory artery

Oral   oral cavity
ox   optic chiasm
Pa   paraventricular hypothalamic nu
Palate   palate
Parietal   parietal bone
PC   paracentral thalamic nucleus
pcer   posterior cerebral artery
pd   predorsal bundle
PDR   posterior dorsal recess of nasal cavity
PDTg   posterodorsal tegmental nu
PF   parafascicular thalamic nucleus
PH   posterior hypothalamic area
Pi   pineal gland
Pk   Purkinje cell layer (cerebellum)
PMR   paramedian raphe nucleus
PN   paranigral nucleus
Pn   pontine nuclei
PnC   pontine reticular nu, caudal part
pnm   pontine migration
PnO   pontine reticular nu, oral part
PPit   posterior lobe of the pituitary
PrC   precommissural nucleus
PrH   prepositus hypoglossal nucleus
PrS   presubiculum

PSph   presphenoid bone
PT   paratenial thalamic nucleus
pyx   pyramidal decussation
R   red nucleus
RI   rostral interstitial nu of mlf
RMg   raphe magnus nucleus
ROb   raphe obscurus nucleus
RPa   raphe pallidus nucleus
RSA   retrosplenial agranular cortex
RSG   retrosplenial granular cortex
RtTg   reticulotegmental nu of the pons
S   subiculum
SC   superior colliculus
scba   superior cerebellar artery
SCh   suprachiasmatic nucleus
sm   stria medullaris of the thalamus
sox   supraoptic decussation
Spinal   spinal cord
SPF   subparafascicular thalamic nu
StHy   striohypothalamic nucleus
SubV   subventricular cortical layer
SuM   supramammillary nucleus
ThyrCart   thyroid cartilage

Tong   tongue
Trachea   trachea
trs   transverse sinus
Tu   olfactory tubercle
VDB   nu vertical limb diagonal band
vert   vertebral artery
vfu   ventral funiculus of the spinal cord
vhc   ventral hippocampal commissure
VM   ventromedial thalamic nucleus
VMH   ventromedial hypothalamic nu
VNO   vomeronasal organ
vno   vomeronasal nerve
VTA   ventral tegmental area (Tsai)
xscp   decussation of sup cerebellar peduncle
ZI   zona incerta

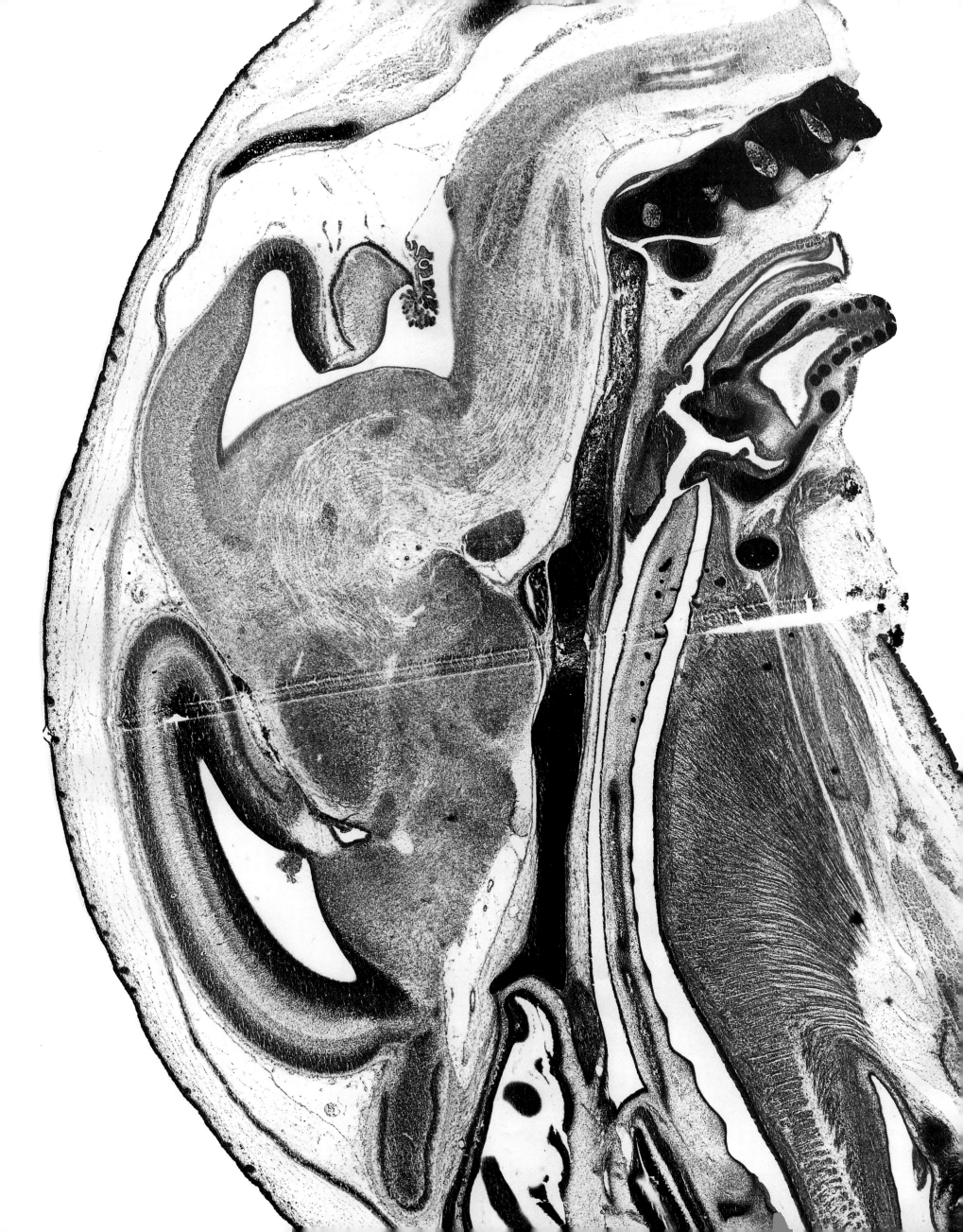

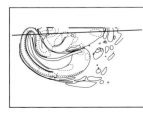

Figure 132
E19 Sagittal 3

1 cortical layer 1
2n optic nerve
4 trochlear nucleus
4V fourth ventricle
7n facial nerve or its root
10 dorsal motor nu of vagus
12 hypoglossal nucleus
A8 A8 dopamine cells
ac anterior commissure
Acb accumbens nucleus
acer anterior cerebral artery
AH anterior hypothalamic nucleus
aica anterior inferior cerebellar artery
AM anteromedial thalamic nucleus
AOP anterior olfactory nu, posterior
APit anterior lobe of the pituitary
Aq aqueduct (Sylvius)
Arc arcuate hypothalamic nucleus
aspina anterior spinal artery
Atlas atlas (C1 vertebra)
Axis axis (C2 vertebra)
BOcc basioccipital bone
BST bed nu of the stria terminalis
CA1 CA1 field of the hippocampus
CA3 CA3 field of the hippocampus
CC central canal
cc corpus callosum
CeF cervical flexure
CG central (periaqueductal) gray
CGPn central gray of the pons
ChP choroid plexus
CL centrolateral thalamic nucleus
cx cortical neuroepi
CxP cortical plate
CxS cortical subplate
DA dorsal hypothalamic area
dfu dorsal funiculus of spinal cord
DG dentate gyrus
Dk nucleus of Darkschewitsch
DM dorsomedial hypothalamic nu
DMTg dorsomedial tegmental area
DpG deep gray layer sup colliculus
DPGi dorsal paragigantocellular nu
DpMe deep mesencephalic nucleus
DpWh deep white layer sup collicul
DR dorsal raphe nucleus
DTg dorsal tegmental nucleus
EGL external germinal layer of Cb
eml external medullary lamina
EndoII endoturbinate II
EndoIII endoturbinate III
EndoIV endoturbinate IV
EPl external plexiform layer of the olf bulb
Eso esophagus
f fornix
Fr frontal cortex
fr fasciculus retroflexus

Gi gigantocellular reticular nucleus
GiA gigantocellular reticular nu, alpha
GiV gigantocellular reticular nu, ventral
Gl glomerular layer of the olf bulb
Gr gracile nucleus
Gu gustatory thalamic nucleus
hi hippocampal formation neuroepi
IC inferior colliculus
ICx intermediate cortical layer
IMLF interstitial nu of the mlf
InG intermed gray layer sup collic
InWh intermed white layer sup coll
IOM inferior olive, medial nucleus
IOPr inferior olive, principal nucleus
IPF interpeduncular fossa
IVF interventricular foramen
LA lateroanterior hypothalamic nu
Larynx larynx
LS lateral septal nucleus
LSD lateral septal nu, dorsal part
LV lateral ventricle
mcp middle cerebellar peduncle
Me5 mesencephalic trigeminal nu
mfb medial forebrain bundle
Mi mitral cell layer of the olf bulb
mlf medial longitudinal fasciculus

MM medial mammillary nu, medial
MoCb molecular layer of cerebellum
MPO medial preoptic nucleus
mt mammillothalamic tract
mtg mammillotegmental tract
Nasal nasal cavity
ne neuroepithelium
obn olfactory bulb neuroepi
Occ occipital cortex
olf olfactory nerve
olfa olfactory artery
Op optic nerve layer sup colliculus
Oral oral cavity
Orb orbital cortex
ox optic chiasm
Pa paraventricular hypothalamic nu
Pal palatine bonepal palatine nerve
Palate palate
Parietal parietal bone
PC paracentral thalamic nucleus
pcb precerebellar nuclei neuroepi
pcer posterior cerebral artery
PDTg posterodorsal tegmental nucleus
PF parafascicular thalamic nucleus
PH posterior hypothalamic area
Pk Purkinje cell layer (cerebellum)

PMD premammillary nu, dorsal part
PN paranigral nucleus
Pn pontine nuclei
PnC pontine reticular nu, caudal part
PnO pontine reticular nu, oral part
PrC precommissural nucleus
PrH prepositus hypoglossal nucleus
PrS presubiculum
PSph presphenoid bone
py pyramidal tract
R red nucleus
RMg raphe magnus nucleus
Ro nucleus of Roller
RRF retrorubral field
RSA retrosplenial agranular cortex
RSG retrosplenial granular cortex
Rt reticular thalamic nucleus
RtTg reticulotegmental nu of the pons
S subiculum
scba superior cerebellar artery
SCh suprachiasmatic nucleus
SHy septohypothalamic nucleus
sm stria medullaris of the thalamus
Sol nucleus of the solitary tract
sox supraoptic decussation
SPF subparafascicular thalamic nu

spt septal neuroepi
Stg strigmoid hypothalamic nucleus
StHy striohypothalamic nucleus
SubV subventricular cortical layer
SuG superfic gray layer sup collic
SuM supramammillary nucleus
Tong tongue
trs transverse sinus
Tu olfactory tubercle
VDB nu vertical limb diagonal band
vert vertebral artery
vfu ventral funiculus of the spinal cord
vhc ventral hippocampal commissure
VL ventrolateral thalamic nucleus
VM ventromedial thalamic nucleus
VMH ventromedial hypothalamic nu
VNO vomeronasal organ
vno nerve of vomeronasal organ
VP ventral pallidum
VTA ventral tegmental area (Tsai)
VTg ventral tegmental nucleus (Gudden)
xscp decussation of the superior cerebellar peduncle
ZI zona incerta
Zo zonal layer of the superior colliculus

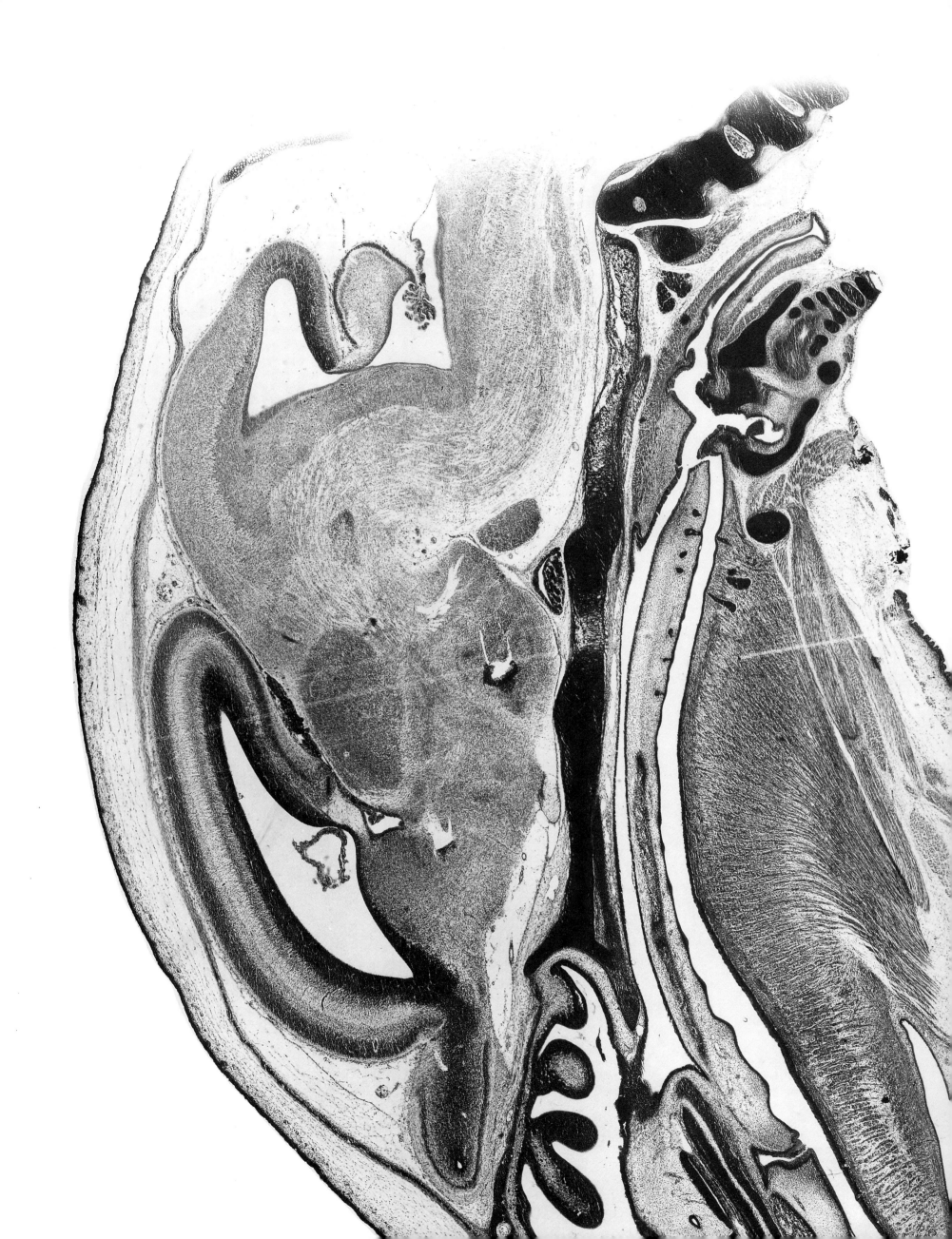

Figure 133
E19 Sagittal 4

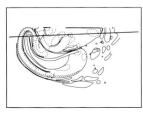

1   cortical layer 1
2n   optic nerve
4   trochlear nucleus
4V   fourth ventricle
6   abducens nucleus
7n   facial nerve or its root
10   dorsal motor nucleus of vagus
12   hypoglossal nucleus
A8   A8 dopamine cells
AB   anterobasal nucleus
ac   anterior commissure
Acb   accumbens nucleus
acer   anterior cerebral artery
aica   anterior inferior cerebellar artery
AH   anterior hypothalamic nucleus
AM   anteromedial thalamic nucleus
AOP   anterior olfactory nu, posterior
APit   anterior lobe of the pituitary
APT   anterior pretectal nucleus
Aq   aqueduct (Sylvius)
Arc   arcuate hypothalamic nucleus
Atlas   atlas (C1 vertebra)
Axis   axis (C2 vertebra)
BOcc   basiooccipital bone
BST   bed nu of the stria terminalis
BSTMA   bed nu of the st, medial division, ant
CA1   CA1 field of the hippocampus
CA3   CA3 field of the hippocampus
cc   corpus callosum
CeF   cervical flexure
CG   central (periaqueductal) gray
ChP   choroid plexus
CL   centrolateral thalamic nucleus
cx   cortical neuroepi
CxP   cortical plate
CxS   cortical subplate
DG   dentate gyrus
DM   dorsomedial hypothalamic nu
DMTg   dorsomedial tegmental area
DPGi   dorsal paragigantocellular nu
DpMe   deep mesencephalic nucleus
DR   dorsal raphe nucleus
DTg   dorsal tegmental nucleus
EGL   external germinal layer of Cb
eml   external medullary lamina
EndoII   endoturbinate II
EndoIII   endoturbinate III
EndoIV   endoturbinate IV
EPl   external plexiform layer of the olf bulb
Fr   frontal cortex
Gi   gigantocellular reticular nucleus
GiA   gigantocellular reticular nu, alpha
GiV   gigantocellular reticular nu, ventral
Gl   glomerular layer of the olf bulb
Gr   gracile nucleus
gr   gracile fasciculus
Gu   gustatory thalamic nucleus

HDB   nu horizon limb diagonal band
hi   hippocampal formation neuroepi
Hyoid   hyoid bone
IC   inferior colliculus
ICx   intermediate cortical layer
IO   inferior olive
IOM   inferior olive, medial nucleus
IVF   interventricular foramen
LA   lateroanterior hypothalamic nu
LD   laterodorsal thalamic nucleus
lfp   longitudinal fasciculus of the pons
LP   lateral posterior thalamic nu
LS   lateral septal nucleus
LSD   lateral septal nu, dorsal part
LV   lateral ventricle
mcp   middle cerebellar peduncle
Me5   mesencephalic trigeminal nu
mfb   medial forebrain bundle
Mi   mitral cell layer of the olf bulb
ML   medial mammillary nu, lateral part
ml   medial lemniscus
mlf   medial longitudinal fasciculus
MoCb   molecular layer of cerebellum
mp   mammillary peduncle

MPO   medial preoptic nucleus
mt   mammillothalamic tract
mtg   mammillotegmental tract
MVe   medial vestibular nucleus
Nasal   nasal cavity
NasoPhar   nasopharyngeal cavity
ne   neuroepithelium
obn   olfactory bulb neuroepi
Oc   olfactory cortex
Occ   occipital bone
olfa   olfactory cavity
Oral   oral cavity
Orb   orbital cortex
OT   nucleus of the optic tract
ox   optic chiasm
Pa   paraventricular hypothalamic nu
Pa4   paratrochlear nucleus
pal   palatine nerve
Palate   palate
Parietal   parietal bone
PC   paracentral thalamic nucleus
pcb   precerebellar nuclei neuroepi
pcer   posterior cerebral artery
PDTg   posterodorsal tegmental nu

PF   parafascicular thalamic nu
Pk   Purkinje cell layer (cerebellum)
PMD   premammillary nu, dorsal part
PMV   premammillary nu, ventral part
Pn   pontine nuclei
PnC   pontine reticular nu, caudal part
PnO   pontine reticular nu, oral part
PPT   posterior pretectal nucleus
PPy   parapyramidal reticular nu
PrH   prepositus hypoglossal nu
PrS   presubiculum
PSph   presphenoid bone
py   pyramidal tract
R   red nucleus
Rathke   Rathke's pouch
RRF   retrorubral field
RSA   retrosplenial agranular cortex
RSG   retrosplenial granular cortex
Rt   reticular thalamic nucleus
RtTg   reticulotegmental nu of the pons
S   subiculum
SC   superior colliculus
scba   superior cerebellar artery
SCGn   superior cervical ganglion

scp   superior cerebellar peduncle
sm   stria medullaris of the thalamus
sox   supraoptic decussation
spt   septal neuroepi
StHy   striohypothalamic nucleus
SubV   subventricular cortical layer
SuM   supramammillary nucleus
sumx   supramammillary decussation
ThyrCart   thyroid cartilage
Tong   tongue
trs   transverse sinus
vert   vertebral artery
vhc   ventral hippocampal commissure
VL   ventrolateral thalamic nucleus
VM   ventromedial thalamic nucleus
VMH   ventromedial hypothalamic nu
VNO   vomeronasal organ
vno   nerve of vomeronasal organ
VP   ventral pallidum
VTA   ventral tegmental area (Tsai)
ZI   zona incerta

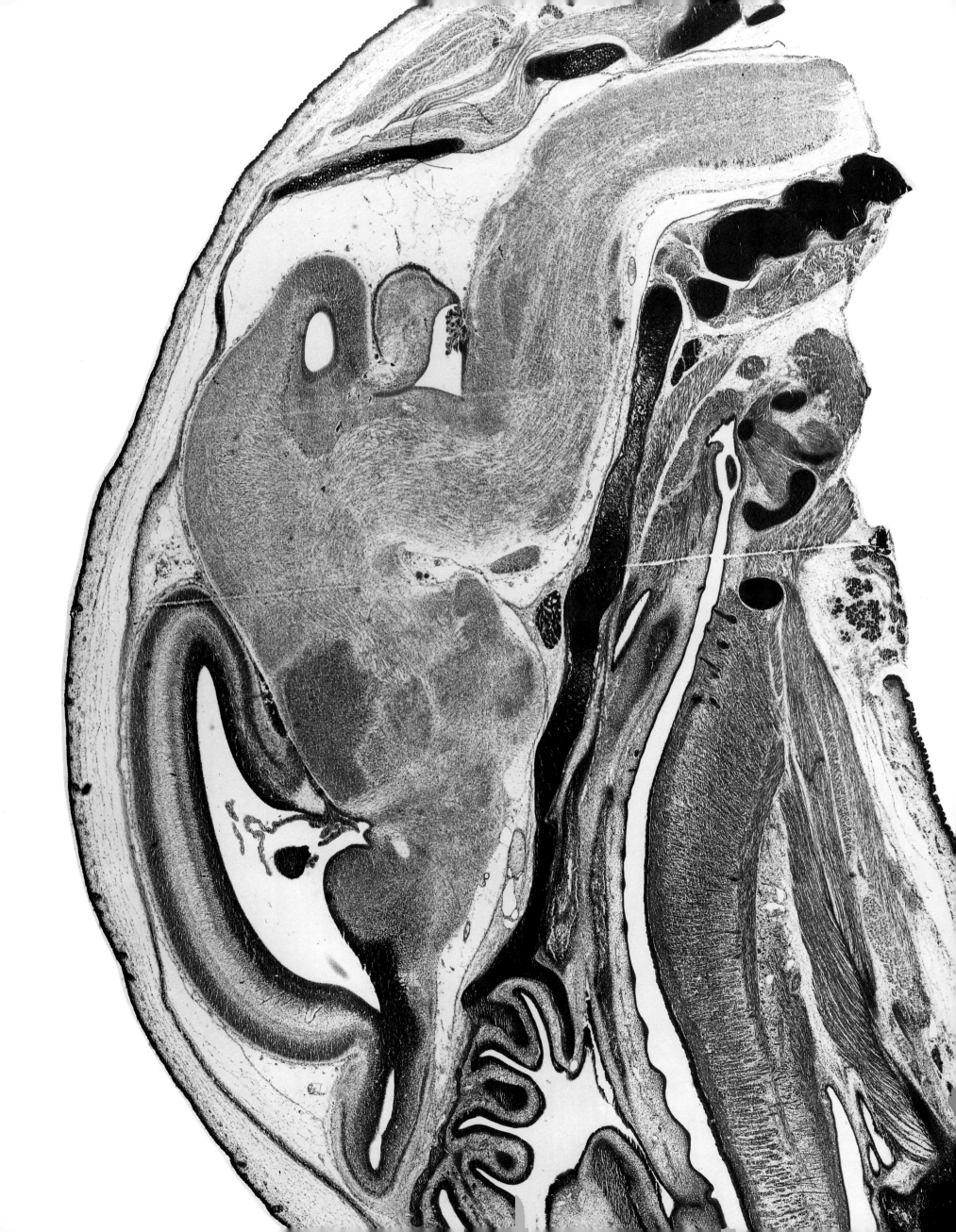

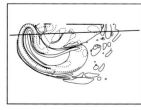

Figure 134
E19 Sagittal 5

1 cortical layer 1
2n optic nerve
4V fourth ventricle
6 abducens nucleus
6n abducens nerve or its root
7n facial nerve or its root
10 dorsal motor nu of vagus
A8 A8 dopamine cells
AB anterobasal nucleus
ac anterior commissure
aca anterior commissure, anterior part
Acb accumbens nucleus
acer anterior cerebral artery
AH anterior hypothalamic nucleus
aica anterior inferior cerebellar artery
AM anteromedial thalamic nucleus
AOP anterior olfactory nu, posterior
APit anterior lobe of the pituitary
APT anterior pretectal nucleus
Aq aqueduct (Sylvius)
Atlas atlas (C1 vertebra)
ATP anterior transitional promontory
BOcc basiooccipital bone
BST bed nu of the stria terminalis
CA1 CA1 field of the hippocampus
CA3 CA3 field of the hippocampus
cav cavernous sinus
cc corpus callosum
CeF cervical flexure
CG central (periaqueductal) gray
CGPn central gray of the pons
ChP choroid plexus
CL centrolateral thalamic nucleus
cpu caudate putamen neuroepi
CrP cribiform plate
cx cortical neuroepi
CxP cortical plate
CxS cortical subplate
dfu dorsal funiculus of spinal cord
DG dentate gyrus
DH dorsal horns of spinal cord
DM dorsomedial hypothalamic nu
DMTg dorsomedial tegmental area
DpG deep gray layer sup colliculus
DpMe deep mesencephalic nucleus
DpWh deep white layer sup colliculus
DR dorsal raphe nucleus
EGL external germinal layer of Cb
eml external medullary lamina
EndoII endoturbinate II
EndoIII endoturbinate III
EndoIV endoturbinate IV
EPI external plexiform layer of olfactory bulb
Eust Eustachian tube
F nucleus of the fields of Forel
Fr frontal cortex
Gem gemini hypothalamic nucleus

Gi gigantocellular reticular nucleus
GiA gigantocellular reticular nu, alpha
GiV gigantocellular reticular nu, ventral
Gl glomerular layer of the olf bulb
Gr gracile nucleus
gr gracile fasciculus
Gu gustatory thalamic nucleus
HDB nu horizon limb diagonal band
hi hippocampal formation neuroepi
Hyoid hyoid bone
IC inferior colliculus
ICx intermediate cortical layer
InG intermed gray layer sup collic
IntZ intermediate zone of spinal gray
InWh intermed white layer sup coll
IOM inferior olive, medial nu
IOPr inferior olive, principal nucleus
IRt intermediate reticular zone
LA lateroanterior hypothalamic nucleus
LD laterodorsal thalamic nucleus
lfp longitudinal fasciculus of the pons
LP lateral posterior thalamic nu
LPO lateral preoptic area
LS lateral septal nucleus
LSD lateral septal nu, dorsal part
LV lateral ventricle
mcp middle cerebellar peduncle

MdD medullary reticular nu, dors
Me5 mesencephalic trigeminal nu
Med medial (fastigial) cerebellar nu
mfb medial forebrain bundle
Mi mitral cell layer of the olf bulb
ML medial mammillary nu, lateral part
ml medial lemniscus
MoCb molecular layer of cerebellum
mp mammillary peduncle
MVe medial vestibular nucleus
Nasal nasal cavity
ne neuroepithelium
obn olfactory bulb neuroepi
Oc occipital cortex
Occ occipital bone
olf olfactory nerve
olfa olfactory artery
Oral oral cavity
Orb orbital cortex
OT nucleus of the optic tract
OV olfactory ventricle
ox optic chiasm
Pa4 paratrochlear nucleus
Pal palatine bone
pal palatine nerve
Parietal parietal bone
PC paracentral thalamic nucleus

pcb precerebellar nuclei neuroepi
pcer posterior cerebral artery
PDR posterior dorsal recess of nasal cavity
PF parafascicular thalamic nucleus
Pk Purkinje cell layer (cerebellum)
PMV premammillary nu, ventral part
Pn pontine nuclei
PnC pontine reticular nu, caudal part
PnO pontine reticular nu, oral part
PPT posterior pretectal nucleus
PPy parapyramidal reticular nu
PR prerubral field
PrS presubiculum
PSph presphenoid bone
py pyramidal tract
R red nucleus
RRF retrorubral field
RSA retrosplenial agranular cortex
RSG retrosplenial granular cortex
Rt reticular thalamic nucleus
RtTg reticulotegmental nucleus of the pons
S subiculum
scba superior cerebellar artery
SCGn superior cervical ganglion
scp superior cerebellar peduncle
sm stria medullaris of the thalamus
Sol nucleus of the solitary tract

sox supraoptic decussation
spa sphenopalatine artery
spt septal neuroepi
SPTg subpeduncular tegmental nu
SpVe spinal vestibular nucleus
StHy striohypothalamic nucleus
SubV subventricular cortical layer
SuG superfic gray layer sup collic
SuM supramammillary nucleus
Thyr thyroid gland
ThyrCart thyroid cartilage
Tong tongue
trs transverse sinus
Tu olfactory tubercle
Tz nucleus of the trapezoid body
vert vertebral artery
vfu ventral funiculus of the spinal cord
VH ventral horn
vhc ventral hippocampal commissure
VL ventrolateral thalamic nucleus
VMH ventromedial hypothalamic nu
VP ventral pallidum
VPM ventral posteromedial thalamic nu
VTA ventral tegmental area (Tsai)
ZI zona incerta

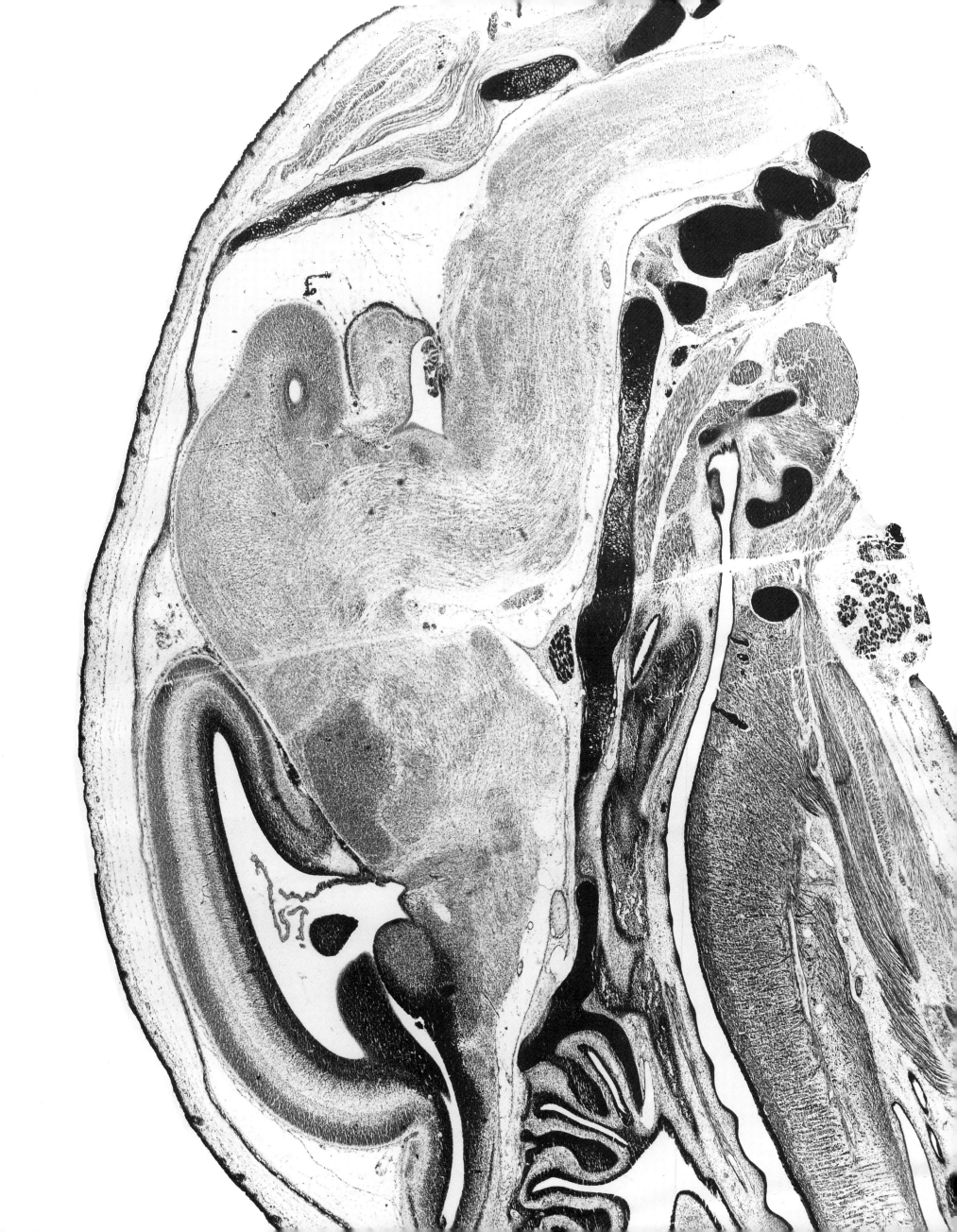

Figure 135
E19 Sagittal 6

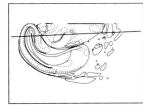

1 cortical layer 1
2n optic nerve
4V fourth ventricle
6 abducens nucleus
6n abducens nerve or its root
7 facial nucleus
7n facial nerve or its root
A8 A8 dopamine cells
ac anterior commissure
aca anterior commissure, anterior part
Acb accumbens nucleus
acer anterior cerebral artery
aica anterior inferior cerebellar artery
AOP anterior olfactory nu, posterior part
APit anterior lobe of the pituitary
APT anterior pretectal nucleus
Aq aqueduct (Sylvius)
Atlas atlas (C1 vertebra)
ATP anterior transitional promontory
AV anteroventral thalamic nucleus
BOcc basioccipital bone
BST bed nu of the stria terminalis
CA1 CA1 field of the hippocampus
CA3 CA3 field of the hippocampus
cav cavernous sinus
CeF cervical flexure
CG central (periaqueductal) gray
CGPn central gray of the pons
ChP choroid plexus
CL centrolateral thalamic nucleus
cpu caudate putamen neuroepi
Cu cuneate nucleus
cu cuneate fasciculus
cx cortical neuroepi
CxP cortical plate
CxS cortical subplate
DG dentate gyrus
DH dorsal horns of spinal cord
DMTg dorsomedial tegmental area
dpal descending palatine artery
DpG deep gray layer sup colliculus
DpMe deep mesencephalic nucleus
DpWh deep white layer sup collicul
EGL external germinal layer of Cb
eml external medullary lamina
EndoII endoturbinate II
EndoIII endoturbinate III
EndoIV endoturbinate IV
EPl external plexiform layer of olf bulb
Eust Eustachian tube
Fr frontal cortex
Frontal frontal bone
Gem gemini hypothalamic nucleus
Gi gigantocellular reticular nucleus
GiV gigantocellular reticular nu, ventral
Gl glomerular layer of the olf bulb
Gu gustatory thalamic nucleus

HDB nu horizon limb diagonal band
hi hippocampal formation neuroepi
Hyoid hyoid bone
IC inferior colliculus
ICx intermediate cortical layer
InG intermed gray layer sup collic
InWh intermed white layer sup coll
IO inferior olive
IPF interpeduncular fossal
IRt intermediate reticular nucleus
LA lateroanterior hypothalamic nu
LC locus coeruleus
LD laterodorsal thalamic nucleus
LDTg laterodorsal tegmental nucleus
lfp longitudinal fasciculus of the pons
LH lateral hypothalamic area
LM lateral mammillary nucleus
LP lateral posterior thalamic nu
LPGi lateral paragigantocellular nu
LPO lateral preoptic area
LS lateral septal nucleus
LV lateral ventricle
Maxilla maxilla
mcp middle cerebellar peduncle
MdD medullary reticular nu, dors
MdV medullary reticular nu, ventral

Me5 mesencephalic trigeminal nu
Med medial (fastigial) cerebellar nu
mfb medial forebrain bundle
Mi mitral cell layer of the olf bulb
ml medial lemniscus
MoCb molecular layer of cerebellum
mp mammillary peduncle
MTu medial tuberal nucleus
MVe medial vestibular nucleus
Nasal nasal cavity
ne neuroepithelium
obn olfactory bulb neuroepi
Oc occipital cortex
Occ occipital bone
olfa olfactory artery
Op optic nerve layer sup colliculus
Oral oral cavity
Orb orbital cortex
OT nucleus of the optic tract
OV olfactory ventricle
ox optic chiasm
P7 perifacial zone
Pal palatine bone
pal palatine nerve
Parietal parietal bone
pcb precerebellar nuclei neuroepi

pcer posterior cerebral artery
PCRt parvocellular reticular nu
PeF periformical nucleus
PF parafascicular thalamic nucleus
Pk Purkinje cell layer (cerebellum)
PMV premammillary nu, ventral part
Pn pontine nuclei
PnC pontine reticular nu, caudal part
PnO pontine reticular nu, oral part
Po posterior thalamic nuclear group
PPT posterior pretectal nucleus
PR prerubral field
PrS presubiculum
PSph presphenoid bone
ptgcn nerve of the pterygoid canal
R red nucleus
RRF retrorubral field
RSG retrosplenial granular cortex
Rt reticular thalamic nucleus
S subiculum
scba superior cerebellar artery
SCGn superior cervical ganglion
scp superior cerebellar peduncle
sm stria medullaris of the thalamus
Sol nucleus of the solitary tract
sol solitary tract

sox supraoptic decussation
spa sphenopalatine artery
SPO superior paraolivary nucleus
spt septal neuroepi
SPTg subpeduncular tegmental nu
SpVe spinal vestibular nucleus
StHy striohypothalamic nucleus
Subl subincertal nucleus
SubV subventricular cortical layer
SuG superfic gray layer sup collic
SuM supramammillary nucleus
Thyr thyroid gland
ThyrCart thyroid cartilage
Tong tongue
trs transverse sinus
Tu olfactory tubercle
ve vestibular area neuroepi
vert vertebral artery
VH ventral horn
vhc ventral hippocampal commissure
VL ventrolateral thalamic nucleus
VP ventral pallidum
VPM ventral posteromedial thalamic nu
VTA ventral tegmental area (Tsai)
ZI zona incerta
Zo zonal layer of the superior colliculus

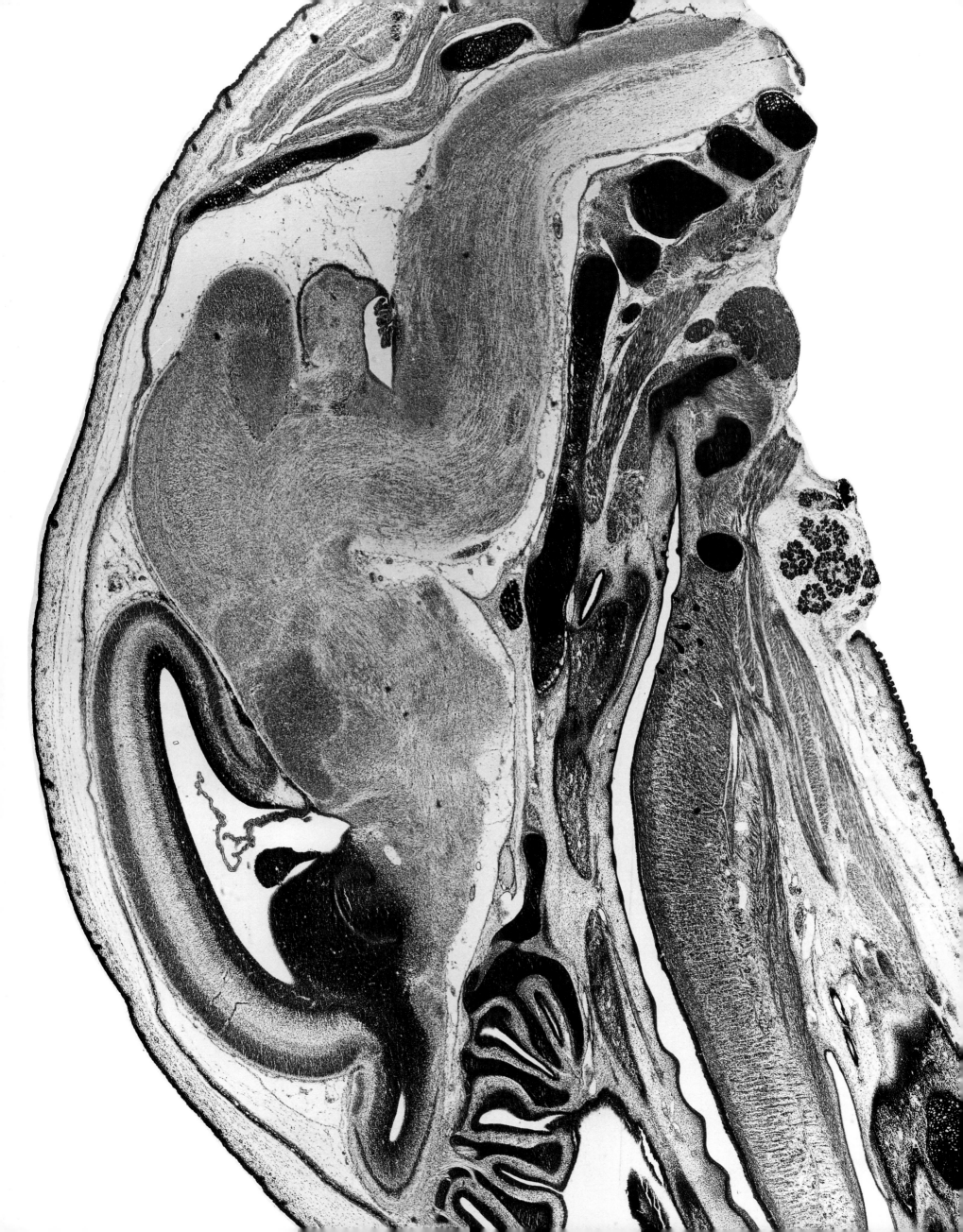

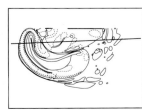

Figure 136
E19 Sagittal 7

1 cortical layer 1
2n optic nerve
4V fourth ventricle
7 facial nucleus
7n facial nerve or its root
A8 A8 dopamine cells
ac anterior commissure
aca anterior commissure, anterior part
Acb accumbens nucleus
acer anterior cerebral artery
aica anterior inferior cerebellar artery
AOB accessory olfactory bulb
AOP anterior olfactory nucleus, posterior
APit anterior lobe of the pituitary
APT anterior pretectal nucleus
Atlas atlas (C1 vertebra)
AV anteroventral thalamic nucleus
Axis axis (C2 vertebra)
BOcc basiooccipital bone
bsc brachium of the superior colliculus
BSph basisphenoid bone
BST bed nu of the stria terminalis
BSTMA bed nu of the st, medial division, ant
CA1 CA1 field of the hippocampus
CA3 CA3 field of the hippocampus
cav cavernous sinus
CeF cervical flexure
CG central (periaqueductal) gray
CGPn central gray of the pons
ChP choroid plexus
CnF cuneiform nucleus
cpu caudate putamen neuroepi
CrP cribiform plate
Cu cuneate nucleus
cu cuneate fasciculus
cx cortical neuroepi
CxP cortical plate
CxS cortical subplate
DG dentate gyrus
DH dorsal horns of spinal cord
dpal descending palatine artery
dpet deep petrosal nerve
DpG deep gray layer sup colliculus
DpMe deep mesencephalic nucleus
DpWh deep white layer sup collicul
EGL external germinal layer of Cb
eml external medullary lamina
EndoII endoturbinate II
EndoIII endoturbinate III
EPI external plexiform layer of the olf bulb
Eth ethmoid bone
Eust Eustachian tube
Fr frontal cortex
Frontal frontal bone
Gi gigantocellular reticular nucleus
GiV gigantocellular reticular nu, vental
Gl glomerular layer of the olf bulb

GPalF greater palatine foramen
Gu gustatory thalamic nucleus
HDB nu horizon limb diagonal band
hi hippocampal formation neuroepi
HL hindlimb area of the cortex
HPtg hamulus of the pterygoid bone
Hyoid hyoid bone
IC inferior colliculus
icap internal capsule
ICGn inferior cervical ganglion
InCo intercollicular nucleus
InG intermed gray layer sup collic
IntZ intermediate zone of spinal gray
InWh intermed white layer sup coll
IO inferior olive
IPF interpeduncular nucleus
IRt intermediate reticular zone
LA lateroanterior hypothalamic nu
LC locus coeruleus
LD laterodorsal thalamic nucleus
LH lateral hypothalamic area
LM lateral mammillary nucleus
LP lateral posterior thalamic nu
LPGi lateral paragigantocellular nu
LPO lateral preoptic area
LV lateral ventricle
Maxilla maxilla
mcp middle cerebellar peduncle

MdD medullary reticular nu, dors
MdV medullary reticular nu, vental
Me5 mesencephalic trigeminal nu
Med medial (fastigial) cerebellar nu
mfb medial forebrain bundle
Mi mitral cell layer of the olf bulb
ml medial lemniscus
MoCb molecular layer of cerebellum
MTu medial tuberal nucleus
MVe medial vestibular nucleus
Nasal nasal cavity
ne neuroepithelium
obn olfactory bulb neuroepi
Oc occipital cortex
Occ occipital bone
olf olfactory nerve
olfa olfactory nerve layer
ON optic nerve layer
Op optic nerve layer sup colliculus
Oral oral cavity
Orb orbital cortex
OT nucleus of the optic tract
OV olfactory ventricle
ox optic chiasm
Pal palatine bone
pal palatine nerve
Parietal parietal bone
pcb precerebellar nuclei neuroepi

pcer posterior cerebral artery
PCRt parvocellular reticular nu
Pk Purkinje cell layer (cerebellum)
PMV premammillary nu, ventral part
Pn pontine nuclei
PnC pontine reticular nu, caudal part
PnO pontine reticular nu, oral part
Po posterior thalamic nuclear group
PPT posterior pretectal nucleus
PR prerubral field
PrS presubiculum
PSph presphenoid bone
ptgn nerve of the pterygoid canal
R red nucleus
RRF retrorubral field
RSA retrosplenial agranular cortex
RSG retrosplenial granular cortex
Rt reticular thalamic nucleus
S subiculum
scba superior cerebellar artery
SCGn superior cervical ganglion
scp superior cerebellar peduncle
sm stria medullaris of the thalamus
SN substantia nigra
SOI superior olive
Sol nucleus of the solitary tract
sox supraoptic decussation
spa sphenopalatine artery

SPFPC subparafascicular thal nu, parvocell
spt septal neuroepi
SPTg subpeduncular tegmental nu
SpVe spinal vestibular nucleus
st stria terminalis
StHy striohypothalamic nucleus
SubC subcoeruleus nucleus
Subl subincertal nucleus
SubV subventricular cortical layer
SuG superfic gray layer sup collic
SuM supramammillary nucleus
Symp sympathetic trunk
Thyr thyroid gland
ThyrCart thyroid cartilage
Tong tongue
trs transverse sinus
Tu olfactory tubercle
vert vertebral artery
vhc ventral hippocampal commissure
VL ventrolateral thalamic nucleus
VP ventral pallidum
VPM ventral posteromedial thalamic nu
VTAR ventral tegmental area, rostral
ZI zona incerta
Zo zonal layer of the superior colliculus

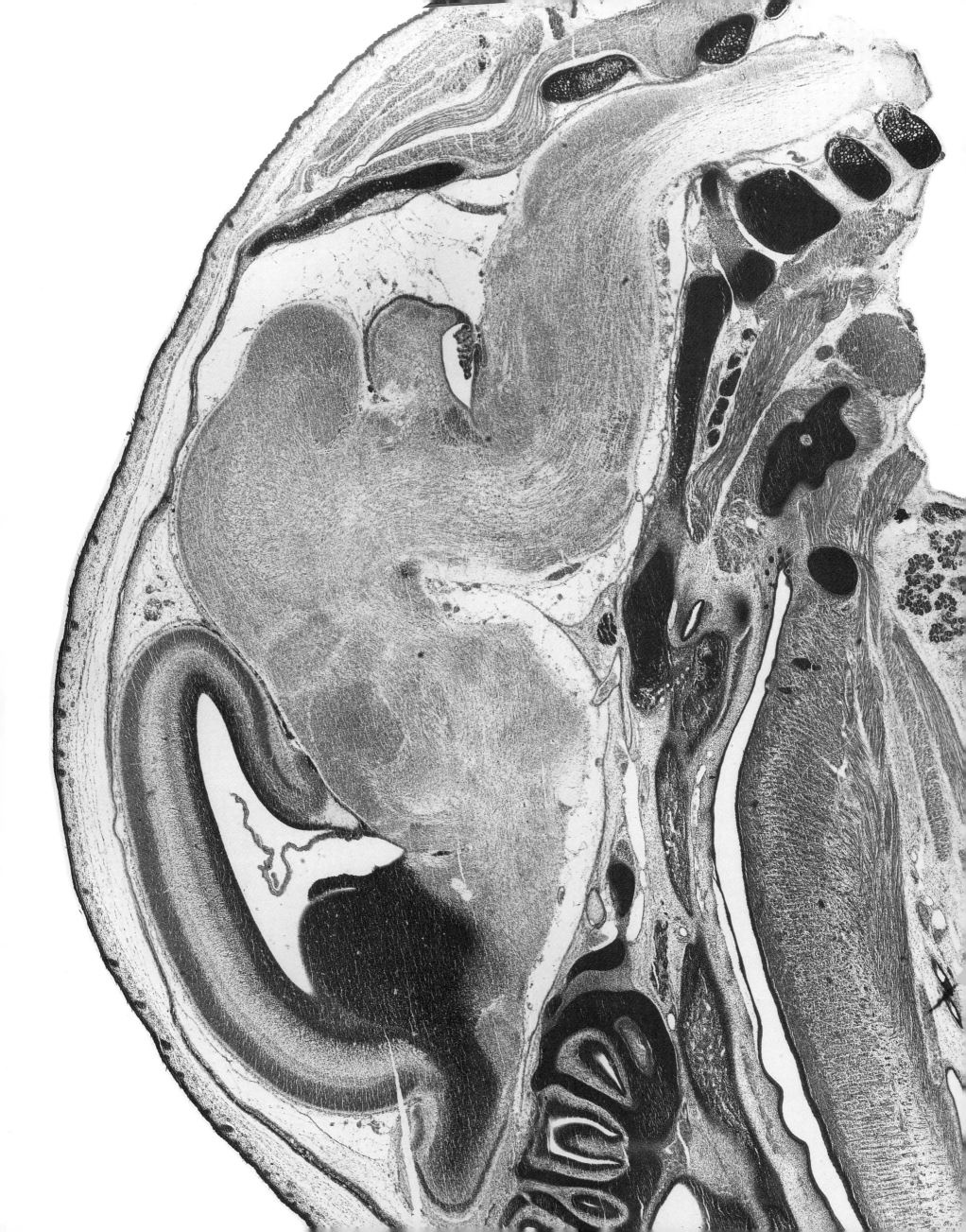

Figure 137
E19 Sagittal 8

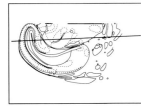

1 cortical layer 1
2n optic nerve
4V fourth ventricle
6n abducens nerve or its root
7 facial nucleus
7n facial nerve or its root
A8 A8 dopamine cells
AB anterobasal nucleus
aca anterior commissure, anterior part
Acb accumbens nucleus
acer anterior cerebral artery
aica anterior inferior cerebellar artery
Amb ambiguus nucleus
AOP anterior olfactory nu, posterior
APit anterior lobe of the pituitary
APT anterior pretectal nucleus
aud auditory neuroepi
AV anteroventral thalamic nucleus
Axis axis (C2 vertebra)
BOcc basioccipital bone
bsc brachium of the superior colliculus
BSph basisphenoid bone
BST bed nu of the stria terminalis
BSTMA bed nu of the st, medial division, ant
CA1 CA1 field of the hippocampus
CA3 CA3 field of the hippocampus
cav cavernous sinus
Cb cerebellum
CeF cervical flexure
CG central (periaqueductal) gray
ChP choroid plexus
CnF cuneiform nucleus
cp cerebral peduncle, basal part
cpu caudate putamen neuroepi
Cu cuneate nucleus
cu cuneate fasciculus
cx cortical neuroepi
CxP cortical plate
CxS cortical subplate
DC dorsal cochlear nucleus
DG dentate gyrus
dpal descending palatine artery
dpet deep petrosal nerve
DpG deep gray layer sup colliculus
DpMe deep mesencephalic nucleus
DpWh deep white layer sup collicul
ECu external cuneate nucleus
EGL external germinal layer of Cb
eml external medullary lamina
EPI external plexiform layer of the olf bulb
Eust Eustachian tube
fi fimbria of the hippocampus
FL forelimb area of the cortex
fmi forceps minor of the corpus callosum
Fr frontal cortex

Frontal frontal bone
Gl glomerular layer of the olf bulb
gpal greater palatine nerve
GPalF greater palatine foramen
hi hippocampal formation neuroepi
HL hindlimb area of the cortex
HPtg hamulus of the pterygoid bone
IC inferior colliculus
icap internal capsule
ictd internal carotid artery
ICx intermed cortical layer
InCo intercollicular nucleus
InG intermed gray layer sup collic
Int interposed cerebellar nucleus
InWh intermed white layer sup coll
IO inferior olive
ipets inferior petrosal sinus
IRt intermediate reticular zone
LC locus coeruleus
LD laterodorsal thalamic nucleus
LH lateral hypothalamic area
ll lateral lemniscus
LP lateral posterior thalamic nu
LPGi lateral paragigantocellular nu
LPO lateral preoptic area
LRec lateral rectus muscle
LV lateral ventricle
Maxilla maxilla
mcp middle cerebellar peduncle
MCPO magnocellular preoptic nu
MdD medullary reticular nu, dors
Me5 mesencephalic trigeminal nu
mfb medial forebrain bundle
Mi mitral cell layer of the olf bulb
MiTg microcellular tegmental nu
ml medial lemniscus
Mo5 motor trigeminal nucleus
MoCb molecular layer of cerebellum
mp mammillary peduncle
MTu medial tuberal nucleus
MVe medial vestibular nucleus
Nasal nasal cavity
ne neuroepithelium
npal nasopalatine nerve
obn olfactory bulb neuroepi
Oc occipital cortex
Occ occipital bone
ON olfactory nerve layer
opt optic tract
OptF optic foramen
Oral oral cavity
Orb orbital cortex
OT nucleus of the optic tract
Pal palatine bone

Parietal parietal bone
PB parabrachial nuclei
pcer posterior cerebral artery
PCRt parvocellular reticular nu
PCRtA parvocell reticular nu, alpha
Pk Purkinje cell layer (cerebellum)
PMV premammillary nu, ventral part
Pn pontine nuclei
PnO pontine reticular nu, oral part
Po posterior thalamic nuclear group
PPT posterior pretectal nucleus
PPTg pedunculopontine tegmental nu
PrS presubiculum
PSph presphenoid bone
ptgcn nerve of the pterygoid canal
RRF retrorubral field
RSA retrosplenial agranular cortex
RSG retrosplenial granular cortex
Rt reticular thalamic nucleus
S subiculum
scba superior cerebellar artery
SCGn superior cervical ganglion
SN substantia nigra
SO superior olive
SOl superior olive
Sol nucleus of the solitary tract
sol solitary tract
SOR supraoptic nu, retrochiasmatic (diffuse) part

sox supraoptic decussation
Sp5C spinal trigeminal nu, caudal part
spa sphenopalatine artery
SphPal sphenopalatine ganglion
SpVe spinal vestibular nucleus
st stria terminalis
str superior thalamic radiation
Su5 supratrigeminal nucleus
SubC subcoeruleus nucleus
SubI subincertal nucleus
SubV subventricular cortical layer
SuG superfic gray layer sup collic
SuM supramammillary nucleus
Symp sympathetic trunk
ThyrCart thyroid cartilage
Tong tongue
trs transverse sinus
Tu olfactory tubercle
v vein (unidentified)
vert vertebral artery
VLL ventral nucleus of the lateral lemniscus
VP ventral pallidum
VPL ventral posterolateral thalamic nucleus
VPM ventral posteromedial thalamic nucleus
VTAR ventral tegmental area, rostral
ZID zona incerta, dorsal part
ZIV zona incerta, ventral part
Zo stratum zonale of superior colliculus

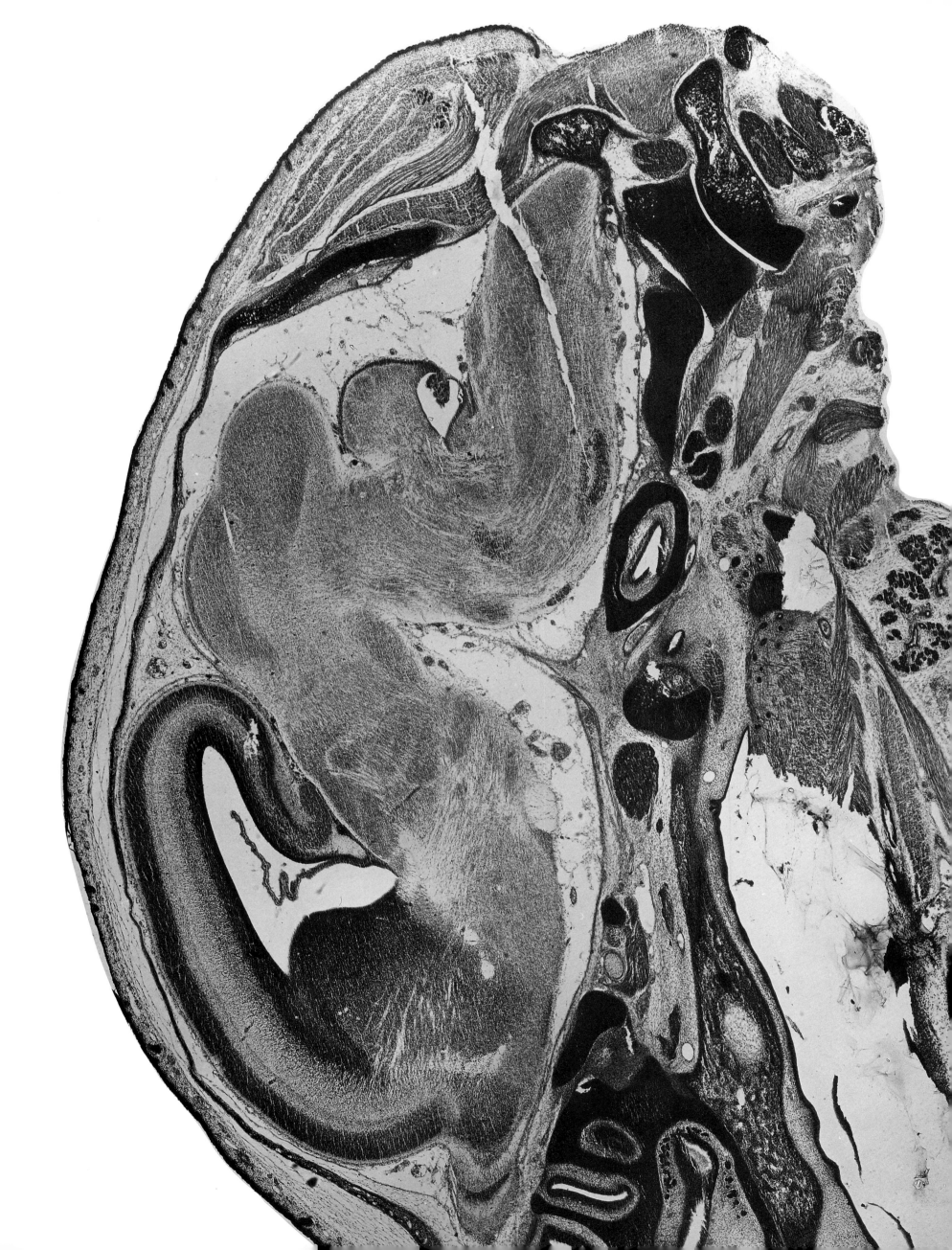

Figure 138
E19 Sagittal 9

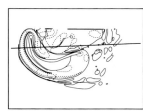

1 cortical layer 1
2n optic nerve
3n oculomotor nerve or its root
4V fourth ventricle
5Gn trigeminal ganglion
5ophth ophthalmic nerve of trigeminal
6n abducens nerve or its root
7 facial nucleus
7n facial nerve or its root
12n hypoglossal nerve or its root
A5 A5 noradrenaline cells
aca anterior commissure, anterior
Acb accumbens nucleus
acer anterior cerebral artery
acp anterior commissure, posterior
aica anterior inferior cerebellar artery
ALF anterior lacerated foramen
Amb ambiguus nucleus
AOP anterior olfactory nu, posterior
ASph alisphenoid bone
Atlas atlas (C1 vertebra)
aud auditory neuroepi
Axis axis (C2 vertebra)
basv basal vein
BIC brachium of inferior colliculus
BOcc basioccipital bone
bsc brachium of the superior colliculus
BST bed nu of the stria terminalis
BSTL bed nu of the st, lateral div
CA1 CA1 field of the hippocampus
CA3 CA3 field of the hippocampus
cav cavernous sinus
Cb cerebellum
CD cochlear duct
ChP choroid plexus
CIC central nu of the inferior colliculus
CnF cuneiform nucleus
cp cerebral peduncle, basal part
CPu caudate putamen (striatum)
cpu caudate putamen neuroepi
cx cortical neuroepi
CxP cortical plate
CxS cortical subplate
DC dorsal cochlear nucleus
DG dentate gyrus
DLG dorsal lateral geniculate nu
dpal descending palatine artery
DRG dorsal root ganglion
ECIC external cortex of the inferior colliculus
ECu external cuneate nucleus
EGL external germinal layer of Cb
eml external medullary lamina
EP entopeduncular nucleus
EPl external plexiform layer of the olf bulb
Eust Eustachian tube
fi fimbria of the hippocampus
FL forelimb area of the cortex

Fr frontal cortex
Frontal frontal bone
Gl glomerular layer of the olf bulb
hi hippocampal formation neuroepi
HL hindlimb area of the cortex
Hyoid hyoid bone
I10Gn inferior vagal (nodose) ganglion
iaud internal auditory artery
icap internal capsule
icp inferior cerebellar peduncle
ictd internal carotid artery
ICx intermediate cortical layer
InCo intercollicular nucleus
Int interposed cerebellar nucleus
ipets inferior petrosal sinus
IRec inferior rectus muscle
IRt intermediate reticular zone
LD laterodorsal thalamic nucleus
LH lateral hypothalamic area
ll lateral lemniscus
LP lateral posterior thalamic nu
LPB lateral parabrachial nucleus
LPO lateral preoptic area
LR4V lateral recess of the 4th ventr
LRec lateral rectus muscle
LRt lateral reticular nucleus
LV lateral ventricle

LVe lateral vestibular nucleus
Maxilla maxilla
MCLH magnocellular nucleus of lateral hypothalamus
mcp middle cerebellar peduncle
MCPO magnocellular preoptic nu
mfb medial forebrain bundle
MG medial geniculate nucleus
Mi mitral cell layer of the olf bulb
MiTg microcellular tegmental nucleus
Mo5 motor trigeminal nucleus
MoCb molecular layer of cerebellum
MRec medial rectus muscle
MTu medial tuberal nucleus
ne neuroepithelium
Oc occipital cortex
Occ occipital bone
opt optic tract
OptF optic foramen
Oral oral cavity
Orb orbital cortex
Pal palatine bone
Parietal parietal bone
PB parabrachial nucleus
pcer posterior cerebral artery
pcoma posterior communicating artery
PCRt parvocellular reticular nu
PCRtA parvocellular reticular nu, alpha

PCTg paracollicular tegmentum
Petrous petrous part of the temp bone
PIL posterior intralaminar thalamic nu
Pir piriform cortex
Pk Purkinje cell layer (cerebellum)
Pn pontine nuclei
pnm pontine migration
PnO pontine reticular nu, oral part
PPTg pedunculopontine tegmental nu
PrS presubiculum
PSph presphenoid bone
Res reservoir of migrating neurons
RR retrorubral area
RSA retrosplenial agranular cortex
Rt reticular thalamic nucleus
S subiculum
SC superior colliculus
scba superior cerebellar artery
SCGn superior cervical ganglion
scp superior cerebellar peduncle
SI substantia innominata
SN substantia nigra
SO supraoptic nucleus
SOR supraoptic nu, retrochiasmatic
sox supraoptic decussation
sp5 spinal trigeminal tract
Sp5C spinal trigeminal nu, caudal part

Sp5I spinal trigeminal nu, interpolar
Sp5O spinal trigeminal nu, oral part
spa sphenopalatine artery
SphPal sphenopalatine ganglion
splex sympathetic plexus
SpVe spinal vestibular nucleus
st stria terminalis
STh subthalamic nucleus
SubB subbrachial nucleus
SubV subventricular cortical layer
SuVe superior vestibular nucleus
Symp sympathetic trunk
Tong tongue
trs transverse sinus
Tu olfactory tubercle
TyBu tympanic bulla
vaf ventral amygdalofugal pathway
vert vertebral artery
VLG ventral lateral geniculate nu
VLL ventral nu of the lateral lemniscus
VP ventral pallidum
VPL ventral posterolateral thalamic nu
ZID zona incerta, dorsal part
ZIV zona incerta, ventral part

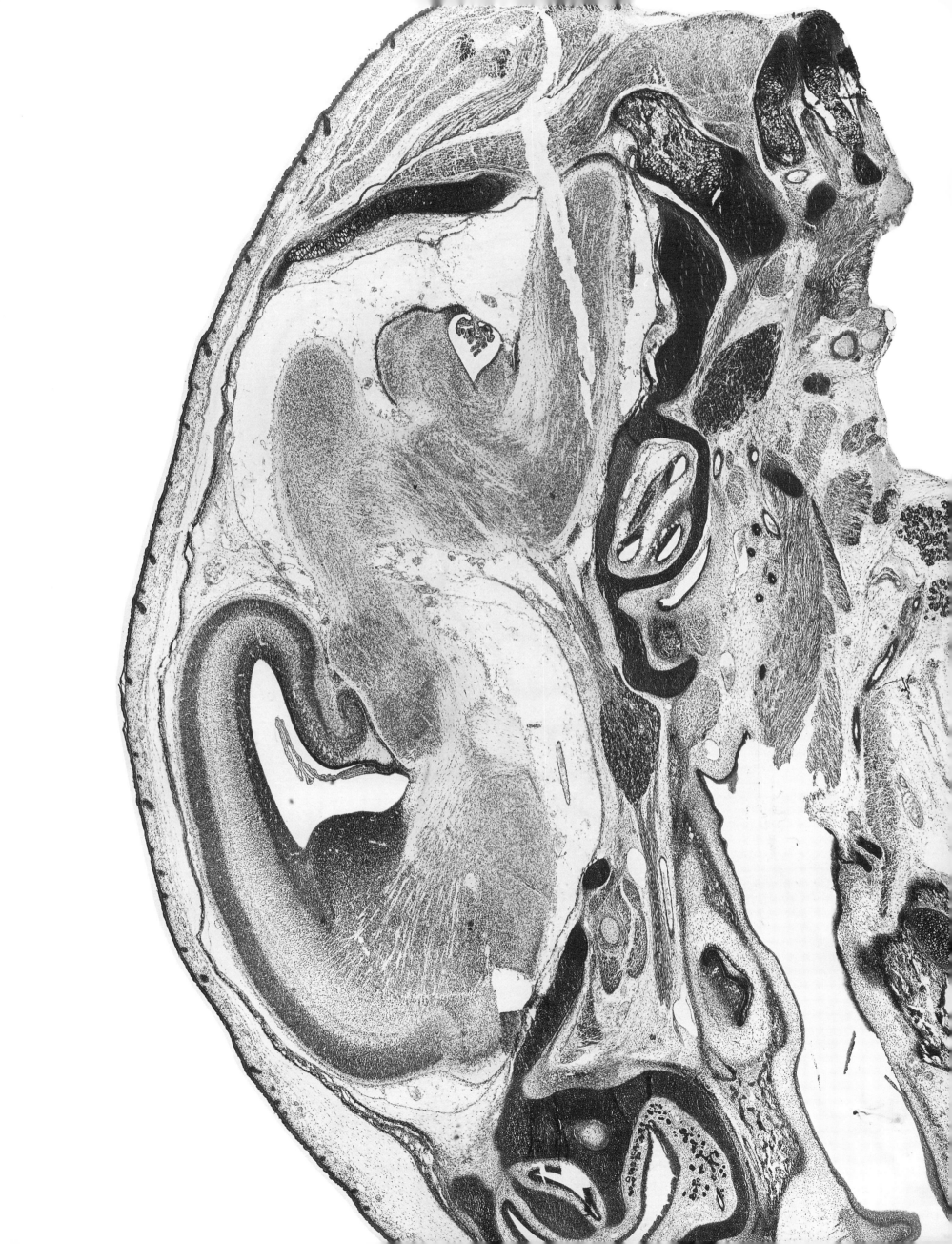

Figure 139
E19 Sagittal 10

1 cortical layer 1
2n optic nerve
3n oculomotor nerve or its root
4n trochlear nerve or its root
5fr frontal branch of trigeminal nerve
5Gn trigeminal ganglion
5max maxillary nerve
5ophth ophthalmic nerve of trigeminal
7n facial nerve or its root
10n vagus nerve
12n hypoglossal nerve or its root
Acb accumbens nucleus
acp anterior commissure, posterior
AI agranular insular cortex
aica anterior inferior cerebellar artery
ALF anterior lacerated foramen
ASph alisphenoid bone
Atlas atlas (C1 vertebra)
Axis axis (C2 vertebra)
BIC nucleus of brachium of inferior colliculus
bic brachium of the inferior colliculus
BOcc basioccipital bone
bsc brachium of the superior colliculus
CA1 CA1 field of the hippocampus
CA3 CA3 field of the hippocampus
cav cavernous sinus
cctd common carotid artery
CD cochlear duct
ChP choroid plexus
CIC central nu of the inferior colliculus
Cil ciliary ganglion
Cl claustrum

cp cerebral peduncle, basal part
CPu caudate putamen (striatum)
cpu caudate putamen neuroepi
CtdB carotid body
cx cortical neuroepi
CxP cortical plate
CxS cortical subplate
DC dorsal cochlear nucleus
DG dentate gyrus
DLG dorsal lateral geniculate nucleus
ECIC external cortex of the inferior colliculus
ectd external carotid artery
EGL external germinal layer of Cb
EnamelO enamel organ (of tooth)
Ethmoid ethmoid bone
fi fimbria of the hippocampus
FL forelimb area of the cortex
Frontal frontal bone
GP globus pallidus
hi hippocampal formation neuroepi
HL hindlimb area of the cortex
Hyoid hyoid bone
I10Gn inferior vagal (nodose) ganglion
iaud internal auditory artery
icap internal capsule

icp inferior cerebellar peduncle
ictd internal carotid artery
ICx intermediate cortical layer
Int interposed cerebellar nucleus
iorb inferior orbital artery
ipets inferior petrosal sinus
IRec inferior rectus muscle
Lat lateral (dentate) cerebellar nu
LD laterodorsal thalamic nucleus
ling lingual nerve
ll lateral lemniscus
lo lateral olfactory tract
LPS levator palpebrae superioris muscle
LR4V lateral recess of the 4th ventr
LRec lateral rectus muscle
LV lateral ventricle
LVe lateral vestibular nucleus
Maxilla maxilla
m5 motor root of the trigeminal nerve
mcp middle cerebellar peduncle
MCPO magnocellular preoptic nucleus
Me medial amygdaloid nucleus
MG medial geniculate nucleus
MiTg microcellular tegmental nucleus

MoCb molecular layer of cerebellum
MPtg medial pterygoid muscle
MRec medial rectus muscle
ne neuroepithelium
Oc occipital cortex
Occ occipital bone
opt optic tract
OptF optic foramen
Oral oral cavity
Orb orbital cortex
Orbit orbital cavity
Par1 parietal cortex, area 1
Parietal parietal bone
PB parabrachial nuclei
PBG parabigeminal nucleus
pcer posterior cerebral artery
PCTg paracollicular tegmentum
Petrous petrous part of the temp bone
Pir piriform cortex
Pk Purkinje cell layer (cerebellum)
pnm pontine migration
Po posterior thalamic nuclear group
PPTg pedunculopontine tegmental nu
Pr5DM principal sensory trigeminal nu, dorsomedial

Pr5VL principal sensory trigeminal nu,
       ventrolateral
PrS presubiculum
PSph presphenoid bone
PSphW presphenoid wing
Res reservoir of migrating neurons
Rt reticular thalamic nucleus
S subiculum
s5 sensory root of the trigeminal nerve
SC superior colliculus
scba superior cerebellar artery
SI substantia innominata
SN substantia nigra
SO supraoptic nucleus
sox supraoptic decussation
sp5 spinal trigeminal tract
Sp5C spinal trigeminal nu, caudal
Sp5I spinal trigeminal nu, interpolar
Sp5O spinal trigeminal nu, oral part
splex sympathetic plexus
SpVe spinal vestibular nucleus
SRec superior rectus muscle
st stria terminalis
str superior thalamic radiation

SubB subbrachial nucleus
SubG subgeniculate nucleus
SubV subventricular cortical layer
SuVe superior vestibular nucleus
Tong tongue
trg germinal trigone
trs transverse sinus
Tu olfactory tubercle
TyBu tympanic bulla
TyC tympanic cavity
v vein (unidentified)
vaf ventral amygdalofugal pathway
vert vertebral artery
VLG ventral lateral geniculate nu
VLL ventral nu of the lateral lemniscus
VP ventral pallidum
VPL ventral posterolateral thalamic nu

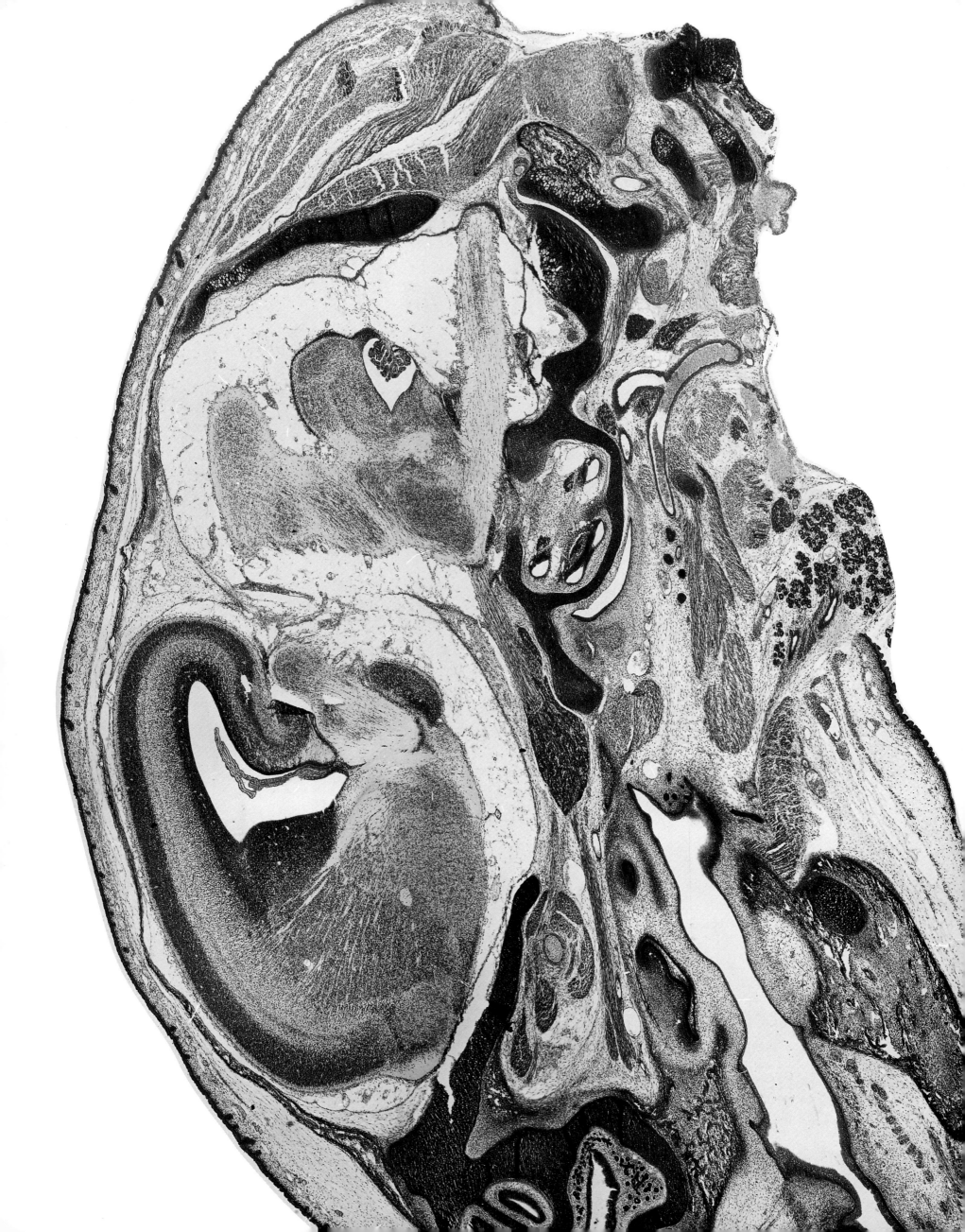

Figure 140
E19 Sagittal 11

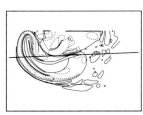

1   cortical layer 1
2n   optic nerve
4n   trochlear nerve or its root
5fr   frontal branch of trigeminal nerve
5Gn   trigeminal ganglion
5max   maxillary nerve
7n   facial nerve or its root
8n   vestibulocochlear nerve
10Gn   vagal ganglion
10n   vagus nerve
12n   hypoglossal nerve or its root
AA   anterior amygdaloid area
aca   anterior commissure, anterior
acp   anterior commissure, posterior
AF   amygdaloid fissure
AI   agranular insular cortex
ALF   anterior lacerated foramen
AOrb   alar orbital bone
ASph   alisphenoid bone
Atlas   atlas (C1 vertebra)
aud   auditory neuroepi
B   basal nucleus of Meynert
bic   brachium of the inferior colliculus
CA1   CA1 field of the hippocampus
CA3   CA3 field of the hippocampus
cav   cavernous sinus
Cb   cerebellum
cctd   common carotid artery
CD   cochlear duct
CGn   cochlear (spiral) ganglion
ChP   choroid plexus
Cl   claustrum
cp   cerebral peduncle, basal part
CPu   caudate putamen (striatum)
cpu   caudate putamen neuroepi
CtdB   carotid body
cx   cortical neuroepi
CxP   cortical plate
CxS   cortical subplate
DC   dorsal cochlear nucleus
DG   dentate gyrus
DLL   dorsal nu of the lateral lemniscus
ECIC   external cortex of the inferior colliculus
ectd   external carotid artery
EGL   external germinal layer of Cb
EnamelO   enamel organ (of tooth)
Ethmoid   ethmoid bone
fi   fimbria of the hippocampus
FL   forelimb area of the cortex
Fr   frontal cortex
Frontal   frontal bone
FStr   fundus striati
GP   globus pallidus
hi   hippocampal formation neuroepi
HL   hindlimb area of the cortex
IAud   internal auditory meatus
icap   internal capsule

icp   inferior cerebellar peduncle
ictd   internal carotid artery
icv   inferior cerebral vein
ICx   intermediate cortical layer
jjug   internal jugular vein
iorb   inferior orbital artery
ipets   inferior petrosal sinus
IRec   inferior rectus muscle
Lat   lateral (dentate) cerebellar nu
ling   lingual nerve
ll   lateral lemniscus
lo   lateral olfactory tract
LPB   lateral parabrachial nucleus
LPS   levator palpebrae superioris muscle
LR4V   lateral recess of the 4th ventr
LRec   lateral rectus muscle
LV   lateral ventricle
m5   motor root of the trigeminal nerve
Maxilla   maxilla
mcer   middle cerebral artery
mcp   middle cerebellar peduncle
Me   medial amygdaloid nucleus
MeAV   medial amyg nu, anteroventral
MG   medial geniculate nucleus

MoCb   molecular layer of cerebellum
MPtg   medial pterygoid muscle
MRec   medial rectus muscle
ne   neuroepithelium
Oc   occipital cortex
oc   olivocerebellar tract
Occ   occipital bone
opt   optic tract
Oral   oral cavity
Orb   orbital cortex
Orbit   orbital cavity
Par1   parietal cortex, area 1
Parietal   parietal bone
PaS   parasubiculum
PBG   parabigeminal nucleus
pcer   posterior cerebral artery
PCTg   paracollicular tegmentum
Petrous   petrous part of the temp bone
Pir   piriform cortex
Pk   Purkinje cell layer (cerebellum)
PMCo   posteromedial cortical amygdaloid nu (C3)
pnm   pontine migration
Pr5   principal sensory trigeminal nu
PrS   presubiculum

PSphW   presphenoid wing
ptgpal   pterygopalatine artery
Res   reservoir of migrating neurons
RF   rhinal fissure
Rt   reticular thalamic nucleus
S   subiculum
s5   sensory root of the trigeminal nerve
scba   superior cerebellar artery
SCGn   superior cervical ganglion
SI   substantia innominata
SOb   superior oblique muscle
sp5   spinal trigeminal tract
Sp5C   spinal trigeminal nu, caudal part
Sp5I   spinal trigeminal nu, interpolar
SRec   superior rectus muscle
st   stria terminalis
SubV   subventricular cortical layer
SuVe   superior vestibular nucleus
trg   germinal trigone
trs   transverse sinus
Tu   olfactory tubercle
TyBu   tympanic bulla
TyC   tympanic cavity
v   vein (unidentified)

VCA   ventral cochlear nu, anterior part
VCP   ventral cochlear nu, posterior part
vert   vertebral artery
VLL   ventral nu of the lateral lemniscus
VP   ventral pallidum
vsc   ventral spinocerebellar tract

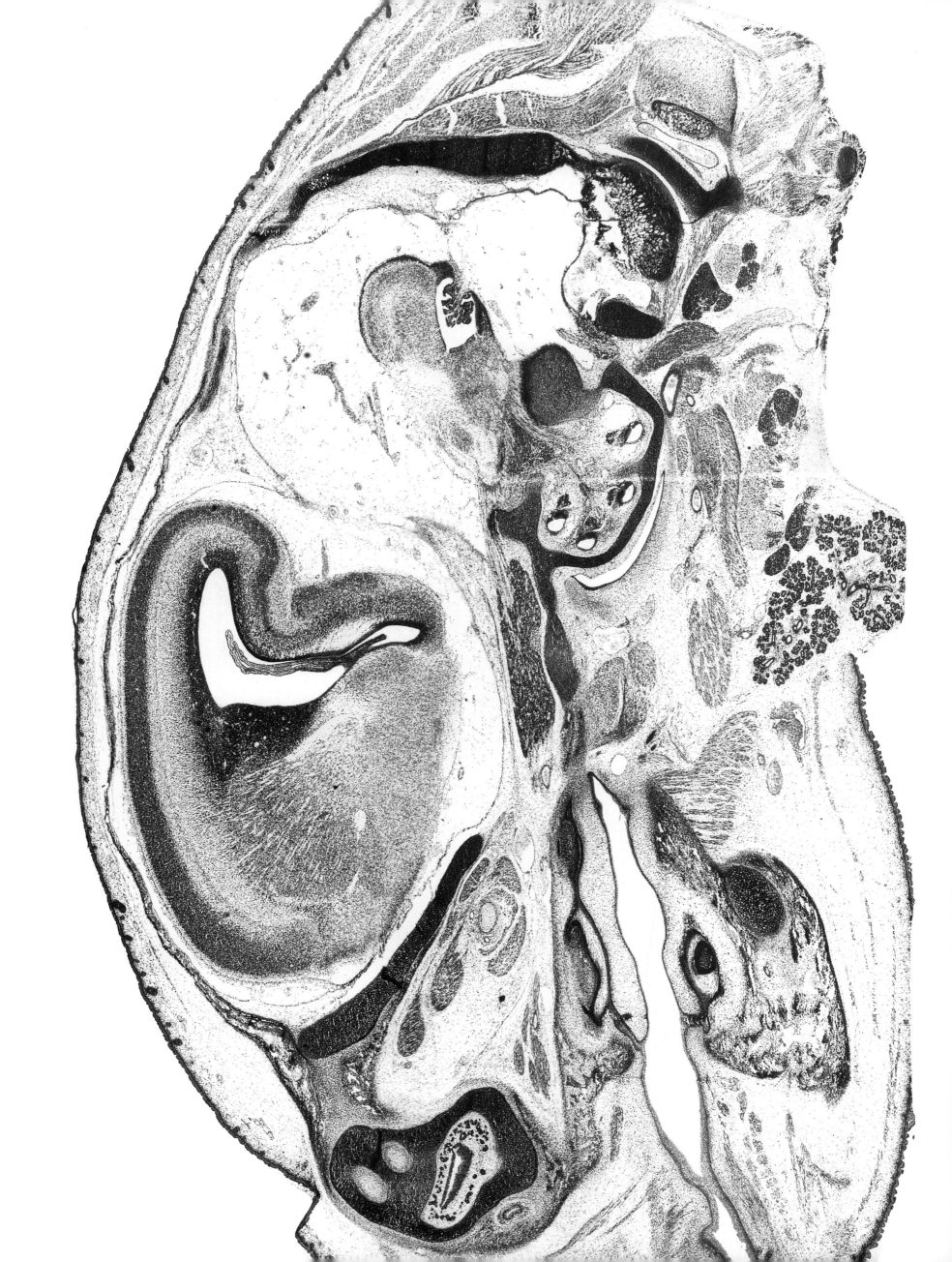

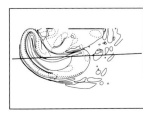

Figure 141
E19 Sagittal 12

1   cortical layer 1
2n   optic nerve
4n   trochlear nerve or its root
5fr   frontal branch of trigeminal nerve
5Gn   trigeminal ganglion
5max   maxillary nerve
8cn   cochlear root of vestibulocochlear nerve
8n   vestibulocochlear nerve
9n   glossopharyngeal nerve
12C   hypoglossal canal
12n   hypoglossal nerve or its root
AA   anterior amygdaloid area
aca   anterior commissure, anterior
Acb   accumbens nucleus
acp   anterior commissure, posterior
AF   amygdaloid fissure
AI   agranular insular cortex
ALF   anterior lacerated foramen
amg   amygdaloid neuroepi
AOrb   alar orbital bone
ASph   alisphenoid bone
Atlas   atlas (C1 vertebra)
aud   auditory neuroepi
B   basal nucleus of Meynert
BAOT   bed nu of the access olf tract
BOcc   basioccipital bone
CA1   CA1 field of the hippocampus
cav   cavernous sinus
Cb   cerebellum
CD   cochlear duct
Ce   central amygdaloid nucleus
CGn   cochlear (spiral) ganglion
ChP   choroid plexus
Cil   ciliary ganglion
Cl   claustrum
CPu   caudate putamen (striatum)
cpu   caudate putamen neuroepi
cx   cortical neuroepi
CxP   cortical plate
CxS   cortical subplate
DC   dorsal cochlear nucleus
DG   dentate gyrus
EGL   external germinal layer of Cb
EnamelO   enamel organ (of tooth)
fi   fimbria of the hippocampus
FL   forelimb area of the cortex
Fr   frontal cortex
Frontal   frontal bone
FStr   fundus striati
GP   globus pallidus
Harderian   Harderian gland
hi   hippocampal formation neuroepi
HiF   hippocampal fissure
HL   hindlimb area of the cortex
I9Gn   inf glossopharyng (petrosal) ganglion
IAud   internal auditory meatus
icap   internal capsule

icv   inferior cerebral vein
ICx   intermediate cortical layer
iorb   inferior orbital artery
IRec   inferior rectus muscle
JugF   jugular foramen
Lat   lateral (dentate) cerebellar nu
ling   lingual nerve
lo   lateral olfactory tract
LOT   nucleus of the lateral olfactory tract
LPS   levator palpebrae superioris muscle
LR4V   lateral recess of the 4th ventr
LRec   lateral rectus muscle
LV   lateral ventricle
m5   motor root of the trigeminal nerve
Mc   Meckel's cartilage
mcer   middle cerebral artery
mcp   middle cerebellar peduncle
MeAV   medial amyg nu, anterovent
MePD   medial amyg nu, posterodors
MePV   medial amyg nu, posterovent
MoCb   molecular layer of cerebellum
MPtg   medial pterygoid muscle
MRec   medial rectus muscle
ne   neuroepithelium
Oc   occipital cortex
Occ   occipital bone

occs   occipital sinus
opha   ophthalmic artery
Oral   oral cavity
Orbit   orbital cavity
Par1   parietal cortex, area 1
Parietal   parietal bone
ParietalP   parietal plate
Petrous   petrous part of the temp bone
Pir   piriform cortex
Pk   Purkinje cell layer (cerebellum)
PMCo   posteromedial cortical amygdaloid nu (C3)
pnm   pontine migration
Pr5   principal sensory trigeminal nu
PrS   presubiculum
PSphW   presphenoid wing
ptgpal   pterygopalatine artery
Res   reservoir of migrating neurons
RF   rhinal fissure
S   subiculum
s5   sensory root of the trigeminal nerve
S9Gn   superior glossopharyngeal ganglion
scba   superior cerebellar artery
SI   substantia innominata
SOb   superior oblique muscle
sp5   spinal trigeminal tract
SRec   superior rectus muscle

st   stria terminalis
SubV   subventricular cortical layer
trg   germinal trigone
trs   transverse sinus
Tu   olfactory tubercle
TyBu   tympanic bulla
TyC   tympanic cavity
v   vein (unidentified)
VCA   ventral cochlear nu, anterior part
VCP   ventral cochlear nu, posterior
VeGn   vestibular ganglion
vert   vertebral artery
VP   ventral pallidum

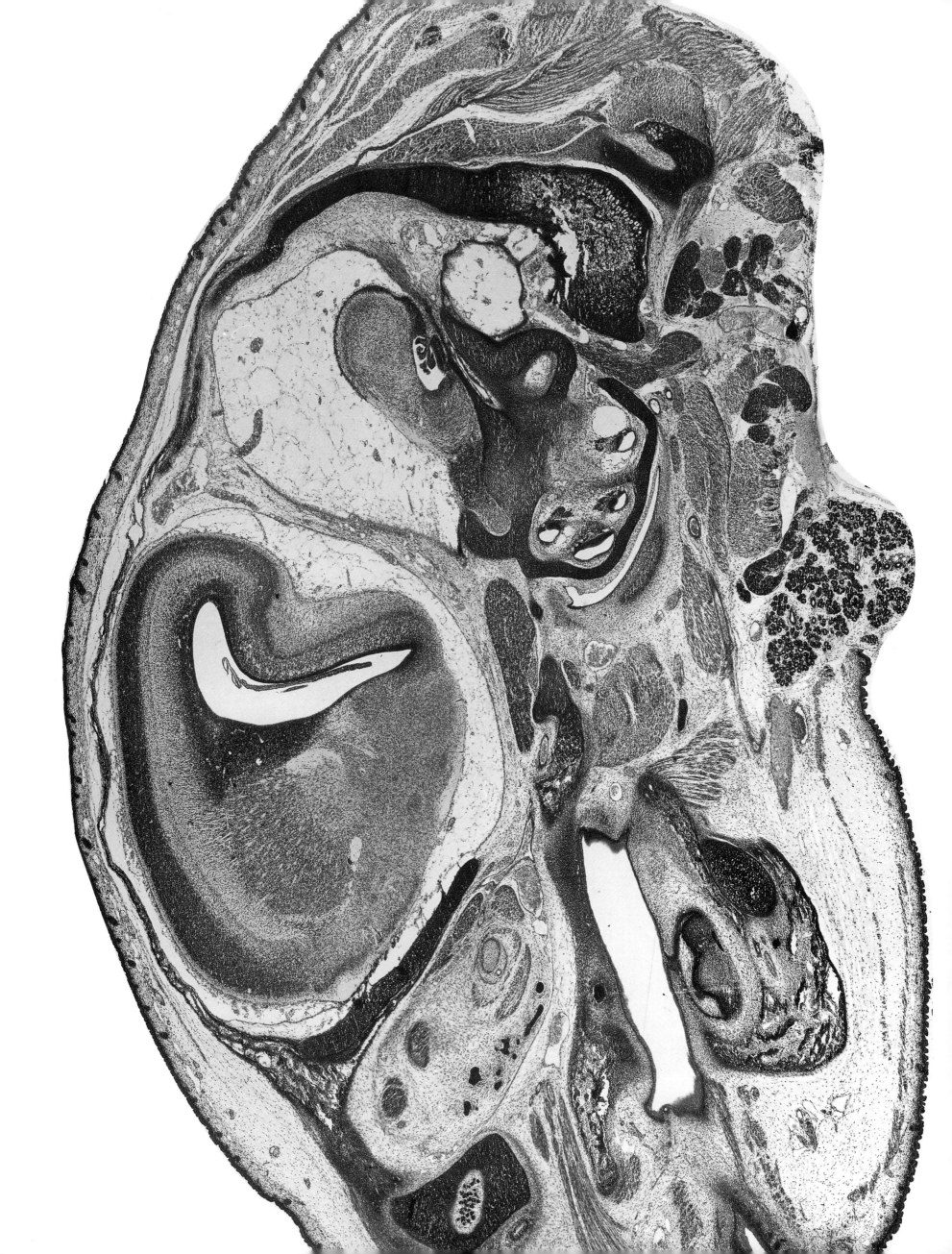

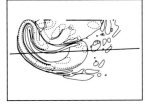

Figure 142
E19 Sagittal 13

1   cortical layer 1
2n   optic nerve
4n   trochlear nerve or its root
5Gn   trigeminal ganglion
5mand   mandibular nerve
8vn   vestibular root of the 8th nerve
AA   anterior amygdaloid area
AAD   anterior amygdaloid area, dors
acer   anterior cerebral artery
ACo   anterior cortical amygdaloid nu
acp   anterior commissure, posterior
AF   amygdaloid fissure
AHi   amygdalohippocampal area
ALF   anterior lacerated foramen
amg   amygdaloid neuroepi
ASph   alisphenoid bone
AStr   amygdalostriatal transition area
Atlas   atlas (C1 vertebra)
aud   auditory neuroepi
B   basal nucleus of Meynert
BOcc   basioccipital bone
CA1   CA1 field of the hippocampus
CA3   CA3 field of the hippocampus
Cb   cerebellum
CD   cochlear duct
Ce   central amygdaloid nucleus
CGn   cochlear (spiral) ganglion
ChP   choroid plexus
Cl   claustrum
CPu   caudate putamen (striatum)
cpu   caudate putamen neuroepi
crhv   caudal rhinal vein
cx   cortical neuroepi
CxP   cortical plate
CxS   cortical subplate
DC   dorsal cochlear nucleus
DEn   dorsal endopiriform nucleus
EGL   external germinal layer of Cb
ELS   endolymphatic sac
EnamelO   enamel organ (of tooth)
FL   forelimb area of the cortex
Fr   frontal cortex
Frontal   frontal bone
FStr   fundus striati
Gen   geniculate ganglion
GP   globus pallidus
gpet   greater petrosal nerve
Harderian   Harderian gland
hi   hippocampal formation neuroepi
HiF   hippocampal fissure
HL   hindlimb area of the cortex
I9Gn   inferior glossopharyngeal (petrosal) ganglion
ialva   inferior alveolar artery
ialvn   inferior alveolar nerve
icap   internal capsule
icv   inferior cerebral vein
ICx   intermediate cortical layer

ipets   inferior petrosal sinus
IRec   inferior rectus muscle
JugF   jugular foramen
Lat   lateral (dentate) cerebellar nu
ling   lingual nerve
lo   lateral olfactory tract
LOT   nucleus of the lateral olfactory tract
LPS   levator palpebrae superioris muscle
LR4V   lateral recess of the 4th ventr
LRec   lateral rectus muscle
LV   lateral ventricle
Maxilla   maxilla
Mc   Meckel's cartilage
mcp   middle cerebellar peduncle
MoCb   molecular layer of cerebellum
MPtg   medial pterygoid muscle
MRec   medial rectus muscle
Oc   occipital cortex
Occ   occipital bone
occs   occipital sinus
opha   ophthalmic artery
ophv   ophthalmic vein
Oral   oral cavity

Orbit   orbital cavity
Otic   otic ganglion
Par1   parietal cortex, area 1
Parietal   parietal bone
Petrous   petrous part of the temp bone
Pir   piriform cortex
Pk   Purkinje cell layer (cerebellum)
PMCo   posteromedial cortical amygdaloid nu (C3)
PrS   presubiculum
psa   posterior superior alveolar artery
PSphW   presphenoid wing
ptgpal   pterygopalatine artery
Res   reservoir of migrating neurons
S   subiculum
S9Gn   superior glossopharyngeal ganglion
scba   superior cerebellar artery
SOb   superior oblique muscle
sp5   spinal trigeminal tract
SRec   superior rectus muscle
SStr   substriatal area
st   stria terminalis
SubV   subventricular cortical layer
trg   germinal trigone

trs   transverse sinus
Tu   olfactory tubercle
TyBu   tympanic bulla
TyC   tympanic cavity
v   vein (unidentified)
VC   ventral cochlear nucleus
VeGn   vestibular ganglion

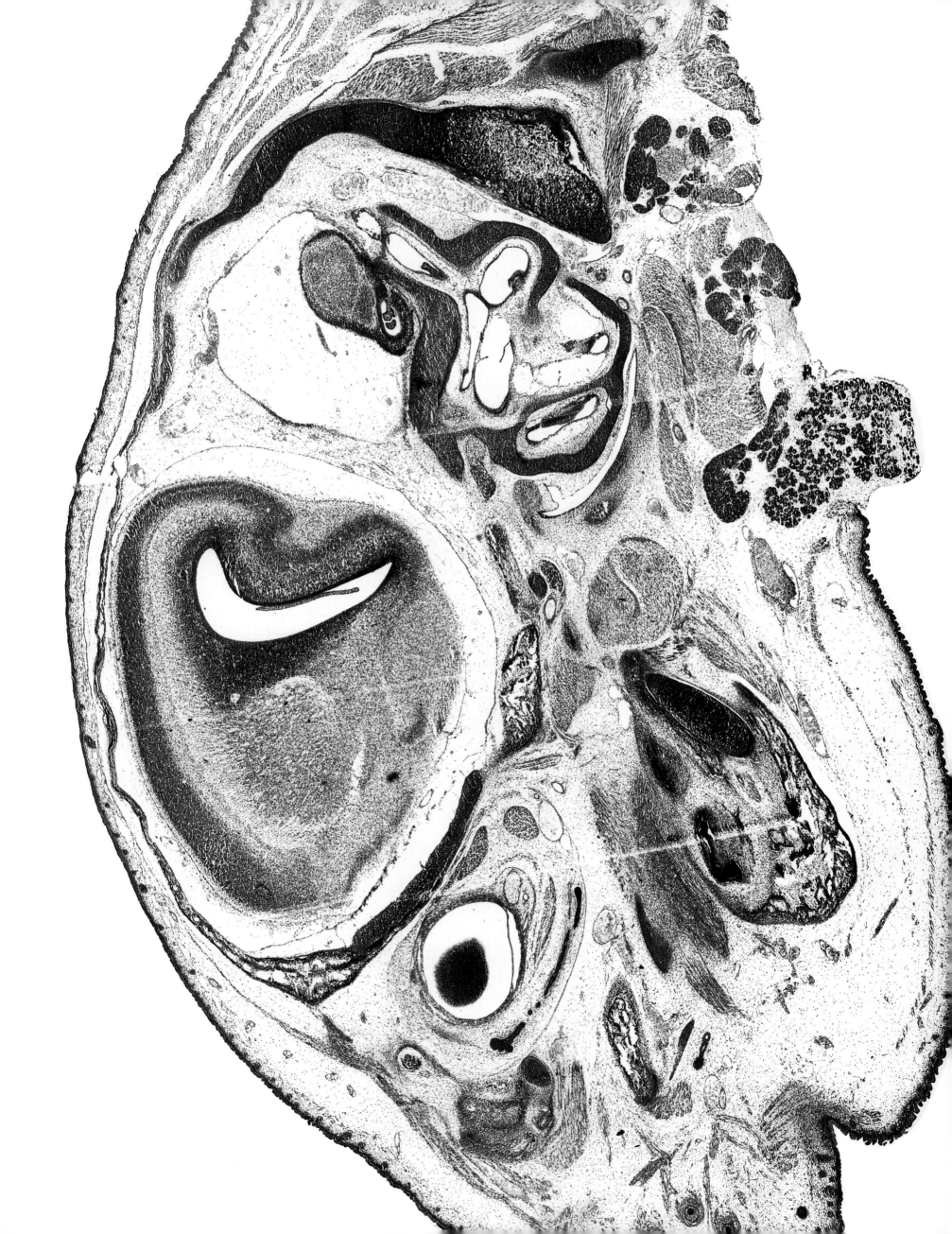

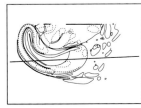

Figure 143
E19 Sagittal 14

1 cortical layer 1
2n optic nerve
5Gn trigeminal ganglion
5mand mandibular nerve
7n facial nerve or its root
8vn vestibular root of the 8th nerve
11n spinal accessory nerve
12n hypoglossal nerve or its root
AA anterior amygdaloid area
AAD anterior amygdaloid area, dors
ACo anterior cortical amygdaloid nu
acp anterior commissure, posterior
AF amygdaloid fissure
AHi amygdalohippocampal area
ALF anterior lacerated foramen
amg amygdaloid neuroepi
Apex apex of the cochlea
ASph alisphenoid bone
AStr amygdalostriatal transition area
Atlas atlas (C1 vertebra)
B basal nucleus of Meynert
BM basomedial amygdaloid nu
buccn buccal nerve
CA1 CA1 field of the hippocampus
CD cochlear duct
Ce central amygdaloid nucleus
CGn cochlear (spiral) ganglion
ChP choroid plexus
Cl claustrum
CPu caudate putamen (striatum)
cx cortical neuroepi
CxP cortical plate
CxS cortical subplate
DEn dorsal endopiriform nucleus
dtn deep temporal nerve
EGL external germinal layer of Cb
eld endolymphatic duct
ELS endolymphatic sac
EnamelO enamel organ (of tooth)
Ent entorhinal cortex
Frontal frontal bone
FStr fundus striati
Gen geniculate ganglion
GP globus pallidus
Harderian Harderian gland
hi hippocampal formation neuroepi
I intercalated nuclei of the amygdala
I9Gn inferior glossopharyngeal (petrosal) ganglion
ialva inferior alveolar artery
ialvn inferior alveolar nerve
icap internal capsule
icv inferior cerebral vein
ICx intermediate cortical layer
Ins insular cortex
IRec inferior rectus muscle
JugF jugular foramen
ling lingual nerve

lo lateral olfactory tract
LOT nucleus of the lateral olfactory tract
LPtg lateral pterygoid
LR4V lateral recess of the 4th ventr
LRec lateral rectus muscle
LV lateral ventricle
Mastoid mastoid process of temporal bone
Mc Meckel's cartilage
mcer middle cerebral artery
MoCb molecular layer of cerebellum
MPtg medial pterygoid muscle
ne neuroepithelium
Oc occipital cortex
Occ occipital bone
occs occipital sinus
opha ophthalmic artery
ophv ophthalmic vein
Otic otic ganglion
Oval oval foramen
Par1 parietal cortex, area 1
Par2 parietal cortex, area 2
Parietal parietal bone
Petrous petrous part of the temp bone

Pig pigment epithelium of retina
Pir piriform cortex
Pk Purkinje cell layer (cerebellum)
PLCo posterolateral cortical amygdaloid nu (C2)
PMCo posteromedial cortical amygdaloid nu (C3)
PSphW presphenoid wing
ptgpal pterygopalatine artery
Res reservoir of migrating neurons
Retina retina
S subiculum
Sacc saccule
scba superior cerebellar artery
SMac saccular macula
SOb superior oblique muscle
SRec superior rectus muscle
SStr substriatal area
st stria terminalis
SubV subventricular cortical layer
Te1 temporal cortex, area 1 (primary auditory cortex)
TensT tensor tympani muscle
trg germinal trigone
trs transverse sinus
Tu olfactory tubercle

TyBu tympanic bulla
TyC tympanic cavity
Utr utricle
v vein (unidentified)
Vent ventricular zone of the retina

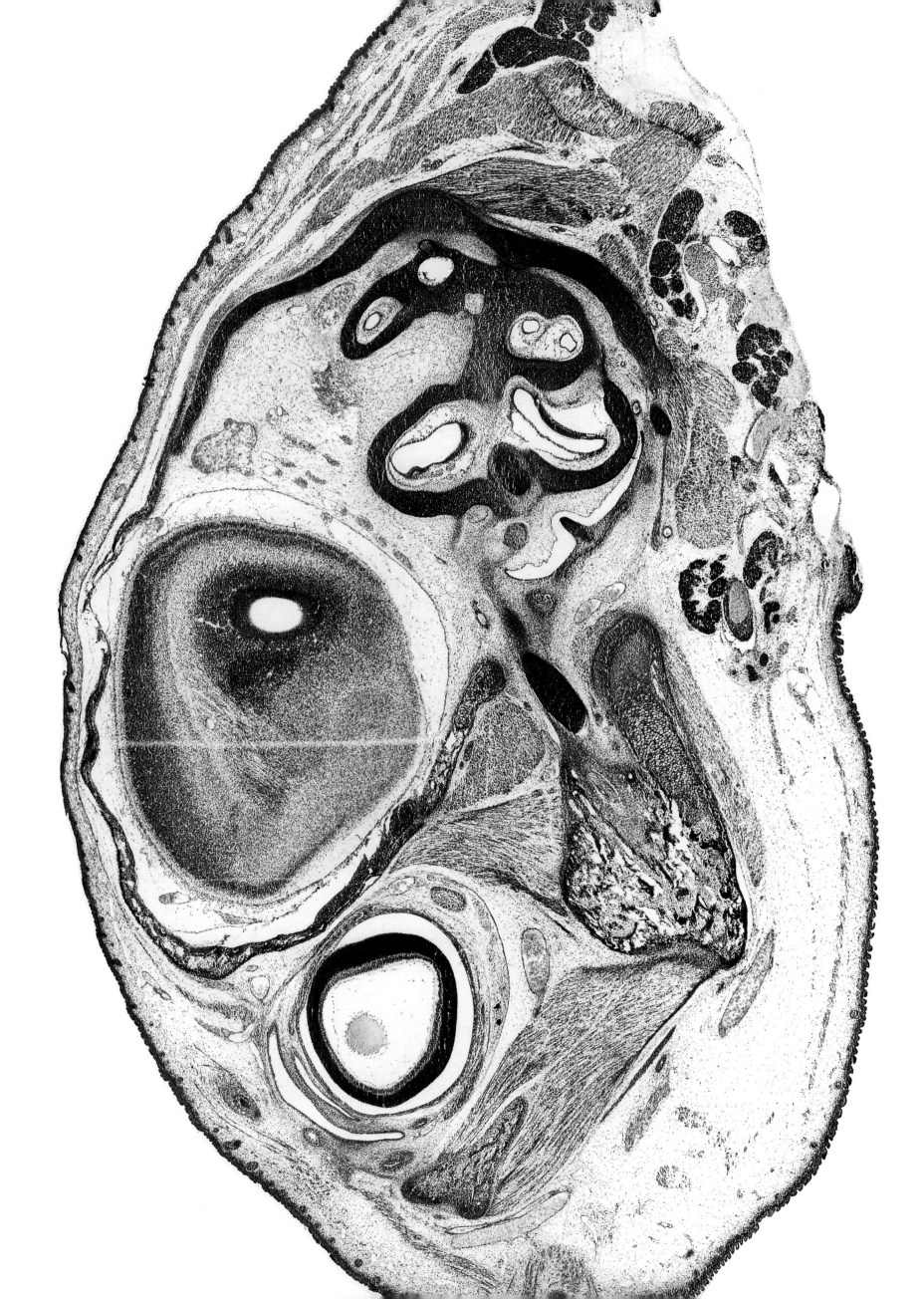

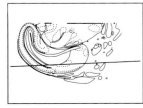

Figure 144
E19 Sagittal 15

1   cortical layer 1
7n   facial nerve or its root
11n   spinal accessory nerve
ACo   anterior cortical amygdaloid nu
acp   anterior commissure, posterior
AHi   amygdalohippocampal area
amg   amygdaloid neuroepi
Ant   anterior semicircular duct
ArcE   arcuate eminence of Petrous
ASph   alisphenoid bone
AStr   amygdalostriatal transition area
aute   auriculotemporal nerve
BL   basolateral amygdaloid nu
BM   basomedial amygdaloid nu
BStapes   base of stapes
CD   cochlear duct
Cl   claustrum
Cornea   cornea
Coronoid   coronoid process of the mandible
CPu   caudate putamen (striatum)
cpu   caudate putamen neuroepi
cty   chorda tympani nerve
cx   cortical neuroepi
CxP   cortical plate
CxS   cortical subplate
DEn   dorsal endopiriform nucleus
dtn   deep temporal nerve
EAM   external auditory meatus
Ent   entorhinal cortex
exorbd   duct of exobital gland
Eyelid   eyelid
Frontal   frontal bone
Gonial   gonial
hi   hippocampal formation neuroepi
Hor   horizontal semicircular duct
HStapes   head of stapes
ialva   inferior alveolar artery
ialvn   inferior alveolar nerve
ICx   intermediate cortical layer
Ins   insular cortex
IOb   inferior oblique muscle
La   lateral amygdaloid nucleus
Lens   lens
ling   lingual nerve
lo   lateral olfactory tract
LPtg   lateral pterygoid
LRec   lateral rectus muscle
LV   lateral ventricle
Mandible   mandible
Masseter   masseter muscle
Mastoid   mastoid process of temp bone
Mc   Meckel's cartilage
n   nerve
ne   neuroepithelium
Occ   occipital bone
occ   occipital artery
OF   optic fiber layer

ophv   ophthalmic vein
Orbit   orbital cavity
Par   parotid gland
Par1   parietal cortex, area 1
Parietal   parietal bone
ParietalP   parietal plate
Petrous   petrous part of the temp bone
PFICv   paraflocular cavity
Pig   pigment epithelium of retina
Pir   piriform cortex
PLCo   posterolateral cortical amygdaloid nu (C2)
Post   posterior semicircular duct
PRh   perirhinal cortex
PSphW   presphenoid wing
ptgpal   pterygopalatine artery
Reichert   Reichert's cartilage
Res   reservoir of migrating neurons
RGn   retinal ganglion cell layer
S   subiculum
SOb   superior oblique muscle
SubV   subventricular cortical layer
Te1   temporal cortex, area 1 (primary auditory cortex)
Te2   temporal cortex, area 2

Temp   temporal muscle
TensT   tensor tympani muscle
trfv   transverse temporal vein
trs   transverse sinus
ty9   tympanic branch of glossopharyngeal nerve
UMac   utricular macula
Utr   utricle
v   vein (unidentified)
Vent   ventricular zone of the retina
Zyg   zygomatic arch

P0 Coronal Section Plan

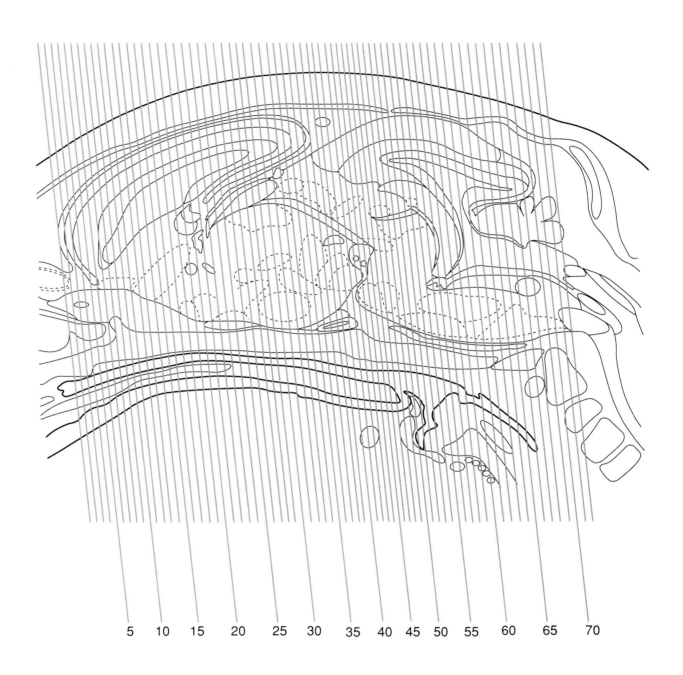

5 10 15 20 25 30 35 40 45 50 55 60 65 70

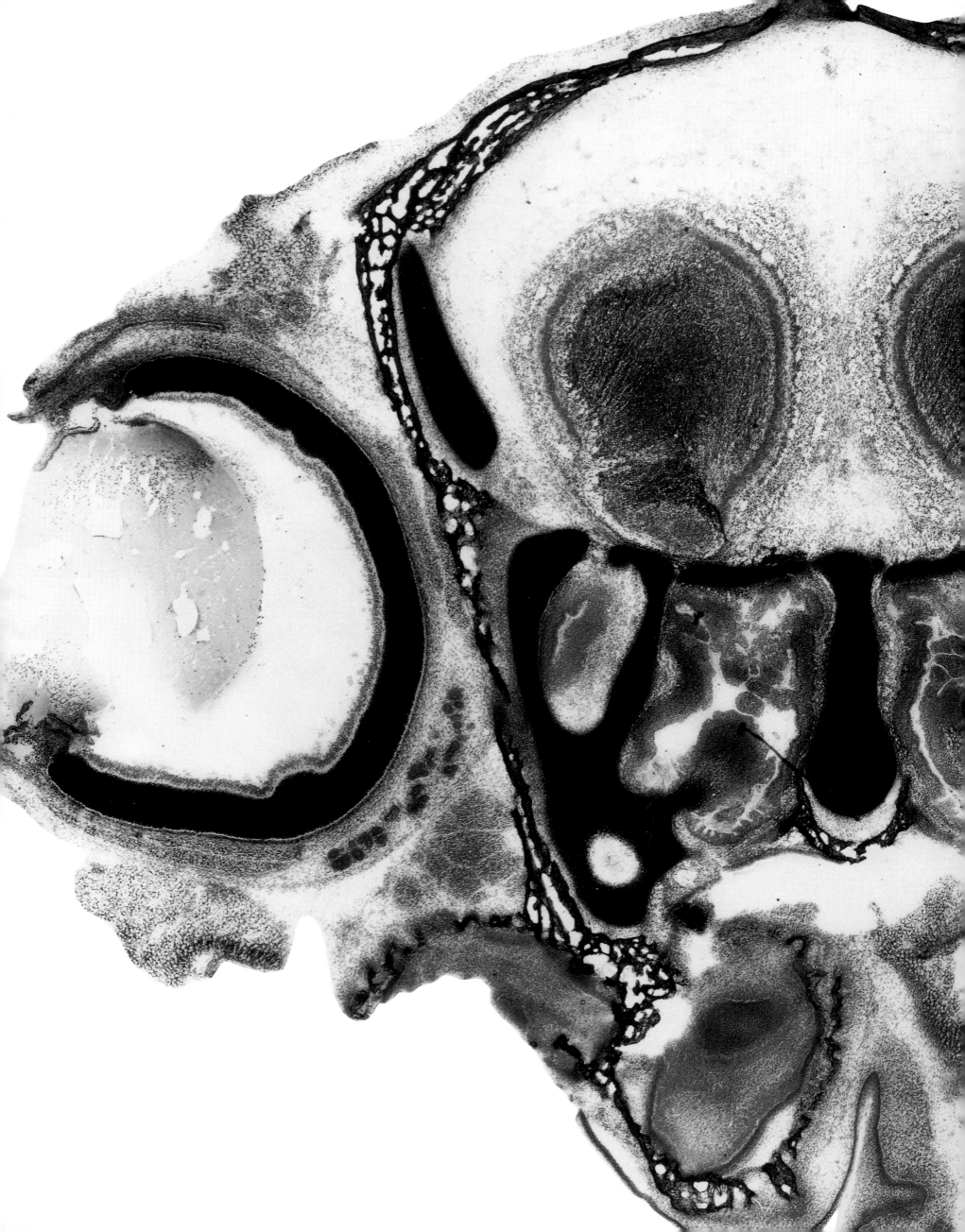

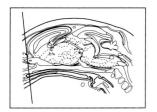

Figure 145
P0-1

AOB   accessory olfactory bulb
EPl   external plexiform layer olf bulb
Ethmoid   ethmoid bone
Frontal   frontal bone
Gl   glomerular layer olfactory bulb
Harderian   Harderian gland
IGr   int granular layer olfactory bulb
InPl   inner plexiform layer of retina
Lens   lens
Maxilla   maxilla
Mi   mitral cell layer olfactory bulb
Nasal   nasal cavity
ne   neuroepithelium
NSpt   nasal septum
obn   olfactory bulb neuroepithelium
OF   optic fiber layer
OV   olfactory ventricle
Pig   pigment epithelium of retina
RGn   retinal ganglion cell layer
SuspLig   suspensory ligament of lens
Vent   ventricular zone of the retina
Vomer   vomer

Frontal

AOB

Mi

IGr

EPl

Gl

obn

● OV

SuspLig

ne

OF

RGn

Lens

InPl

Vent

Nasal

Pig

NSpt

Harderian   Ethmoid

Vomer

Maxilla

4   3   2   1   0

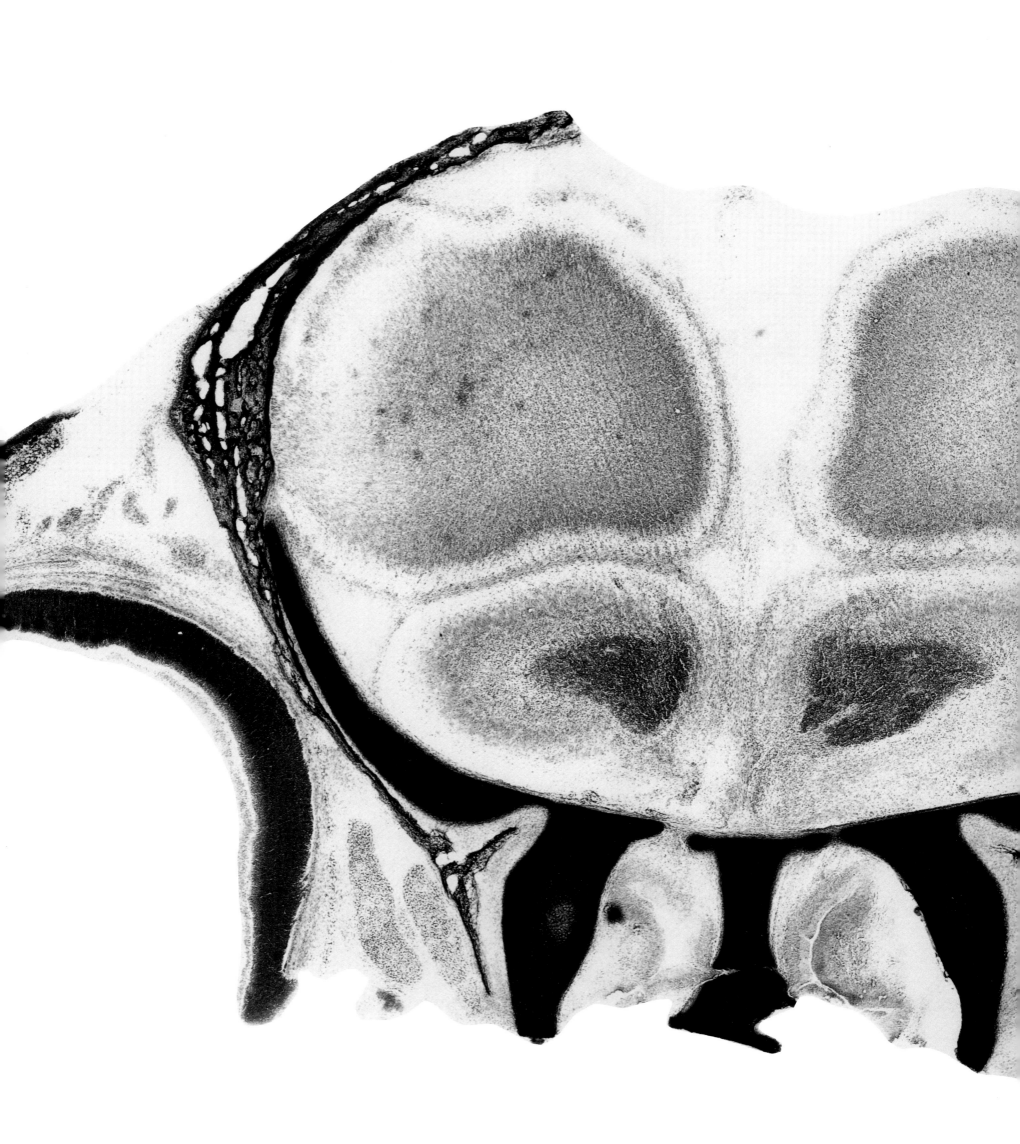

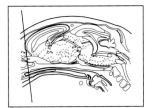

Figure 146
P0-2

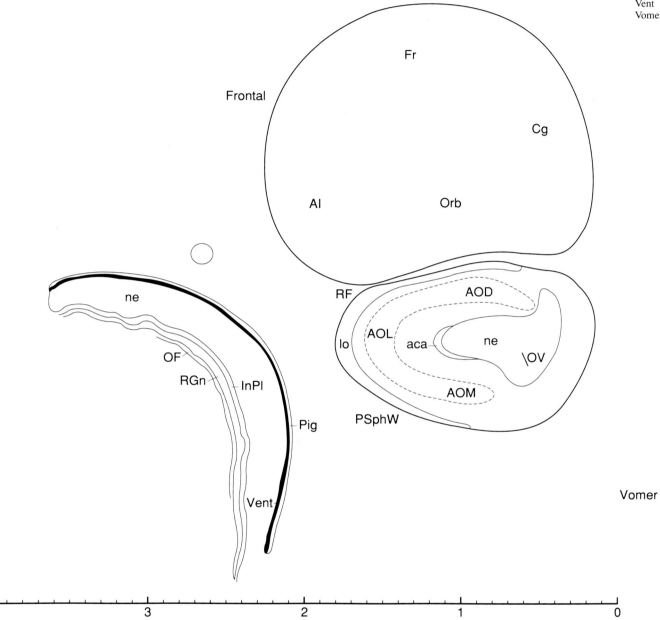

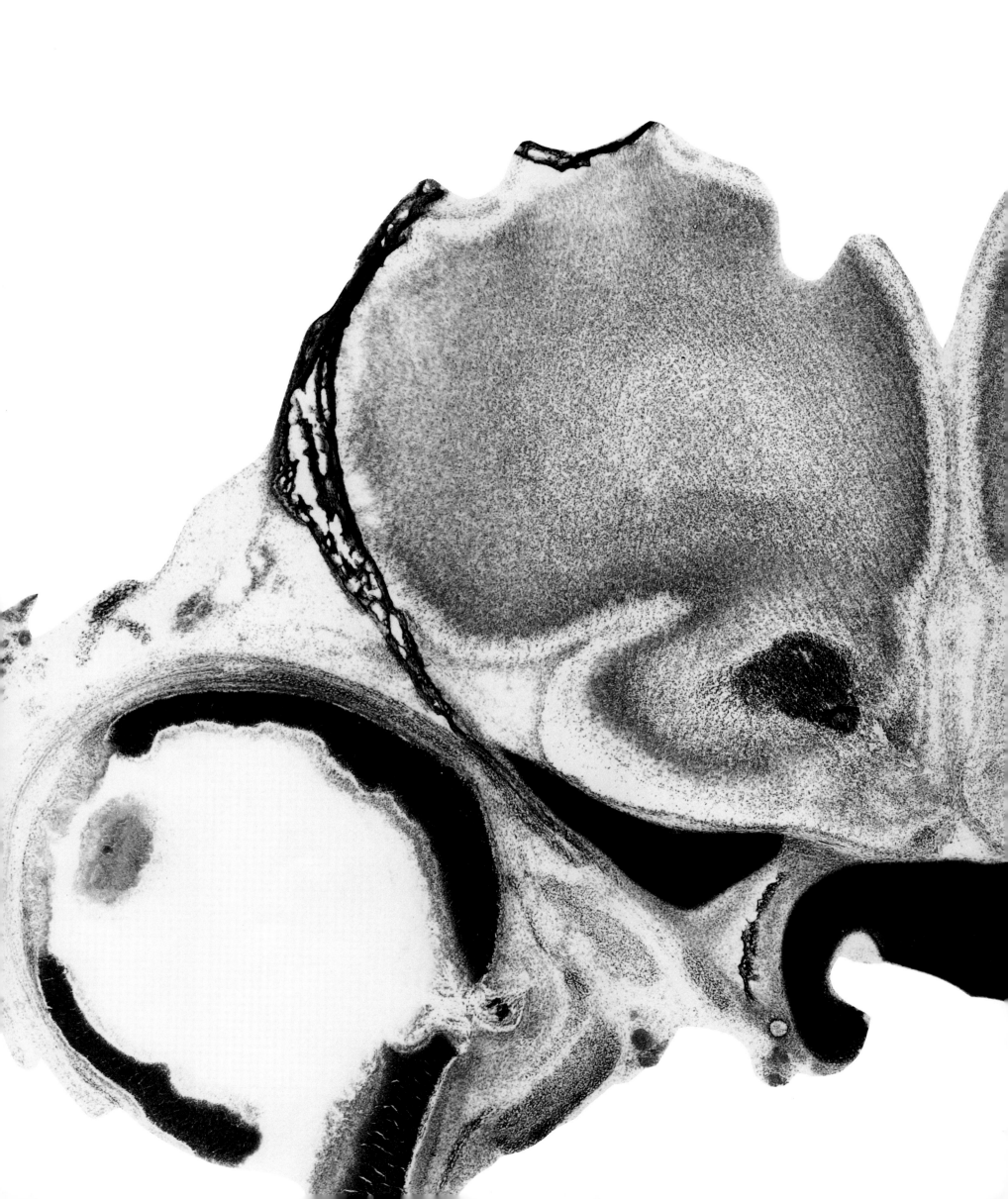

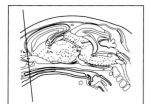

Figure 147
P0-3

2n   optic nerve
aca   anterior commissure, anterior
AI   agranular insular Cx
AO   anterior olfactory nu
Cg   cingulate cortex
DTr   dorsal transition zone
Fr   frontal cortex
Frontal   frontal bone
InPl   inner plexiform layer of retina
Lens   lens
lo   lateral olfactory tract
ne   neuroepithelium
OF   optic fiber layer
Orb   orbital cortex
OV   olfactory ventricle
Pig   pigment epithelium of retina
Pir   piriform cortex
PSph   presphenoid bone
PSphW   presphenoid wing
RF   rhinal fissure
RGn   retinal ganglion cell layer
TT   tenia tecta
Vent   ventricular zone of the retina

Fr

Frontal

Cg

AI

Orb

DTr

TT

RF

Pir

ne

OV

aca

AO

InPl

ne

lo

Lens

Vent

PSphW

OF

Pig

PSph

RGn

2n

4   3   2   1   0

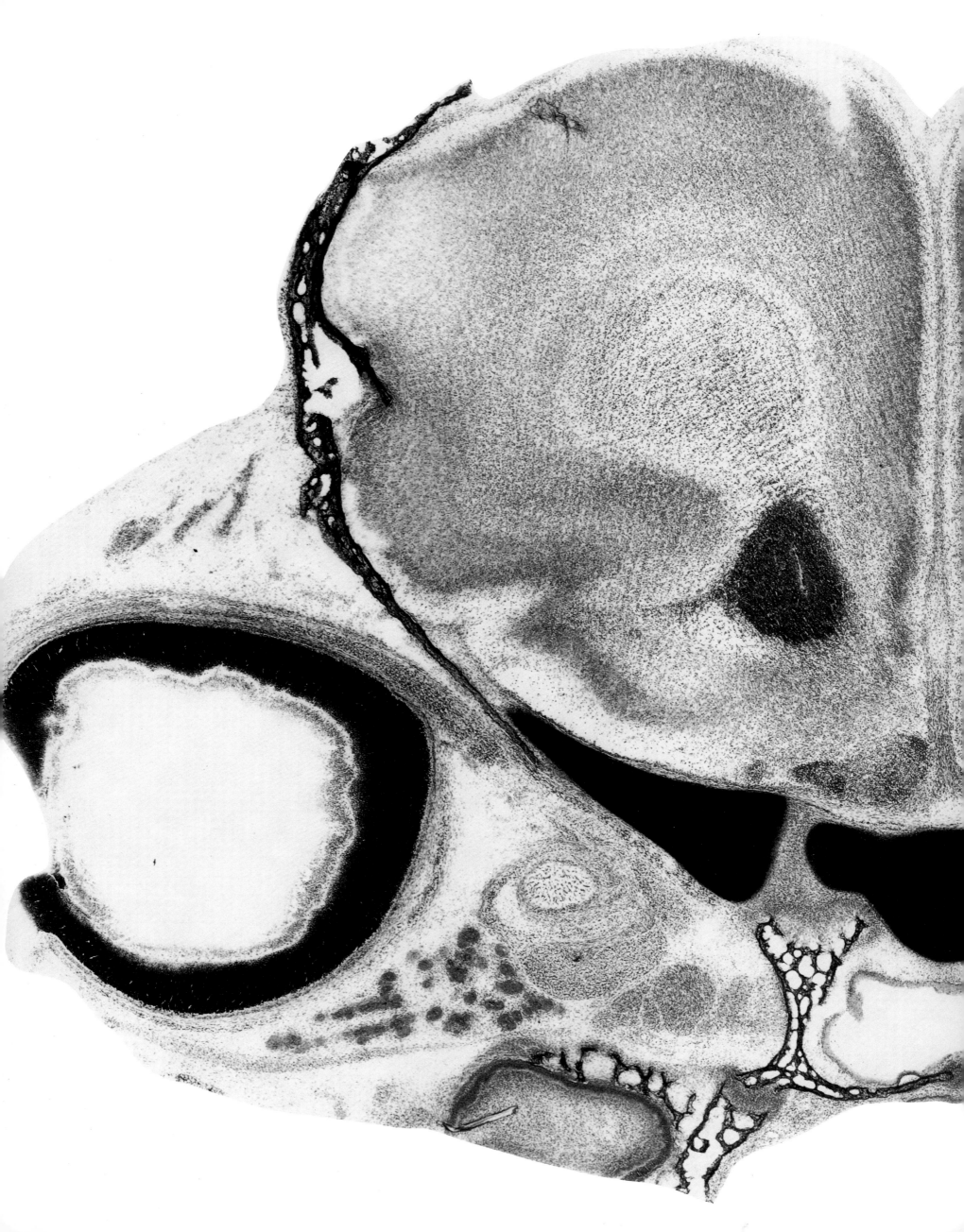

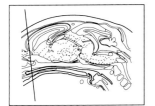

Figure 148
P0-4

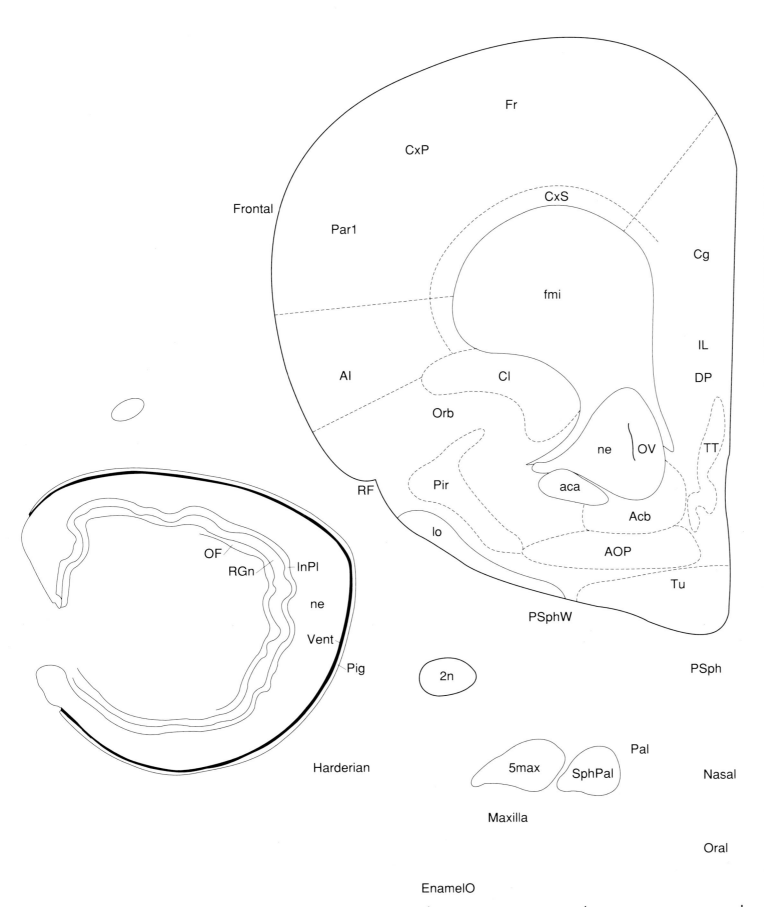

2n   optic nerve
5max   maxillary nerve
aca   anterior commissure, anterior
Acb   accumbens nu
AI   agranular insular Cx
AOP   anterior olfactory nu, posterior
Cg   cingulate cortex
Cl   claustrum
CxP   cortical plate
CxS   cortical subplate
DP   dorsal peduncular cortex
EnamelO   enamel organ (of tooth)
fmi   forceps minor corpus callosum
Fr   frontal cortex
Frontal   frontal bone
Harderian   Harderian gland
IL   infralimbic cortex
InPl   inner plexiform layer of retina
lo   lateral olfactory tract
Maxilla   maxilla
Nasal   nasal cavity
ne   neuroepithelium
OF   optic fiber layer
Oral   oral cavity
Orb   orbital cortex
OV   olfactory ventricle
Pal   palatine bone
Par1   parietal cortex, area 1
Pig   pigment epithelium of retina
Pir   piriform cortex
PSph   presphenoid bone
PSphW   presphenoid wing
RF   rhinal fissure
RGn   retinal ganglion cell layer
SphPal   sphenopalatine ganglion
TT   tenia tecta
Tu   olfactory tubercule
Vent   ventricular zone of the retina

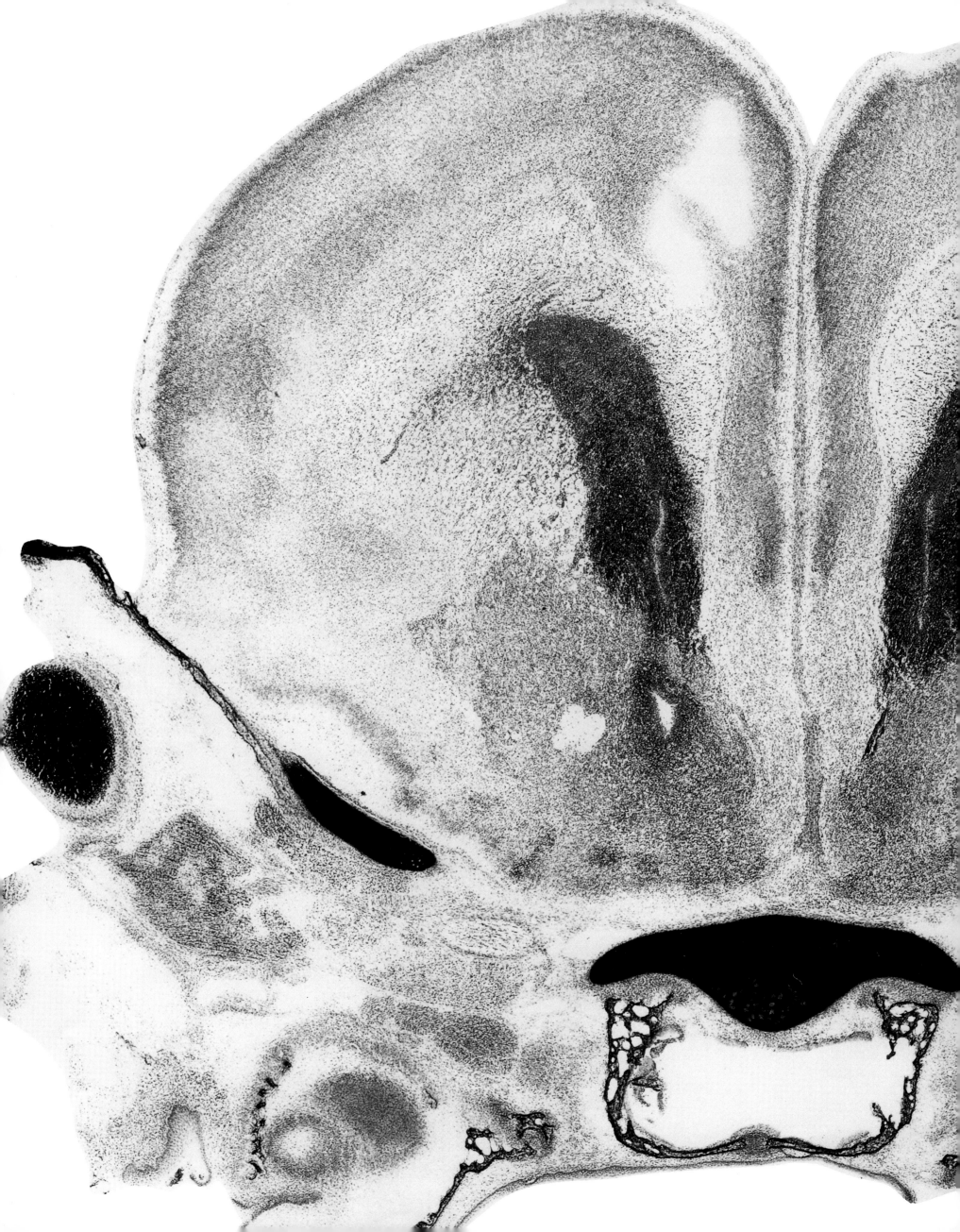

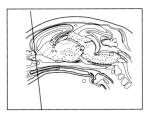

Figure 149
P0-5

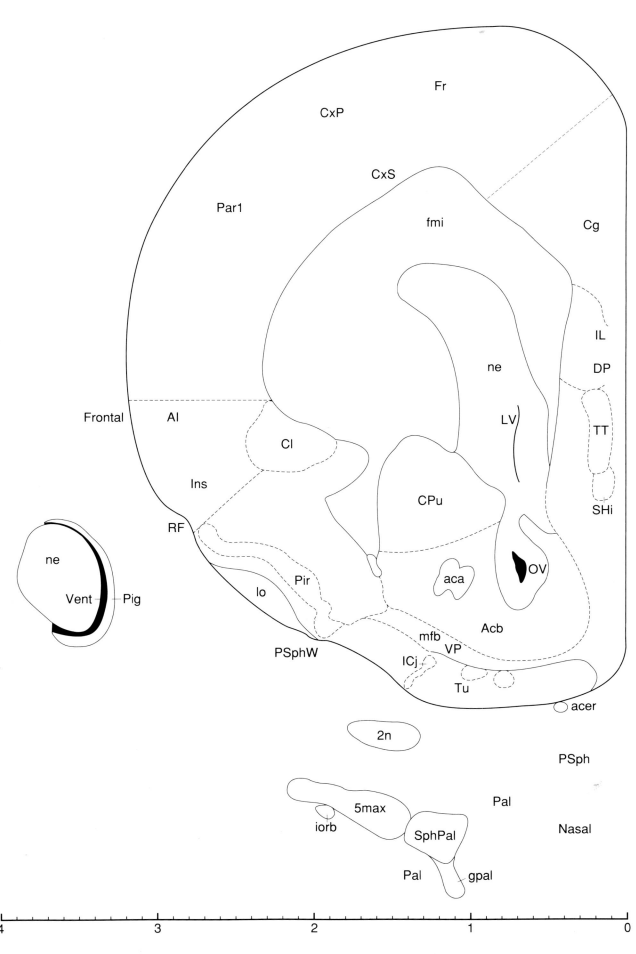

2n   optic nerve
5max   maxillary nerve
aca   anterior commissure, anterior
Acb   accumbens nu
acer   anterior cerebral artery
AI   agranular insular Cx
Cg   cingulate cortex
Cl   claustrum
CPu   caudate putamen
CxP   cortical plate
CxS   cortical subplate
DP   dorsal peduncular cortex
fmi   forceps minor corpus callosum
Fr   frontal cortex
Frontal   frontal bone
gpal   greater palatine nerve
ICj   islands of Calleja
IL   infralimbic cortex
Ins   insular cortex
iorb   inferior orbital artery
lo   lateral olfactory tract
LV   lateral ventricle
mfb   medial forebrain bundle
Nasal   nasal cavity
ne   neuroepithelium
OV   olfactory ventricle
Pal   palatine bone
Par1   parietal cortex, area 1
Pig   pigment epithelium of retina
Pir   piriform cortex
PSph   presphenoid bone
PSphW   presphenoid wing
RF   rhinal fissure
SHi   septohippocampal nu
SphPal   sphenopalatine ganglion
TT   tenia tecta
Tu   olfactory tubercule
Vent   ventricular zone of the retina
VP   ventral pallidum

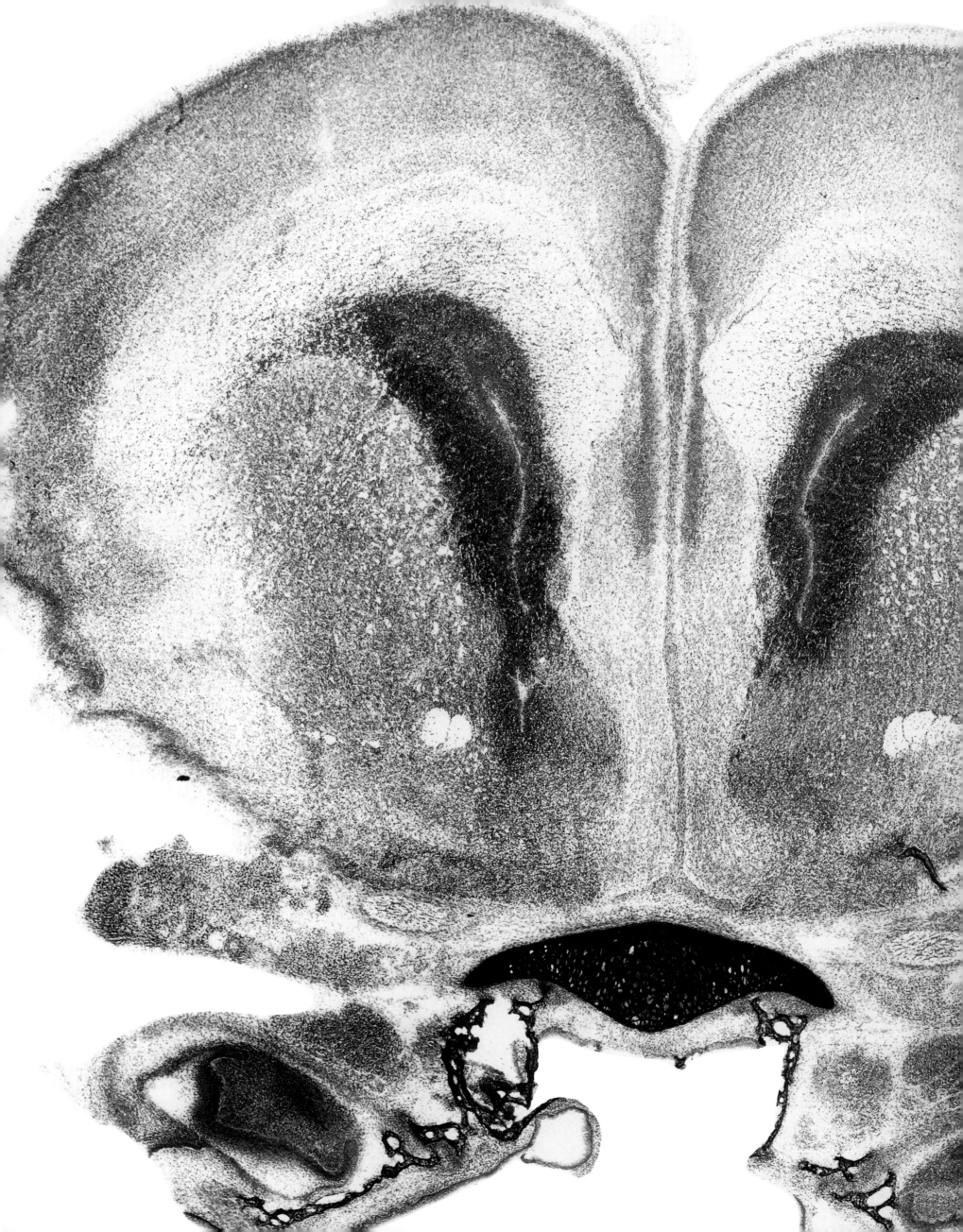

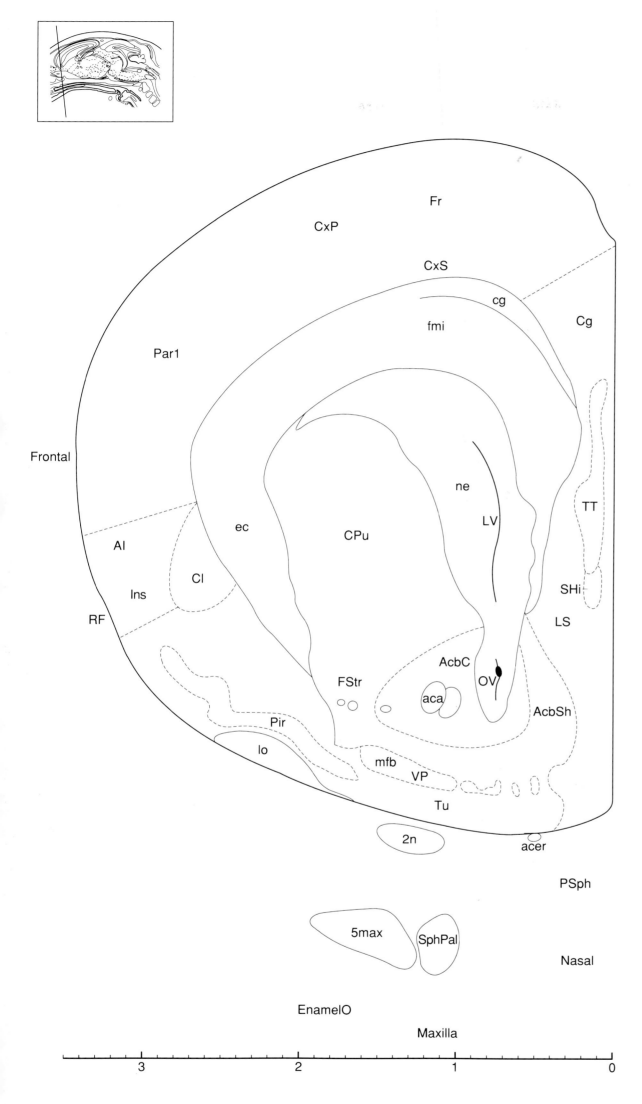

Figure 150
P0-6

2n   optic nerve
5max   maxillary nerve
aca   anterior commissure, anterior
AcbC   accumbens nu, core
AcbSh   accumbens nu, shell
acer   anterior cerebral artery
AI   agranular insular Cx
Cg   cingulate cortex
cg   cingulum
Cl   claustrum
CPu   caudate putamen
CxP   cortical plate
CxS   cortical subplate
ec   external capsule
EnamelO   enamel organ (of tooth)
fmi   forceps minor corpus callosum
Fr   frontal cortex
Frontal   frontal bone
FStr   fundus striati
Ins   insular cortex
lo   lateral olfactory tract
LS   lateral septal nu
LV   lateral ventricle
Maxilla   maxilla
mfb   medial forebrain bundle
Nasal   nasal cavity
ne   neuroepithelium
OV   olfactory ventricle
Par1   parietal cortex, area 1
Pir   piriform cortex
PSph   presphenoid bone
RF   rhinal fissure
SHi   septohippocampal nu
SphPal   sphenopalatine ganglion
TT   tenia tecta
Tu   olfactory tubercule
VP   ventral pallidum

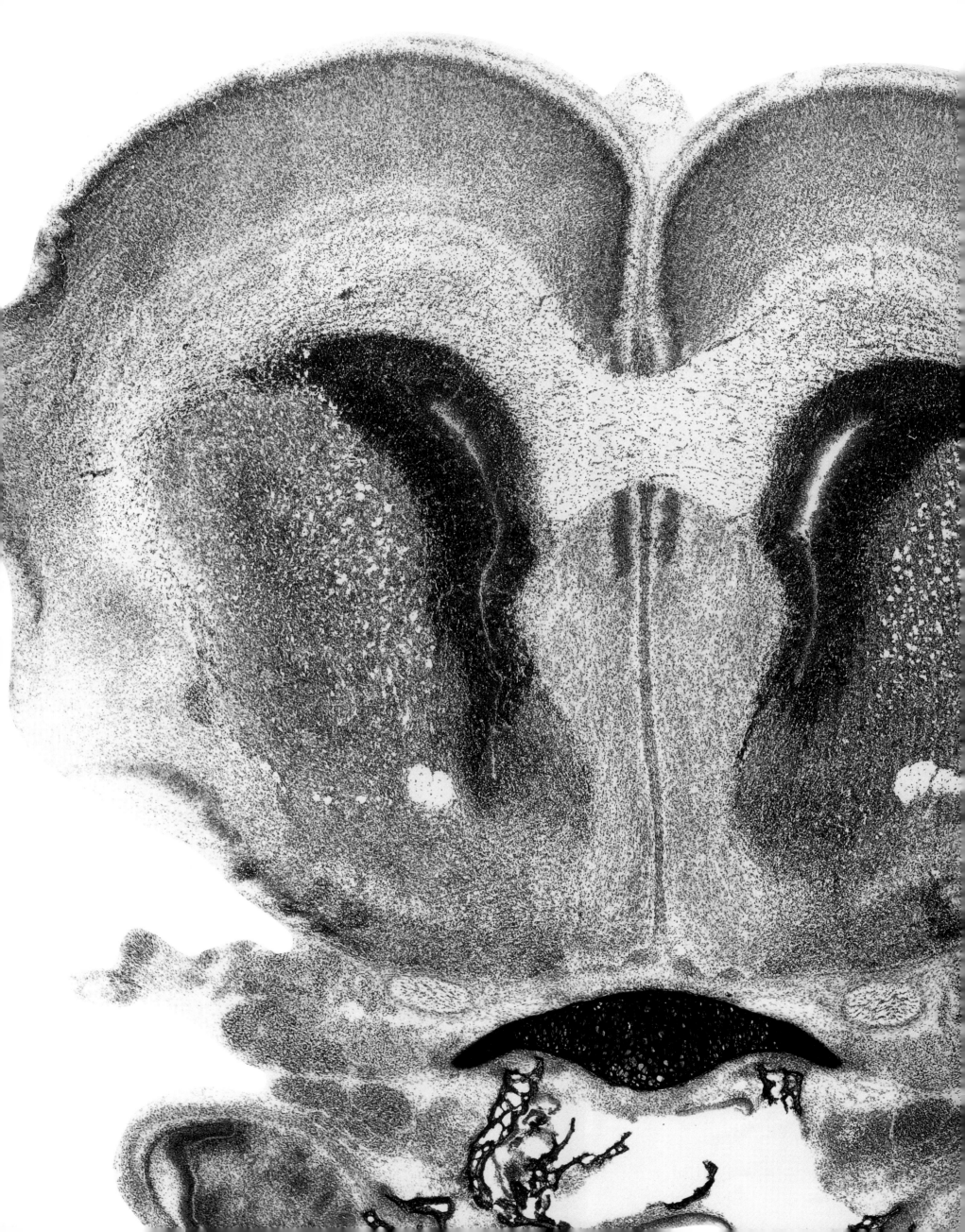

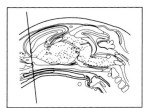

Figure 151
P0-7

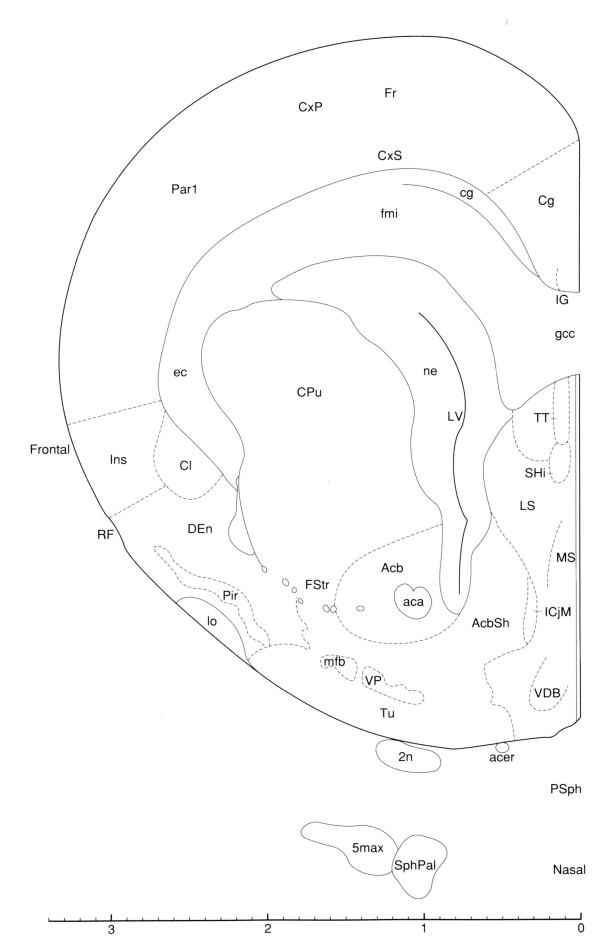

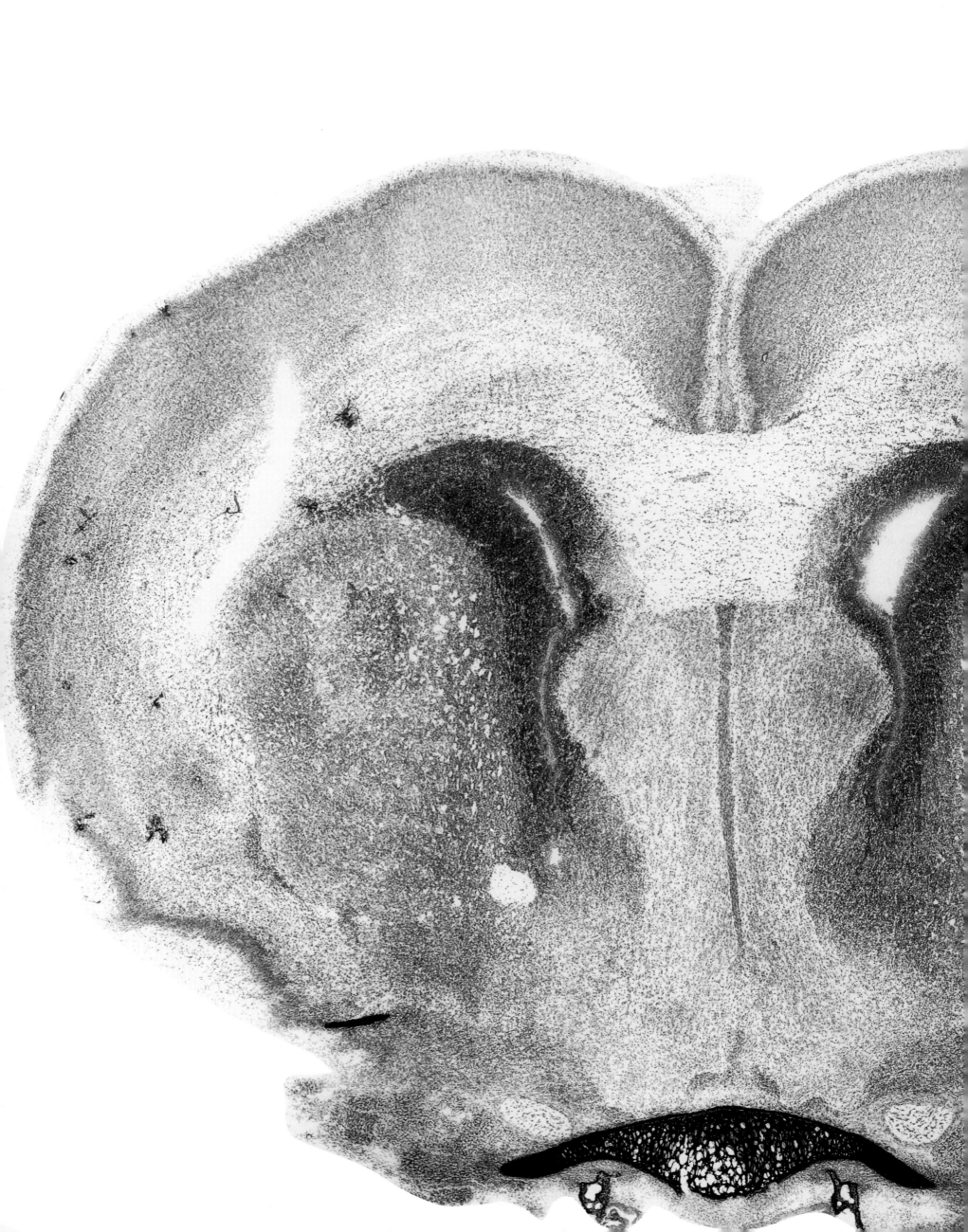

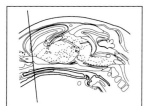

Figure 152
P0-8

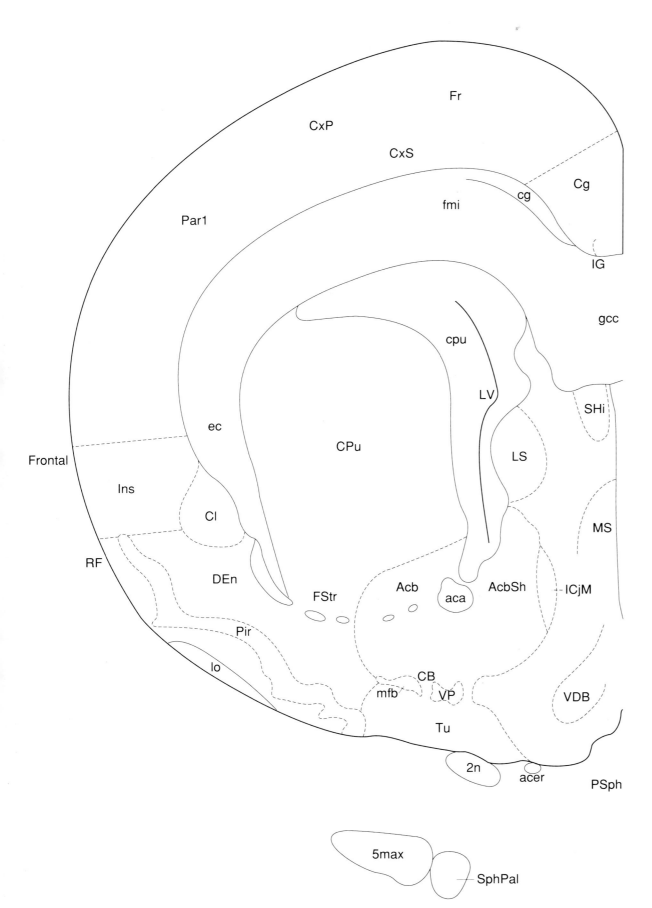

2n   optic nerve
5max   maxillary nerve
aca   anterior commissure, anterior
Acb   accumbens nu
AcbSh   accumbens nu, shell
acer   anterior cerebral artery
CB   cell bridges ventral striatum
Cg   cingulate cortex
cg   cingulum
Cl   claustrum
CPu   caudate putamen
cpu   caudate putamen neuroepithelium
CxP   cortical plate
CxS   cortical subplate
DEn   dorsal endopiriform nu
ec   external capsule
fmi   forceps minor corpus callosum
Fr   frontal cortex
Frontal   frontal bone
FStr   fundus striati
gcc   genu corpus callosum
ICjM   islands of Calleja, major island
IG   indusium griseum
Ins   insular cortex
lo   lateral olfactory tract
LS   lateral septal nu
LV   lateral ventricle
mfb   medial forebrain bundle
MS   medial septal nu
Par1   parietal cortex, area 1
Pir   piriform cortex
PSph   presphenoid bone
RF   rhinal fissure
SHi   septohippocampal nu
SphPal   sphenopalatine ganglion
Tu   olfactory tubercule
VDB   nu vertical limb diagonal band
VP   ventral pallidum

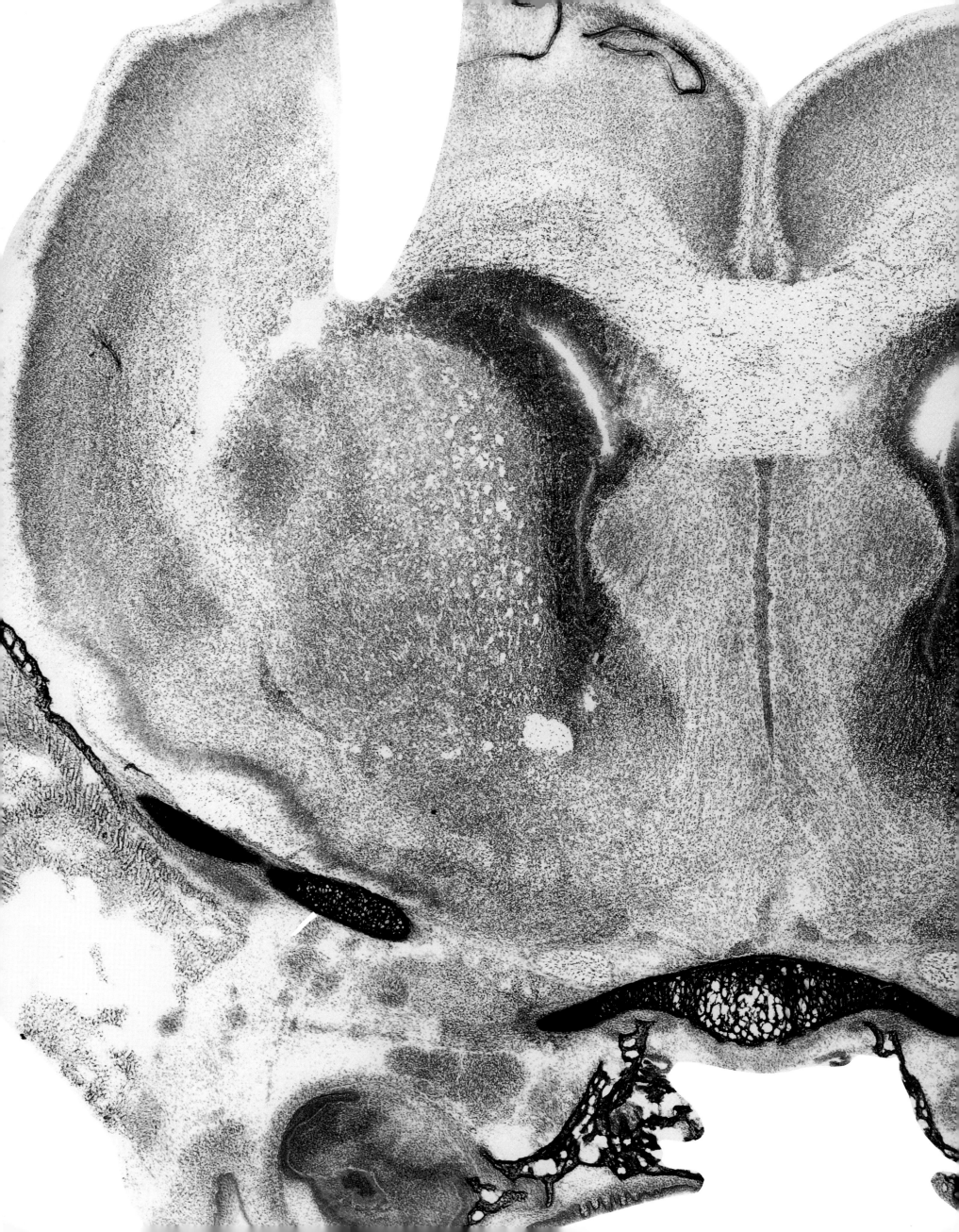

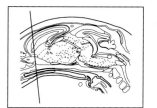

Figure 153
P0-9

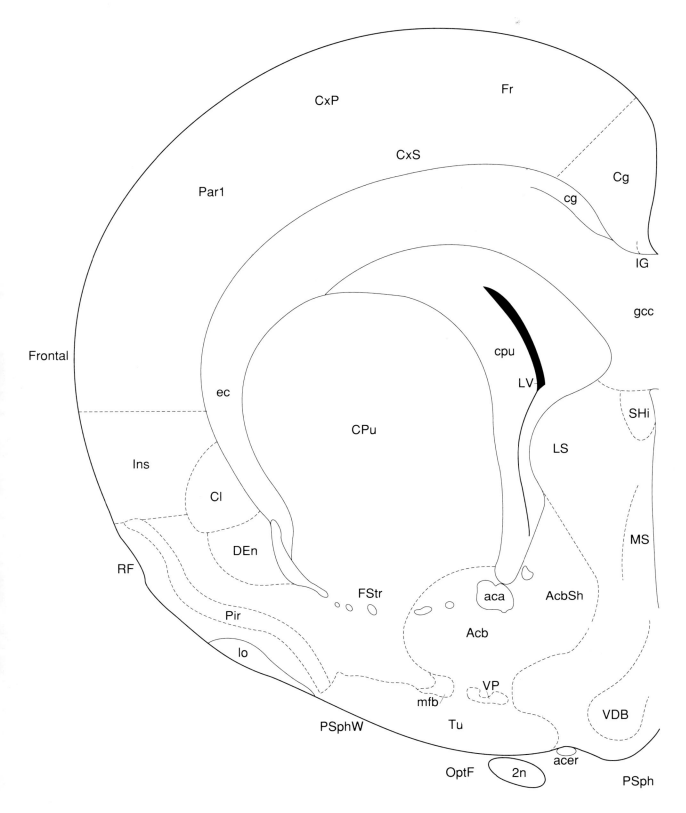

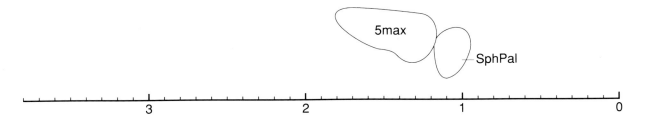

2n   optic nerve
5max   maxillary nerve
aca   anterior commissure, anterior
Acb   accumbens nu
AcbSh   accumbens nu, shell
acer   anterior cerebral artery
Cg   cingulate cortex
cg   cingulum
Cl   claustrum
CPu   caudate putamen
cpu   caudate putamen neuroepithelium
CxP   cortical plate
CxS   cortical subplate
DEn   dorsal endopiriform nu
ec   external capsule
Fr   frontal cortex
Frontal   frontal bone
FStr   fundus striati
gcc   genu corpus callosum
IG   indusium griseum
Ins   insular cortex
lo   lateral olfactory tract
LS   lateral septal nu
LV   lateral ventricle
mfb   medial forebrain bundle
MS   medial septal nu
OptF   optic foramen
Par1   parietal cortex, area 1
Pir   piriform cortex
PSph   presphenoid bone
PSphW   presphenoid wing
RF   rhinal fissure
SHi   septohippocampal nu
SphPal   sphenopalatine ganglion
Tu   olfactory tubercule
VDB   nu vertical limb diagonal band
VP   ventral pallidum

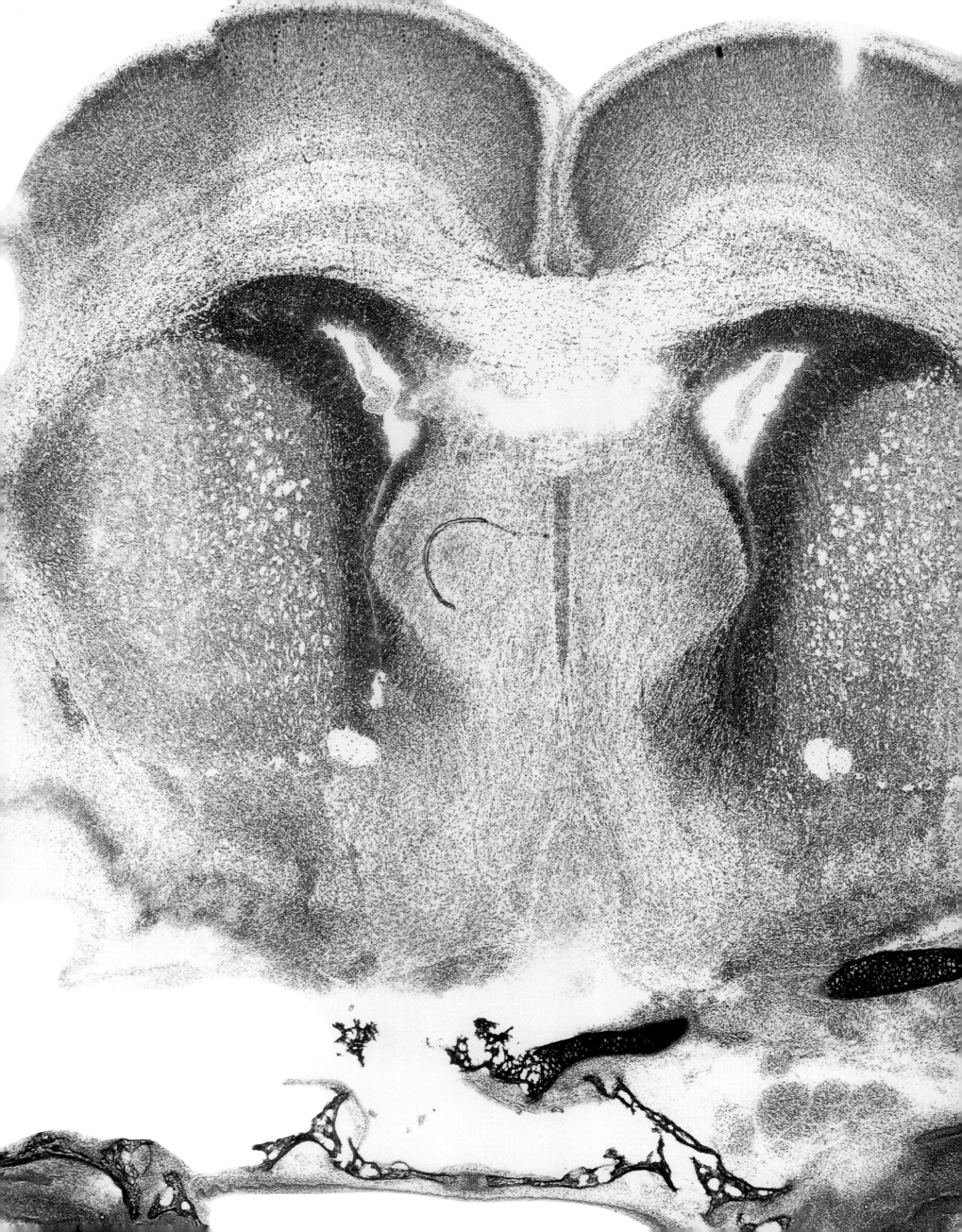

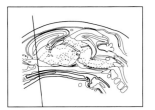

Figure 154
P0-10

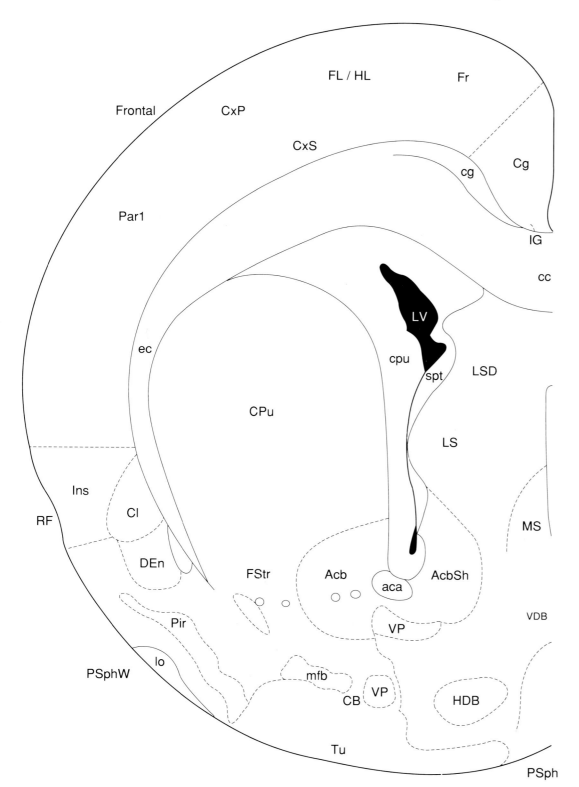

aca   anterior commissure, anterior
Acb   accumbens nu
AcbSh   accumbens nu, shell
CB   cell bridges ventral striatum
cc   corpus callosum
Cg   cingulate cortex
cg   cingulum
Cl   claustrum
CPu   caudate putamen
cpu   caudate putamen neuroepithelium
CxP   cortical plate
CxS   cortical subplate
DEn   dorsal endopiriform nu
ec   external capsule
FL   forelimb area of cortex
Fr   frontal cortex
Frontal   frontal bone
FStr   fundus striati
HDB   nu horiz limb diagonal band
HL   hindlimb area of cortex
IG   indusium griseum
Ins   insular cortex
lo   lateral olfactory tract
LS   lateral septal nu
LSD   lateral septal nu, dorsal
LV   lateral ventricle
mfb   medial forebrain bundle
MS   medial septal nu
Par1   parietal cortex, area 1
Pir   piriform cortex
PSph   presphenoid bone
PSphW   presphenoid wing
RF   rhinal fissure
spt   septal neuroepithelium
Tu   olfactory tubercule
VDB   nu vertical limb diagonal band
VP   ventral pallidum

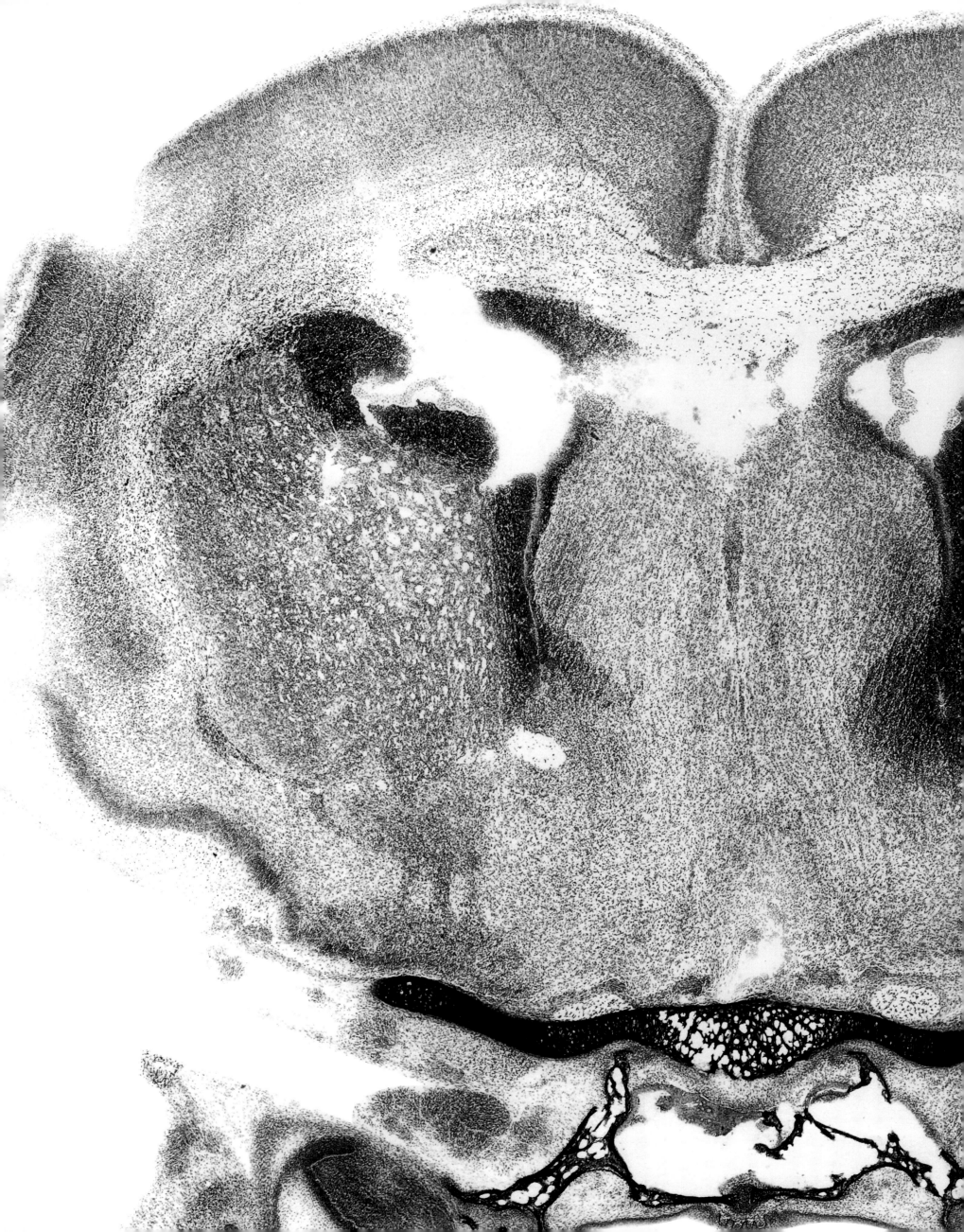

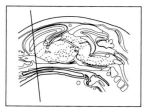

Figure 155
P0-11

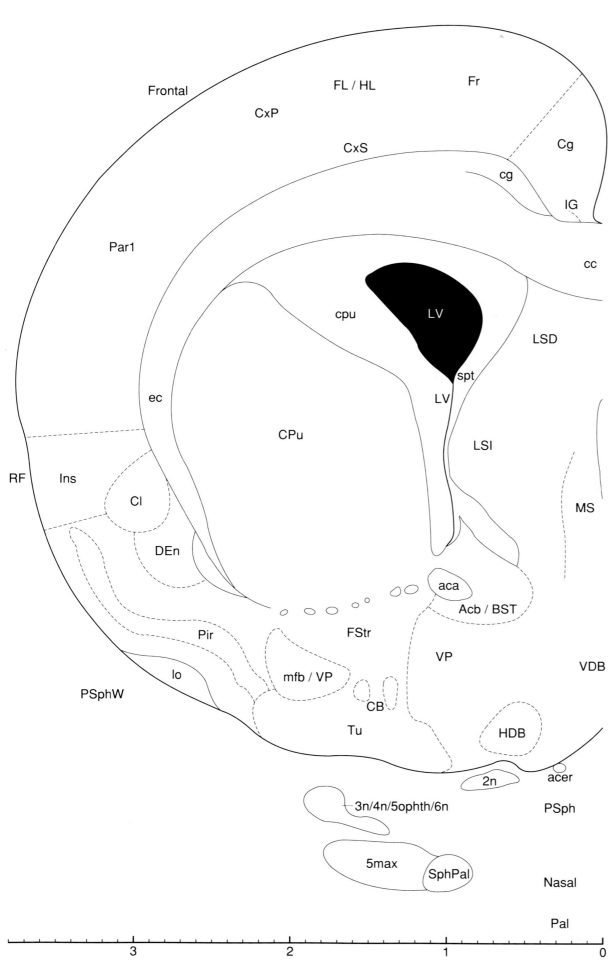

2n   optic nerve
3n   oculomotor nerve or its root
4n   trochlear nerve or its root
5max   maxillary nerve
5ophth   ophthalmic n of trigem
6n   abducens nerve or its root
aca   anterior commissure, anterior
Acb   accumbens nu
acer   anterior cerebral artery
BST   bed nu stria terminalis
CB   cell bridges ventral striatum
cc   corpus callosum
Cg   cingulate cortex
cg   cingulum
Cl   claustrum
CPu   caudate putamen
cpu   caudate putamen neuroepithelium
CxP   cortical plate
CxS   cortical subplate
DEn   dorsal endopiriform nu
ec   external capsule
FL   forelimb area of cortex
Fr   frontal cortex
Frontal   frontal bone
FStr   fundus striati
HDB   nu horiz limb diagonal band
HL   hindlimb area of cortex
IG   indusium griseum
Ins   insular cortex
lo   lateral olfactory tract
LSD   lateral septal nu, dorsal
LSI   lateral septal nu, intermediate
LV   lateral ventricle
mfb   medial forebrain bundle
MS   medial septal nu
Nasal   nasal cavity
Pal   palatine bone
Par1   parietal cortex, area 1
Pir   piriform cortex
PSph   presphenoid bone
PSphW   presphenoid wing
RF   rhinal fissure
SphPal   sphenopalatine ganglion
spt   septal neuroepithelium
Tu   olfactory tubercule
VDB   nu vertical limb diagonal band
VP   ventral pallidum
VP   ventral pallidum

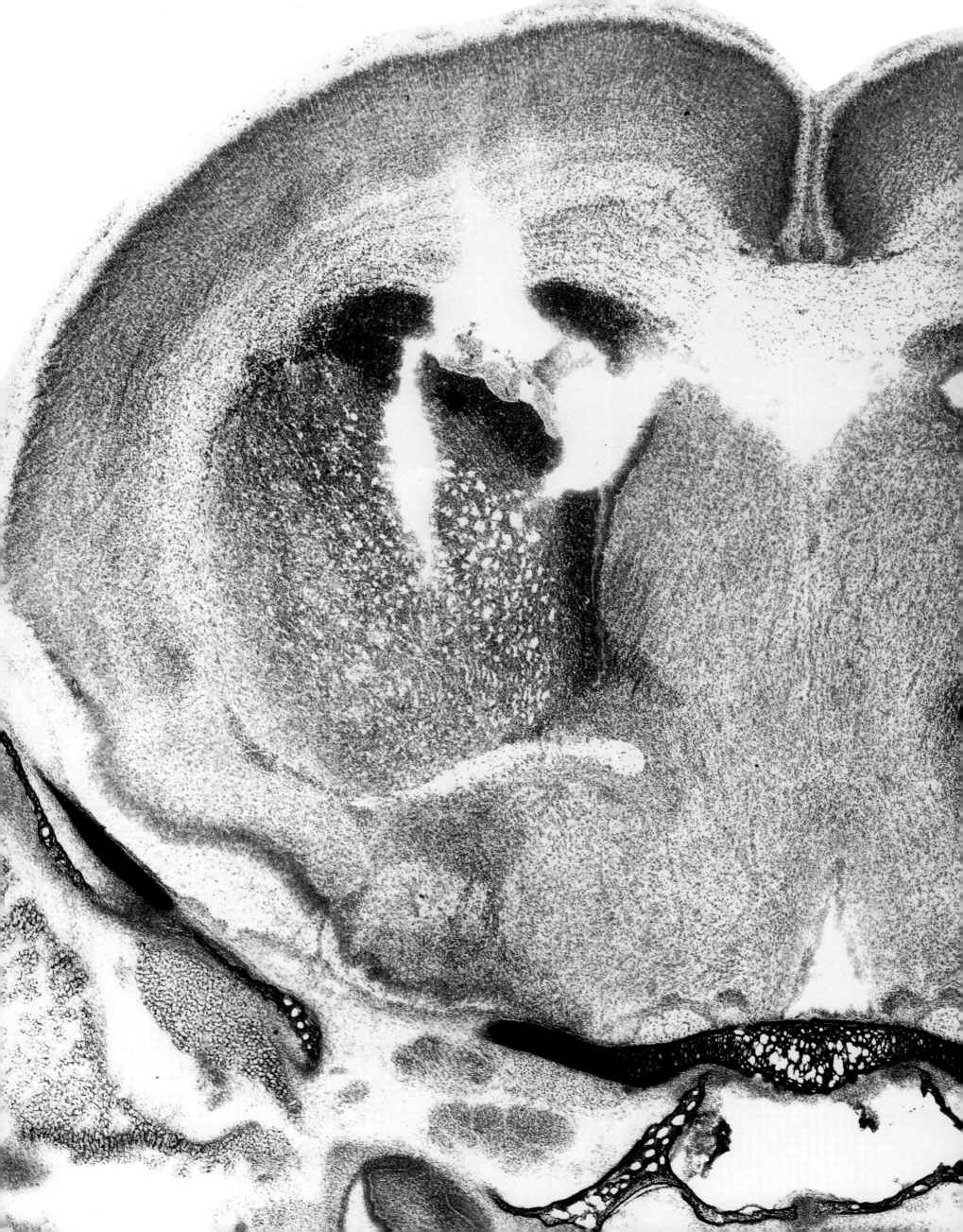

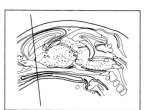

Figure 156
P0-12

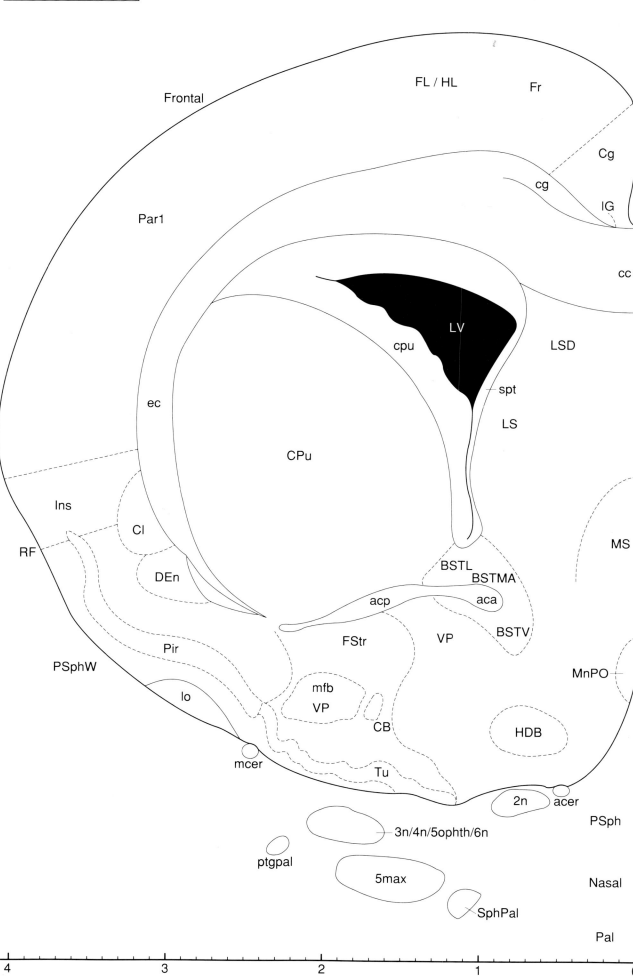

2n   optic nerve
3n   oculomotor nerve or its root
4n   trochlear nerve or its root
5max   maxillary nerve
5ophth   ophthalmic n of trigem
6n   abducens nerve or its root
aca   anterior commissure, anterior
acer   anterior cerebral artery
acp   anterior commissure, posterior
BSTL   bed nu st, lateral div
BSTMA   bed nu st, med div, ant
BSTV   bed nu st, ventral div
CB   cell bridges ventral striatum
cc   corpus callosum
Cg   cingulate cortex
cg   cingulum
Cl   claustrum
CPu   caudate putamen
cpu   caudate putamen neuroepithelium
DEn   dorsal endopiriform nu
ec   external capsule
FL   forelimb area of cortex
Fr   frontal cortex
Frontal   frontal bone
FStr   fundus striati
HDB   nu horiz limb diagonal band
HL   hindlimb area of cortex
IG   indusium griseum
Ins   insular cortex
lo   lateral olfactory tract
LS   lateral septal nu
LSD   lateral septal nu, dorsal
LV   lateral ventricle
mcer   middle cerebral artery
mfb   medial forebrain bundle
MnPO   median preoptic nu
MS   medial septal nu
Nasal   nasal cavity
Pal   palatine bone
Par1   parietal cortex, area 1
Pir   piriform cortex
PSph   presphenoid bone
PSphW   presphenoid wing
ptgpal   pterygopalatine artery
RF   rhinal fissure
SphPal   sphenopalatine ganglion
spt   septal neuroepithelium
Tu   olfactory tubercule
VP   ventral pallidum

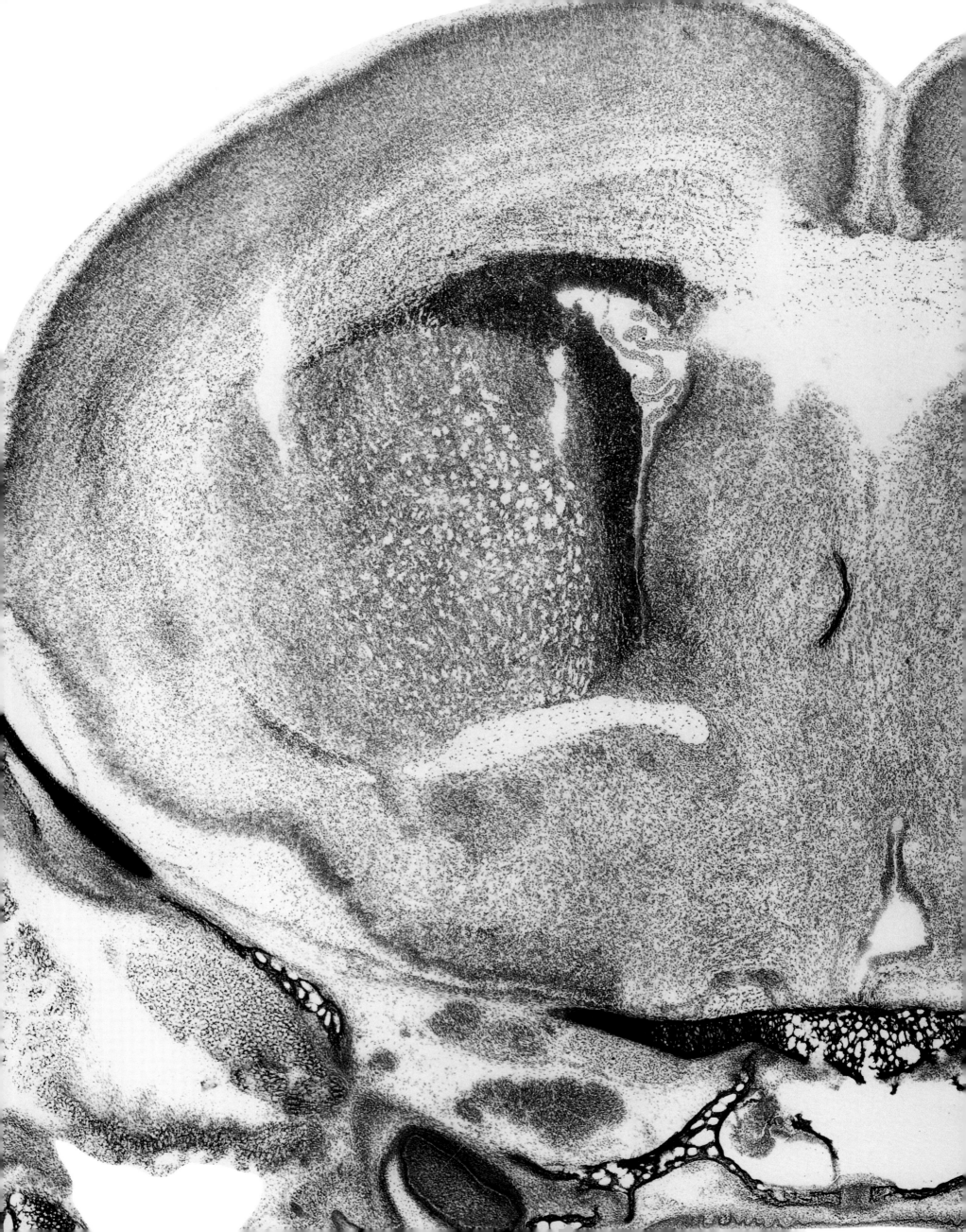

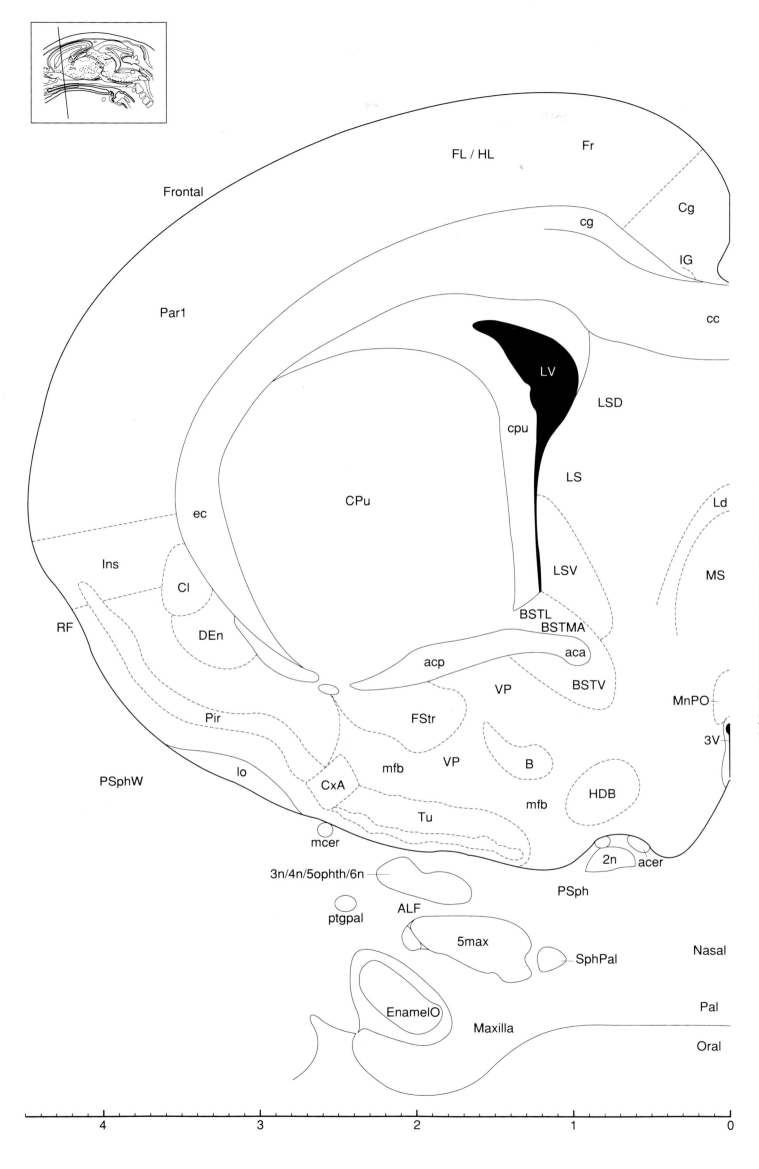

Figure 157
P0-13

2n   optic nerve
3n   oculomotor nerve or its root
3V   3rd ventricle
4n   trochlear nerve or its root
5max   maxillary nerve
5ophth   ophthalmic n of trigem
6n   abducens nerve or its root
aca   anterior commissure, anterior
acer   anterior cerebral artery
acp   anterior commissure, posterior
ALF   anterior lacerated foramen
B   basal nu Meynert
BSTL   bed nu st, lateral div
BSTMA   bed nu st, med div, ant
BSTV   bed nu st, ventral div
cc   corpus callosum
Cg   cingulate cortex
cg   cingulum
Cl   claustrum
CPu   caudate putamen
cpu   caudate putamen neuroepithelium
CxA   cortex-amygdala transition zone
DEn   dorsal endopiriform nu
ec   external capsule
EnamelO   enamel organ (of tooth)
FL   forelimb area of cortex
Fr   frontal cortex
Frontal   frontal bone
FStr   fundus striati
HDB   nu horiz limb diagonal band
HL   hindlimb area of cortex
IG   indusium griseum
Ins   insular cortex
Ld   lambdoid septal zone
lo   lateral olfactory tract
LS   lateral septal nu
LSD   lateral septal nu, dorsal
LSV   lateral septal nu, ventral
LV   lateral ventricle
Maxilla   maxilla
mcer   middle cerebral artery
mfb   medial forebrain bundle
MnPO   median preoptic nu
MS   medial septal nu
Nasal   nasal cavity
Oral   oral cavity
Pal   palatine bone
Par1   parietal cortex, area 1
Pir   piriform cortex
PSph   presphenoid bone
PSphW   presphenoid wing
ptgpal   pterygopalatine artery
RF   rhinal fissure
SphPal   sphenopalatine ganglion
Tu   olfactory tubercule
VP   ventral pallidum

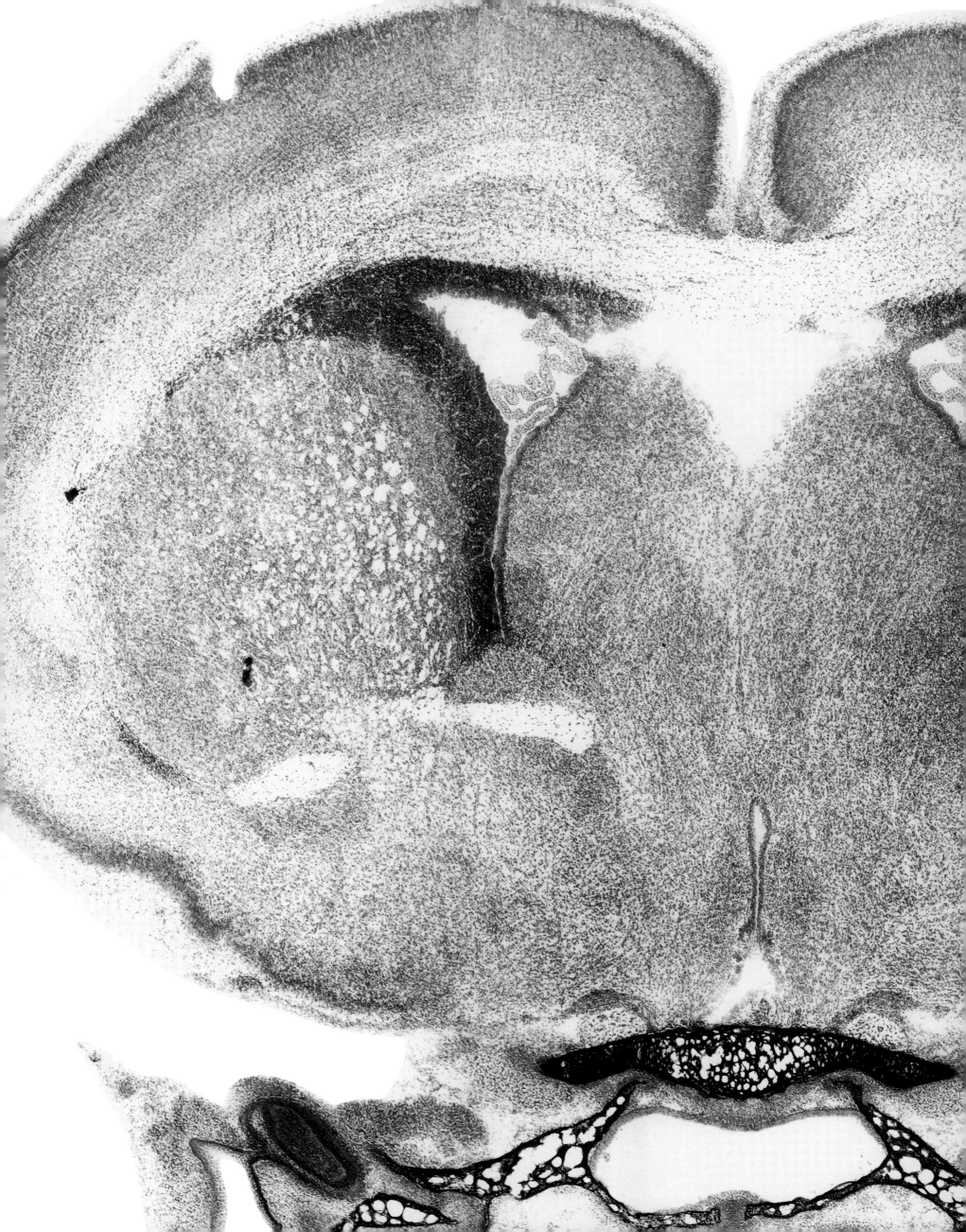

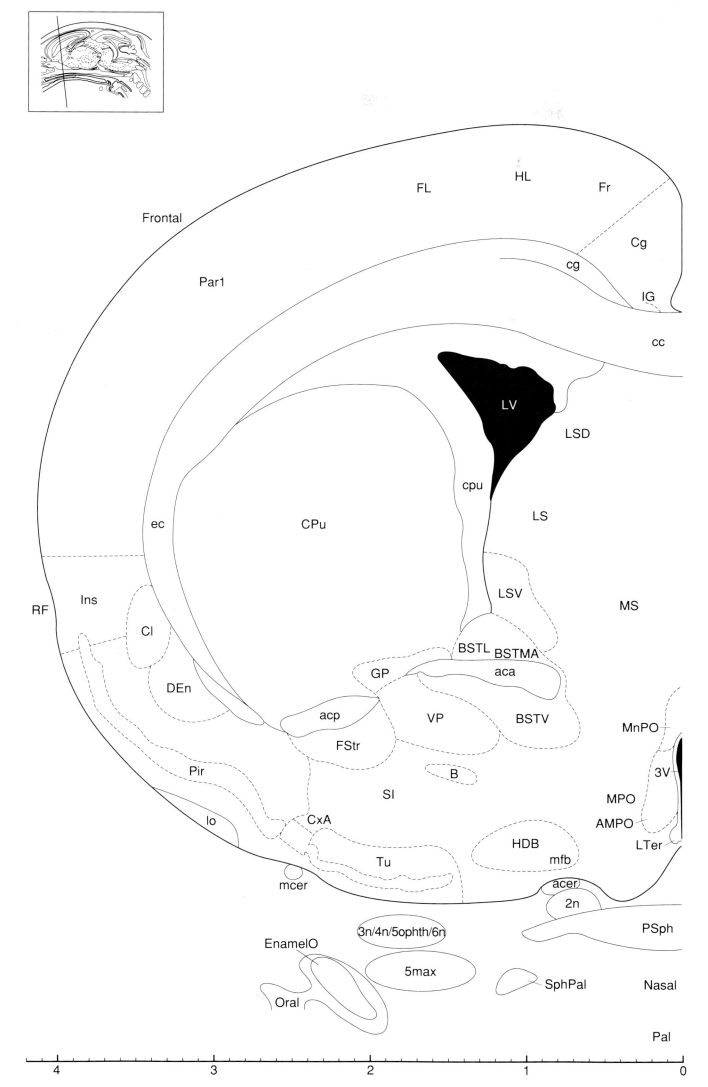

Figure 158
P0-14

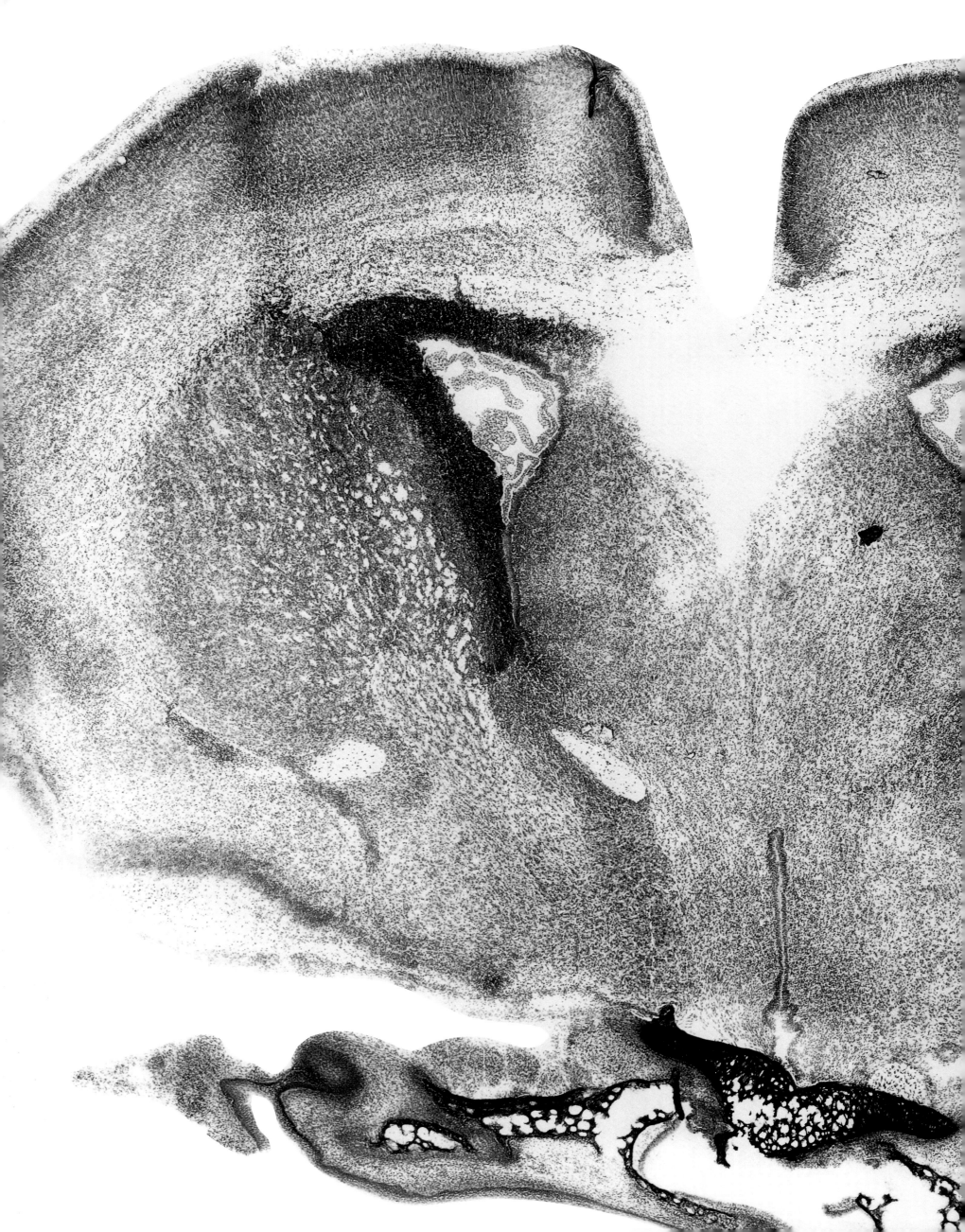

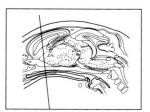

Figure 159
P0-15

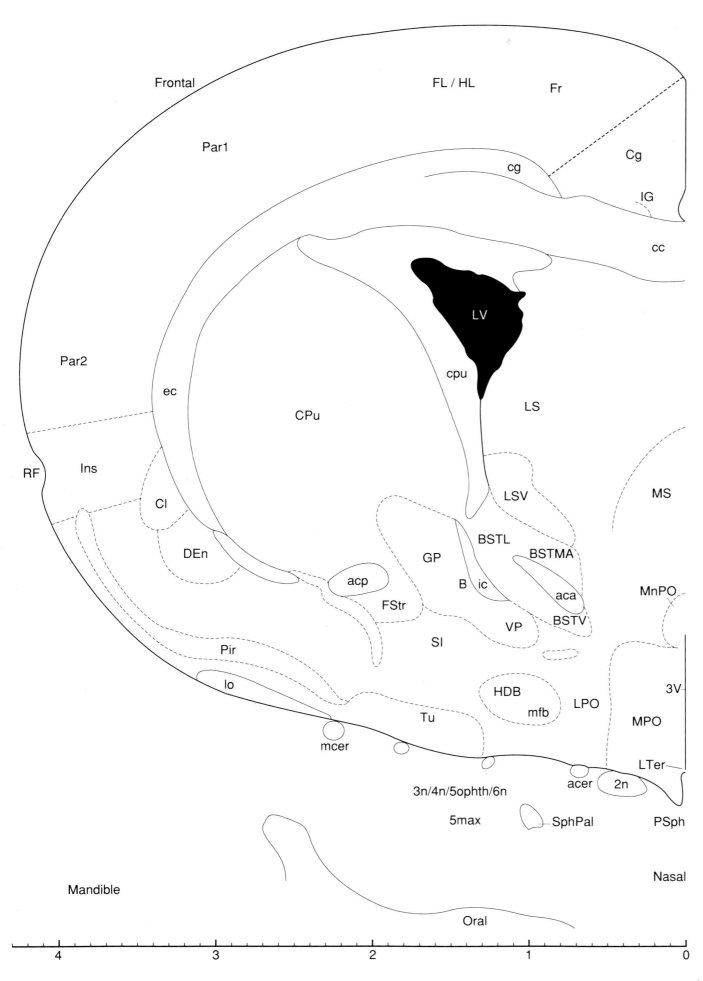

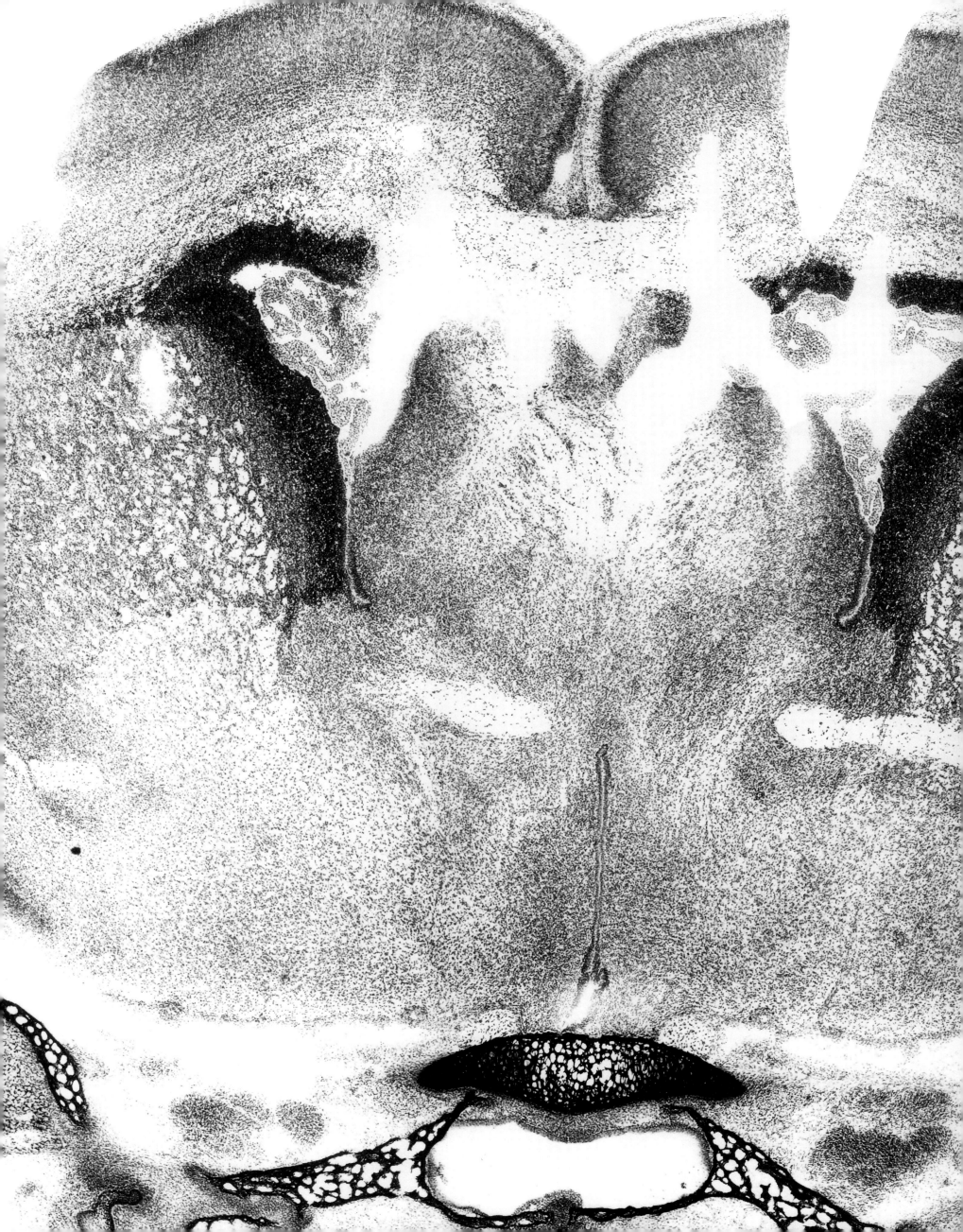

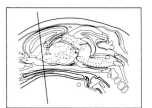

Figure 160
P0-16

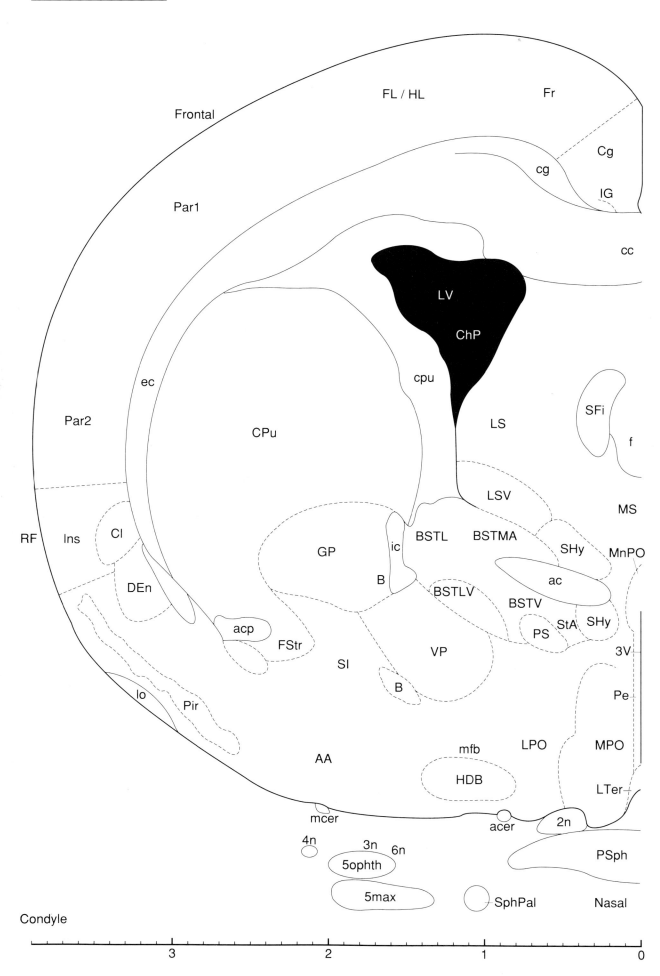

2n   optic nerve
3n   oculomotor nerve or its root
3V   3rd ventricle
4n   trochlear nerve or its root
5max   maxillary nerve
5ophth   ophthalmic n of trigem
6n   abducens nerve or its root
AA   anterior amygdaloid area
ac   anterior commissure
acer   anterior cerebral artery
acp   anterior commissure, posterior
B   basal nu Meynert
BSTL   bed nu st, lateral div
BSTLV   bed nu st, lat div, vent
BSTMA   bed nu st, med div, ant
BSTV   bed nu st, ventral div
cc   corpus callosum
Cg   cingulate cortex
cg   cingulum
ChP   choroid plexus
Cl   claustrum
Condyle   condyloid process of mandib
CPu   caudate putamen
cpu   caudate putamen neuroepithelium
DEn   dorsal endopiriform nu
ec   external capsule
f   fornix
FL   forelimb area of cortex
Fr   frontal cortex
Frontal   frontal bone
FStr   fundus striati
GP   globus pallidus
HDB   nu horiz limb diagonal band
HL   hindlimb area of cortex
ic   inferior colliculus neuroepithelium
IG   indusium griseum
Ins   insular cortex
lo   lateral olfactory tract
LPO   lateral preoptic area
LS   lateral septal nu
LSV   lateral septal nu, ventral
LTer   lamina terminalis
LV   lateral ventricle
mcer   middle cerebral artery
mfb   medial forebrain bundle
MnPO   median preoptic nu
MPO   medial preoptic nu
MS   medial septal nu
Nasal   nasal cavity
Par1   parietal cortex, area 1
Par2   parietal cortex, area 2
Pe   periventricular hypoth nu
Pir   piriform cortex
PS   parastrial nu
PSph   presphenoid bone
RF   rhinal fissure
SFi   septofimbrial nu
SHy   septohypothalamic nu
SI   substantia innominata
SphPal   sphenopalatine ganglion
StA   strial part preoptic area
VP   ventral pallidum

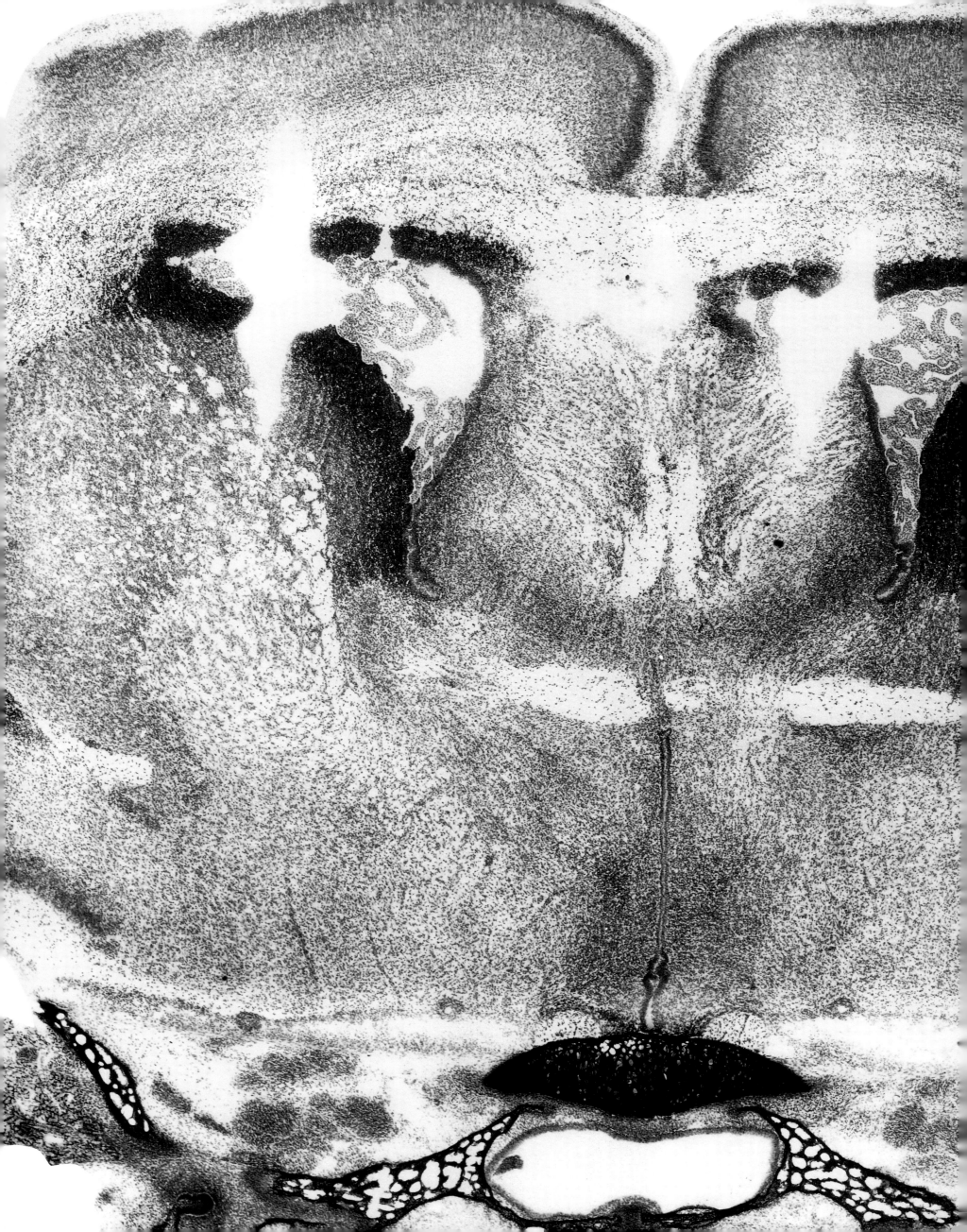

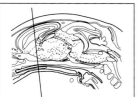

Figure 161
P0-17

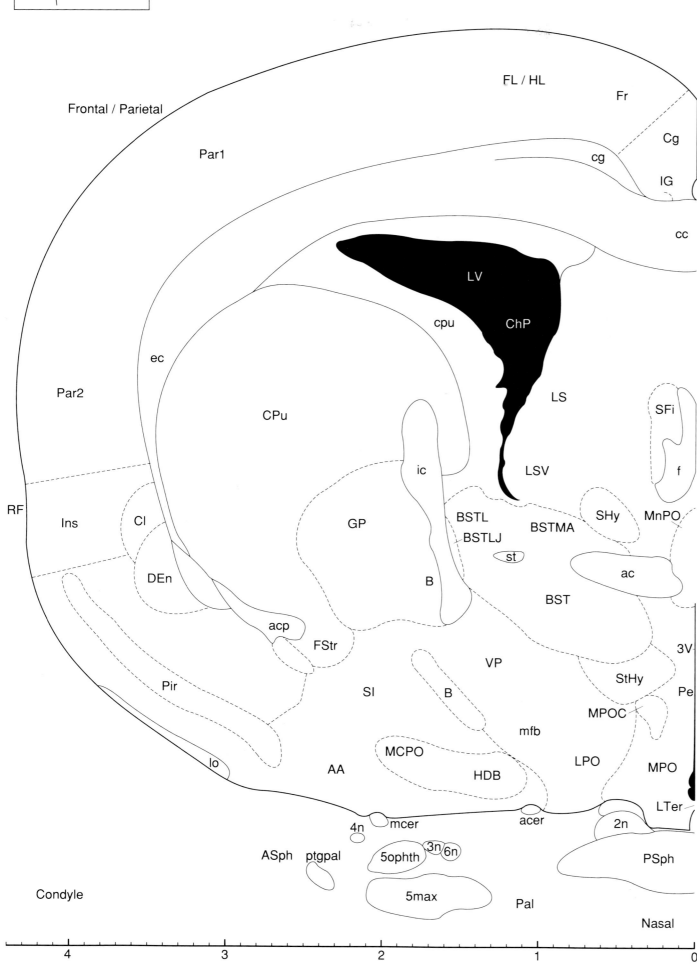

Frontal / Parietal

FL / HL

Fr

Par1

Cg

cg

IG

cc

LV

cpu

ChP

ec

LS

SFi

Par2

CPu

f

RF

ic

LSV

Ins

Cl

GP

BSTL

BSTMA

SHy

MnPO

DEn

BSTLJ

st

ac

B

BST

acp

FStr

3V

Pir

VP

StHy

SI

B

Pe

MPOC

mfb

MPO

MCPO

LPO

AA

HDB

lo

4n mcer

acer

2n

LTer

ASph ptgpal

3n 6n

5ophth

PSph

Condyle

5max

Pal

Nasal

4   3   2   1   0

2n   optic nerve
3n   oculomotor nerve or its root
3V   3rd ventricle
4n   trochlear nerve or its root
5max maxillary nerve
5ophth ophthalmic n of trigem
6n   abducens nerve or its root
AA   anterior amygdaloid area
ac   anterior commissure
acer anterior cerebral artery
acp   anterior commissure, posterior
ASph alisphenoid bone
B   basal nu Meynert
BST   bed nu stria terminalis
BSTL bed nu st, lateral div
BSTLJ bed nu st, lat, juxtacapsular
BSTMA bed nu st, med div, ant
cc   corpus callosum
Cg   cingulate cortex
cg   cingulum
ChP   choroid plexus
Cl   claustrum
Condyle condyloid process of mandib
CPu   caudate putamen
cpu   caudate putamen neuroepithelium
DEn   dorsal endopiriform nu
ec   external capsule
f   fornix
FL   forelimb area of cortex
Fr   frontal cortex
Frontal frontal bone
FStr fundus striati
GP   globus pallidus
HDB   nu horiz limb diagonal band
HL   hindlimb area of cortex
ic   inferior colliculus neuroepithelium
IG   indusium griseum
Ins   insular cortex
lo   lateral olfactory tract
LPO   lateral preoptic area
LS   lateral septal nu
LSV   lateral septal nu, ventral
LTer lamina terminalis
LV   lateral ventricle
mcer middle cerebral artery
MCPO magnocellular preoptic nu
mfb   medial forebrain bundle
MnPO median preoptic nu
MPO   medial preoptic nu
MPOC medial preoptic nu, central
Nasal nasal cavity
Pal   palatine bone
Par1 parietal cortex, area 1
Par2 parietal cortex, area 2
Parietal parietal bone
Pe   periventricular hypoth nu
Pir   piriform cortex
PSph presphenoid bone
ptgpal pterygopalatine artery
RF   rhinal fissure
SFi   septofimbrial nu
SHy   septohypothalamic nu
SI   substantia innominata
st   stria terminalis
StHy striohypothalamic nu
VP   ventral pallidum

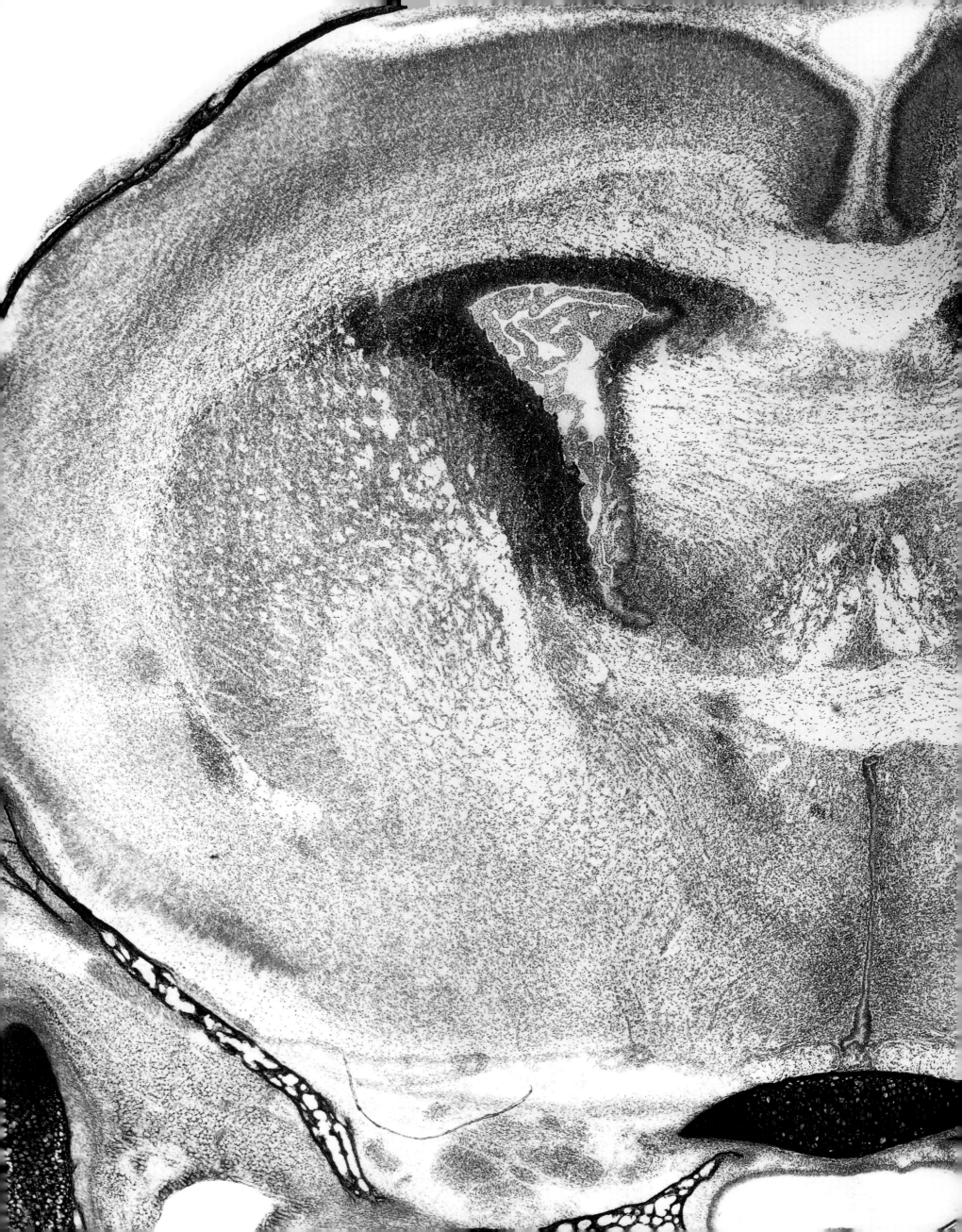

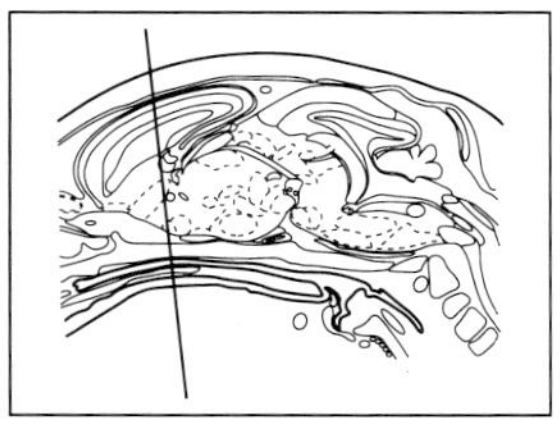

Figure 162
P0-18

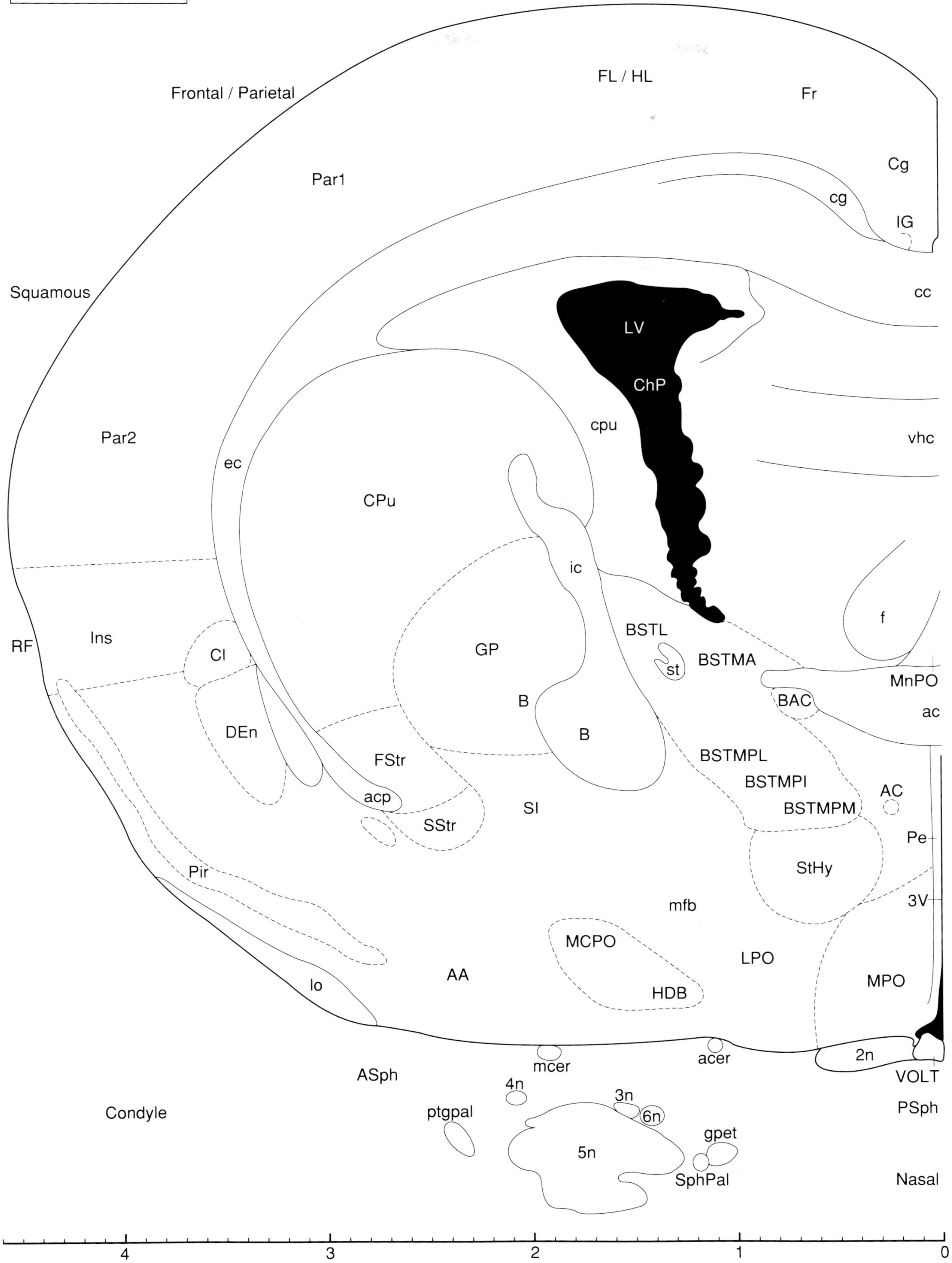

2n   optic nerve
3n   oculomotor nerve or its root
3V   3rd ventricle
4n   trochlear nerve or its root
5n   trigeminal nerve
6n   abducens nerve or its root
AA   anterior amygdaloid area
AC   anterior commissural nu
ac   anterior commissure
acer   anterior cerebral artery
acp   anterior commissure, posterior
ASph   alisphenoid bone
B   basal nu Meynert
BAC   bed nu anterior commissure
BSTL   bed nu st, lateral div
BSTMA   bed nu st, med div, ant
BSTMPI   bed nu st, med div, posteroin
BSTMPL   bed nu st, med, posterolat
BSTMPM   bed nu st, med, posteromed
cc   corpus callosum
Cg   cingulate cortex
cg   cingulum
ChP   choroid plexus
Cl   claustrum
Condyle   condyloid process of mandib
CPu   caudate putamen
cpu   caudate putamen neuroepithelium
DEn   dorsal endopiriform nu
ec   external capsule
f   fornix
FL   forelimb area of cortex
Fr   frontal cortex
Frontal   frontal bone
FStr   fundus striati
GP   globus pallidus
gpet   greater petrosal nerve
HDB   nu horiz limb diagonal band
HL   hindlimb area of cortex
ic   inferior colliculus neuroepithelium
IG   indusium griseum
Ins   insular cortex
lo   lateral olfactory tract
LPO   lateral preoptic area
LV   lateral ventricle
mcer   middle cerebral artery
MCPO   magnocellular preoptic nu
mfb   medial forebrain bundle
MnPO   median preoptic nu
MPO   medial preoptic nu
Nasal   nasal cavity
Par1   parietal cortex, area 1
Par2   parietal cortex, area 2
Parietal   parietal bone
Pe   periventricular hypoth nu
Pir   piriform cortex
PSph   presphenoid bone
ptgpal   pterygopalatine artery
RF   rhinal fissure
SI   substantia innominata
SphPal   sphenopalatine ganglion
Squamous   squamous part of temp
SStr   substriatal area
st   stria terminalis
StHy   striohypothalamic nu
vhc   ventral hip commissure
VOLT   vascular organ lamina terminal

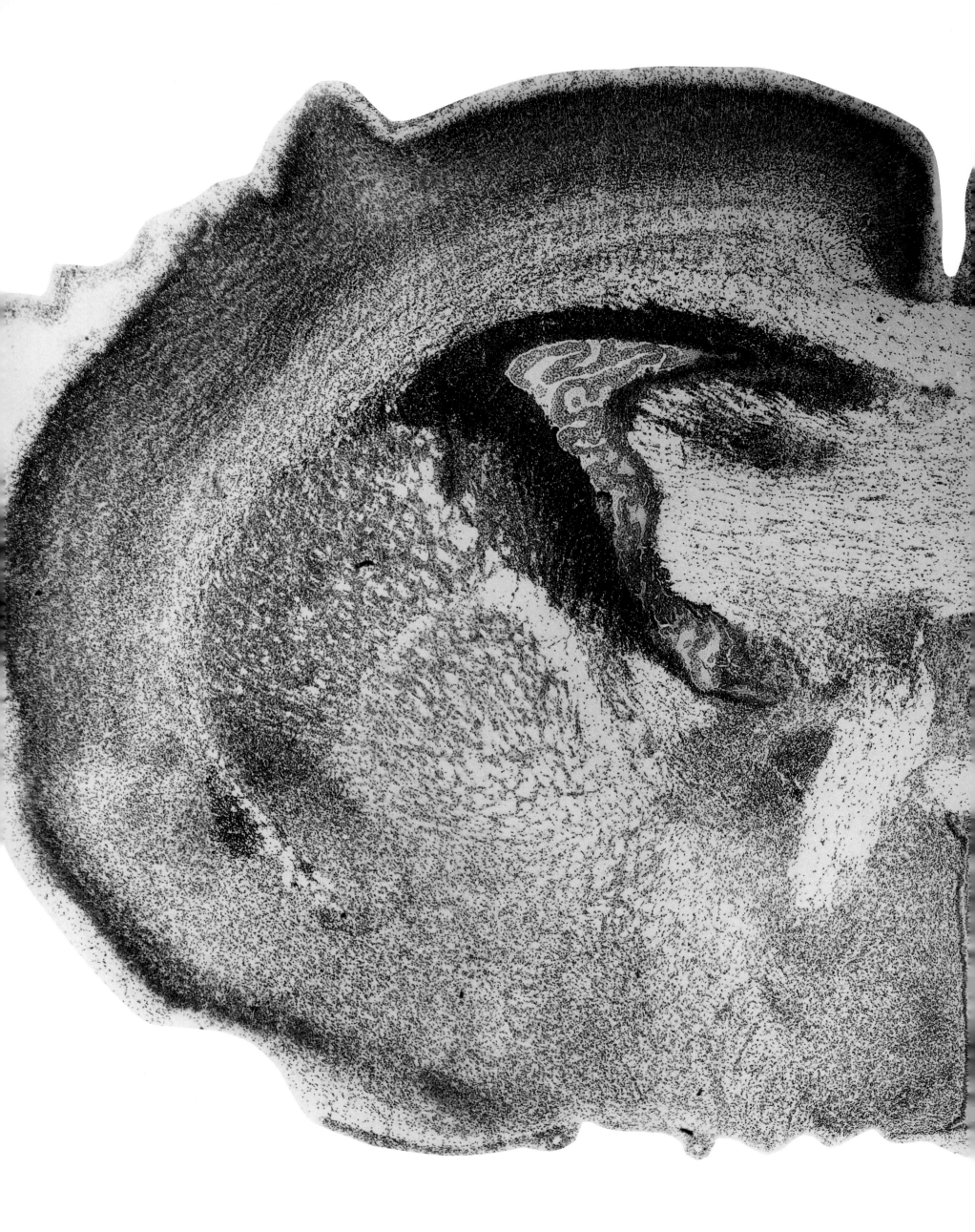

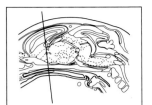

Figure 163
P0-19

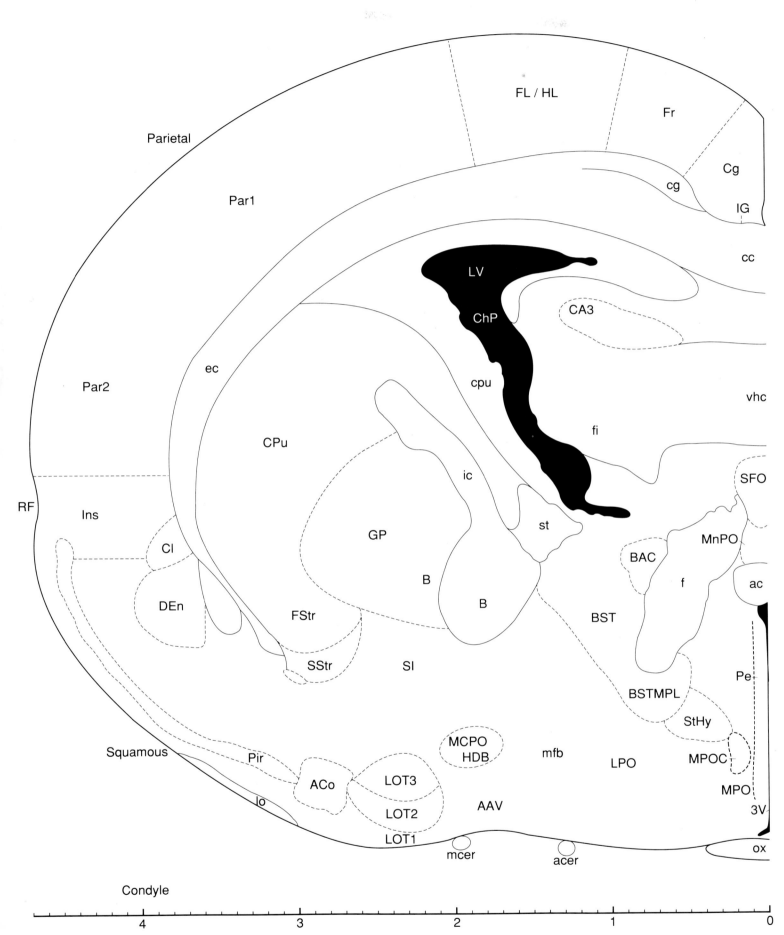

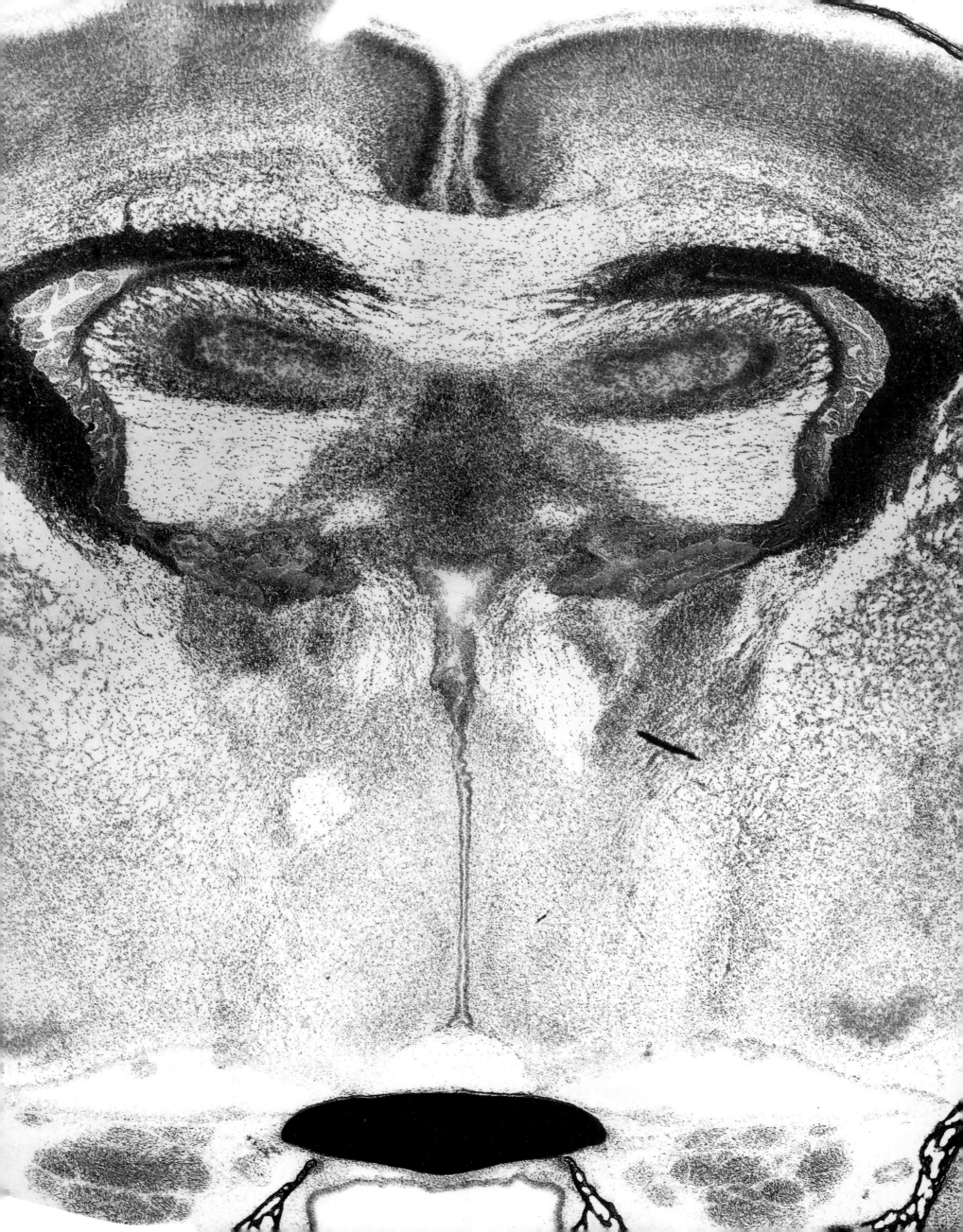

Figure 164
P0-20

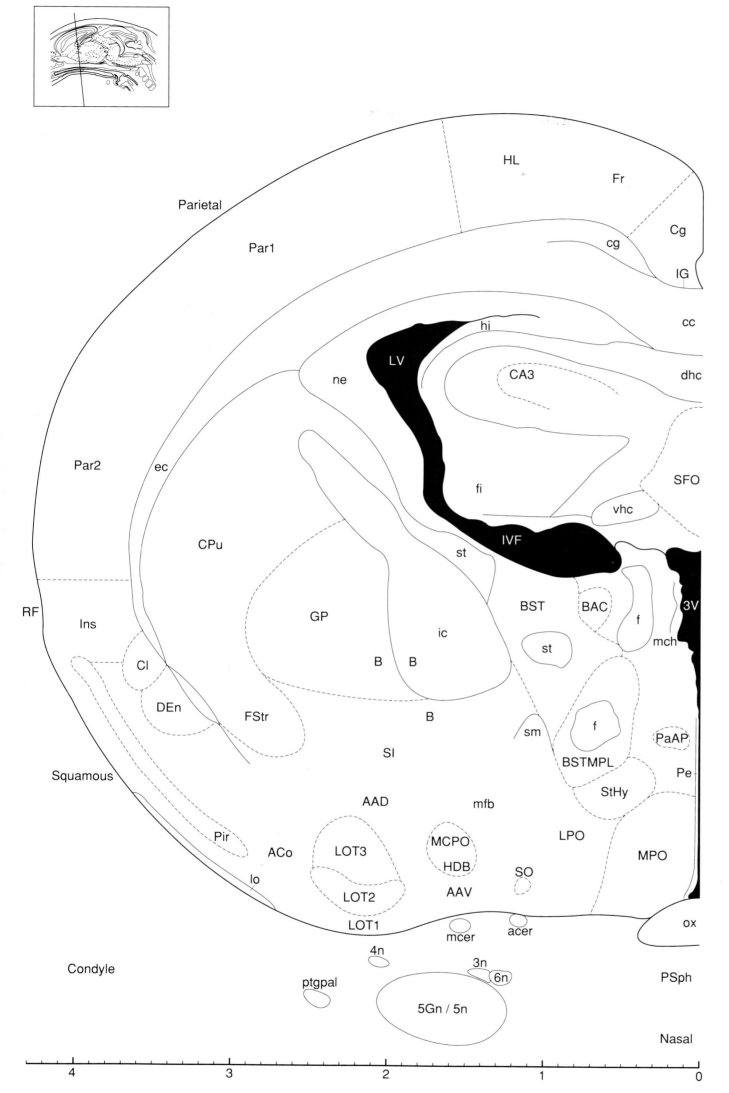

3n   oculomotor nerve or its root
3V   3rd ventricle
4n   trochlear nerve or its root
5Gn trigeminal ganlion
5n   trigeminal nerve
6n   abducens nerve or its root
AAD   anterior amygdaloid area, dorsal
AAV   anterior amygdaloid area, ventral
acer   anterior cerebral artery
ACo   anterior cortical amygdaloid nu
B   basal nu Meynert
BAC   bed nu anterior commissure
BST   bed nu stria terminalis
BSTMPL   bed nu st, med, posterolat
CA3   CA3 field of the hippocampus
cc   corpus callosum
Cg   cingulate cortex
cg   cingulum
Cl   claustrum
Condyle   condyloid process of mandib
CPu   caudate putamen
DEn   dorsal endopiriform nu
dhc   dorsal hippocampal commissure
ec   external capsule
f   fornix
fi   fimbria hippocampus
Fr   frontal cortex
FStr   fundus striati
GP   globus pallidus
HDB   nu horiz limb diagonal band
hi   hippocampal formation neuroepi
HL   hindlimb area of cortex
ic   inferior colliculus neuroepithelium
IG   indusium griseum
Ins   insular cortex
IVF   interventricular foramen
lo   lateral olfactory tract
LOT1   lat olf nu, layer 1
LOT2   lat olf nu, layer 2
LOT3   lat olf nu, layer 3
LPO   lateral preoptic area
LV   lateral ventricle
mcer   middle cerebral artery
mch   med corticohypothalamic tr
MCPO   magnocellular preoptic nu
mfb   medial forebrain bundle
MPO   medial preoptic nu
Nasal   nasal cavity
ne   neuroepithelium
ox   optic chiasm
PaAP   paravent hypoth nu, ant parvo
Par1   parietal cortex, area 1
Par2   parietal cortex, area 2
Parietal   parietal bone
Pe   periventricular hypoth nu
Pir   piriform cortex
PSph   presphenoid bone
ptgpal   pterygopalatine artery
RF   rhinal fissure
SFO   subfornical organ
SI   substantia innominata
sm   stria medullaris thalami
SO   supraoptic nu
Squamous   squamous part of temp
st   stria terminalis
StHy   striohypothalamic nu
vhc   ventral hip commissure

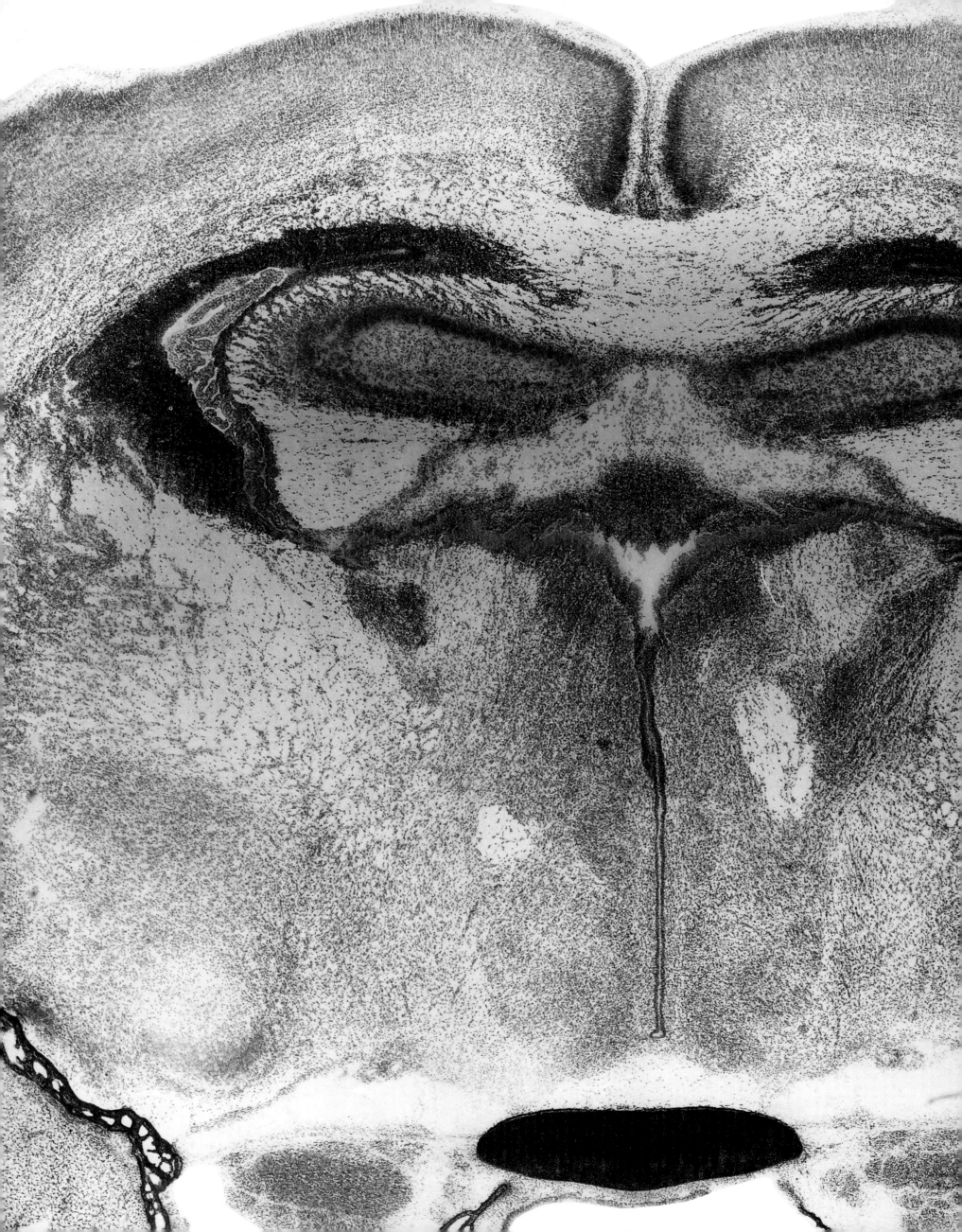

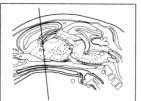

Figure 165
P0-21

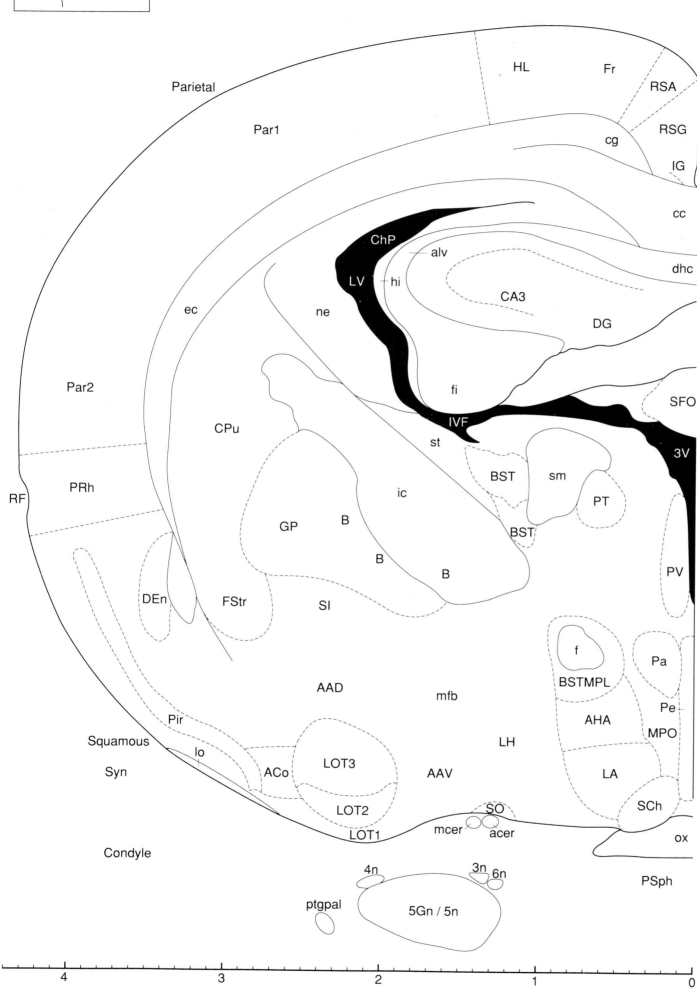

Parietal
Par1
HL   Fr
RSA
RSG
cg
IG
cc
ChP
alv
dhc
LV   hi
ec
ne   CA3
DG
fi
SFO
st
IVF
3V
BST   sm
ic   PT
GP   B   BST
B
B   PV
f   Pa
BSTMPL
AAD   mfb   Pe
AHA   MPO
Pir   LH
Squamous   LOT3   LA   SCh
Syn   ACo   AAV
LOT2   SO
LOT1   mcer   acer   ox
Condyle   PSph
4n   3n 6n
ptgpal   5Gn / 5n

Par2
CPu
RF
PRh
DEn   FStr   SI

4   3   2   1   0

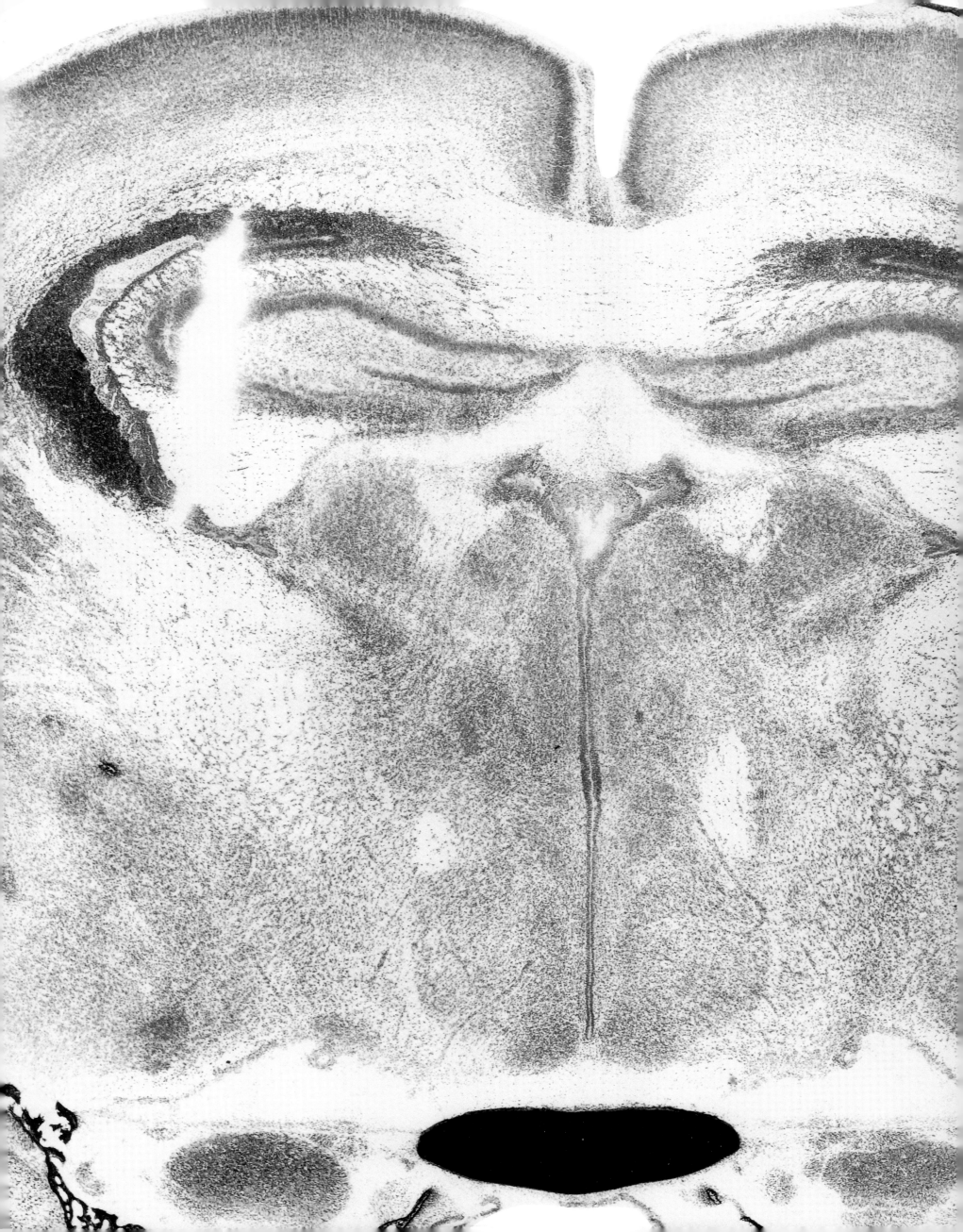

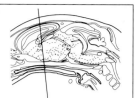

Figure 166
P0-22

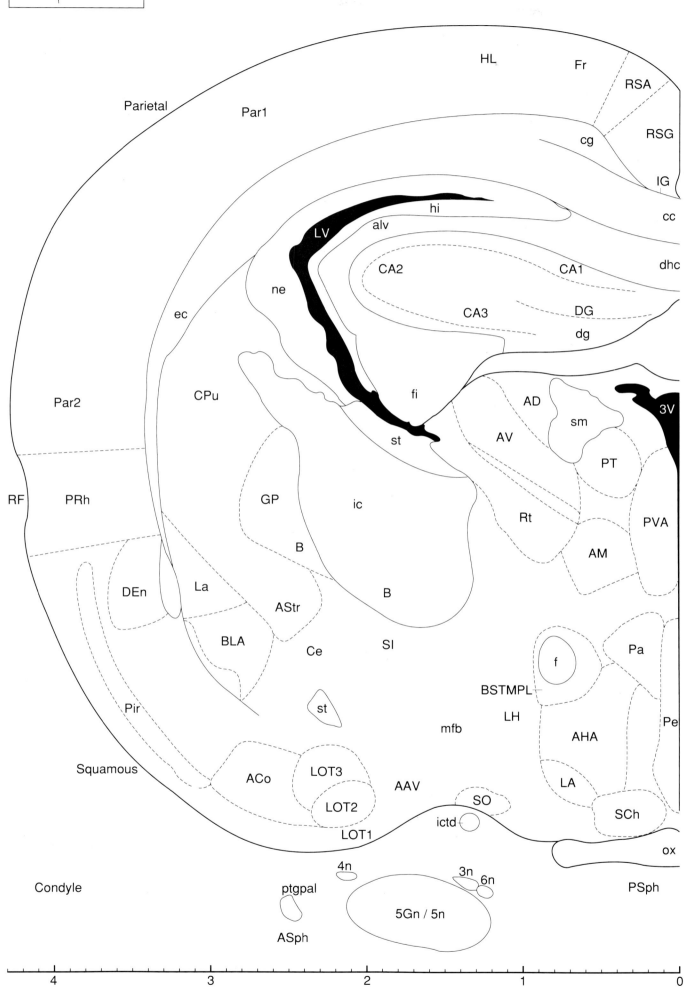

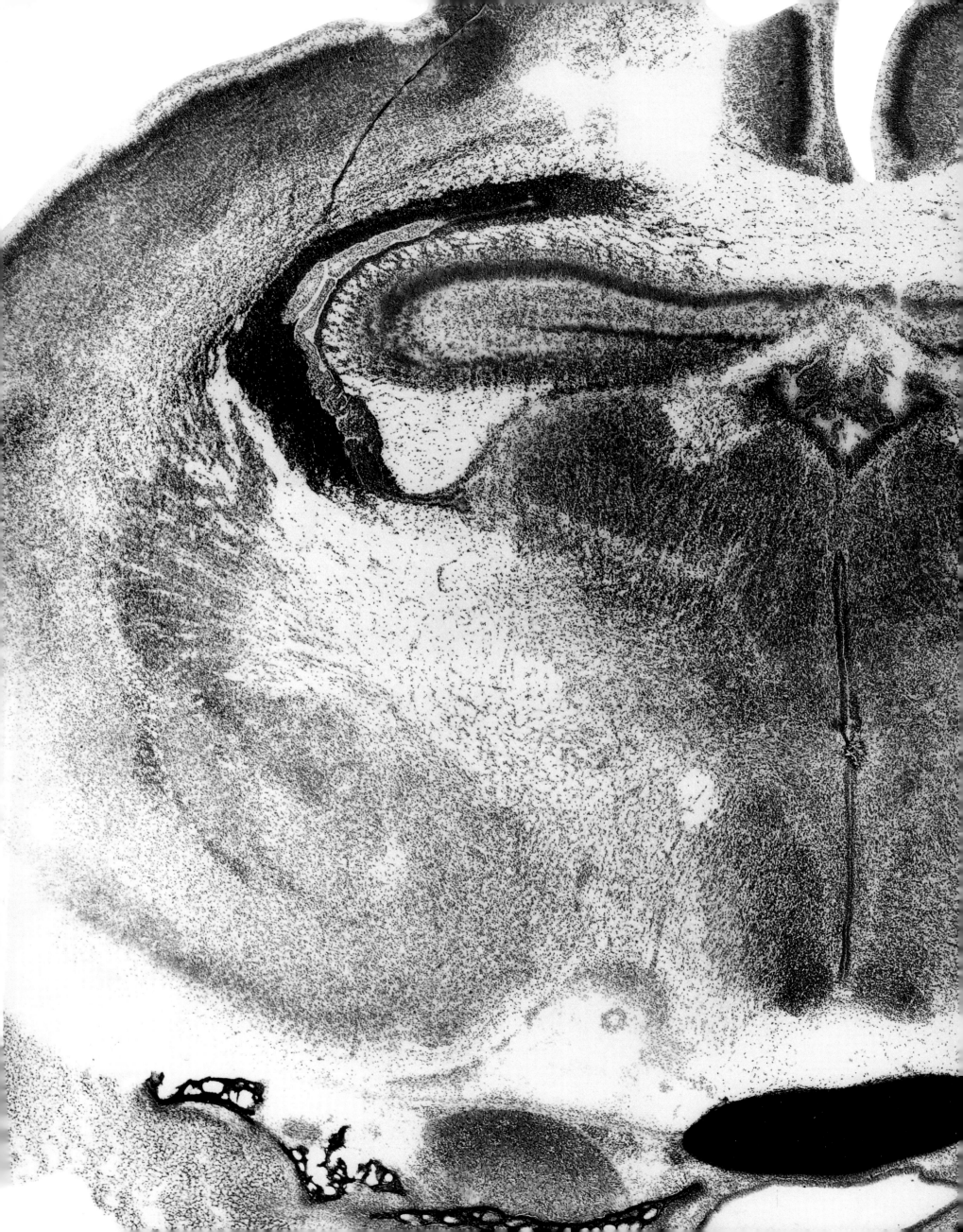

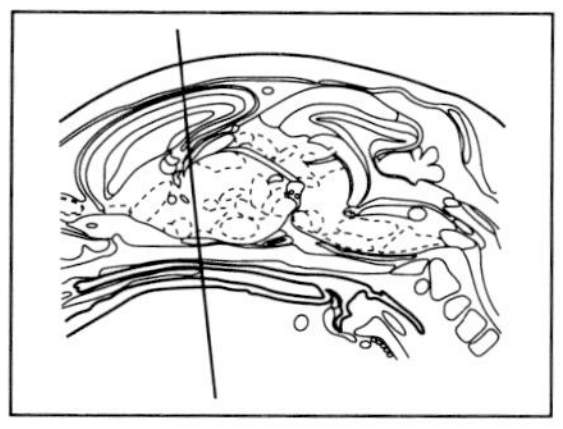

Figure 167
P0-23

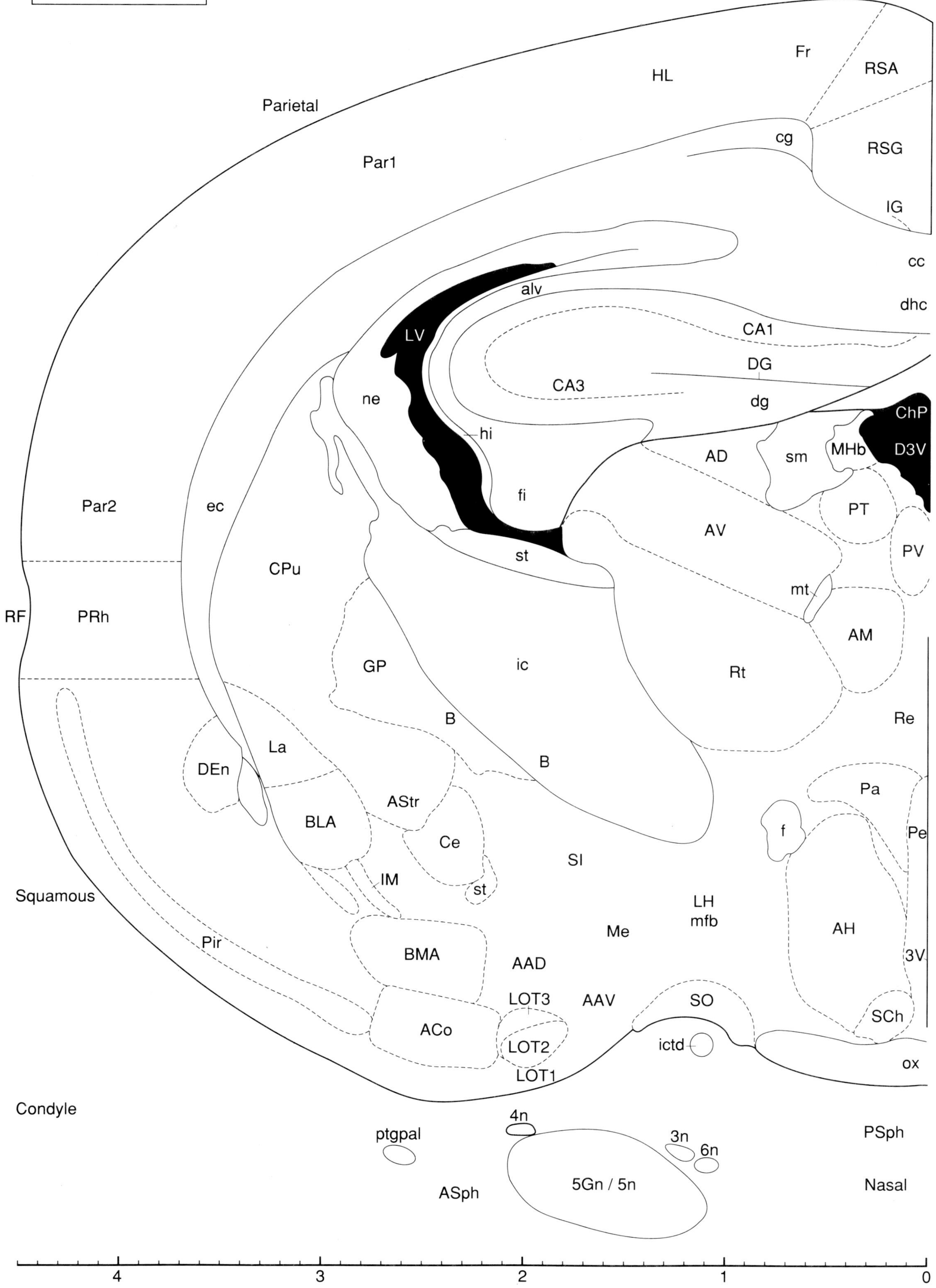

3n   oculomotor nerve or its root
3V   3rd ventricle
4n   trochlear nerve or its root
5Gn   trigeminal ganlion
5n   trigeminal nerve
6n   abducens nerve or its root
AAD   anterior amygdaloid area, dorsal
AAV   anterior amygdaloid area, ventral
ACo   anterior cortical amygdaloid nu
AD   anterodorsal thal nu
AH   anterior hypoth nu
alv   alveus of hippocampus
AM   anteromedial thal nu
ASph alisphenoid bone
AStr amygdalostriatal transition area
AV   anteroventral thal nu
B   basal nu Meynert
BLA   basolateral amygdaloid nu, ant
BMA   basomedial amygdaloid nu, ant
CA1   CA1 field of the hippocampus
CA3   CA3 field of the hippocampus
cc   corpus callosum
Ce   central amygdaloid nu
cg   cingulum
ChP   choroid plexus
Condyle   condyloid process of mandib
CPu   caudate putamen
D3V   dorsal third ventricle
DEn   dorsal endopiriform nu
DG   dentate gyrus
dg   dentate gyrus neuroepithelium
dhc   dorsal hippocampal commissure
ec   external capsule
f   fornix
fi   fimbria hippocampus
Fr   frontal cortex
GP   globus pallidus
hi   hippocampal formation neuroepi
HL   hindlimb area of cortex
ic   inferior colliculus neuroepithelium
ictd internal carotid artery
IG   indusium griseum
IM   intercalated amygdaloid nu, main
La   lateral amygdaloid nu
LH   lateral hypothalamic area
LOT1 lat olf nu, layer 1
LOT2 lat olf nu, layer 2
LOT3 lat olf nu, layer 3
LV   lateral ventricle
Me   medial amygdaloid nu
mfb   medial forebrain bundle
MHb   medial habenular nu
mt   mammillothalamic tract
Nasal nasal cavity
ne   neuroepithelium
ox   optic chiasm
Pa   paraventricular hypoth nu
Par1 parietal cortex, area 1
Par2 parietal cortex, area 2
Parietal   parietal bone
Pe   periventricular hypoth nu
Pir   piriform cortex
PRh   perirhinal cortex
PSph presphenoid bone
PT   paratenial thal nu
ptgpal pterygopalatine artery
PV   paraventricular thal nu
Re   reuniens thal nu
RF   rhinal fissure
RSA   retrosplenial agranular Cx
RSG   retrosplenial granular Cx
Rt   reticular thal nu
SCh   suprachiasmatic nu
SI   substantia innominata
sm   stria medullaris thalami
SO   supraoptic nu
Squamous   squamous part of temp
st   stria terminalis

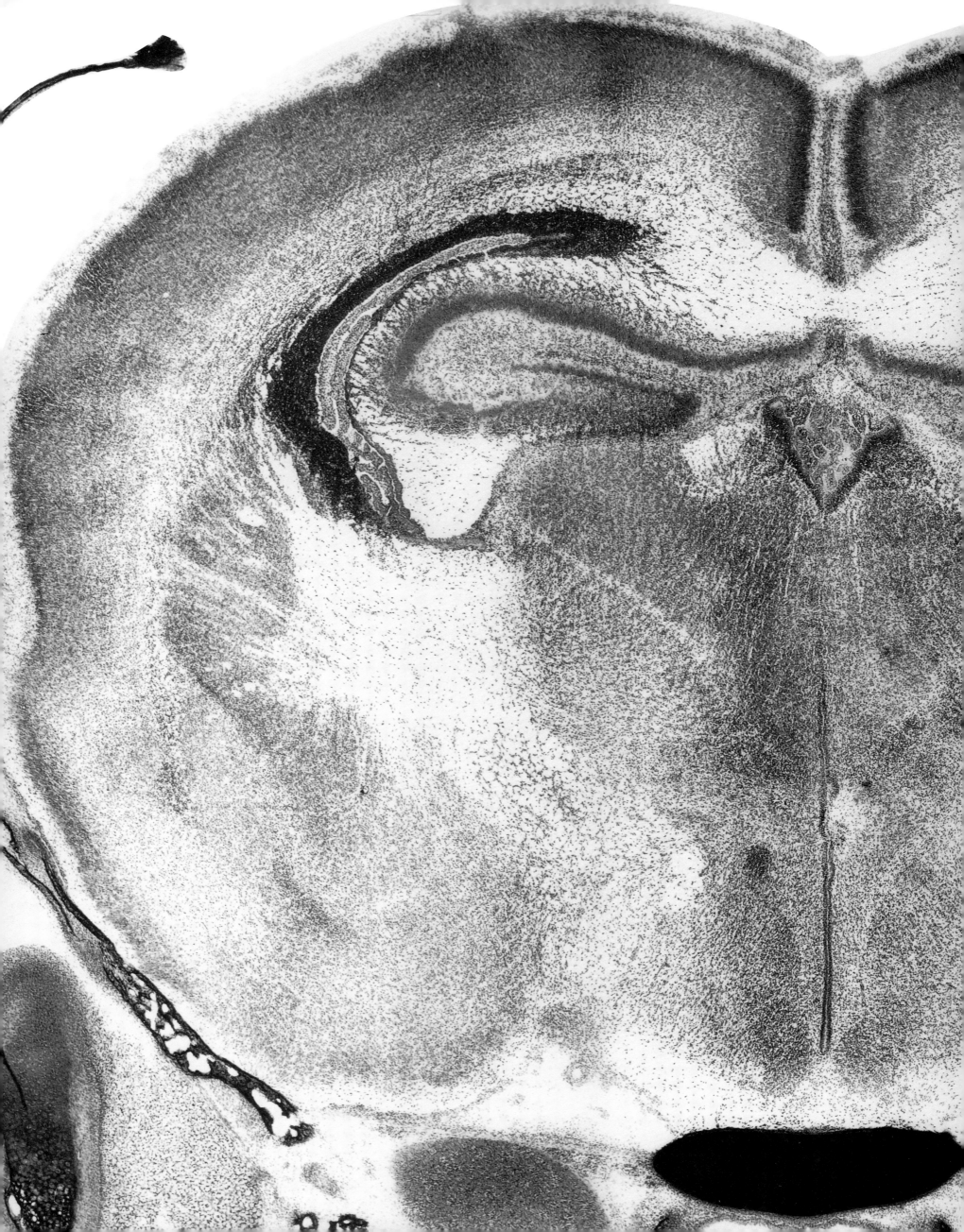

Figure 168
P0-24

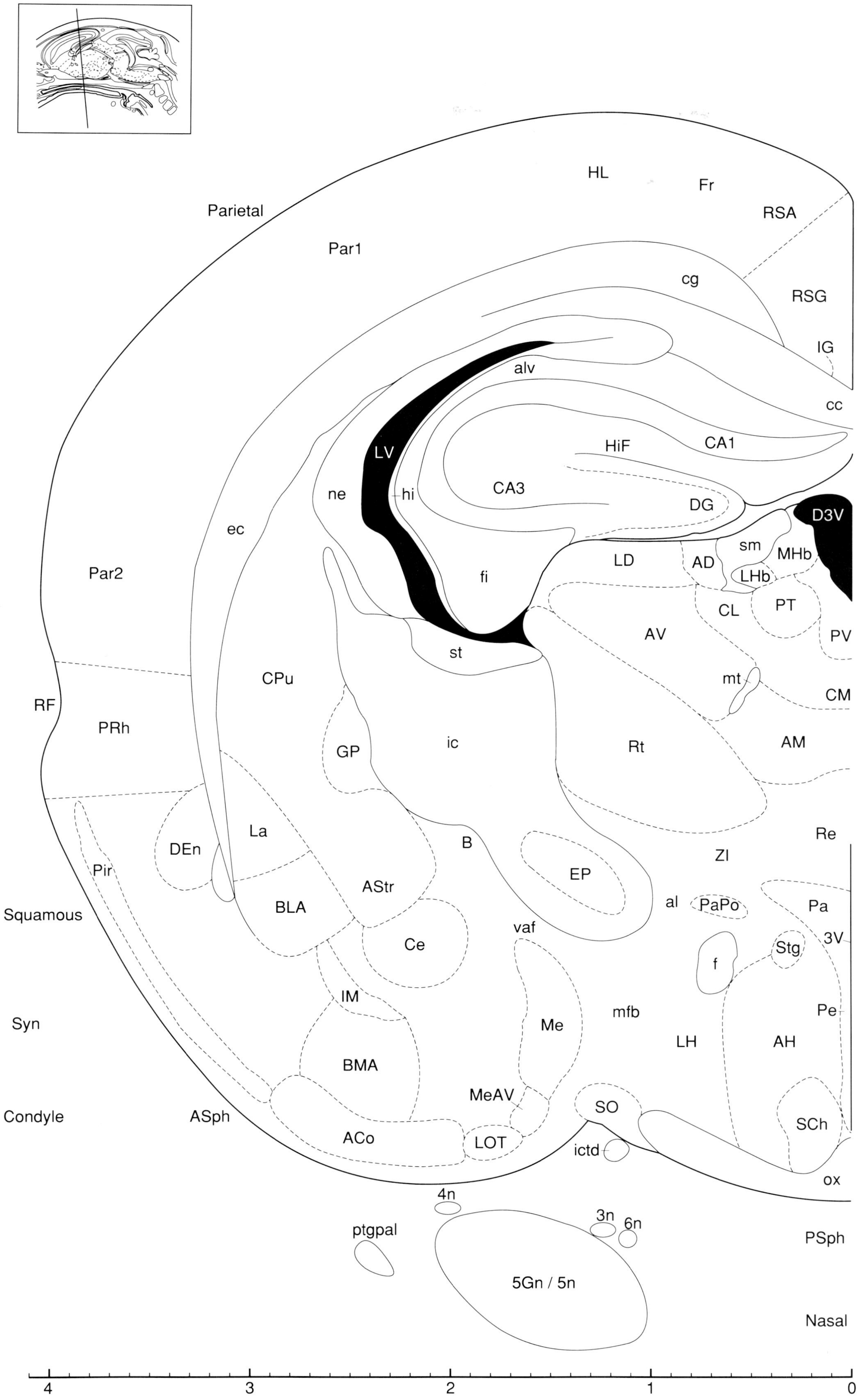

3n   oculomotor nerve or its root
3V   3rd ventricle
4n   trochlear nerve or its root
5Gn   trigeminal ganlion
5n   trigeminal nerve
6n   abducens nerve or its root
ACo   anterior cortical amygdaloid nu
AD   anterodorsal thal nu
AH   anterior hypoth nu
al   ansa lenticularis
alv   alveus of hippocampus
AM   anteromedial thal nu
ASph   alisphenoid bone
AStr   amygdalostriatal transition area
AV   anteroventral thal nu
B   basal nu Meynert
BLA   basolateral amygdaloid nu, ant
BMA   basomedial amygdaloid nu, ant
CA1   CA1 field of the hippocampus
CA3   CA3 field of the hippocampus
cc   corpus callosum
Ce   central amygdaloid nu
cg   cingulum
CL   centrolateral thal nu
CM   central medial thal nu
Condyle   condyloid process of mandib
CPu   caudate putamen
D3V   dorsal third ventricle
DEn   dorsal endopiriform nu
DG   dentate gyrus
ec   external capsule
EP   entopeduncular nu
f   fornix
fi   fimbria hippocampus
Fr   frontal cortex
GP   globus pallidus
hi   hippocampal formation neuroepi
HiF   hippocampal fissure
HL   hindlimb area of cortex
ic   inferior colliculus neuroepithelium
ictd   internal carotid artery
IG   indusium griseum
IM   intercalated amygdaloid nu, main
La   lateral amygdaloid nu
LD   laterodorsal thal nu
LH   lateral hypothalamic area
LHb   lateral habenular nu
LOT   nu lateral olfactory tract
LV   lateral ventricle
Me   medial amygdaloid nu
MeAV   medial amyg nu, anterventral
mfb   medial forebrain bundle
MHb   medial habenular nu
mt   mammillothalamic tract
Nasal   nasal cavity
ne   neuroepithelium
ox   optic chiasm
Pa   paraventricular hypoth nu
PaPo   paraventricular hypoth nu, post
Par1   parietal cortex, area 1
Par2   parietal cortex, area 2
Parietal   parietal bone
Pe   periventricular hypoth nu
Pir   piriform cortex
PRh   perirhinal cortex
PSph   presphenoid bone
PT   paratenial thal nu
ptgpal   pterygopalatine artery
PV   paraventricular thal nu
Re   reuniens thal nu
RF   rhinal fissure
RSA   retrosplenial agranular Cx
RSG   retrosplenial granular Cx
Rt   reticular thal nu
SCh   suprachiasmatic nu
sm   stria medullaris thalami
SO   supraoptic nu
Squamous   squamous part of temp
st   stria terminalis
Stg   stigmoid hypoth nu
Syn   synovial cavity
vaf   ventral amygdalofugal pathway
ZI   zona incerta

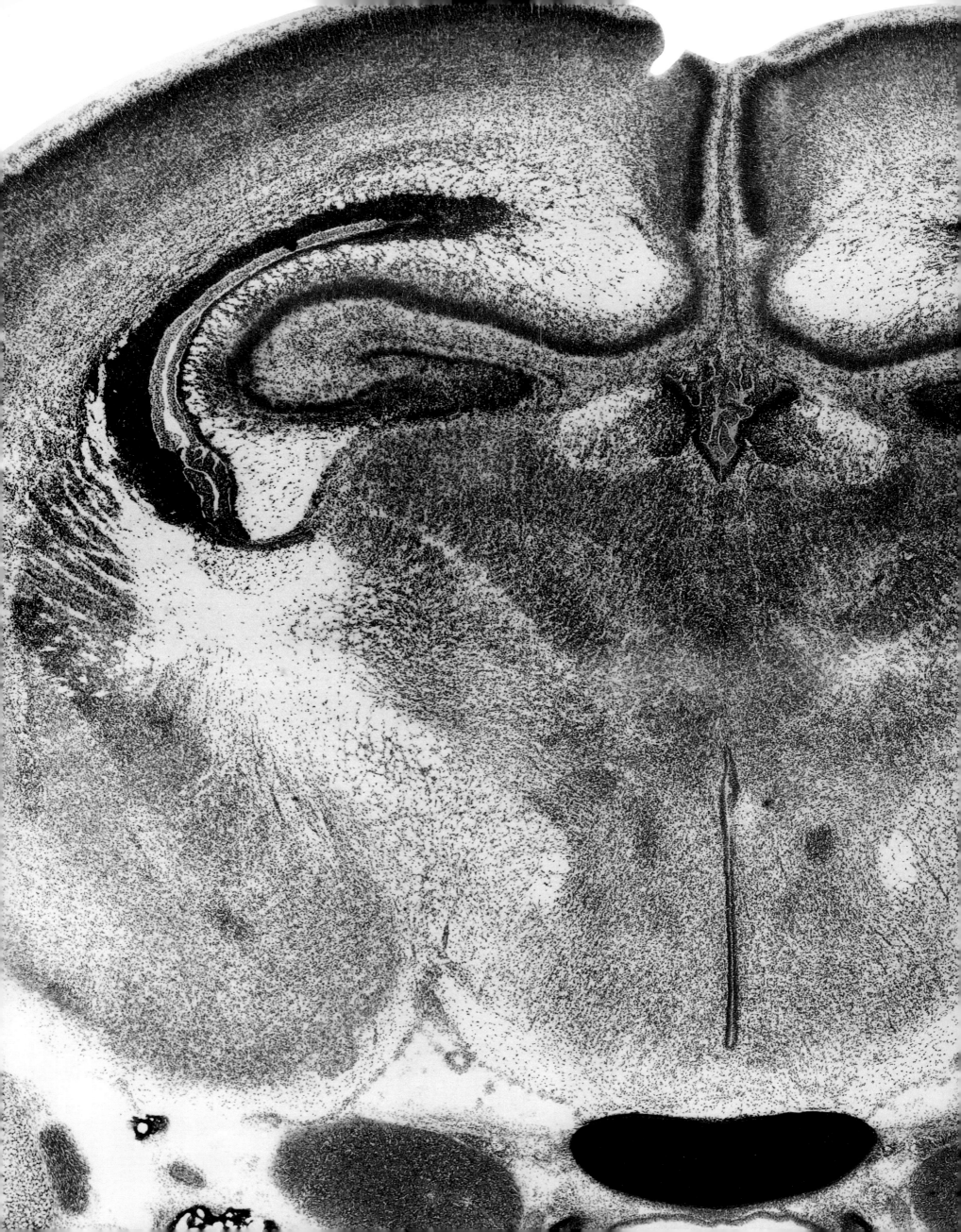

Figure 169
P0-25

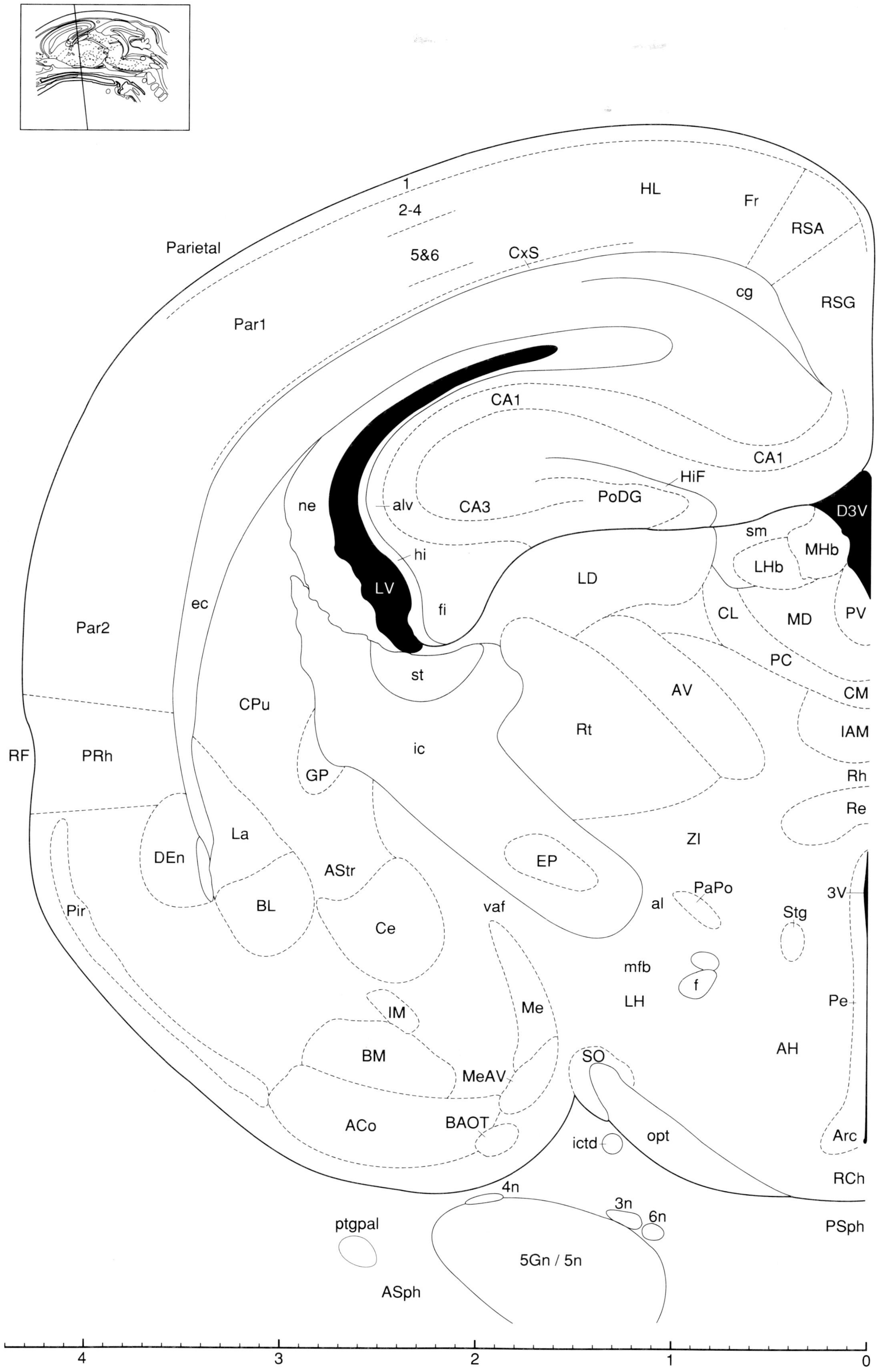

1 cortical layer 1
2-4 cortical layers
3n oculomotor nerve or its root
3V 3rd ventricle
4n trochlear nerve or its root
5&6 cortical layers
5Gn trigeminal ganlion
5n trigeminal nerve
6n abducens nerve or its root
ACo anterior cortical amygdaloid nu
AH anterior hypoth nu
al ansa lenticularis
alv alveus of hippocampus
Arc arcuate hypoth nu
ASph alisphenoid bone
AStr amygdalostriatal transition area
AV anteroventral thal nu
BAOT bed nu accessory olfactory tr
BL basolateral amygdaloid nu
BM basomedial amygdaloid nu
CA1 CA1 field of the hippocampus
CA3 CA3 field of the hippocampus
Ce central amygdaloid nu
cg cingulum
CL centrolateral thal nu
CM central medial thal nu
CPu caudate putamen
CxS cortical subplate
D3V dorsal third ventricle
DEn dorsal endopiriform nu
ec external capsule
EP entopeduncular nu
f fornix
fi fimbria hippocampus
Fr frontal cortex
GP globus pallidus
hi hippocampal formation neuroepi
HiF hippocampal fissure
HL hindlimb area of cortex
IAM interanteromedial thal nu
ic inferior colliculus neuroepithelium
ictd internal carotid artery
IM intercalated amygdaloid nu, main
La lateral amygdaloid nu
LD laterodorsal thal nu
LH lateral hypothalamic area
LHb lateral habenular nu
LV lateral ventricle
MD mediodorsal thal nu
Me medial amygdaloid nu
MeAV medial amyg nu, anterventral
mfb medial forebrain bundle
MHb medial habenular nu
ne neuroepithelium
opt optic tract
PaPo paraventricular hypoth nu, post
Par1 parietal cortex, area 1
Par2 parietal cortex, area 2
Parietal parietal bone
PC paracentral thal nu
Pe periventricular hypoth nu
Pir piriform cortex
PoDG polymorph layer dentate gyrus
PRh perirhinal cortex
PSph presphenoid bone
ptgpal pterygopalatine artery
PV paraventricular thal nu
RCh retrochiasmatic area
Re reuniens thal nu
RF rhinal fissure
Rh rhomboid thal nu
RSA retrosplenial agranular Cx
RSG retrosplenial granular Cx
Rt reticular thal nu
sm stria medullaris thalami
SO supraoptic nu
st stria terminalis
Stg stigmoid hypoth nu
vaf ventral amygdalofugal pathway
ZI zona incerta

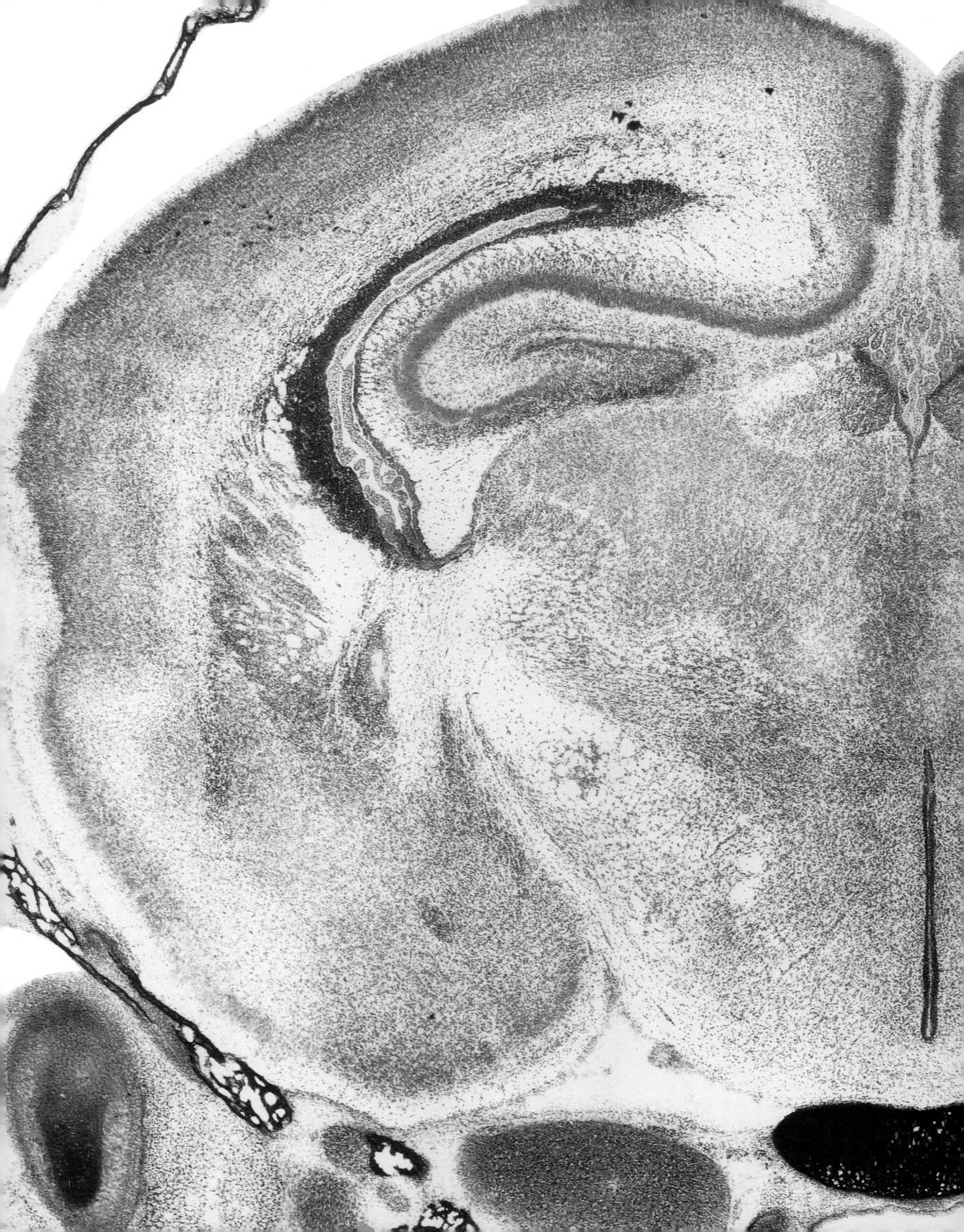

Figure 170
P0-26

3n   oculomotor nerve or its root
3V   3rd ventricle
4n   trochlear nerve or its root
5Gn   trigeminal ganlion
5mand mandibular nerve
5n   trigeminal nerve
6n   abducens nerve or its root
ACo   anterior cortical amygdaloid nu
AH   anterior hypoth nu
AL   nu ansa lenticularis
alv   alveus of hippocampus
Arc   arcuate hypoth nu
ASph alisphenoid bone
AStr amygdalostriatal transition area
BAOT bed nu accessory olfactory tr
BL   basolateral amygdaloid nu
BM   basomedial amygdaloid nu
BSTIA bed nu st, intraamyg div
CA1   CA1 field of the hippocampus
CA3   CA3 field of the hippocampus
Ce   central amygdaloid nu
cg   cingulum
CL   centrolateral thal nu
CM   central medial thal nu
Condyle condyloid process of mandib
CPu   caudate putamen
D3V   dorsal third ventricle
DEn   dorsal endopiriform nu
DLG   dorsal lateral geniculate nu
ec   external capsule
EP   entopeduncular nu
f   fornix
fi   fimbria hippocampus
Fr   frontal cortex
hi   hippocampal formation neuroepi
HiF   hippocampal fissure
HL   hindlimb area of cortex
IAM   interanteromedial thal nu
ic   inferior colliculus neuroepithelium
ictd internal carotid artery
IM   intercalated amygdaloid nu, main
La   lateral amygdaloid nu
LD   laterodorsal thal nu
LH   lateral hypothalamic area
LHb   lateral habenular nu
LV   lateral ventricle
MD   mediodorsal thal nu
Me   medial amygdaloid nu
MeAV medial amyg nu, anterventral
mfb   medial forebrain bundle
MHb   medial habenular nu
mt   mammillothalamic tract
ne   neuroepithelium
opt   optic tract
Par1 parietal cortex, area 1
Par2 parietal cortex, area 2
Parietal parietal bone
PC   paracentral thal nu
Pe   periventricular hypoth nu
Pir   piriform cortex
PLCo posterolateral cortical amyg nu
PoDG polymorph layer dentate gyrus
PRh   perirhinal cortex
PSph presphenoid bone
ptgpal pterygopalatine artery
PV   paraventricular thal nu
RCh   retrochiasmatic area
Re   reuniens thal nu
RF   rhinal fissure
Rh   rhomboid thal nu
RSA   retrosplenial agranular Cx
RSG   retrosplenial granular Cx
Rt   reticular thal nu
sm   stria medullaris thalami
sox   supraoptic decussation
Squamous squamous part of temp
st   stria terminalis
VL   ventrolateral thal nu
VM   ventromedial thal
ZI   zona incerta

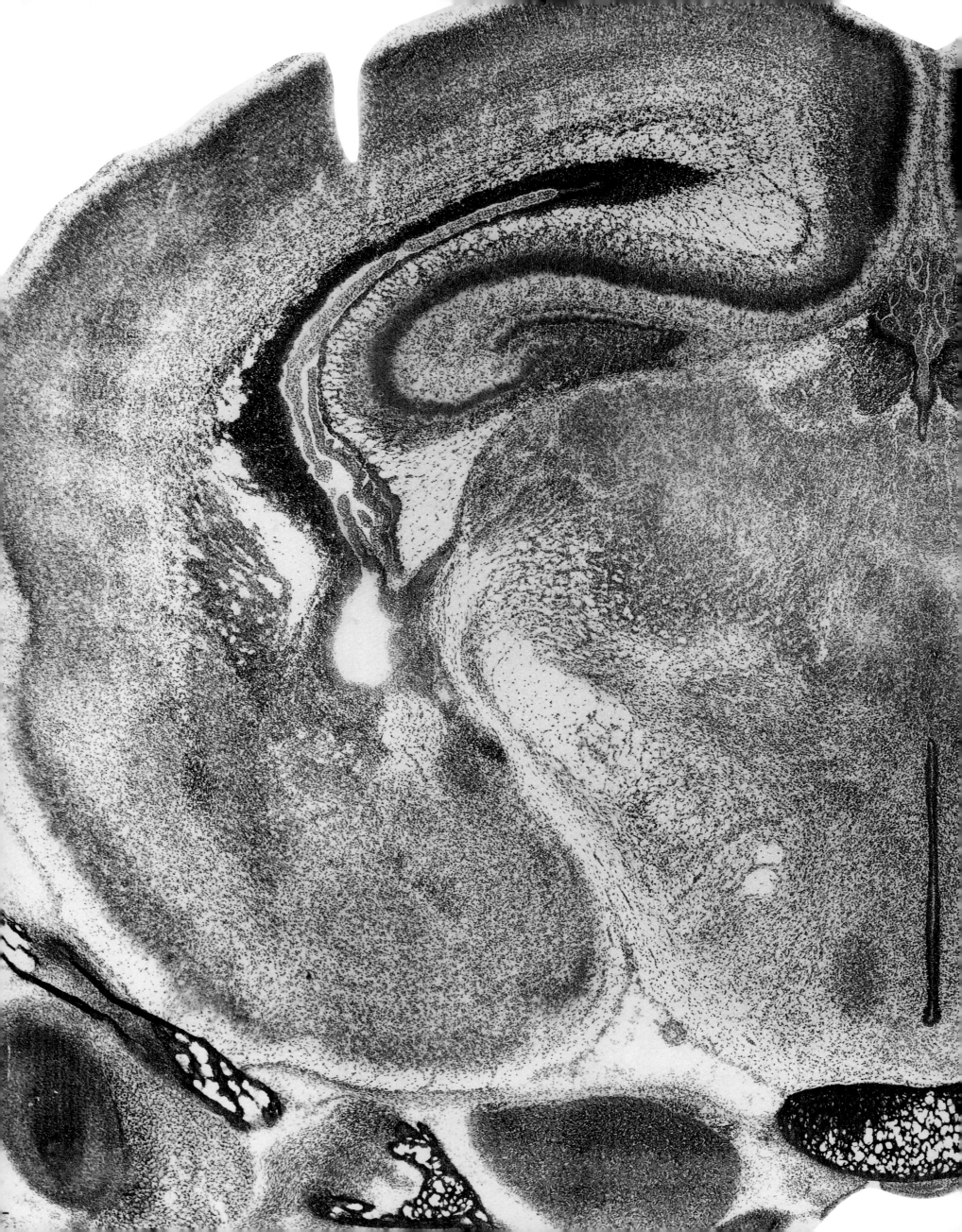

Figure 171
P0-27

HL
Fr
Parietal
Par1
RSA
RSG
Or   Py
Rad
CA1
LMol
HiF
D3V
GrDG
Mol
PoDG
sm
MHb
CA3
alv
LHb
Par2
LD
LP
PV
ne
hi
DLG
CL
MD
ic
LV
fi
PC
VLG
CM
VL
PRh
RF
VPL
ec
CPu
Rt
G
ne
mt
Rh
ic
VM
La
Re
AStr
st
DEn
Ce
MePD
ZI
BL
EP
BSTIA
3V
IM
mfb
Pir
BLV
LH
f
AH
Pe
BM
Me
VMH
MeAV
Squamous
PCo
ACo
ictd   sox
Arc
Condyle
ASph
4n
3n
Oval
5mand
ASph
6n
PSph
5Gn / 5n
Oval
ptgpal

4   3   2   1   0

3n   oculomotor nerve or its root
3V   3rd ventricle
4n   trochlear nerve or its root
5Gn   trigeminal ganlion
5mand   mandibular nerve
5n   trigeminal nerve
6n   abducens nerve or its root
ACo   anterior cortical amygdaloid nu
AH   anterior hypoth nu
alv   alveus of hippocampus
Arc   arcuate hypoth nu
ASph   alisphenoid bone
AStr   amygdalostriatal transition area
BL   basolateral amygdaloid nu
BLV   basolateral amygdaloid nu, vent
BM   basomedial amygdaloid nu
BSTIA   bed nu st, intraamyg div
CA1   CA1 field of the hippocampus
CA3   CA3 field of the hippocampus
Ce   central amygdaloid nu
CL   centrolateral thal nu
CM   central medial thal nu
Condyle   condyloid process of mandib
CPu   caudate putamen
D3V   dorsal third ventricle
DEn   dorsal endopiriform nu
DLG   dorsal lateral geniculate nu
ec   external capsule
EP   entopeduncular nu
f   fornix
fi   fimbria hippocampus
Fr   frontal cortex
G   gelatinosus thal nu
GrDG   granular layer dentate gyrus
hi   hippocampal formation neuroepi
HiF   hippocampal fissure
HL   hindlimb area of cortex
ic   inferior colliculus neuroepithelium
ictd   internal carotid artery
IM   intercalated amygdaloid nu, main
La   lateral amygdaloid nu
LD   laterodorsal thal nu
LH   lateral hypothalamic area
LHb   lateral habenular nu
LMol   lacunosum moleculare layer hip
LP   lateral posterior thal nu
LV   lateral ventricle
MD   mediodorsal thal nu
Me   medial amygdaloid nu
MeAV   medial amyg nu, anterventral
MePD   medial amyg nu, posterodorsal
mfb   medial forebrain bundle
MHb   medial habenular nu
Mol   molecular layer dentate gyrus
mt   mammillothalamic tract
ne   neuroepithelium
opt   optic tract
Or   oriens layer hippocampus
Oval   oval foramen
Par1   parietal cortex, area 1
Par2   parietal cortex, area 2
Parietal   parietal bone
PC   paracentral thal nu
PCo   posterior cortical amygdaloid nu
Pe   periventricular hypoth nu
Pir   piriform cortex
PoDG   polymorph layer dentate gyrus
PRh   perirhinal cortex
PSph   presphenoid bone
ptgpal   pterygopalatine artery
PV   paraventricular thal nu
Py   pyramidal cell layer hippocampus
Rad   stratum radiatum hippocampus
Re   reuniens thal nu
RF   rhinal fissure
Rh   rhomboid thal nu
RSA   retrosplenial agranular Cx
RSG   retrosplenial granular Cx
Rt   reticular thal nu
sm   stria medullaris thalami
sox   supraoptic decussation
Squamous   squamous part of temp
st   stria terminalis
VL   ventrolateral thal nu
VLG   ventral lateral geniculate nu
VM   ventromedial thal
VMH   ventromedial hypoth nu
VPL   ventral posterolat thal nu
ZI   zona incerta

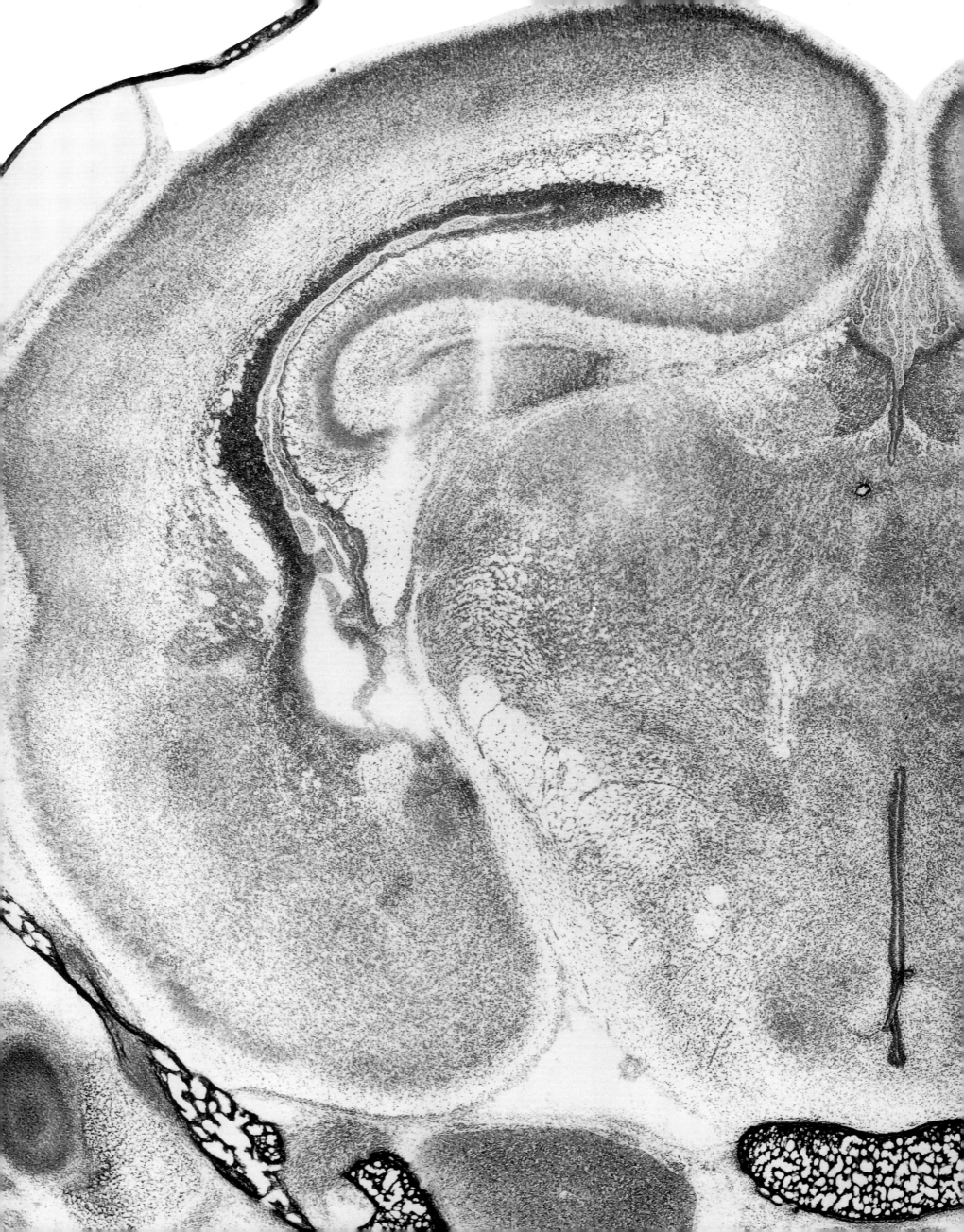

Figure 172
P0-28

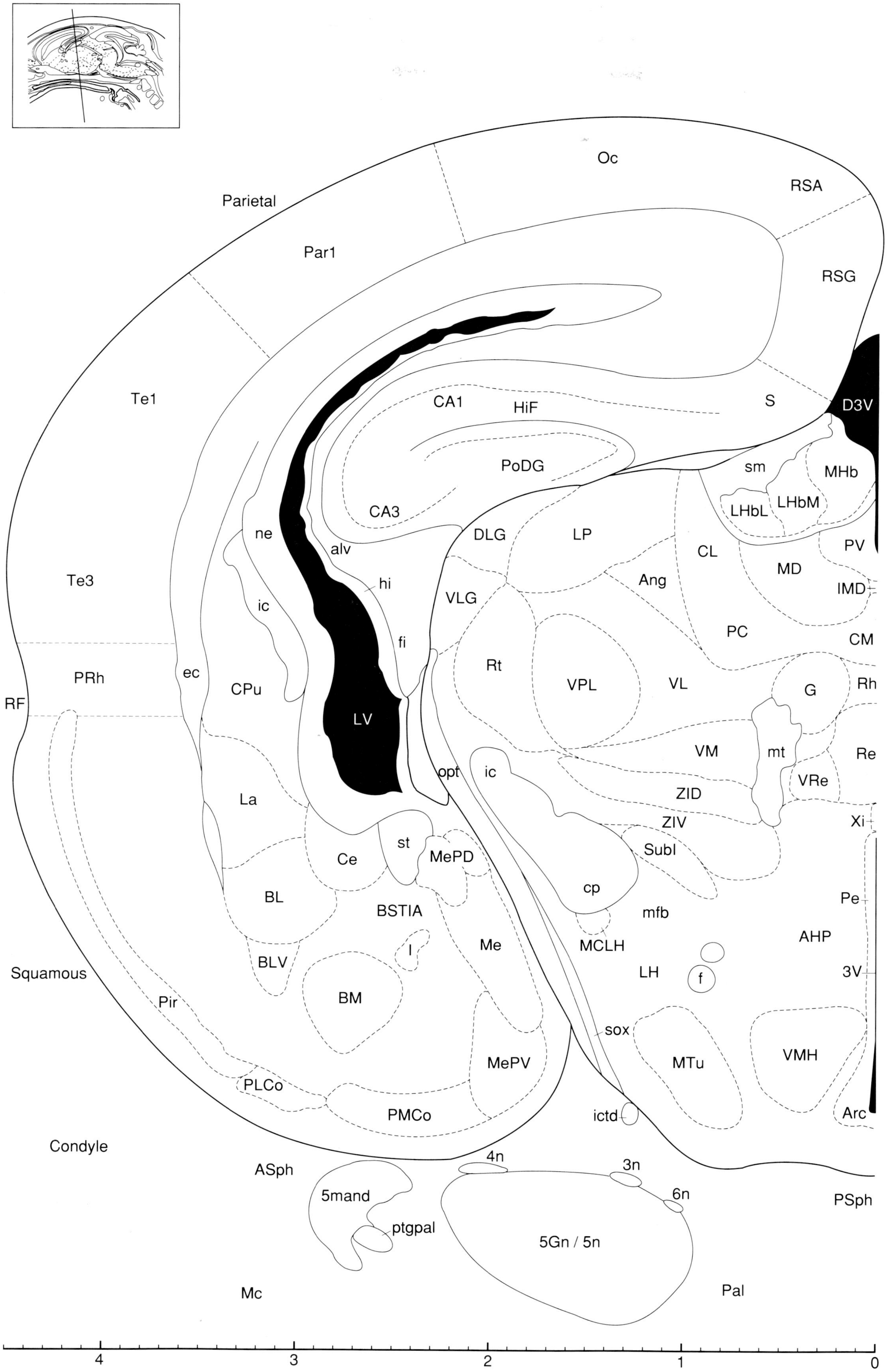

3n   oculomotor nerve or its root
3V   3rd ventricle
4n   trochlear nerve or its root
5Gn   trigeminal ganlion
5mand   mandibular nerve
5n   trigeminal nerve
6n   abducens nerve or its root
AHP   anterior hypoth area, posterior
alv   alveus of hippocampus
Ang   angular thal nu
Arc   arcuate hypoth nu
ASph   alisphenoid bone
BL   basolateral amygdaloid nu
BLV   basolateral amygdaloid nu, vent
BM   basomedial amygdaloid nu
BSTIA   bed nu st, intraamyg div
CA1   CA1 field of the hippocampus
CA3   CA3 field of the hippocampus
Ce   central amygdaloid nu
CL   centrolateral thal nu
CM   central medial thal nu
Condyle   condyloid process of mandib
cp   cerebral peduncle, basal
CPu   caudate putamen
D3V   dorsal third ventricle
DLG   dorsal lateral geniculate nu
ec   external capsule
f   fornix
fi   fimbria hippocampus
G   gelatinosus thal nu
hi   hippocampal formation neuroepi
HiF   hippocampal fissure
I   intercalated nuclei amygdala
ic   inferior colliculus neuroepithelium
ictd   internal carotid artery
IMD   intermediodorsal thal nu
La   lateral amygdaloid nu
LH   lateral hypothalamic area
LHbL   lateral habenular nu, lateral
LHbM   lateral habenular nu, medial
LP   lateral posterior thal nu
LV   lateral ventricle
Mc   Meckel's cartilage
MCLH   magnocellular nu lat hypoth
MD   mediodorsal thal nu
Me   medial amygdaloid nu
MePD   medial amyg nu, posterodorsal
MePV   medial amyg nu, posteroventr
mfb   medial forebrain bundle
MHb   medial habenular nu
mt   mammillothalamic tract
MTu   medial tuberal nu
ne   neuroepithelium
Oc   occipital cortex
opt   optic tract
Pal   palatine bone
Par1   parietal cortex, area 1
Parietal   parietal bone
PC   paracentral thal nu
Pe   periventricular hypoth nu
Pir   piriform cortex
PLCo   posterolateral cortical amyg nu
PMCo   posteromed cortical amyg nu
PoDG   polymorph layer dentate gyrus
PRh   perirhinal cortex
PSph   presphenoid bone
ptgpal   pterygopalatine artery
PV   paraventricular thal nu
Re   reuniens thal nu
RF   rhinal fissure
Rh   rhomboid thal nu
RSA   retrosplenial agranular Cx
RSG   retrosplenial granular Cx
Rt   reticular thal nu
S   subiculum
sm   stria medullaris thalami
sox   supraoptic decussation
Squamous   squamous part of temp
st   stria terminalis
SubI   subincertal nu
Te1   temporal cortex, area 1
Te3   temporal cortex, area 3
VL   ventrolateral thal nu
VLG   ventral lateral geniculate nu
VM   ventromedial thal
VMH   ventromedial hypoth nu
VPL   ventral posterolat thal nu
VRe   ventral reuniens thal nu
Xi   xiphoid thal nu
ZID   zona incerta, dorsal
ZIV   zona incerta, ventral

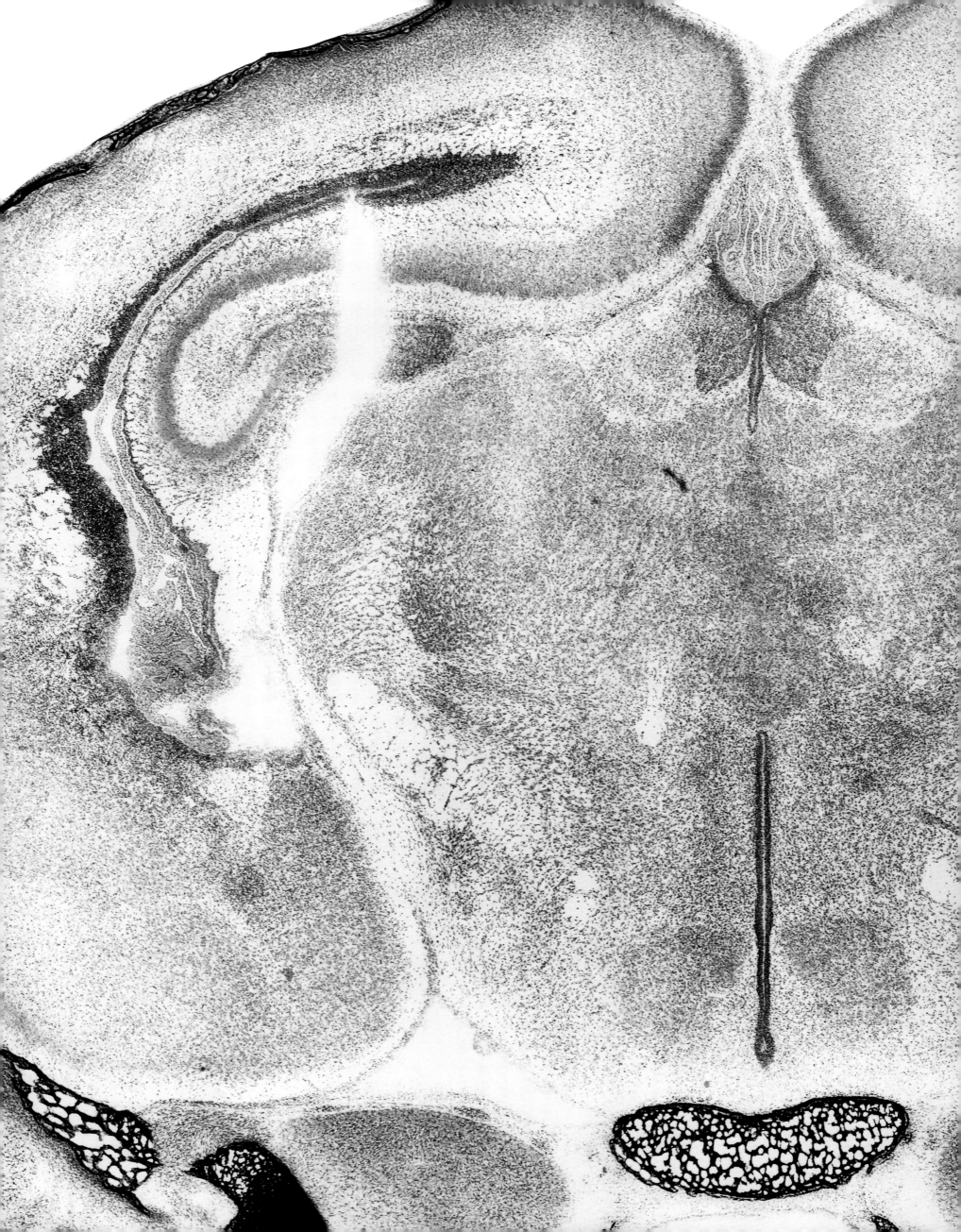

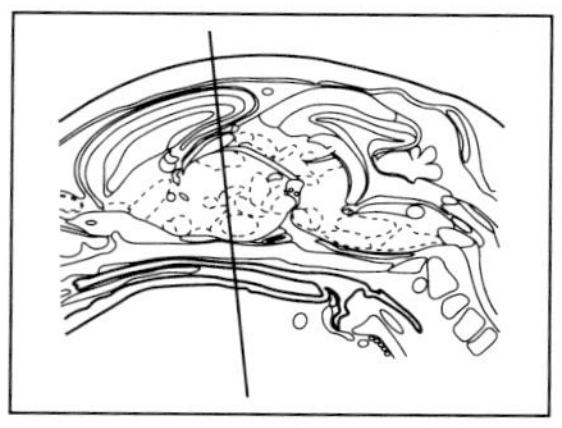

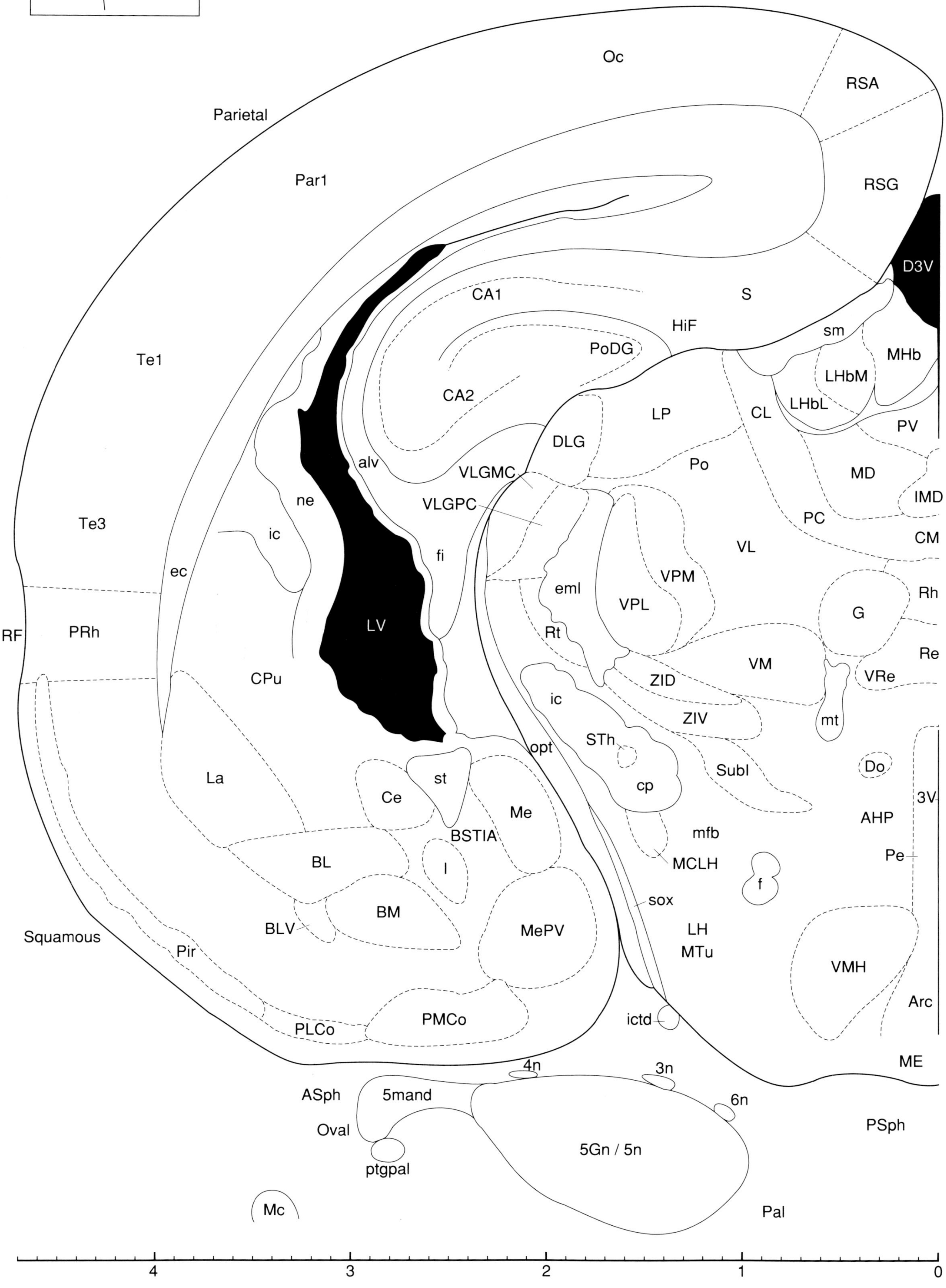

Figure 173
P0-29

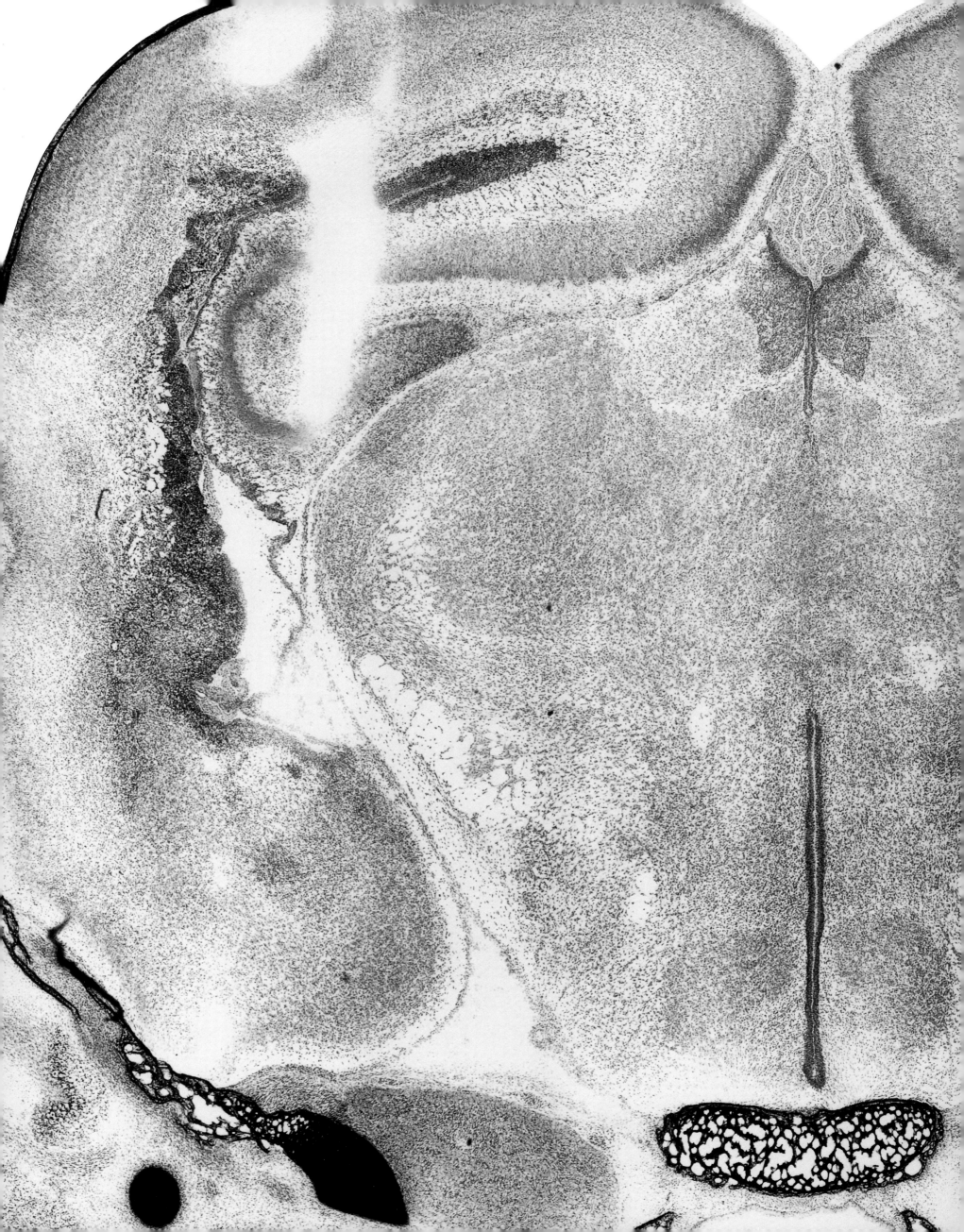

Figure 174
P0-30

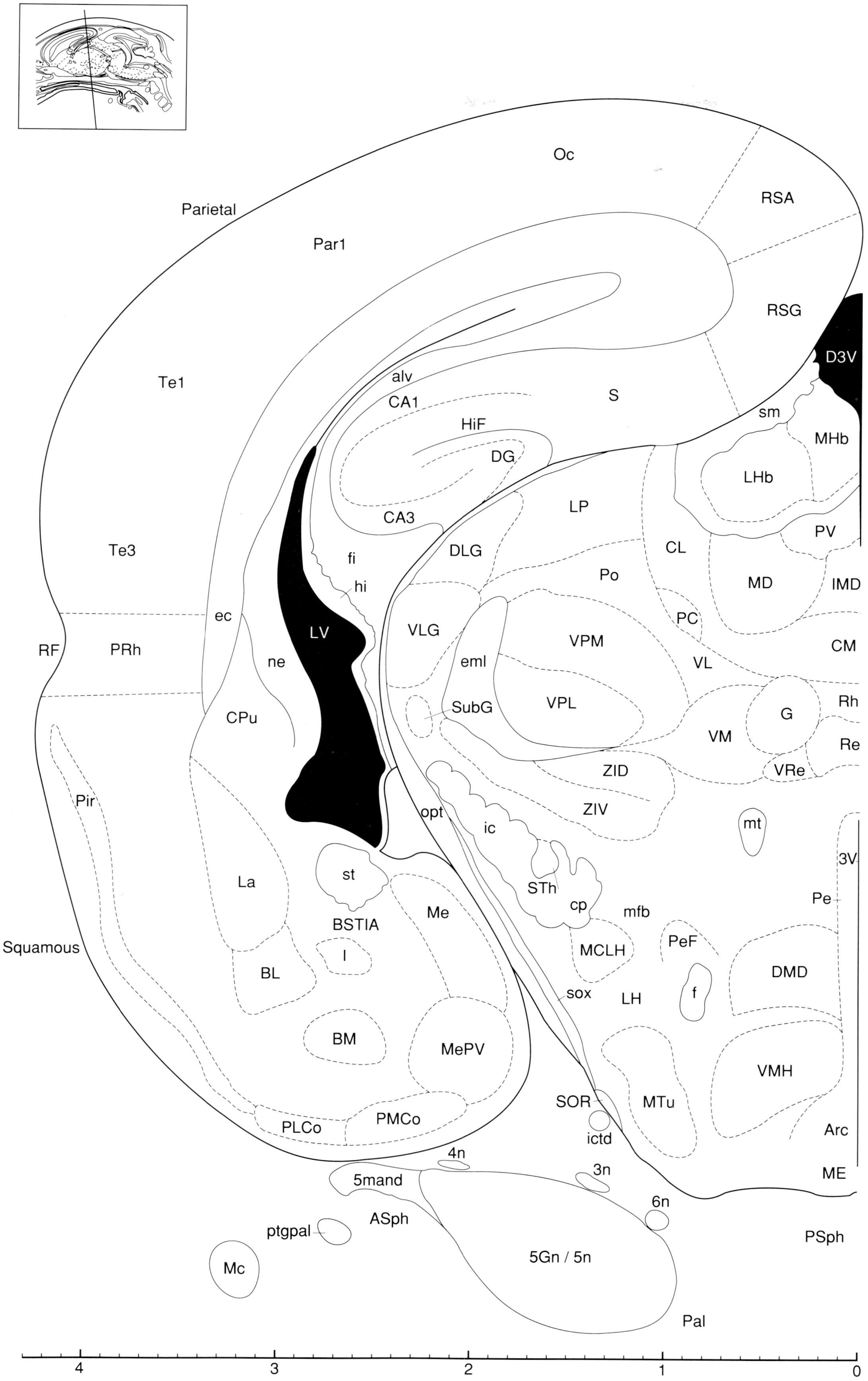

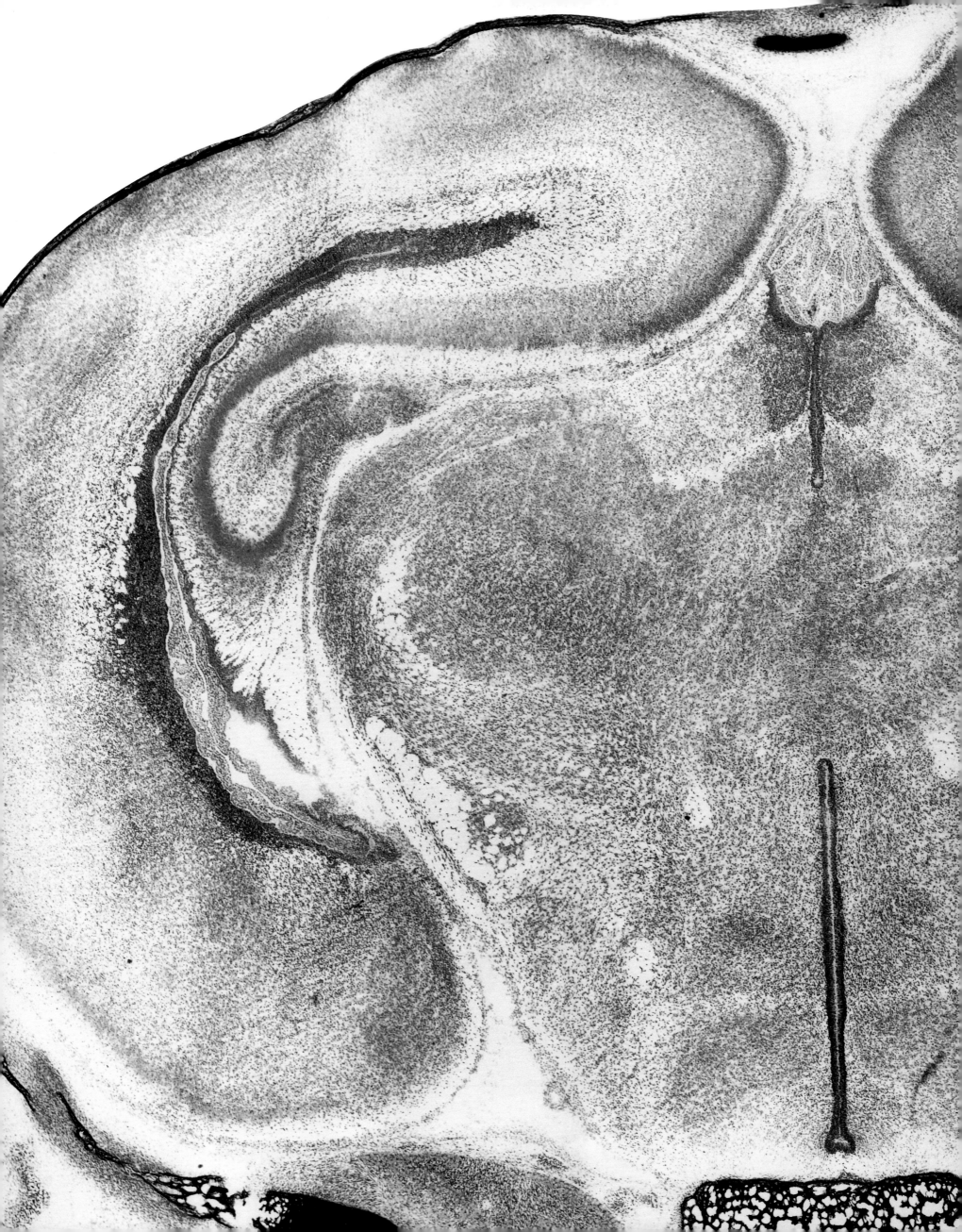

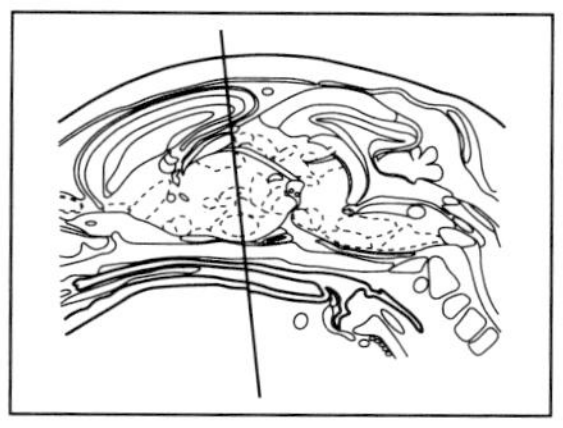

Figure 175
P0-31

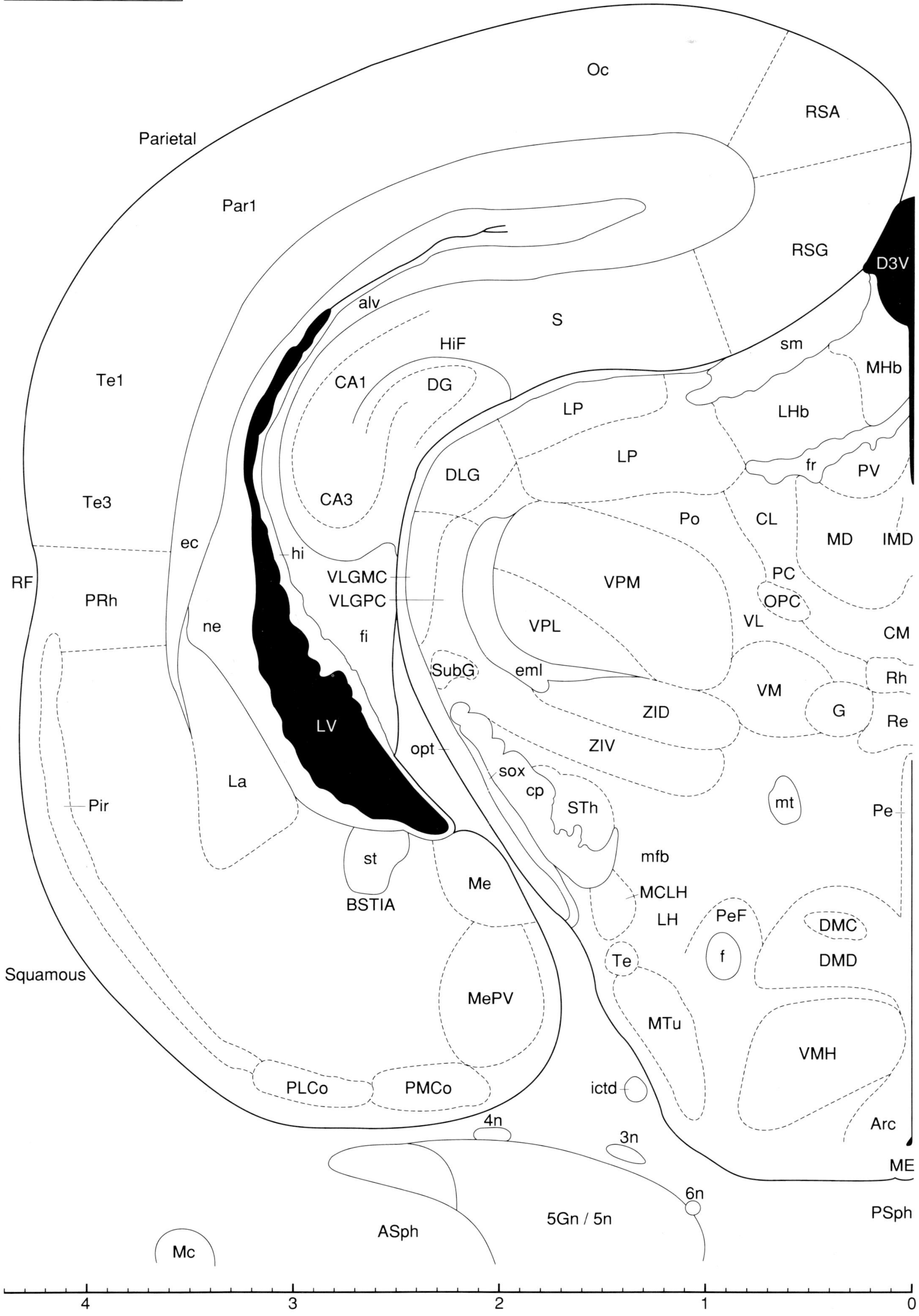

3n   oculomotor nerve or its root
4n   trochlear nerve or its root
5Gn   trigeminal ganlion
5n   trigeminal nerve
6n   abducens nerve or its root
alv   alveus of hippocampus
Arc   arcuate hypoth nu
ASph   alisphenoid bone
BSTIA   bed nu st, intraamyg div
CA1   CA1 field of the hippocampus
CA3   CA3 field of the hippocampus
CL   centrolateral thal nu
CM   central medial thal nu
cp   cerebral peduncle, basal
D3V   dorsal third ventricle
DG   dentate gyrus
DLG   dorsal lateral geniculate nu
DMC   dorsomed hypoth nu, compact
DMD   dorsomed hypoth nu, diffuse
ec   external capsule
eml   external medullary lamina
f   fornix
fi   fimbria hippocampus
fr   fasciculus retroflexus
G   gelatinosus thal nu
hi   hippocampal formation neuroepi
HiF   hippocampal fissure
ictd   internal carotid artery
IMD   intermediodorsal thal nu
La   lateral amygdaloid nu
LH   lateral hypothalamic area
LHb   lateral habenular nu
LP   lateral posterior thal nu
LV   lateral ventricle
Mc   Meckel's cartilage
MCLH   magnocellular nu lat hypoth
MD   mediodorsal thal nu
ME   median eminence
Me   medial amygdaloid nu
MePV   medial amyg nu, posteroventr
mfb   medial forebrain bundle
MHb   medial habenular nu
mt   mammillothalamic tract
MTu   medial tuberal nu
ne   neuroepithelium
Oc   occipital cortex
OPC   oval paracentral thal nu
opt   optic tract
Par1   parietal cortex, area 1
Parietal parietal bone
PC   paracentral thal nu
Pe   periventricular hypoth nu
PeF   perifornical nu
Pir   piriform cortex
PLCo   posterolateral cortical amyg nu
PMCo   posteromed cortical amyg nu
Po   posterior thal nuclear group
PRh   perirhinal cortex
PSph   presphenoid bone
PV   paraventricular thal nu
Re   reuniens thal nu
RF   rhinal fissure
Rh   rhomboid thal nu
RSA   retrosplenial agranular Cx
RSG   retrosplenial granular Cx
S   subiculum
sm   stria medullaris thalami
sox   supraoptic decussation
Squamous squamous part of temp
st   stria terminalis
STh   subthalamic nu
SubG   subgeniculate nu
Te   terete hypoth nu
Te1   temporal cortex, area 1
Te3   temporal cortex, area 3
VL   ventrolateral thal nu
VLGMC   vent lat genicul nu, magnoce
VLGPC   vent lat genicul nu, parvocell
VM   ventromedial thal
VMH   ventromedial hypoth nu
VPL   ventral posterolat thal nu
VPM   ventral posteromedial thal nu
ZID   zona incerta, dorsal
ZIV   zona incerta, ventral

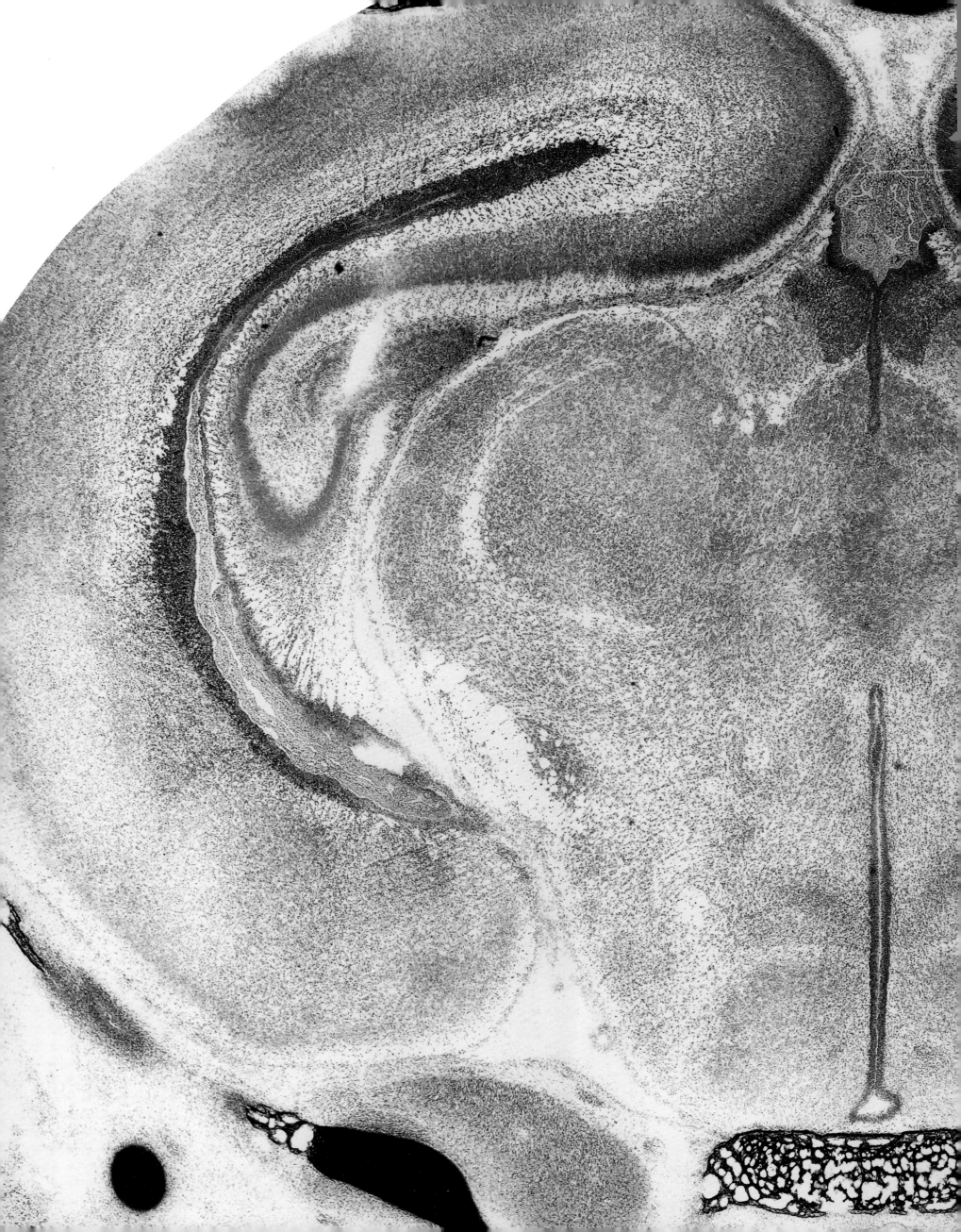

Figure 176
P0-32

Oc

RSA

Parietal

Par1

RSG

Te1

D3V

alv

S

HiF

MHb

sm

PoDG

LP

LHb

CA1

LP

CL

DLG

fr

PV

Te3

IGL

Po

PF

CA3

MD

IMD

RF

PRh

ne

PC

VPM

VL

OPC

CM

eml

VPL

VLGMC

ml

VM

SubG

VLGPC

ZID

G

Rh

hi

fi

ZIV

Re

La

LV

sox

opt

cp

STh

mt

Pe

mfb

MCLH

Pir

LH

Squamous

Te

f

DMC

DMD

3V

MTu

MePV

VMH

ictd

PLCo

PMCo

4n

3n

Arc

ptgpal

5mand

ME

6n

Mc

ASph

5Gn / 5n

PSph

Otic

Pal

4   3   2   1   0

3n   oculomotor nerve or its root
3V   3rd ventricle
4n   trochlear nerve or its root
5Gn   trigeminal ganlion
5mand mandibular nerve
5n   trigeminal nerve
6n   abducens nerve or its root
alv   alveus of hippocampus
Arc   arcuate hypoth nu
ASph alisphenoid bone
CA1   CA1 field of the hippocampus
CA3   CA3 field of the hippocampus
CL   centrolateral thal nu
CM   central medial thal nu
cp   cerebral peduncle, basal
D3V   dorsal third ventricle
DLG   dorsal lateral geniculate nu
DMC   dorsomed hypoth nu, compact
DMD   dorsomed hypoth nu, diffuse
eml   external medullary lamina
f   fornix
fi   fimbria hippocampus
fr   fasciculus retroflexus
G   gelatinosus thal nu
hi   hippocampal formation neuroepi
HiF   hippocampal fissure
ictd internal carotid artery
IGL   intergeniculate leaf
IMD   intermediodorsal thal nu
La   lateral amygdaloid nu
LH   lateral hypothalamic area
LHb   lateral habenular nu
LP   lateral posterior thal nu
LV   lateral ventricle
Mc   Meckel's cartilage
MCLH magnocellular nu lat hypoth
MD   mediodorsal thal nu
ME   median eminence
MePV medial amyg nu, posteroventr
mfb   medial forebrain bundle
MHb   medial habenular nu
ml   medial lemniscus
mt   mammillothalamic tract
MTu   medial tuberal nu
ne   neuroepithelium
Oc   occipital cortex
OPC   oval paracentral thal nu
opt   optic tract
Otic otic ganglion
Pal   palatine bone
Par1 parietal cortex, area 1
Parietal parietal bone
PC   paracentral thal nu
Pe   periventricular hypoth nu
PF   parafascicular thal nu
Pir   piriform cortex
PLCo posterolateral cortical amyg nu
PMCo posteromed cortical amyg nu
Po   posterior thal nuclear group
PoDG polymorph layer dentate gyrus
PRh   perirhinal cortex
PSph presphenoid bone
ptgpal pterygopalatine artery
PV   paraventricular thal nu
Re   reuniens thal nu
RF   rhinal fissure
Rh   rhomboid thal nu
RSA   retrosplenial agranular Cx
RSG   retrosplenial granular Cx
S   subiculum
sm   stria medullaris thalami
sox   supraoptic decussation
Squamous squamous part of temp
STh   subthalamic nu
SubG subgeniculate nu
Te   terete hypoth nu
Te1   temporal cortex, area 1
Te3   temporal cortex, area 3
VL   ventrolateral thal nu
VLGMC vent lat genicul nu, magnoce
VLGPC vent lat genicul nu, parvocell
VM   ventromedial thal
VMH   ventromedial hypoth nu
VPL   ventral posterolat thal nu
VPM   ventral posteromedial thal nu
ZID   zona incerta, dorsal
ZIV   zona incerta, ventral

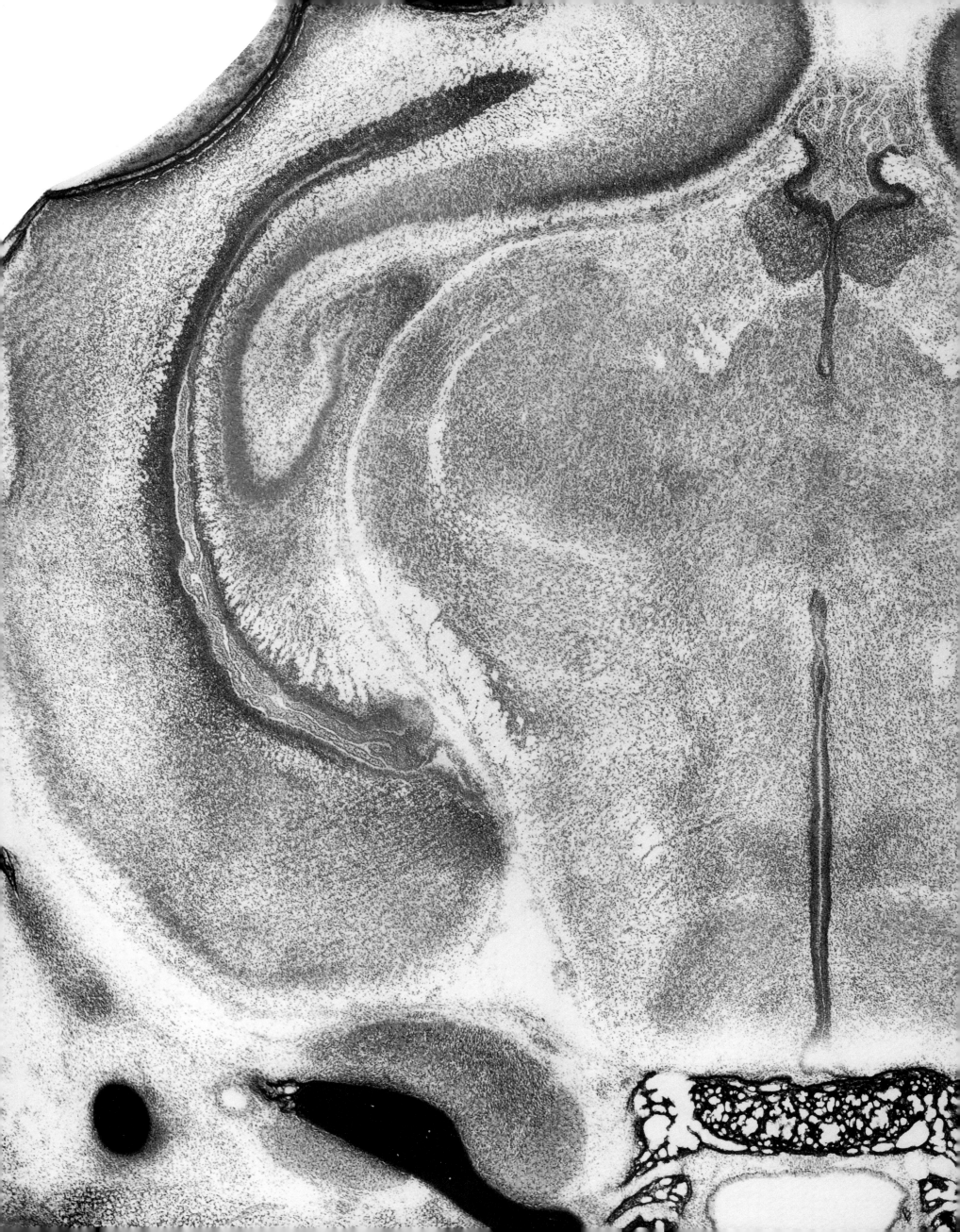

Figure 177
P0-33

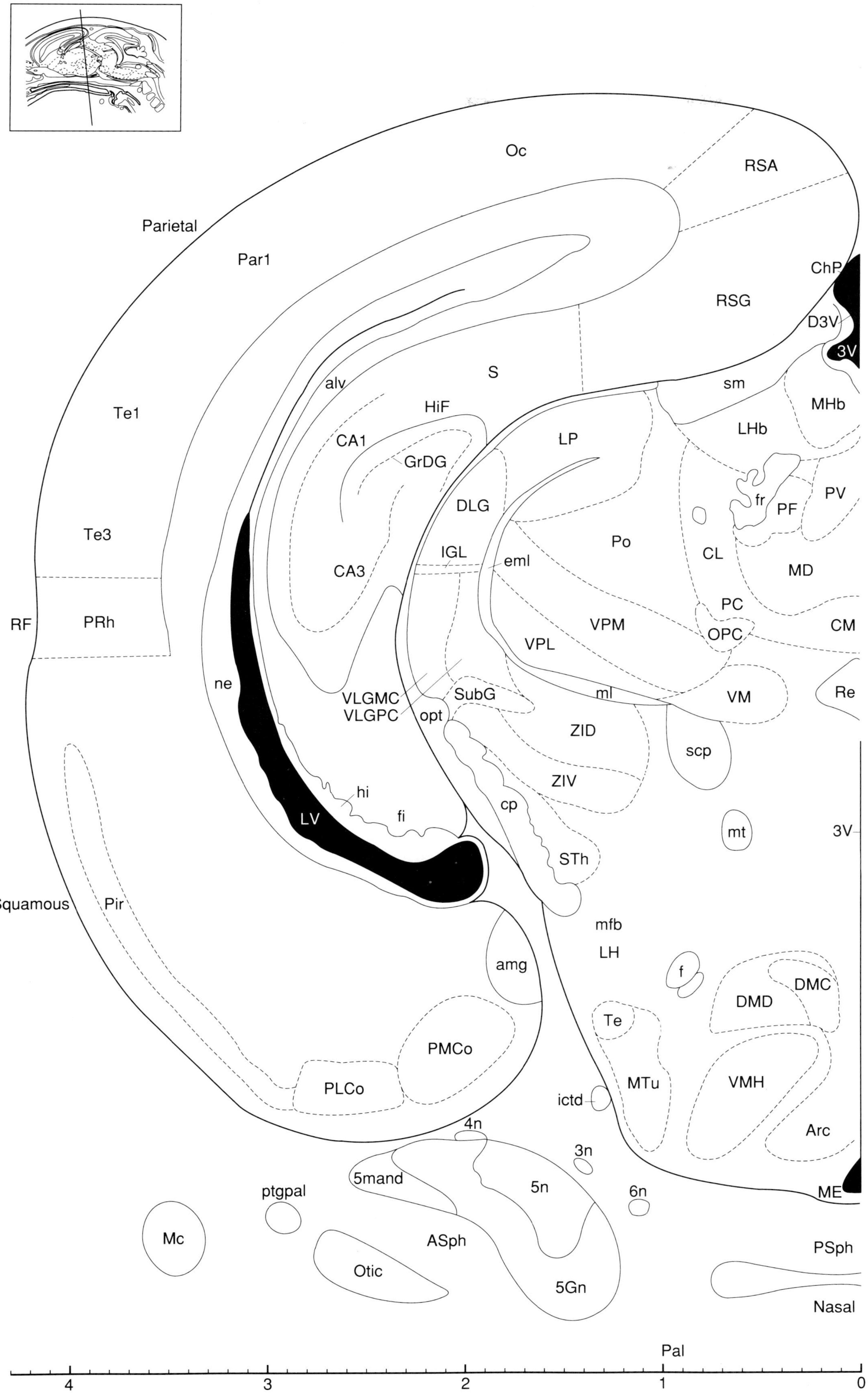

3n   oculomotor nerve or its root
3V   3rd ventricle
4n   trochlear nerve or its root
5Gn   trigeminal ganglion
5mand   mandibular nerve
5n   trigeminal nerve
6n   abducens nerve or its root
alv   alveus of hippocampus
amg   amygdaloid neuroepithelium
Arc   arcuate hypoth nu
ASph   alisphenoid bone
CA1   CA1 field of the hippocampus
CA3   CA3 field of the hippocampus
ChP   choroid plexus
CL   centrolateral thal nu
CM   central medial thal nu
cp   cerebral peduncle, basal
D3V   dorsal third ventricle
DLG   dorsal lateral geniculate nu
DMC   dorsomed hypoth nu, compact
DMD   dorsomed hypoth nu, diffuse
eml   external medullary lamina
f   fornix
fi   fimbria hippocampus
fr   fasciculus retroflexus
GrDG   granular layer dentate gyrus
hi   hippocampal formation neuroepi
HiF   hippocampal fissure
ictd   internal carotid artery
IGL   intergeniculate leaf
LH   lateral hypothalamic area
LHb   lateral habenular nu
LP   lateral posterior thal nu
LV   lateral ventricle
Mc   Meckel's cartilage
MD   mediodorsal thal nu
ME   median eminence
mfb   medial forebrain bundle
MHb   medial habenular nu
ml   medial lemniscus
mt   mammillothalamic tract
MTu   medial tuberal nu
Nasal   nasal cavity
ne   neuroepithelium
Oc   occipital cortex
OPC   oval paracentral thal nu
opt   optic tract
Otic   otic ganglion
Pal   palatine bone
Par1   parietal cortex, area 1
Parietal   parietal bone
PC   paracentral thal nu
PF   parafascicular thal nu
Pir   piriform cortex
PLCo   posterolateral cortical amyg nu
PMCo   posteromed cortical amyg nu
Po   posterior thal nuclear group
PRh   perirhinal cortex
PSph   presphenoid bone
ptgpal   pterygopalatine artery
PV   paraventricular thal nu
Re   reuniens thal nu
RF   rhinal fissure
RSA   retrosplenial agranular Cx
RSG   retrosplenial granular Cx
S   subiculum
scp   superior cerebellar peduncle
sm   stria medullaris thalami
Squamous   squamous part of temp
STh   subthalamic nu
SubG   subgeniculate nu
Te   terete hypoth nu
Te1   temporal cortex, area 1
Te3   temporal cortex, area 3
VLGMC   vent lat genicul nu, magnoce
VLGPC   vent lat genicul nu, parvocell
VM   ventromedial thal
VMH   ventromedial hypoth nu
VPL   ventral posterolat thal nu
VPM   ventral posteromedial thal nu
ZID   zona incerta, dorsal
ZIV   zona incerta, ventral

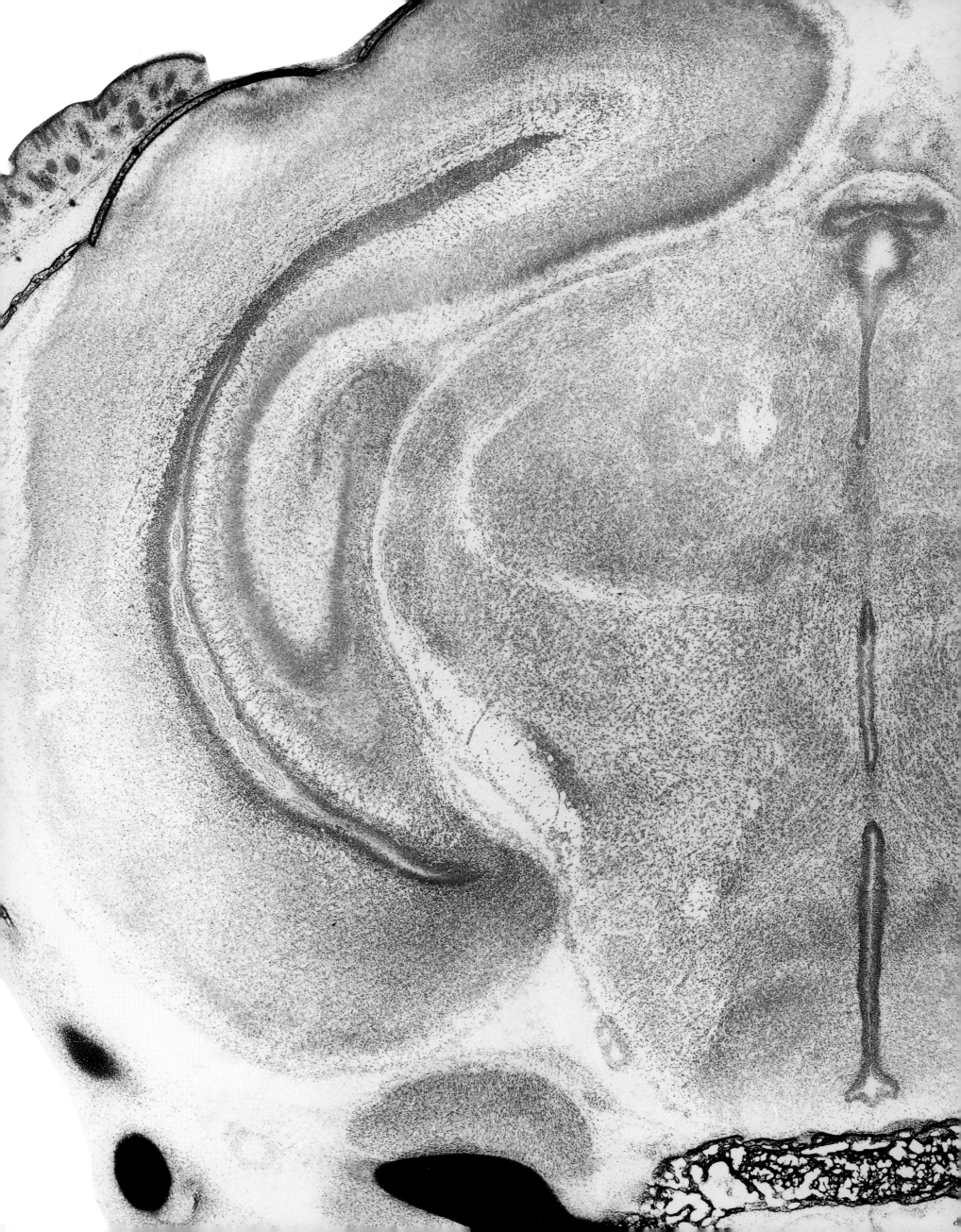

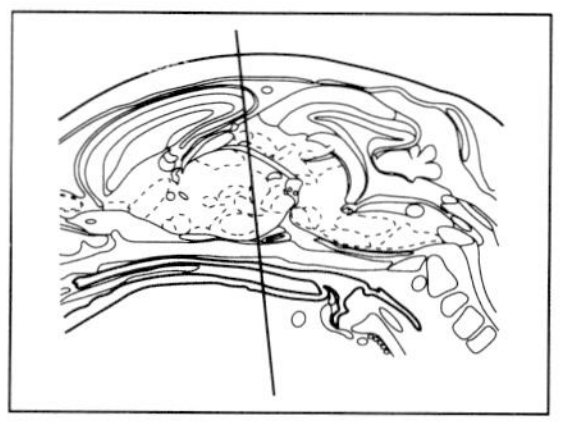

Figure 178
P0-34

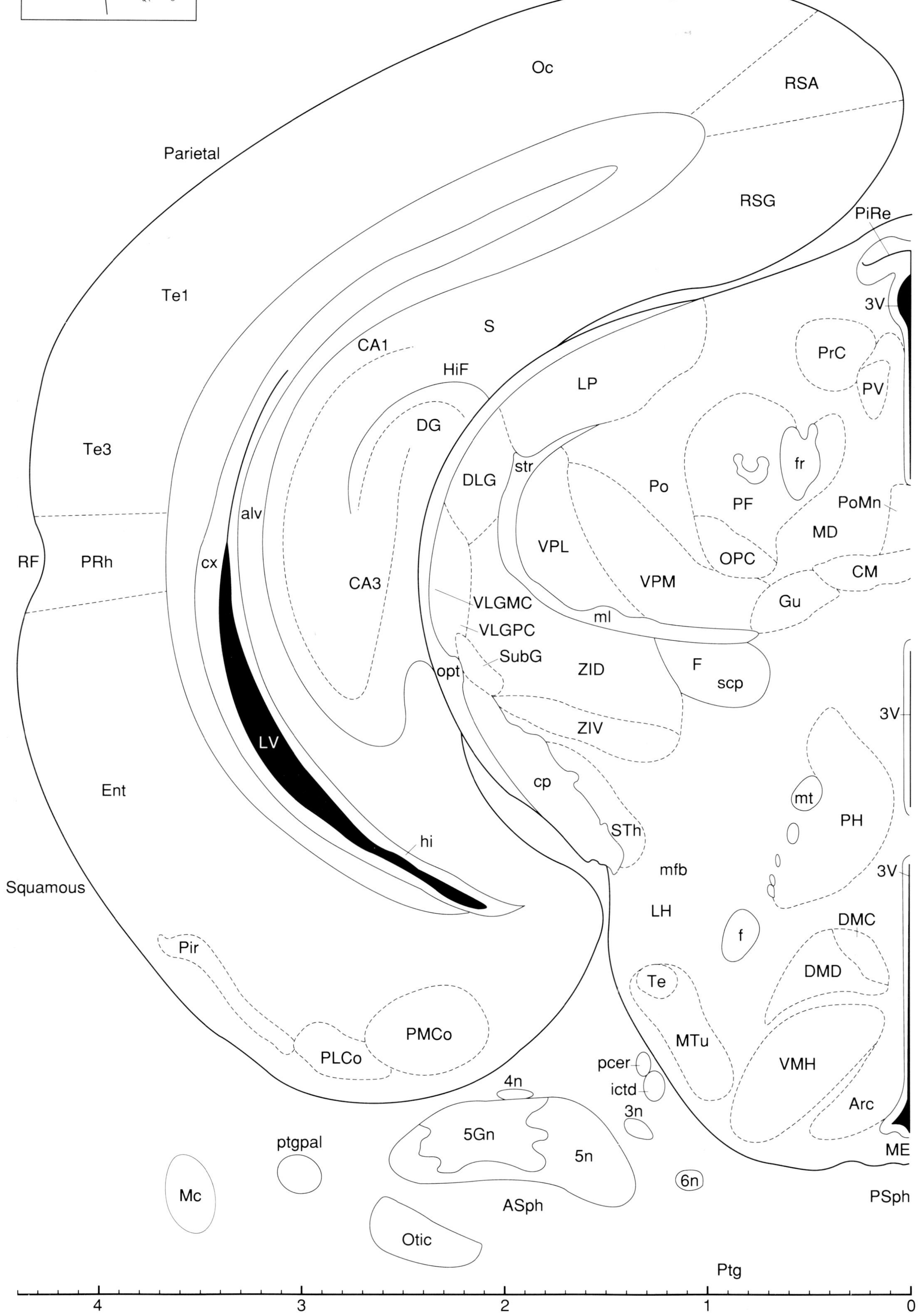

3n oculomotor nerve or its root
3V 3rd ventricle
4n trochlear nerve or its root
5Gn trigeminal ganglion
5n trigeminal nerve
6n abducens nerve or its root
alv alveus of hippocampus
Arc arcuate hypoth nu
ASph alisphenoid bone
CA1 CA1 field of the hippocampus
CA3 CA3 field of the hippocampus
CM central medial thal nu
cp cerebral peduncle, basal
cx cortical neuroepithelium
DG dentate gyrus
DLG dorsal lateral geniculate nu
DMC dorsomed hypoth nu, compact
DMD dorsomed hypoth nu, diffuse
Ent entorhinal cortex
F nu fields of Forel
f fornix
fr fasciculus retroflexus
Gu gustatory thalamic nu
hi hippocampal formation neuroepi
HiF hippocampal fissure
ictd internal carotid artery
LH lateral hypothalamic area
LP lateral posterior thal nu
LV lateral ventricle
Mc Meckel's cartilage
MD mediodorsal thal nu
ME median eminence
mfb medial forebrain bundle
ml medial lemniscus
mt mammillothalamic tract
MTu medial tuberal nu
Oc occipital cortex
OPC oval paracentral thal nu
opt optic tract
Otic otic ganglion
Parietal parietal bone
pcer posterior cerebral artery
PF parafascicular thal nu
PH posterior hypoth area
Pir piriform cortex
PiRe pineal recess of 3V
PLCo posterolateral cortical amyg nu
PMCo posteromed cortical amyg nu
Po posterior thal nuclear group
PoMn posteromedian thal nu
PrC precommissural nu
PRh perirhinal cortex
PSph presphenoid bone
Ptg pterygoid bone
ptgpal pterygopalatine artery
PV paraventricular thal nu
RF rhinal fissure
RSA retrosplenial agranular Cx
RSG retrosplenial granular Cx
S subiculum
scp superior cerebellar peduncle
Squamous squamous part of temp
STh subthalamic nu
str superior thal radiation
SubG subgeniculate nu
Te terete hypoth nu
Te1 temporal cortex, area 1
Te3 temporal cortex, area 3
VLGMC vent lat genicul nu, magnoce
VLGPC vent lat genicul nu, parvocell
VMH ventromedial hypoth nu
VPL ventral posterolat thal nu
VPM ventral posteromedial thal nu
ZID zona incerta, dorsal
ZIV zona incerta, ventral

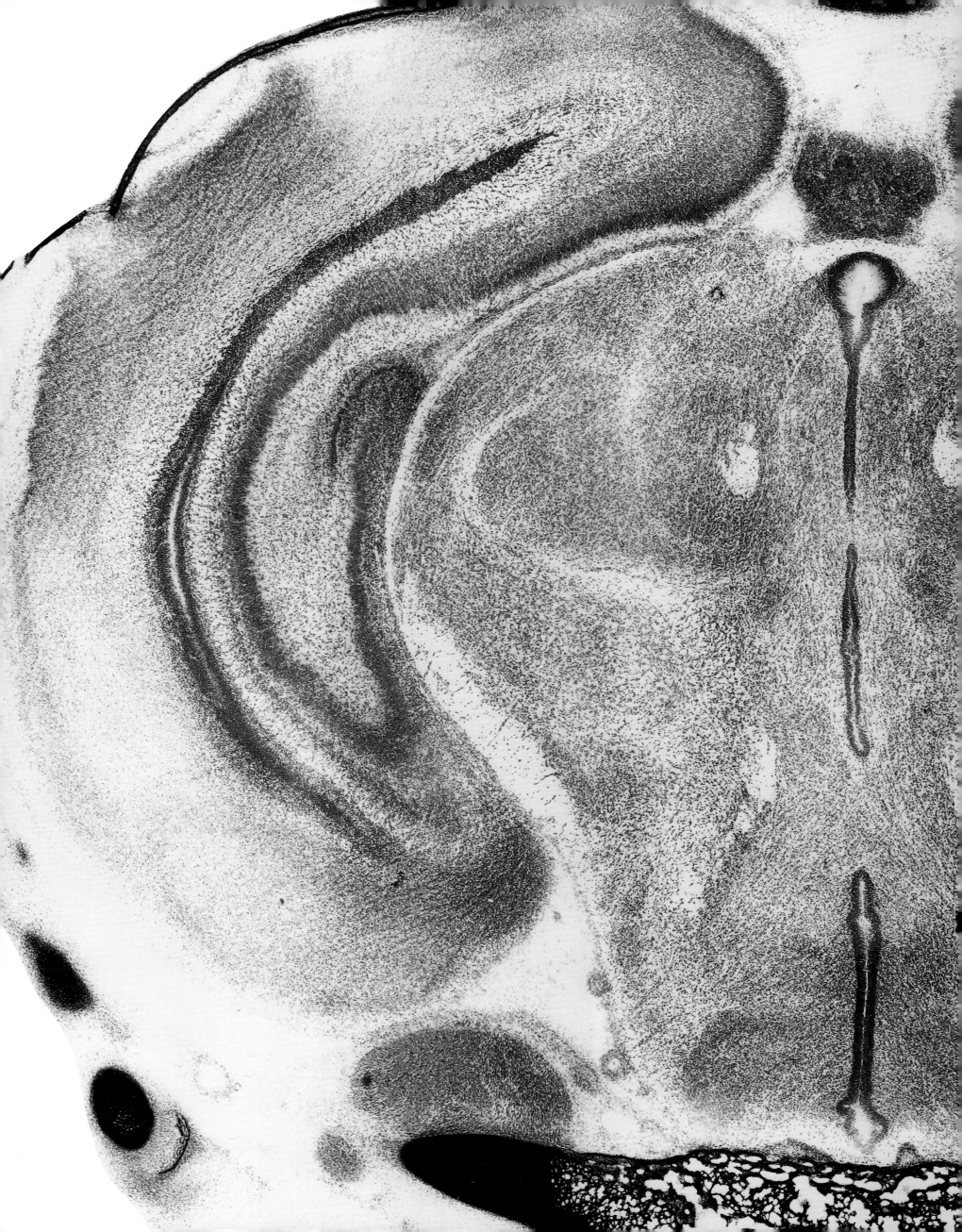

Figure 179
P0-35

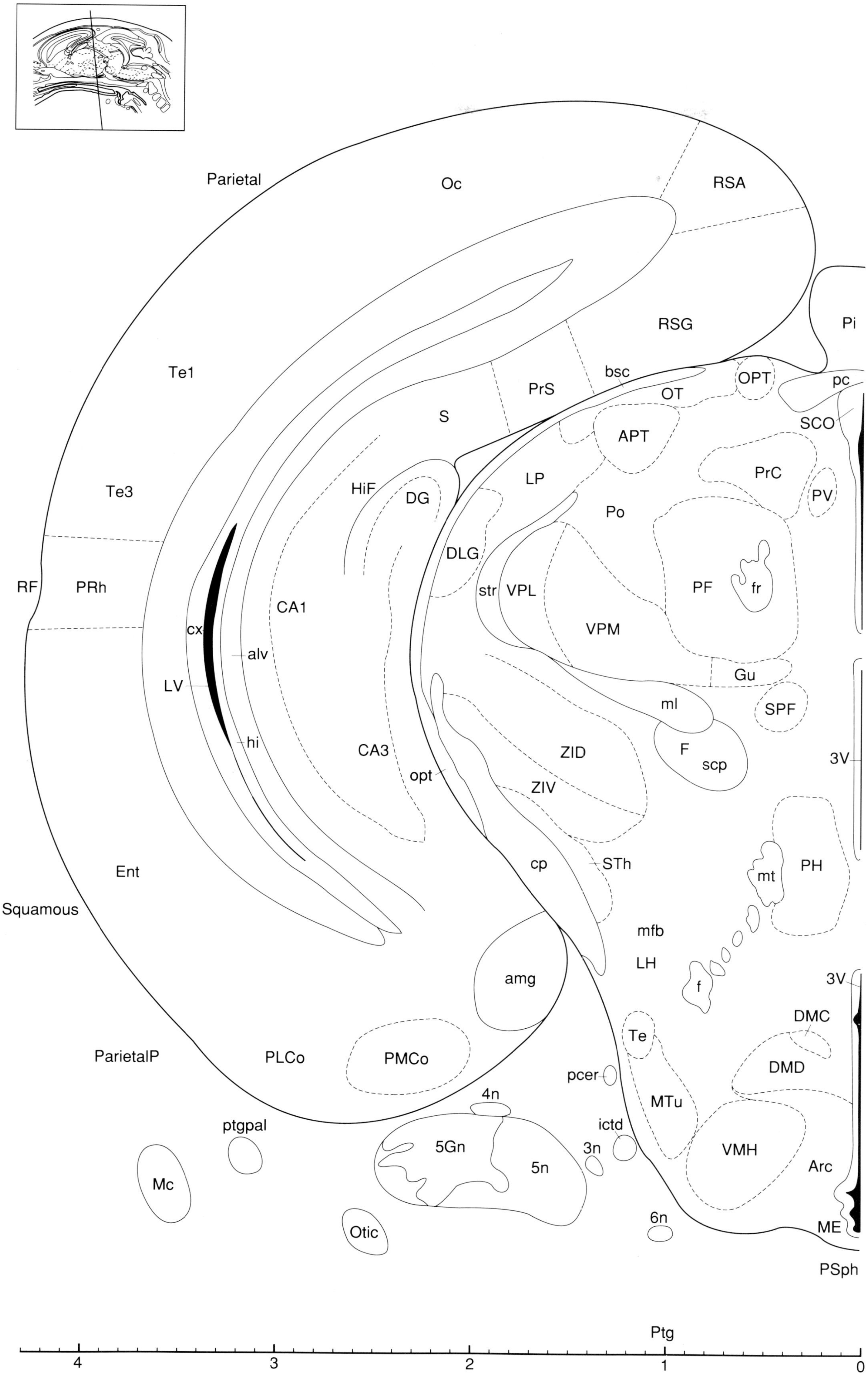

3n   oculomotor nerve or its root
3V   3rd ventricle
4n   trochlear nerve or its root
5Gn   trigeminal ganglion
5n   trigeminal nerve
6n   abducens nerve or its root
alv   alveus of hippocampus
amg   amygdaloid neuroepithelium
APT   anterior pretectal nu
Arc   arcuate hypoth nu
bsc   brachium superior colliculus
CA1   CA1 field of the hippocampus
CA3   CA3 field of the hippocampus
cp   cerebral peduncle, basal
cx   cortical neuroepithelium
DG   dentate gyrus
DLG   dorsal lateral geniculate nu
DMC   dorsomed hypoth nu, compact
DMD   dorsomed hypoth nu, diffuse
Ent   entorhinal cortex
F   nu fields of Forel
f   fornix
fr   fasciculus retroflexus
Gu   gustatory thalamic nu
hi   hippocampal formation neuroepi
HiF   hippocampal fissure
ictd internal carotid artery
LH   lateral hypothalamic area
LP   lateral posterior thal nu
LV   lateral ventricle
Mc   Meckel's cartilage
ME   median eminence
mfb   medial forebrain bundle
ml   medial lemniscus
mt   mammillothalamic tract
MTu   medial tuberal nu
Oc   occipital cortex
OPT   olivary pretectal nu
opt   optic tract
OT   nu optic tract
Otic otic ganglion
Parietal   parietal bone
ParietalP   parietal plate
pc   posterior commissure
pcer posterior cerebral artery
PF   parafascicular thal nu
PH   posterior hypoth area
Pi   pineal gland
PLCo posterolateral cortical amyg nu
PMCo posteromed cortical amyg nu
Po   posterior thal nuclear group
PrC   precommissural nu
PRh   perirhinal cortex
PrS   presubiculum
PSph presphenoid bone
Ptg   pterygoid bone
ptgpal pterygopalatine artery
PV   paraventricular thal nu
RF   rhinal fissure
RSA   retrosplenial agranular Cx
RSG   retrosplenial granular Cx
S   subiculum
SCO   subcommissural organ
scp   superior cerebellar peduncle
SPF   subparafascicular thal nu
Squamous   squamous part of temp
STh   subthalamic nu
str   superior thal radiation
Te   terete hypoth nu
Te1   temporal cortex, area 1
Te3   temporal cortex, area 3
VMH   ventromedial hypoth nu
VPL   ventral posterolat thal nu
VPM   ventral posteromedial thal nu
ZID   zona incerta, dorsal
ZIV   zona incerta, ventral

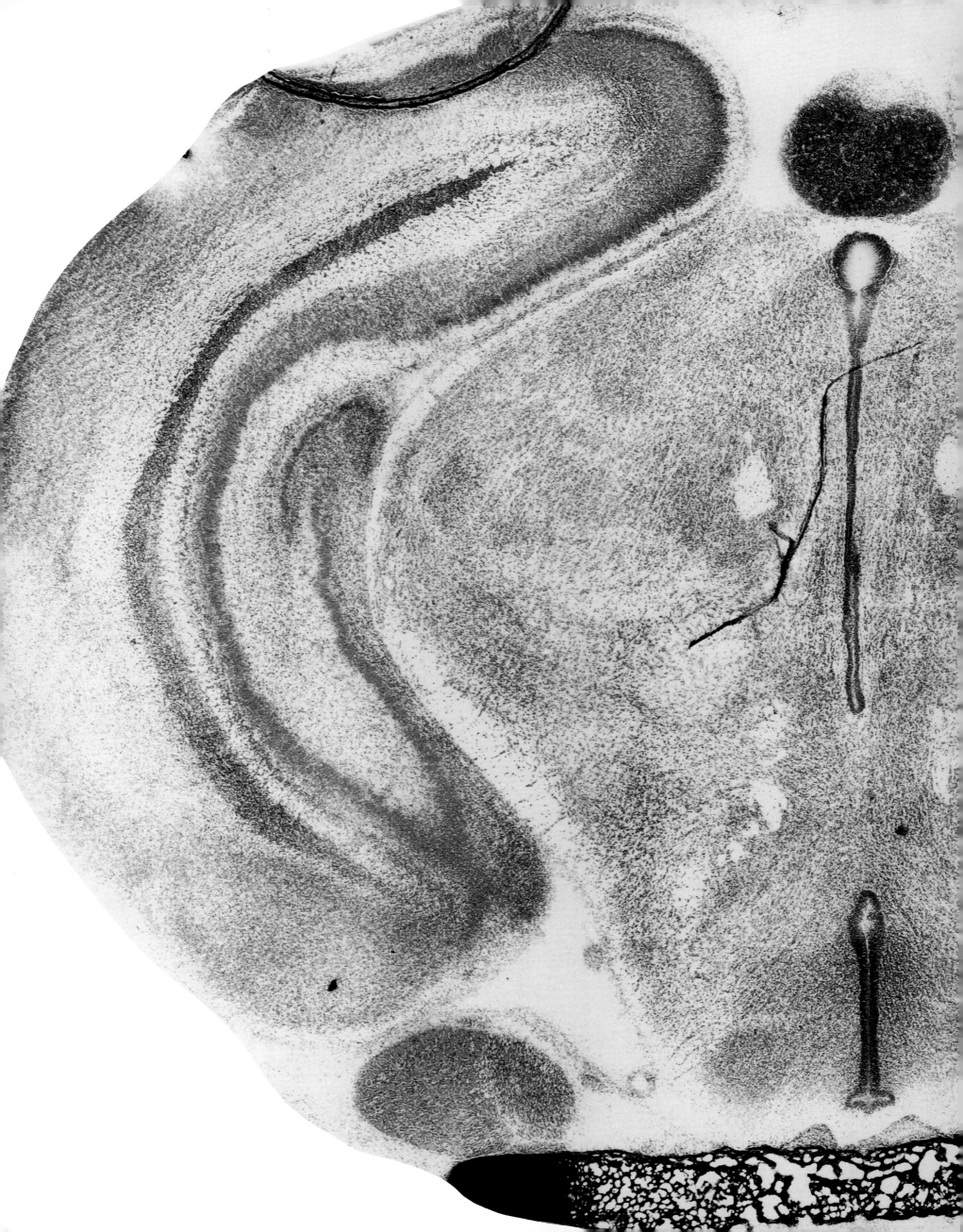

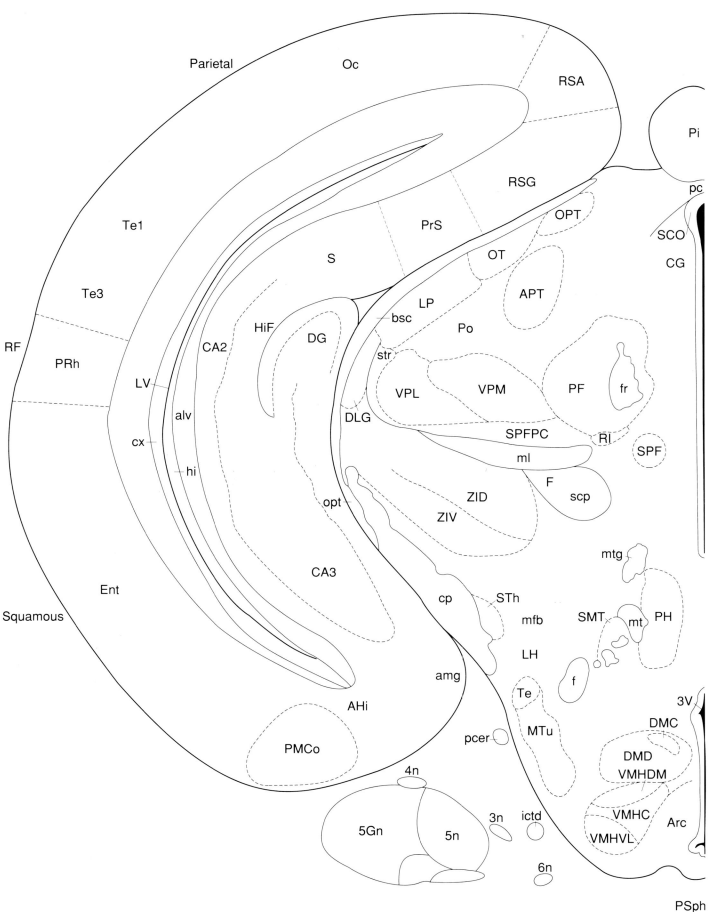

Figure 180
P0-36

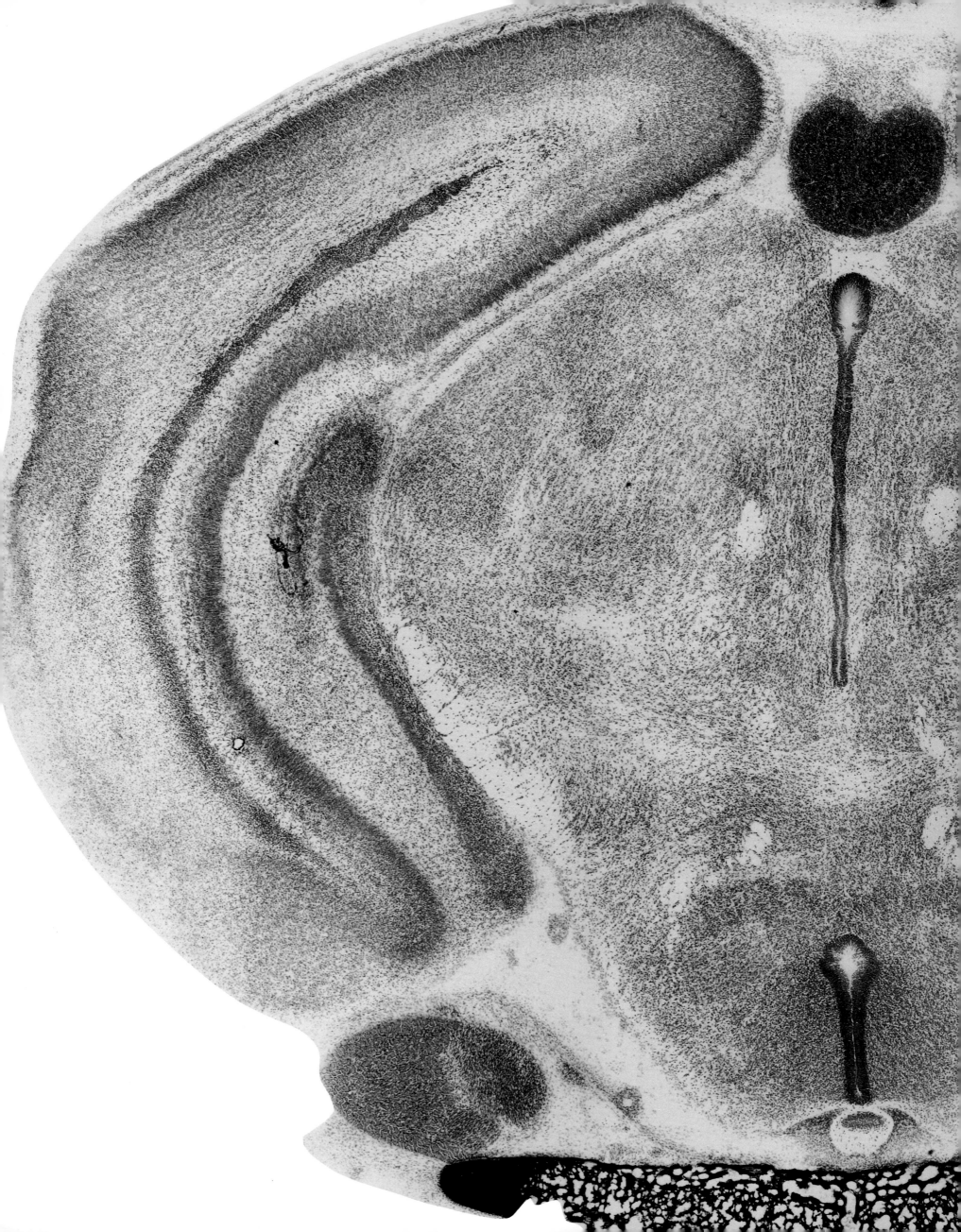

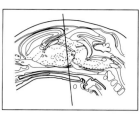

Figure 181
P0-37

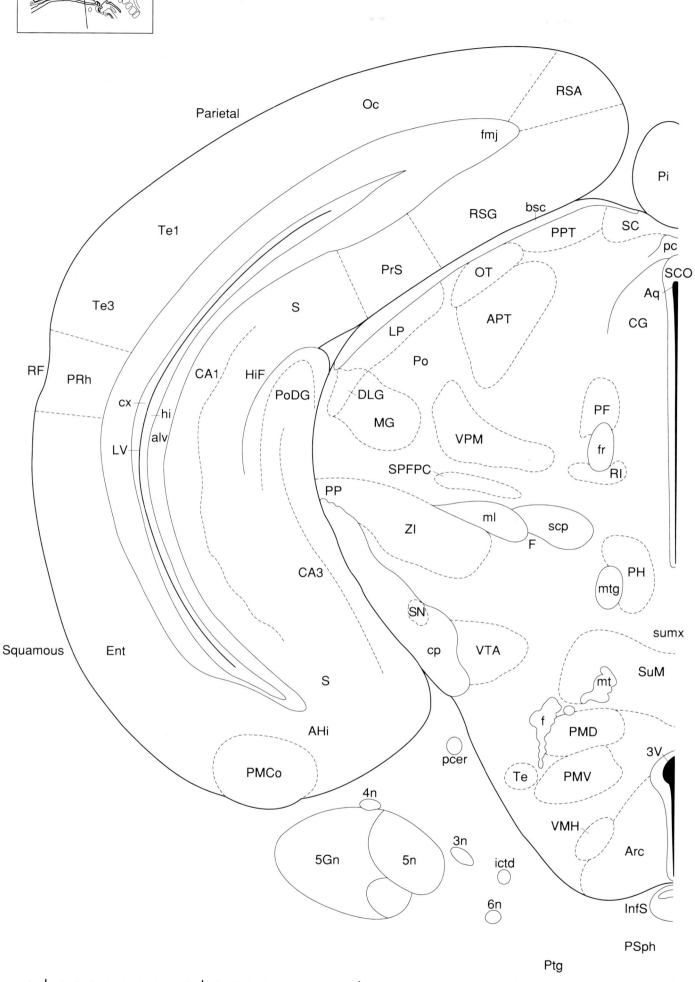

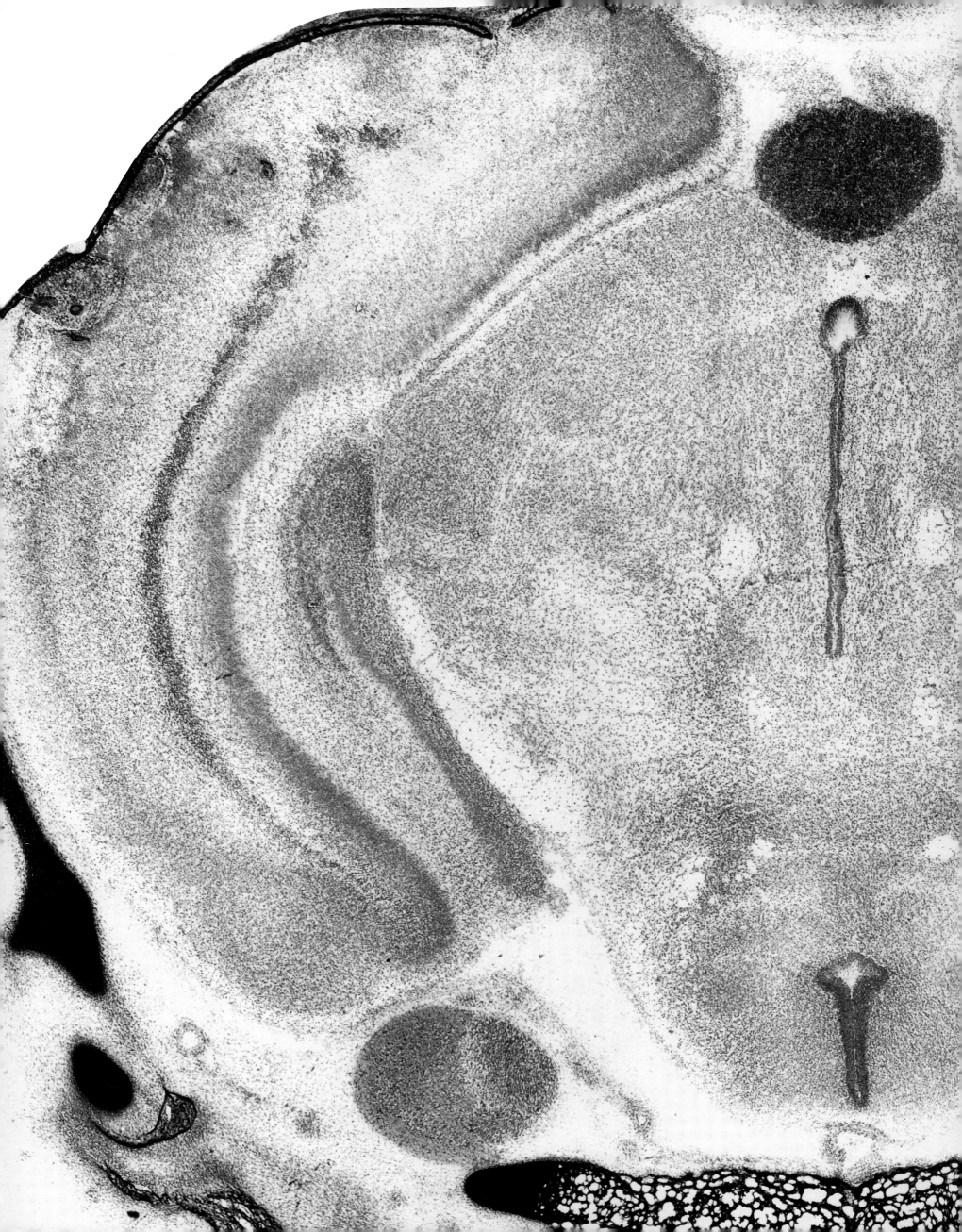

Figure 182
P0-38

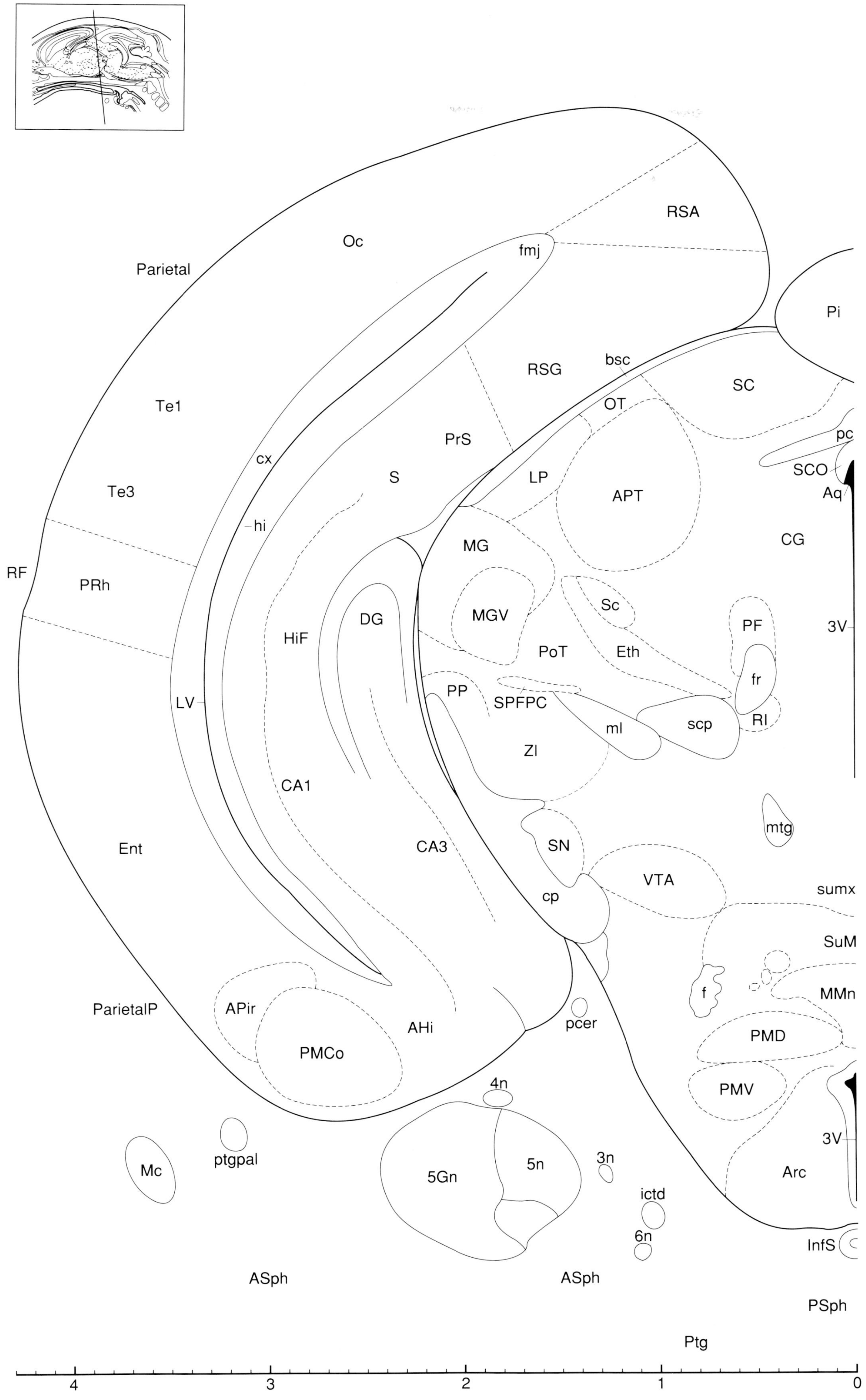

RSA
Oc
fmj
Parietal
Pi
Te1
bsc
RSG
SC
PrS
OT
cx
S
pc
LP
SCO
Te3
hi
APT
Aq
PRh
MG
CG
RF
MGV
Sc
DG
PF
HiF
PoT
Eth
3V
LV
PP
SPFPC
fr
ml
scp
RI
ZI
CA1
mtg
CA3
SN
Ent
VTA
sumx
cp
ParietalP
APir
SuM
f
AHi
pcer
MMn
PMCo
4n
PMD
Mc
ptgpal
5Gn
5n
3n
Arc
PMV
ictd
3V
6n
InfS
ASph
ASph
PSph
Ptg

4   3   2   1   0

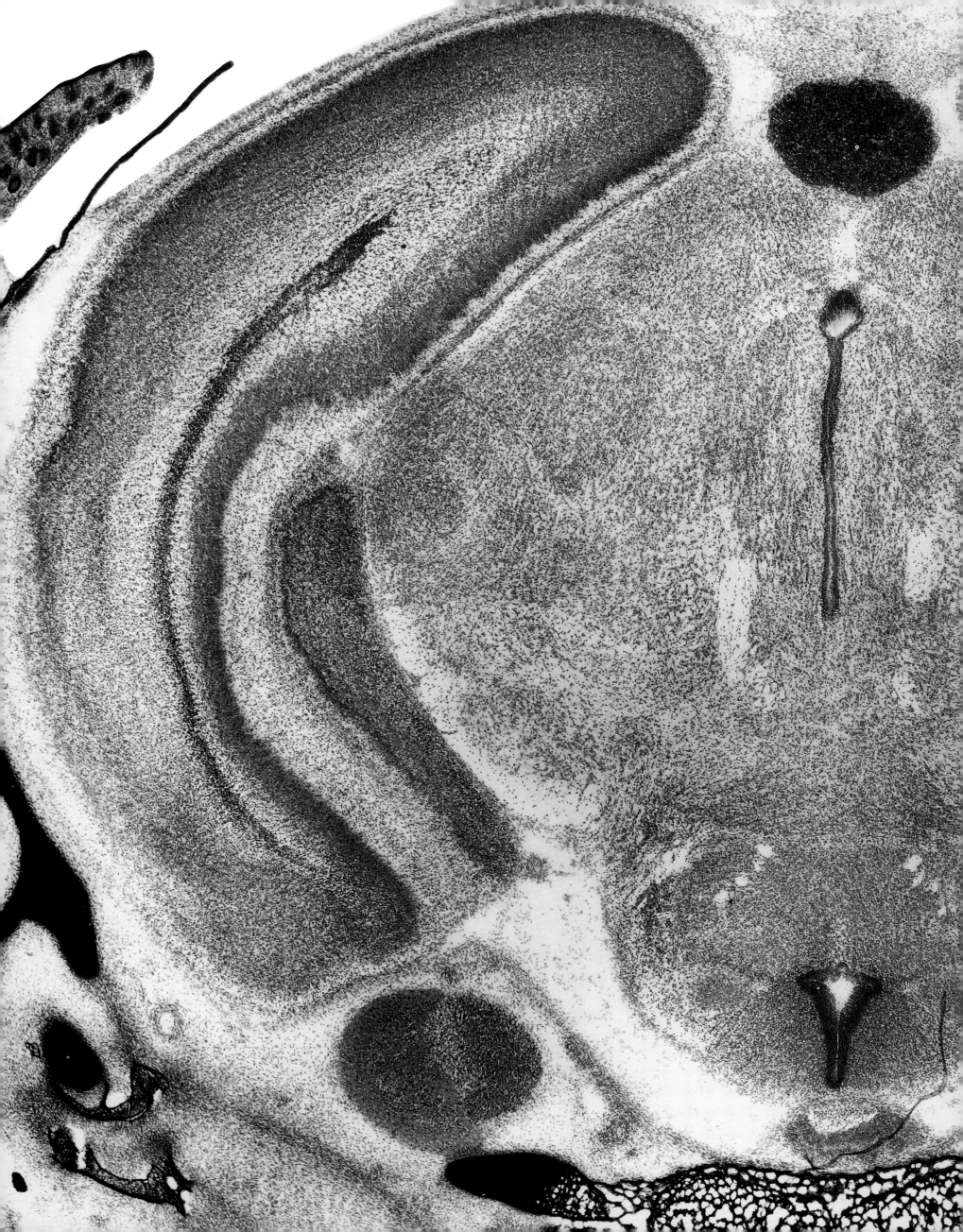

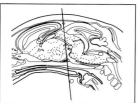

Figure 183
P0-39

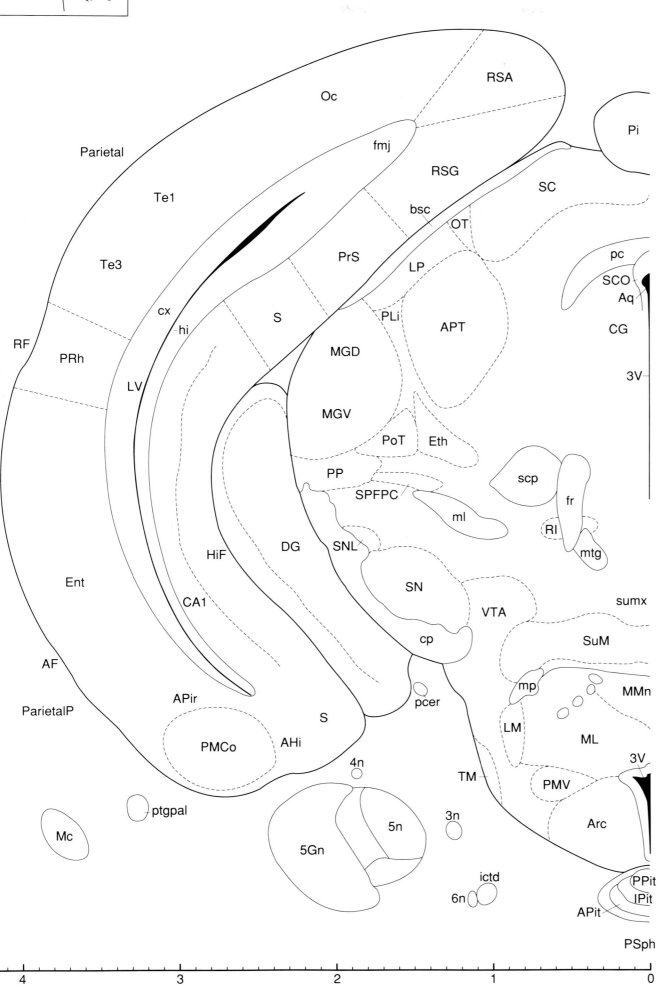

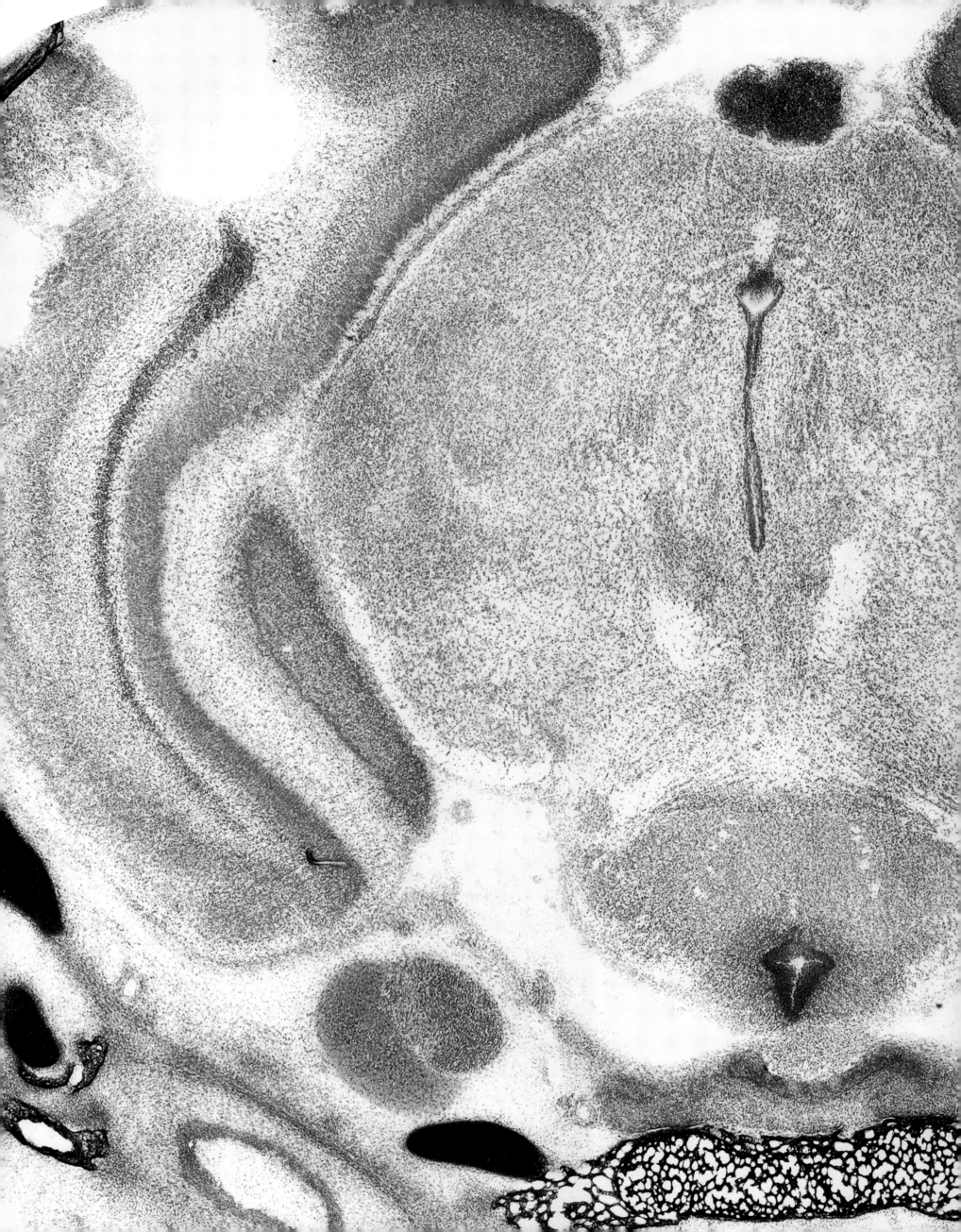

Figure 184
P0-40

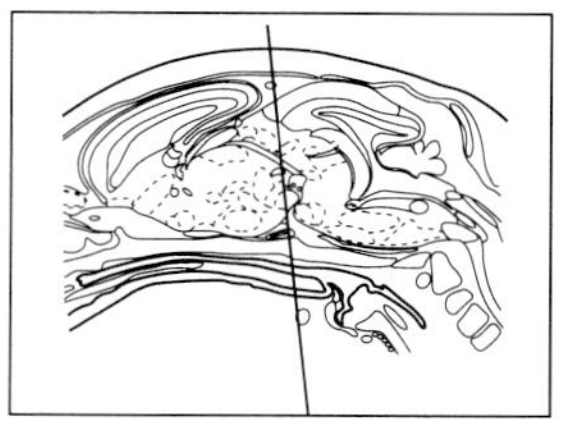

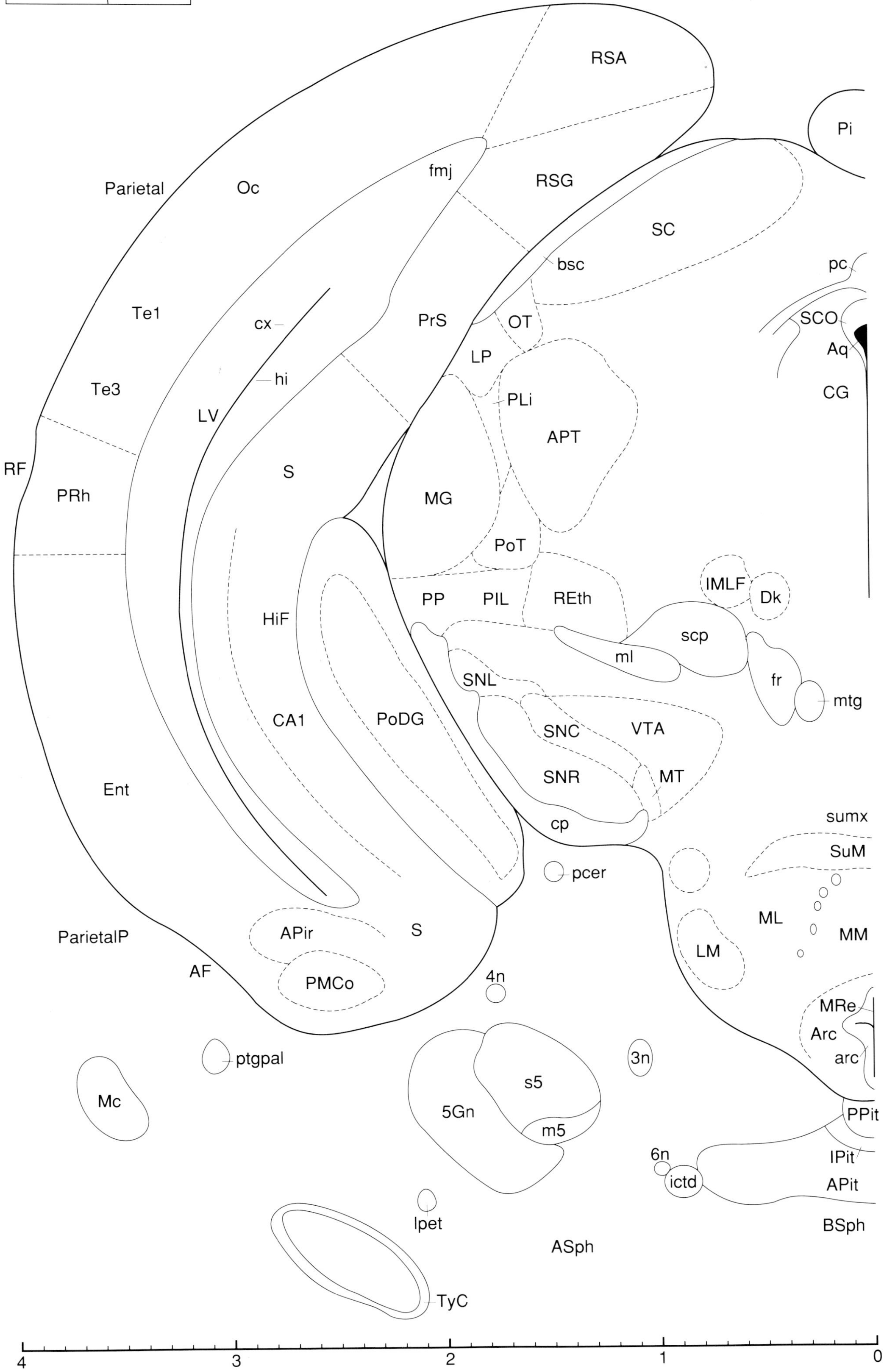

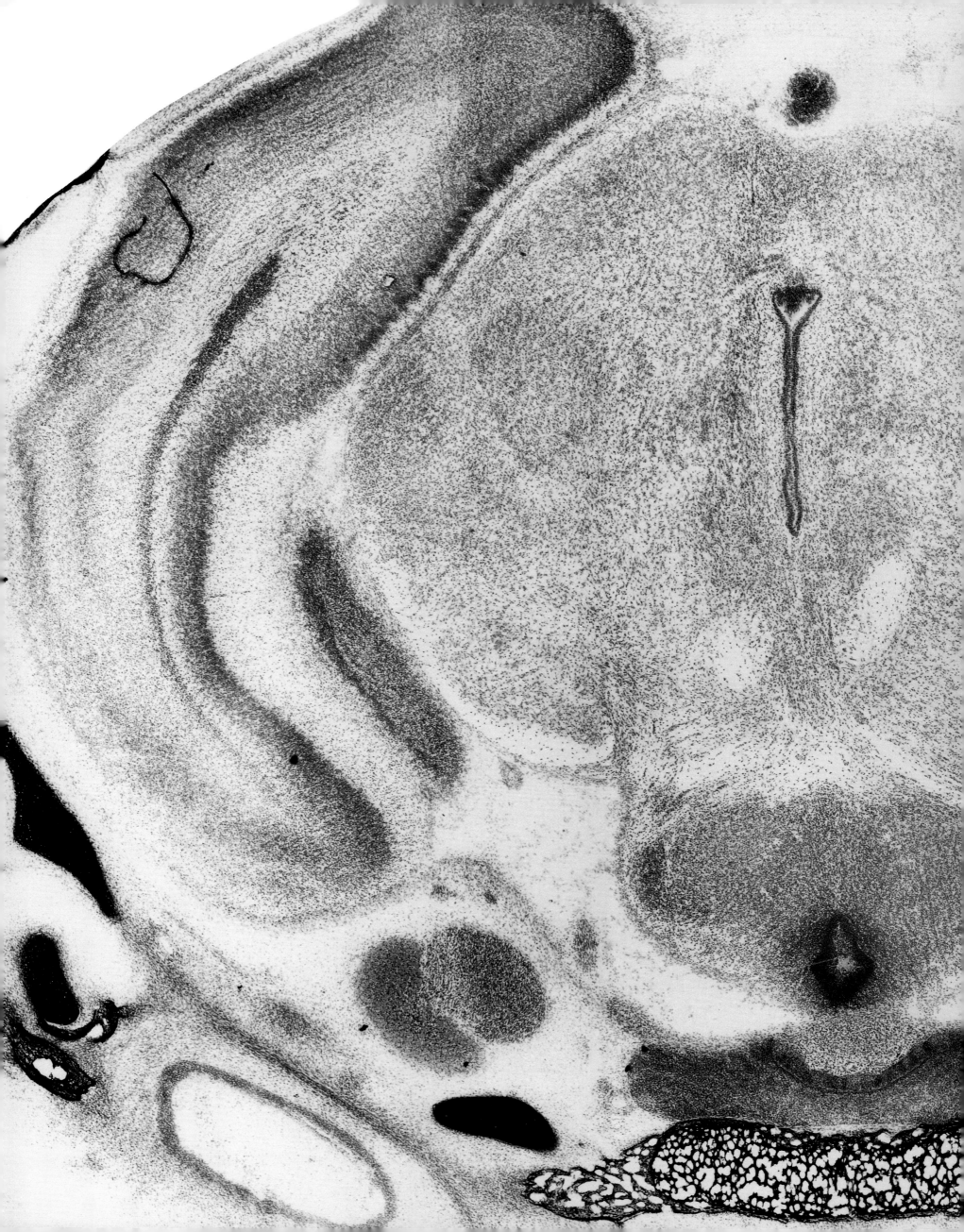

Figure 185
P0-41

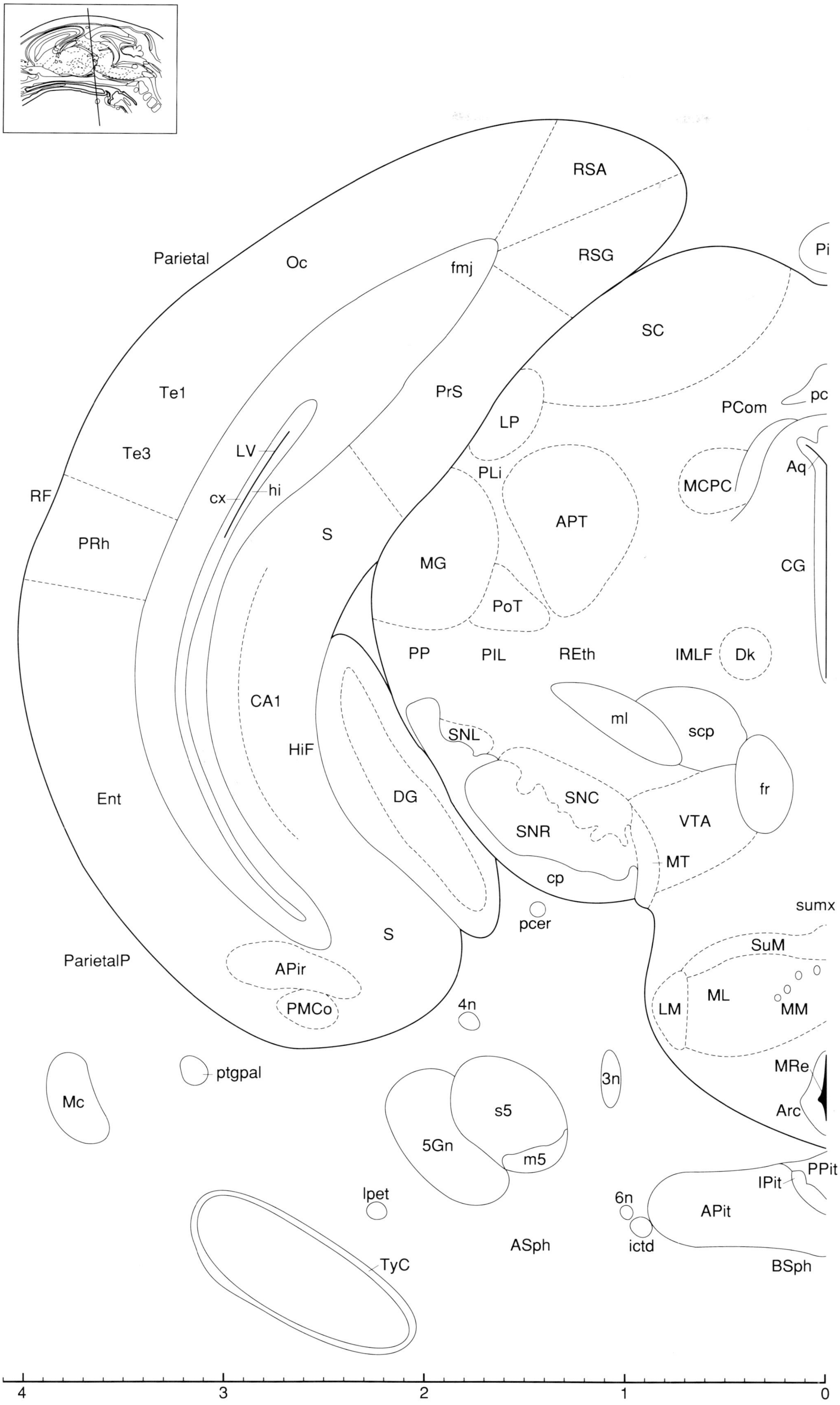

3n   oculomotor nerve or its root
4n   trochlear nerve or its root
5Gn   trigeminal ganglion
6n   abducens nerve or its root
APir amygdalopiriform transition
APit anterior lobe pituitary
APT   anterior pretectal nu
Aq   aqueduct
Arc   arcuate hypoth nu
ASph alisphenoid bone
BSph basisphenoid bone
CA1   CA1 field of the hippocampus
CG   central gray
cp   cerebral peduncle, basal
cx   cortical neuroepithelium
DG   dentate gyrus
Dk   nucleus Darkschewitsch
Ent   entorhinal cortex
fmj   forceps major corpus callosum
fr   fasciculus retroflexus
hi   hippocampal formation neuroepi
HiF   hippocampal fissure
ictd internal carotid artery
IMLF interstitial nu mlf
IPit intermediate lobe pituitary
LM   lateral mammillary nu
LP   lateral posterior thal nu
lpet lesser petrosal nerve
LV   lateral ventricle
m5   motor root trigeminal nerve
Mc   Meckel's cartilage
MCPC magnocellular nu post com
MG   medial geniculate nu
ML   medial mammillary nu, lateral
ml   medial lemniscus
MM   medial mammillary nu, medial
MRe   mammillary recess 3rd ventricle
MT   med terminal nu accessory optic tr
Oc   occipital cortex
Parietal   parietal bone
ParietalP parietal plate
pc   posterior commissure
pcer posterior cerebral artery
PCom nu posterior commissure
Pi   pineal gland
PIL   posterior intralaminar thal nu
PLi   posterior limitans thal nu
PMCo posteromed cortical amyg nu
PoT   posterior thal nu gr, triangular
PP   peripeduncular nu
PPit posterior lobe pituitary
PRh   perirhinal cortex
PrS   presubiculum
ptgpal pterygopalatine artery
REth retroethmoid nu
RF   rhinal fissure
RSA   retrosplenial agranular Cx
RSG   retrosplenial granular Cx
S   subiculum
s5   sensory root trigeminal nerve
SC   superior colliculus
scp   superior cerebellar peduncle
SNC   substantia nigra, compact
SNL   substantia nigra, lateral
SNR   substantia nigra, reticular
SuM   supramammillary nu
sumx supramammillary decussation
Te1   temporal cortex, area 1
Te3   temporal cortex, area 3
TyC   tympanic cavity
VTA   ventral tegmental area

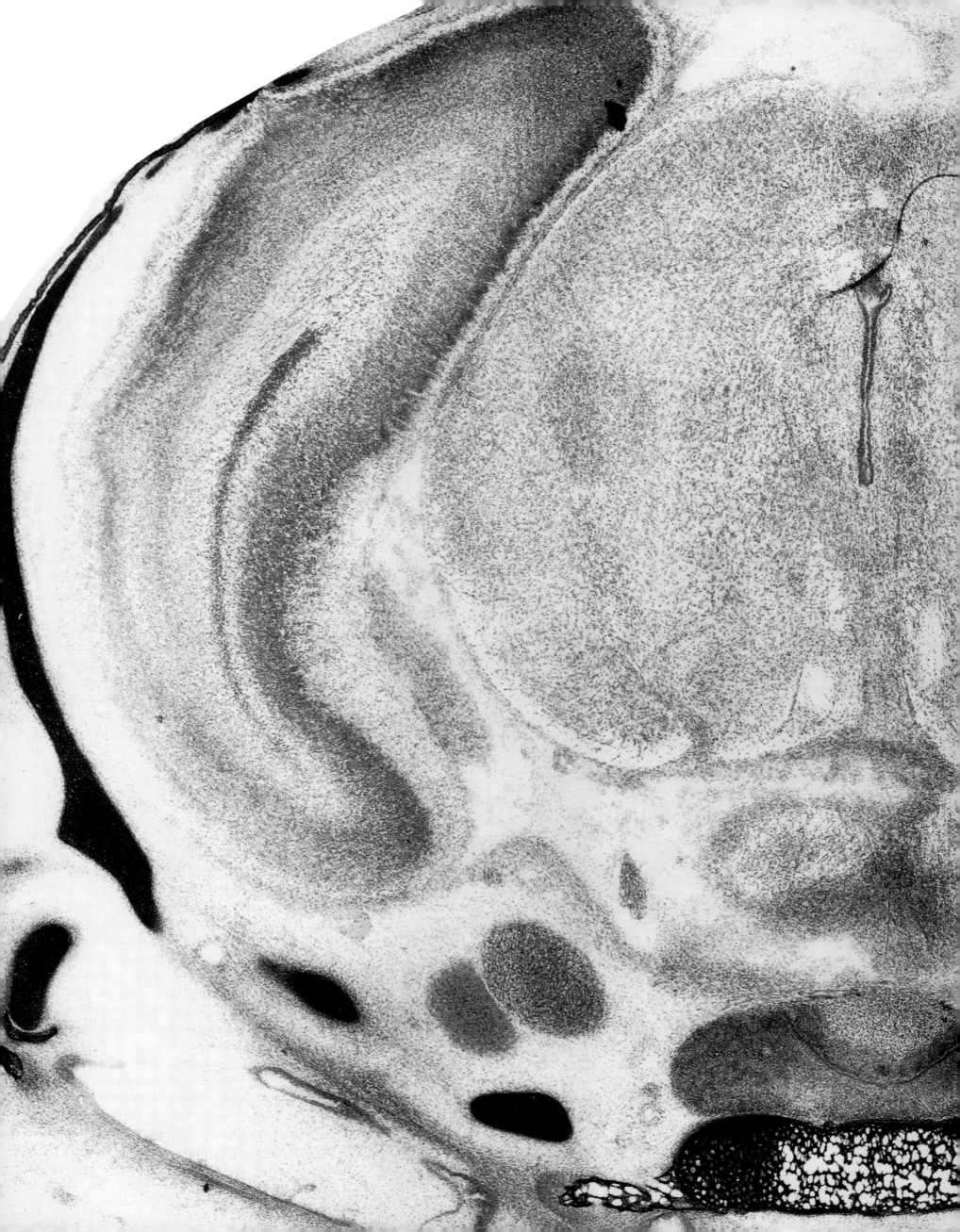

Figure 186
P0-42

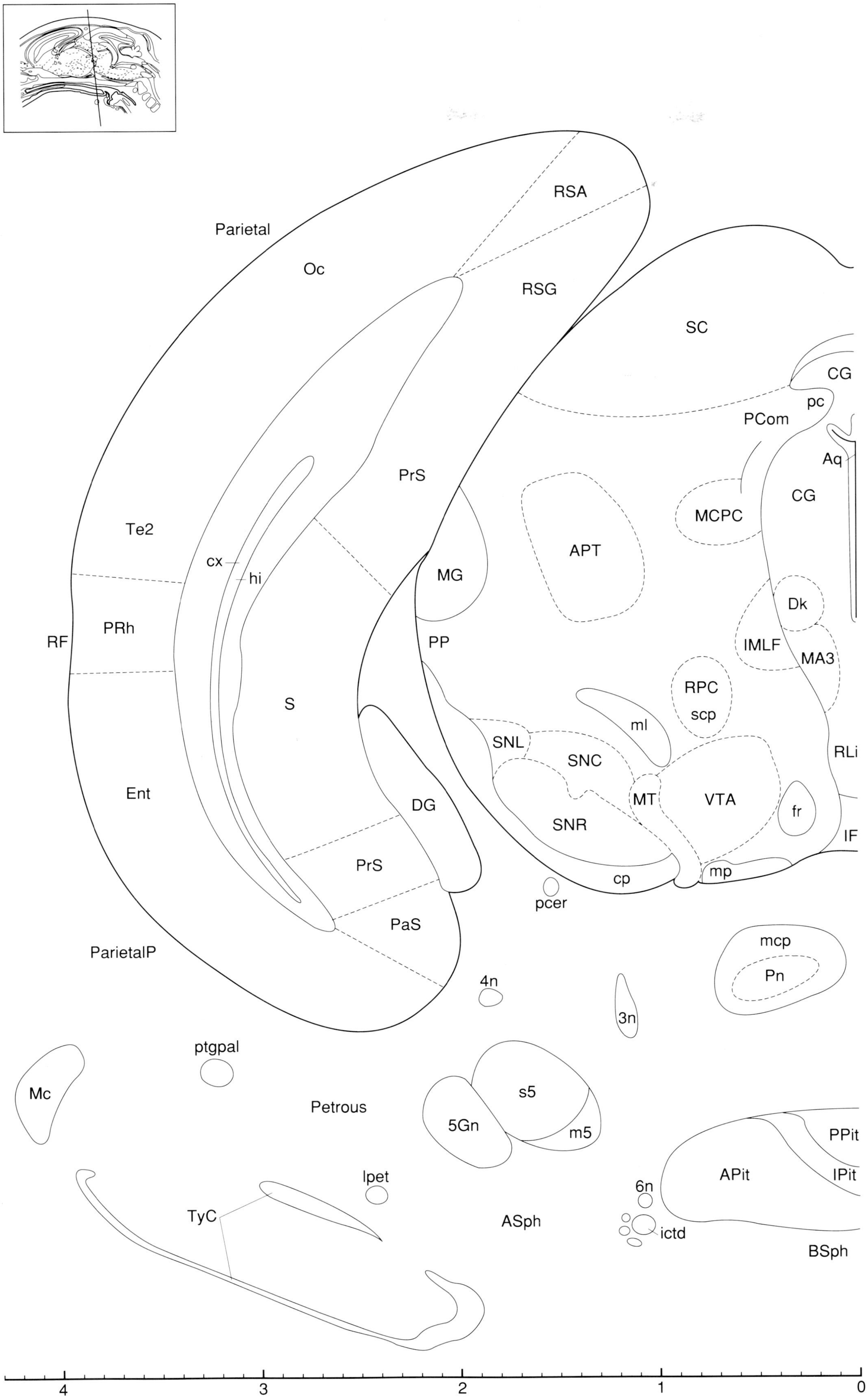

3n   oculomotor nerve or its root
4n   trochlear nerve or its root
5Gn trigeminal ganglion
6n   abducens nerve or its root
APit anterior lobe pituitary
APT anterior pretectal nu
Aq   aqueduct
ASph alisphenoid bone
BSph basisphenoid bone
CG   central gray
cp   cerebral peduncle, basal
cx   cortical neuroepithelium
DG   dentate gyrus
Dk   nucleus Darkschewitsch
Ent   entorhinal cortex
fr   fasciculus retroflexus
hi   hippocampal formation neuroepi
ictd   internal carotid artery
IF   interfascicular nucleus
IMLF   interstitial nu mlf
IPit   intermediate lobe pituitary
lpet   lesser petrosal nerve
m5   motor root trigeminal nerve
MA3   medial acs oculomotor nu
Mc   Meckel's cartilage
mcp   middle cerebellar peduncle
MCPC   magnocellular nu post com
MG   medial geniculate nu
ml   medial lemniscus
mp   mammillary peduncle
MT   med terminal nu accessory optic tr
Oc   occipital cortex
Parietal   parietal bone
ParietalP   parietal plate
PaS   parasubiculum
pc   posterior commissure
pcer   posterior cerebral artery
PCom   nu posterior commissure
Petrous   petrous part, temporal bone
Pn   pontine nuclei
PP   peripeduncular nu
PPit   posterior lobe pituitary
PRh   perirhinal cortex
PrS   presubiculum
ptgpal   pterygopalatine artery
RF   rhinal fissure
RLi   rostral linear nu raphe
RPC   red nu, parvocellular
RSA   retrosplenial agranular Cx
RSG   retrosplenial granular Cx
S   subiculum
s5   sensory root trigeminal nerve
SC   superior colliculus
scp   superior cerebellar peduncle
SNC   substantia nigra, compact
SNL   substantia nigra, lateral
SNR   substantia nigra, reticular
Te2   temporal cortex, area 2
TyC   tympanic cavity
VTA   ventral tegmental area

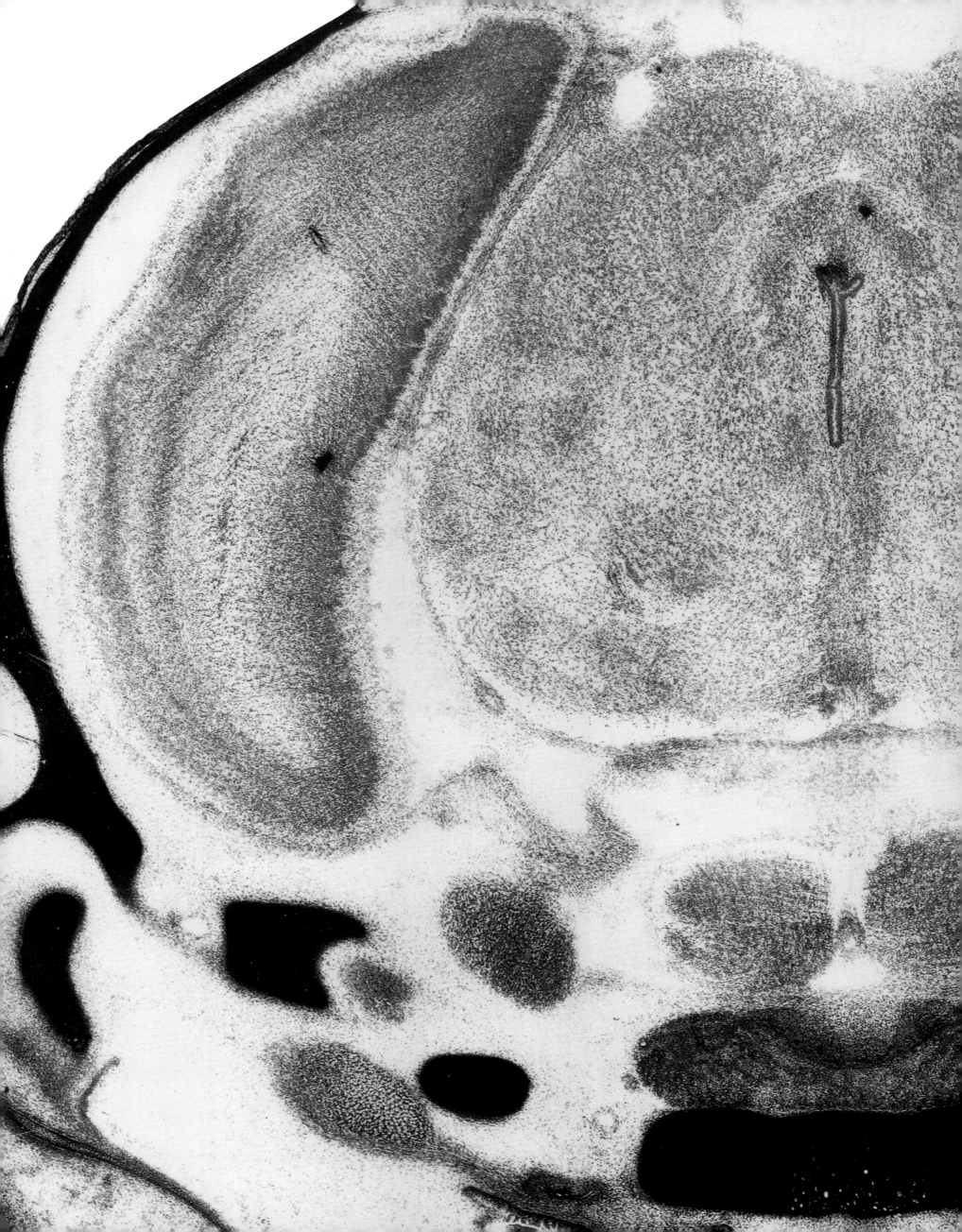

P0-43

3n   oculomotor nerve or its root
4n   trochlear nerve or its root
6n   abducens nerve or its root
APit   anterior lobe pituitary
APT   anterior pretectal nu
Aq   aqueduct
bas   basilar artery
BSph   basisphenoid bone
CG   central gray
cp   cerebral peduncle, basal
csc   commissure superior colliculus
Ctd   carotid canal
Dk   nucleus Darkschewitsch
DpG   deep gray layer of superior coll
DpMe   deep mesencephalic nu
DpWh   deep white layer of super coll
EAM   external auditory meatus
Ent   entorhinal cortex
Eust   Eustachian tube
fr   fasciculus retroflexus
Gen   geniculate ganglion
ictd   internal carotid artery
IF   interfascicular nucleus
IMLF   interstitial nu mlf
InWh   intermediate white layer sup col
IP   interpeduncular nu
IPit   intermediate lobe pituitary
m5   motor root trigeminal nerve
MA3   medial acs oculomotor nu
Mc   Meckel's cartilage
mcp   middle cerebellar peduncle
MG   medial geniculate nu
ml   medial lemniscus
Oc   occipital cortex
Parietal   parietal bone
ParietalP   parietal plate
PaS   parasubiculum
PBP   parabrachial pigmented nu
pcer   posterior cerebral artery
Petrous   petrous part, temporal bone
PN   paranigral nu
Pn   pontine nuclei
PP   peripeduncular nu
PPit   posterior lobe pituitary
PRh   perirhinal cortex
PrS   presubiculum
ptgpal   pterygopalatine artery
R   red nu
RF   rhinal fissure
RLi   rostral linear nu raphe
RSA   retrosplenial agranular Cx
RSG   retrosplenial granular Cx
s5   sensory root trigeminal nerve
SC   superior colliculus
scp   superior cerebellar peduncle
SNC   substantia nigra, compact
SNL   substantia nigra, lateral
SNR   substantia nigra, reticular
Te2   temporal cortex, area 2
TensT   tensor tympani muscle
TyC   tympanic cavity
VTA   ventral tegmental area

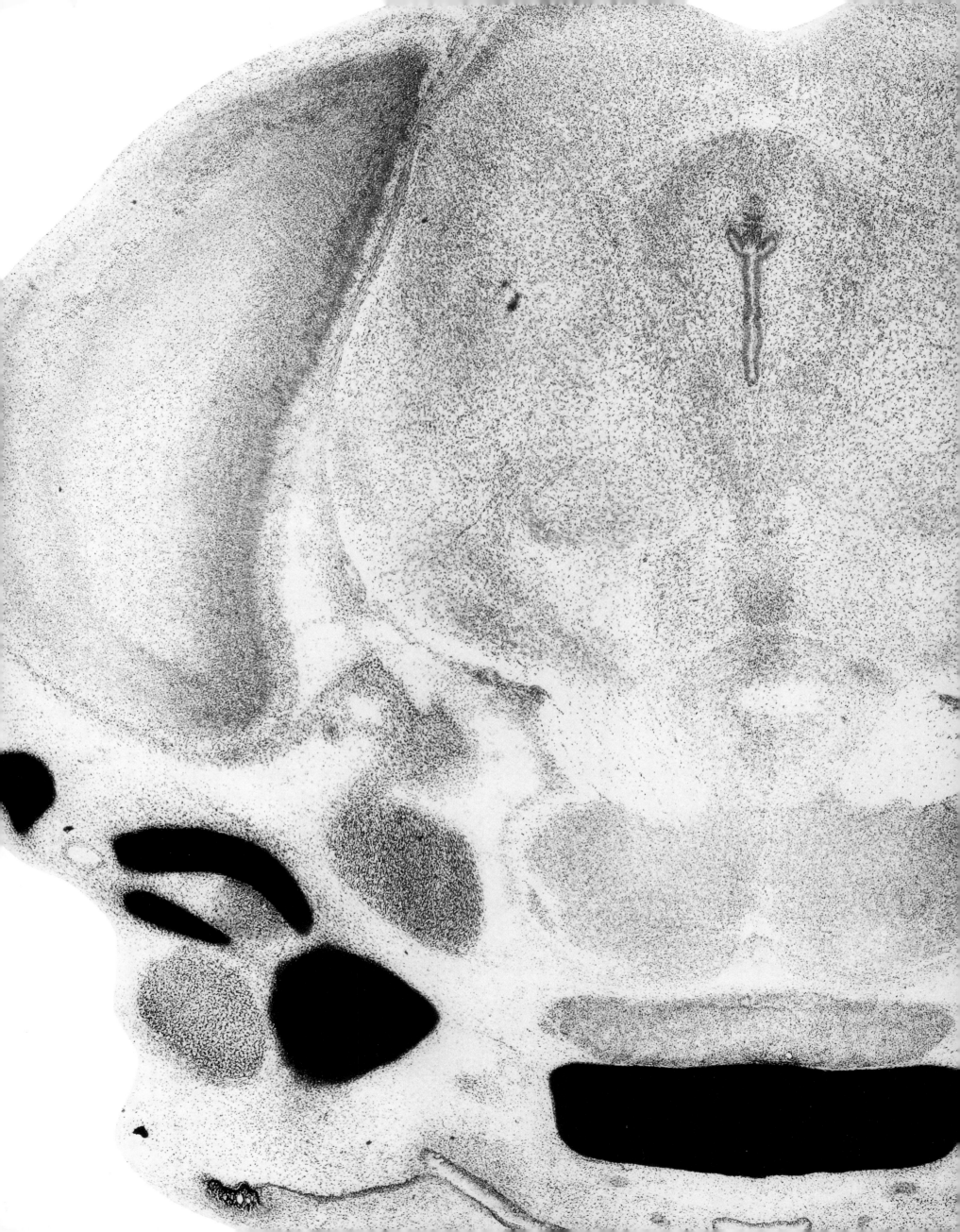

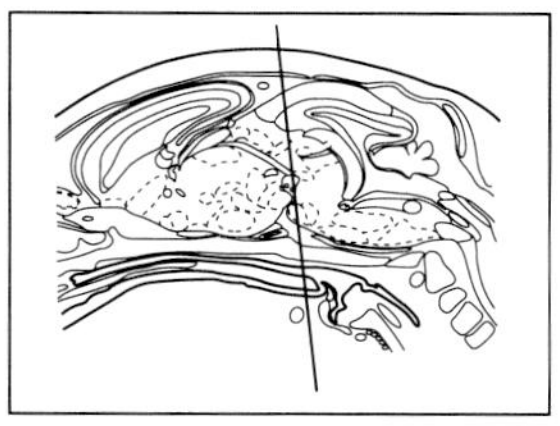

Figure 188
P0-44

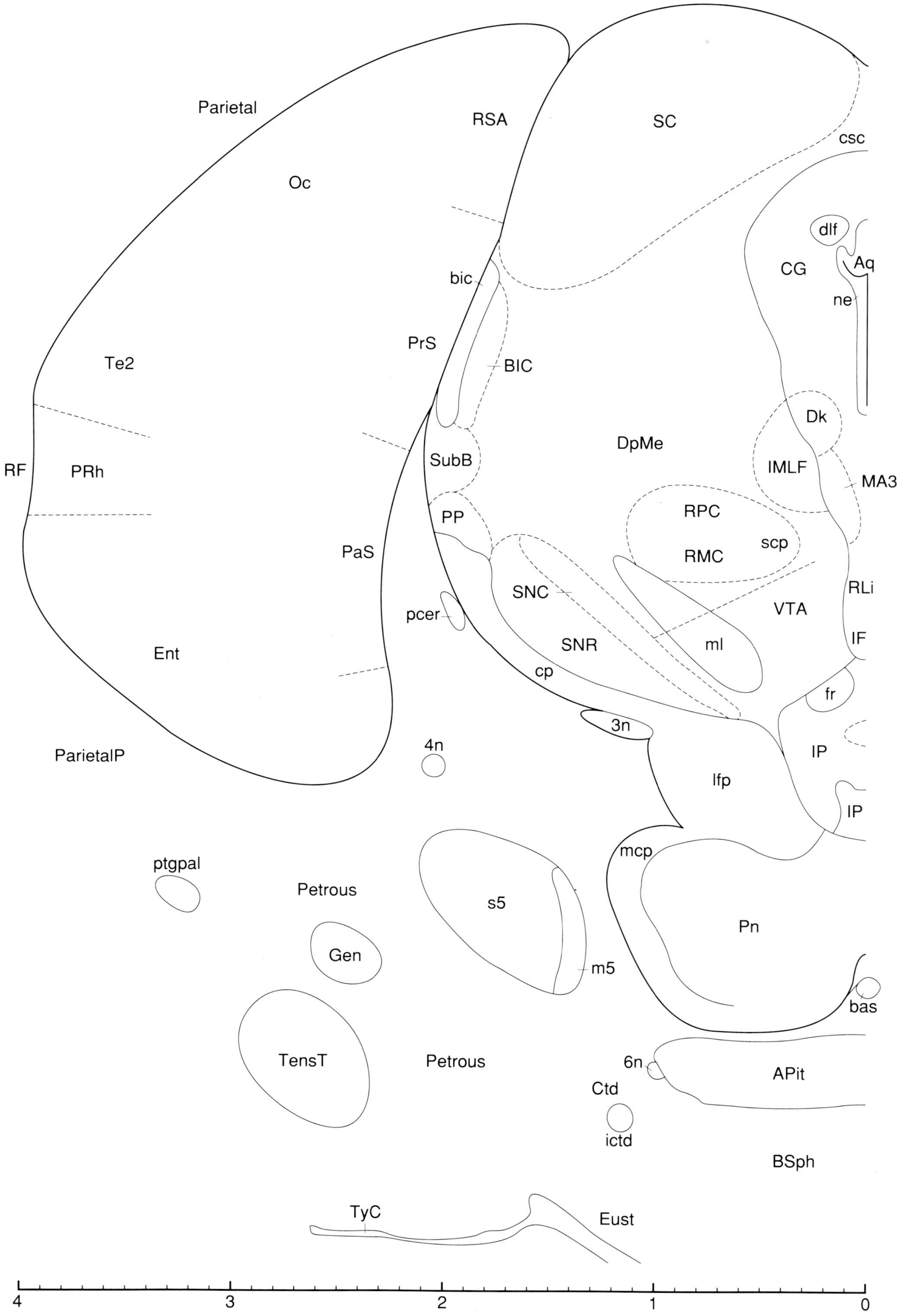

3n   oculomotor nerve or its root
4n   trochlear nerve or its root
6n   abducens nerve or its root
APit   anterior lobe pituitary
Aq   aqueduct
bas   basilar artery
BIC   nu brachium inferior colliculus
bic   brachium inferior colliculus
BSph   basisphenoid bone
CG   central gray
cp   cerebral peduncle, basal
csc   commissure superior colliculus
Ctd   carotid canal
Dk   nucleus Darkschewitsch
dlf   dorsal longitudinal fasciculus
DpMe   deep mesencephalic nu
Ent   entorhinal cortex
Eust   Eustachian tube
fr   fasciculus retroflexus
Gen   geniculate ganglion
ictd   internal carotid artery
IF   interfascicular nucleus
IMLF   interstitial nu mlf
IP   interpeduncular nu
lfp   longitudinal fasciculus pons
m5   motor root trigeminal nerve
MA3   medial acs oculomotor nu
mcp   middle cerebellar peduncle
ml   medial lemniscus
ne   neuroepithelium
Oc   occipital cortex
Parietal   parietal bone
ParietalP   parietal plate
PaS   parasubiculum
pcer   posterior cerebral artery
Petrous   petrous part, temporal bone
Pn   pontine nuclei
PP   peripeduncular nu
PRh   perirhinal cortex
PrS   presubiculum
ptgpal   pterygopalatine artery
RF   rhinal fissure
RLi   rostral linear nu raphe
RMC   red nu, magnocellular
RPC   red nu, parvocellular
RSA   retrosplenial agranular Cx
s5   sensory root trigeminal nerve
SC   superior colliculus
scp   superior cerebellar peduncle
SNC   substantia nigra, compact
SNR   substantia nigra, reticular
SubB   subbrachial nu
Te2   temporal cortex, area 2
TensT   tensor tympani muscle
TyC   tympanic cavity
VTA   ventral tegmental area

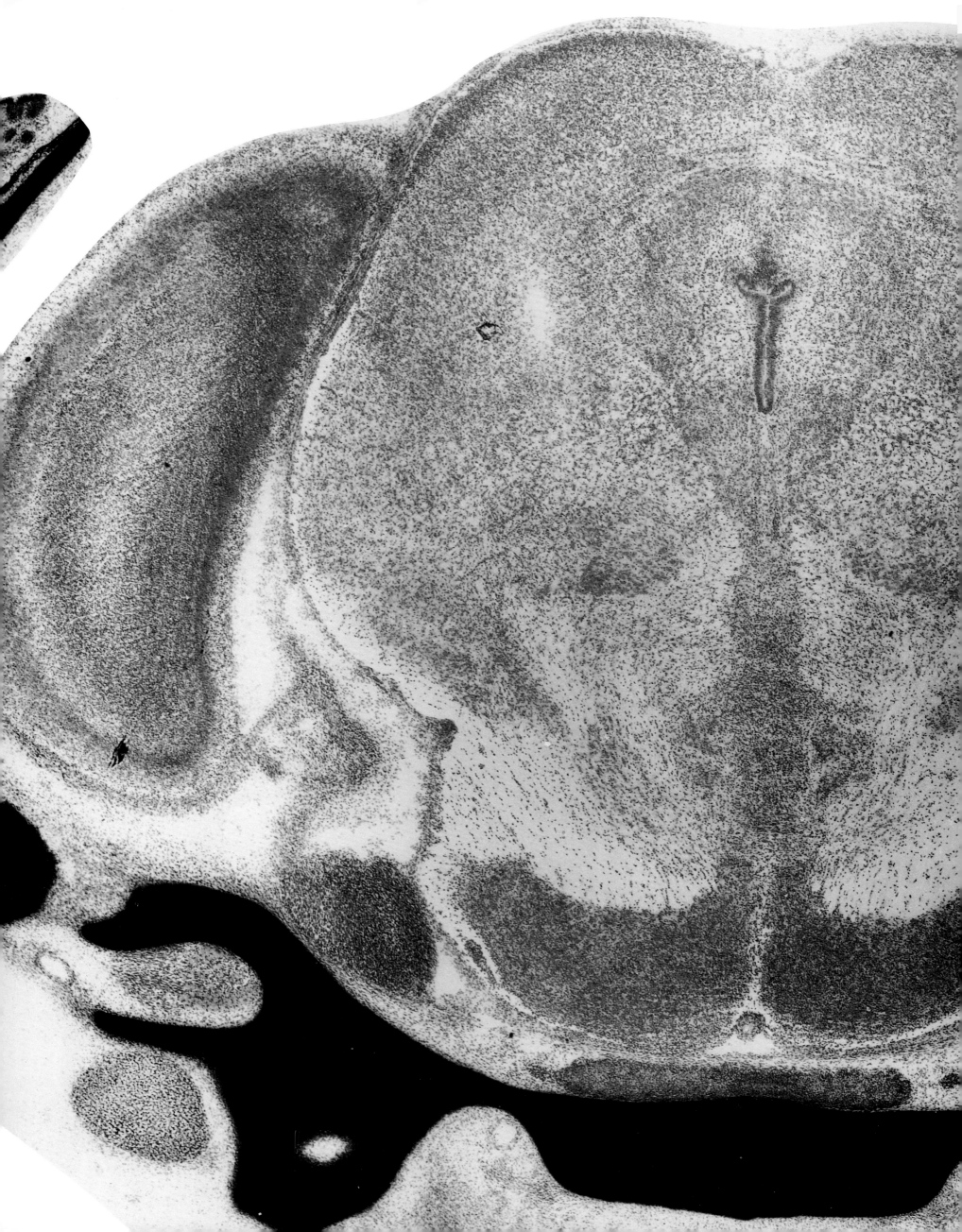

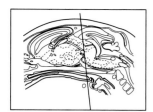

Figure 189
P0-45

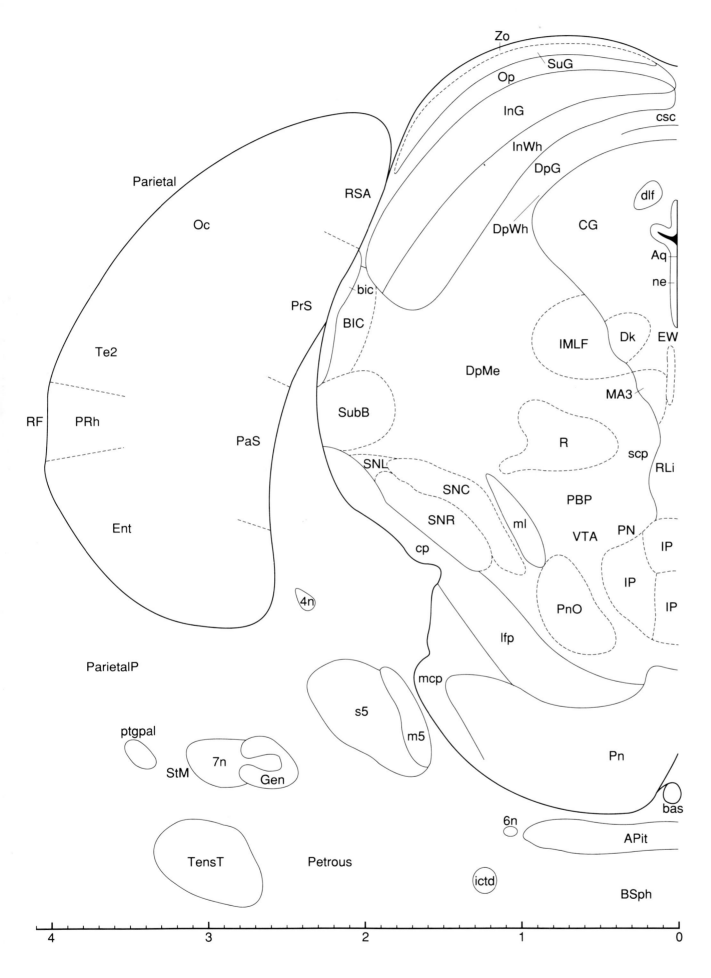

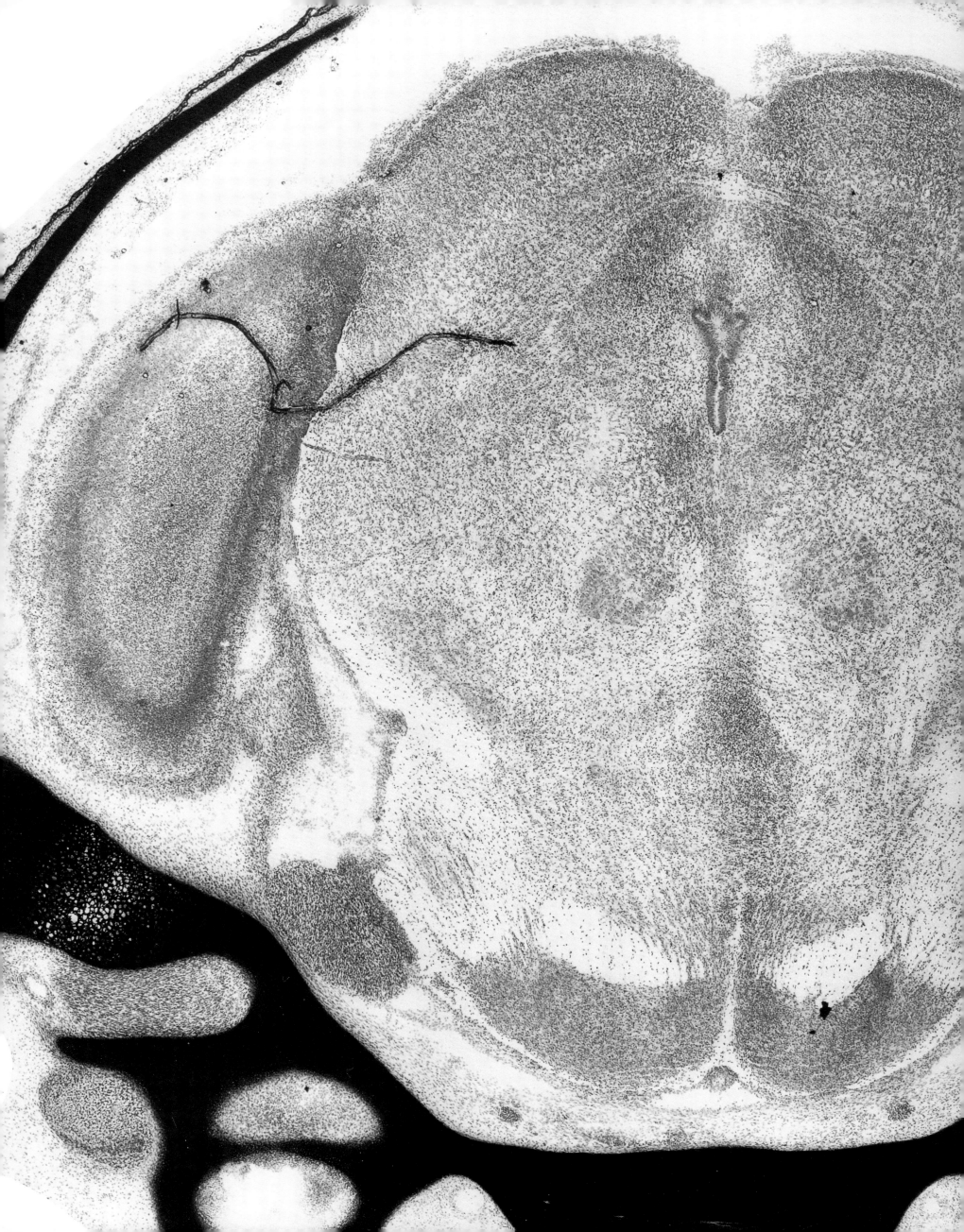

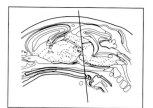

Figure 190
P0-46

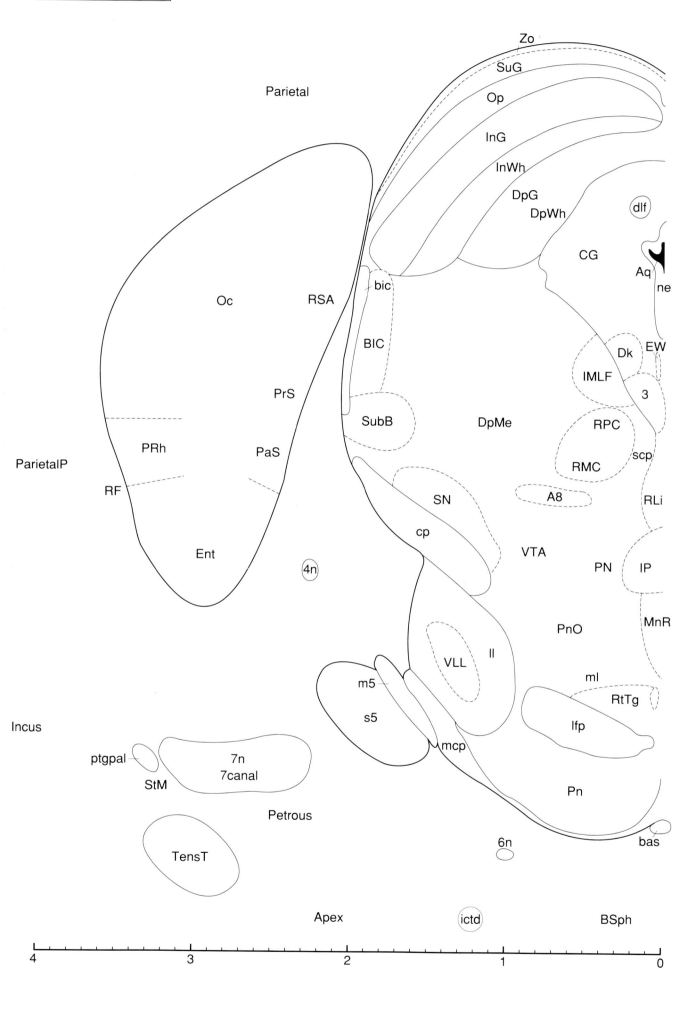

3 oculomotor nu
4n trochlear nerve or its root
6n abducens nerve or its root
7canal canal of the facial nerve
7n facial nerve or its root
A8 A8 dopamine cells
Apex apex of the cochlea
Aq aqueduct
bas basilar artery
BIC nu brachium inferior colliculus
bic brachium inferior colliculus
BSph basisphenoid bone
CG central gray
cp cerebral peduncle, basal
Dk nucleus Darkschewitsch
dlf dorsal longitudinal fasciculus
DpG deep gray layer of superior coll
DpMe deep mesencephalic nu
DpWh deep white layer of super coll
Ent entorhinal cortex
EW Edinger-Westphal nu
ictd internal carotid artery
IMLF interstitial nu mlf
Incus incus (ossicle)
InG intermediate gray layer sup col
InWh intermediate white layer sup col
IP interpeduncular nu
lfp longitudinal fasciculus pons
ll lateral lemniscus
m5 motor root trigeminal nerve
mcp middle cerebellar peduncle
ml medial lemniscus
MnR median raphe nu
ne neuroepithelium
Oc occipital cortex
Op optic nerve layer superior collicul
Parietal parietal bone
ParietalP parietal plate
PaS parasubiculum
Petrous petrous part, temporal bone
PN paranigral nu
Pn pontine nuclei
PnO pontine reticular nu, oral
PRh perirhinal cortex
PrS presubiculum
ptgpal pterygopalatine artery
RF rhinal fissure
RLi rostral linear nu raphe
RMC red nu, magnocellular
RPC red nu, parvocellular
RSA retrosplenial agranular Cx
RtTg reticulotegmental nu pons
s5 sensory root trigeminal nerve
scp superior cerebellar peduncle
SN substantia nigra
StM sternomastoid muscle
SubB subbrachial nu
SuG superficial gray layer sup col
TensT tensor tympani muscle
VLL ventral nu lateral lemniscus
VTA ventral tegmental area
Zo zonal layer of the superior coll

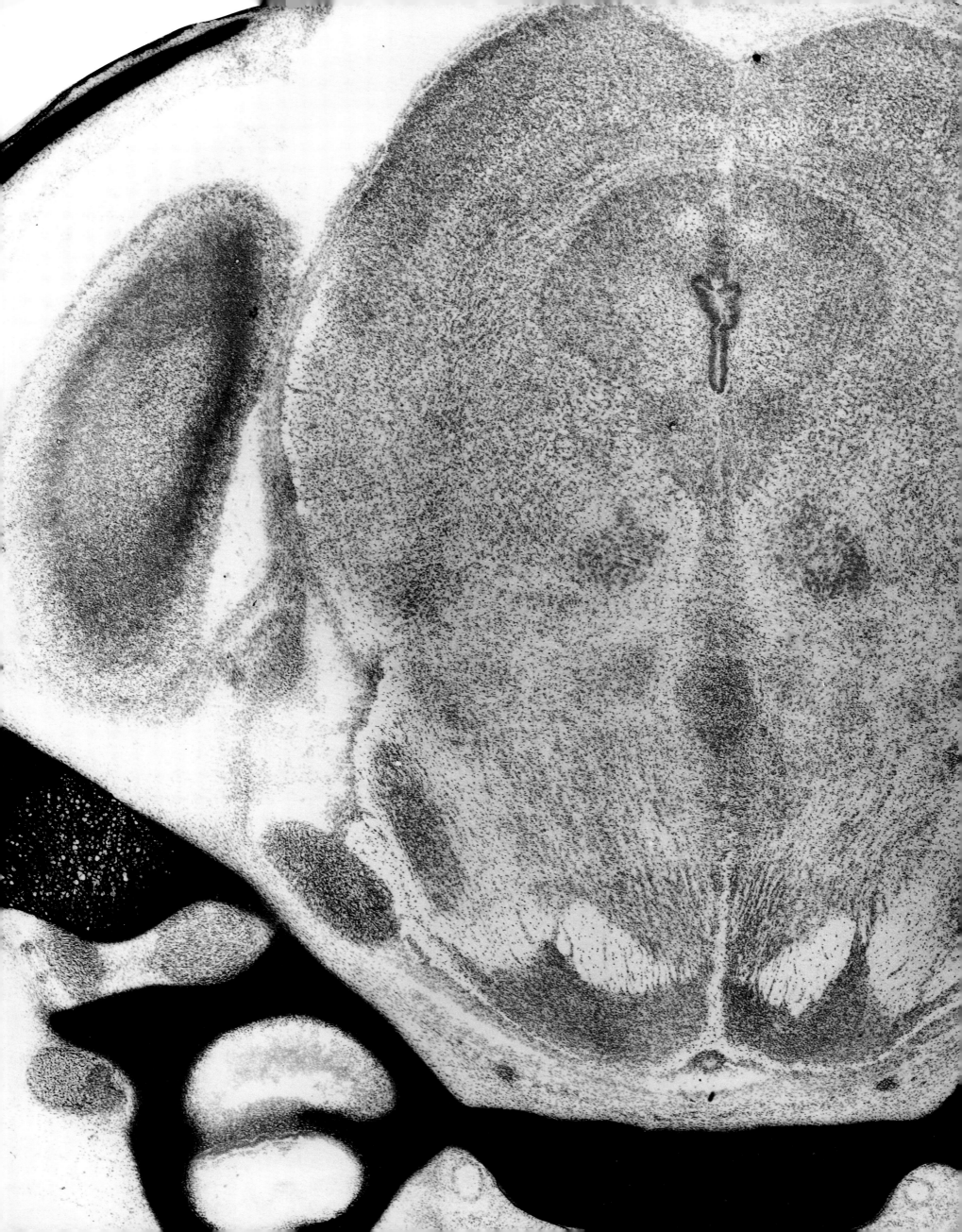

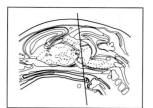

Figure 191
P0-47

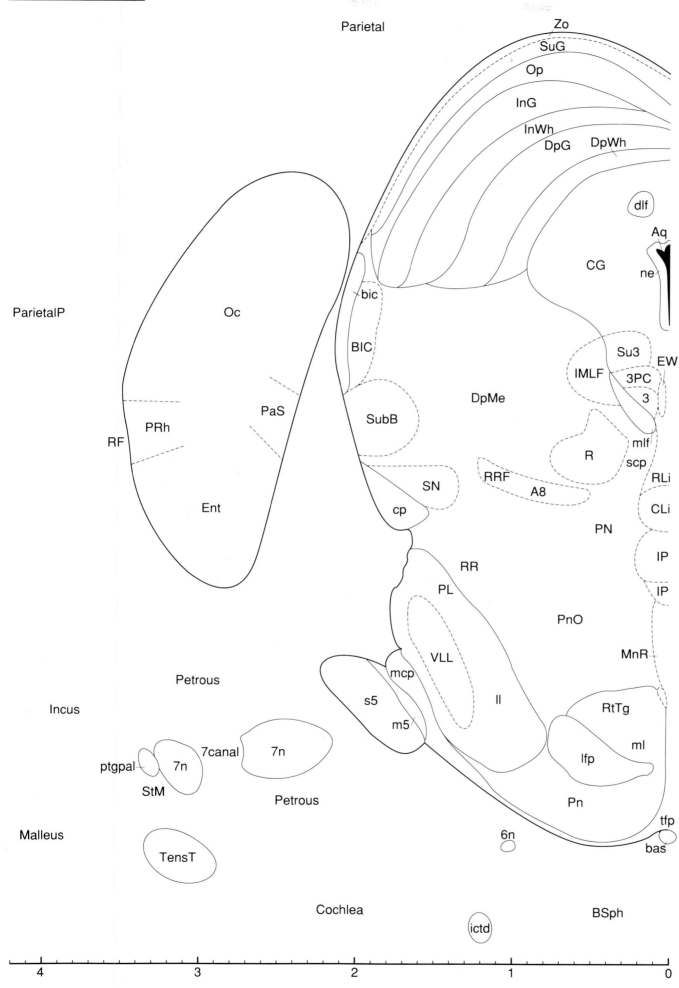

3   oculomotor nu
3PC   oculomotor nu, parvocellular
6n   abducens nerve or its root
7canal   canal of the facial nerve
7n   facial nerve or its root
A8   A8 dopamine cells
Aq   aqueduct
bas   basilar artery
BIC   nu brachium inferior colliculus
bic   brachium inferior colliculus
BSph basisphenoid bone
CG   central gray
CLi   caudal linear nu raphe
Cochlea   cochlea
cp   cerebral peduncle, basal
dlf   dorsal longitudinal fasciculus
DpG   deep gray layer of superior coll
DpMe deep mesencephalic nu
DpWh deep white layer of super coll
Ent   entorhinal cortex
EW   Edinger-Westphal nu
ictd internal carotid artery
IMLF interstitial nu mlf
Incus   incus (ossicle)
InG   intermediate gray layer sup col
InWh intermediate white layer sup col
IP   interpeduncular nu
lfp   longitudinal fasciculus pons
ll   lateral lemniscus
m5   motor root trigeminal nerve
Malleus   malleus (ossicle)
mcp   middle cerebellar peduncle
ml   medial lemniscus
mlf   medial longitudinal fasciculus
MnR   median raphe nu
ne   neuroepithelium
Oc   occipital cortex
Op   optic nerve layer superior collicul
Parietal   parietal bone
ParietalP   parietal plate
PaS   parasubiculum
Petrous   petrous part, temporal bone
PL   paralemniscal nu
PN   paranigral nu
Pn   pontine nuclei
PnO   pontine reticular nu, oral
PRh   perirhinal cortex
ptgpal   pterygopalatine artery
R   red nu
RF   rhinal fissure
RLi   rostral linear nu raphe
RR   retrorubral nu
RRF   retrorubral field
RtTg reticulotegmental nu pons
s5   sensory root trigeminal nerve
scp   superior cerebellar peduncle
SN   substantia nigra
StM   sternomastoid muscle
Su3   supraoculomotor central gray
SubB subbrachial nu
SuG   superficial gray layer sup col
TensT   tensor tympani muscle
tfp   transverse fibers pons
VLL   ventral nu lateral lemniscus
Zo   zonal layer of the superior coll

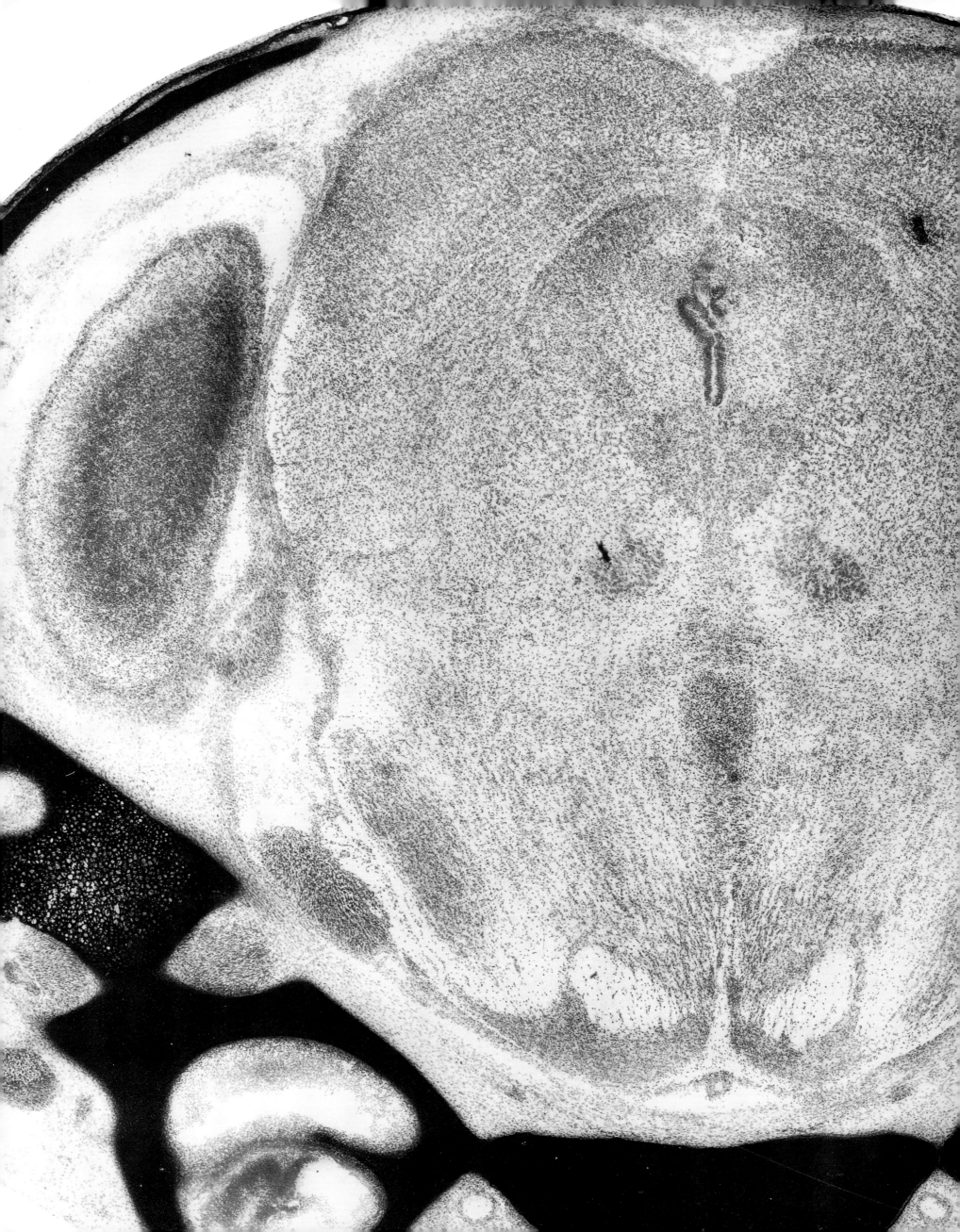

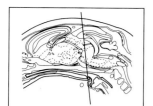

Figure 192
P0-48

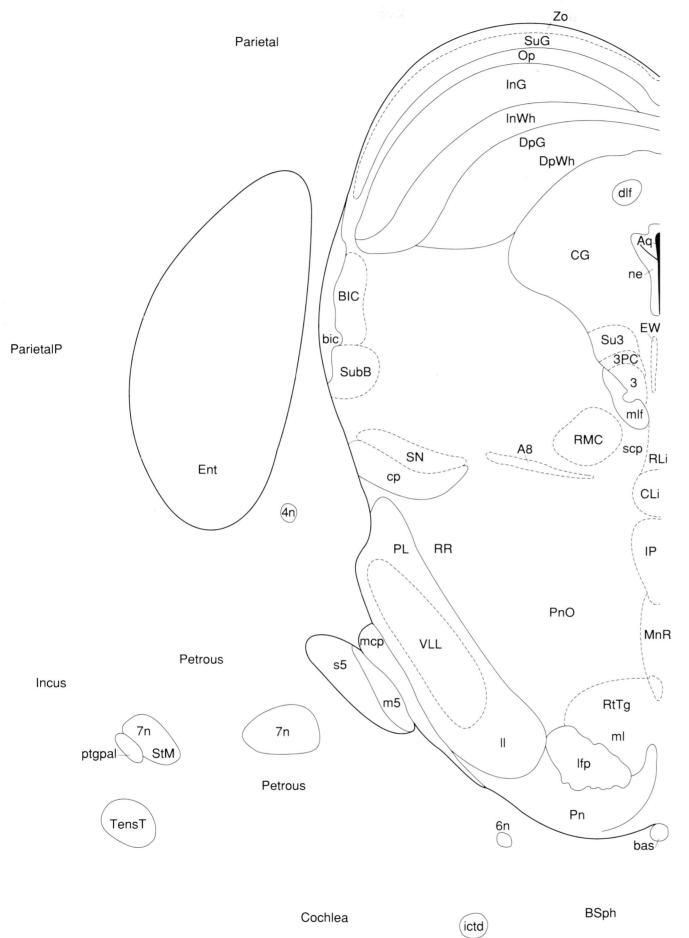

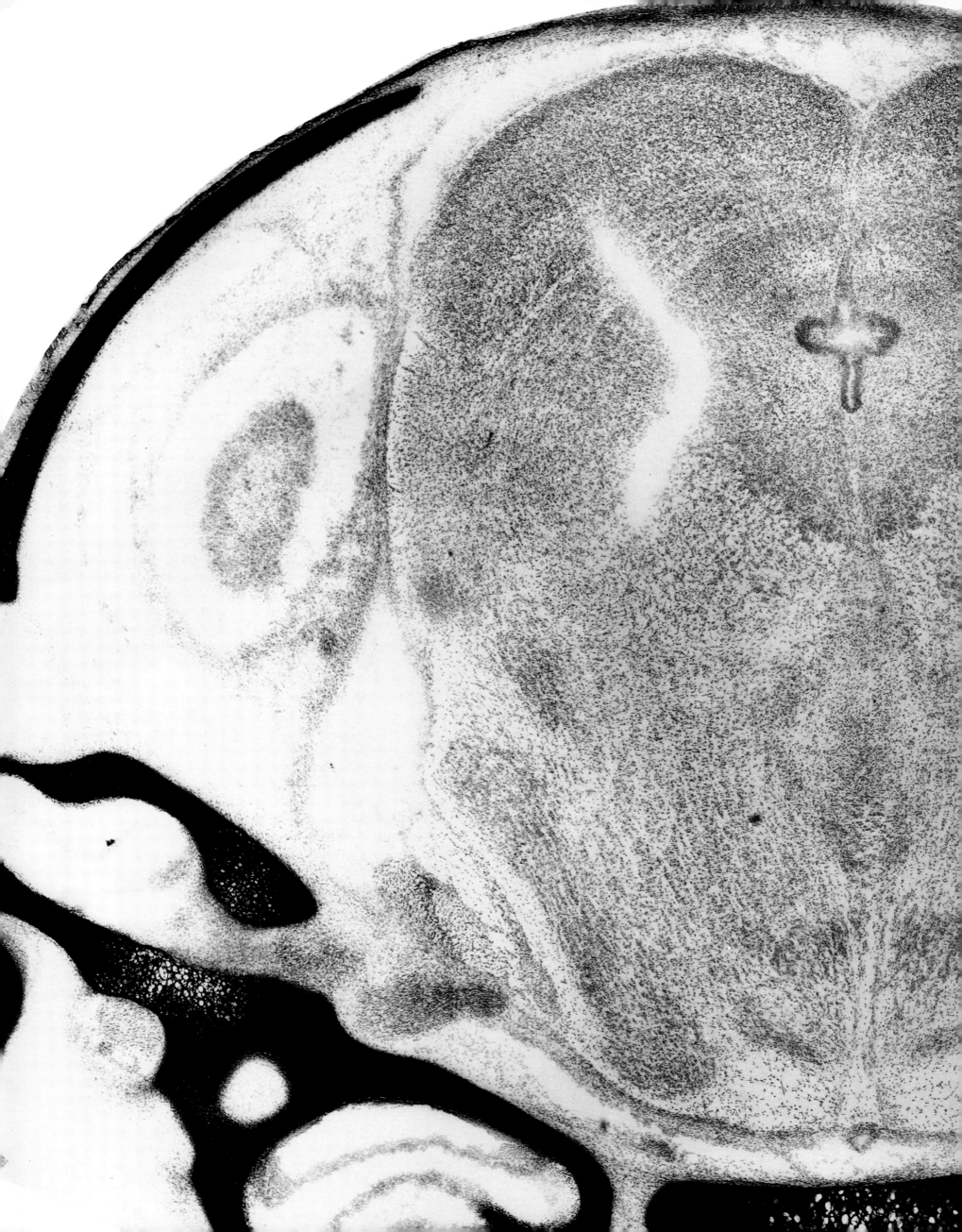

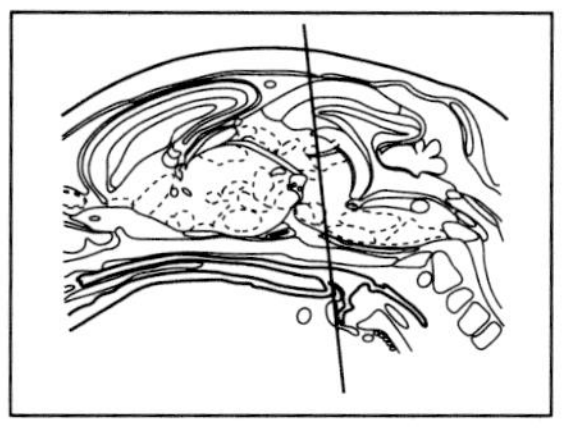

Figure 193
P0-49

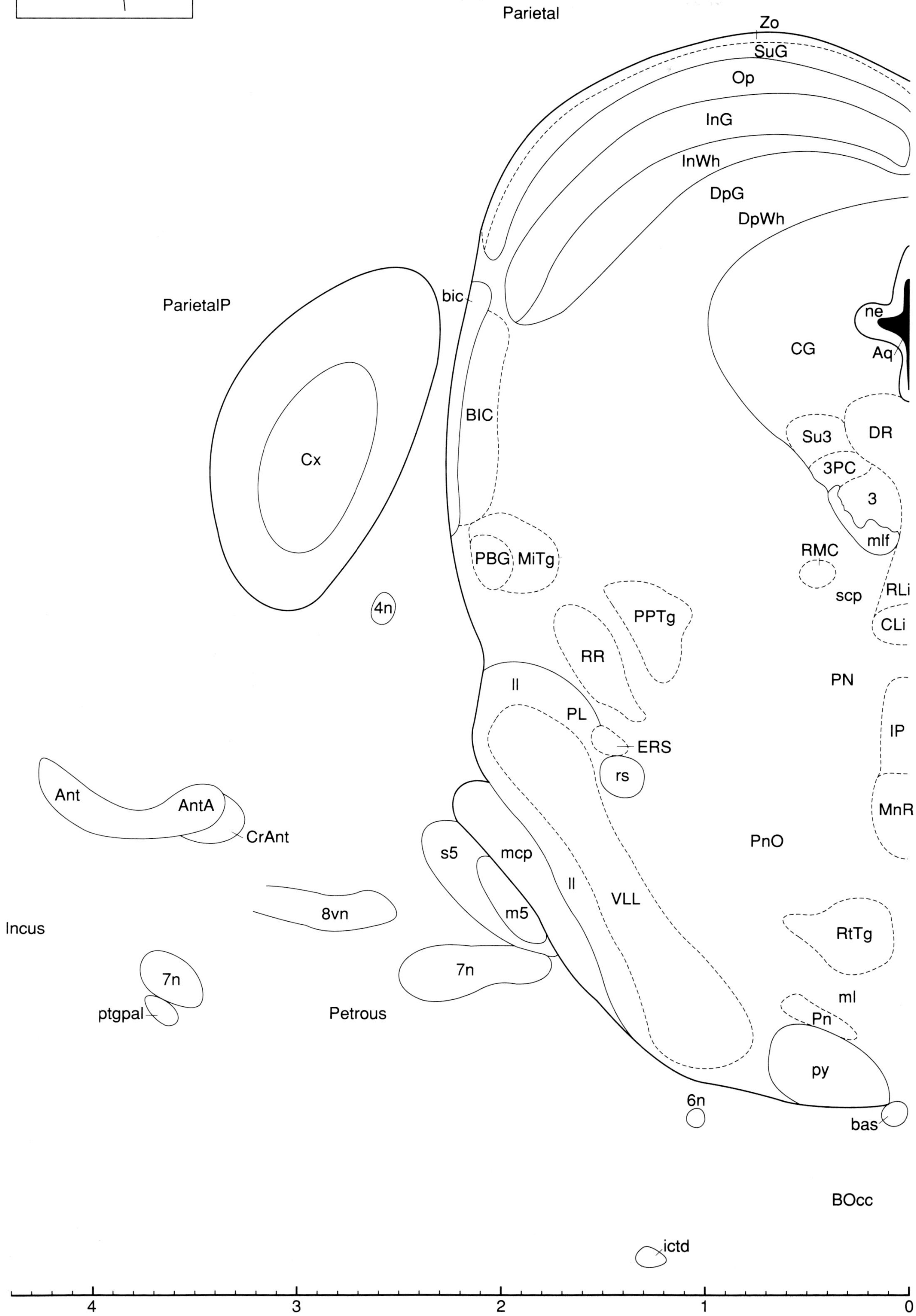

3   oculomotor nu
3PC   oculomotor nu, parvocellular
4n   trochlear nerve or its root
6n   abducens nerve or its root
7n   facial nerve or its root
8vn   vestibular root vestibulococh n
Ant   anterior semicircular duct
AntA   ampulla of ant semicircular duct
Aq   aqueduct
bas   basilar artery
BIC   nu brachium inferior colliculus
bic   brachium inferior colliculus
BOcc   basioccipital bone
CG   central gray
CLi   caudal linear nu raphe
CrAnt   crista of anterior ampulla
Cx   cerebral cortex
DpG   deep gray layer of superior coll
DpWh   deep white layer of super coll
DR   dorsal raphe nu
ERS   epirubrospinal nu
ictd   internal carotid artery
Incus   incus (ossicle)
InG   intermediate gray layer sup col
InWh   intermediate white layer sup col
IP   interpeduncular nu
ll   lateral lemniscus
m5   motor root trigeminal nerve
mcp   middle cerebellar peduncle
MiTg   microcellular tegmental nu
ml   medial lemniscus
mlf   medial longitudinal fasciculus
MnR   median raphe nu
ne   neuroepithelium
Op   optic nerve layer superior collicul
Parietal   parietal bone
ParietalP   parietal plate
PBG   parabigeminal nu
Petrous   petrous part, temporal bone
PL   paralemniscal nu
PN   paranigral nu
Pn   pontine nuclei
PnO   pontine reticular nu, oral
PPTg   pedunculopontine tegmental nu
ptgpal   pterygopalatine artery
py   pyramidal tract
RLi   rostral linear nu raphe
RMC   red nu, magnocellular
RR   retrorubral nu
rs   rubrospinal tract
RtTg   reticulotegmental nu pons
s5   sensory root trigeminal nerve
scp   superior cerebellar peduncle
Su3   supraoculomotor central gray
SuG   superficial gray layer sup col
VLL   ventral nu lateral lemniscus
Zo   zonal layer of the superior coll

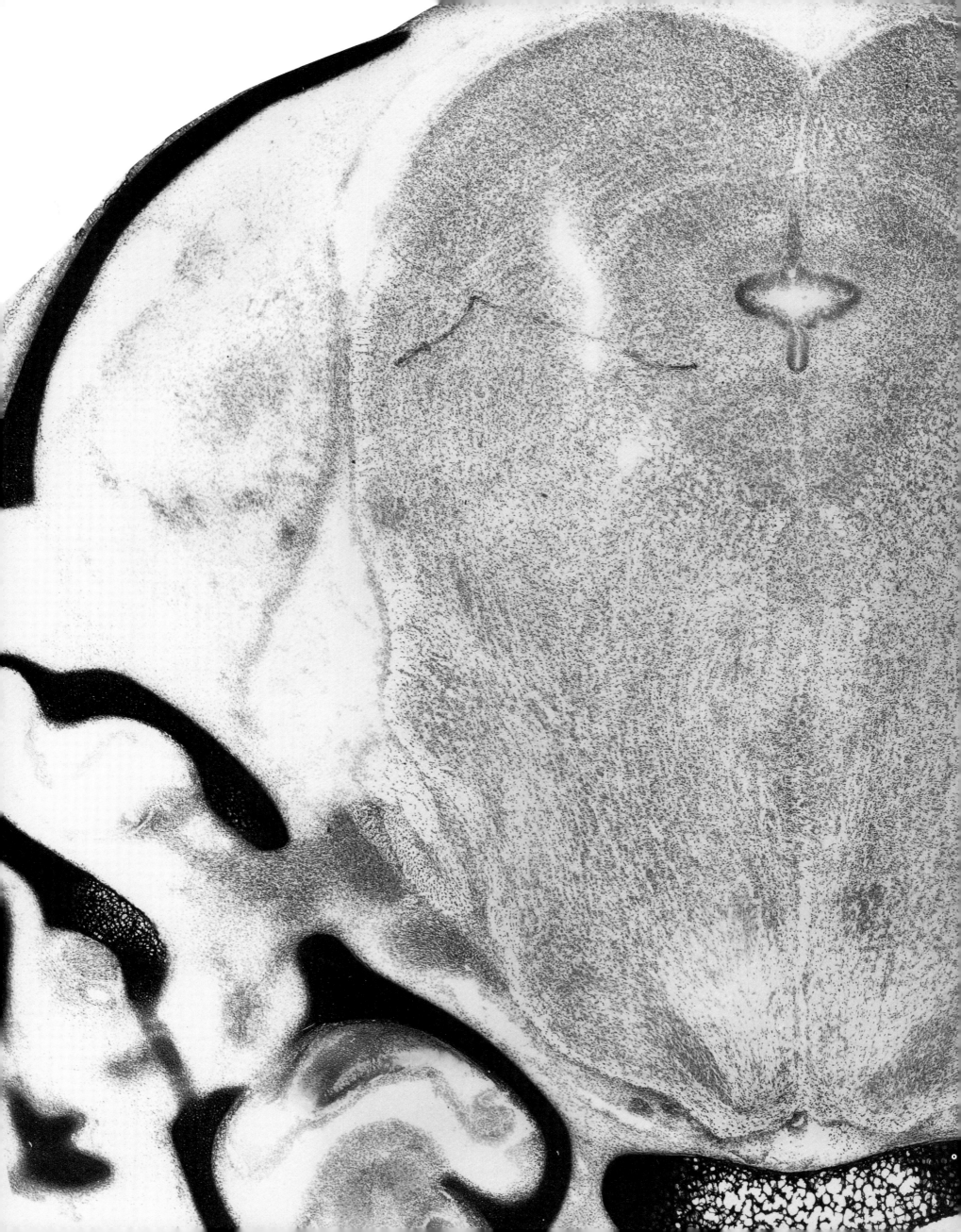

Figure 194
P0-50

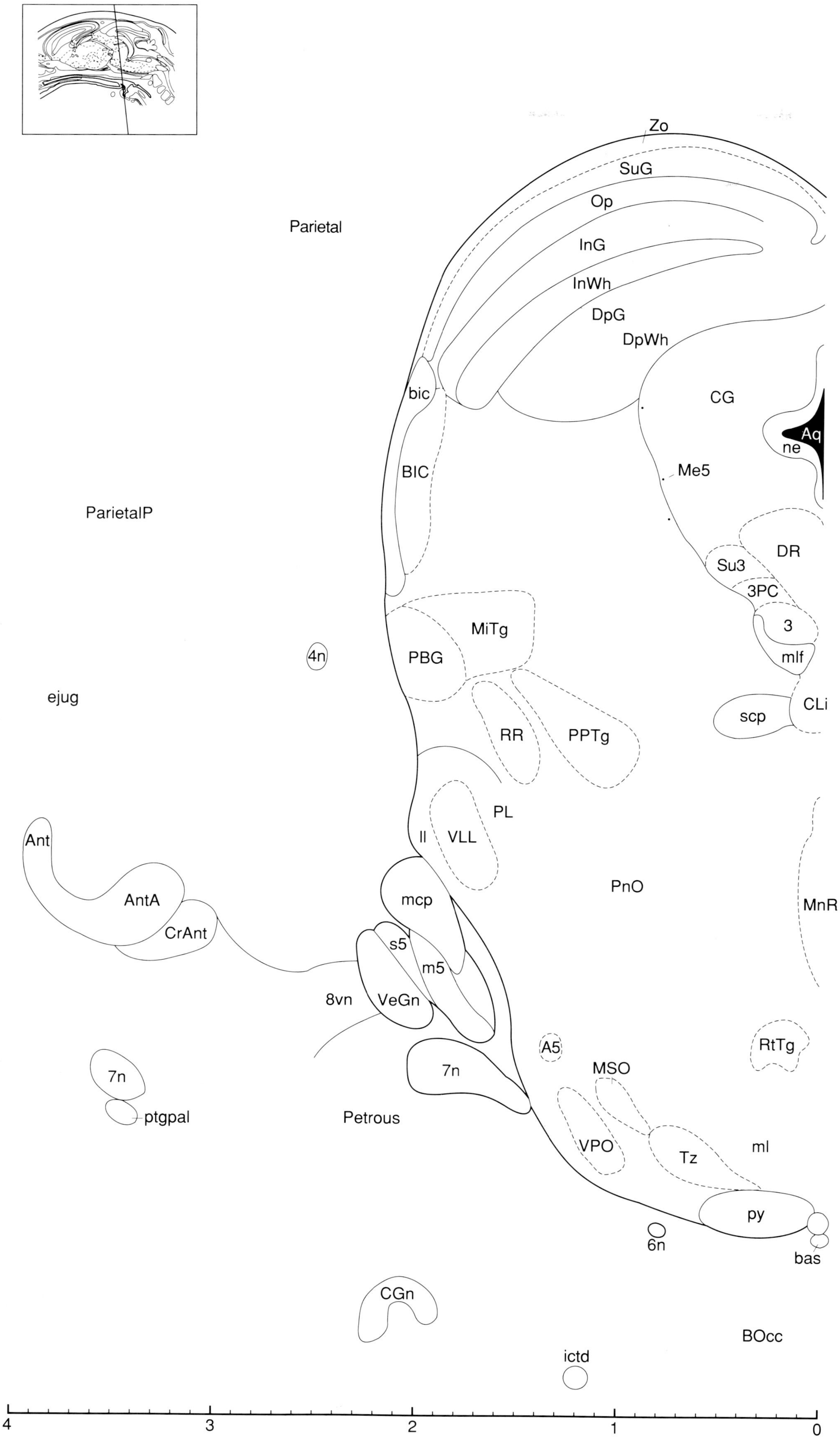

Zo
SuG
Op
InG
Parietal
InWh
DpG
DpWh
bic
CG
BIC
Aq
ne
ParietalP
Me5
DR
Su3
3PC
MiTg
3
PBG
mlf
4n
ejug
scp
CLi
RR
PPTg
PL
Ant
ll
VLL
AntA
PnO
CrAnt
mcp
MnR
s5
m5
8vn
VeGn
A5
RtTg
7n
MSO
7n
ptgpal
Petrous
VPO
Tz
ml
py
6n
bas
CGn
BOcc
ictd

4   3   2   1   0

3   oculomotor nu
3PC   oculomotor nu, parvocellular
4n   trochlear nerve or its root
6n   abducens nerve or its root
7n   facial nerve or its root
8vn   vestibular root vestibulococh n
A5   A5 noradrenaline cells
Ant   anterior semicircular duct
AntA   ampulla of ant semicircular duct
Aq   aqueduct
bas   basilar artery
BIC   nu brachium inferior colliculus
bic   brachium inferior colliculus
BOcc   basioccipital bone
CG   central gray
CGn   cochlear (spiral) ganglion
CLi   caudal linear nu raphe
CrAnt   crista of anterior ampulla
DpG   deep gray layer of superior coll
DpWh   deep white layer of super coll
DR   dorsal raphe nu
ejug   external jugular vein
ictd   internal carotid artery
InG   intermediate gray layer sup col
InWh   intermediate white layer sup col
ll   lateral lemniscus
m5   motor root trigeminal nerve
mcp   middle cerebellar peduncle
Me5   mesencephalic trigeminal nu
MiTg   microcellular tegmental nu
ml   medial lemniscus
mlf   medial longitudinal fasciculus
MnR   median raphe nu
MSO   medial superior olive
ne   neuroepithelium
Op   optic nerve layer superior collicul
Parietal   parietal bone
ParietalP   parietal plate
PBG   parabigeminal nu
Petrous   petrous part, temporal bone
PL   paralemniscal nu
PnO   pontine reticular nu, oral
PPTg   pedunculopontine tegmental nu
ptgpal   pterygopalatine artery
py   pyramidal tract
RR   retrorubral nu
RtTg   reticulotegmental nu pons
s5   sensory root trigeminal nerve
scp   superior cerebellar peduncle
Su3   supraoculomotor central gray
SuG   superficial gray layer sup col
Tz   nu trapezoid body
VeGn   vestibular ganglion
VLL   ventral nu lateral lemniscus
VPO   ventral periolivary nuclei
Zo   zonal layer of the superior coll

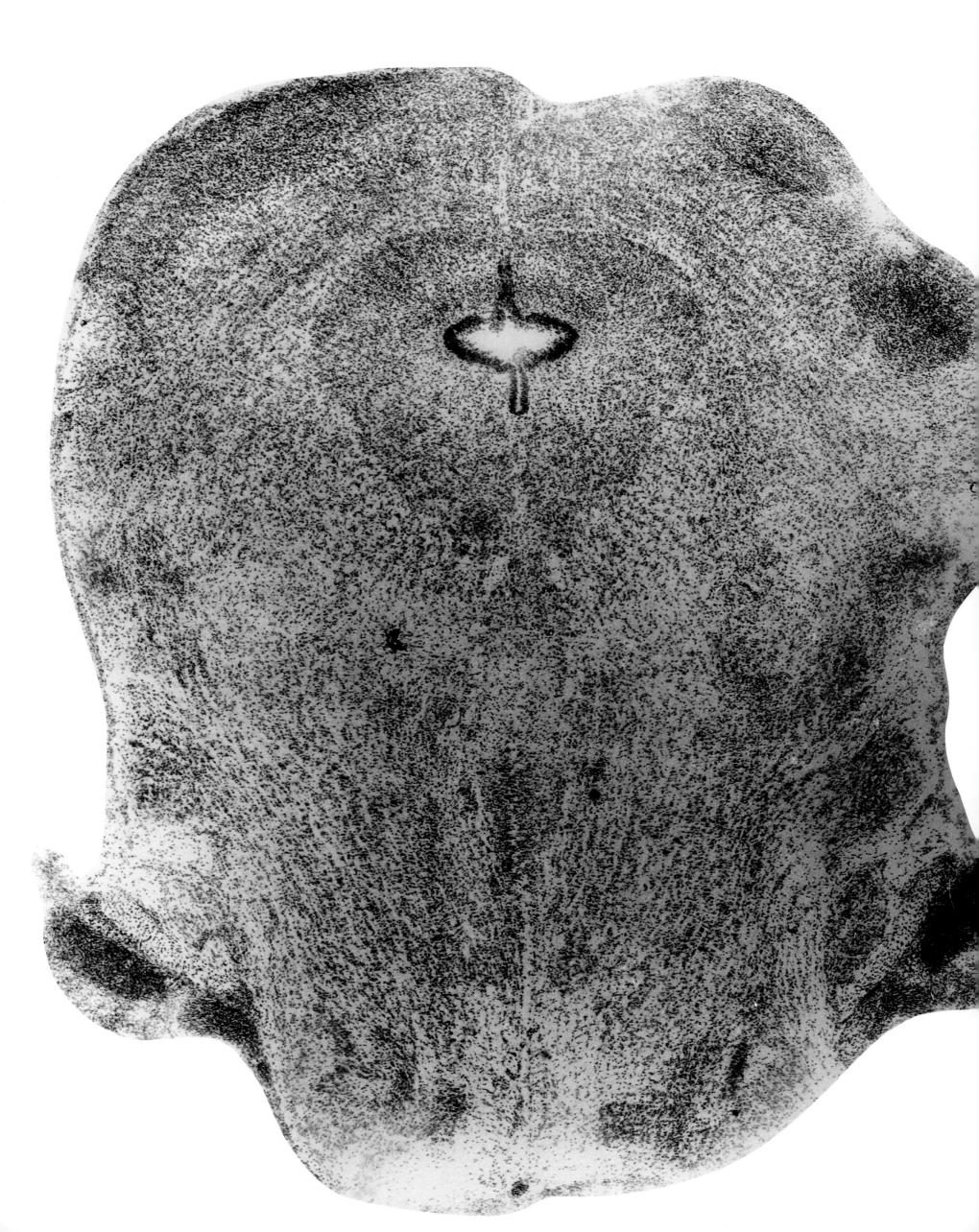

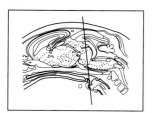

Figure 195
P0-51

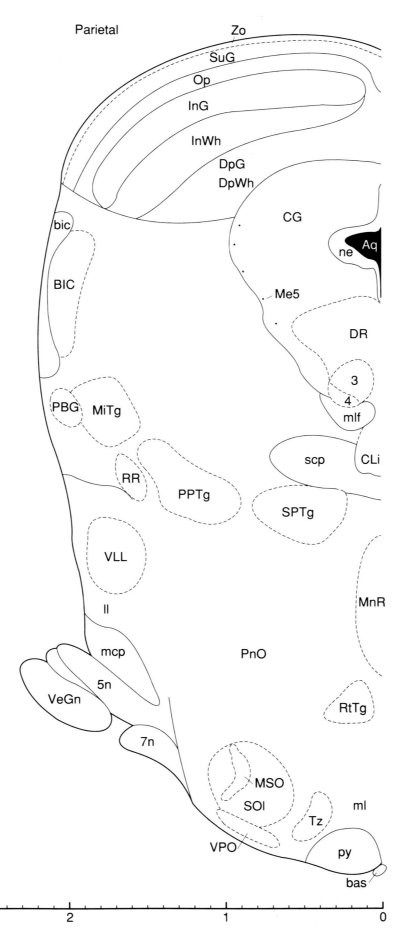

Parietal

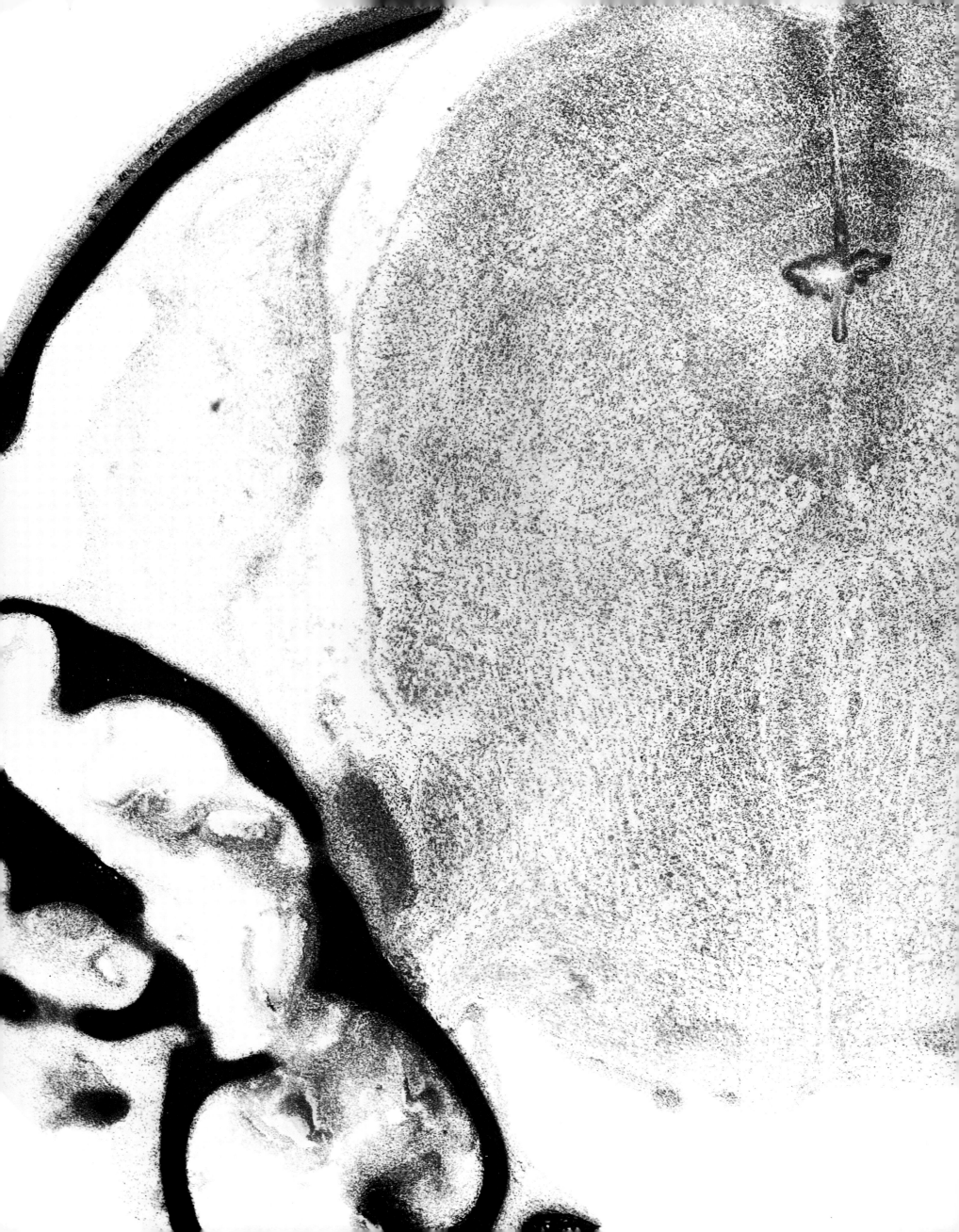

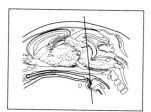

Figure 196
P0-52

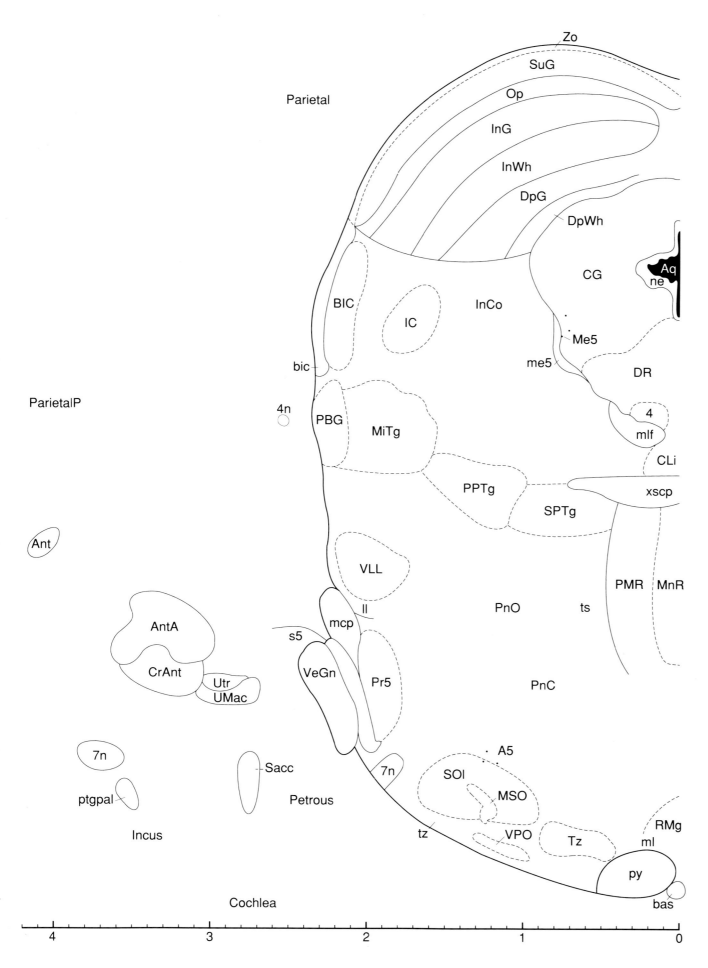

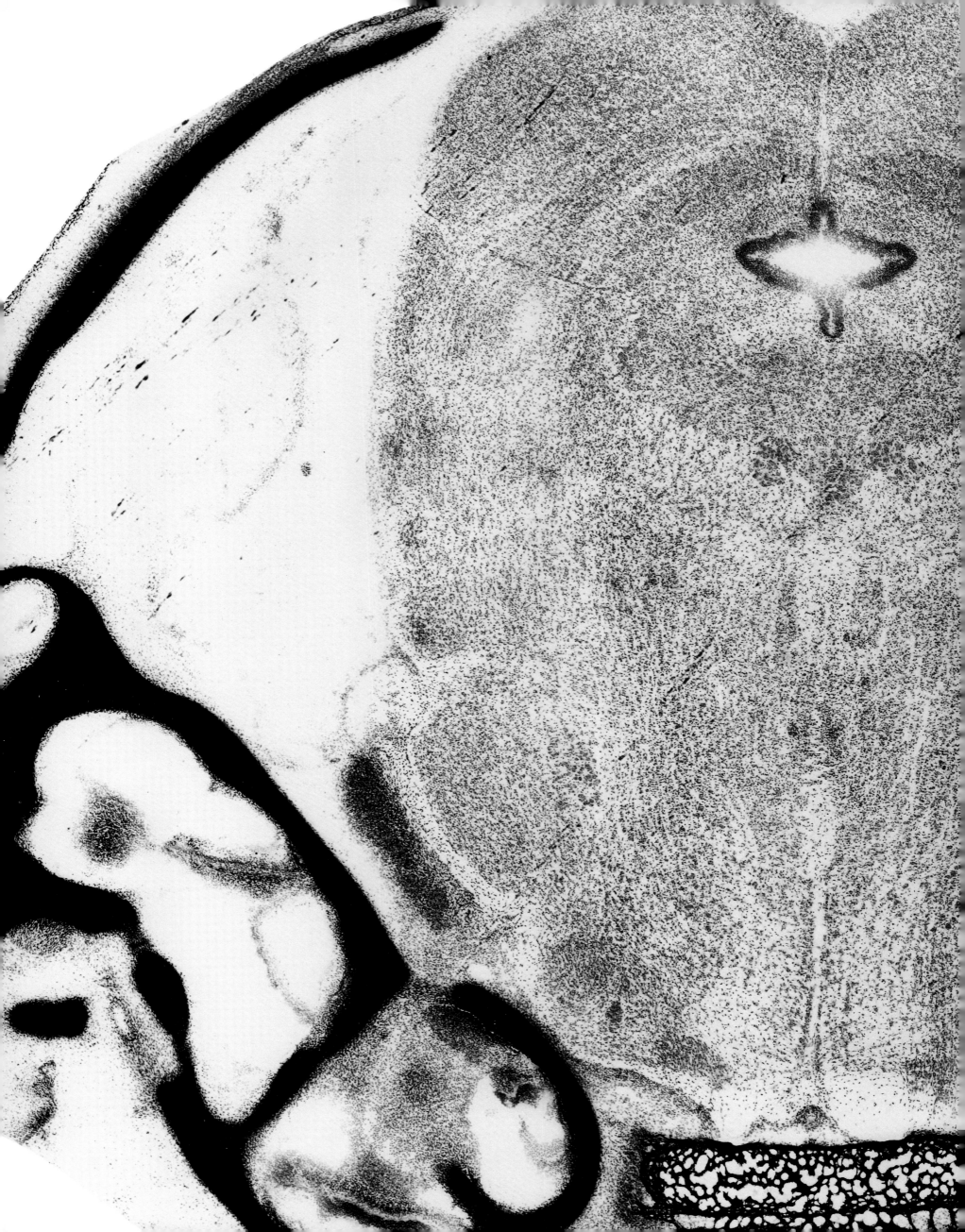

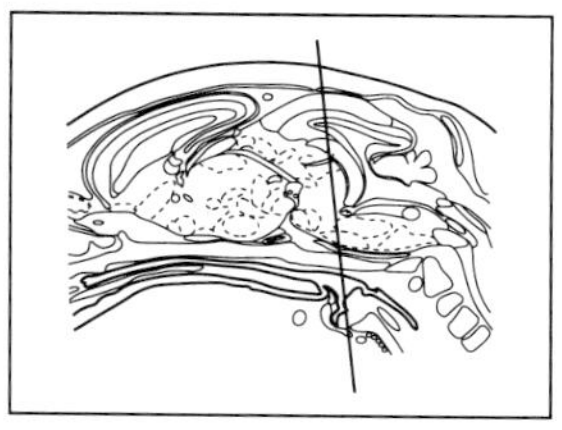

Figure 197
P0-53

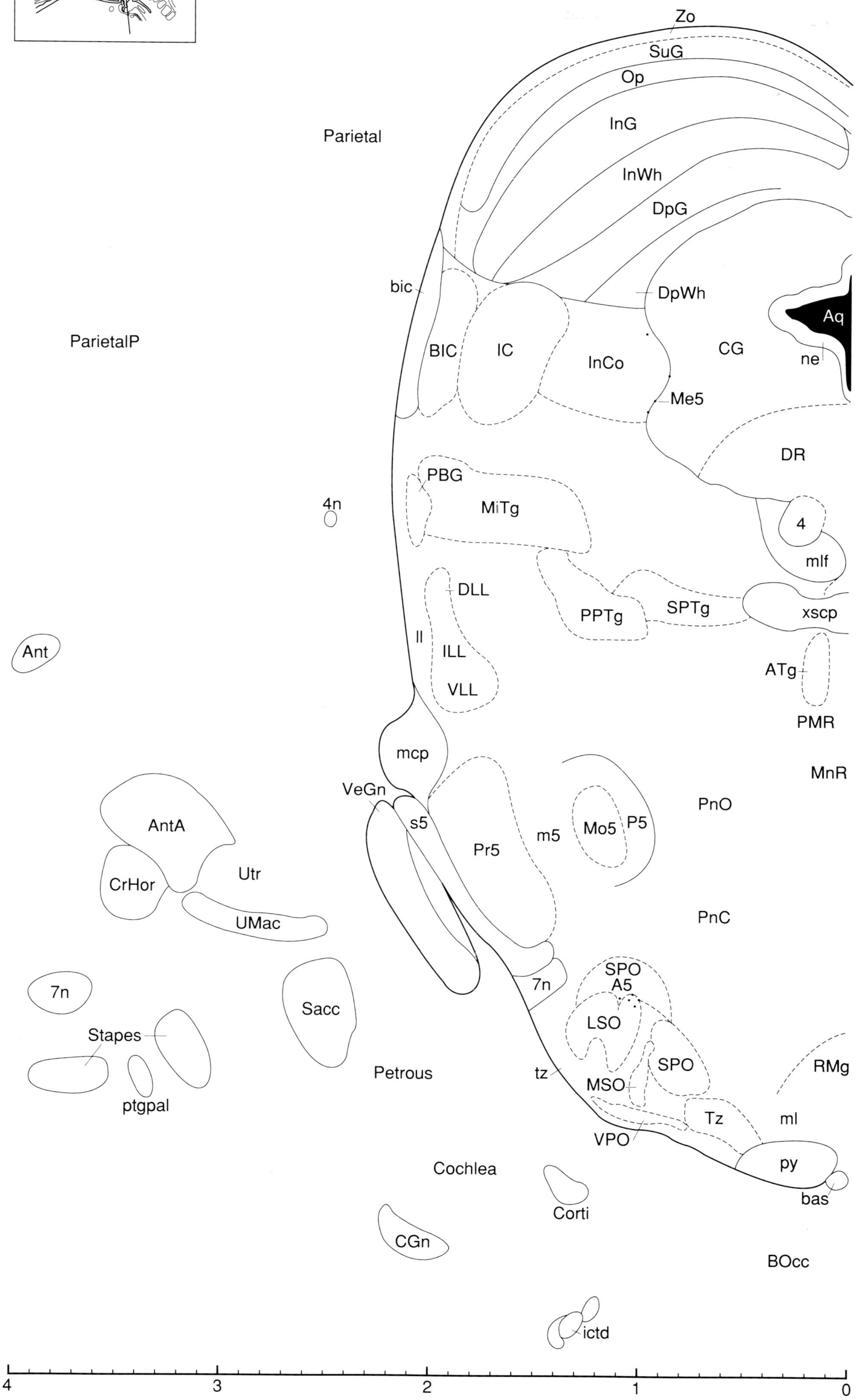

4   trochlear nu
4n   trochlear nerve or its root
7n   facial nerve or its root
A5   A5 noradrenaline cells
Ant anterior semicircular duct
AntA ampulla of ant semicircular duct
Aq   aqueduct
ATg anterior tegmental nu
bas basilar artery
BIC nu brachium inferior colliculus
bic brachium inferior colliculus
BOcc basioccipital bone
CG   central gray
CGn cochlear (spiral) ganglion
Cochlea   cochlea
Corti   organ of Corti
CrHor   crista of horizontal ampulla
DLL dorsal nu lateral lemniscus
DpG deep gray layer of superior coll
DpWh deep white layer of super coll
DR   dorsal raphe nu
IC   inferior colliculus
ictd internal carotid artery
ILL intermediate nu lat lemniscus
InCo intercollicular nu
InG intermediate gray layer sup col
InWh intermediate white layer sup col
ll   lateral lemniscus
LSO lateral superior olive
m5   motor root trigeminal nerve
mcp middle cerebellar peduncle
Me5 mesencephalic trigeminal nu
MiTg microcellular tegmental nu
ml   medial lemniscus
mlf medial longitudinal fasciculus
MnR median raphe nu
Mo5 motor trigeminal nu
MSO medial superior olive
ne   neuroepithelium
Op   optic nerve layer superior collicul
P5   peritrigeminal zone
Parietal   parietal bone
ParietalP parietal plate
PBG parabigeminal nu
Petrous   petrous part, temporal bone
PMR paramedian raphe
PnC pontine reticular nu, caudal
PnO pontine reticular nu, oral
PPTg pedunculopontine tegmental nu
Pr5 principal sensory trigeminal nu
ptgpal   pterygopalatine artery
py   pyramidal tract
RMg raphe magnus nu
s5   sensory root trigeminal nerve
Sacc saccule
SPO superior paraolivary nu
SPTg subpeduncular tegmental nu
Stapes   stapes (ossicle)
SuG superficial gray layer sup col
Tz   nu trapezoid body
tz   trapezoid body
UMac utricular macula
Utr utricle
VeGn vestibular ganglion
VLL ventral nu lateral lemniscus
VPO ventral periolivary nuclei
xscp decussation sup Cb peduncle
Zo   zonal layer of the superior coll

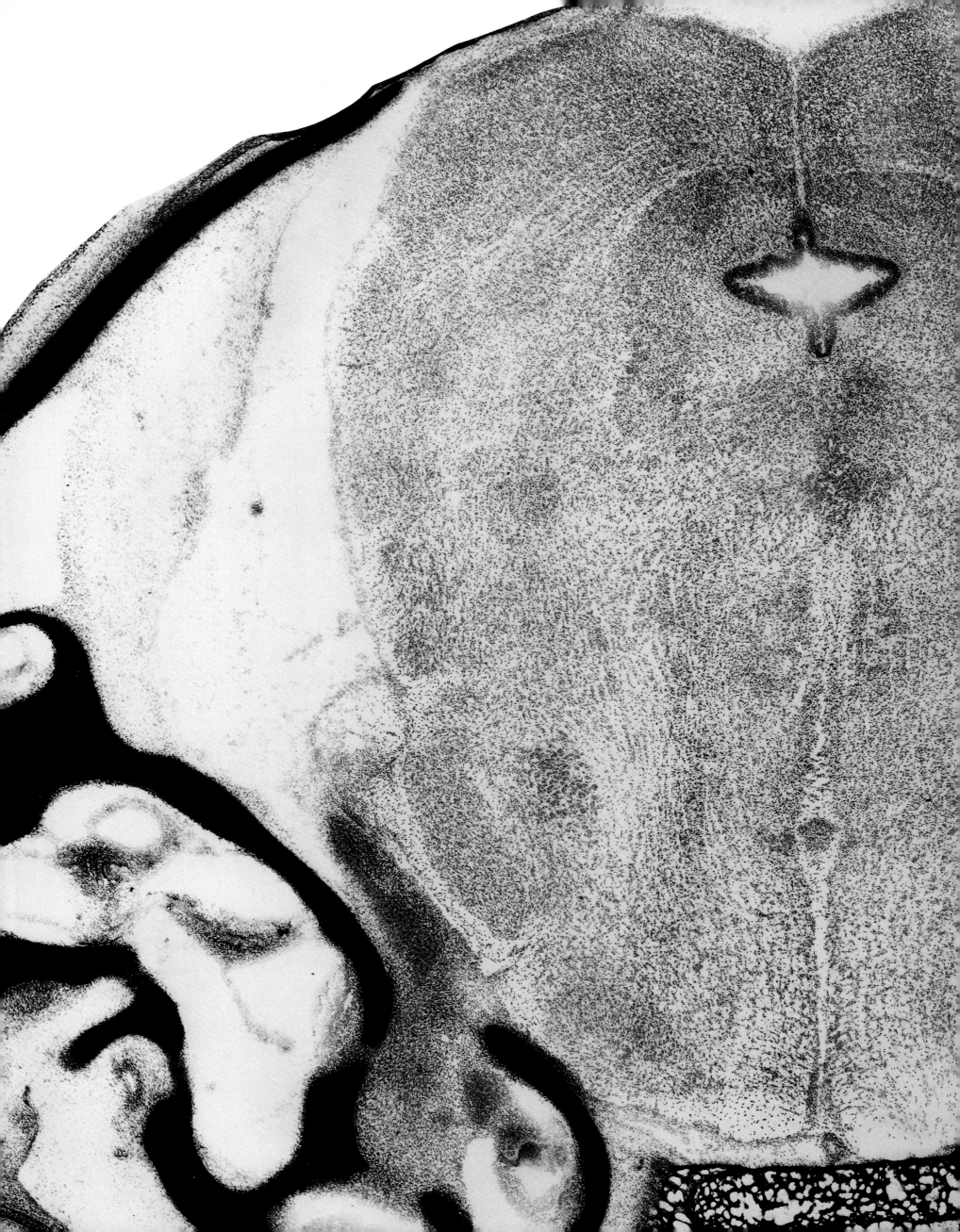

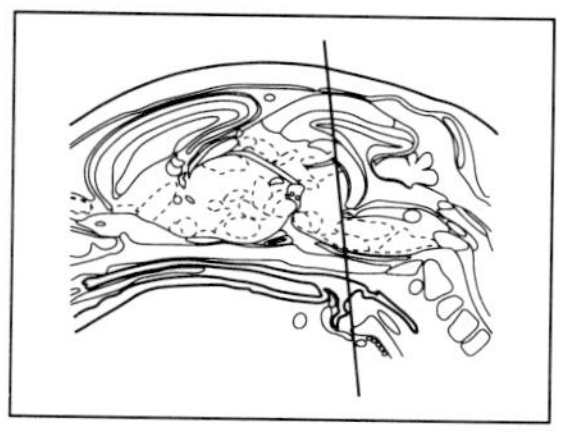

Figure 198
P0-54

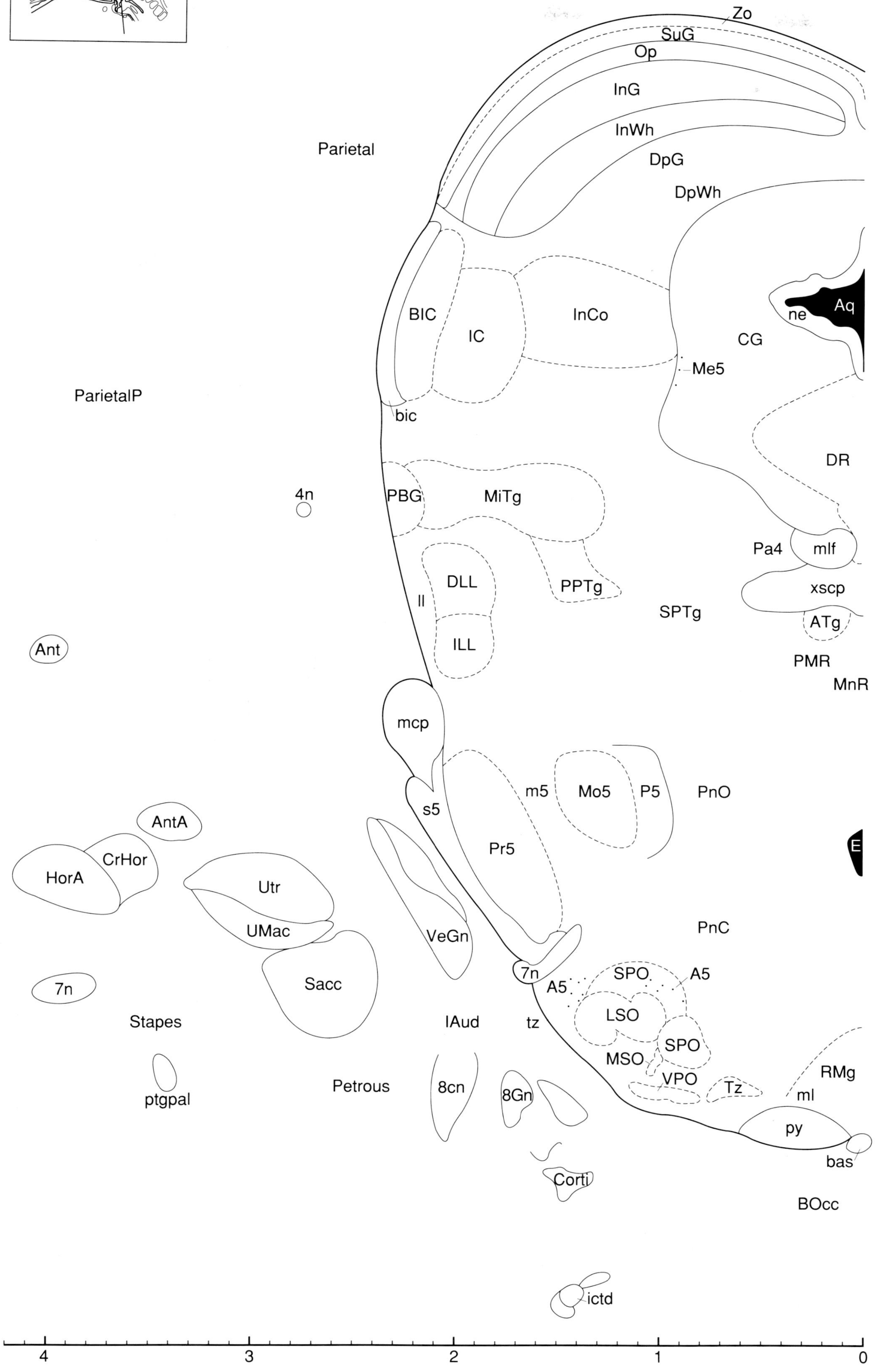

Zo
SuG
Op
InG
InWh
DpG
DpWh

Parietal

BIC
IC
InCo
CG
ne
Aq
Me5

bic

ParietalP

DR

4n

PBG
MiTg

Pa4
mlf

DLL
PPTg
xscp

ll
ATg

ILL
SPTg

Ant
PMR
MnR

mcp

s5
m5
Mo5
P5
PnO

AntA

CrHor
Pr5

HorA

Utr

UMac
VeGn

PnC

Sacc

7n
A5
SPO
A5

7n
A5

Stapes
LSO

IAud
tz
SPO

Petrous
MSO
RMg

ptgpal
8cn
8Gn
VPO
Tz
ml

py

bas

Corti

BOcc

ictd

4n   trochlear nerve or its root
7n   facial nerve or its root
8cn   cochlear root vestibulococh nerve
8Gn   ganglion of vestibulocochlear n
A5   A5 noradrenaline cells
Ant   anterior semicircular duct
AntA   ampulla of ant semicircular duct
Aq   aqueduct
ATg   anterior tegmental nu
bas   basilar artery
BIC   nu brachium inferior colliculus
bic   brachium inferior colliculus
BOcc   basioccipital bone
CG   central gray
Corti   organ of Corti
CrHor   crista of horizontal ampulla
DLL   dorsal nu lateral lemniscus
DpG   deep gray layer of superior coll
DpWh   deep white layer of super coll
DR   dorsal raphe nu
E   ependyma and subependymal layer
HorA   ampulla of hor semicircular duct
IAud   internal auditory meatus
IC   inferior colliculus
ictd   internal carotid artery
ILL   intermediate nu lat lemniscus
InCo   intercollicular nu
InG   intermediate gray layer sup col
InWh   intermediate white layer sup col
ll   lateral lemniscus
LSO   lateral superior olive
m5   motor root trigeminal nerve
mcp   middle cerebellar peduncle
Me5   mesencephalic trigeminal nu
MiTg   microcellular tegmental nu
ml   medial lemniscus
mlf   medial longitudinal fasciculus
MnR   median raphe nu
Mo5   motor trigeminal nu
MSO   medial superior olive
ne   neuroepithelium
Op   optic nerve layer superior collicul
P5   peritrigeminal zone
Pa4   paratrochlear nu
Parietal   parietal bone
ParietalP   parietal plate
PBG   parabigeminal nu
Petrous   petrous part, temporal bone
PMR   paramedian raphe
PnC   pontine reticular nu, caudal
PnO   pontine reticular nu, oral
PPTg   pedunculopontine tegmental nu
Pr5   principal sensory trigeminal nu
ptgpal   pterygopalatine artery
py   pyramidal tract
RMg   raphe magnus nu
s5   sensory root trigeminal nerve
Sacc   saccule
SPO   superior paraolivary nu
SPTg   subpeduncular tegmental nu
Stapes   stapes (ossicle)
SuG   superficial gray layer sup col
Tz   nu trapezoid body
tz   trapezoid body
UMac   utricular macula
Utr   utricle
VeGn   vestibular ganglion
VPO   ventral periolivary nuclei
xscp   decussation sup Cb peduncle
Zo   zonal layer of the superior coll

4   3   2   1   0

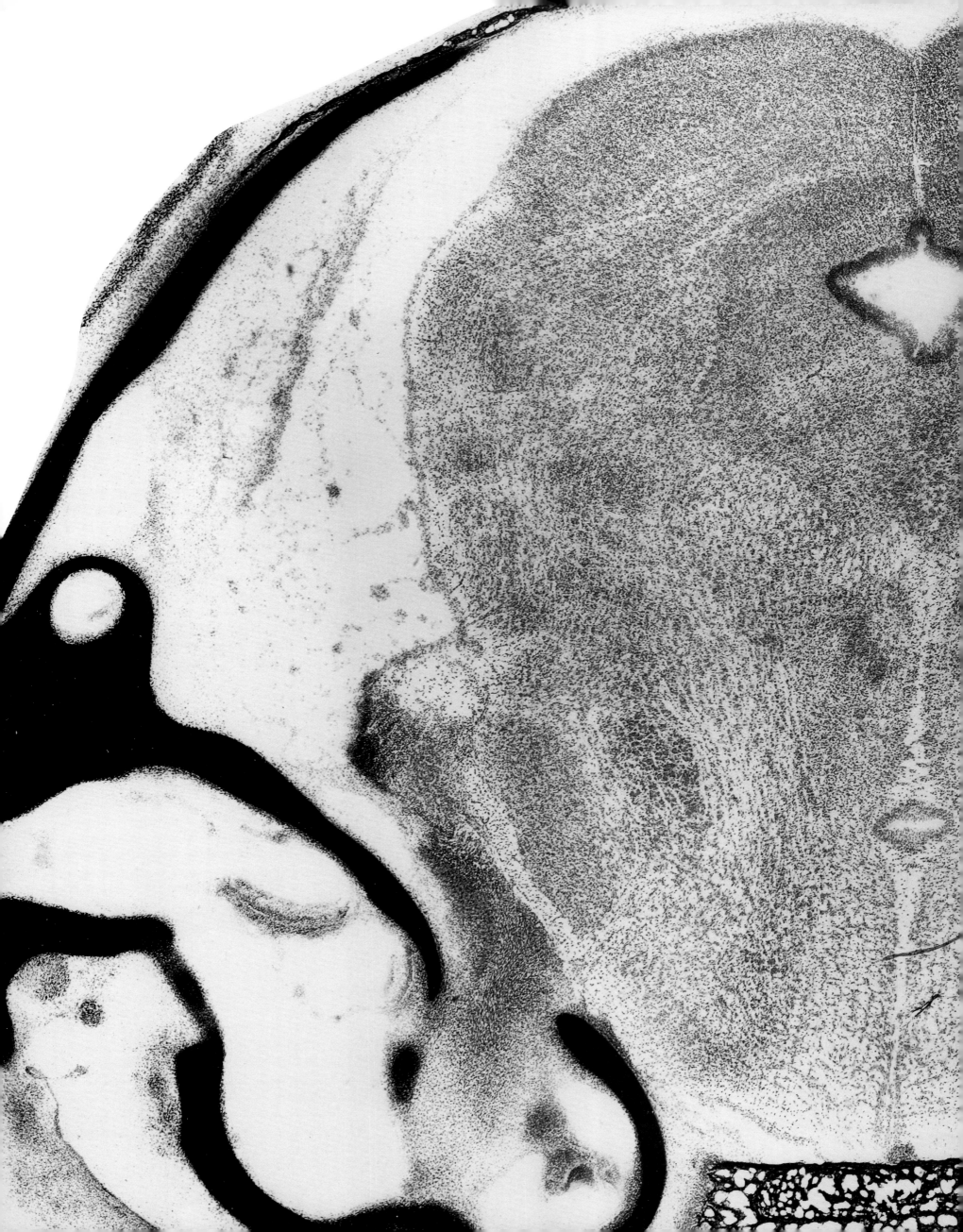

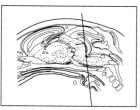

Figure 199
P0-55

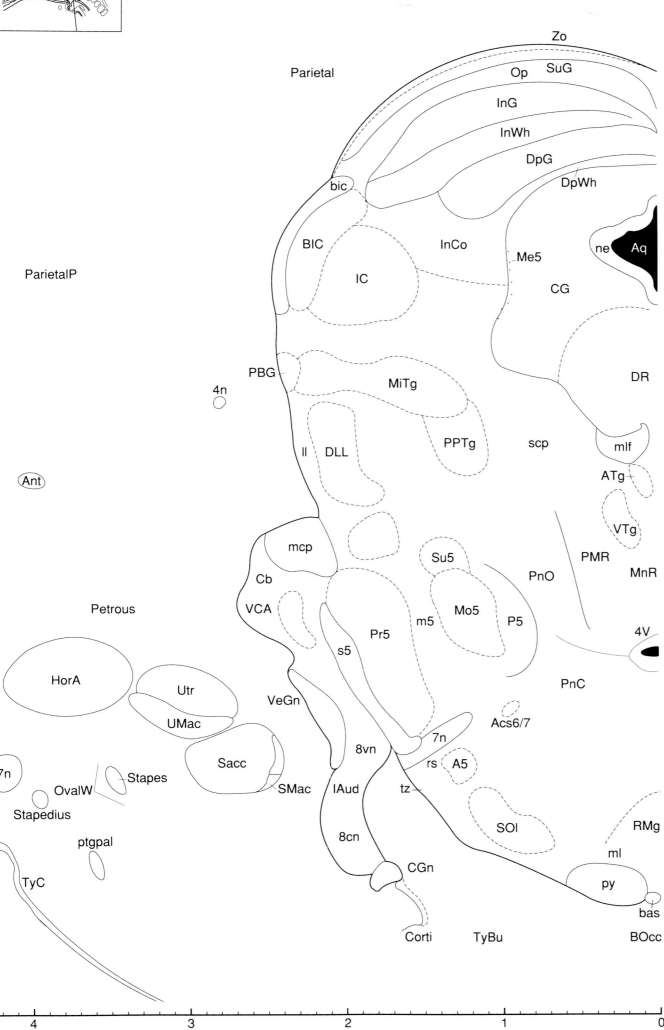

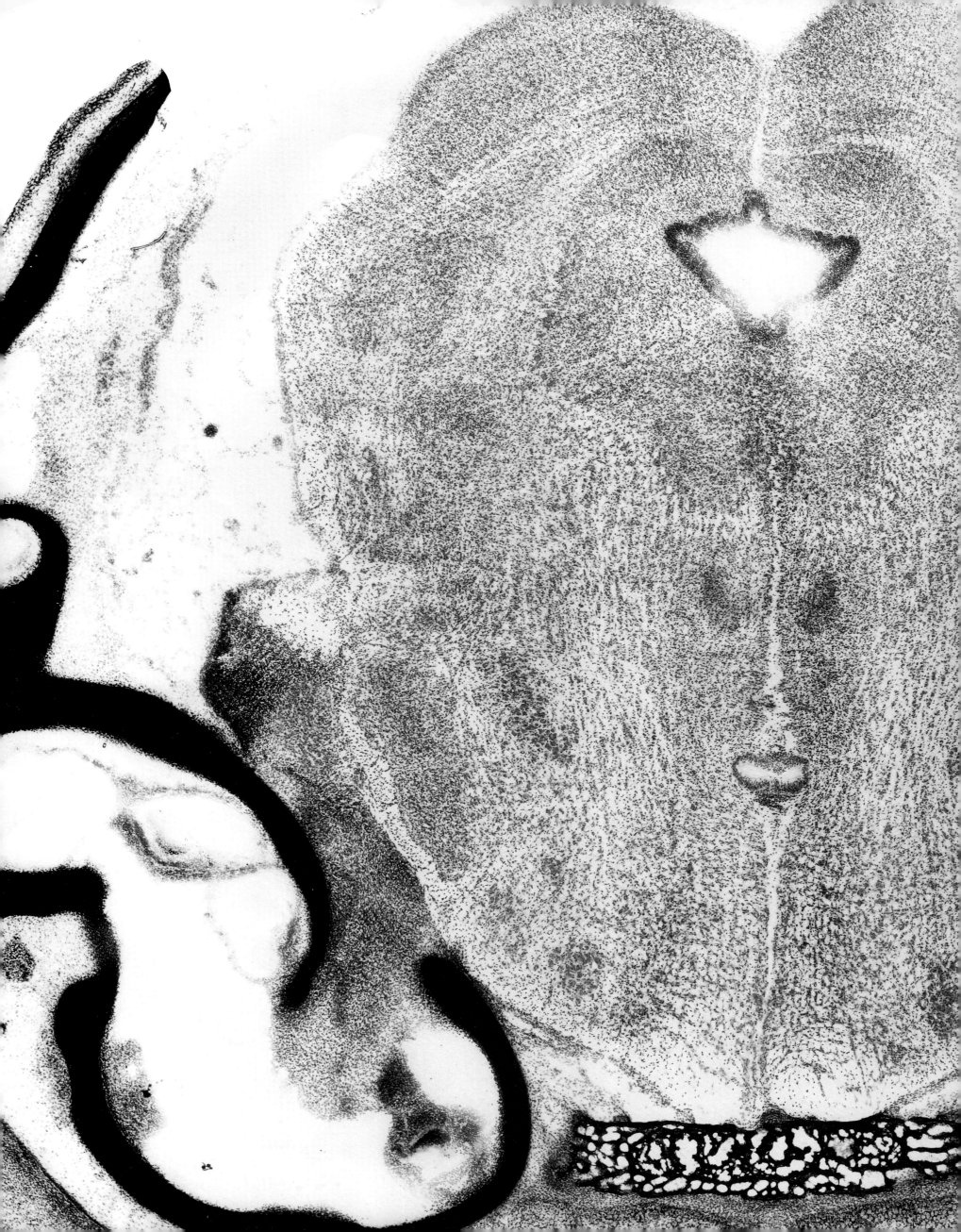

Figure 200
P0-56

Parietal

Zo

SuG

InG   Op

InWh

DpG

bic

DpWh

BIC

ne   Aq

IC

CG

Me5

DR

4n

MiTg   CnF

scp

ll   DLL   PPTg

mlf

PMR

MnR

mcp

VTg

Ant

LR4V

Pr5DM   Su5

Cb

Mo5   SubC

DMTg

VCA

Acs5

m5   PnO

Petrous

P5

HorA

s5

4V

Utr

8vn   Pr5VL

UMac

VeGn   7n   Acs6/7

Sacc   PnC

7n   A5

SMac   IAud   P7   LPGi   Gi

Stapedius   7

ptgpal   GiA

8cn   tz   RMg

TyC

ml

CGn   py

bas

Corti   BOcc

ictd

4   3   2   1   0

4n   trochlear nerve or its root
4V   4th ventricle
7   facial nu
7n   facial nerve or its root
8cn   cochlear root vestibulococh nerve
8vn   vestibular root vestibulococh n
A5   A5 noradrenaline cells
Acs5   accessory trigeminal nu
Acs6   accessory abducens nu
Acs7   accessory facial nu
Ant   anterior semicircular duct
Aq   aqueduct
bas   basilar artery
BIC   nu brachium inferior colliculus
bic   brachium inferior colliculus
BOcc   basioccipital bone
Cb   cerebellum
CG   central gray
CGn   cochlear (spiral) ganglion
CnF   cuneiform nu
Corti   organ of Corti
DLL   dorsal nu lateral lemniscus
DMTg   dorsomedial tegmental area
DpG   deep gray layer of superior coll
DpWh   deep white layer of super coll
DR   dorsal raphe nu
Gi   gigantocellular reticular nucleus
GiA   gigantocell reticular nu, alpha
HorA   ampulla of hor semicircular duct
IAud   internal auditory meatus
IC   inferior colliculus
ictd   internal carotid artery
InG   intermediate gray layer sup col
InWh   intermediate white layer sup col
ll   lateral lemniscus
LPGi   lateral paragigantocellular nu
LR4V   lateral recess 4th ventricle
m5   motor root trigeminal nerve
mcp   middle cerebellar peduncle
Me5   mesencephalic trigeminal nu
MiTg   microcellular tegmental nu
ml   medial lemniscus
mlf   medial longitudinal fasciculus
MnR   median raphe nu
Mo5   motor trigeminal nu
ne   neuroepithelium
Op   optic nerve layer superior collicul
P5   peritrigeminal zone
P7   perifacial zone
Parietal   parietal bone
ParietalP   parietal plate
Petrous   petrous part, temporal bone
PMR   paramedian raphe
PnC   pontine reticular nu, caudal
PnO   pontine reticular nu, oral
PPTg   pedunculopontine tegmental nu
Pr5DM   princ sens trigem nu, dorsome
Pr5VL   princ sens trigem nu, ventrolat
ptgpal   pterygopalatine artery
py   pyramidal tract
RMg   raphe magnus nu
s5   sensory root trigeminal nerve
Sacc   saccule
scp   superior cerebellar peduncle
SMac   saccular macula
Stapedius   stapedius muscle
Su5   supratrigeminal nu
SubC   subcoeruleus nu
SuG   superficial gray layer sup col
TyC   tympanic cavity
tz   trapezoid body
UMac   utricular macula
Utr   utricle
VCA   ventral cochlear nu, anterior
VeGn   vestibular ganglion
VTg   ventral tegmental nu
Zo   zonal layer of the superior coll

ParietalP

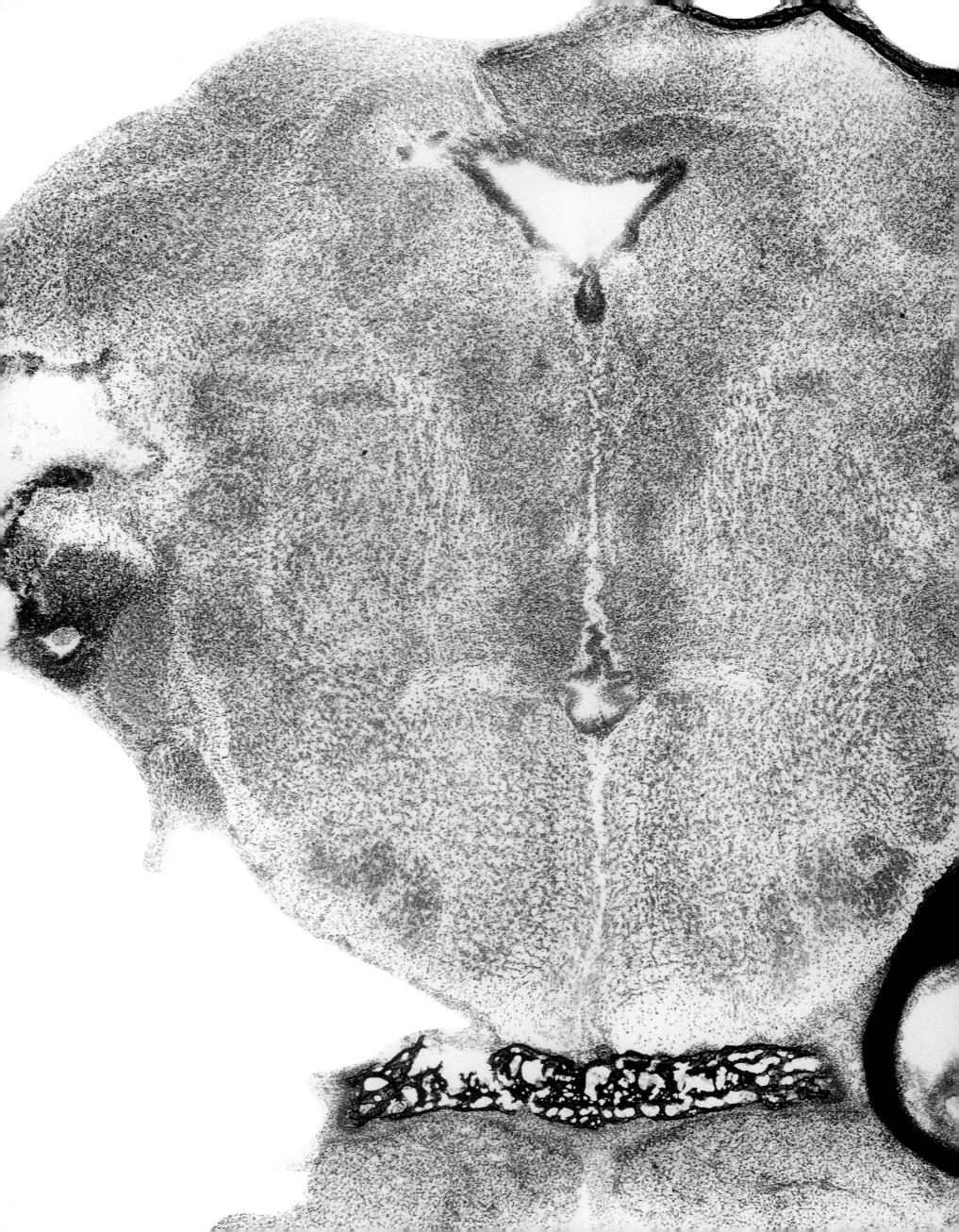

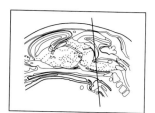

Figure 201
P0-57

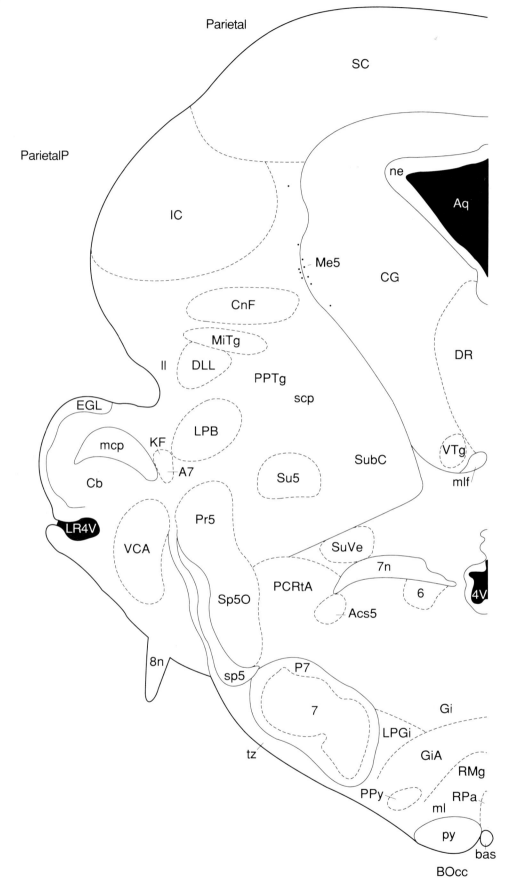

Parietal

ParietalP

SC

IC

ne

Aq

Me5

CG

CnF

MiTg

DR

ll

DLL

PPTg

scp

EGL

LPB

VTg

mcp

KF

A7

SubC

mlf

Cb

Su5

LR4V

Pr5

VCA

SuVe

7n

Sp5O

PCRtA

6

4V

Acs5

8n

sp5

P7

7

Gi

LPGi

tz

GiA

RMg

PPy

RPa

ml

py

bas

BOcc

4   3   2   1   0

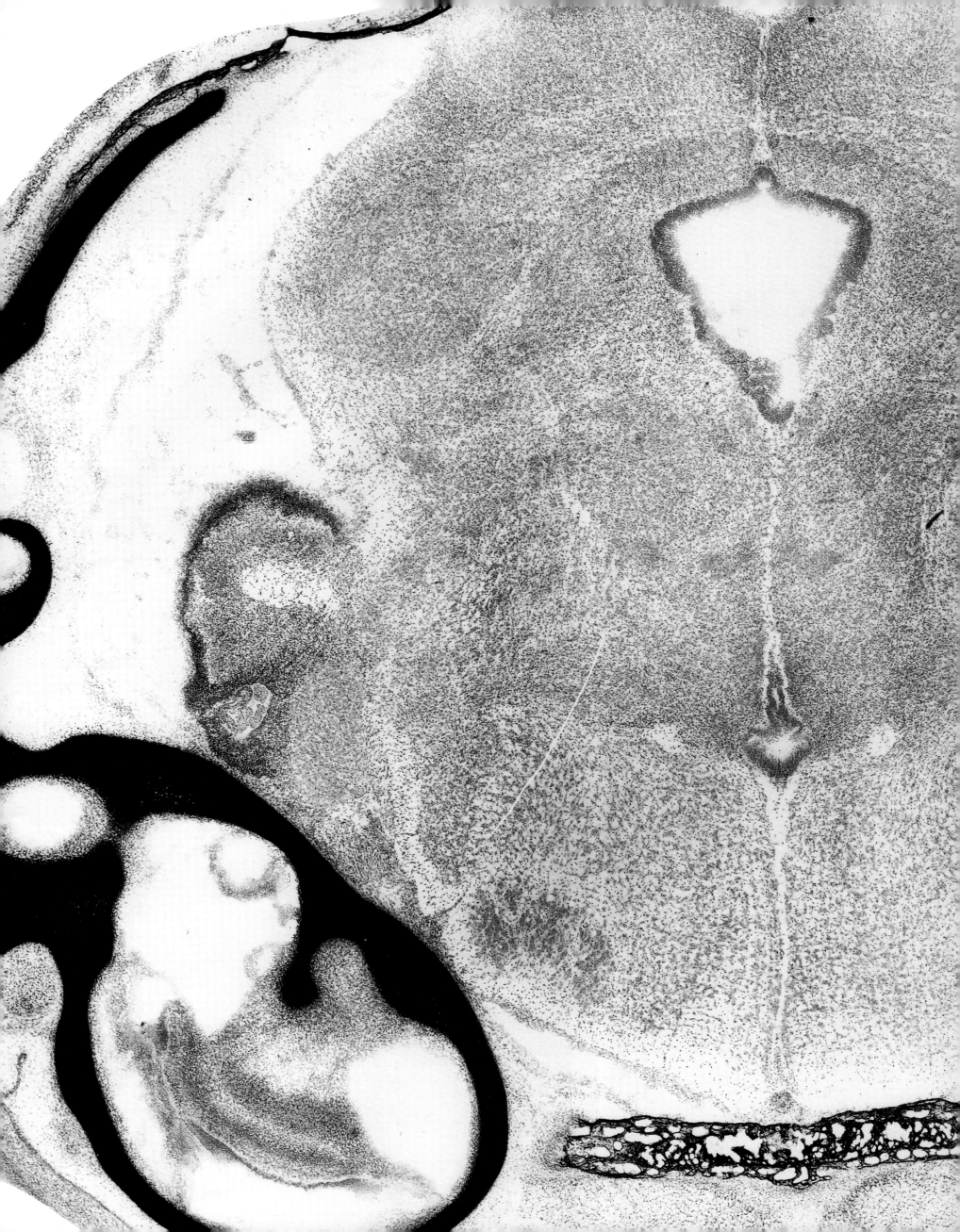

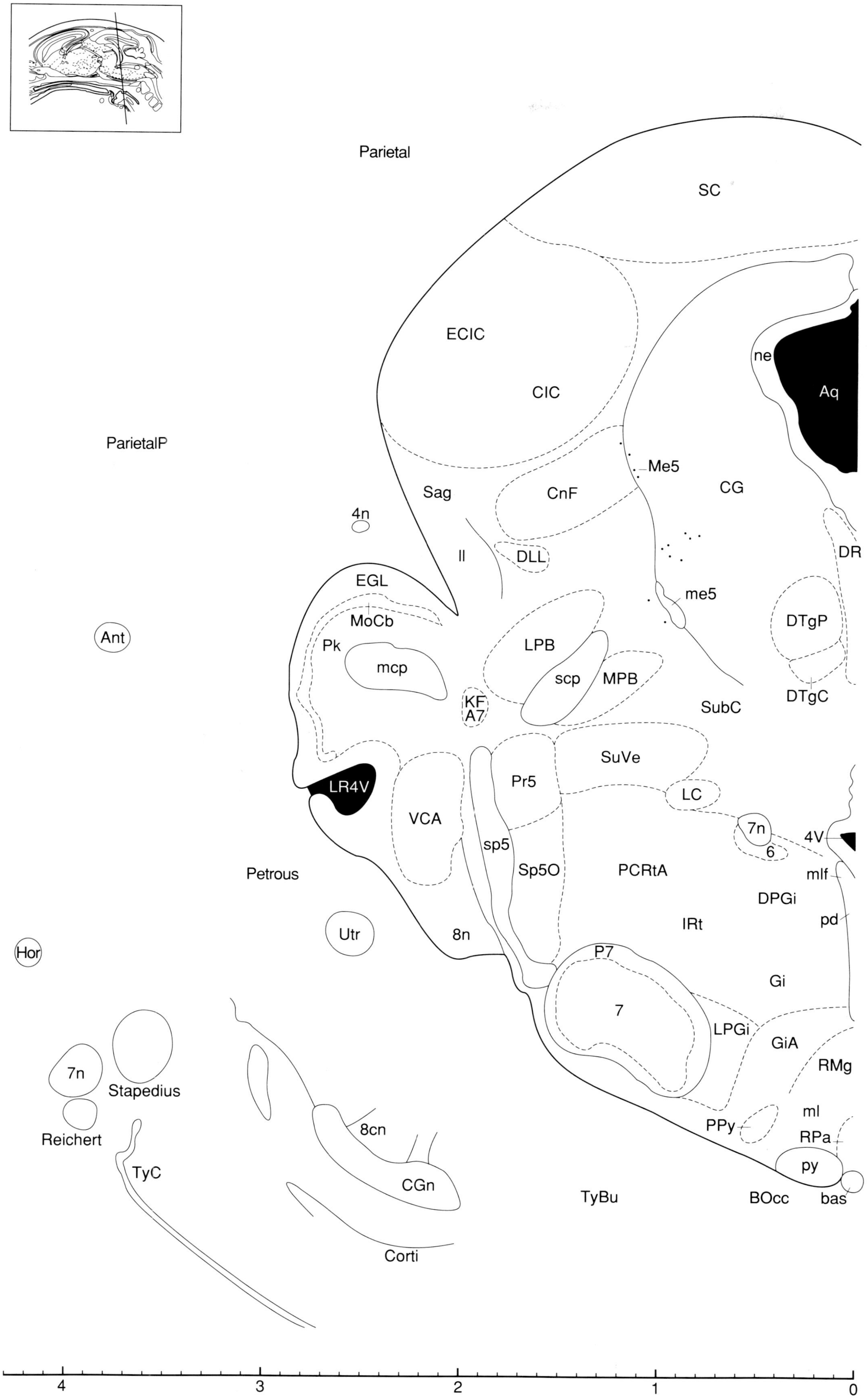

Figure 202

P0-58

Parietal

SC

ECIC

CIC

ne

Aq

ParietalP

Sag

Me5

CnF

CG

4n

ll

DLL

me5

DR

EGL

MoCb

DTgP

Pk

LPB

Ant

mcp

scp

MPB

DTgC

KF
A7

SubC

LR4V

SuVe

VCA

Pr5

LC

7n

4V

Petrous

sp5

6

Sp5O

PCRtA

mlf

Utr

8n

IRt

pd

Hor

DPGi

P7

Gi

7

LPGi

GiA

RMg

7n

Stapedius

ml

PPy

RPa

Reichert

8cn

py

TyC

CGn

TyBu

BOcc

bas

Corti

4	3	2	1	0

4n   trochlear nerve or its root
4V   4th ventricle
6   abducens nu
7   facial nu
7n   facial nerve or its root
8cn   cochlear root vestibulococh nerve
8n   vestibulocochlear nerve
A7   A7 noradrenaline cells
Ant   anterior semicircular duct
Aq   aqueduct
bas   basilar artery
BOcc   basioccipital bone
CG   central gray
CGn   cochlear (spiral) ganglion
CIC   central nu inferior colliculus
CnF   cuneiform nu
Corti   organ of Corti
DLL   dorsal nu lateral lemniscus
DPGi   dorsal paragigantocellular nu
DR   dorsal raphe nu
DTgC   dorsal tegmental nu, central
DTgP   dorsal tegmental nu, pericentral
ECIC   external cortex inf colliculus
EGL   external germinal layer of cb
Gi   gigantocellular reticular nucleus
GiA   gigantocell reticular nu, alpha
Hor   horizontal semicircular duct
IRt   intermediate reticular zone
KF   Kölliker-Fuse nu
LC   locus coeruleus
ll   lateral lemniscus
LPB   lateral parabrachial nu
LPGi   lateral paragigantocellular nu
LR4V   lateral recess 4th ventricle
mcp   middle cerebellar peduncle
Me5   mesencephalic trigeminal nu
me5   mesencephalic trigeminal tract
ml   medial lemniscus
mlf   medial longitudinal fasciculus
MoCb   molecular layer of cerebellum
MPB   medial parabrachial nu
ne   neuroepithelium
P7   perifacial zone
Parietal   parietal bone
ParietalP   parietal plate
PCRtA   parvocellular ret nu, alpha
pd   predorsal bundle
Petrous   petrous part, temporal bone
Pk   Purkinje cell layer (cerebellum)
PPy   parapyramidal reticular nu
Pr5   principal sensory trigeminal nu
py   pyramidal tract
Reichert   Reichert's cartilage
RMg   raphe magnus nu
RPa   raphe pallidus nu
Sag   sagulum nu
SC   superior colliculus
scp   superior cerebellar peduncle
sp5   spinal trigeminal tract
Sp5O   spinal trigeminal nu, oral
Stapedius   stapedius muscle
SubC   subcoeruleus nu
SuVe   superior vestibular nu
TyBu   tympanic bulla
TyC   tympanic cavity
Utr   utricle
VCA   ventral cochlear nu, anterior

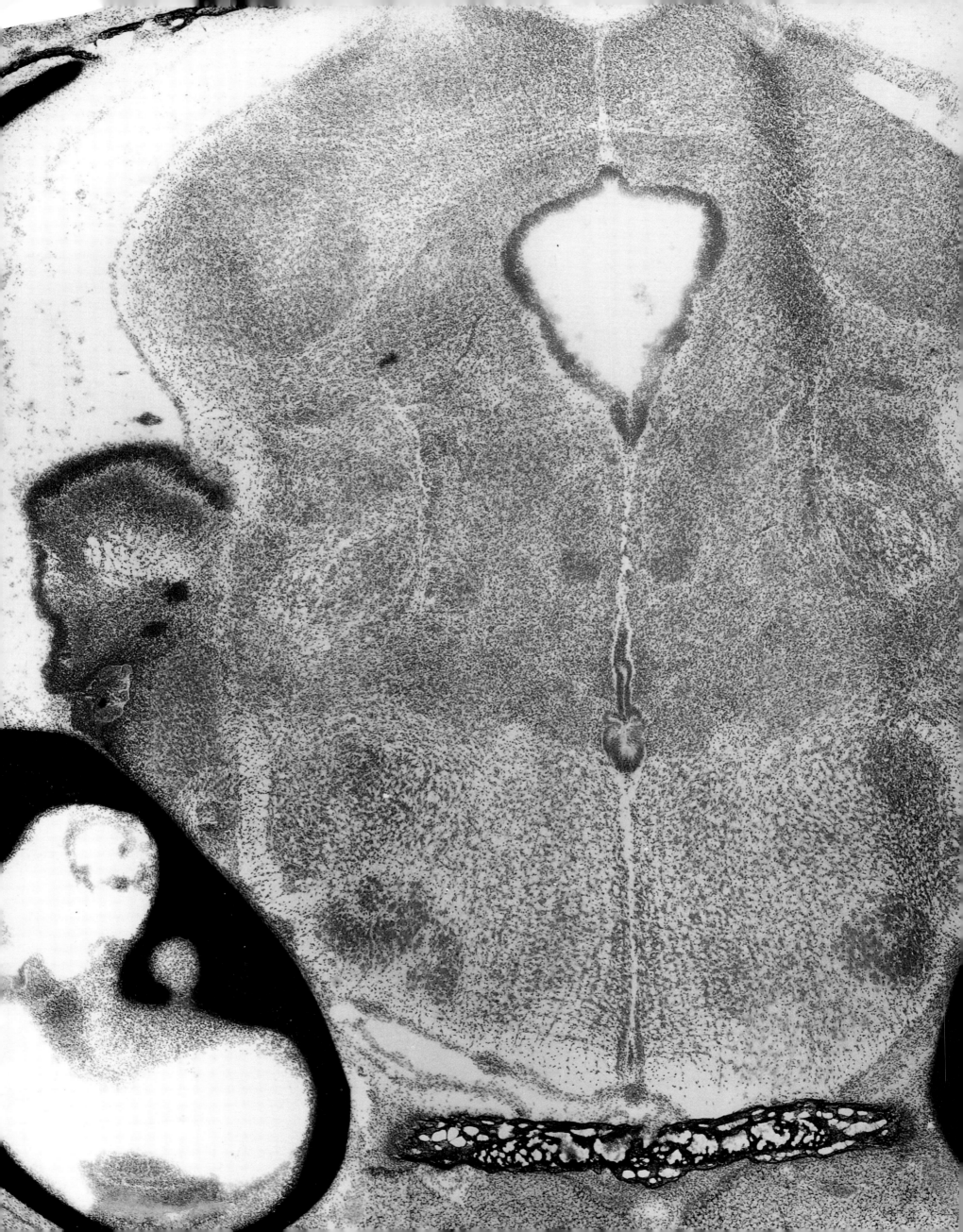

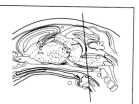

Figure 203
P0-59

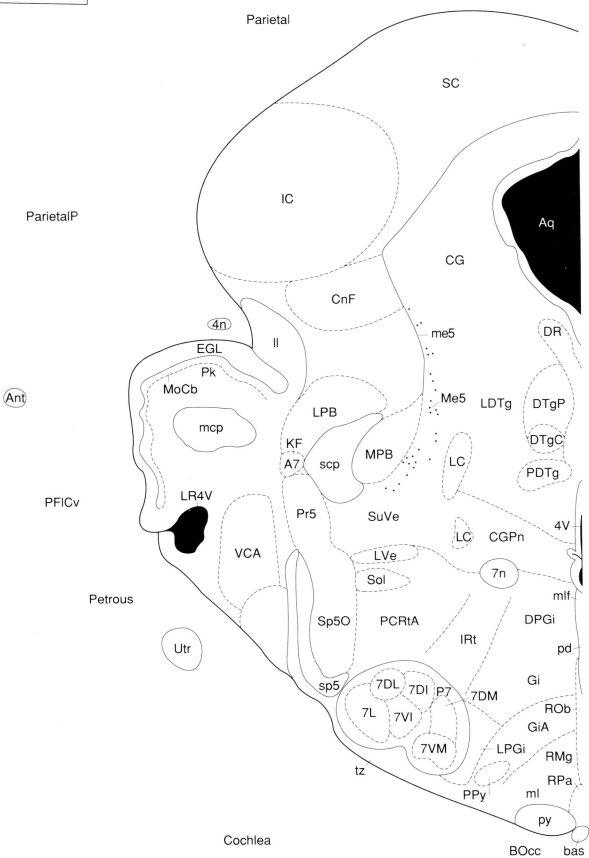

4n   trochlear nerve or its root
4V   4th ventricle
7DI   facial nu, dorsal intermed subnu
7DL   facial nu, dorsolateral subnu
7DM   facial nu, dorsomedial subnu
7L   facial nu, lateral subnu
7n   facial nerve or its root
7VI   facial nu, ventral intermed subnu
7VM   facial nu, ventromedial subnu
A7   A7 noradrenaline cells
Ant   anterior semicircular duct
Aq   aqueduct
bas   basilar artery
BOcc   basioccipital bone
CG   central gray
CGPn   central gray, pons
CnF   cuneiform nu
Cochlea   cochlea
DPGi   dorsal paragigantocellular nu
DR   dorsal raphe nu
DTgC   dorsal tegmental nu, central
DTgP   dorsal tegmental nu, pericentral
EGL   external germinal layer of cb
Gi   gigantocellular reticular nucleus
GiA   gigantocell reticular nu, alpha
IC   inferior colliculus
ictd   internal carotid artery
IRt   intermediate reticular zone
KF   Kölliker-Fuse nu
LC   locus coeruleus
LDTg   laterodorsal tegmental nu
ll   lateral lemniscus
LPB   lateral parabrachial nu
LPGi   lateral paragigantocellular nu
LR4V   lateral recess 4th ventricle
LVe   lateral vestibular nu
mcp   middle cerebellar peduncle
Me5   mesencephalic trigeminal nu
me5   mesencephalic trigeminal tract
ml   medial lemniscus
mlf   medial longitudinal fasciculus
MoCb   molecular layer of cerebellum
MPB   medial parabrachial nu
P7   perifacial zone
Parietal   parietal bone
ParietalP   parietal plate
PCRtA   parvocellular ret nu, alpha
pd   predorsal bundle
PDTg   posterior dorsal tegmental nu
Petrous   petrous part, temporal bone
PFlCv   parafloccular cavity
Pk   Purkinje cell layer (cerebellum)
PPy   parapyramidal reticular nu
Pr5   principal sensory trigeminal nu
py   pyramidal tract
RMg   raphe magnus nu
ROb   raphe obscurus nu
RPa   raphe pallidus nu
SC   superior colliculus
scp   superior cerebellar peduncle
Sol   nu solitary tract
sp5   spinal trigeminal tract
Sp5O   spinal trigeminal nu, oral
SuVe   superior vestibular nu
tz   trapezoid body
Utr   utricle
VCA   ventral cochlear nu, anterior

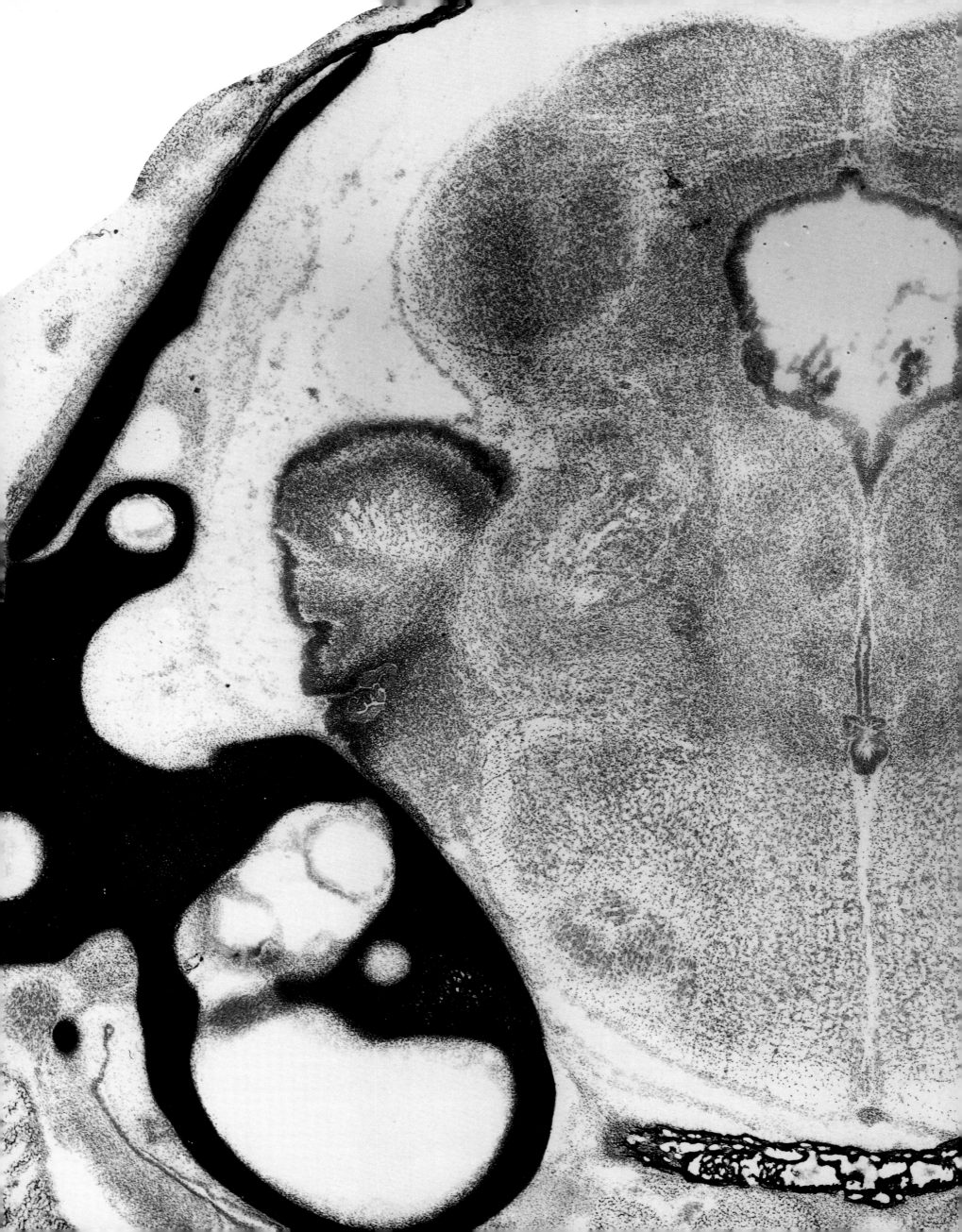

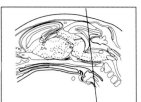

Figure 204
P0-60

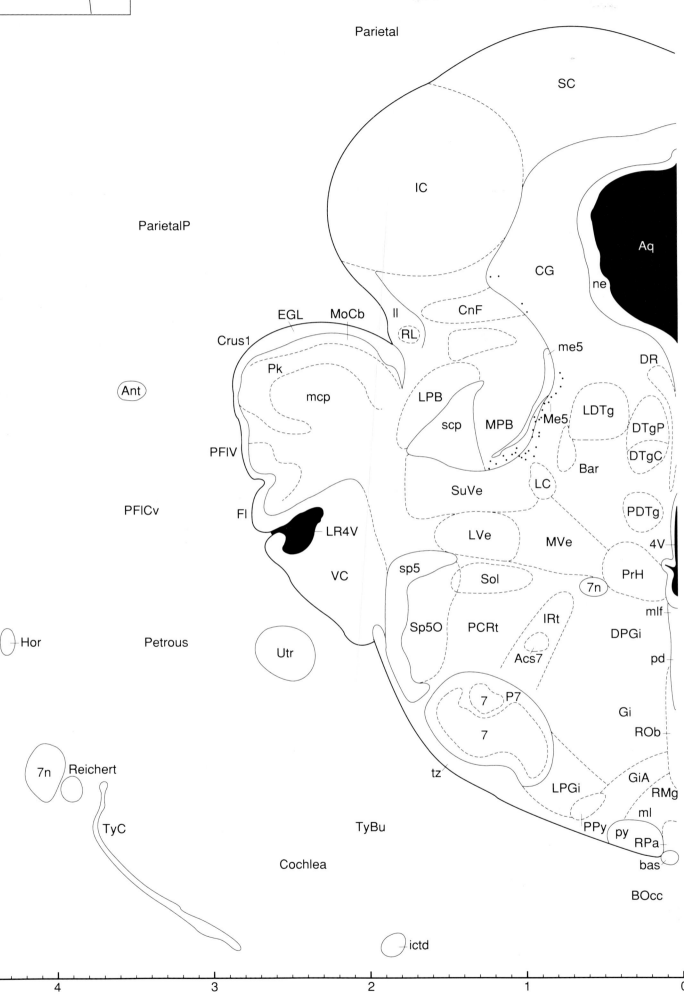

Parietal

SC

IC

ParietalP

Aq

CG

ne

CnF

EGL   MoCb   ll

Crus1   RL

me5

DR

Pk

Ant

mcp   LPB

LDTg

scp   MPB   Me5

DTgP

PFlV

DTgC

Bar

SuVe   LC

PDTg

PFlCv   Fl

LR4V   LVe

MVe

4V

VC   sp5

Sol   PrH

7n

Hor   Petrous   Sp5O   PCRt   IRt   mlf

Utr   Acs7   DPGi

pd

7   P7

Gi

7   ROb

7n   Reichert   tz   GiA   RMg

ml

TyC   LPGi   PPy   py   RPa

TyBu   bas

Cochlea   BOcc

ictd

4   3   2   1   0

4V   4th ventricle
7   facial nu
7n   facial nerve or its root
Acs7   accessory facial nu
Ant   anterior semicircular duct
Aq   aqueduct
Bar   Barrington's nu
bas   basilar artery
BOcc   basioccipital bone
CG   central gray
CnF   cuneiform nu
Cochlea   cochlea
Crus1   crus 1 ansiform lobule
DPGi   dorsal paragigantocellular nu
DR   dorsal raphe nu
DTgC   dorsal tegmental nu, central
DTgP   dorsal tegmental nu, pericentral
EGL   external germinal layer of cb
Fl   flocculus
Gi   gigantocellular reticular nucleus
GiA   gigantocell reticular nu, alpha
Hor   horizontal semicircular duct
IC   inferior colliculus
ictd   internal carotid artery
IRt   intermediate reticular zone
LC   locus coeruleus
LDTg   laterodorsal tegmental nu
ll   lateral lemniscus
LPB   lateral parabrachial nu
LPGi   lateral paragigantocellular nu
LR4V   lateral recess 4th ventricle
LVe   lateral vestibular nu
mcp   middle cerebellar peduncle
Me5   mesencephalic trigeminal nu
me5   mesencephalic trigeminal tract
ml   medial lemniscus
mlf   medial longitudinal fasciculus
MoCb   molecular layer of cerebellum
MPB   medial parabrachial nu
MVe   medial vestibular nu
ne   neuroepithelium
P7   perifacial zone
Parietal   parietal bone
ParietalP   parietal plate
PCRt   parvocellular reticular nu
pd   predorsal bundle
PDTg   posterior dorsal tegmental nu
Petrous   petrous part, temporal bone
PFlCv   paraflocular cavity
PFlV   paraflocculus, ventral part
Pk   Purkinje cell layer (cerebellum)
PPy   parapyramidal reticular nu
PrH   prepositus hypoglossal nu
py   pyramidal tract
Reichert   Reichert's cartilage
RL   retrolemniscal nu
RMg   raphe magnus nu
ROb   raphe obscurus nu
RPa   raphe pallidus nu
SC   superior colliculus
scp   superior cerebellar peduncle
Sol   nu solitary tract
sp5   spinal trigeminal tract
Sp5O   spinal trigeminal nu, oral
SuVe   superior vestibular nu
TyBu   tympanic bulla
TyC   tympanic cavity
tz   trapezoid body
Utr   utricle
VC   ventral cochlear nu

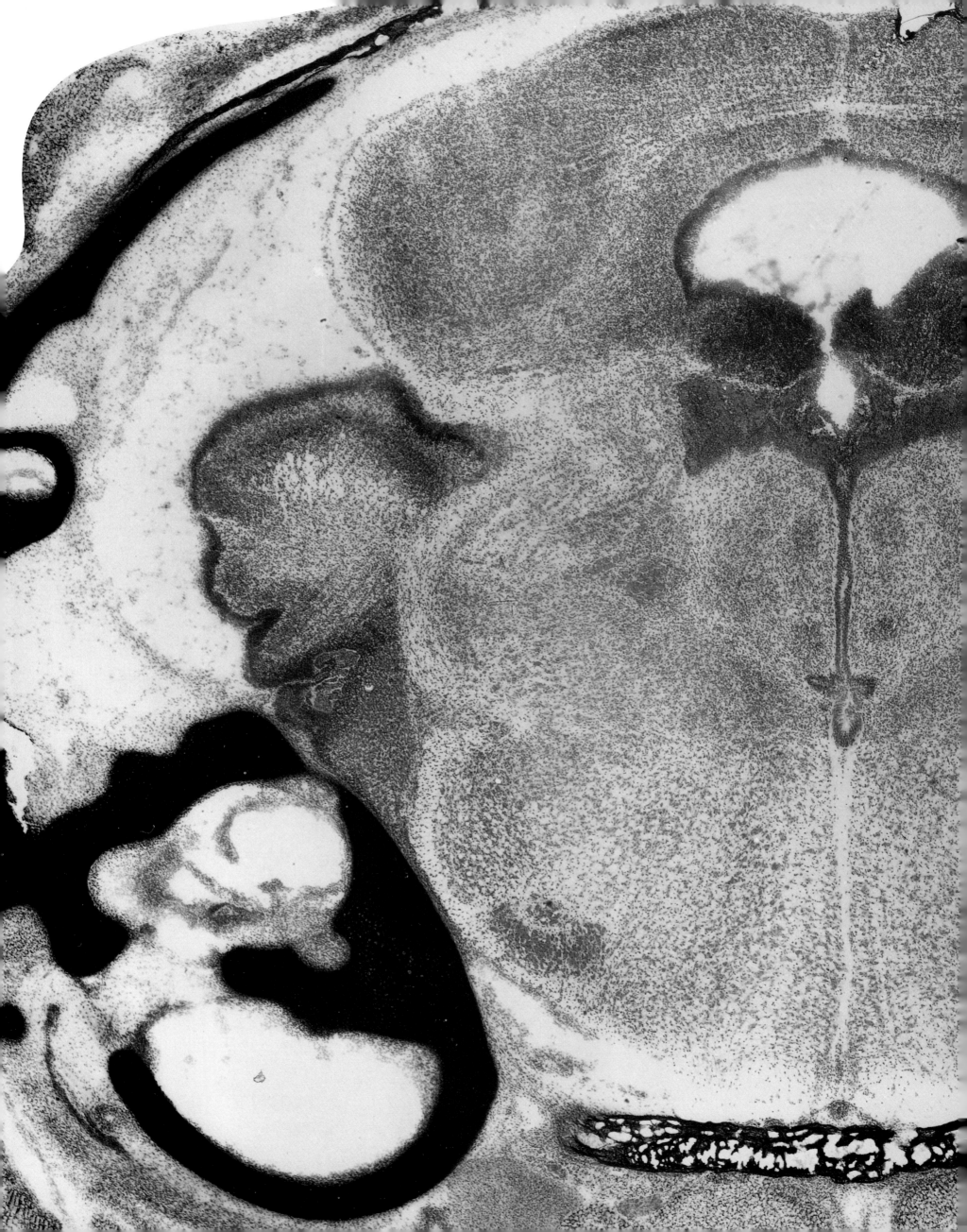

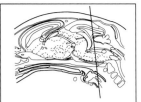

Figure 205
P0-61

Parietal

SC

CG

IC

Aq

ParietalP

ne

PCTg

ne

EGL

MoCb

Crus1

Pk

mcp

me5

DR

Me5

icp

MPB

DTg

Ant

scp

LC

LDTg

PFlV

LPB

Sph

SuVe

PDTg

PFlCv

MVe

Fl

LVe

MVeV

PrH

LR4V

4V

VC

icp

sp5

sol

mlf

Sol

Petrous

pd

Sp5O

PCRt

IRt

DPGi

Utr

Gi

Hor

7 P7

ROb

7

Reichert

LPGi

GiA

tz

RMg

ml

RPa

Cochlea

py

bas

ptgpal

BOcc

ictd

| 4 | 3 | 2 | 1 | 0 |

4V   4th ventricle
7   facial nu
Ant   anterior semicircular duct
Aq   aqueduct
bas   basilar artery
BOcc   basioccipital bone
CG   central gray
Cochlea   cochlea
Crus1   crus 1 ansiform lobule
DPGi   dorsal paragigantocellular nu
DR   dorsal raphe nu
DTg   dorsal tegmental nu
EGL   external germinal layer of cb
Fl   flocculus
Gi   gigantocellular reticular nucleus
GiA   gigantocell reticular nu, alpha
Hor   horizontal semicircular duct
IC   inferior colliculus
icp   inferior cerebellar peduncle
ictd   internal carotid artery
IRt   intermediate reticular zone
LC   locus coeruleus
LDTg   laterodorsal tegmental nu
LPB   lateral parabrachial nu
LPGi   lateral paragigantocellular nu
LR4V   lateral recess 4th ventricle
LVe   lateral vestibular nu
mcp   middle cerebellar peduncle
Me5   mesencephalic trigeminal nu
me5   mesencephalic trigeminal tract
ml   medial lemniscus
mlf   medial longitudinal fasciculus
MoCb   molecular layer of cerebellum
MPB   medial parabrachial nu
MVe   medial vestibular nu
MVeV   medial vestibular nu, ventral
ne   neuroepithelium
P7   perifacial zone
Parietal   parietal bone
ParietalP   parietal plate
PCRt   parvocellular reticular nu
PCTg   paracollicular tegmentum
pd   predorsal bundle
PDTg   posterior dorsal tegmental nu
Petrous   petrous part, temporal bone
PFlCv   paraflocular cavity
PFlV   paraflocculus, ventral part
Pk   Purkinje cell layer (cerebellum)
PrH   prepositus hypoglossal nu
ptgpal   pterygopalatine artery
py   pyramidal tract
Reichert   Reichert's cartilage
RMg   raphe magnus nu
ROb   raphe obscurus nu
RPa   raphe pallidus nu
SC   superior colliculus
scp   superior cerebellar peduncle
Sol   nu solitary tract
sol   solitary tract
sp5   spinal trigeminal tract
Sp5O   spinal trigeminal nu, oral
Sph   sphenoid nu
SuVe   superior vestibular nu
tz   trapezoid body
Utr   utricle
VC   ventral cochlear nu

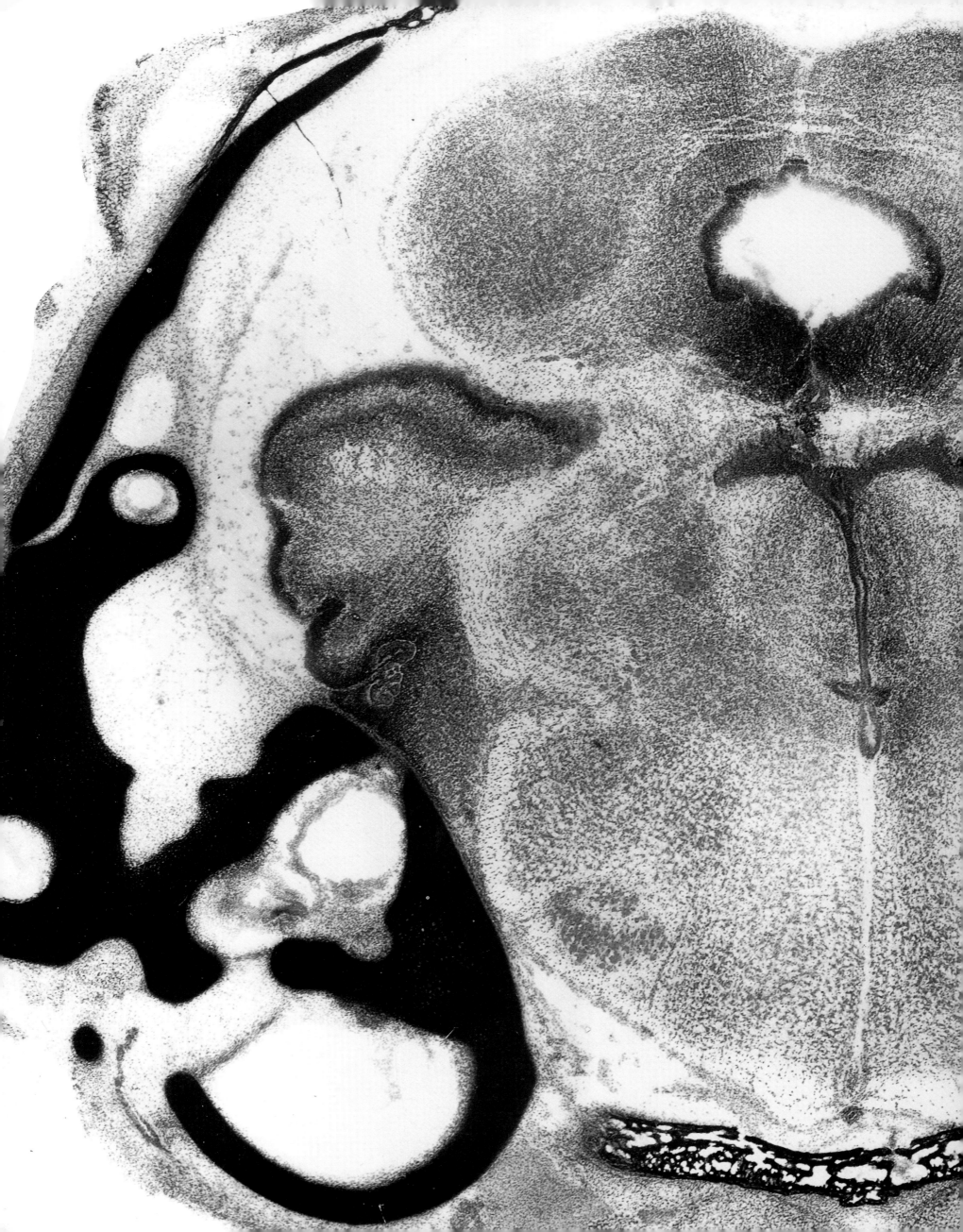

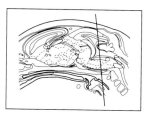

Figure 206
P0-62

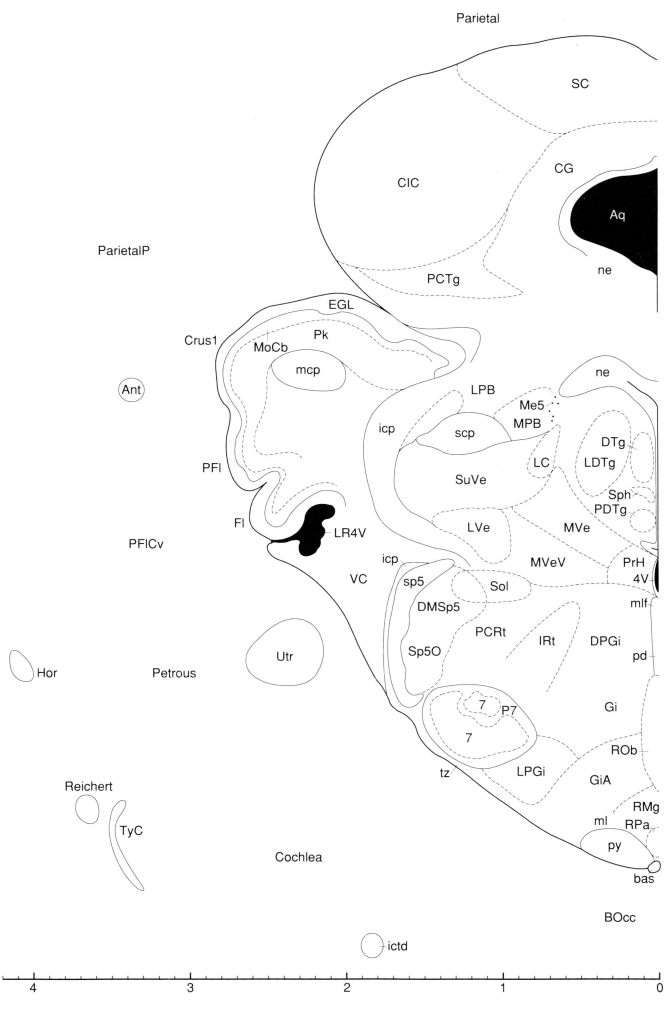

4V   4th ventricle
7   facial nu
Ant   anterior semicircular duct
Aq   aqueduct
bas   basilar artery
BOcc   basioccipital bone
CG   central gray
CIC   central nu inferior colliculus
Cochlea   cochlea
Crus1   crus 1 ansiform lobule
DMSp5   dorsomedial spinal trig nu
DPGi   dorsal paragigantocellular nu
DTg   dorsal tegmental nu
EGL   external germinal layer of cb
Fl   flocculus
Gi   gigantocellular reticular nucleus
GiA   gigantocell reticular nu, alpha
Hor   horizontal semicircular duct
icp   inferior cerebellar peduncle
ictd   internal carotid artery
IRt   intermediate reticular zone
LC   locus coeruleus
LDTg   laterodorsal tegmental nu
LPB   lateral parabrachial nu
LPGi   lateral paragigantocellular nu
LR4V   lateral recess 4th ventricle
LVe   lateral vestibular nu
mcp   middle cerebellar peduncle
Me5   mesencephalic trigeminal nu
ml   medial lemniscus
mlf   medial longitudinal fasciculus
MoCb   molecular layer of cerebellum
MPB   medial parabrachial nu
MVe   medial vestibular nu
MVeV   medial vestibular nu, ventral
ne   neuroepithelium
P7   perifacial zone
Parietal   parietal bone
ParietalP   parietal plate
PCRt   parvocellular reticular nu
PCTg   paracollicular tegmentum
pd   predorsal bundle
PDTg   posterior dorsal tegmental nu
Petrous   petrous part, temporal bone
PFl   paraflocculus
PFlCv   paraflocular cavity
Pk   Purkinje cell layer (cerebellum)
PrH   prepositus hypoglossal nu
py   pyramidal tract
Reichert   Reichert's cartilage
RMg   raphe magnus nu
ROb   raphe obscurus nu
RPa   raphe pallidus nu
SC   superior colliculus
scp   superior cerebellar peduncle
Sol   nu solitary tract
sp5   spinal trigeminal tract
Sp5O   spinal trigeminal nu, oral
Sph   sphenoid nu
SuVe   superior vestibular nu
TyC   tympanic cavity
tz   trapezoid body
Utr   utricle
VC   ventral cochlear nu

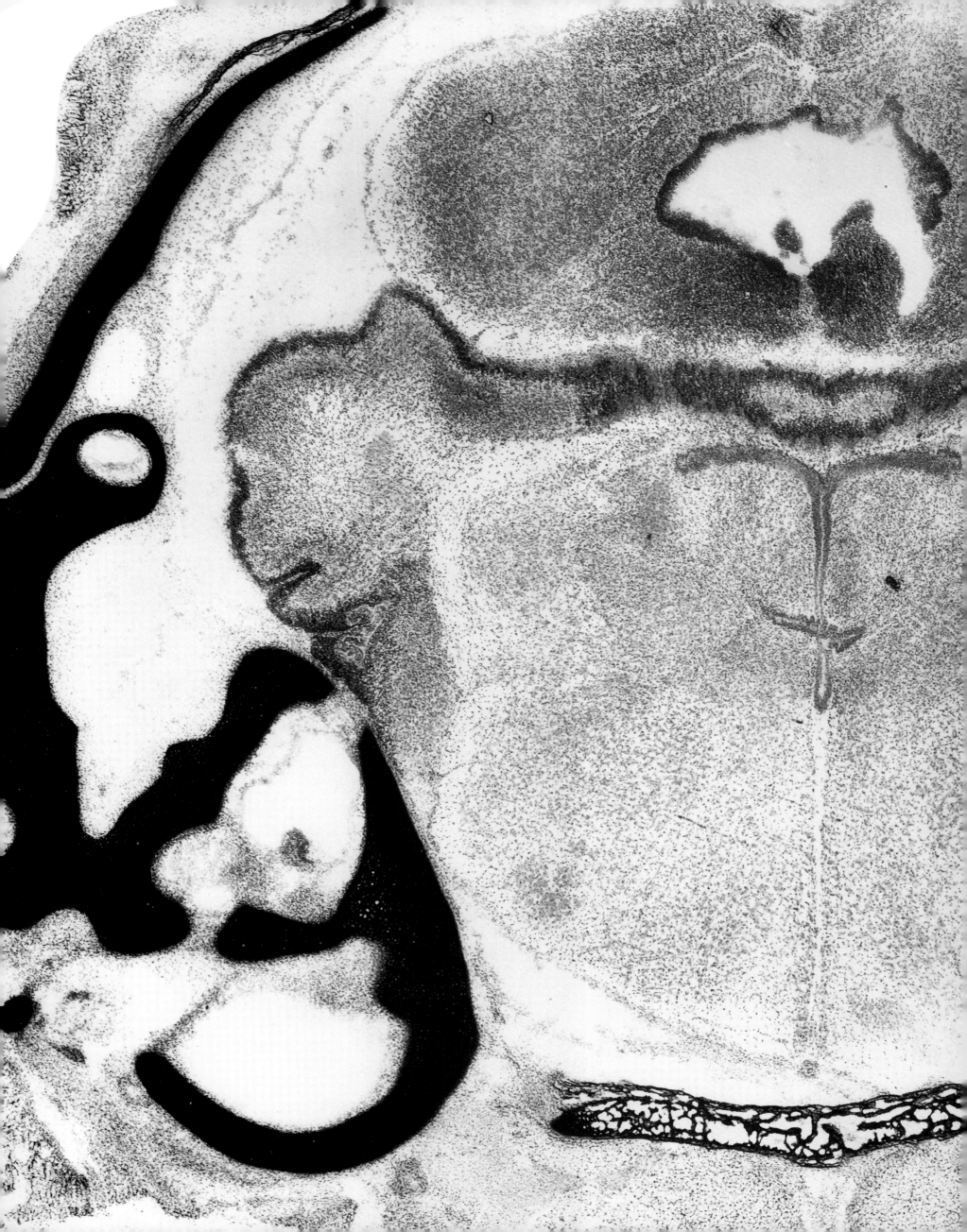

Figure 207

P0-63

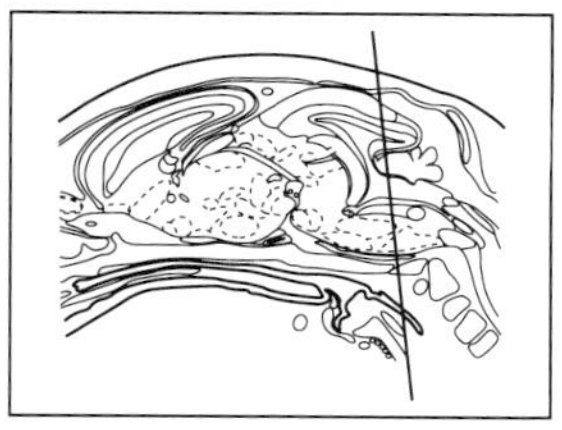

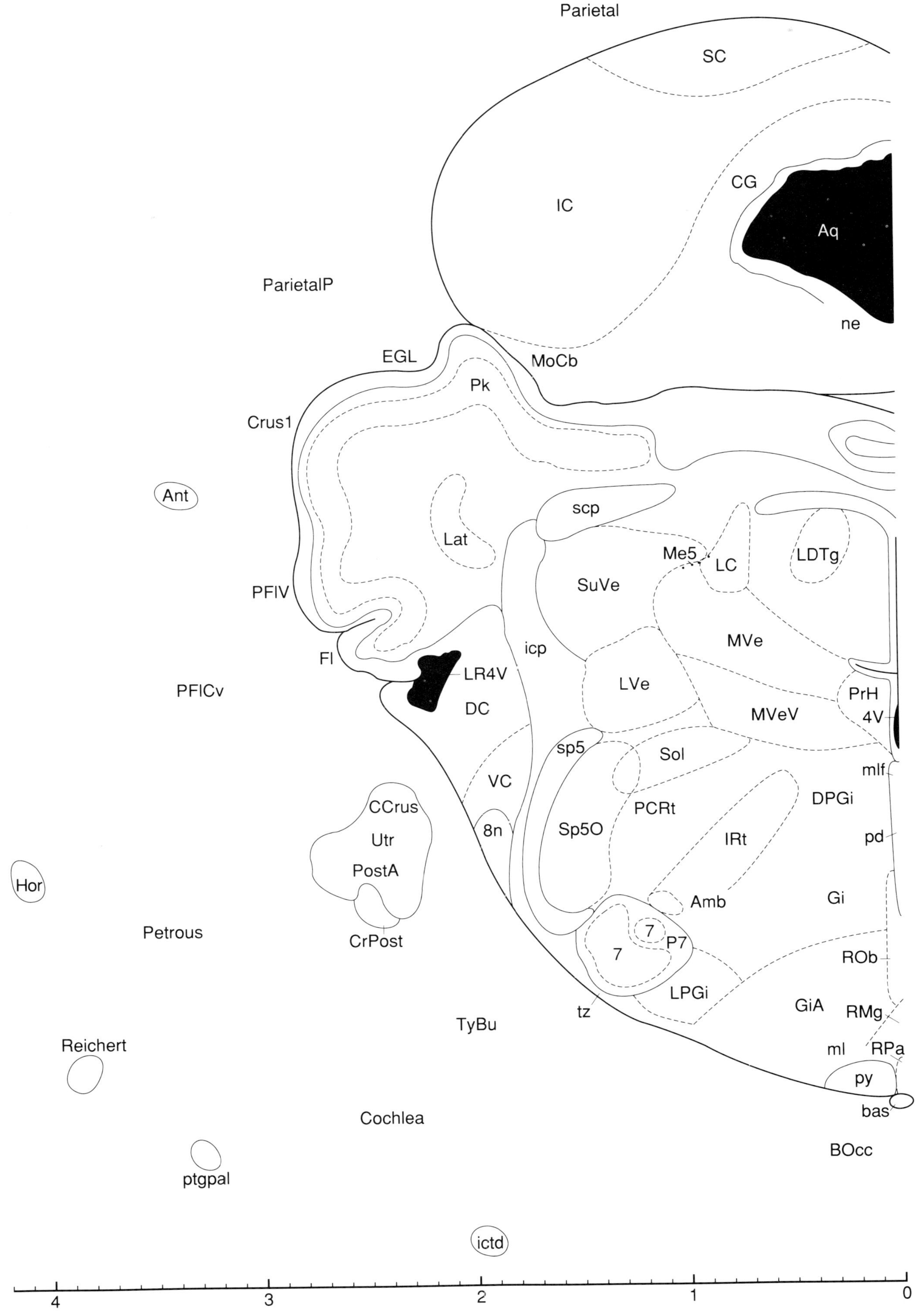

Parietal
SC
IC
CG
Aq
ne
ParietalP
EGL
MoCb
Pk
Crus1
Ant
Lat
scp
Me5
LC
LDTg
SuVe
PFlV
MVe
Fl
icp
LR4V
PFlCv
DC
LVe
PrH
MVeV
4V
sp5
Sol
mlf
VC
CCrus
PCRt
DPGi
Utr
8n
Sp5O
IRt
pd
PostA
Hor
Amb
Gi
Petrous
CrPost
7
P7
7
ROb
LPGi
GiA
RMg
tz
TyBu
Reichert
ml
RPa
py
bas
Cochlea
BOcc
ptgpal
ictd
4
3
2
1
0

4V 4th ventricle
7 facial nu
8n vestibulocochlear nerve
Amb ambiguus nu
Ant anterior semicircular duct
Aq aqueduct
bas basilar artery
BOcc basioccipital bone
CCrus common crus of ant and post s
CG central gray
Cochlea cochlea
CrPost crista of posterior ampulla
Crus1 crus 1 ansiform lobule
DC dorsal cochlear nu
DPGi dorsal paragigantocellular nu
EGL external germinal layer of cb
Fl flocculus
Gi gigantocellular reticular nucleus
GiA gigantocell reticular nu, alpha
Hor horizontal semicircular duct
IC inferior colliculus
icp inferior cerebellar peduncle
ictd internal carotid artery
IRt intermediate reticular zone
Lat lateral cerebellar nu
LC locus coeruleus
LDTg laterodorsal tegmental nu
LPGi lateral paragigantocellular nu
LR4V lateral recess 4th ventricle
LVe lateral vestibular nu
Me5 mesencephalic trigeminal nu
ml medial lemniscus
mlf medial longitudinal fasciculus
MoCb molecular layer of cerebellum
MVe medial vestibular nu
MVeV medial vestibular nu, ventral
ne neuroepithelium
P7 perifacial zone
Parietal parietal bone
ParietalP parietal plate
PCRt parvocellular reticular nu
pd predorsal bundle
Petrous petrous part, temporal bone
PFlCv parafloccular cavity
PFlV paraflocculus, ventral part
Pk Purkinje cell layer (cerebellum)
PostA ampulla of post semicirc duct
PrH prepositus hypoglossal nu
ptgpal pterygopalatine artery
py pyramidal tract
Reichert Reichert's cartilage
RMg raphe magnus nu
ROb raphe obscurus nu
RPa raphe pallidus nu
SC superior colliculus
scp superior cerebellar peduncle
Sol nu solitary tract
sp5 spinal trigeminal tract
Sp5O spinal trigeminal nu, oral
SuVe superior vestibular nu
TyBu tympanic bulla
tz trapezoid body
Utr utricle
VC ventral cochlear nu

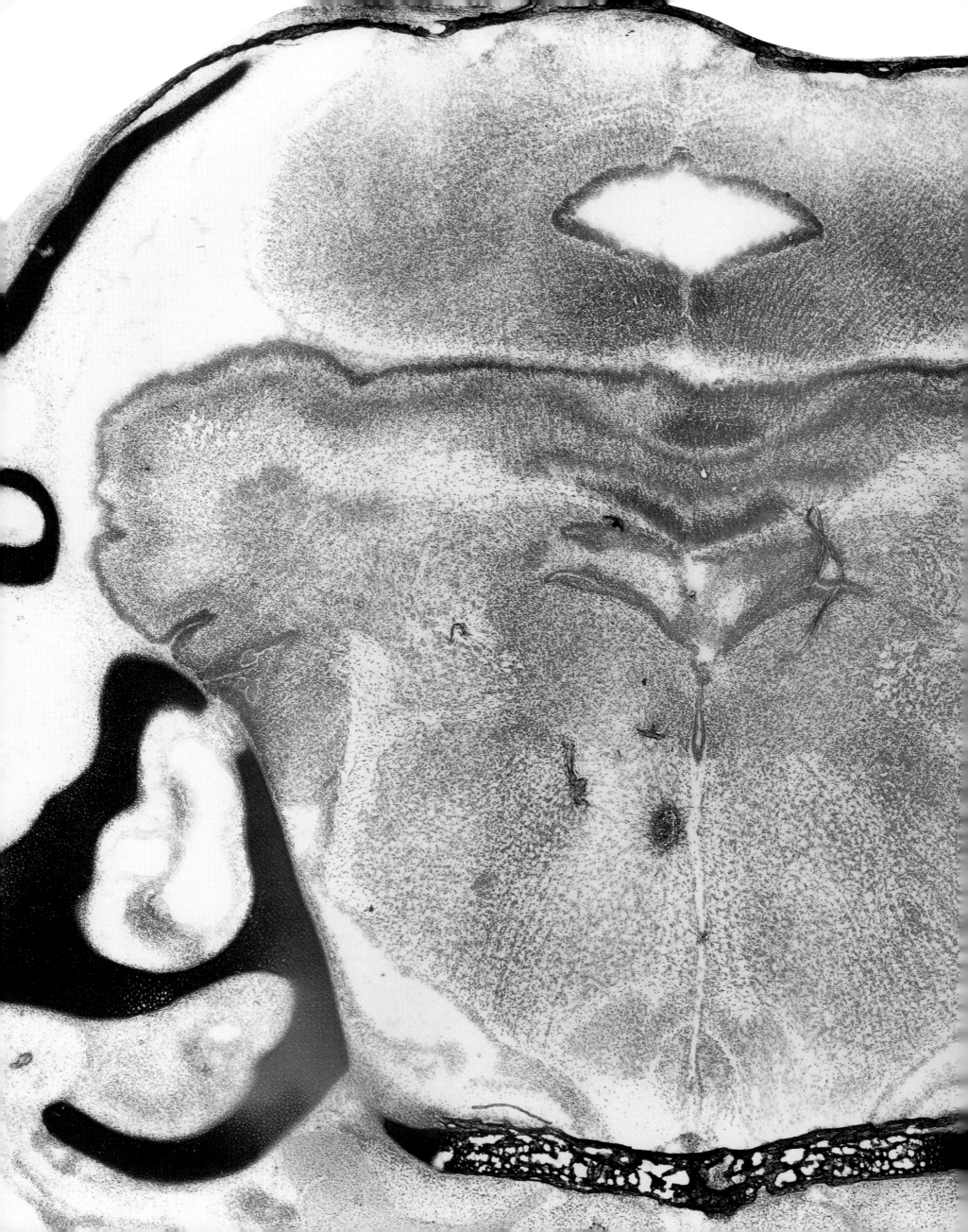

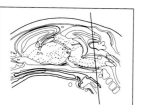

Figure 208
P0-64

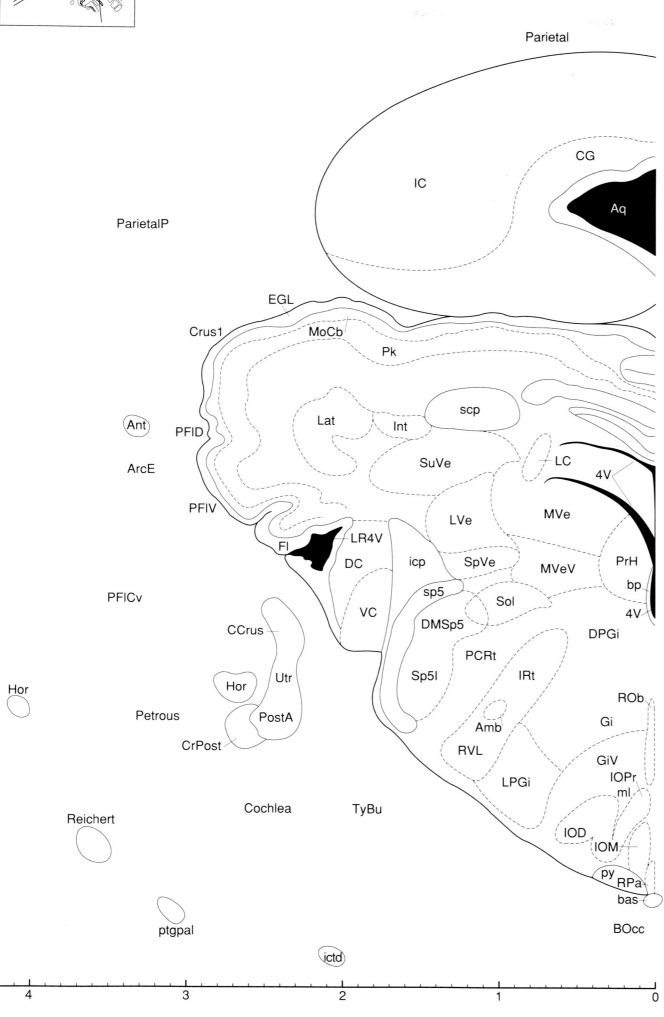

4V   4th ventricle
Amb   ambiguus nu
Ant   anterior semicircular duct
Aq   aqueduct
ArcE   arcuate eminence
bas   basilar artery
BOcc   basioccipital bone
bp   basal plate neuroepithelium
CCrus   common crus of ant and post s
CG   central gray
Cochlea   cochlea
CrPost   crista of posterior ampulla
Crus1   crus 1 ansiform lobule
DC   dorsal cochlear nu
DMSp5   dorsomedial spinal trig nu
DPGi   dorsal paragigantocellular nu
EGL   external germinal layer of cb
Fl   flocculus
Gi   gigantocellular reticular nucleus
GiV   gigantocell reticular nu, vent
Hor   horizontal semicircular duct
IC   inferior colliculus
icp   inferior cerebellar peduncle
ictd   internal carotid artery
Int   interposed cerebellar nu
IOD   inferior olive, dorsal nu
IOM   inferior olive, medial nu
IOPr   inferior olive, principal nu
IRt   intermediate reticular zone
Lat   lateral cerebellar nu
LC   locus coeruleus
LPGi   lateral paragigantocellular nu
LR4V   lateral recess 4th ventricle
LVe   lateral vestibular nu
ml   medial lemniscus
MoCb   molecular layer of cerebellum
MVe   medial vestibular nu
MVeV   medial vestibular nu, ventral
Parietal   parietal bone
ParietalP   parietal plate
PCRt   parvocellular reticular nu
Petrous   petrous part, temporal bone
PFlCv   parafloccular cavity
PFlD   paraflocculus, dorsal part
PFlV   paraflocculus, ventral part
Pk   Purkinje cell layer (cerebellum)
PostA   ampulla of post semicirc duct
PrH   prepositus hypoglossal nu
ptgpal   pterygopalatine artery
py   pyramidal tract
Reichert   Reichert's cartilage
ROb   raphe obscurus nu
RPa   raphe pallidus nu
RVL   rostroventrolateral retic nu
scp   superior cerebellar peduncle
Sol   nu solitary tract
sp5   spinal trigeminal tract
Sp5I   spinal trigem nu, interpolar
SpVe   spinal vestibular nu
SuVe   superior vestibular nu
TyBu   tympanic bulla
Utr   utricle
VC   ventral cochlear nu

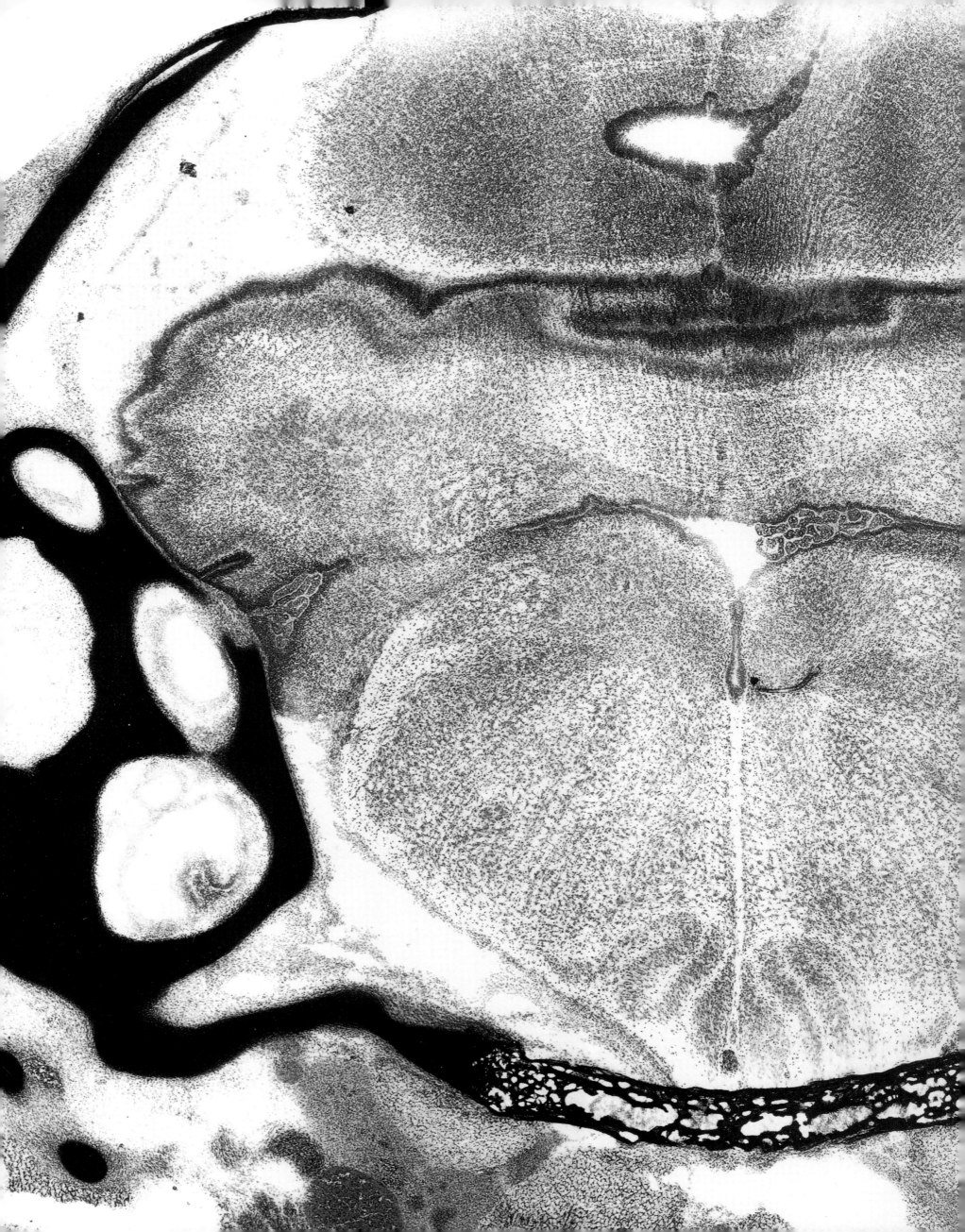

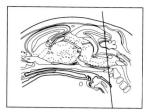

Figure 209
P0-65

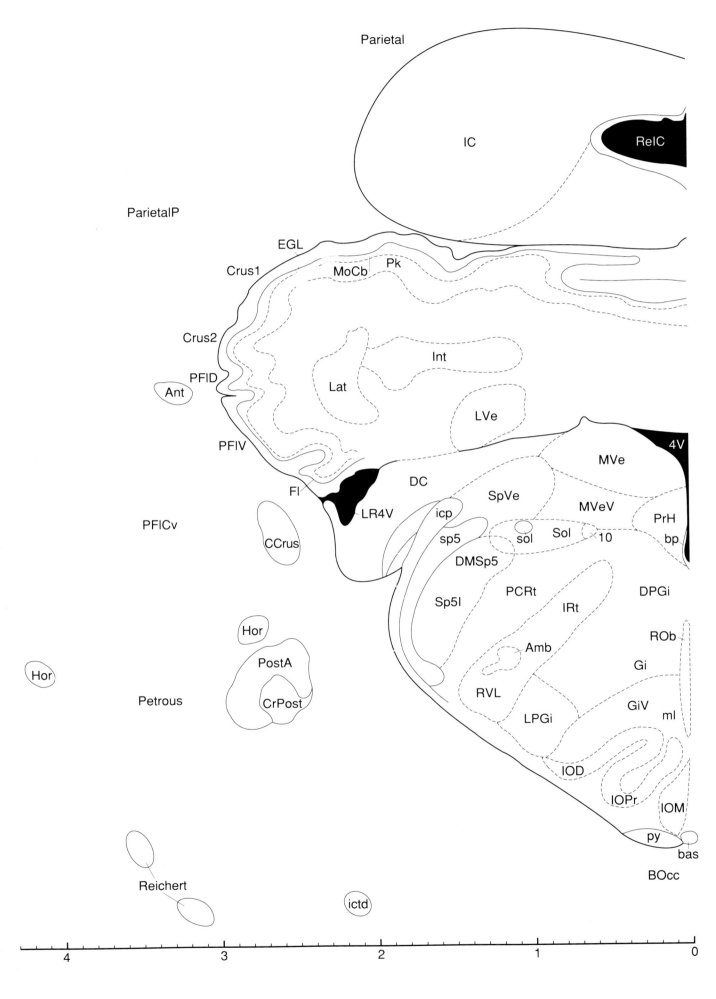

4V   4th ventricle
10   dorsal motor nu vagus
Amb   ambiguus nu
Ant   anterior semicircular duct
bas   basilar artery
BOcc   basioccipital bone
bp   basal plate neuroepithelium
CCrus   common crus of ant and post s
CrPost   crista of posterior ampulla
Crus1   crus 1 ansiform lobule
Crus2   crus 2 ansiform lobule
DC   dorsal cochlear nu
DMSp5   dorsomedial spinal trig nu
DPGi   dorsal paragigantocellular nu
EGL   external germinal layer of cb
Fl   flocculus
Gi   gigantocellular reticular nucleus
GiV   gigantocell reticular nu, vent
Hor   horizontal semicircular duct
IC   inferior colliculus
icp   inferior cerebellar peduncle
ictd   internal carotid artery
Int   interposed cerebellar nu
IOD   inferior olive, dorsal nu
IOM   inferior olive, medial nu
IOPr   inferior olive, principal nu
IRt   intermediate reticular zone
Lat   lateral cerebellar nu
LPGi   lateral paragigantocellular nu
LR4V   lateral recess 4th ventricle
LVe   lateral vestibular nu
ml   medial lemniscus
MoCb   molecular layer of cerebellum
MVe   medial vestibular nu
MVeV   medial vestibular nu, ventral
Parietal   parietal bone
ParietalP   parietal plate
PCRt   parvocellular reticular nu
Petrous   petrous part, temporal bone
PFlCv   parafloccular cavity
PFlD   paraflocculus, dorsal part
PFlV   paraflocculus, ventral part
Pk   Purkinje cell layer (cerebellum)
PostA   ampulla of post semicirc duct
PrH   prepositus hypoglossal nu
py   pyramidal tract
ReIC   recess inferior colliculus
Reichert   Reichert's cartilage
ROb   raphe obscurus nu
RVL   rostroventrolateral retic nu
Sol   nu solitary tract
sol   solitary tract
sp5   spinal trigeminal tract
Sp5I   spinal trigem nu, interpolar
SpVe   spinal vestibular nu

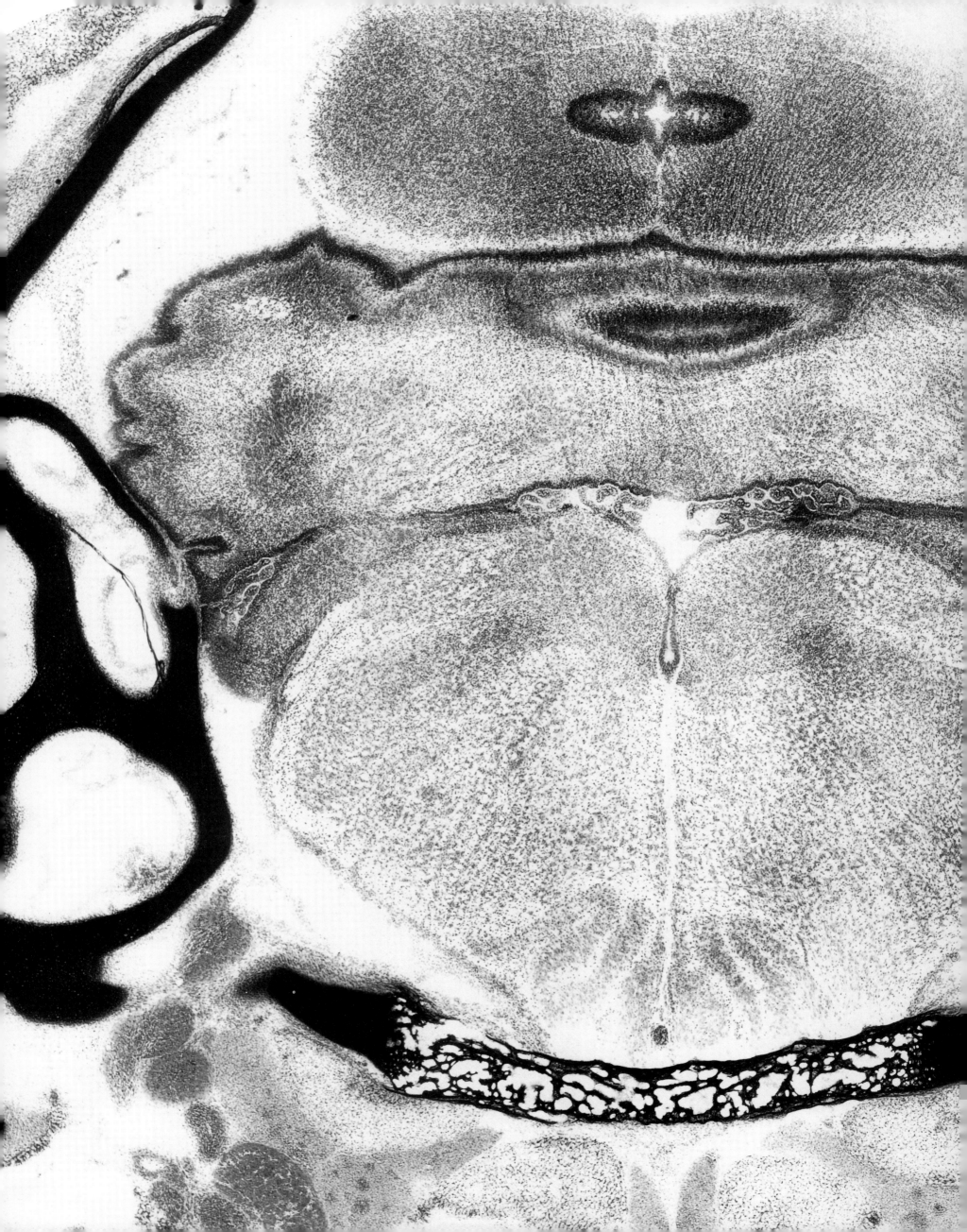

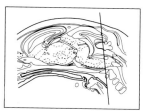

Figure 210
P0-66

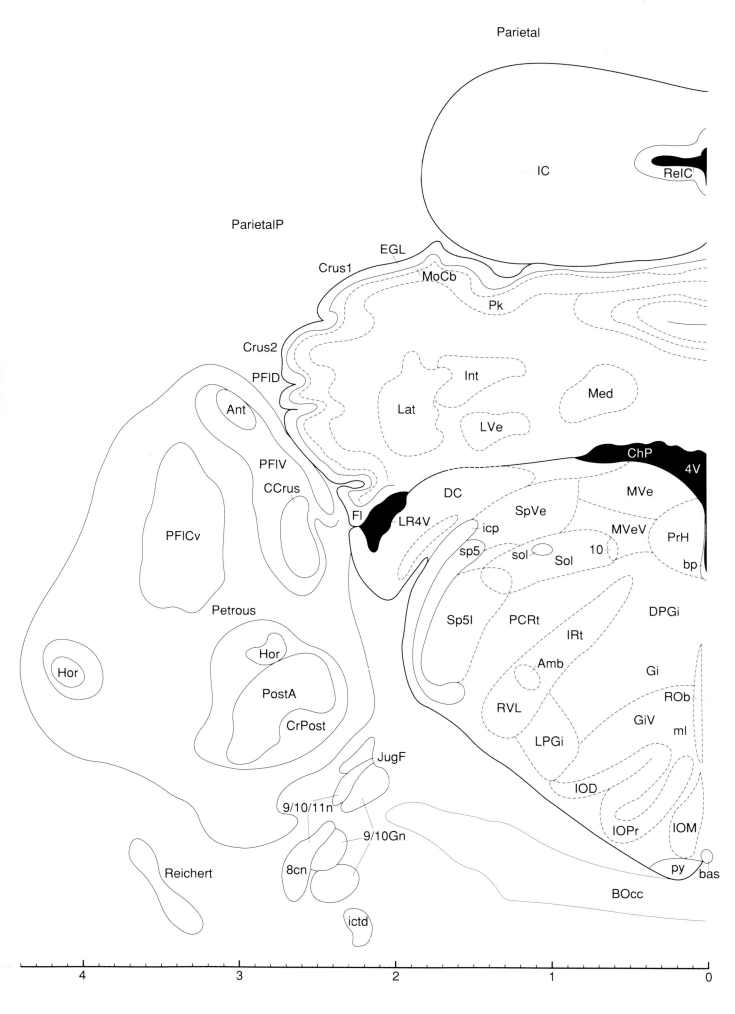

4V   4th ventricle
8cn   cochlear root vestibulococh nerve
9Gn   glossopharyngeal ganglion
9n   glossopharyngeal nerve
10   dorsal motor nu vagus
10Gn   vagal ganglion
10n   vagus nerve
11n   spinal accessory nerve
Amb   ambiguus nu
Ant   anterior semicircular duct
bas   basilar artery
BOcc   basioccipital bone
bp   basal plate neuroepithelium
CCrus   common crus of ant and post s
ChP   choroid plexus
CrPost   crista of posterior ampulla
Crus1   crus 1 ansiform lobule
Crus2   crus 2 ansiform lobule
DC   dorsal cochlear nu
DPGi   dorsal paragigantocellular nu
EGL   external germinal layer of cb
Fl   flocculus
Gi   gigantocellular reticular nucleus
GiV   gigantocell reticular nu, vent
Hor   horizontal semicircular duct
IC   inferior colliculus
icp   inferior cerebellar peduncle
ictd   internal carotid artery
Int   interposed cerebellar nu
IOD   inferior olive, dorsal nu
IOM   inferior olive, medial nu
IOPr   inferior olive, principal nu
IRt   intermediate reticular zone
JugF   jugular foramen
Lat   lateral cerebellar nu
LPGi   lateral paragigantocellular nu
LR4V   lateral recess 4th ventricle
LVe   lateral vestibular nu
Med   medial cerebellar nu
ml   medial lemniscus
MoCb   molecular layer of cerebellum
MVe   medial vestibular nu
MVeV   medial vestibular nu, ventral
Parietal   parietal bone
ParietalP   parietal plate
PCRt   parvocellular reticular nu
Petrous   petrous part, temporal bone
PFlCv   paraflocular cavity
PFlD   paraflocculus, dorsal part
PFlV   paraflocculus, ventral part
Pk   Purkinje cell layer (cerebellum)
PostA   ampulla of post semicirc duct
PrH   prepositus hypoglossal nu
py   pyramidal tract
ReIC   recess inferior colliculus
Reichert   Reichert's cartilage
ROb   raphe obscurus nu
RVL   rostroventrolateral retic nu
Sol   nu solitary tract
sol   solitary tract
sp5   spinal trigeminal tract
Sp5I   spinal trigem nu, interpolar
SpVe   spinal vestibular nu

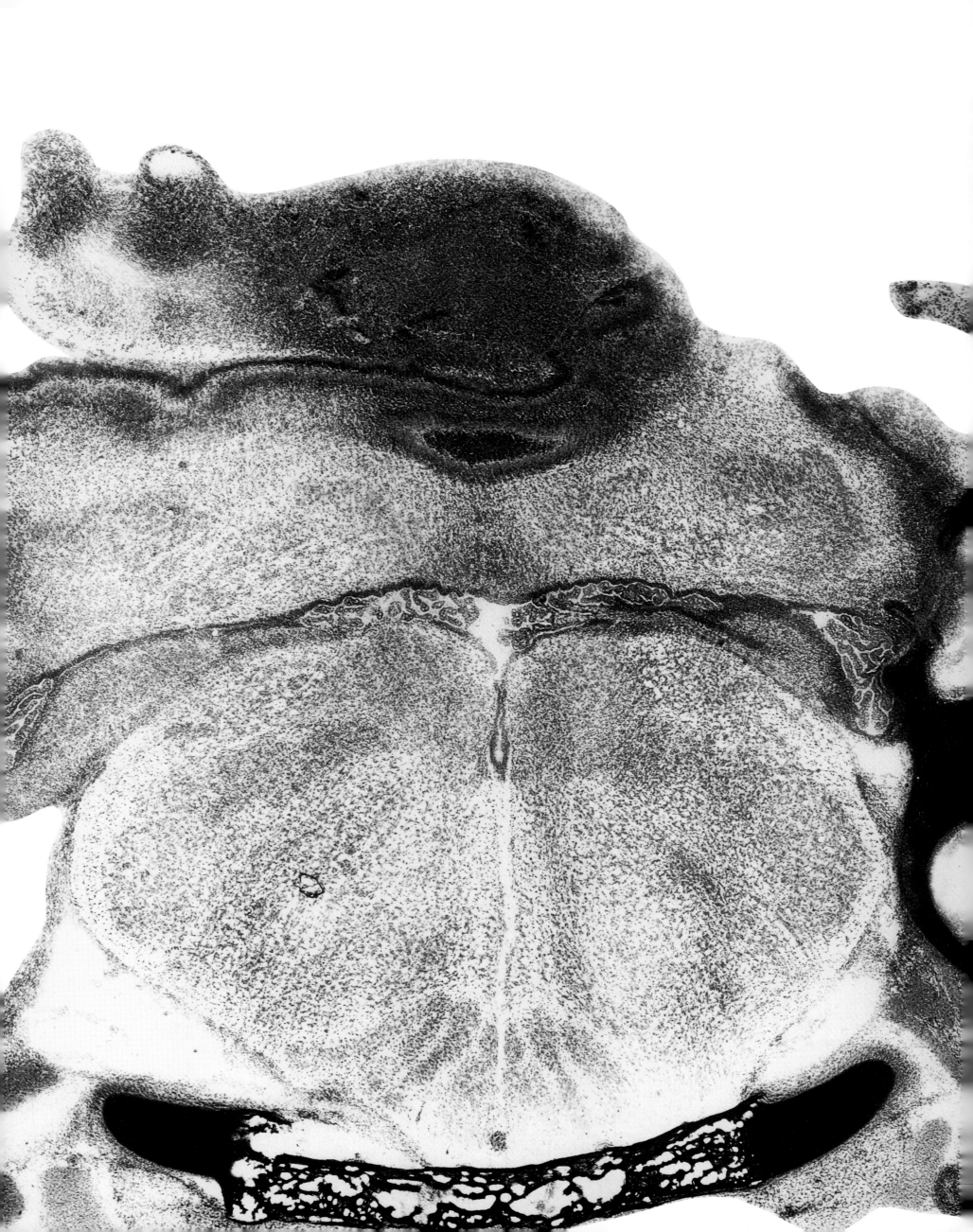

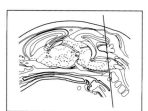

Figure 211
P0-67

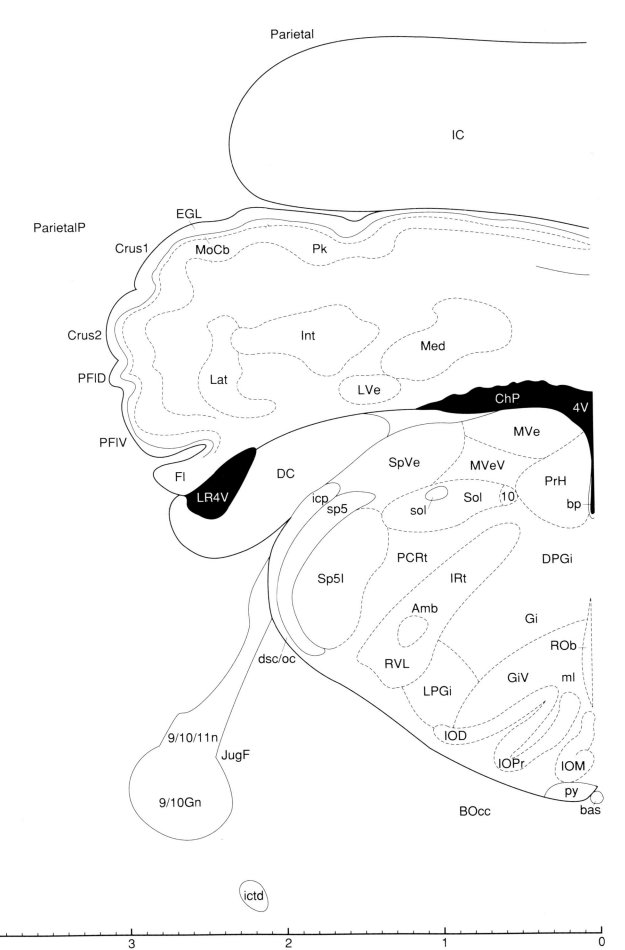

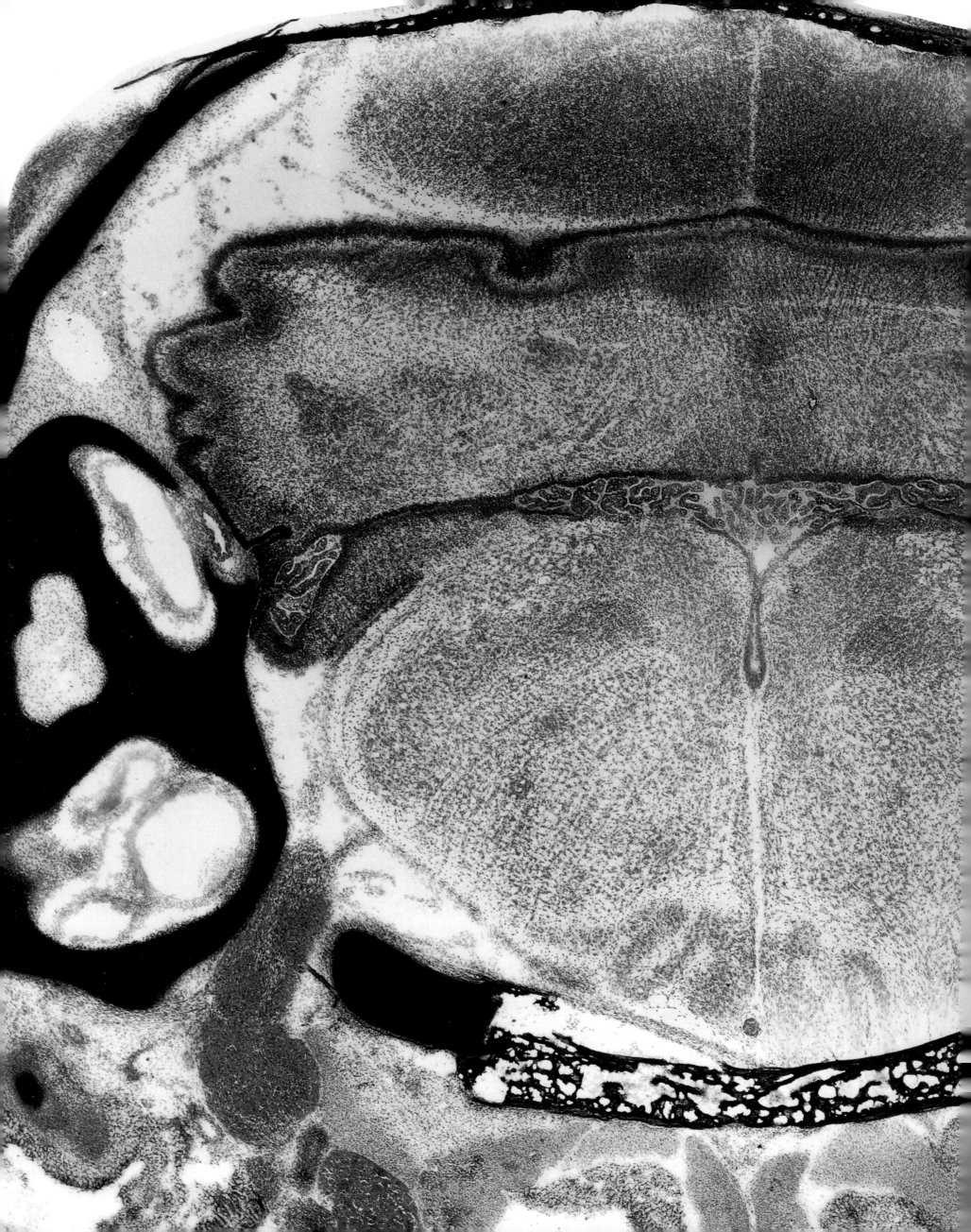

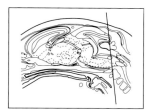

Figure 212
P0-68

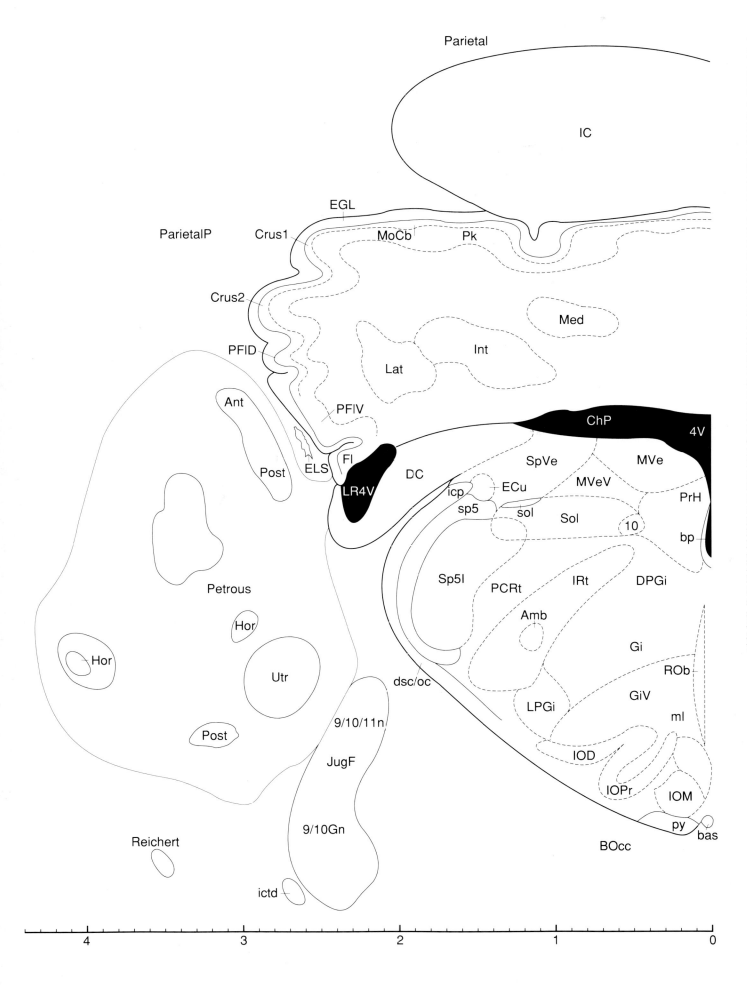

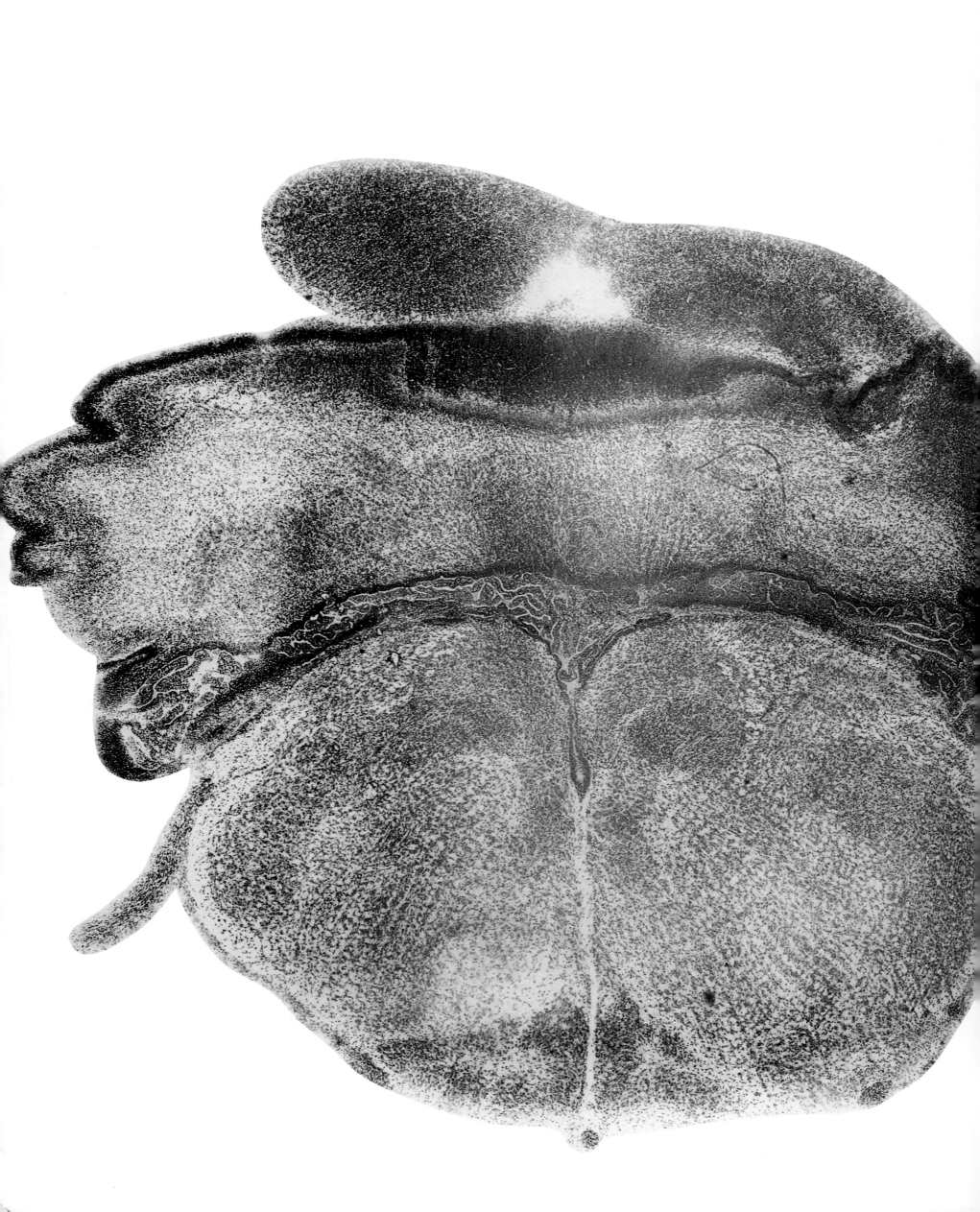

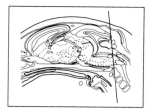

Figure 213
P0-69

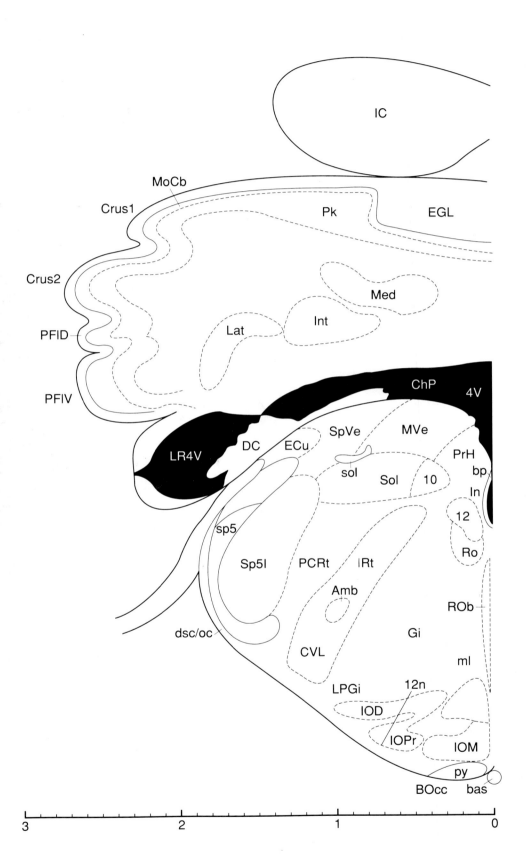

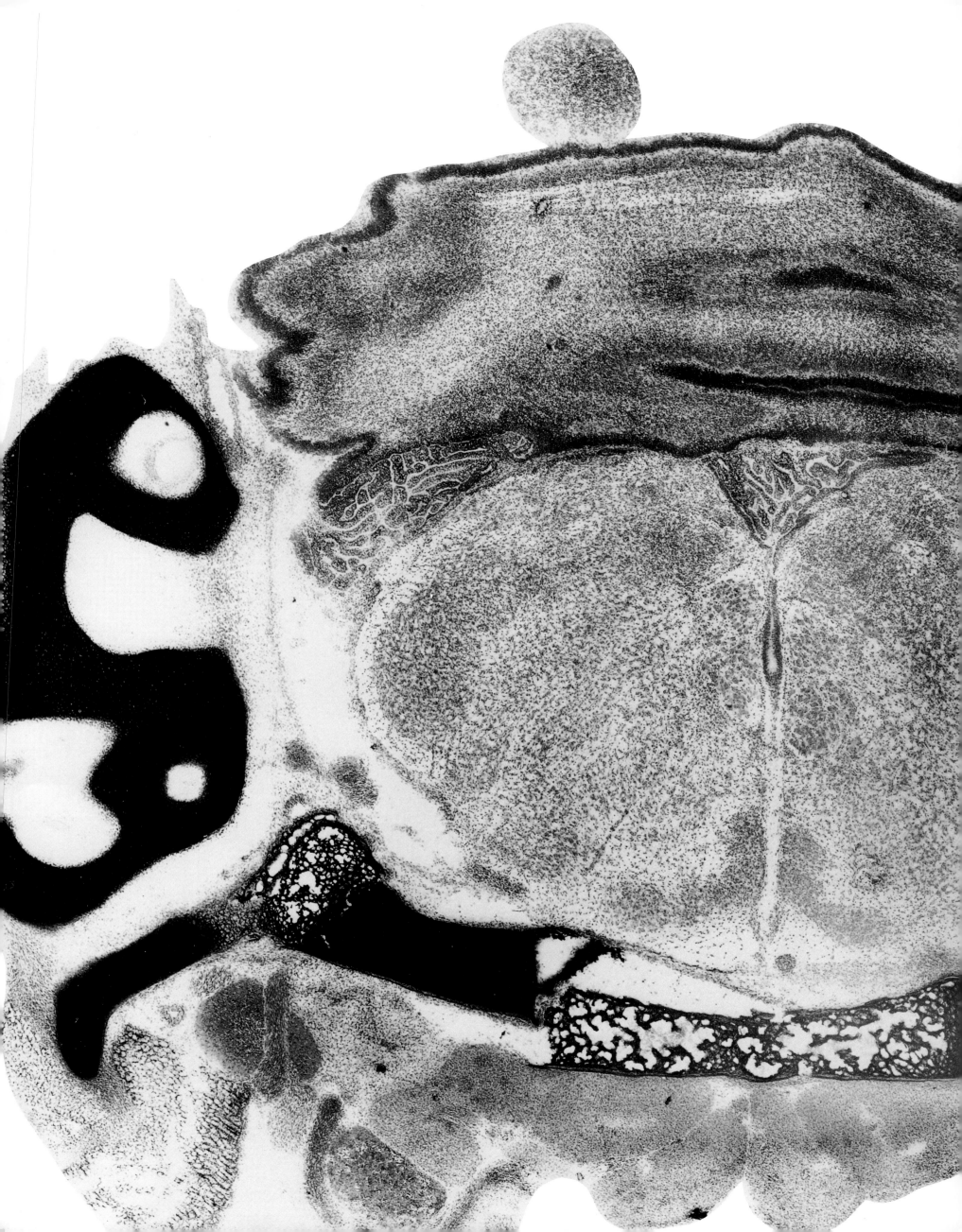

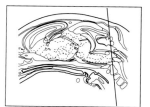

Figure 214

P0-70

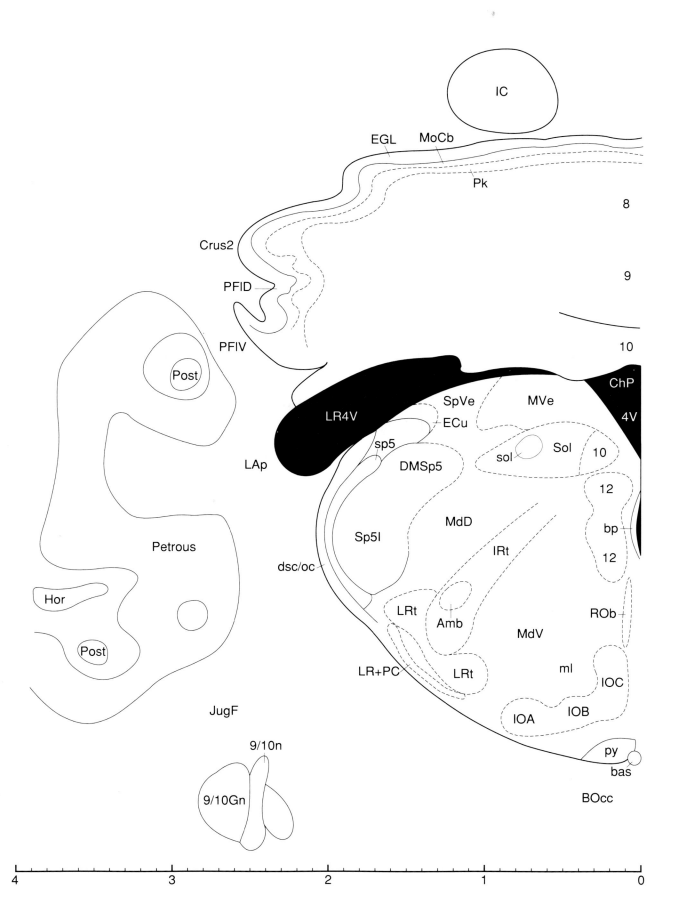

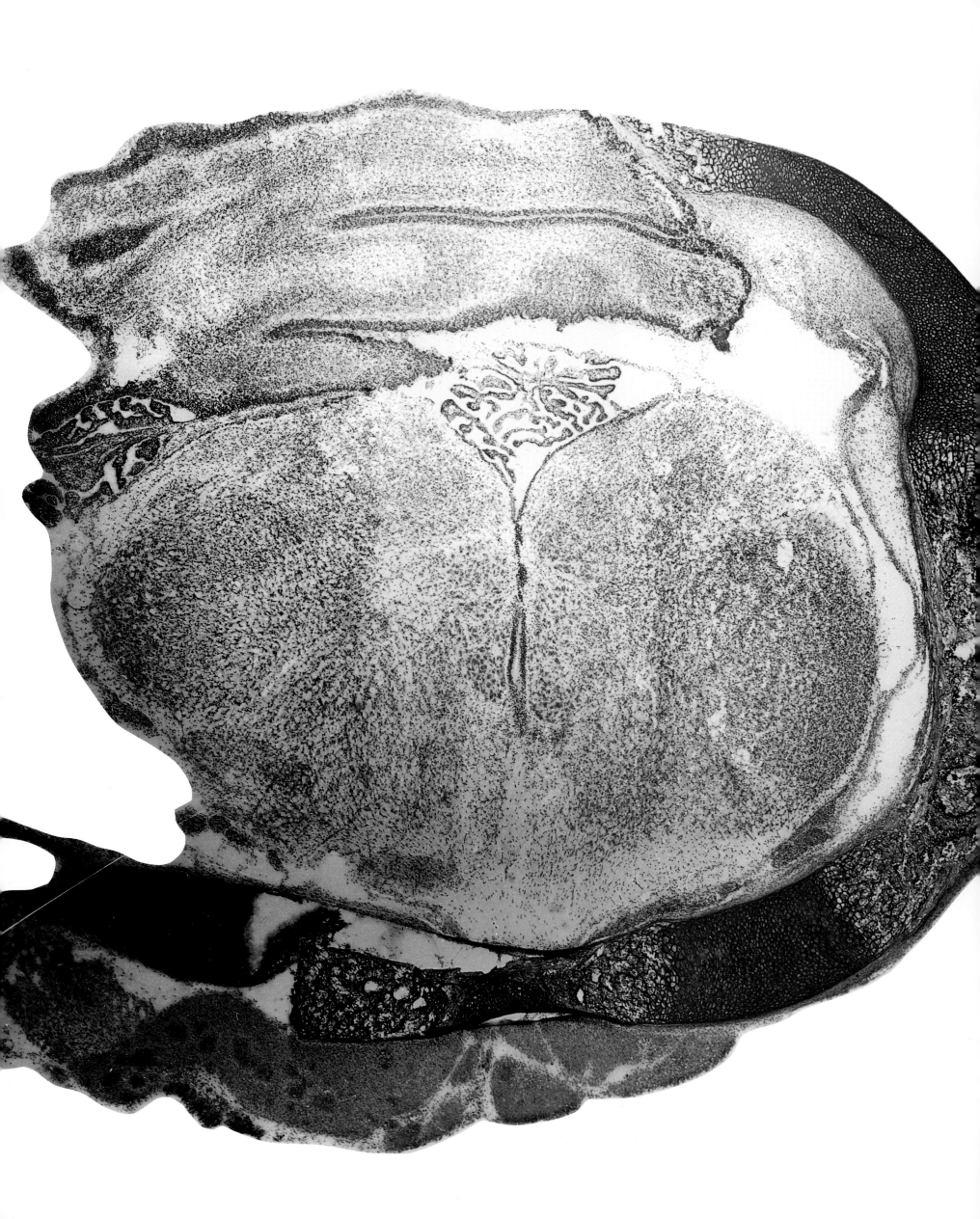

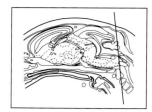

Figure 215
P0-71

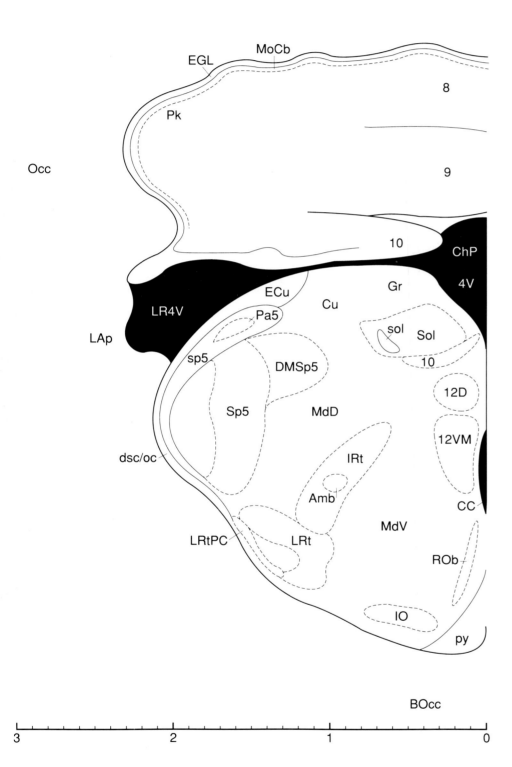

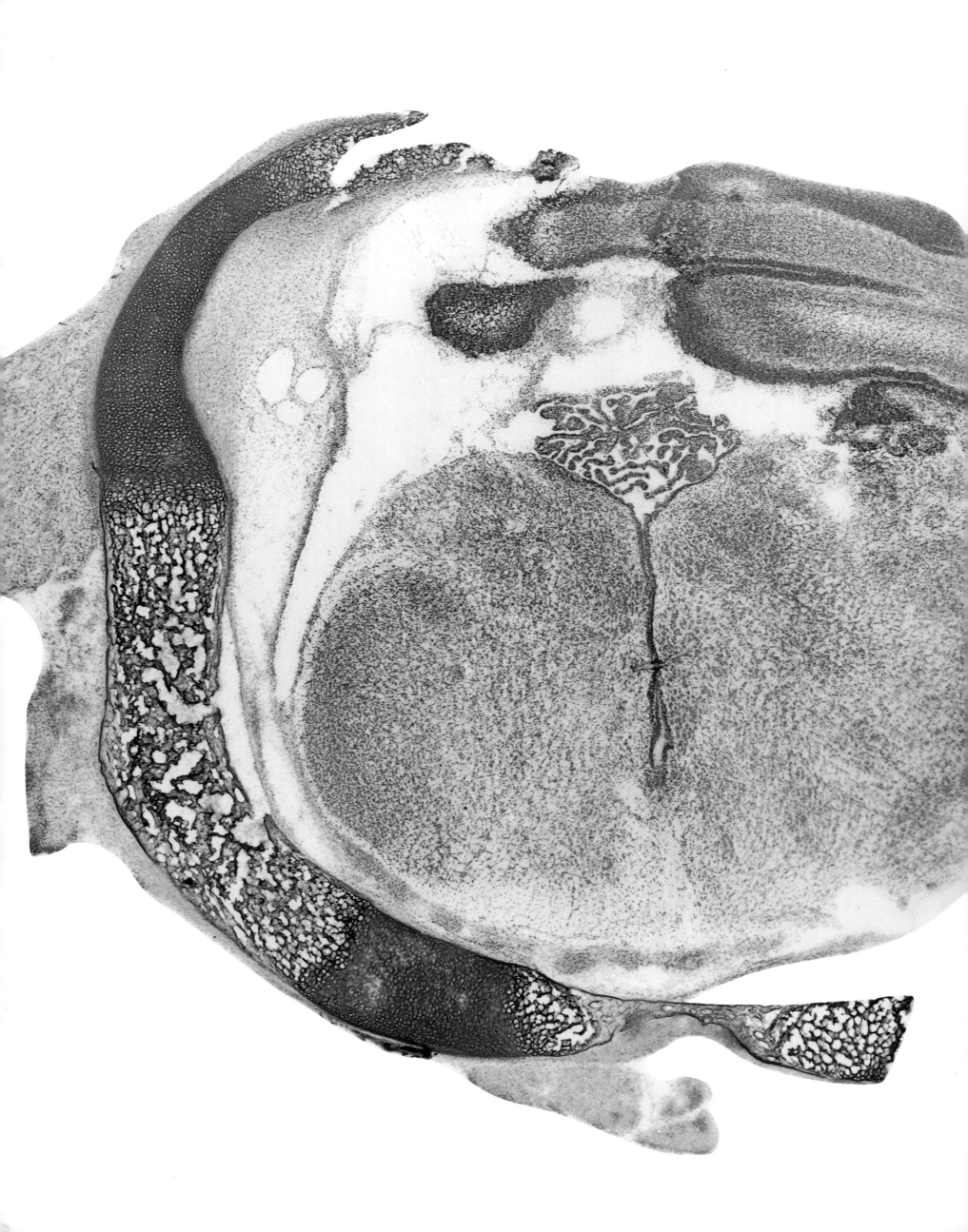

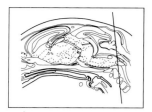

Figure 216
P0-72

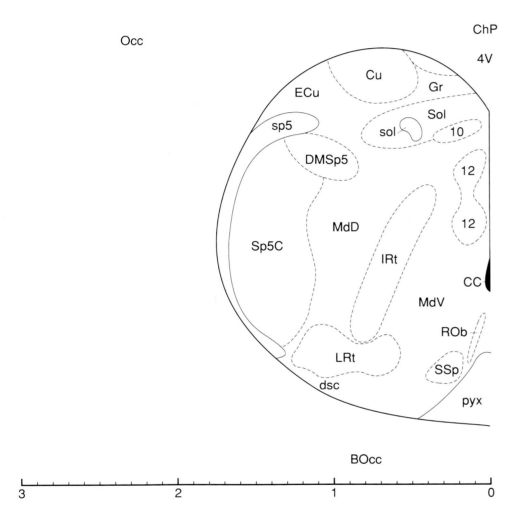

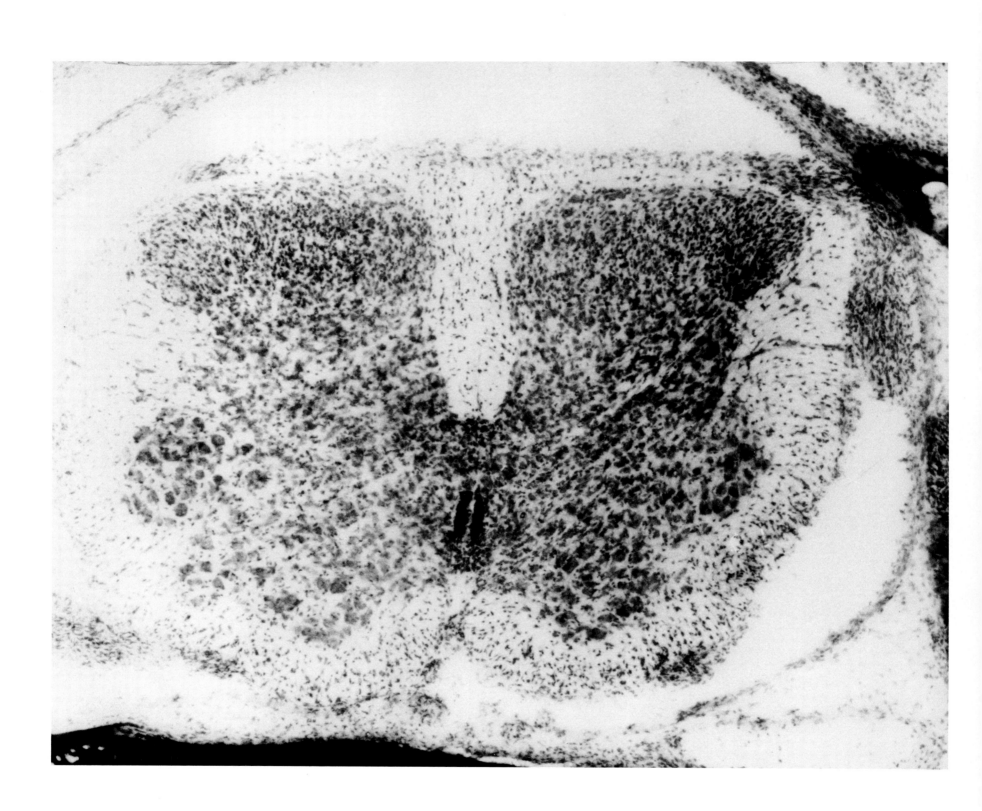

Figure 217
P0-Spinal Cord, Cervical

1-10   layers of the spinal cord
cu   cuneate fasciculus
dcs   dorsal corticospinal tract
dfu   dorsal funiculus of spinal cord
dl   dorsolateral fasciculus Sp cord
gr   gracile fasciculus
IML   intermediolateral cell column
lfu   lateral funiculus spinal cord
LSp   lateral spinal nu
vfu   ventral funiculus spinal cord
VMnF   vent median fissure spinal cord

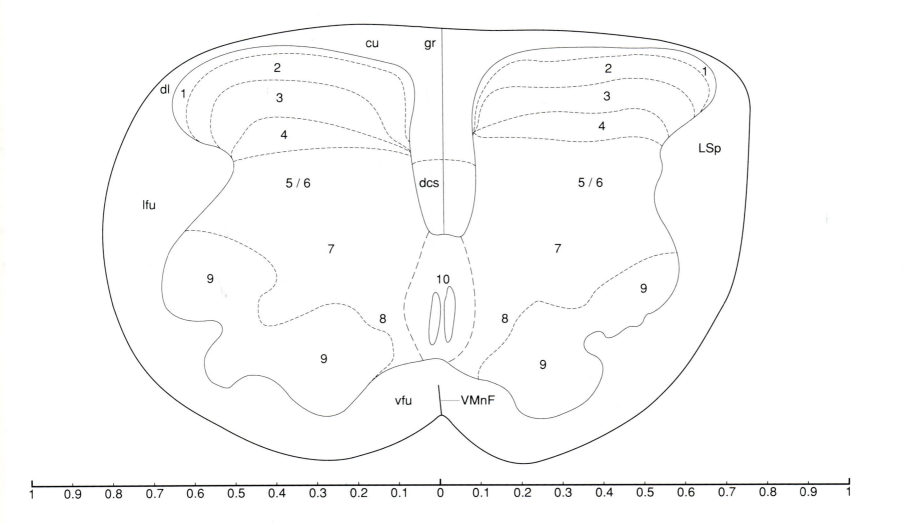

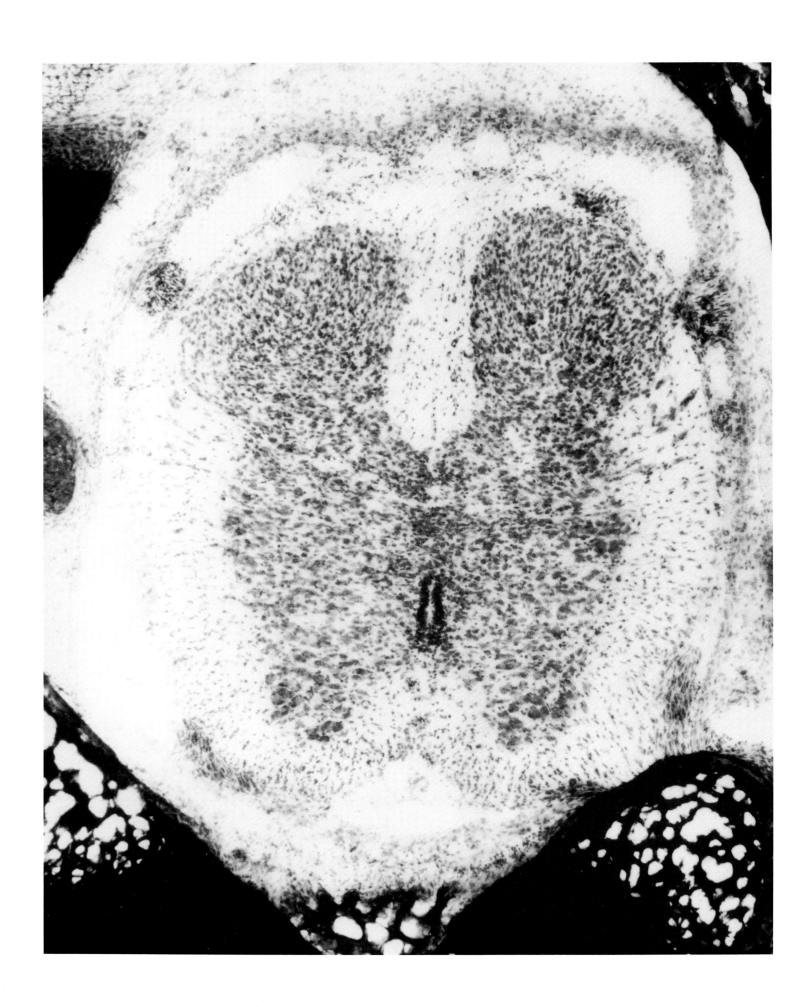

Figure 218
P0-Spinal Cord, Thoracic

1-10 layers of the spinal cord
cu   cuneate fasciculus
dcs   dorsal corticospinal tract
dfu   dorsal funiculus of spinal cord
dl   dorsolateral fasciculus Sp cord
gr   gracile fasciculus
IML   intermediolateral cell column
lfu   lateral funiculus spinal cord
LSp   lateral spinal nu
vfu   ventral funiculus spinal cord
VMnF vent median fissure spinal cord

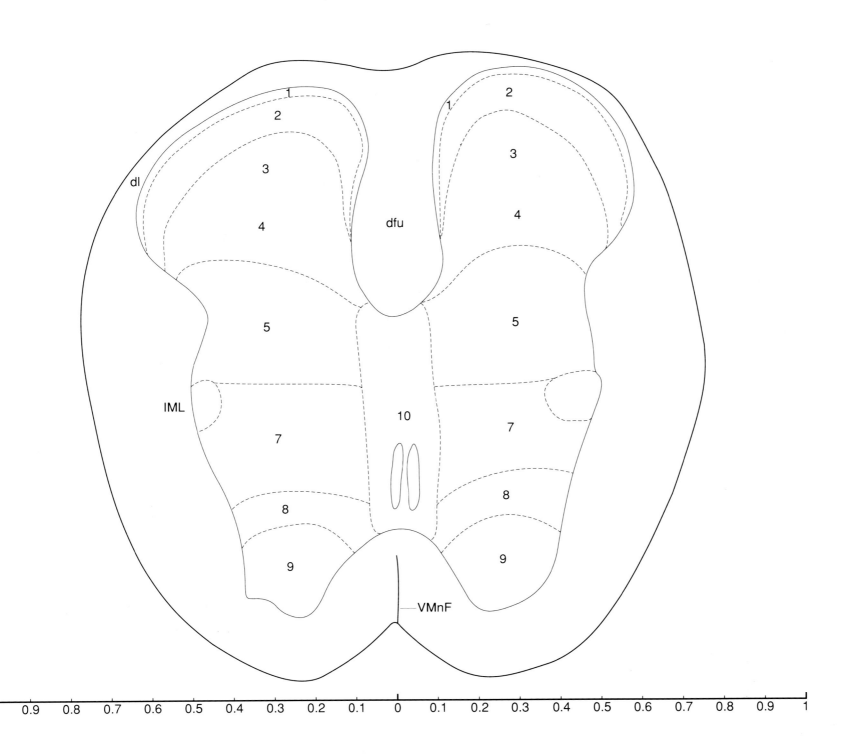